FIBRINOGEN, THROMBOSIS, COAGULATION, AND FIBRINOLYSIS

ADVANCES IN EXPERIMENTAL MEDICINE AND BIOLOGY

Recent Volumes in this Series

FIBRINOGEN, THROMBOSIS, COAGULATION, AND FIBRINOLYSIS

Edited by

Chung Yuan Liu

Institute of Biomedical Sciences
Academia Sinica
Taipei, Taiwan, Republic of China

and

Shu Chien

University of California, San Diego
La Jolla, California

SPRINGER SCIENCE+BUSINESS MEDIA, LLC

Library of Congress Cataloging-in-Publication Data

International Scientific Symposium on Fibrinogen, Thrombosis, Coagulation, and Fibrinolysis (1989 : Taipei, Taiwan)
Fibrinogen, thrombosis, coagulation, and fibrinolysis / edited by Chung Yuan Liu and Shu Chien.
p. cm. -- (Advances in experimental medicine and biology ; v. 281)
"Proceedings of the International Scientific Symposium on Fibrinogen, Thrombosis, Coagulation, and Fibrinolysis, held August 30-September 1, 1989, in Taipei, Taiwan, Republic of China"--T.p. verso.
Includes bibliographical references and indexes.
ISBN 978-1-4613-6697-3 ISBN 978-1-4615-3806-6 (eBook)
DOI 10.1007/978-1-4615-3806-6
1. Blood--Coagulation--Congresses. 2. Fibrinogen--Congresses. 3. Fibrinolysis--Congresses. 4. Thrombosis--Congresses. I. Liu, Chung Yuan, 1934- . II. Chien, Shu. III. Title. IV. Series.
[DNLM: 1. Blood Coagulation--physiology--congresses. 2. Coronary Thrombosis--drug therapy--congresses. 3. Endocardium--cytology--congresses. 4. Fibrinogen--physiology--congresses. 5. Myocardial Infarction--drug therapy--congresses. 6. Thrombolytic Therapy--congresses. W1 AD559 v. 281 / WH 310 I5975f 1989]
QP93.5.I538 1989
612.1'15--dc20
DNLM/DLC
for Library of Congress 90-14299
CIP

Proceedings of the International Scientific Symposium on Fibrinogen, Thrombosis, Coagulation, and Fibrinolysis, held August 30–September 1, 1989, in Taipei, Taiwan, Republic of China

ISBN 978-1-4613-6697-3

PREFACE

The International Scientific Symposium on Fibrinogen, Thrombosis, Coagulation, and Fibrinolysis was held in Academia Sinica, Taipei, Taiwan, Republic of China, on August 30 — September 1, 1989. This Symposium has provided a forum for the free exchange of information in this important and rapidly advancing research field. This proceedings volume provides a published record of 46 papers presented at the Symposium. The sponsors have exerted no influence on the scientific opinions or positions of the participants in the Symposium. It is hoped that this Symposium will stimulate further worldwide cooperation and collaboration in these vital fields for the benefit of all human kind.

This volume is composed of four parts. The first part consists of 8 papers on **Fibrinogen and Fibrin:** Biochemistry, Molecular Biology, and Physiology. The second part contains 16 papers on **Coagulation and Fibrinolysis:** Biochemistry, Molecular Biology, and Physiology. The third part has 10 papers on **Cardiovascular Cell Biology:** Biochemistry, Molecular Biology, and Physiology. The fourth part comprises 12 papers on **Clinical Studies of the Cardiovascular System:** Thrombotic and Bleeding Disorders and Thrombolytic Therapy. The Author Index with addresses of all contributors and the Subject Index of all 46 papers are arranged at the end of this volume.

On behalf of the Symposium Organizing Committee, it is indeed our pleasure to sincerely thank all of the participants for their participation in the Symposium. We appreciate very much the valuable contributions by the distinguished scientists coming from many nations around the world and the Republic of China as speakers and chairpersons of scientific sessions in this Symposium and as members of Symposium Committees. Their participation has assured the high quality of the Symposium.

We would like to acknowledge with sincere gratitude the support and encouragement by the sponsors of this Symposium: Academia Sinica, National Science Council, Department of Health, and fourteen Scientific Societies, and the co-sponsors: the Foundation for Biomedical Sciences and the China Committee for Scientific and Scholarly Cooperation with U.S.A. We also wish to express our appreciation for the generous support from several government agencies and private sectors of the Republic of China toward this Symposium.

The organization of the Symposium was initiated about one year before the 60th anniversary of Academia Sinica of the Republic of China (June 9, 1988), and the Symposium was held about one year after the Anniversary. We have dedicated this Symposium to Academia Sinica as a part of its 60th anniversary celebration, and we would also like to dedicate this proceedings volume to this anniversary.

To our colleagues in the scientific and medical communities of the Republic of China and the staff members of the Institute of Biomedical Sciences (IBMS) of Academia Sinica we wish to extend our deepest appreciation for their marvelous efforts, total dedication and wonderful cooperation and for giving us a most rewarding and memorable Symposium.

We would like particularly to acknowledge the important contributions by Ms. Jenny Chen and Ms. Maggie Shih of IBMS for their assistance in planning for this proceedings volume, by Ms. Grace Han of Enjoy Enterprise Co., Ltd. in Taipei for her help in retyping the entire volume, and by the staff of Plenum Publishing Corporation in New York for their suggestions, comments and assistance in general for the organization and publication of this volume.

Shu Chien, Ph.D., M.D.
Chairman

Chung Yuan Liu, Ph.D.
Secretary General

Symposium Organizing Committee
July 30, 1990

CONTENTS

PART I

FIBRINOGEN AND FIBRIN
Biochemistry, Molecular Biology, and Physiology

PART II

COAGULATION AND FIBRINOLYSIS
Biochemistry, Molecular Biology, and Physiology

A COAGULATION

B FIBRINOLYSIS

PART III

CARDIOVASCULAR CELL BIOLOGY
Biochemistry, Molecular Biology, and Physiology

A PLATELETS

B ENDOTHELIAL CELLS AND VESSEL WALL

PART IV

CLINICAL STUDIES OF THE CARDIOVASCULAR SYSTEM
Thrombotic and Bleeding Disorders and Thrombolytic Therapy

NATIVE FIBRIN GEL NETWORKS AND FACTORS INFLUENCING THEIR FORMATION IN HEALTH AND DISEASE

B. Blombäck [a,b], D. Banerjee[b], K. Carlsson[c], A. Hamsten[d],
B. Hessel[a], R. Procyk[b], A. Silveira[a], and L. Zacharski[e]

[a]Karolinska Institutet, S-10401 Stockholm, Sweden,
[b]The New York Blood Center, New York, NY 10021, USA,
[c]The Royal Institute of Technology, S-10444 Stockholm (Sweden),
[d]King Gustaf V Research Institute, Karolinska Hospital, S-10401 Stockholm and
[e]Veterans Administration Hospital, White River Junction, VT 05001, USA

SUMMARY

Hydrated fibrin gels were studied by confocal laser 3D microscopy, liquid permeation and turbidity. The gels from normal fibrinogen were found to be composed of straight rod-like fiber elements which sometimes originated from denser nodes. In gels formed at increasing thrombin or fibrinogen concentrations, the gel networks became tighter and the porosity decreased. The fiber strands also became shorter. Gel porosity of the network decreased dramatically in gels formed at increasing ionic strengths. Shortening of the fibers were observed and fiber swelling occurred at ionic strength above 0.24.

Albumin and dextran, when present in the gel forming system, affected the formation of more porous structures with strands of larger mass-length ratio and fiber thickness. This type of gels were also formed in plasma. Albumin and lipoproteins may be among the determinants for the formation of this type of gel structure in plasma.

Gels formed when factor XIIIa instead of thrombin was used as catalyst for gelation showed a completely different structure in which lumps of polymeric material were held together by a network of fine fiber strands.

Our studies have also shown that the methodologies employed may be useful in studies of gel structures in certain dysfibrinogenemias as well as in other diseases. We give examples of two patients with abnormal fibrinogen and of patients with ischaemic heart disease.

INTRODUCTION

Extensive knowledge of the structure of fibrinogen and the formation and properties of fibrin polymers has been accumulated during the past decades through studies employing chemical and physicochemical methods (1, 2, 3).

The fibrinogen molecule is an asymmetric rod-shaped dimer (about 10x45 nm). On activation, the fibrinogen molecules polymerize by end-to-end and side-to-side association there by forming protofibrils. The protofibrils associate to form fibrin fibers and the latter join into bundles of larger widths. Protofibrils were suggested to be twisted in a helical fashion (4, 5, 6). The twisting is possibly a reflection of an indigenous screw symmetry in the fibrinogen molecule itself. Fibrin fibers, also twisted, were seen to grow to a limiting size (7, 8). The limitation in growth was explained as a consequence of stretching of the protofibrils near the surface of the fiber. When the amount of energy necessary to stretch

Fibrinogen, Thrombosis, Coagulation, and Fibrinolysis, Edited by
C. Y. Liu and S. Chien, Plenum Press, New York

a protofibril exceeded the energy available from bonding, the lateral growth would cease (8).

The fibrin gel network has been studied by a variety of physicochemical methods. Light scattering, liquid permeation, viscoelastic measurements and studies by light microscopy, electron microscopy (EM) and scanning electron microscopy (SEM) have all been used to characterize the overall fibrin gel network architecture. Ferry and Morrison (9) investigated the mechanical and optical properties of fibrin gels classifying matrices as ranging between two extremes: fine (thin network and small liquid spaces) and coarse (built up by thick bundles and wide liquid spaces). Carr *et al.* (10, 11) determined the average fiber thickness in fibrin gels from turbidity and liquid permeation experiments. Shah *et al.* (12) showed with these techniques that clots from plasma contained significantly thicker fibers in contrast to gels formed from purified fibrinogen. By liquid permeation of gels formed under a variety of conditions it was shown that kinetic factors (*i.e.* the concentration of thrombin) determined the porosity of fibrin gels (13). In light microscopy (14-16) formation of needle-like structures after addition of thrombin to fibrinogen or plasma were observed. These needles eventually formed networks. In electron microscopic studies (17-21) of compacted, dehydrated fibrins, irregular networks of fibers or bundles of fibers of varying widths were observed. In contrast to these images, scanning electron microscopy of crack surfaces in freeze-dried fibrin gel samples revealed ordered network structures alternating with less ordered ones (16). Müller *et al.* (20) identified tri-functional branch points in the networks which were created by the twisting of fibers around each other. It therefore seems that fibrin fibers are both individually twisted but they are also twisted around each other at branch points (8).

Many factors can affect fibrin formation and gel structure. Calcium ions are bound to fibrinogen (22) and their removal hampers the clotting of fibrinogen and leads to gross structural changes (23). There are three high affinity binding sites for calcium and several low affinity binding sites (22). The high affinity binding sites are important for the integrity of the molecule and the low affinity sites are most likely important in the assembly of the network structure (24). Other effects on gel formation are seen in the presence of albumin and dextran. These compounds shorten the thrombin induced clotting time of fibrinogen, but do not affect the activation stage (25, 26). Increase in strand width and pore size occurs in the presence of dextran or albumin (26-28)). Water exclusion has been offered as an explanation for the effect of dextran by most investigators, but interaction of dextran with fibrinogen has also been raised as a possibility (29).

A significant factor determining the gel structure is the fibrinogen molecule itself, and modification of its structural domains will lead to abnormal polymerization and/or formation of fibrin gels with different physical properties. A unique type of modification occurs naturally in patients with congenital dysfibrinogenemia (for Review, see Ref. 30). Certain mutations or deletions at or in the vicinity of the thrombin susceptible bonds in the Aα- and Bβ-chains of fibrinogen hamper polymerization and gelation. In some patients, *e.g.* fibrinogen Detroit (31) and Rouen (32), the defects are associated with bleeding symptoms, while in others, *e.g.* fibrinogen New York (33) and fibrinogen Chapel Hill (34), thromboembolic manifestations occur.

Fibrinogen is an acute phase protein and temporarily increased levels in blood are consequences of inflammatory reactions. The blood level of fibrinogen is also determined by genetic factors (35). Epidemiological studies have shown that persistently elevated levels of fibrinogen in blood carries with it a strongly increased risk for development of ischacmic heart disease (IHD) (36-40). The occlusive event in IHD is in most cases a fibrin clot. However, it is unclear whether the mass of fibrin including cellular components or rather the architecture of the fibrin gel plays the determining role in this event.

In terms of function, the gelation of fibrinogen by thrombin is usually considered to be the only haemostatic pathway *in vivo*. This concept may be an oversimplification, since we and others have shown that polymerization and gelation of fibrinogen do not necessarily require activation by thrombin (41-44). In one of the non-thrombin pathways, polymerization and gelation of fibrinogen is accomplished by factor XIII catalysis, leading to formation of covalently crosslinked oligomers which eventually become incorporated into gel structures (fibrinogenin) (44-45). A second thrombin independent pathway involves fibronectin in addition to fibrinogen (45-46). In this reaction crosslinked heterooligomers (heteronectin) are formed. At present, the physiological importance of these alternative clotting pathways is not clear.

In the present work, part of which was published elsewhere (47), we have studied the hydrated fibrin or fibrinogenin network structures by confocal laser microscopy, which allows three-dimensional imaging of a specimen. We have, furthermore, employed liquid permeation and turbidity measurements for quantitation of features in the structure. We have demonstrated the potential usefulness of these techniques in studies of fibrin gels from patients with dysfibrinogenemia and IHD.

MATERIAL AND METHODS

Materials

When not otherwise stated the materials used were described in a recent report (47). Factor XIII was prepared as previously described (48). Human albumin was obtained from KABI AB, Stockholm, Sweden and Dextran T40 from Pharmacia, Uppsala, Sweden. Blood from normal individuals and patients was collected in citrate (1 part 0.13 M sodium citrate + 9 parts of blood) and plasma prepared by centrifugation (2000-3000 xg for 20 min). Platelet-poor plasma was prepared by recentrifugation of the plasma at 10000 xg for 20 min.

Lipoprotein-deficient plasma was prepared from normal blood collected in acid citrate dextrose (ACD) solution. After centrifugation, an aliquot of the plasma was adjusted to a density of 1.215 with solid KBr and subsequently centrifuged at 214000 xg for 23 h at 4-10°C (49). The floated lipids and lipoproteins were removed from the top and the remainder of the plasma plus the pellet resuspended and dialyzed extensively against TNE-buffer (0.05 M Tris, 0.1 M NaCl, 1 mM EDTA, pH 7.4). The lipoprotein containing plasma was also dialyzed against TNE-buffer. The fibrinogen concentration was determined in both samples and the lipoprotein containing plasma sample was diluted with TNE-buffer to the same fibrinogen concentration as the lipoprotein free plasma. The osmolarity of both samples was identical after dialysis.

Flow measurements

The construction of the microchambers shown in Fig. 1 and the gel formation in the chambers were as described previously (47). The clotting procedure of fibrinogen solutions containing albumin or dextran were essentially the same as for fibrinogen without these additions. Plasma samples were dialyzed against TNE-buffer containing trasylol (at 5 KIE/mL) and subsequently recalcified with 1 M $CaCl_2$ to a final concentration of 20 mM and then clotted with human thrombin (at 0.18 NIH units/ml). The microchambers containing the gels were placed in holders and the gels equilibrated with deareated buffer (0.02 M Tris-imidazole, 0.1 M NaCl, pH 7.4) (13) containing trasylol (at 5 KIE/mL), alanine (at 1 mM) and [3 H]-alanine (at 0.1 μ Ci/mL). Subsequently, buffer was percolated through the gel for a certain time at five different hydrostatic pressures. The radioactivity (dpm) of the effluent was determined by liquid scintillation counting. The volume was

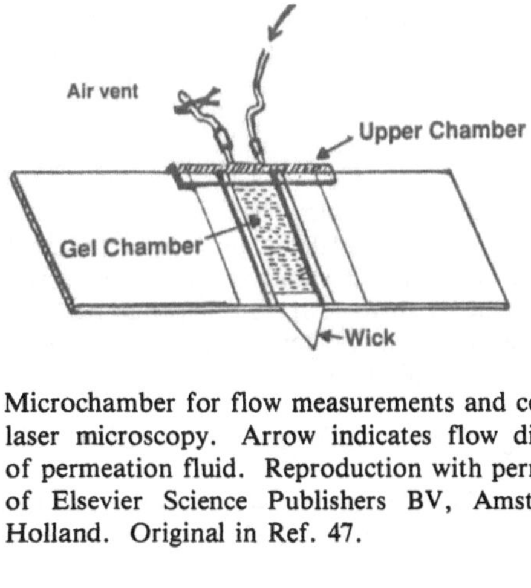

Fig. 1. Microchamber for flow measurements and confocal laser microscopy. Arrow indicates flow direction of permeation fluid. Reproduction with permission of Elsevier Science Publishers BV, Amsterdam, Holland. Original in Ref. 47.

calculated from the specific radioactivity of the permeant. The flow rates ($L/cm^2/h$) plotted against the pressure (in $dyne/cm^2$) fit a straight line. The dyne value at zero flow was subtracted from the pressure applied and the resulting value of presssure was used in calculation of the permeability (or Darcy) constant (K_s), as previously described (47).

The fiber diameter, d, was calculated from the K_s values using the equation of Signer *et al.* (50). In this calculation one must know the fractional volume of the fibers in the gel. For our calculations we derived the fractional volume of the dehydrated fiber from the protein concentration and the density (51) of the protein. In order to obtain the fractional volume for the hydrated fiber, we assumed a fiber hydration of 6 g/g protein (52), since this hydration value gave in general the best agreement with the turbidometric and to some extent with the microscopic analyses. For further explanations see Reference 47.

Analysis of the gel structures by microscopy

The gels used in the flow studies were labeled by percolation of 1.6 mM fluorescein isothiocyanate (FITC) in buffer (0.01 M Tris, 0.1 M NaCl, 1 mM EDTA, pH 8.0) through the gels. FITC should be in contact with the entire gel matrix for about 30 to 60 min. The gels were washed extensively by slow percolation of buffer and then used for microscopic analysis. The confocal microscope scanner used in this study, PHOIBOS, was developed at the Royal Institute of Technology in Stockholm, Sweden (53). To illuminate the specimen a focused beam of light from an argon laser was scanned in a raster pattern. The excitation wavelength used in this study was 488 nm and the emission wavelength approximately 530 nm. By introducing a small aperture in front of the detector (a photomul-
calculated from the specific radioactivity of the permeant. The flow rates ($L/cm^2/h$) plotted against the pressure (in $dyne/cm^2$) fit a straight line. The dyne value at zero flow was
The laser beam caused bleaching of the specimen to a varying extent. In most experiments we therefore percolated a p-phenylenediamine solution (1.0 mg/mL) in buffer (0.011 M sodium phosphate, 0.02 M NaCl, 0.003 M KCl, 0.2 M sodium bicarbonate, pH 8.0) through the gel prior to scanning (54). Scanning and data sampling were controlled by a microprocessor, which also transferred the data to a host computer (Compaq 386/20). The host computer produced stereo images of the gel structure corresponding to a ± 10° tilt relative to the optical axis. The average width of the strands in the recorded images was measured by recording the light intensity profiles along cross-sections of the strands. The width of each strand was then calculated as the distance between points where the intensity value was half of the maximum value, *i.e.* the full-width half-maximum (FWHM) value was calculated.

Clotting time (Ct) and wavelength dependence of turbidity

For accurate determination of C_t. the fibrinogen-thrombin mixture was allowed to gel in a cuvette in a spectrophotometer and the development of turbidity with time was monitored at 450 nm. The tangent of the sigmoidal curve was extrapolated to the base line and the corresponding time was defined as clotting time (13). The $C/\tau \bullet \lambda^3$ was plotted against $1/\lambda^2$, where C is the fibrin concentration (in g/cm^3), τ the turbidity (per cm) and λ the wavelength (in cm) (29, 55). In the plots, the intercept on the y-axis gives the mass-length ratio (μ) and the ratio of slope over intercept represents the radius square of the hydrated fibers.

Fibrin concentration

In most of the experiments employing purified fibrinogen-thrombin systems, fibrin was determined by clotting a sample of fibrinogen under the same conditions as was used in the permeability and turbidity experiments. The clots were synerized and the fibrin content determined as described earlier (56). In plasma, the samples were diluted before clotting in order to avoid occlusion of proteins during syneresis (56). The value for fibrin thus obtained was used in the calculations.

Determination of crosslinking of fibrin chains by SDS-gel electrophoresis

The gels formed during the experiments were solubilized in the presence of reducing reagent (dithiothreitol) and used for SDS-gel electrophoresis as described previously (57).

Table 1. Influence of Thrombin Concentration on K_s, Mass-Length Ratio and Average Fiber Diameter as Measured by Permeability and Turbidity

		Permeability			Turbidity	
Thrombin Units/mL[a]	C_t s	K_s $cm^2 \cdot 10^9$	μ $\cdot 10^{-12}$	ϕ μm	μ $\cdot 10^{-12}$	ϕ μm
A 0.08	315	6.9	24.1	0.161	25.5	0.130
0.15	183	5.1	17.7	0.138	22.1	0.114
0.23	168	5.3	18.3	0.140	22.2	0.116
1.04	51	2.6	9.0	0.098	13.7	0.084
B 0.05	531	10.6	36.5	0.198	29.6	0.135
0.21	225	10.4	36.2	0.197	23.0	0.121
0.52	96	5.0	17.4	0.134	17.6	0.104
1.04	60	3.0	10.6	0.107	14.8	0.096
2.02	33	2.7	9.2	0.099	12.5	0.093

Fibrinogen (1,15 mg/mL) was clotted with thrombin at different thrombin concentrations and ionic strength 0,21. The data are from two different experiments (A and B). C_t, clotting time, μ, fiber mass-length ratio in dalton/cm. ϕ, average fiber diameter. a. NIH units. For further explanations, see Materials, Methods and Text.

RESULTS

Gel structure at different thrombin concentrations

In these studies fibrinogen was clotted at concentrations of thrombin ranging from 0.05-2.0 NIH unit/mL and at an ionic strength of 0.21. All gel matrices were fully cross-linked.

As shown in Table 1, the K_s values were inversely related to the thrombin concentration (13, 47, 58). This means that the higher the thrombin concentration the tighter gel structures with smaller pores were formed. The mass-length ratio and the average fiber diameter as calculated from permeability also decreased with increasing thrombin concentrations. Calculations based on the turbidometric data showed a similar decrease.

3D reconstructions from confocal microscopy data of the gels showed that the gel porosity decreased with increasing thrombin concentrations which was in agreement with the permeability data (cf. Table 1). The photos, displayed as stereo pairs in Fig. 2A and B, show the architecture at a low and high thrombin concentration, respectively. The fiber strands are straight, cylindrical, rod-like structures at all thrombin concentrations. Many strands protrude perpendicularly to the plane of the flow chambers throughout the entire volume element examined (usually 30 μm) The length of the fibers appears to decrease with increasing thrombin concentration and the networks seem to become more regular and achieve a honeycomb-like structure. As judged form the fluorescence intensities there appears to be a variation in strand width. The fiber strands cross each other at numerous places. Branching fibers are occasionally seen. Some strands in the gel network emerge from node structures in the solution and proceed from these points in different directions in space until they encounter strands from other nodes thus creating a noded architecture. Usually, more than three strands protrude from each node. The number of nodes per volume element appears to increase with increasing thrombin concentration. The average fiber diameter measured in these images did not show any statistically significant change with increasing thrombin concentrations. Therefore, we suggest that the strand widths at the different thrombin concentrations are below the resolving power (about 0.2-0.3 μm) of the microscope.

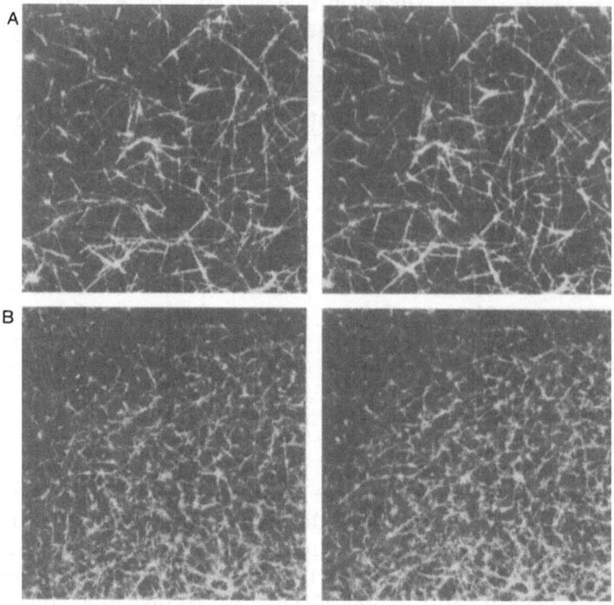

Fig. 2. 3D microscope pictures of fibrin gel networks at different thrombin concentrations. The stereo pairs shown are computer reconstructions made from confocal laser microscope recordings. Gel formation: in TNE buffer, pH 7.4, containing $CaCl_2$ (at 20 mM) of ionic strength 0.21, and at different thrombin concentrations. Temperature, 24°C. Fibrinogen, 1.15 mg/mL. Permeation: with Tris-imidazole-NaCl buffer (see Materials and Methods). A) Thrombin, 0.05 NIH units/mL. B) Thrombin, 2.0 NIH units/mL. Dimensions: 100×100 μm.

Table 2. Influence of Fibrinogen Concentration on K_s, Mass-Length Ratio and Average Fiber Diameter as Measured by Permeability and Turbidity

		Permeability			Turbidity	
Fibrinogen mg/mL	C_t s	K_s $cm^2 \cdot 10^9$	μ $\cdot 10^{-12}$	ϕ μm	μ $\cdot 10^{-12}$	ϕ μm
A 4.0	48	2.5	47.9	0.223	218.7	0.699
3.5	47	3.6	57.7	0.248	96.8	0.407
3.0	55	3.2	41.5	0.209	98.5	0.388
2.5	57	8.8	89.4	0.309	82.7	0.336
2.0	53	8.7	64.8	0.263	65.5	0.274
1.5	67	12.4	62.2	0.257	57.4	0.217
1.0	67	—	—	—	44.0	0.179
0.5	69	106.8	118.5	0.356	40.1	0.184
B 1.9	165	2.7	18.2	0.140	20.7	0.122
0.9	135	4.4	11.5	0.111	15.4	0.098
0.5	165	24.4	24.9	0.163	12.9	0.085

Fibrinogen at different concentrations was clotted with thrombin at 35 NIH units/mL and ionic strength 0.16 (in A) and 0.21 (in B). C_t, clotting time. μ, fiber mass-length ratio in dalton/cm. ϕ, average fiber diameter. For further explanations, see Materials, Methods, and Text.

Gel structure at varying concentrations of fibrinogen

Fibrinogen at concentrations between 4.0 and 0.5 mg/mL was clotted with thrombin at two ionic strengths. Under all conditions, crosslinking of the chains of fibrin was complete.

The analysis by permeation (Table 2) showed that the K_s values of the gels increased logarithmically with decreasing fibrin concentrations (13, 47, 58). The calculated fiber mass-length ratio and fiber diameter did not change significantly with decreasing fibrinogen concentration between 4 and 1 mg/mL. However, at the lowest fibrin concentration (0.5 mg/mL) the strand width seemed to increase (Table 2). The turbidity measurements of the gels at concentrations below 2.0-2.5 mg/mL were in reasonable agreement with the permeability data, although widening of the fibers at the lowest concentration was not seen.

Confocal microscopy showed that the porosity of the gel matrix increased with decreasing fibrin concentrations in agreement with the permeability data. The fiber strands and the microscope images had the appearance of straight rods. The average length of the fibers appeared to decrease with increasing fibrinogen concentration. There seemed to be some variation in the width of the fiber strands seen at each fibrin concentration. The statistical analysis gave a significant (at the 90% confidence level) increase in average

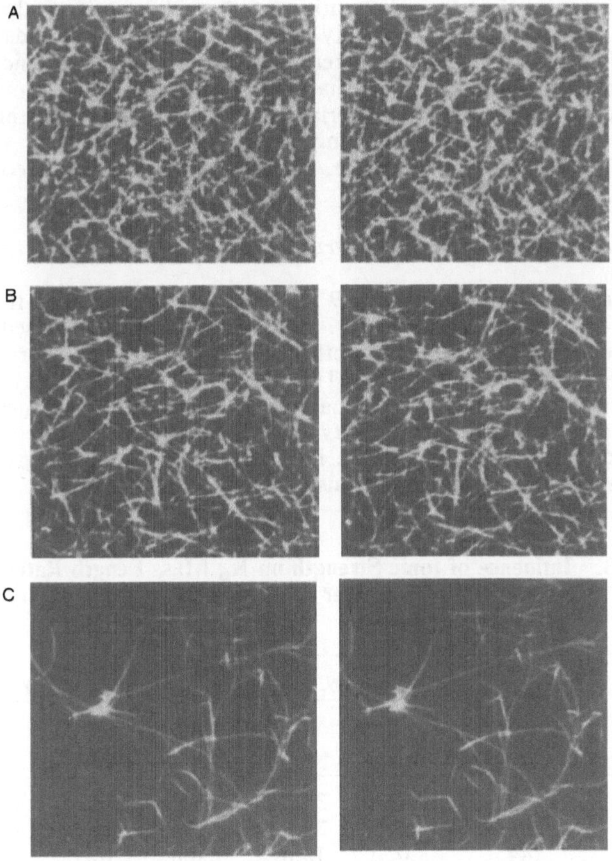

Fig. 3. 3D microscope pictures of fibrin gel networks at different fibrinogen concentrations. Gel formation: in TNE-buffer, pH 7.4, containing $CaCl_2$ (at 20 mM) of ionic strength 0.16. Temperature, 24°C. Thrombin, 0.35 NIH units/mL. Permeation: same as described in legend to Fig. 2. A) Fibrinogen, 3.5 mg/mL. B) Fibrinogen, 1.5 mg/mL. C) Fibrinogen, 0.5 mg/mL. Stereo pairs are shown. Dimensions: 100×100 μm. The gels were treated with *p*-phenylenediamine.

fiber diameter at the lowest fibrinogen concentration, which was also suggested by the permeability data. Like in the previous experiment several strands seemed to emerge from node structures in the solution and the number of nodes appeared to decrease with decreasing fibrinogen concentration. In Fig. 3A-C are shown pictures of the gel matrices at three different fibrinogen concentrations. At the lowest fibrin concentration a few bent and collapsed structures are seen (Fig. 3C), probably arising from perturbation in this experiment of the gel structure caused by the flow of permeant through the extremely porous structure. In other experiments (not shown) unperturbed porous gels were demonstrated.

Gel structure at different ionic strengths

In a third type of experiment, the fibrinogen solution was clotted with thrombin at different ionic strengths. All of the gel matrices were fully crosslinked.

The permeability analysis showed that the porosity of the gels, as measured by their K_s values, decreased with increasing ionic strength (Table 3). The fiber mass-length ratio and average strand width decreased in concert with K_s. The turbidity studies of these gels showed a similar decrease in mass-length ratio with increasing ionic strength (Table 3). The strand width as measured by turbidity also decreased up to an ionic strength of about 0.24. In contrast, however, in the turbidity measurements at higher ionic strengths an apparent widening of the fibers was observed.

Confocal microscopy showed a diminution of the gel porosity with increasing ionic strenghts in agreement with the permeability measurements. The strands became shorter and the number of nodes increased with increasing ionic strength. At ionic strengths above 0.24 the fibers swelled. They lost their sharpness and appeared as thin structures surrounded by a fuzzy zone of particulate material. At ionic strength 0.28, this fuzzy material covers almost the whole visual field and only faint fiber structures are visible. At ionic strength 0.4 even less stuctures were visible. Fig. 4A-C shows the stereo pictures of the network at three different ionic strengths.

Effect of albumin and dextran on fibrin gel structure

In these experiments fibrinogen (at 1.9 mg/mL) was clotted with thrombin (at 0.35 NIH units/mL) at an ionic strength of 0.21. Clotting was also performed in the presence of albumin and Dextran T40 (Table 4). Both albumin and dextran shortened the clotting time of the fibrinogen thrombin mixture. This shortening was not, as was the case with increasing thrombin concentrations, accompanied by a decrease in permeability and fiber dimensions. On the contrary, the permeability and turbidity studies showed that the presence of these compounds resulted in an increase in porosity, fiber mass-length ratio and fiber width. The microscope images supported this conclusion (Fig. 5, A-C).

Table 3. Influence of Ionic Strength on K_s, Mass-Length Ratio and Average Fiber Diameter as Measured by Permeability and Turbidity

		Permeability			Turbidity	
Ionic strength	C_t s	K_S $cm^2 \cdot 10^9$	μ $\cdot 10^{-12}$	ϕ μm	μ $\cdot 10^{-12}$	ϕ μm
0.15	69	10.9	38.0	0.197	38.3	0.183
0.19	84	6.5	22.6	0.154	20.6	0.136
0.22	144	3.0	10.6	0.106	10.7	0.085
0.24	144	0.9	3.2	0.059	2.1	0.094
0.26	204	0.3	1.1	0.033	1.0	0.125
0.28	204	0.3	0.9	0.028	0.6	—

Fibrinogen (1.15 mg/mL) was clotted with thrombin at 0.35 NIH units/mL and at different ionic strengths. C_t, clotting time. μ, fiber mass-length ratio in dalton/cm. ϕ, average fiber diameter. For further explanation, see Material, Methods and Text.

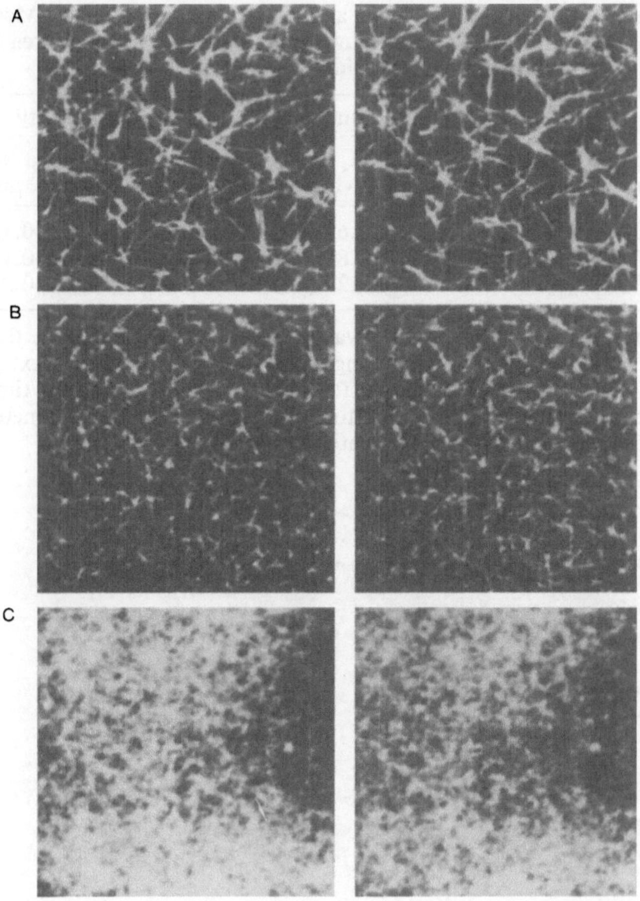

Fig. 4. 3D microscope pictures of fibrin gels at different ionic strengths. Gel formation: in TNE-buffer, pH 7.4, containing $CaCl_2$ (at 20 mM) of different ionic strengths. The desired ionic strength between 0.15-0.28 was achieved by adding sodium chloride. Temperature, 23-24°C. Fibrinogen, 1.15 mg/mL. Thrombin, 0.35 NIH units/mL. Permeation: same as described in legend to Fig. 2. A) Ionic strength, 0.15. B) Ionic strength, 0.24. C) Ionic strength, 0.28. The gels were treated with *p*-phenylenediamine. Stereo pairs are shown. Dimensions: 100×100 μm.

Gel structure in the alternative pathway of clotting — Fibrinogenin gels

Factor XIIIa can catalyze gelation of fibrinogen without participation of thrombin (44, 45). The permeability characteristics of factor XIIIa induced gels were somewhat different from those of thrombin induced gels (compare Table 5 with Table 2). This became more apparent when plotting K_s against the fibrinogen concentration. In this plot the exponent of the logarithmic function relating K_s to fibrinogen concentration had a larger negative value for the fibrinogenin gels (-2.93 vs. -1.77). As a consequence, the calculated fiber mass-length ratio and fiber diameter as calculated by permeability increased with decreasing fibrinogen concentration. On the contrary, the turbidity measurements of the fibrinogenin gels showed a small decrease in these parameters with decreasing fibrinogen concentration (Table 5).

The microscope pictures of the gels at two different fibrinogen concentrations clearly showed the difference between these gels. This is exemplified at one fibrinogen concentration in Fig 6. The factor XIIIa induced gels are composed of lumps of polymeric material

Table 4. Influence of Albumin and Dextran on K_s. Fiber Mass-length Ratio and Average Fiber Diameter as Measured by Permeability and Turbidity

			Permeability			Turbidity	
Albumin mg/mL	Dextran mg/mL	C_t s	K_s $cm^2 \cdot 10^9$	μ $\cdot 10^{-12}$	ϕ μm	μ $\cdot 10^{-12}$	ϕ μm
0	0	162	2.6	17.0	0.136	24.5	0.135
37	0	78	9.8	66.5	0.267	39.7	0.194
0	9.3	63	8.2	55.6	0.243	69.2	0.252

Fibrinogen (at 1.87 mg/mL) was clotted with thrombin at 0.35 NIH units/mL and ionic strength 0.21. Albumin and Dextran were dialyzed against TNE-buffer before use. C_t, clotting time. μ, fiber mass-length ratio in dalton/cm. ϕ, average fiber diameter. For further explanation, see Materials, Methods and Text.

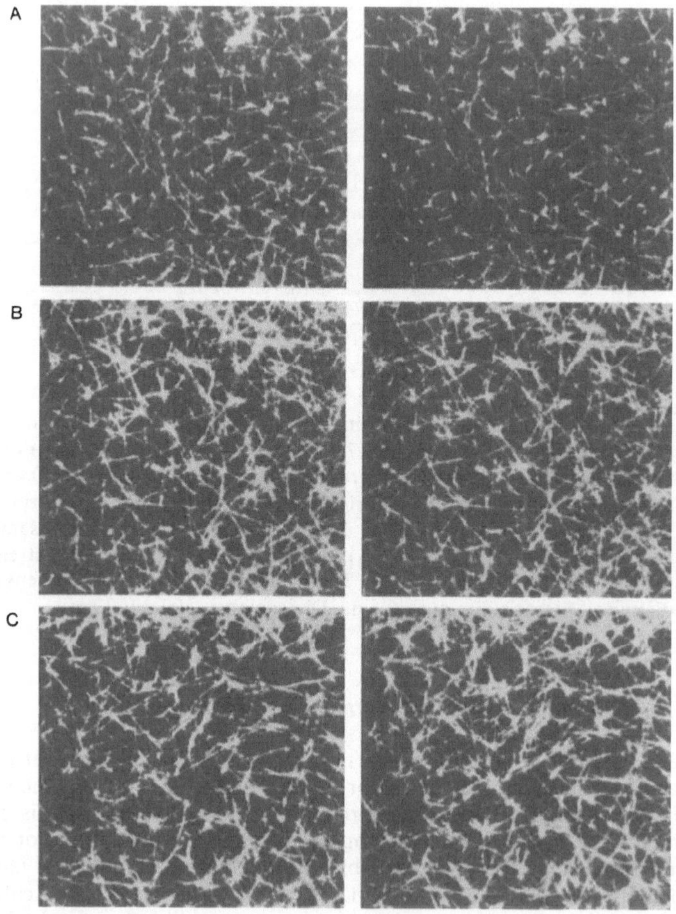

Fig. 5. 3D microscope pictures of fibrin gels formed in the absence and presence of albumin and dextran. A) Gel formation: in TNE-buffer, pH 7.4, containing $CaCl_2$ (at 20 mM) of ionic strength 0.21. Temperature, 24°C. Thrombin, 0.35 NIH units/mL. Fibrinogen, 1.87 mg/mL. B) Gel formation in presence of albumin (at 37 mg/mL). C) Gel formation in presence of dextran (at 9.3 mg/mL). Permeation: same as described in legend to Fig. 2. Stereo pairs are shown. Dimensions: 100 × 100 μm.

Table 5. FXIII Induced Gelation — Porosity, Fiber Mass-Length Ratio and Fiber Dimensions as Measured by Permeability and Turbidity

		Permeability			Turbidity	
Fbg mg/mL	C_t s	K_s $cm^2 \cdot 10^9$	μ $\cdot 10^{-12}$	ϕ μm	μ $\cdot 10^{-12}$	ϕ μm
3.00	405	1.4	17.6	0.137	35.1	0.291
2.50	345	1.8	18.1	0.138	30.2	0.274
1.50	435	11.2	56.1	0.245	27.9	0.268
1.20	201	16.5	60.7	0.255	24.7	0.248

Fibrinogen in TNE-buffer was clotted with factor XIII_a (at 0.4 units/mL) in the presence of $CaCl_2$ (at 20 mM), dithiothreitol (at 0.5 mM), hirudin (at 0.1 ATU/mL) and at ionic strength 0.21. Factor XIII (at 8 units/mL) was activated by thrombin (at 4 units/mL) for 60 min at room temperature. Thrombin was quenched by addition of Hirudin (at 16 ATU/mL). C_t, clotting time. μ, fiber mass-length ratio in dalton/cm. ϕ, average fiber diameter. For further explanations, see Materials, Methods and Text.

A

B

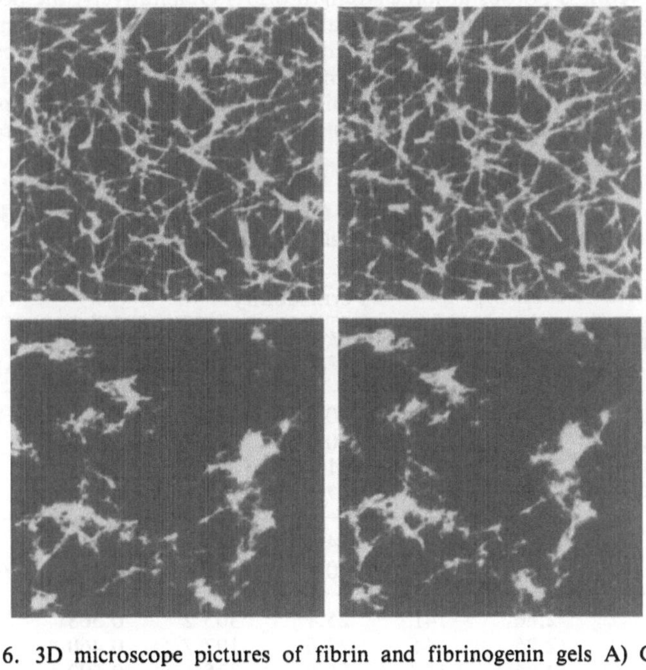

Fig. 6. 3D microscope pictures of fibrin and fibrinogenin gels A) Gel formation in presence of thrombin: in TNE-buffer, pH 7.4, containing $CaCl_2$ (at 20 mM) of ionic strength 0.15 at 24°C. Thombin, 0.35 NIH units/mL. Fibrinogen, 1.5 mg/mL. B) Gel formation in presence of factor XIIIa: in TNE-buffer, pH 7.4, containing $CaCl_2$ (at 20 mM) of ionic strength 0.16 at 24°C. Fibrinogen: 1.2 mg/mL. Factor XIIIa: 0.4 units/mL. Dithiothreitol: 0.5 mM. Factor XIII was activated as described in legend to Table 5. Permeation for A and B: same as described in legend to Fig. 2. Stereo pairs are shown. Dimensions: 100 × 100 μm. The gels were treated with *p*-phenylenediamine.

which are held together by an extremely fine network of fibers which in many places are not readily visible in the pictures. The calculation of fiber mass-length ratios based on turbidity and permeability data rests on the assumption of rod-like structures in the network. The microscope pictures show that this condition may not apply for the alternative gels. Therefore, the accuracy of the values for average mass-length ratios and average fiber dimensions as shown in Table 5 may be questioned.

Fibrin gel structures formed in normal plasma by thrombin

Citrated normal plasma, containing residual platelets (at about 10000/μm^3), and citrated normal platelet-poor plasma was clotted with thrombin (at 0.2 NIH units/mL) in the presence of added calcium chloride (at 20 mM) and at an ionic strength of 0.21. Complete conversion of fibrinogen appeared to occur under these conditions. The results obtained are shown in Table 6, A and B. As judged by permeability the gels formed in plasma (at comparable clotting times and fibrinogen concentrations) had much larger K_s and fiber mass-length ratios than those formed in a pure system (compare Table 2B and Table 6). It is obvious that the turbidity measurements in plasma gave lower values for fiber mass-length ratios and fiber diameters than the permeability measurements. However, the variation in values obtained by turbidity was much larger than the corresponding parameters measured by permeability (Table 6).

Examples of the normal plasma fibrin network as observed by microscopy is shown in Fig. 9A. The apparent porosity is larger than expected for a fibrin network formed in a pure thrombin-fibrinogen system at similar fibrinogen concentration and ionic strength. For the purpose of illustration one may compare Fig. 9A and Fig. 3A. (It should be noted that the experiment with isolated fibrinogen was performed at ionic strength 0.16. The network would be even less porous at ionic strength 0.21.) Qualitatively, the porosity observed by microscopy is in agreement with the permeability data. Also the fiber diameters measured in the micrscope agree with the permeability data, *i.e.* 0.44 and 0.52 μm, respectively.

The difference in results obtained by permeability and turbidity measurements in plasma is remarkable, since after addition of albumin (which is the major protein component in plasma) to the purified fibrinogen-thrombin system the two methods agreed rather well

Table 6. Fibrin Gels Formed in Plasma — Porosity, Fiber Mass-Length Ratio and Fiber Dimensions as Measured by Permeability and Turbidity

				Permeability			Turbidity	
	Plasma	Fbg mg/mL	C_t s	K_s $cm^2 \cdot 10^9$	μ $\cdot 10^{-12}$	ϕ μm	μ $\cdot 10^{-12}$	ϕ μm
A	B-1	2.74	102	20.7	237.8	0.505	88.2	0.741
	B-2	2.79	110	20.6	242.6	0.509	42.3	0.317
	B-3	2.07	102	31.1	243.5	0.511	19.8	0.317
	B-4	2.24	123	27.2	236.6	0.503	32.8	0.300
Mean	—	—	—	24.9	240.1	0.51	45.8	0.42
SD, %	—	—	—	20.8	1.4	0.7	65.0	51.3
B	B-1	2.86	141	25.1	305.2	0.568	285.5	0.741
	B-2	2.25	144	22.6	197.7	0.460	85.9	0.317
	B-3	2.36	144	35.9	335.6	0.599	72.9	0.300
	B-5	2.30	177	25.8	233.0	0.500	49.7	0.237
Mean	—	—	—	27.3	267.9	0.53	123.5	0.400
SD, %	—	—	—	21.5	23.7	11.9	88.3	57.9

Citrated blood was in A centrifuged at 2000 xg for 20 min. In B the plasma was recentrifuged at 10000 xg for 20 min to obtain platelet poor plasma. The plasmas were dialyzed against TNE-buffer containing trasylol (at 5 KIE/mL). The plasmas were recalcified and clotted with thrombin. Fbg, fibrinogen. C_t, clotting time. μ, fiber mass-length ratio in dalton/cm. ϕ, fiber diameter. For further explanation, see Materials, Methods and Text.

Table 7. Fibrin Gels Formed in Plasma — Effect of Removal of Lipoproteins on Permeability and Turbidity Parameters of Fibrin Gels

Plasma	Fbg mg/mL	C_t s	Permeability K_s $cm^2 \cdot 10^9$	μ $\cdot 10^{-12}$	ϕ μm	Turbidity μ $\cdot 10^{-12}$	ϕ μm
Normal plasma	2.37	45	27.81	261.5	0.529	18.1	0.166
Lipoprotein free	2.37	75	14.9	139.6	0.387	21.1	0.184

Citrated blood was centrifuged at 2000 xg for 20 min. To an aliquot of the plasma, KBr was added and the lipids and lipoproteins floated by centrifugation (see Materials and Methods). Both samples were dialyzed against TNE-buffer. After recalcification (at 20 mM $CaCl_2$), the plasma was clotted with thrombin (at 0.2 NIH units/mL) and at ionic strength 0.21. Fbg, fibrinogen, C_t, clotting time, μ, fiber mass-length ratio in dalton/cm. ϕ, fiber diameter. For further explanations, see Material, Methods and Text.

(*cf.* Table 4). The large porosity and fiber dimensions of plasma clots, as measured by permeability, can therefore only partly be explained by the presence of albumin in plasma. Most likely other components in plasma are also exerting an effect in the formation of porous fibrin gels in the plasma medium. In preliminary experiments we have shown that lipoproteins may be of importance, since permeability measurements of fibrin gels formed in lipoprotein-free plasma show significantly smaller K_s and fiber mass-length ratios than lipoprotein containing plasma (Table 7). The turbidity measurements, did not show any significant difference between these samples in regard of fiber mass-length ratios and fiber diameters (Table 7).

Fibrin gel structure in dysfibrinogenemia

We have begun an investigation of the fibrin gel structure in two patients with dysfibrinogenemia. One patient with dysfibrinogenemia Tampere (investigated in cooperation with dr Takala *et al.* (59)) suffered recurrent thromboembolic episodes. The other patient with dysfibrinogenemia Laconia (investigated in cooperation with dr Zacharski) also suffered thromboembolic episodes. Plasma and the isolated fibrinogen fraction of both patients yield fibrin gels with thrombin which are more transparent than normal ones (Fig. 7). In Table 8 are shown permeability and turbidity data for fibrin gels formed from purified fibrinogen in the presence of thrombin. In the permeability studies, the gels formed from

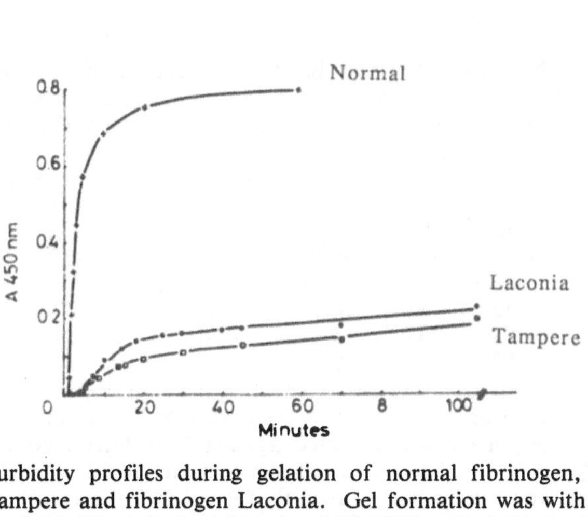

Fig. 7. Turbidity profiles during gelation of normal fibrinogen, fibrinogen Tampere and fibrinogen Laconia. Gel formation was with fibrinogen (at 1.5/mL) and thrombin (at 0.35 NIH units/mL) in TNE-buffer containing $CaCl_2$ (at 20 mM) and at ionic strength 0.21.

Table 8. Fibrin Gels Formed from Purified Fibrinogen Tampere and Fibrinogen Laconia — Porosity, Fiber Mass-Length Ratio and Fiber Diameter as Measured by Permeability and Turbidity

Fbg	Fbg mg/mL	C_t s	Permeability K_s $cm^2 \cdot 10^9$	Permeability μ $\cdot 10^{-12}$	Permeability ϕ μm	Turbidity μ $\cdot 10^{-12}$	Turbidity ϕ μm
Normal	2.00	114	3.5	26.1	0.167	37.7	0.156
Laconia	2.00	174	0.6	4.5	0.069	11.8	0.211
Tampere	2.00	177	1.0	7.2	0.087	12.6	0.240

Fibrinogen in TNE-buffer was clotted with thrombin (at 0.2 NIH units/mL) in the presence of $CaCl_2$ (at 20 mM) and at ionic strength 0.21. Fbg, fibrinogen. C_t, clotting time. μ, fiber mass-length ratio in dalton/cm. ϕ, fiber diameter. For further explanation, see Materials, Methods and Text.

the abnormal fibrinogens were much less porous and the fiber mass-length ratio and fiber diameters were also much smaller than for normal gels. The turbidity measurements showed about the same relative decrease in fiber mass-length ratio but the average fiber diameter was increased in the abnormal gels.

When citrated plasma was clotted with thrombin in presence of excess calcium the difference in clot structure was also obvious according to permeability measurements (Table 9). The turbidity data showed relatively less difference between the normal and abnormal samples in regard of fiber mass-length ratios. In comparison with normal plasma the fiber diameter showed only a small diminution in fibrinogen Tampere and in the case of fibrinogen Laconia there was actually an increase in fiber diameter.

The microscope pictures of the gels from the abnormal fibrinogens confirm that the gels formed from isolated fibrinogen (Fig. 8) or in plasma (Fig. 9) are less porous than fibrin gels from normal fibrinogen or plasma. It should also be noted that the gels from the abnormal fibrinogens (especially from fibrinogen Laconia) in comparison with normal fibrinogen display strands of a knotted and somewhat fuzzy appearance. There seems to be a larger number of node-like structures and these have often an irregular lump-like appearance (somewhat resembling the lumps in fibrinogenin gels). There also appears to be a larger variation in strand widths with some of the finest protruding from knots on the

Table 9. Fibrin Gels Formed in Normal and Dysfibrinogenemic Plasma — Porosity, Fiber Mass-Length Ratio and Fiber Diameters as Measured by Permeability and Turbidity

Plasma	Fbg mg/mL	C_t s	Permeability K_s $cm^2 \cdot 10^9$	Permeability μ $\cdot 10^{-12}$	Permeability ϕ μm	Turbidity μ $\cdot 10^{-12}$	Turbidity ϕ μm
Normal	2.83	81	24.1	289.0	0.555	22.5	0.167
Tampere	1.94	137	5.8	41.4	0.210	13.6	0.152
Laconia	3.80	97	0.1	1.6	0.040	1.7	0.267

Citrated blood was centrifuged at 2000 xg for 20 min and plasma secured. The plasma was dialyzed against TNE-buffer containing trasylol (at 5 KIE/mL). After recalcification (at 20 mM $CaCl_2$), the plasma was clotted with thrombin (at 0.2 NIH units/mL) at an ionic strength of 0.21. Fbg, fibrinogen. C_t, clotting time. μ, fiber mass-length ratio in dalton/cm. ϕ, fiber diameter. For further explanation, see Materials, Methods and Text.

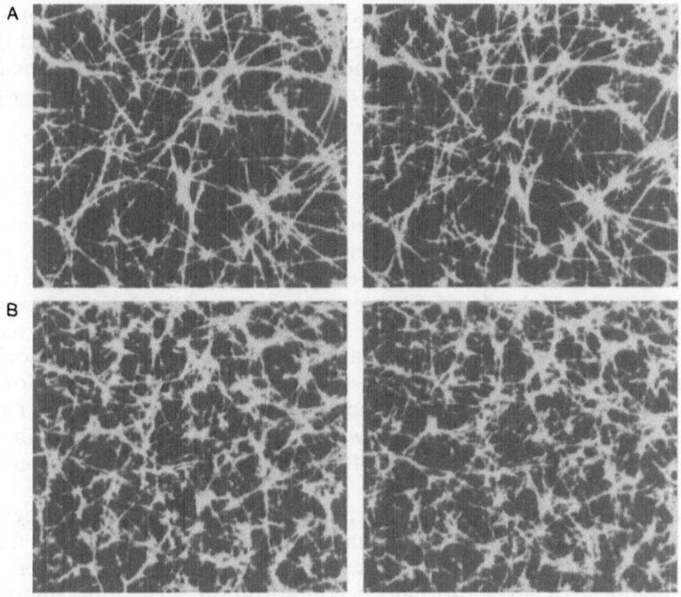

Fig. 8. 3D microscope pictures of fibrin gels formed from normal fibrinogen and fibrinogen Tampere. Gel formation: in TNE-buffer, pH 7.4, containing $CaCl_2$ (at 20 mM) of ionic strength 0.21 at 23°C. Fibrinogen, 1.0 mg/mL. Thrombin, 0.18 NIH units/mL. Permeation: same as described in legend to Fig. 2. A) Normal fibrinogen. B) Fibrinogen Tampere. Stereo pairs are shown. Dimensions: 100×100 μm. The gels were treated with *p*-phenylenediamine.

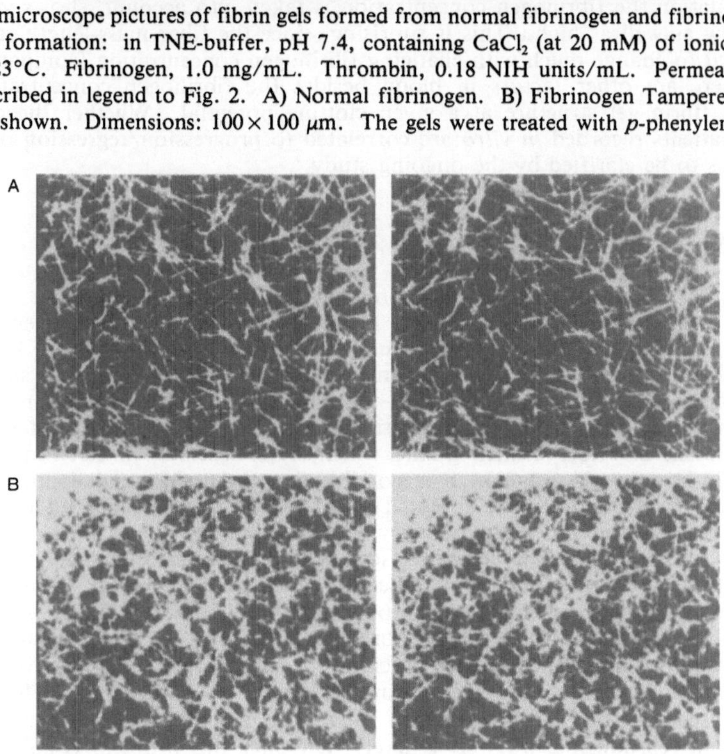

Fig. 9. 3D microscope pictures of fibrin gels formed in normal and dysfibrinogenemic plasma. Gel formation: plasma was dialyzed against TNE-buffer, pH 7.4, containing trasylol (at 5 KIE/mL). $CaCl_2$ was added (at 20mM) and clotting induced by adding thrombin (at 0.18 NIH units/mL). Ionic strength, 0.21. Temperature, 24°C. Permeation: same as described in legend to Fig. 2. A) Normal plasma. Fibrinogen, 2.8 mg/mL. Clotting time, 81 s. B) Plasma of dysfibrinogenemia Laconia. Fibrinogen, 3.8 mg/mL. Clotting time, 97 s. Stereo pairs are shown. Dimensions: 100×100 μm. The gels were treated with *p*-phenylenediamine.

strands. The average strand widths in plasma gels as measured in the microscope, were not significantly thinner in the dysfibrinogenemias (*i.e.* 0.44 μm for normal and 0.41 and 0.46 for Tampere and Laconia, respectively). Since the porosity (K_s) was drastically decreased this means that there must also exist an extremely fine network of fibers which is either not visible or being over-estimated in the microscope analysis.

Fibrin gel structure in ischaemic heart disease

A prospective cohort study on disease progression/regression in individuals with a first heart infarction before the age of 45 was begun in Stockholm in 1980 (60, 61). Some of the individuals in this study have elevated levels of fibrinogen in plasma. It is known from other studies that increased levels of fibrinogen in plasma show an indepent association with development of ischaemic heart disease (IHD) (36, 38-40). We have recently started a study on the fibrin gel structure in the plasma of the patients of the Stockholm study. We want to know whether there exists an association between the formation of a certain type of gel structure *in vitro* and disease progression/regression *in vivo.* Table 10 shows permeability data on a few of the patients with various fibrinogen levels in their plasma. In these experiments citrated plasma was clotted with thrombin in the presence of calcium and at ionic strength 0.21.

The gel structure in the patients shows remarkable differences with regard to K_s and fiber mass-length ratio. In general there is an inverse relationship between fibrinogen concentration and the porosity of the gels as reflected in the K_s values. However, the Ks values fluctuate considerably at a given fibrinogen concentration. The fiber mass-length ratio, in which calculation the fibrinogen concentration is taken into account, shows values both abnormally low as well as high. This is suprising, since the fiber mass-length ratio would not be expected to change much with changing fibrinogen concentration (compare Table 2) Obviously, there are other factors in blood besides the fibrinogen concentration which determine the fibrin gel structure at a given clotting potential. Whether the differences between the patients recorded *in vitro* are correlated to progression/regression of the IHD disease remains to be clarified by the ongoing study.

DISCUSSION

The network structures, formed either in a purified fibrinogen-thrombin system or in plasma were, as observed in the microscope, composed of straight rod-like elements which often came together at denser nodes. The nature of the nodes is still elusive. It is unlikely that these nodes are the result of fiber crossings or fiber branch points, since such features are in most instances easily discriminated in the images. In general, three or more fibers protrude from these nodes in different directions in space. For this reason it is less likely that they represent the fiber branch points which were observed in electron-microscopic picutures of fibrin networks (17-21). It is possible that the nodes result from the twisting of fibers around each other as was observed by Müller *et al.* (20), and this could explain the apparent tri- or higher order functionality of the nodes.

The permeation and turbidometric analyses of the gels provide quantitative information on the porosity of the gels, the fiber mass-length ratio and the average strand width and thus supplement the qualitative information obtained by microscopy. There is good agreement between the K_s values, which are proportional to the surface area available for flow, and the assessment of apparent porosity observed by microscopy. Quantitative assessment of the pore size, node density, and the existing geometries by microscopy suffers from the difficulty to discriminate between objects of interest, *e.g.* nodes, and the environment in an unbiased and reproducible way. It is likely that once this problem is solved, the unbiased counting rules for particle density in biological tissues as suggested by Howard *et al.* (62) can be used for quantitation of nodes in the network. The optical resolution applied in this work and in a previous communication (47) is insufficient to resolve strands at or below the dimensions suggested by the permeability and turbidity data for fibers formed in a pure fibrinogen-thrombin system (Table 1-3). However, most of the fibers which are formed in plasma are much wider and these can most likely be measured in the images. The use of antifading agents may also make it possible to use a smaller pixel size (*i.e.* 0.1 μm instead of the present 0.2 μm) in future analyses of fiber dimensions. This will allow measurements of somewhat thinner fibers than is now possible.

Table 10. Permeability Parameters of Fibrin Gels Formed in Plasma of Patients with Ischaemic Heart Disease

Plasma	Fbg mg/mL	K_s cm^2•10^9	μ •10^{-12}	ϕ μm
Normal	2.46 ± 0.3	26.9 ± 5.9	266.6 ± 55.1	0.533 ± 0.05
H-204	2.50	16.8	169.6	0.426
H-210	2.60	14.4	154.0	0.405
H-286	2.60	20.4	218.3	0.484
H-304	4.60	8.7	202.2	0.466
H-358	4.60	13.0	303.7	0.570
H-228	5.70	10.4	326.1	0.591
H-162	7.00	9.2	384.0	0.642

Citrated plasma was prepared and clotted with thrombin as described in Table 9. Fbg, fibrinogen, μ, fiber mass-length ratio in Dalton/cm. ϕ, fiber diameter. For further explanation, see Materials, Methods and Text.

The structures we observe by confocal microscopy of hydrated gel structures show similarities but also many dissimilarities with network structures observed by electron microscopy (17-21). Electron microscopy requires the removal of water and handling procedures that inevitably compact and/or distort fibrin gel structures, which usually contain more than 99% of water. Therefore, what were once originally straight rod-like elements will inevitably become bent, twisted, and coalesced structures. The gel networks we visualize are supported by the water captured in the network and therefore most likely represent native structures. In agreement with results obtained by electron microscopy (19, 21), the networks we observe appear to have a considerable variation in strand widths, as based on the variation in fluorescence intensities. Weisel *el al.* (7, 8) reported that a rather homogeneous population of fibers existed in fibrin gel matrices. In dried fibrin preparations the fibers had an average diameter of about 100 nm. Bundles of fibers of varying widths were also observed. The fact that most of the fiber structures we visualize have dimensions below the optical resolution of the microscope makes it impossible to estimate the actual strand widths in the network. However, it appears that there are at least two classes of fibers with different diameters, which may represent the bimodal fiber system described by Shah *et al.* (19). It is also possible that the fainter structures represent the twisted fibers shown by Weisel (8) and that the brighter fibers represent the bundles.

Our present and previous (47) studies show that, at a constant ionic strength and calcium ion concentration, the gel structure is determined by kinetic factors, *i.e.* the clotting potential (thrombin concentration) and the fibrinogen concentration. Increasing the thrombin concentration leads to decrease in porosity, fiber mass-length ratio and strand width as calculated both from permeation and turbidity data. Our analyses show that the porosity of the gels decreases logarithmically with increasing fibrinogen concentration. However, there are only small changes in mass-length ratio and fiber width. The concentration range of fibrinogen in our study covers both concentrations below and above the K_m for the fibrinogen-thrombin interaction with regard to release of fibrinopeptide A (FPA) (*i.e.*, 3-10 μM). On increasing the fibrinogen concentration one would expect the affinity between thrombin and fibrinogen to increase. This would therefore have the same effect as increasing the thrombin concentration, *i.e.* the porosity, fiber mass-length ratio and fiber width would decrease. The fiber mass-length ratio and the fiber width are, however, less affected since the fibrinogen concentration was also increasing. The permeability and microscopic images support this interpretation, which is also in reasonable agreement with the turbidity data in a limited concentration range of fibrinogen.

The permeation and turbidity measurements showed that in changes in ionic strength have a dramatic and interesting effect on the gel structure. In the ionic strength range 0.15-0.24, the porosity and fiber mass-length ratio decreased with increasing ionic strength and this was also supported by the microscope pictures. The effect of ionic strength was similar to that seen when the thrombin concentration was increased. This may suggest an increase in thrombin activity with increasing ionic strength in this range. However, previous studies (63) suggested that the activation of fibrinogen by thrombin rather decreased with increasing ionic strength. When the ionic strength was increased above 0.24, the microscope pictures and turbidity data suggested that swelling of the fiber strands occurred. The fibers became shorter and less compact, the contours acquired a fuzzy appearance, giving the impression of an increase in strand width. The swelling is most likely caused by a decrease in protein-protein interaction which results in a loosening up of the fibers with otherwise little effect on the noded architecture. The strand width as deduced from permeability measurements did not indicate any swelling at ionic strengths above 0.24. However, strand widths as derived from permeability data are mathematical transformations of K_s, with an assumption of a constant degree of hydration. This is most likely not the case, since Voter *et al.* (64) have demonstrated an appreciable increase in water content of fibrin fibers with increasing ionic strength. The decrease in flow at increasing ionic strength is thus due not only to the increase in the tightness of the fiber network, as was the case with inceasing thrombin concentration, but also to the swelling of fibers, which will dramatically restrict the surface available to flow. Our findings regarding the gel structure at high ionic strengths appear contrary to the classical concept by Ferry *et al.* and reiterated in later reports (9, 10, 65, 66). This concept states that at high ionic strength the width of the fibers decreases and their length increases.

There are also factors that have an effect on gel formation without interfering with the kinetics, *i.e.* the activation of fibrinogen. As was shown here, such compounds are albumin and dextran. The shortening of the clotting time in presence of these compounds is not due to faster release of fibrinopeptides (25, 26) but to the fact that a coarser network is formed and less activation is needed for a coarse network to form than for a fine one (13). Albumin and specially dextran may exert their effects by exclusion of water from the reacting molecules as was shown for other enzyme-substrate systems in presence of dextran (67). We assume that fibrinogen and to some extent thrombin are excluded from the water phases containing albumin or dextran. This leaves less space in a given volume element where the thrombin-fibrinogen reaction can occur and consequently a coarser network structure is formed. The gel structure can, as we have shown previously (24, 68), also be influenced by binding of components to fibrinogen and/or fibrin during the polymerization and gelation process. Example of such components are fibronectin and calcium ions.

As judged from permeability data, much coarser gel structures were formed in plasma in comparison with those in a purified fibrinogen-thrombin system. This is partly explained by the presence of albumin in plasma. However, since the gels formed after addition of albumin to purified fibrinogen only partly mimicks the plasma clots, other components in plasma also appear to play a role in the development of the coarse plasma clot. Fibronectin is incorporated in the fibrin network and is therefore certainly among those factors influencing the strand structure of a plasma clot (68). Fibronectin, however, is most likely not responsible for the large porosity of the plasma clot, since it has only a moderate effect on the permeability of fibrin gels (68). As we have shown here also lipoproteins may play a role in the development of the clot structure in plasma and there may be other components, so far unknown, which participate in the development of the plasma clot architecture.

It is puzzling that the fiber mass-length ratios and fiber dimensions in plasma, as calculated from turbidity data, gave different values from those obtained by permeation measurements. Furthermore, removal of lipoproteins affected the permeation measurements but not the turbidity data. These discrepancies are the more remarkable, since the measured increase in fiber coarseness after addition of albumin or dextran to a purified fibrinogen-thrombin system is about the same with both methods (*cf.* Table 4). It is possible that the larger fiber mass-length ratios obtained from permeability data of plasma clots are artificially high, which would be the case if the fiber hydration was lower in plasma gels than in gels from purified fibrinogen. However, this is somewhat unlikely since addition of large amounts of albumin and dextran (which would influence the water distribution in the system) affected permeability and turbidity parameters in a similar fashion. Also, recalculation of the fiber mass-length ratio using the lower hydration value determined for fibrin (2 g/g protein) (69) still gave values of fiber dimensions larger than those obtained by turbidity measurements. Since the average strand widths as calculated from permeability data agreed with those obtained by microscopy, we suggest that the theory for fiber dimensions based on turbidity data (29, 55) may not apply to a plasma clot where not only the fibrin network is present but also a complex mixture of plasma proteins and other constituents.

The propensity to develop abnormal fibrin gel structures in dysfibrinogenemic plasma *in vitro* was associated with a thromboembolic disease *in vivo*. It is possible that there exists a causal connection between the two phenomena. The formation of the abnormal gel structures *in vitro* is most likely due to an inherent abnormality in the fibrinogen molecule leading to abnormal polymerization and gelation. The microscope pictures of the abnormal fibrin gels displayed a grave disturbance in clot architecture. Extremely tight network structures are seen in which coarse knotted fiber strands coexist with finer strands. We hypothesize that these abnormal clot structures are thrombogenic when developed in the vasculature *in vivo*. The heterogeneity in fiber sizes may explain why the measurements of fiber dimensions by permeability and turbidity (Table 8) gave so different results. In a heterogeneous mixture of fine and coarse fibers the turbidity measurements will overestimate the proportion of coarse strands (19), whereas in permeability measurement the fine networks is the limiting factor for flow and this will be reflected in the estimation of average strand width (19).

Recent epidemiological studies have shown that moderatly increased levels of fibrinogen and/or factor VII activity carries with it an increased risk for development of IHD or stroke (36-40). Both factors showed in several studies an independent association with IHD (36, 38-40). In one study (70) the red cell sedimentation rate (ESR) appeared to be a risk factor for IHD. The increased ESR also indirectly links fibrinogen to the process. Our finding that the structure of the fibrin gel is determined by the clotting potential and the fibrinogen concentration acting in concert, poses the question whether the common denominator for factor VII activity (clotting potential) and fibrinogen concentration as risk factors in IHD is the architecture of an established clot network rather than the mass of fibrin deposited. In a fibrinogen-thrombin system it is the water kept in the gel network that has the volume filling capacity and the resiliency of the gel structure would determine the occlusive power of such a structure. Tight network structures with small pores and thin fiber strands are rigid and brittle and rather break than give off the water trapped in the clot. An increase in the clotting potential or fibrinogen concentration favours formation of these types of clots as shown in this report. Elevated levels of factor VII activity and/or fibrinogen *in vivo* might favour formation of these tight, rigid and space filling clots at sites of lesions in the vasculature and thereby increase the risk for thromboembolic diseases. The above considerations are the rationale behind our ongoing IHD-study in which the gel structure formed under standardized conditions in the plasma of the patients is studied. The preliminary data suggest that in contrast to healthy individuals a great variation in fibrin gel structures may be found in the patients and this trail of research is therefore worthwhile to continue. Our hypothesis is that patients with a propensity to form tight and rigid gel strucures *in vitro* may be at greater risk for progression of IHD *in vivo*. This may be associated not only with an increase in fibrinogen concentration and clotting potential. There exist several modulators in plasma, which at a given clotting potential and fibrinogen concentration may tilt gel formation either towards tight and rigid gels or coarse and plastic ones. The latter clots may be less thrombogenic since they would give way in flow and coalesce with the vessel wall.

The present dogma in haemostasis is that thrombin is solely responsible for clot formation. Copley (71-73) has in rheological studies demonstrated formation of fibrinogen gels at interfaces and proposed a physiological significance for a lining on the endothelium of fibrin and/or fibrinogen gels. Other studies (41-45) have also shown that gel formation can occur without participation of thrombin. In our studies, factor XIII (especially factor XIIIa) was shown to induce gelation of fibrinogen (44, 45). In this report we demonstrate that the fibrinogen gels formed in presence of factor XIIIa have a completely different architecture as compared to those formed with thrombin. The physiological or pathophysiological role of this alternative clotting pathway is elusive. In instances when the thrombin clotting potential is high this pathway is certainly of minor importance *in vivo*. However, the alternative pathway may be of importance when the thrombin clotting potential is low, provided that an excess of activated factor XIII is available. Thrombin can activate factor XIII and in fact early fibrin polymers facilitate this activation (74, 75). There are also other enzymes which have been shown to activate factor XIII, one being the intracellular enzyme calpain (76), which may be released from damaged cells. In dysfibrinogenemia Aarhus, thrombin induced clotting is almost completely abolished, still no bleeding manifestations occur *in vivo*. Factor XIII induced gelation of fibrinogen Aarhus is normal or almost normal (77). It is possible that the alternative pathway in this case is responsible for the observed unimpaired haemostasis of the patient with dysfibrinogenemia Aarhus.

In conclusion our studies have shown that studies of fibrin gel structure by a combination of several physico-chemical techniques under a variety of conditions can give detailed information on this structure. Furthermore, the architectural quality of the fibrin gel may have clinical correlates.

ACKNOWLEDGMENT

This work was supported by grants from the National Institutes of Health, Bethesda (HL27279-09), The Swedish Medical Research Council (B89-13X-02475-22C and B89-19X-08691-01), The Swedish Heart-Lung Foundation and The Tornspiran Foundation, Stockholm, Sweden. We thank Dr. Paul Appleton, Lakes Region General Hospital, Laconia, New Hampshire, for providing the plasma samples of dysfibrinogenemia Laconia. The excellent technical assistance by Lisbeth Therkildsen, Gerd Sjöberg and Birgitta Strimme is gratefully acknowledged.

REFERENCES

1. R. F. Dollittle, Fibrinogen and Fibrin, *Ann. Rev. Biochem.,* **53:**195-229 (1984).
2. B. Blombäck, Fibrinogen to fibrin — An overview, *in:* "Fibrinogen-Structural Variants and Interactions, A. Henschen, B. Hessel, J. McDonagh and T. Saldeen, eds., Walter de Gruyter and Co., Berlin-New York. (1985).
3. G. R. Crabtree, The molecular biology of fibrinogen, *in:* "The Molecular Basis of Blood Diseases", G. Stamatoyannopoulos, A. W. Nienhuis, P. Leder and P. W. Majerus, eds., W.B. Saunders Co., Philadelphia (1987).
4. W. Krakow, G. F. Endres, B. M. Siegel and H. A. Scheraga, An electron microscopic investigation of the polymerization of bovine fibrin monomer, *J. Mol. Biol.,* **71:**95-103 (1972).
5. R. C. Williams, Morphology of fibrinogen monomers and of fibrin protofibrils, *Ann. N.Y. Acad. Sci.,* **408:**180-193 (1983).
6. J. Hermans, Models of fibrin, *Proc. Natl. Acad. Sci. USA,* **76:**1189-1193 (1979).
7. J. W. Weisel, Fibrin assembly. Lateral aggregation and the role of the two pairs of fibrinopeptides, *Biophys. J.,* **50:**1079-1093 (1986).
8. J. W. Weisel, C. Nagaswami, and L. Makowski, Twisting of fibrin fibers limits their radial growth, *Proc. Natl. Acad. Sci. USA,* **84:**8991-8995 (1987).
9. J. D. Ferry and P. R. Morrison, Preparation and properties of serum and plasma proteins — VIII. The conversion of human fibrinogen to fibrin under various conditions, *J. Am. Chem. Soc.,* **69:**388-400 (1947).
10. M. E. Carr, Jr., L. L. Shen and J. Hermans, Mass-length ratio of fibrin fibers from gel permeation and light scattering, *Biopolymers,* **16:**1-15 (1977).
11. G. A. Shah, C. H. Nair and D. P. Dhall, Physiological studies on fibrin network structure, *Thromb. Res.,* **40:**181-188 (1985).
12. G. A. Shah, C. H. Nair and D. P. Dhall, Comparison of fibrin networks in plasma and fibrinogen solution, *Thromb. Res.,* **45:**257-264 (1987).
13. B. Blombäck and M. Okada, Fibrin gel structure and clotting time, *Thromb. Res.,* **25:**51-70 (1982).
14. W. H. Howell, The clotting of blood as seen with the ultramicroscope, *Am. J. Physiol.,* **35:**143-149 (1914).
15. J. M. Buchanan, L. B. Chen, T. Hamazaki, E. Lenk and D. F. Waugh, The early development of fibrin clot structure, *in:* "Chemistry and Biology of thrombin", R. L. Lundblad, J. W. Fenton, II and K. G. Mann, eds., Ann Arbor Science Publishers Inc., New York (1977).
16. B. Blombäck, M. Okada, B. Forslind, and U. Larsson, Fibrin gels as biological filters and interfaces, *Biorheology,* **21:**93-104 (1984).
17. C. v. Z. Hawn and K. R. Porter, The fine structure of clots formed from purified bovine fibrinogen and thrombin: A study with the electron microscope, *J. Exp. Med.,* **86:**285-292 (1947).
18. K. R. Porter, and C. v. Z. Hawn, Sequences in the formation of clots from purified bovine fibrinogen and thrombin: A study with the electron microscope, *J. Exp. Med.,* **90:**225-232 (1949).

19. G. A. Shah, I. A. Ferguson, T. Z. Dhall and D. P. Dhall, Polydispersion in the diameter of fibers in fibrin networks: Consequences on the measurement of mass-length ratio by permeability and turbidity, *Biopolymers,* **21**:1037-1047 (1982).
20. M. F. Müller, H. Ris. and J. D. Ferry, Electron microscopy of fine fibrin clots and fine and coarse fibrin films, *J. Mol. Biol.,* **174**:369-384 (1984).
21. M. W. Mosesson, J. P. DiOrio, M. F. Müller, J. R. Shainoff, K. R. Siebenlist, D. L. Amrani, G. A. Homandberg, J. Soria, C. Soria, and M. Samama, Studies on the ultrastructure of fibrin lacking fibrinopeptide B (β-fibrin), *Blood,* **69**:1073-1081 (1987).
22. G. Marguerie, G. Chagniel, and M. Suscillon, The binding of calcium to bovine fibrinogen, *Biochim. Biophys. Acta,* **490**:94-103 (1977).
23. B. Blombäck, M. Blombäck, T. C. Laurent and H. Pertoft, Effect of EDTA on fibrinogen, *Biochim. Biophys. Acta,* **127**:560-562 (1966).
24. M. Okada and B. Blombäck, Calcium and fibrin gel structure, *Thromb. Res.,* **29**:269-280 (1983).
25. U. Abildgaard, Acceleration of fibrin polymerization by dextran and ficoll. Interaction with calcium and plasma proteins, *Scand. J. Clin. Lab. Invest.,* **18**:518-524 (1966).
26. M. Okada, B. Blombäck, and M. Block, Effect of albumin and dextran on fibrin gel structure, *Thromb. Haemostas.,* **50**:185 (1983).
27. M. E. Carr, and D. A. Gabriel, The effect of dextran 70 on the structure of plasma-derived fibrin gels, *J. Lab. Clin. Med.,* **96**:985-993 (1980).
28. O. Tangen, K. O. Wik, I. A. M. Almquist, K.-E. Arfors and H.C. Hint, Effects of dextran on the structure and plasmin-induced lysis of human fibrin, *Thromb. Res.,* **1**:487-492 (1972).
29. M. E. Carr and D. A. Gabriel, Dextran-induced changes in fibrin fiber size and density based on wavelength dependence of gel turbidity, *Macro-molecules,* **13**:1473-1477 (1980).
30. C. Southan, Molecular and genetic abnormalities of fibrinogen, *in:* Fibrinogen, Fibrin Stabilisation and Fibrinolysis J. L. Francis, ed., E. Horwood Ltd., Chichester, England (1988).
31. M. Blombäck, B. Blombäck, E. F. Mammen and A. S. Prasad, Fibrinogen Detroit — A molecular defect in the *N*-terminal disulphide knot of human fibrinogen? *Nature,* **218**:134-137 (1968).
32. F. Ni, Y. Konishi, L. D. Bullock, M. N. Rivetna, and H. A. Scheraga, High-resolution NMR studies of fibrinogen-like peptides in solution: Structural basis for the bleeding disorder caused by a single mutation of Gly(12) to Val(12) in the Aα chain of human fibrinogen Rouen, *Biochemistry,* **28**:3106-3119 (1989).
33. C. Y. Liu, J. A. Koehn and F. J. Morgan, Characterization of fibrinogen New York 1, *J. Biol. Chem.,* **260**:4390-4396 (1985).
34. N. Carrell, D. A. Gabriel, P. M. Blatt, M. E. Carr, and J. McDonagh, Hereditary dysfibrinogenemia in a patient with thrombotic disease, *Blood,* **62**:439-447 (1983).
35. S. E. Humphries, M. Cook, M. Dubowitz, Y. Stirling and T. W. Meade, Role of genetic variation at the fibrinogen locus in determination of plasma fibrinogen concentrations, *The Lancet,* **i**:1452-1455 (1987).
36. T. W. Meade, S. Mellows, M. Brozovic, G. J. Miller, R. R. Chakrabarti, W. R. S. North, A. P. Haines, Y. Stirling, J. D. Imeson and S. G. Thompson, Haemostatic function and ischaemic heart disease: Principal results of the Northwick Park heart study, *The Lancet,* **ii**:533-537 (1986).
37. L. Wilhelmsen, K. Svärdsudd, K. Korsan-Bengtsen, B. Larsson, L. Welin and G. Tibblin, Fibrinogen as a risk factor for stroke and myocardial infarction, *New Engl. J. Med.,* **311**:501-505 (1984).
38. M. C. Stone and and J. M. Thorp, Plasma fibrinogen — A major coronary risk factor, *J. Royal College of Gen. Practitioners,* **35**:565-569 (1985).
39. H. L. Markowe, M. G. Marmot, M. J. Shipley, C. J. Bulpitt, T. W. Meade, Y. Stirling, M. V. Vickers and A. Semmence, Fibrinogen: a possible link between social class and coronary heart disease, *Brit. Med. J.,* **291**:1312-1314 (1985).
40. W. B. Kannel, W. P. Castelli and S. L. Meeks, Fibrinogen and cardiovascular disease, *J. Am. Coll. Card.,* **5**:517 (1985).
41. J. (Brunner) Lorand, T. Urayama and L. Lorand, Transglutaminase as a blood clotting enzyme, *Biochem. Biophys. Res. Commun.,* **23**:828-834 (1966).
42. B. Ly, P. Kierulf and E. Jakobsen, Stabilization of soluble fibrin/fibrinogen complexes by fibrin stabilizing factor (FSF), *Thromb. Res.,* **4**:509-522 (1974).

43. H. Kanaide and J. R. Shainoff, Cross-linking of fibrinogen and fibrin by fibrin-stabilizing factor (factor XIIIa), *J. Lab. Clin. Med.*, **85:**574-597 (1975).
44. B. Blombäck, R. Procyk, L. Adamson and B. Hessel, FXIII induced gelation of human fibrinogen — An alternative thiol enhanced, thrombin independent pathway, *Thromb. Res.*, **37:**613-628 (1985).
45. R. Procyk and B. Blombäck, Factor XIII-induced crosslinking in solutions of fibrinogen and fibronectin, *Biochim. Biophys. Acta,* **967:**304-313. (1988).
46. R. Procyk, L. Adamson, M. Block and B. Blombäck, Factor XIII catalyzed formation of fibrinogen-fibronectin oligomers — A thiol enhanced process, *Thromb. Res.*, **40:**833-852 (1985).
47. B. Blombäck, K. Carlsson, B. Hessel, A. Liljeborg, R. Procyk and N. Åslund, Native fibrin gel networks observed by 3D microscopy, permeation and turbidity, *Biochim. Biophys. Acta,* **997:**96-110 (1989).
48. L. Lorand and T. Gotoh, Fibrinoligase — The fibrin stabilizing factor system, *Methods in Enzymology,* **19:**770-782 (1970).
49. M. S. Brown, S. E. Dana and J. L. Goldstein, Regulation of 3-hydroxy-3-methylglutaryl coenzyme A reductase activity in cultured human fibroblasts, *J. Biol. Chem.*, **249:**789-796 (1974).
50. R. Signer and H. Egli, Sedimentation von Makromolekülen und Durchströmung von Gelen, *Recueil,* **69:**45-58 (1950).
51. H. A. Scheraga and M. Laskowski, Jr., The fibrinogen-fibrin conversion, *Adv. Protein Chem.*, **12:**1-131 (1957).
52. G. Marguerie and H. B. Stuhrmann, A neutron small-angle scattering study of bovine fibrinogen, *J. Mol. Biol.*, **102:**143-156 (1976).
53. K. Carlsson and A. Liljeborg, A confocal laser microscope scanner for digital recording of optical serial sections, *J. Microscopy,* **153:**171-180 (1989).
53a. T. Wilson and C. J. R. Sheppard, Theory and Practice of Scanning optical microscopy, Academic Press, London (1984).
54. J. L. Platt and A. F. Michael, Retardation of fading and enhancement of intensity of immunofluorescence by *p*-Phenylenediamine, *J. Histochem. Cytochem.*, **31:**840-842 (1983).
55. M. E. Carr, Jr. and J. Hermans, Size and density of fibrin fibers from turbidity, *Macromolecules,* **11:**46-50 (1978).
56. B. Blombäck and M. Blombäck, Purification of human and bovine fibrinogen, *Arkiv Kemi,* **10:**415-443 (1957).
57. U. K. Laemmli, Cleavage of structural proteins during the assembly of the head of bacteriophage T4, *Nature,* **227:**680-685 (1970).
58. M. Okada and B. Blombäck, Factors influencing fibrin gel structure studied by flow measurements, *Ann. N.Y. Acad. Sci.*, **408:**233-253 (1983).
59. T. Takala, H. Oksa, V. Rasi and R. Tuimala, Dysfibrinogenemia associated with thrombosis and third trimester fetal loss. A case report, *Thromb. Res.*, In press (1990).
60. A. Hamsten, U. de Faire, G. Walldius, G. Dahlén, A. Szamosi, C. Landou, M. Blomäck and B. Wiman, Plasminogen activator inhibitor in plasma: Risk factor for recurrent myocardial infarction, *The Lancet,* **ii:** 3-9 (1987).
61. A. Hamsten, M. Blombäck, B. Wiman, J. Svensson, A. Szamosi, U. de Faire and L. Mettinger, Haemostatic function in myocardial infarction, *Brit. Heart J.*, **55:**58-66 (1986).
62. V. Howard, S. Reid, A. Baddeley and A. Boyde, Unbiased estimation of particle density in the tandem scanning reflected light microscope, *J. Microscopy,* **138:**203-212 (1985).
63. B. Blombäck, Studies on the action of thrombic enzymes on bovine fibrinogen as measured by *N*-terminal analysis, *Arkiv Kemi,* **12:**321-335 (1958).
64. W. A. Voter, C. Lucaveche and H. P. Erickson, Concentration of protein in fibrin fibers and fibrinogen polymers determined by refractive index matching, *Biopolymers,* **25:**2375-2384 (1986).
65. G. W. Nelb, C. Gerth, J. D. Ferry and L. Lorand, Rheology of fibrin clots. III. Shear creep and creep recovery of fine ligated and coarse unligated clots, *Biophys. Chem.*, **5:**377-387 (1976).
66. R. R. Hantgan and J. Hermans, Assembly of fibrin — A light scattering study, *J. Biol. Chem.*, **254:**11272-11281 (1979).
67. T. C. Laurent, Enzyme reactions in polymer media, *Eur. J. Biochem.*, **21:**498-506 (1971).

68. M. Okada, B. Blombäck, M.-D. Chang and B. Horowitz, Fibronectin and fibrin gel structure, *J. Biol. Chem.*, **260**:1811-1820 (1985).
69. U. Larsson, R. Rigler, B. Blombäck, K. Mortensen and R. Bauer, Polymerisation of fibrinogen to fibrin studied by time-resolved small angel neutron scattering, *in:* Springer Series in Biophysics, Structure, Dynamics and Function of Biomolecules, A. Ehrenberg, R. Rigler, A. Gräslund and L. Nilsson, eds., Springer Verlag, Heidelberg (1987).
70. L. A. Carlson, L. E. Böttiger and P. E. Ahfeldt, Risk factors for myocardial infarction in the Stockholm prospective study, *Acta Med. Scand.*, **206**:351-360 (1979).
71. A. L. Copley, The endoendothelial fibrin(ogenin) lining and its physiological significance, *Biorheology,* **25**:377-399 (1988).
72. A. L. Copley, Perihemorheology: The bridge between the vessel-blood organ and the organs it penetrates, *Biorheology,* **26**:377-388 (1989).
73. Copley, A. L., The endo-endothelial fibrin lining. A historical account, *in:*The endoendothelial fibrin lining. Symposium of the XII Eur. Conf. on Microcirculation, Jerusalem, Israel, Sept (1982), A. L. Copley, ed., Pergamon Press, New York-Oxford, *Thromb. Res.* Suppl. V (1983).
74. S. D. Lewis, L. Lorand, J. W. Fenton, II and J. A. Shafer, Catalytic competence of human α- and γ-thrombin in the activation of fibrinogen and factor XIII, *Biochemistry,* **26**:7597-7603 (1987).
75. C. S. Greenberg, K. E. Achyuthan, S. Rajagopalan and S. V. Pizzo, Characterization of the fibrin polymer structure that accelerates thrombin cleavage of plasma factor XIII, *Arch. Biochem. Biophys.*, **262**:142-148 (1988).
76. Y. Ando, S. Imamura, Y. Yamagata, A. Kitahara, H. Saji, T. Murachi and R. Kannagi, Platelet factor XIII is activated by calpain, *Biochem. Biophys. Res. Commun.*, **144**:484-490 (1987).
77. B. Hessel, L. Adamson, R. Procyk, L. Therkildsen, S. Stenbjerg, B. Blombäck, Fibrinogen Aarhus and factor XIII induced polymerization and gel formation, *Brit. J. Haematology,* **66**:355-361 (1987).

THE STRUCTURE AND EVOLUTION OF VERTEBRATE FIBRINOGEN: A COMPARISON OF THE LAMPREY AND MAMMALIAN PROTEINS

Russell F. Doolittle

Center for Molecular Genetics M-034
University of California, San Diego
La Jolla, CA 92093, USA

ABSTRACT

The blood plasmas of all vertebrate animals contain a six-chained fibrinogen molecule that is polymerized into fibrin upon the thrombin-catalyzed removal of fibrinopeptides. In all cases, also, the polymerization reaction is inhibited by Gly-Pro-Arg-ending peptides. The complete amino acid sequences of human, rat and lamprey fibrinogens are known, permitting an assessment of just which sequence features are essential for polymerization. To an extent, the same approach can also be applied to the associated phenomena of fibrin cross-linking by factor XIII, plasminogen and plasminogen activator binding, and vessel wall-fibrinogen interactions.

INTRODUCTION

The coagulation of vertebrate blood is the result of a complex interplay of cells (including mammalian platelets) and a variety of extracellular proteins. The obvious function of the process is the prevention of blood loss when the circulatory system is damaged. For the past 30 years, along with my students, I have been trying to track down where and how this complicated process evolved. From the outset, it was clear that the process is fundamentally similar in all true vertebrates. Lampreys, which along with hagfish are the most primitive vertebrates, have a typical clotting system centered around the thrombin-catalyzed conversion of fibrinogen to fibrin. Lamprey thrombin generation itself is the result of the interplay of thrombocytes and tissue factor, calcium ions, and a family of calcium-binding procoagulants that includes prothrombin (Doolittle et al, 1962; Doolittle & Surgenor, 1962).

Moreover, lamprey fibrinogen, like its mammalian equivalents, is composed of three pairs of non-identical polypeptide chains covalently bound by an array of disulfide bonds ($\alpha_2 \beta_2 \gamma_2$). As in mammals, the soluble lamprey protein is transformed into a spontaneously polymerizing unit by the thrombin-catalyzed release of peptides from the amino-terminal portions of the α and β chains (Doolittle, 1965). Moreover, lamprey fibrin is covalently crosslinked by a transglutaminase with the properties of Factor XIII, and fibrinolysis appears to occur in an "ordinary" way. It is clear, then, that most of the key features of blood coagulation were already in place 450 million years when lampreys and mammals last had a common ancestor.

We have used this information to good advantage in tracing the roots of vertebrate fibrinogen back to invertebrates, a subject we discuss in detail elsewhere (Xu & Doolittle, 1990). In the meantime, we can learn a great deal about essential features of vertebrate

Table 1. Functional Comparison of Human and Lamprey Fibrinogens

Similarities

- Both clotted as a result of thrombin-catalyzed release of fibrinopeptides (A and B).
- Polymerization inhibited by Gly-Pro-Arg peptides.
- Factor XIII crosslinking inhibited by same family of amines.

Differences

- In lampreys, fibrinopeptide A and B released at same rate.
- Lamprey fibrinogen readily clotted by exclusive release of fibrinopeptide B (at room temperature).
- Lamprey fibrin formation measurably inhibited by Gly-Val-Arg peptides.

fibrinogen and its conversion to fibrin by a rigorous comparison of the significantly diverged lamprey and mammalian molecules.

General Features of Vertebrate Fibrinogens

Vertebrate fibrinogens are large proteins, their molecular weights ranging between 320,000 and 400,000. Invariably, the γ chain is the smallest subunit, an apparent molecular weight of 47,000 $\pm$ 1,000 being found in all the many cases examined (Doolittle, 1973). β chains are slightly more variable, apparent molecular weights of 55,000-62,000 being observed on gels. Some of the larger β chains are known to have long fibrinopeptides, and, in at least one case, a fibrinopeptide-attached carbohydrate cluster adds materially to the molecular weight (Doolittle & Cottrell, 1974). Past surveys indicate that α chains vary considerably from species to species. Among mammals, rat α chains are small (58,000) and horse α chains are large (80,000). As for other vertebrates, lamprey α chains exhibit apparent molecular weights greater than 100,000 (Doolittle & Wooding, 1974; Murtaugh et al, 1974).

Functional Aspects

Fibrinopeptide Release and Polymerization. From a functional point of view, the most interesting feature of lamprey fibrinogen is that it is fully clotted by mammalian thrombins (Table 1), even though these enzymes only release the lamprey fibrinopeptide B (Doolittle, 1965). In this respect, the situation is similar to the more recently reported observation that mammalian fibrinogens can be clotted by a copperhead snake venom enzyme that only releases the fibrinopeptide B, albeit in these cases only at temperatures below 14°C (Shainoff & Dardik, 1979). Moreover, lamprey thrombin, which ordinarily releases both fibrinopeptides A and B at about the same rate, can remove fibrinopeptide A from lamprey fibrin generated by treatment with mammalian thrombin (Doolittle, 1965). As might be anticipated, lamprey thrombin is more effective in clotting lamprey fibrinogen than mammalian types, and mammalian thrombins in turn clot mammalian fibrinogens more speedily than they do lamprey.

Fibrin Stabilization. In mammals, fibrin is stabilized by the Factor XIII-catalyzed introduction of covalent crosslinks between neighboring units in the polymer. In particular, a reciprocal set of isopeptide bonds is formed between a lysine donor and a glutamine acceptor near the carboxy-termini of abutting γ chain (Chen & Doolittle, 1971). If the disulfide bonds in fibrin are reduced and subunit interactions disrupted, virtually all the constituent γ subunits are found in the form of covalently bond dimers, indicating that in the fibrin polymer all γ chains are oriented in the same way with their carboxy-terminal segments anti-parallel in adjacent units. Glutamyl-lysine isopeptides are also incorporated between fibrin α chains. The incorporation is slower in this case, and the connections are such that large multimeric arrays become involved (McKee et al, 1967).

In the lamprey, γ-γ dimers form under the same conditions and to the same extent as is observed in mammalian systems, regardless of whether lamprey or mammalian thrombin is used to clot the Factor XIII-containing fibrinogen. Furthermore, fluorescent or radio-

```
Human     YVATRDNccILDERFGSYcPTTcGIADFLSTYQTKVDKDLQSLEDILHQVENKTSEVKQL
Lamprey    QVRDLKQcSNDPEFGRYcPTTcGVADVLSKYAKGVDEDSSFIDSVLTQLAAKHGIVEGN

Human     IKAIQLTYNPDESSKPNMIDAATLKSRKMLEEIMKYEASILTHDSSIRYLQEIYNSNNQK
Lamprey   VNIVNEDVRITRDEAQIIKDSGQKTVQKILEEVRILEQIGVSHDAQIQELSEMWRVNQQF

Human     IVNLKEKVAQLEAQcQEPcKDTV  QIHDITGKDcQDIANKGAKQSGLYFIKPLKANQQF
Lamprey   VTRLQQQLVDIRQTcSRPcQDTTANKISPITGKDcQQVVDNGGKDSGLYYIKPLKAKQPF

Human     LVYcEIDGSGNGWTVFQKRLDGSVDFKKNWIQYKEGFGHLSPTGTTEFWLGNEKIHLIST
Lamprey   LVFcEI ENGNGWTVIQHRHDGSVNFTRDWV3YREGFGYLAPTLTTEFWLGNEKIHLLTG

Human     QSAIPYALRVELEDWNGRTSTADYAMFKVGPEADKYRLTYAYFAGGDAGDAFDGFDFGDD
Lamprey   QQA   YRLRIDLTDWENTHRYADYGHFKLTPESDEYRLFYSMYLDGDAGNAFDGFDFGDD

Human     PSDKFFTSHNGMQFSTWDNDNDKFEGNcAEQDGSGWWMNKcHAGHLNGVYYQGGTYSKAS
Lamprey   PQDKFYTTHLGMLFSTPERDNDKYEGScAEQDGSGWWMNRcHAGHLNGKYYFGGNYRKTD

Human     TPNGYDNGIIWATWKTRWYSMKKTTMKIIPFNRLTIGEGQQHHLGGAKQAGDV
Lamprey   VEFPYDDGIIWATWHDRWYSLKMTTMKLLPMGRDLSGHGGQQQSKGNSRGDN
```

Fig. 1. Comparison of human and lamprey fibrinogen γ-chain sequences. The chains have 205 identities at the 406 aligned positions (50.5%). Arrows denote known crosslink acceptors; bars denote crosslink donors. From Strong et al (1985).

active substitute donors are readily incorporated and can be localized to the carboxy-terminus of the γ chain (Jue & Doolittle, 1985). When purified fibrinogen is used as a starting material, the crosslinking of α chains is sluggish, but the clotting of fresh plasma results in the same multimeric crosslinking observed in mammals.

Structural Aspects

Sequence Comparison. The sequences of lamprey and human γ chains are just about 50% identical (Fig. 1), as are the β chains (Fig. 2). In both cases there are regions of near identity and others were the similarity is very low. The homology between β and γ chains (Takagi and Doolittle, 1975; Henschen and Lottspeich, 1977; Watt et al, 1978) continues to serve as an aid to our structural interpretation of the molecule, particularly in the major terminal domains (fragments D), and the more so with the lamprey sequences in hand.

The lamprey α chain is unusually long at 961 amino acids (Fig. 3), and the correspondence between lamprey and human α chains is quite low, partly because of some usually genetic intrusions that have led to differently repeated regions. Nonetheless, the amino-terminal portions of these α chains can be readily aligned (Fig. 4).

Disulfide Bond Arrangement. The disulfide bond arrangement in mammalian and lamprey fibrinogens is very similar, 26 of the 29 disulfides present in the human molecule being situated at exactly the same positions in lamprey. The lamprey molecule lacks one of the two bonds that hold the two halves of the molecule together in a dimer. It also lacks the α chain intrachain bonds. The fact that all 26 cysteines in the three lamprey fibrinogen chains are at exactly the same locations as in mammalian fibrinogens has important implications, the foremost of which is that the general folding of the constituent polypeptide chains must be the same. Thus, the "disulfide rings" that hold the three chains together on either end of the "coiled-coils" (Doolittle et al, 1977a) are highly conserved, whereas the "coiled-coils" themselves only maintain similar sidechain polarities (Fig. 5).

Asn-Linked Carbohydrate. Previous surveys by gel electrophoresis have revealed that the β and γ chains of mammalian fibrinogens invariably yield positive results for carbohydrate but α chains do not. These observations have been confirmed in all those cases where

```
Human    GHRPLDK KREEAPSLRPAPPPISGGGYRARPAKAAATQKKVERKAPDAGGcLHADPDLG
Lamprey  GVRPLPSGTRVRRPPLR  HRRLAPGAVMSRDPPASPRPQEAQKAIRDEGGcMLPESDLG

Human    VLcPTGcQLQEALLQQERPIRNSVDELNNNVEAVSQTSSSSFQYMYLLKDLWQKRQKQVK
Lamprey  VLcPTGcELREELLKQRDPVRYKISMLKQNLTYFINSFDRMASDSNTLKQNVQTLRRRLN

Human    DNENVVNEYSSELEKHQLYIDETVNSNIPTNLRVLRSILENLRSKIQKLESDVSAQMEYc
Lamprey  SRSSTHVNAQKEIENRYKEVKIRIESTVAGSLRSMKSVLEHLRAKMQRMEEAIKTQKELc

Human    RTPcTVScNIPVVSGKEcEEIIRKGGETSEMYLIQPDSSVKPYRVYcDMNTENGGWTVIQ
Lamprey  SAPcTVNcRVPVVSGMHcEDIYRNGGRTSEAYYIQPDLFSEPYKVFcDMESHGGGWTVVQ

Human    NRQDGSVDFGRKWDPYKQGFGNVATNTDGKNYcGLPGEYWLGNDKISQLTRMGPTELLIE
Lamprey  NRVDGSSNFARDWNTYKAEFGNIA FGNGKSIcNIPGEYWLGTKTVHQLTKQHTQQVLFD

Human    MEDWKGDKVKAHYGGFTVQNEANKYQISVNKYRGTAGNALMDGASQLMGENRTMTIHNGM
Lamprey  MSDWEGSSVYAQYASFRPENEAQGYRLWVEDYSGNAGNALLEGATQLMGDNRTMTIHNGM

Human    FFSTYDRDNDGWLTSDPRKQcSKEDGGGWWYNRcHAANPNGRYYWGGQYTWDMAKHGTDD
Lamprey  QFSTFDRDNDNWNPGDPTKHcSREDAGGWWYNRcHAANPNGRYYWGGIYTKEQADYGTDD

Human    GVVWMNWKGSWYSMRKMSMKIRPFFPQQ
Lamprey  GVVWMNWKGSWYSMRQMAMKLRPKWP
```

Fig. 2. Comparison of human and lamprey fibrin B-chain sequences. The chains have 218 identities at the 438 aligned positions (49.8%). From Bohonus et al. (1986).

amino acid analysis has been conducted, amino sugars being evident in β and γ chains but not in α chains. In human fibrinogen, it is known that all the carbohydrate is asparagine-linked and can be removed by appropriate endoglycosidases (Langer et al, 1987).

In lampreys, also, the β and γ chains contain carbohydrate and α chains do not. Moreover, all the carbohydrate is found in peptides containing the asparaginyl consensus signal required for Asn-linked carbohydrate clusters. The distribution of these carbohydrates is significantly different than in mammals, however. Thus, the γ-chain lacks the well known attachment situated on the "coiled-coils" at human residue γ-52 (Fig. 5). Instead, it has a cluster at γ-203. The lamprey β chain has three carbohydrate attachments: one on its fibrinopeptide (Doolittle & Cottrell, 1974), one at a point on the "coiled-coils" at residue β88, and a third at β348 the same as is found in mammals (e.g., human β350). It seems significant that β-Asn 88 exists at about the same location in the "coiled-coil" as γ Asn-52 (Fig. 5). Interestingly, although neither lamprey nor human α chains contain carbohydrate, both have the tripeptidyl consensus sequence characteristic of asparginyl-linked sugars. Other workers (Henschen et al, 1983) have assumed that the absence of carbohydrate from the human α chain consensus position is because proline is a neighboring residue, the sequences in question being N-P-S(S) and N-V-S(P). That this explanation cannot be the whole story is provided by the fact that in the lamprey α chain the unoccupied site is N-G-T(T). To further complicate explanations, it has been reported that in chicken fibrinogen, carbohydrate is indeed present in some α chains (Greininger et al, 1984).

Polymerization Sites. Although lamprey fibrin forms readily upon the exposure of the β-chain Gly-Val-Arg sequence, the polymerization is significantly more sensitive to Gly-Pro-Arg-type inhibitors than it is to Gly-Val-Arg-derivatives. Nonetheless, the data make it clear that mammalian thrombins can attack the β-chain arginyl-glycine bond in lamprey fibrinogen without first releasing the fibrinopeptide A. They also underscore the fact that it is not necessary to expose the α-chain Gly-Pro-Arg sequence for polymerization to occur. It should be borne in mind that about half the amino acids are exactly the same

```
LAMPREY FIBRINOGEN ALPHA CHAIN

      ↓
  1   DDISLRGPRLTEQRSAGQGScASATADLcVHGDWGRKcPNGcRMQGLMSHAEKDIGKRIG

 61   DLTERLARLGRLYTQVHTDFRAVSDTSGQTLNEHNELEVRYSEVLRELERRIIHLQRRIN

121   MQLQQLTLLQHNIKTQVSQILRVEVDIDVALRAcKGScARYLEYRLDKEKNLQLEKAASY

181   IANLKFERFEEVVVEETLNRRVETSSHAFQPTHGQGTPQPGHGTHSLSATSSITSAPNFV

241   PHRQPTYVDHGRLSNPNQVAHSASSSSTHTSSSSSPSQPVSPDSAFPLPGSNTGTSEWDF

301   NFHDESTPGNGPRDEAAASSSALSPSTASHHTATSTTSFSSGTSGKDVAPLGTGVTHDGG

361   VRTSGSLMDGGSSDTGTGGVSKTTTFTGSAQGGSWSTGGSTATNTGSAQGGSWSTGGRTE

421   PNTGSGQGGSWGTGGRTEPNTGSGQGGSWGTGGRTEPNTGSGQGGSWGTGGRTEPNTGSA

481   QGGSWGTGGRTEPNTGSAQGGSWGTGGRTEPNTGSAQGGSWSTGGRTEPNTGSAKGGSWG

541   TGGRTEPNTGSAKGGSWSTGGRTEPNTGSAKGGSWGTGGRTEPNTGSAQGGSWGTGGRTE

601   PNTGSAQGGSWGTGGRTEPNTGSAQGGSWGTGGRTEPNTGSAQGGSWGTGGRTEPNTGSA

661   QGGSWSTGGRTEPNTGSGQGGSWGTGGRTEPNTGSGQGGSWSTGGRTEPNTGSGQGGSWG

721   TGGRTEPNTGSAQGGSWGTGGRTEPNTGSAQGGSWGTGGSTATNTGSAQGGGGYAAGGTG

781   AQTGSGSTSTHSAHSASGGMSSLDMLPALPDFGTWDMPDHSDIFSRRRVSTSSTTSSSSG

841   GGHAGAAAGGGGDGASRFGSLFTTDFGPEFHEEFRSMLPGASRLSSSSSSSTRSTSSTSG

901   GKVVTESVVTKVLSNGTTITHHTKHVSTSDGTGAASDGVSPLLTGRKTKAARSRRAKATRP
```

Fig. 3. Complete 961-amino acid sequence of lamprey fibrinogen α chain. The arrow denotes thrombin cleavage point for releae of six-residue fibrinopeptide A. From Wang et al. (1989).

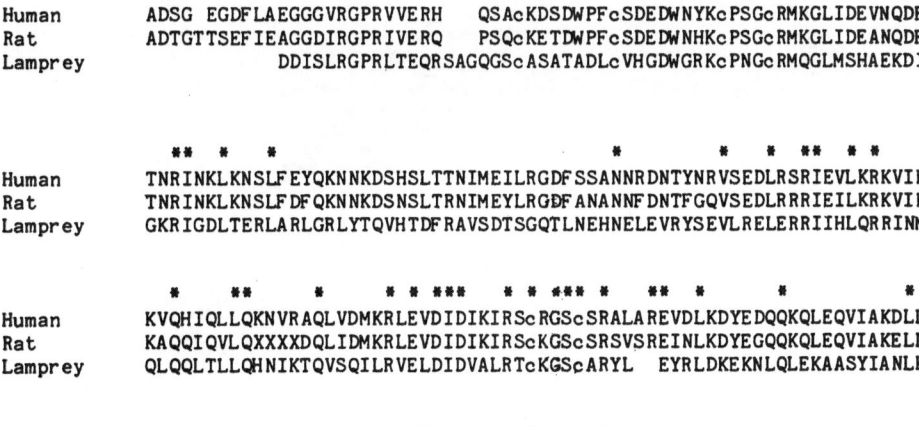

```
          ** *    *       ****  *        *       *   **  *** **** **      *
Human     ADSG EGDFLAEGGGVRGPRVVERH   QSAcKDSDWPFcSDEDWNYKcPSGcRMKGLIDEVNQDF
Rat       ADTGTTSEFIEAGGDIRGPRIVERQ   PSQcKETDWPFcSDEDWNHKcPSGcRMKGLIDEANQDF
Lamprey              DDISLRGPRLTEQRSAGQGScASATADLcVHGDWGRKcPNGcRMQGLMSHAEKDI

            **  *   *                            *        *   *  **  * *
Human     TNRINKLKNSLFEYQKNNKDSHSLTTNIMEILRGDFSSANNRDNTYNRVSEDLRSRIEVLKRKVIE
Rat       TNRINKLKNSLFDFQKNNKDSNSLTRNIMEYLRGDFANANNFDNTFGQVSEDLRRRIEILKRKVIE
Lamprey   GKRIGDLTERLARLGRLYTQVHTDFRAVSDTSGQTLNEHNELEVRYSEVLRELERRIIHLQRRINM

            *    **     *     * * ***   * * *** *   **  *      *          *
Human     KVQHIQLLQKNVRAQLVDMKRLEVDIDIKIRScRGScSRALAREVDLKDYEDQQKQLEQVIAKDLL
Rat       KAQQIQVLQXXXXDQLIDMKRLEVDIDIKIRScKGScSRSVSREINLKDYEGQQKQLEQVIAKELL
Lamprey   QLQQLTLLQHNIKTQVSQILRVELDIDVALRTcKGScARYL  EYRLDKEKNLQLEKAASYIANLK

                                 *        *     *
Human     PSRDRQHLPLIKMKPVPDLVPGNFKSQLQKVPPEWKALTDMPQMR
Rat       PAKDRQYLPAIKMSPVPDLVPGSFKSQLQEGPPEWKALTEMRQMR
Lamprey   FERFEEVVVEETLNRRVETSSHAFQPTHGQGTPQPGHGTHSLSAT
```

Fig. 4. Computer alignment of amino-terminal segments of human, rat and lamprey fibrinogen α chains. Asterisks (*) denote positions identical in all three species. From Wang et al. (1989).

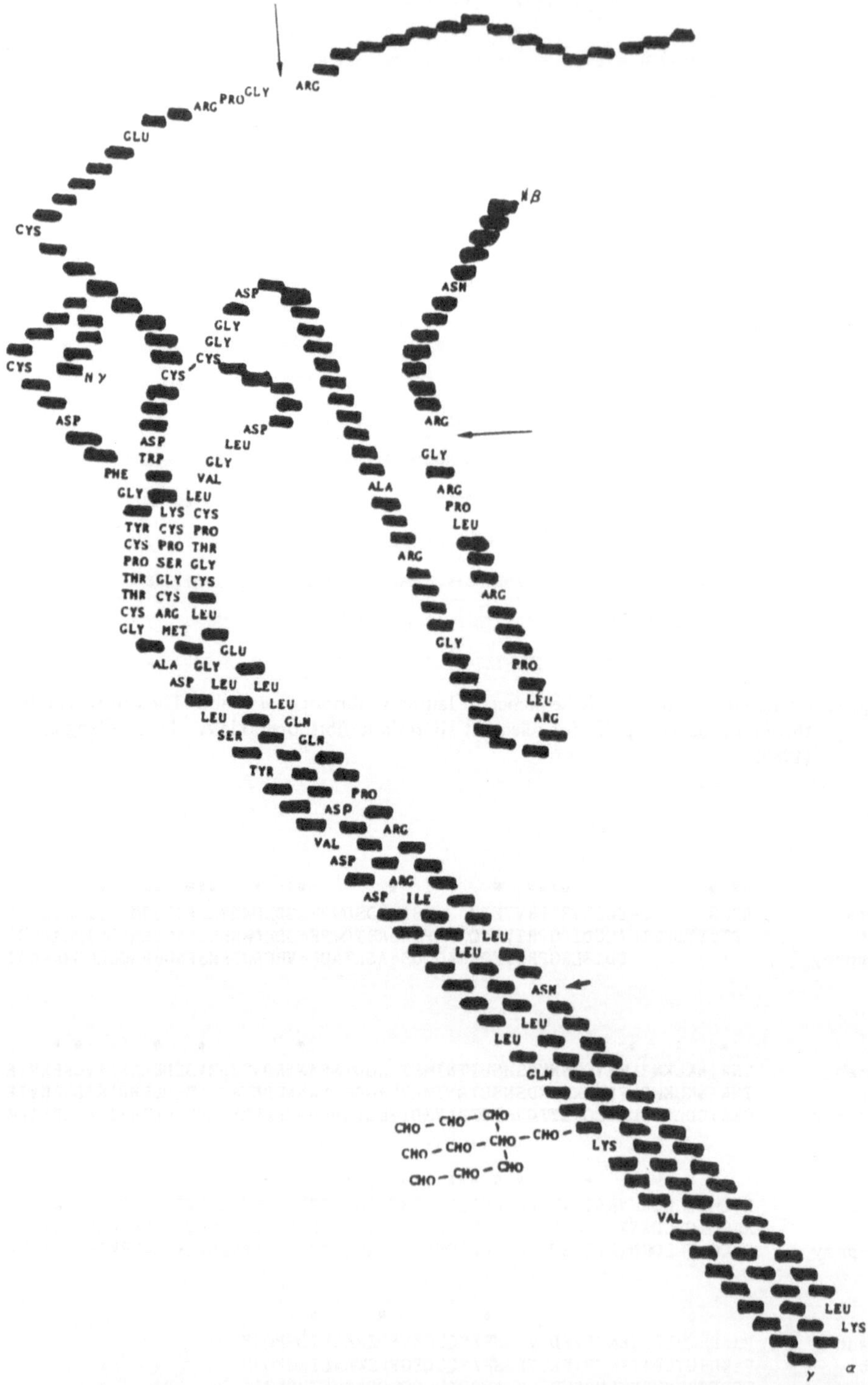

Fig. 5. Two-dimensional arrangement of amino-terminal sequences of human fibrinogen. Blackened residues denote differences between human and lamprey fibrinogens. Long arrows indicate thrombin cleavage points. The carbohydrate cluster (CHO-etc.) on the human γ chain does not occur in lampreys, but a carbohydrate cluster does occur in the lamprey β chain at the Asn denoted with a short arrow.

in this region of lamprey and mammalian fibrinogens (Fig. 5). The most likely reason that mammalian thrombins do not remove lamprey fibrinopeptide B is the presence of a leucine adjacent to the critical Arg-Gly target, a position most often occupied by a valine, or less often isoleucine (Fig. 3).

As for the complementary sites with which the Gly-Pro-Arg "knobs" interact, a consideration of those residues in the terminal domain that are conserved in lampreys and humans allows us to discount certain regions as candidates. Thus, an early notion (Olexa & Budzynski, 1981) about the carboxy-terminal 38 residues containing the site seems quite unreasonable in the light of the large amount of amino acid replacement in this region (Fig. 6). Rather, the conservation pattern is more in line with other observations that embrace a much large portion of the γ chain (e.g., Varadi & Scheraga, 1986). Previously, we suggested γ chain residues 318-352 (human numbering) or 363-385 as likely candidate regions (Strong et al, 1985). More recently, affinity-labeling experiments have implicated the cyanogen bromide peptide that stretches from γ-res. 337-379 (Shimizu & Doolittle, 1989). It is also interesting to note in passing that most variant human fibrinogens with impaired polymerization (e.g., Miyata et al, 1989) involve residues that are conserved between humans and lampreys.

The question arises, is the exposed Gly-Val-Arg filling a specific complementary site on the terminal domain of fibrinogen? Or, is it fitting uncomfortably into the Gly-Pro-Arg complementary site, a hole thought to be situated somewhere in the carboxy-terminal third of the γ chain? Binding studies on both lamprey and human fibrinogen have shown that Gly-Pro-Arg and its derivatives can bind to both sites, but the inverse, that Gly-Val-Arg-peptides can bind to both sites, is not firmly established, even though these peptides are weak inhibitors of lamprey fibrin formation. In contrast, Gly-His-Arg peptides, which correspond to the amino-terminus of mammalian fibrin β chains, do not inhibit polymerization in either system (Laudano & Doolittle, 1980), even though Gly-His-Arg peptides compete for the same lamprey site ordinarily occupied by Gly-Val-Arg (Laudano et al, 1983). That the final orientation of monomeric units is the same in fibrins A and B is attested to by the fact that full sets of γ-γ dimers accrue in either case.

The Free-Swimming Appendages. Of all the structural differences between lamprey and human fibrinogens, the most bizarre has to do with those highly exposed regions of the α chain called "free-swimming appendages." These polar protuberances are readily removed by a wide variety of proteases. Their irregular structure is revealed by the fact that the isolated material fails to exhibit α helix or β structure by circular dichroism (Takagi & Doolittle, 1975). In this regard, the parent molecule after their removal is much richer in such organized structure (Huseby et al, 1970).

In human fibrinogen, a large portion of the appendage is composed of a series of imperfect 13-residue repeats (Doolittle et al, 1979); the repeats are rich in glycine and serine, and most of them have a tryptophan residue. Lamprey α chains also have a highly repeated sequence in this region; this repeat, also, is rich in glycine and serine and contains one tryptophan per repeat. Astonishingly, the lamprey repeat, of which there are 23, is 18 residues long (Fig. 7). Since there is no way a 13-residue repeat can be converted into an 18-residue one, or vice-versa, these two structures have evolved independently.

The fact that the two kinds of repeats are composed of similar amino acids in similar arrays suggests that some fundamentally advantageous property is common to them. In this regard, tryptophan is a common occupant of the fourth position of β-turns. Given this meagre toe-hold, we have reasoned that the repeated regions in both molecules may involve an accordion structure composed of a series of tight turns that interact locally with themselves or with α chains on other molecules (Fig. 8). Evidence has been reported that implicates these regions in the polymerization process itself (Medved' et al, 1985). Given the similar nature of fibrin formation in all cases, the species variation in length for these appendages is nothing short of remarkable (Fig. 9).

Crosslinking Sites. Although lamprey fibrin is reciprocally crosslinked by glutamyl-lysine isopeptide bonds located at γ-chain carboxy termini, the sequence around the acceptor site is significantly different from that observed in mammals, the participating glutamine and lysine being only four residues apart istead of eight (Fig. 1). It is interesting that in either case the joined sidechains would be situated on the appropriate faces of antiparallel α helices (Doolittle, 1973).

The α chain crosslinking situation remains mysterious. In human fibrinogen, two glutamine acceptors have been pinpointed at residues α 328 and 366 (Cottrell et al, 1979). In fact, there is only one other glutamine residue in the entire carboxy-terminal half of

Fig. 6. Two-dimensional rendering of the Fragment D portion of human Fibrinogen. Blackened residues denote differences between human and lamprey fibrinogens.

	Lamprey 18-Res. Repeat	Human 13-Res. Repeat
	...LMDGGSSDTGTGGVSKTTT	...PGGNEITRGGSTSYGTGSE
1	FTGSAQGGSWSTGGSTAT	TESPRNPSSAGSW
2	NTGSAQGGSWSTGGRTEP	NSGSSGPGSTGNR
3	NTGSGQGGSWGTGGRTEP	NPGSSGTGSGATW
4	NTGSGQGGSWGTGGRTEP	KPGSSGPGSTGSW
5	NTGSGQGGSWGTGGRTEP	NSGSSGTGSTGNQ
6	NTGSAQGGSWGTGGRTEP	NPGSPRPGSTGTW
7	NTGSAQGGSWGTGGRTEP	NPGSSERGSAGHW
8	NTGSAQGGSWSTGGRTEP	TSESSVSGSTGQW
9	NTGSAKGGSWGTGGRTEP	HSESGSFRPDSPG
10	NTGSAKGGSWSTGGRTEP	SGNARPNDPNW
11	NTGSAKGGSWGTGGRTEP	
12	NTGSAQGGSWGTGGRTEP	
13	NTGSAQGGSWGTGGRTEP	
14	NTGSAQGGSWGTGGRTEP	
15	NTGSAQGGSWGTGGRTEP	
16	NTGSAQGGSWSTGGRTEP	
17	NTGSGQGGSWGTGGRTEP	
18	NTGSGQGGSWSTGGRTEP	
19	NTGSGQGGSWGTGGRTEP	
20	NTGSAQGGSWGTGGRTEP	
21	NTGSAQGGSWGTGGSTAT	
(22)	NTGSAQGGGGYAAGGTGA	
(23)	QTGSGSTSTHSAHSASGG	
	MSSLDMLPALPDFGTWDMPDHS...	GTFEEVSGNVSPGTRREYHT...

Fig. 7. Comparison of lamprey α-chain 18-residue repeat (left) with human α-chain 13-residue repeat (right) the lamprey sequence begins at upper left with Leu-367 and ends with Ser-821. The human sequence shown begins with Pro-245 and ends with Thr-411 at lower right. From Wang et al. (1989).

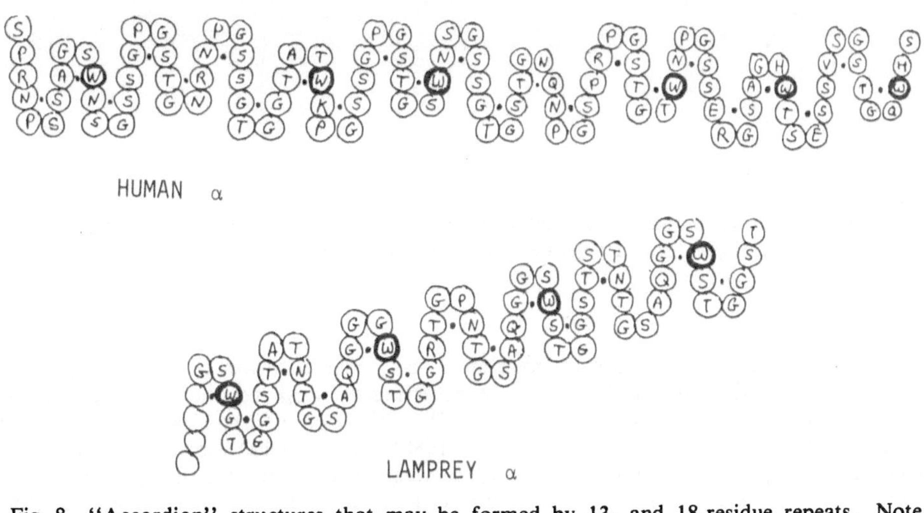

Fig. 8. "Accordion" structures that may be formed by 13- and 18-residue repeats. Note that tryptophan (ω) frequently occupies the fourth position of the β-turns.

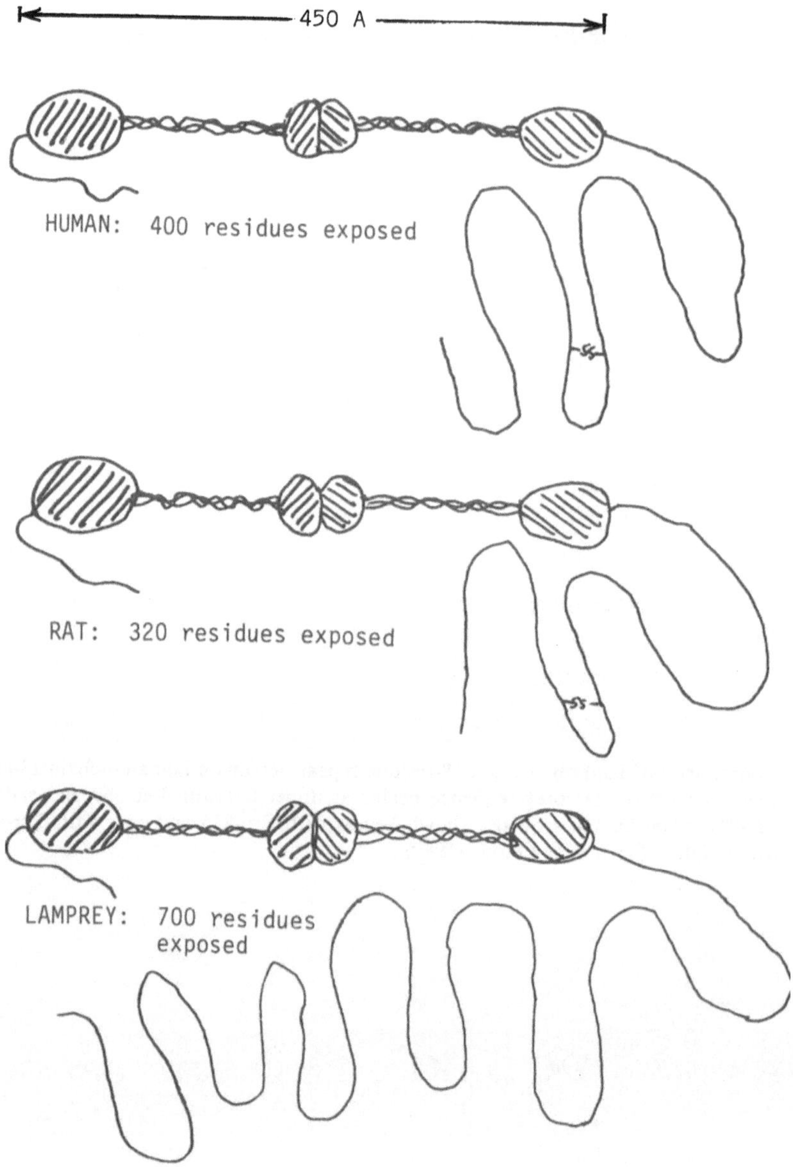

Fig. 9. Exposed α-chain free-swimming appendages in human, rat and lamprey fibrinogens drawn to scale. Note absence of disulfide bond in lamprey segment.

the human α chain, and it occurs at position α-563. Identification of the α-chain lysine donors has proved difficult, although early experiments localized them to the carboxy-terminal hundred residues (Doolittle et al, 1977b). Further efforts aimed at unravelling the situation have been reported (Sobel et al, 1988). The elucidation of the rat α-chain sequence from its cDNA (Crabtree et al, 1985) was revealing in that none of the corresponding glutamine acceptors is conserved between rat and human, even though rat fibrin α chains became fully crosslinked. Moreover, the rat α chain is the shortest known so far (Fig. 9).

Glutamine availability is surely not the limiting aspect in lamprey α chains, since 19 of the 23 repeats have a glutamine residue (Fig. 7), although none exist among the last 180 carboxy-terminal residues. That region of the chain, like mammalian α chains, does contain a substantial number of lysines, however. In any event, genuine questions still remain about α-chain crosslinking. The exact identification of participatory residues is also critical to our understanding of how α chains are finally situated in the fibrin polymer.

Other Interaction Sites. As is well known, mammalian fibrinogen binds to platelets and is an essential co-factor in several platelet reactions. Platelets are peculiar to mammals, of course, but infra-mammalian vertebrates have a white cell equivalent that participates in clotting (Doolittle & Surgenor, 1962). It can be anticipated that lamprey fibrinogen interacts with these and other lymphoid cells.

In mammals, the sequence Arg-Gly-Asp ("RGD") has been suggested as playing a role in fibrinogen binding to platelets (Peerschke & Galanakis, 1987; Timmons et al, 1989). The sequence occurs twice in the α chains of the human and rat molecules, albeit in only one of these cases are they are equivalent positions. The RGD sequence does not occur at all in the lamprey α chain. It does occur at the very carboxy-terminus of the lamprey γ chain (Fig. 1). This may be a mere coincidence, however, especially when it is realized that this sequence ought to occur randomly once in about every 5000 residues and is well known to occur in many proteins that do not bind to cells at all.

As for other interactive sites, human fibrin is well known to bind a plasminogen activator (e.g., Ichinose et al, 1986). In this regard, lysine α-157 (human numbering) has been identified as particularly critical (Voskuilen et al, 1987). As it happens, there is no lysine in this particular region of the lamprey α chain, the residue in question being a leucine instead (Fig. 4). No information about the binding of plasminogen activators to lamprey fibrin is available, however, and it would be misleading to read too much into this difference.

In mammals, also, it is known that the 30-residue disulfide loops on the distal portions of γ chains can become linked through a mixed disulfide to α_1 antitrypsin (Laurell & Thulin, 1975). The absence of this loop from lamprey α chains precludes such a phenomenon in lamprey fibrinogen.

Minor Forms of Fibrinogen. In humans and other mammals, circulating fibrinogen occurs in major and minor forms (Wolfenstein-Todel et al, 1981), the latter the product of alternative exon splicing during the maturation of γ-chain mRNA (Crabtree & Kant, 1982). We have not observed a minor form of lamprey fibrinogen. Moreover, the genomic sequence of the lamprey γ chain has revealed that alternative splicing of the sort found in mammalian systems is not possible (Pan & Doolittle, in preparation).

CONCLUSIONS

The overall structures of lamprey and mammalian fibrinogens are quite similar, even though each has some unique features. The exposure of Gly-Xxx-Arg-ending "knobs" by the thrombin-catalyzed release of amino-terminal peptides is the key to fibrin formation in both situations. Exactly where the "holes" into which those knobs fit during polymerization are located is still not known exactly, but we can be sure that their architecture must be virtually the same in lamprey and mammals. That being the case, we can rule out certain sectors of the β and γ chains as not being intimately involved in these interactions, and can make some conjectures about just which sections participate.

ACKNOWLEDGMENTS

I am grateful to Karen Anderson for assistance in the preparation of this manuscript and to many of my past co-workers who have contributed to the lamprey-human fibrinogen project over the years. Most of our work has been supported by the National Heart, Lung and Blood Institute.

REFERENCES

Bohonus, V., Doolittle, R. F., Pontes, M., and Strong, D. D. 1986, Complementary DNA sequence of lamprey fibrinogen β chain. *Biochemistry,* **25**:6512-6516.

Chen, R., and Doolittle, R. F., 1971, γ-γ Cross-linking sites in human and bovine fibrin. *Biochemistry,* **10**:4486-4491.

Cottrell, B. A., Strong, D. D., Watt, K. W. K., and Doolittle, R. F., 1979, Amino acid sequence studies on the α-chain of human fibrinogen. Exact location of cross-linking acceptor sites. *Biochemistry,* **18:**5405-5410.

Crabtree, G. R., and Kant, J. A., 1982, Organization of the rat γ-fibrinogen gene: alternative mRNA splice patterns produce the γA and γB (gamma') chains of fibrinogen. *Cell,* **31:**159-166.

Crabtree, G. R., Comeau, C. M., Fowlkes, D. M., Fornace, A. J., Jr., Malley, J. D., and Kant, J. A., 1985, Evolution and structure of the fibrinogen genes. Random insertion of introns or selective loss? *J. Mol. Biol.,* **185:**1-19.

Doolittle, R. F., 1965, Difference in the clotting of lamprey fibrinogen by lamprey and bovine thrombins. *Biochem. J.,* **94:**735-741.

Doolittle, R. F., 1973, Structural aspects of the fibrinogen-fibrin conversion. *Advances in Protein Chemistry,* **27:**1-109.

Doolittle, R. F. and Surgenor, D. M., 1962, Blood coagulation in fish. *Am. J. Physiol.,* **203:**964-970.

Doolittle, R. F. and Cottrell, B. A., 1974, Lamprey fibrinopeptide is a glycopeptide. *Biochem. Biophys. Res. Commun.,* **60:**1090-1096.

Doolittle, R. F., and Wooding, G. L., 1974, The subunit structure of lamprey fibrinogen and fibrin. *Biochim. Biophys. Acta,* **271:**277-282.

Doolittle, R. F., Oncley, J. L. and Surgenor, D. M., 1962, Species differences in the interaction of thrombin and fibrinogen. *J. Biol. Chem.,* **237:**3123-3127.

Doolittle, R. F., Cassman, K. G., Cottrell, B. A., Friezner, S. J., and Takagi, T., 1977a, Amino acid sequence studies on the α-chain of human fibrinogen. The covalent structure of the α-chain portion of fragment D. *Biochemistry,* **16:**1710-1715.

Doolittle, R. F., Cassman, K. G., Cottrell, B. A., and Friezner, S. J., 1977b, Amino acid sequence studies on the α-chain of human fibrinogen. Isolation and characterization of two linked α-chain cyanogen bromide fragments from fully cross-linked fibrin. *Biochemistry,* **16:**1715-1719.

Doolittle, R. F., Watt, K. W. K., Cottrell, B. A., Strong, D. D., and Riley, M., 1979a., The amino acid sequence of the α-chain of human fibrinogen *Nature,* **280:**464-468.

Greininger, G., Plant, P. W. and Kossoff, H. S., 1984, Glycosylation of A α chains in chicken fibrinogen. *Biochemistry,* **23:**5888-5892.

Henschen, A., and Lottspeich, F., 1977, Sequence homology between β-chain and γ-chain in human fibrin. *Thrombosis Research,* **11:**869-880.

Henschen, A., Lottspeich, F., Kehl, M. and Southan, C., 1983, Covalent structure of fibrinogen. *Ann. N.Y. Acad. Sci.,* **408:**28-43.

Huseby, R. M. Mosesson, M. W. and Murray, M., 1970, Studies of the amino acid composition and conformation of human fibrinogen: comparison of fractions I-4 and I-8. *Physiol. Chem. & Physics,* **2:**374-384.

Ichinose, A., Takio, K., and Fujikawa, K., 1986, Localization of the binding site of tissue-type plasminogen activator to fibrin. *J. Clin. Invest.,* **78:**163-169.

Jue, R. A., and Doolittle, R. F., 1985, Determination of the relative positions of amino acids by partial specific cleavages of end-labeled proteins. *Biochemistry,* **24:**162-170.

Langer, B. G., Hong, S. K., Schmelzer, C. H., and Bell, W. R., 1987, Deglycosylation of a native, protease-sensitive glycoprotein by peptide *N*-glycosidase F without protease inhibitors. *Anal. Biochem.,* **166:**212-217.

Laudano, A. P., and Doolittle, R. F., 1980, Studies on synthetic peptides that bind to fibrinogen and prevent fibrin polymerization. Structural requirements, numbers of binding sites and species differences. *Biochemistry,* **19:**1013-1019.

Laudano, A. P., Cottrell, B. A. and Doolittle, R. F., 1983, Synthetic peptides modeled on fibrin polymerization sites. *Ann. N.Y. Acad. Sci.,* **408:**315-329.

Laurell, C.-B., and Thulin, E., 1975, Complexes in human plasma between α_1-antitrypsin and IgA, and α_1-antitrypsin and fibrinogen. *Scand. J. Immunol.,* **4:**Suppl. 2, 7-12.

McKee, P. A., Mattock, P., and Hill, R. L., 1970, Subunit structure of human fibrinogen, soluble fibrin, and cross-linked insoluble fibrin. *Proc. Natl. Acad. Sci., USA,* **66:**738-774.

Medved', L., Gorkun, O. V., Manyakov, V. F., and Belitser, V. A., 1985, The role of fibrinogen α C-domains in the fibrin assembly process. *FEBS Lett.,* **181:**109-112.

Miyata, T., Furukawa, K., Iwanaga, S., Takamatsu, J., and Saito, H., 1989, Fibrinogen Nagoya, a replacement of glutamine-329 by arginine in the γ-chain that impairs the polymerization of fibrin monomer. *J. Biochem.,* **105:**10-14.

Murtaugh, P. A., Halver, J. E., Lewis, M. S. and Gladner, J. A., 1974, Cross-linking reactions of lamprey fibrinogen and fibrin. *Biochim. Biophys. Acta,* **359:**415-420.

Olexa, S. A., and Budzynski, A. Z., 1981, Localization of a fibrin polymerization site. *J. Biol. Chem.,* **256:**3544-3549.

Peerschke, E. I. B., and Galanakis, D. K., 1987, The synthetic RGDS peptide inhibits the binding of fibrinogen lacking intact α chain carboxy terminal sequences to human blood platelets. *Blood,* **69:**950-952.

Shainoff, J. R., and Dardik, B. N., 1979, Fibrinopeptide B and aggregation of fibrinogen. *Science,* **204:**200-202.

Shimizu, A. and Doolittle, R. F., 1989, Identification of fibrin polymerization site by photoaffinity labeling. *XII Intern. Cong. Thromb. Haem., Tokyo* (abstract).

Sobel, J. H. Thibodeau, C. A., and Canfield, R. E., 1988, Early alpha chain crosslinking in human fibrin preparations. *Thrombosis & Haemostasis,* **60:**153-159.

Strong, D. D., Moore, M., Cottrell, B. A., Bohonus, V. L. Pontes, M., Evans, B., Riley, M., and Doolittle, R. F., 1985, Lamprey fibrinogen γ chain: cloning, cDNA sequencing, and general characterization. *Biochemistry,* **24:**92-101.

Takagi, T., and Doolittle, R. F., 1975, Amino acid sequence studies on the α-chain of human fibrinogen: location of four plasmin attack points and a covalent crosslinking sites. *Biochemistry,* **14:**5149-5156.

Takagi, T., and Doolittle, R. F., 1975, Amino acid sequence of the carboxy-terminal cyanogen bromide peptide of the human fibrinogen ß chain: homology with the corresponding γ-chain peptide and presence in fragment D. *Biochim. Biophys. Acta,* **386:**617-622.

Timmons, S., Bednarek, M. A., Kloczewiak, M., and Hawiger, J., 1989, Antiplatelet "hybrid" peptides analogous to receptor recognition domains on γ and α chains of human fibrinogen. *Biochemistry,* **28:**2919-2923.

Varadi, A., and Scheraga, H. A., 1986, Localization of segments essential for polymerization and for calcium binding in the γ-chain of human fibrinogen. *Biochemistry,* **25:**519-528.

Voskuilen, M., Vermond, A., Veeneman, G. H., van Boom, J. H., Klasen, E. A., Zegers, and N. D., Nieuwenhuizen, W., 1987, Fibrinogen lysine residue A α157 plays a crucial role in the fibrin-induced acceleration of plasminogen activation, catalyzed by tissue-type plasminogen activator. *J. Biol. Chem.,* **262:**5944-5946.

Wang, Y. Z., Patterson, J., Gray, J. E., Yu, C., Cottrell, B. A., Shimizu, A., Graham, D., Riley, M., and Doolittle, R. F., 1989, Complete sequence of the lamprey fibrinogen α chain. *Biochemistry,* **28:**9801-9806.

Watt, K. W. K., Takagi, T., and Doolittle, R. F., 1978, Amino acid sequence of the ß-chain of human fibrinogen: homology with the γ-chain. *Proc. Natl. Acad. Sci., USA,* **75:**1731-1735.

Wolfenstein-Todel, C., and Mosesson, M. W., 1981, Carboxy-terminal amino acid sequence of a human fibrinogen γ-chain variant (γ'). *Biochemistry,* **20:**6146-6149.

Xu, X. and Doolittle, R. F., 1990, Presence of a fibrinogen-like sequence in an echinoderm. *Proc. Natl. Acad. Sci., USA,* in press.

NUCLEOTIDE SEQUENCES OF THE THREE GENES CODING FOR HUMAN FIBRINOGEN

Dominic W. Chung, Jeff E. Harris, and Earl W. Davie

Department of Biochemistry
University of Washington
Seattle, WA 98195, USA

INTRODUCTION

Fibrinogen is synthesized in the liver by hepatic parenchymal cells and is secreted into the circulation (1). Hepatic synthesis of fibrinogen is constitutive but the rate can be modulated by a number of physiological and nonphysiological factors. The three chains of fibrinogen are encoded by distinct species of mRNA that are derived from the expression of three single copy genes (2, 3). Present evidence indicates that the three genes of human fibrinogen are linked and are located in a region that extends approximately 45 kb on chromosome 4q23-q32 (4). The genes are arranged in the order of γ-Aα-Bβ. The γ and Aα genes are transcribed in the same direction while the Bβ gene is transcribed in the opposite direction.

Gene Isolation

Three recombinant lambda phage genomic libraries were screened by the plaque hybridization technique of Benton and Davis as modified by Woo (5) using cDNAs as hybridization probes (6, 7, 8). These libraries include a fetal liver library kindly provided by Maniatis (9), a white cell library obtained from Clontech, and a fibroblast library provided by Yoshitake (10). Initially, clones for the Bβ gene were identified using a bovine cDNA as hybridization probe (7). Recombinant phage containing overlapping segments of the three human fibrinogen genes were isolated and plaque purified. They were mapped by the method of Rackwitz et al. using radiolabelled oligonucleotides specific for the cohesive ends of lambda phage (11). Contiguous and overlapping fragments of the three genes were subcloned into various plasmid vectors. The complete DNA sequence of the three genes was determined by the dideoxy chain terminator method of Sanger (12). Subcloned gene fragments were digested with exonuclease Bal 31 for varying durations to generate a series of overlapping deletions. Sequences were determined from the Bal 31 deleted ends after cloning these fragments into appropriate sites of M13mp18 or M13mp19. Over 85% of the sequence was determined on both strands of the gene.

Overall Organization of the Three Genes

The genes for the Aα, Bβ, and γ chains are 5.4, 8.2, and 8.4 kilobases in length. The Aα gene consists of 5 exons, whereas the Bβ gene has 8 exons and the γ gene has 10 exons. Comparisons of the amino acid sequences of the three chains suggested that they are derived from a common ancestral gene and that the time of divergence between the Bβ and the γ gene was approximately 600 million years ago, and that the divergence of the Aα gene occurred over 1000 million years ago (13). The organization of the three genes provides additional evidence of such evolutionary relatedness. The location of 2 introns in each of the three genes occurs at conserved positions. As shown in Fig. 1, an intron, with a type 0 splice junction, interrupts the coding region in each of the three genes at precisely 3 residues prior to the first cysteine residues in the proximal disulfide ring. A second conserved intron, with type I splice junction, occurs in the middle of the coiled-coil region. The sequences of these conserved introns, however, are different in length, and show no significant identity. The conservation of the location of these introns relative to the coding sequence, and the conservation of splice junction types provide further evidence of the evolutionary relatedness of the three genes.

A-alpha Gene

The Aα gene is located in the middle of the fibrinogen gene cluster. It is located at the 3' end of the γ gene, and both the Aα and γ genes are transcribed in the same direction. The sequence of the Aα gene is shown in Fig. 2. Three single nucleotide differences with the cDNA sequence have been observed. These differences (nucleotides 4325, 4328, and 4805) occur at the third base of the respective codons and do not change the amino acids encoded. The organization of the gene is unique in that the majority of the primary translation product (amino acid residues 153-625) is encoded in one large exon (exon V, Fig. 2). This exon contains the tandem repeats unique to the Aα chain (residues 270-374). Each of these repeats contains the sequence of Gly-Ser-Ser, which is encoded by the nucleotide sequence GGG-AGC-TCT. This sequence contains the recognition sequence of the restriction endonuclease Sst I. These repeat sequences, however, are unrelated to the human Sst I family of repeats. Another unique feature of the Aα gene is that it contains a segment of 100 nucleotides in intron C in which the sequence is composed of exclusively pyrimidine residues (nucleotides 2893-2994), and over 70% of which is composed of T residues. This region is partially duplicated in the 3' flanking region of the Aα gene in the form of an inverted repeat (GA rich sequence, nucleotides 5738-5779). These regions are highly homologous to regions of the apolipoprotein C III, beta actin pseudogene, and ribosomal spacer sequences. To a less extent, significant homology is also apparent with the poly A region of the Alu family of repeats. Thus far, no function has been attributed to these sequences. This segment of the Aα gene is absent in the rat Aα gene. Fig. 3 shows an analysis of the human and rat Aα gene sequences (14) in the form of a dot matrix comparison, in which a dot represents a segment with 60% or higher identity. This comparison examines gene sequences that are much closer related in evolutionary time. A major difference between the two genes is in the size of intron A; the rat gene contains an apparent insert of approximately 1000 base pairs. In addition to the exons, introns A and C show significant conservation of sequences.

```
                Intron B Type 0                         Intron C type I
                     ▼                                        ▼
A-alpha         D W N Y K C P S G C R M K ..........F S S A N N R D N .....
                40        45        49                100

                Intron B Type 0                         Intron C type I
                     ▼                                        ▼
B-beta          D L G V L C P T G C Q L Q ..........Q K Q V K D N E N .....
                71        76        80                131

                Intron B Type 0                         Intron C type I
                     ▼                                        ▼
gamma           R F G S Y C P T T C G I A ..........E S S K P N M I D .....
                14        19        23                74
```

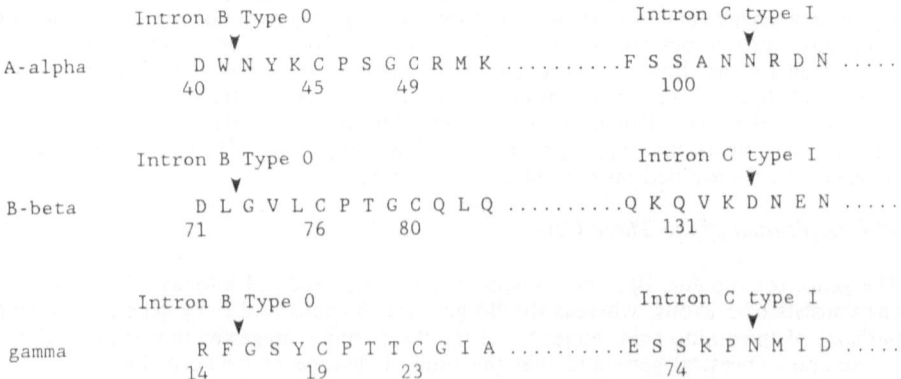

Fig. 1. Locations of conserved introns and their splice junction types in the three genes coding for the Aα, Bß, and γ Chains of human fibrinogen.

```
                                  -19Met Phe Ser Met Arg Ile Val Cys Leu Val Leu Ser Val Val Gly Thr Ala Trp-2
GTCTAGGAGC CAGCCCCACC CTTAGAAAAG   ATG TTT TCC ATG AGG ATC GTC TGC CTA GTT CTA AGT GTG GTG GGC ACA GCA TGG       84
][ GTATGGCCCT TTTCATTTTT TCTTCTTGCT TTCTCTCTGG TGTTTATTCC ACAAAGAGCC TGGAGGTCAG AGTCTACCTG CTCTATGTCC TGACAC   180
ACTCTTAGCT TTATGACCCC AGGCCTGGGA GGAAATTTCC TGGGTGGGCT TGACACCTCA AGAATACAGG GTAATATGAC ACCAAGAGGA AGATCTTAGA   280
TGGATGAGAG TGTACAACTA CAAGGGAAAC TTTAGCATCT GTCATTCAGT CTTACCACAT TTTGTTTTGT TTTGTTTTAA AAAGGGCAAG AATTATTTGC   380
CATCCTTGTA CCTATAAAGC CTTGGTGCAT TATAATGCTA GTTAATGGAA TAAAACATTT TATGGTAAGA TTTGTTTTCT TTAGTTATTA ATTTCTTGCT   480
ACTTGTCCAT AATAAGCAGA ACTTTTAGTG TTAGTACAGT TTTGCTGAAA GGTTATTGTT GTGTTTGTCA AGACAGAAGA AAAAGCAAAC GAATTATCTT   580
TGGAAATATC TTTGCAGTAT CAGAAGAGAT TAGTTAGTAA GGCAATACGC TTTTCCGCAG TAATGGTATT CTTTTAAATT ATGAATCCAT CTCTAAAGGT   680
TACATAGAAA CTTGAAGGAG AGAGGAACAT TCAGTTAAGA TAGTCTAGGT TTTTCTACTG AAGCAGCAAT TACAGGAGAA AGAGCTCTAC AGTAGTTTTC   780
AACTTTCTGT CTGCAGTCAT TAGTAAAAAT GAAAAGGTAA AATTTAACTG ATTTTATAGA TTCAAATAAT TTTCCTTTTA GGATGGATTC TTTAAAACTC   880
CTAATATTTA TCAAATGCTT ATTTAAGTGT CACACACAGT TAAGAAATTT GTACACCTTG TCTCCTTTAA TTCTCATAAC AACTCCATAA AATGGGTCCT   980
AGGATTTCCA TTTGAAGATA AGAAACCTGA AGCTTGCCGA AGCCCTGTGT CTGCTCTCCT TAATCTCTGT GAGAGTGCCA TCTCTTCCTG GGGACTTGTA  1080
                                                                                      -1Thr Ala Asp Ser Gly Glu
GGCATGCCAC TGTCTCCTCT TCTGGCTAAC ATTGCTGTTG CTCTCTTTTG TGTATGTGAA TGAATCTTTA AAG ]| ACT GCA GAT AGT GGT GAA  1171

Gly Asp Phe Leu Ala Glu Gly Gly Gly Val Arg Gly Pro Arg Val Val Glu Arg His Gln Ser Ala Cys Lys Asp Ser Asp
GGT GAC TTT CTA GCT GAA GGA GGA GGC GTG CGT GGC CCA AGG GTT GTG GAA AGA CAT CAA TCT GCC TGC AAA GAT TCA GAC   1252

Trp Pro Phe Cys Ser Asp Glu Asp Trp41
TGG CCC TTC TGC TCT GAT GAA GAC TGG  |[  GTAAGCAGTC AGCGGGGGAA GCAGGAGATT CCTTCCCTCT GATGCTAGAG GGGCTCACAG   1339
GCTGACCTGA TTGGTCCCAG AAACTTTTTT AAATAGAAAA TAATTGAATA GTTACCTACA TAGCAAATAA AGAAAAGGAA CCTACTCCCA AGAGCACTGT  1439
TTATTTACCT CCCCAACTCT GGATCATTAG TGGGTGAACA GACAGGATTT CAGTTGCATG CTCAGGCAAA ACCAGGCTCC TGAGTATTGT GGCCTCAATT  1539
TCCTGGCACC TATTTATGGC TAAGTGGACC CTCATTCCAG AGTTTCTCTG CGACCTCTAA CTAGTCCTCT TACCTACTTT TAAGCCAACT TATCTGGAAG  1639
AGAAAGGGTA GGAAGAAATG GGGGCTGCAT GGAAACATGC AAAATTATTC TGAATCTGAG AGATAGATCC TTACTGTAAT TTTCTCCCTT CACTTTCAG ]| 1738
42Asn Tyr Lys Cys Pro Ser Gly Cys Arg Met Lys Gly Leu Ile Asp Glu Val Asn Gln Asp Phe Thr Asn Arg Ile Asn Lys
  AAC TAC AAA TGC CCT TCT GGC TGC AGG ATG AAA GGG TTG ATT GAT GAA GTC AAT CAA GAT TTT ACA AAC AGA ATA AAT AAG  1819

Leu Lys Asn Ser Leu Phe Glu Tyr Gln Lys Asn Asn Lys Asp Ser His Ser Leu Thr Thr Asn Ile Met Glu Ile Leu Arg
CTC AAA AAT TCA CTA TTT GAA TAT CAG AAG AAC AAT AAG GAT TCT CAT TCG TTG ACC ACT AAT ATA ATG GAA ATT TTG AGA   1900

Gly Asp Phe Ser Ser Ala Asn102
GGC GAT TTT TCC TCA GCC AAT A    |[ GTAAGTATT ACATATTTAC TTCTTTGACT TTATAACAGA AACAACAAAA ATCCTAAATA AATATGATAT   1991
CCGCTTATAT CTATGACAAT TTCATCCCAA AGTACTTAGT GTAGAAACAC ATACCTTCAT AATATCCCTG AAAATTTTAA GAGGGAGCTT TTGTTTTCGT   2091
TATTTTTTCA AAGTAAAAGA TGTTAACTGA GATTGTTTAA GGTCACAAAA TAAGTCAGAA TTTTGGATTA AAACAAGAAT TTAAATGTGT TCTTTTCAAC   2191
AGTATATACT GAAAGTAGGA TGGGTCAGAC TCTTTGAGTT GATATTTTTG TTTCTGCTTT GTAAAGGTGA AAACTGAGAG GTCAAGGAAC TTGTTCAAAG   2291
ACACAGAGCT GGGAATTCAA CTCCCAGACT CCACTGAGCT GATTAGGTAG ATTTTTAAAT TTAAAATATA GGGTCAAGCT ACGTCATTCT CACAGTCTAC   2391
TCATTAGGGT TAGGAAACAT TGCATTCACT CTGGGCATGG ACAGCGAGTC TAGGGAGTCC TCAGTTTCTC AAGTTTTGCT TTGCCTTTTT ACACCTTCAC   2491
AAACACTTGA CATTTAAAAT CAGTGATGCC AACACTAGCT GGCAAGTGAG TGATCCTGTT GACCCAAAAC AGCTTAGGAA CCATTTCAAA TCTATAGAGT   2591
TAAAAAGAAA AGCTCATCAG TAAGAAAATC CAATATGTTC AAGTCCCTTG ATTAAGGATG TTATAAAATA ATTGAAATGC AATCAAACCA ACTATTTTAA   2691
CTCCAAATTA CACCTTTAAA ATTCCAAAGA AAGTTCTTCT TCTATATTTC TTTGGGATTA CTAATTGCTA TTAGGACATC TTAACTGGCA TTCATGGAAG   2791
GCTGCAGGGC ATAACATTAT CCAAAAGTCA AATGCCCCAT AGGTTTTGAA CTCACAGATT AAACTGTAAC CAAAATAAAA TTAGGCATAT TTACAAGCTA   2891
GTTTCTTTCT TTCTTTTTTC TCTTTCTTTC TTTCTTTCTT TCTTTCTTTC TTTCTTTCTT TCTTTCTTTC TTTCTCCTTC CTTCCTTTCT TCCTTTCTTT   2991
                                                                          103Asn Arg Asp Asn Thr Tyr Asn Arg Val Ser
TTTGCTGGCA ATTACAGACA AATCACTCAG CAGCTACTTC AATAACCATA TTTTCGATTT CAG ]|  AC CGT GAT AAT ACC TAC AAC CGA GTG TCA 3083

Glu Asp Leu Arg Ser Arg Ile Glu Val Leu Lys Arg Lys Val Ile Glu Lys Val Gln His Ile Gln Leu Leu Gln Lys Asn
GAG GAT CTG AGA AGC AGA ATT GAA GTC CTG AAG CGC AAA GTC ATA GAA AAA GTA CAG CAT ATC CAG CTT CTG CAG AAA AAT   3164

Val Arg Ala Gln Leu Val Asp Met Lys Arg Leu Glu Val152
GTT AGA GCT CAG TTG GTT GAT ATG AAA CGA CTG GAG GT  |[ AAGTATGTG GCTGTGGTCC CGAGTGTCCT TGTTTTTGAG TAGAGGGAAA   3251
AGGAAGGCGA TAGTTATGCA CTGAGTGTCT ACTATATGCA GAGAAAAGTG TTATATCCAT CATCTACCTA AAAGTAGGTA TTATTTTCCT CACTCCACAG  3351
TTGAAGAAAA AAAAATTCAG AGATATTAAG TAAATTTTCC AACGTACATA GATAGTAATT CAAAGCAATG TTCAGTCCCT GTCTATTCCA AGCCATTACA  3451
TCACCACACC TCTGAGCCCT CAGCCTGAGT TCACCAAGGA TCATTTAATT AGCGTTTCCT TTGAGAGGGA ATAGCACCTT ACTCTTGATC CATTCTGAGG  3551
CTAAGATGAA TTAAACAGCA TCCATTGCTT ATCCTGGCTA GCCCTGCAAT ACCCAACATC TCTTCCACTG AGGGTGCTCG ATAGGCAGAA AACAGAGAAT  3651
ATTAAGTGGT AGGTCTCCGA GTCAAAAAAA ATGAAACCAG TTTCCAGAAG GAAAATTAAC TACCAGGAAC TCAATAGACG TAGTTTATGT ATTTGTATCT  3751
                                            153Asp Ile Asp Ile Lys Ile Arg Ser Cys Arg Gly Ser Cys Ser Arg Ala
ACATTTTCTC TTTATTTTTC TCCCCTCTCT CTAGGT   ]| G GAC ATT GAT ATT AAG ATC CGA TCT TGT CGA GGG TCA TGC AGT AGG GCT    3836

Leu Ala Arg Glu Val Asp Leu Lys Asp Tyr Glu Asp Gln Gln Lys Gln Leu Glu Gln Val Ile Ala Lys Asp Leu Leu Pro
TTA GCT CGT GAA GTA GAT CTG AAG GAC TAT GAA GAT CAG CAG AAG CAA CTT GAA CAG GTC ATT GCC AAA GAC TTA CTT CCC   3917

Ser Arg Asp Arg Gln His Leu Pro Leu Ile Lys Met Lys Pro Val Pro Asp Leu Val Pro Gly Asn Phe Lys Ser Gln Leu
TCT AGA GAT AGG CAA CAC TTA CCA CTG ATA AAA ATG AAA CCA GTT CCA GAC TTG GTT CCC GGA AAT TTT AAG AGC CAG CTT   3998

Gln Lys Val Pro Pro Glu Trp Lys Ala Leu Thr Asp Met Pro Gln Met Arg Met Glu Leu Glu Arg Pro Gly Gly Asn Glu
CAG AAG GTA CCC CCA GAG TGG AAG GCA TTA ACA GAC ATG CCG CAG ATG AGA ATG GAG TTA GAG AGA CCT GGT GGA AAT GAG   4079

Ile Thr Arg Gly Gly Ser Thr Ser Tyr Gly Thr Gly Ser Glu Thr Glu Ser Pro Arg Asn Pro Ser Ser Ala Gly Ser Trp
ATT ACT CGA GGA GGC TCC ACC TCT TAT GGA ACC GGA TCA GAG ACG GAA AGC CCC AGG AAC CCT AGC AGT GCT GGA AGC TGG   4160

Asn Ser Gly Ser Ser Gly Pro Gly Ser Thr Gly Asn Arg Asn Pro Gly Ser Ser Gly Thr Gly Gly Thr Ala Thr Trp Lys
AAC TCT GGG AGC TCT GGA CCT GGA AGT ACT GGA AAC CGA AAC CCT GGG AGC TCT GGG ACT GGA GGG ACT GCA ACC TGG AAA   4241

Pro Gly Ser Ser Gly Pro Gly Ser Ala Gly Ser Trp Asn Ser Gly Ser Ser Gly Thr Gly Ser Thr Gly Asn Gln Asn Pro
CCT GGG AGC TCT GGA CCT GGA AGT GCT GGA AGC TGG AAC TCT GGG AGC TCT GGA ACT GGA AGT ACT GGA AAC CAA AAC CCT   4322

Gly Ser Pro Arg Pro Gly Ser Thr Gly Thr Trp Asn Pro Gly Ser Ser Glu Arg Gly Ser Ala Gly His Trp Thr Ser Glu
GGG AGC CCT AGA CCT GGT AGT ACC GGA ACC TGG AAT CCT GGC AGC TCT GAA CGC GGA AGT GCT GGG CAC TGG ACC TCT GAG   4403

Ser Ser Val Ser Gly Ser Thr Gly Gln Trp His Ser Glu Ser Gly Ser Phe Arg Pro Asp Ser Pro Gly Ser Gly Asn Ala
AGC TCT GTA TCT GGT AGT ACT GGA CAA TGG CAC TCT GAA TCT GGA AGT TTT AGG CCA GAT AGC CCA GGC TCT GGG AAC GCG   4484
```

Fig. 2. Nucleotide sequence of the gene coding for the Aα chain of human fibrinogen. Three single base differences with the cDNA sequence in exon V are underlined (nucleotides 4325, 4328, and 4805).

```
Arg Pro Asn Asn Pro Asp Trp Gly Thr Phe Glu Glu Val Ser Gly Asn Val Ser Pro Gly Thr Arg Arg Glu Tyr His Thr
AGG CCT AAC AAC CCA GAC TGG GGC ACA TTT GAA GAG GTG TCA GGA AAT GTA AGT CCA GGG ACA AGG AGA GAG TAC CAC ACA      4565

Glu Lys Leu Val Thr Ser Lys Gly Asp Lys Glu Leu Arg Thr Gly Lys Glu Lys Val Thr Ser Gly Ser Thr Thr Thr Thr
GAA AAA CTG GTC ACT TCT AAA GGA GAT AAA GAG CTC AGG ACT GGT AAA GAG AAG GTC ACC TCT GGT AGC ACA ACC ACC ACG      4646

Arg Arg Ser Cys Ser Lys Thr Val Thr Lys Thr Val Ile Gly Pro Asp Gly His Lys Glu Val Thr Lys Glu Val Val Thr
CGT CGT TCA TGC TCT AAA ACC GTT ACT AAG ACT GTT ATT GGT CCT GAT GGT CAC AAA GAA GTT ACC AAA GAA GTG GTG ACC      4727

Ser Glu Asp Gly Ser Asp Cys Pro Glu Ala Met Asp Leu Gly Thr Leu Ser Gly Ile Gly Thr Leu Asp Gly Phe Arg His
TCC GAA GAT GGT TCT GAC TGT CCC GAG GCA ATG GAT TTA GGC ACA TTG TCT GGC ATA GGT ACT CTG GAT GGG TTC CGC CAT      4808

Arg His Pro Asp Glu Ala Ala Phe Phe Asp Thr Ala Ser Thr Gly Lys Thr Phe Pro Gly Phe Phe Ser Pro Met Leu Gly
AGG CAC CCT GAT GAA GCT GCC TTC TTC GAC ACT GCC TCA ACT GGA AAA ACA TTC CCA GGT TTC TTC TCA CCT ATG TTA GGA      4889

Glu Phe Val Ser Glu Thr Glu Ser Arg Gly Ser Glu Ser Gly Ile Phe Thr Asn Thr Lys Glu Ser Ser Ser His His Pro
GAG TTT GTC AGT GAG ACT GAG TCT AGG GGC TCA GAA TCT GGC ATC TTC ACA AAT ACA AAG GAA TCC AGT TCT CAT CAC CCT      4970

Gly Ile Ala Glu Phe Pro Ser Arg Gly Lys Ser Ser Ser Tyr Ser Lys Gln Phe Thr Ser Ser Thr Ser Tyr Asn Arg Gly
GGG ATA GCT GAA TTC CCT TCC CGT GGT AAA TCT TCA AGT TAC AGC AAA CAA TTT ACT AGT AGC ACG AGT TAC AAC AGA GGA      5051

Asp Ser Thr Phe Glu Ser Lys Ser Tyr Lys Met Ala Asp Glu Ala Gly Ser Glu Ala Asp His Glu Gly Thr His Ser Thr
GAC TCC ACA TTT GAA AGC AAG AGC TAT AAA ATG GCA GAT GAG GCC GGA AGT GAA GCC GAT CAT GAA GGA ACA CAT AGC ACC      5132

Lys Arg Gly His Ala Lys Ser Arg Pro Val Arg Gly Ile His Thr Ser Pro Leu Gly Lys Pro Ser Leu Ser Pro ***
AAG AGA GGG CAT GCT AAA TCT CGC CCT GTC AGA GGT ATC CAC ACT TCT CCT TTG GGG AAG CCT TCC CTG TCC CCC TAG          5210

ACTAAGTTAA ATATTTCTGC ACAGTGTTCC CATGGCCCCT TGCATTTCCT TCTTAACTCT CTGTTACACG TCATTGAAAC TACACTTTTT TGGTCTGTTT   5310
TTGTGCTAGA CTGTAAGTTC CTTGGGGGCA GGGCCTTTGT CTGTCTCATC TCTGTATTCC CAAATGCCTA ACAGTACAGA GCCATGACTC AATAAATACA   5410
TGTTAAATGG ATGAATG↓AAT TCCTCTGAAA CTCTATTTGA GCTTATTTAG TCAAATTCTT TCACTATTCA AAGTGTGTGC TATTAGAATT GTCACCCAAC  5510
TGATTAATCA CATTTTTAGT ATGTGTCTCA GTTGACATTT AGGTCAGGCT AAATACAAGT TGTGTTAGTA TTAAGTGAGC TTAGCTACCT GTACTGGTTA   5610
CTTGCTATTA GTTTGTGCAA GTAAAATTCC AAATACATTT GAGGAAAATC CCCTTTGCAA TTTGTAGGTA TAAATAACCG CTTATTTGCA TAAGTTCTAT   5710
CCCACTGTAA GTGCATCCTT TCCCTATGGA GGGAAGGAAA GGAGGAAGAA AGAAAGGAAG GGAAAGAAAC AGTATTTGCC TTATTTAATC TGAGCCGTGC   5810
CTATCTTTGT AAAGTTAAAT GAGAATAACT TCTTCCAACC AGCTTAATTT TTTTTTTAGA CTGTGATGAT GTCCTCCAAA CACATCCTTC AGGTACCCAA   5910
AGTGGCATTT TCAATATCAA GCTATCCGGA TCC                                                                           5943
```

Fig. 2 (Continued).

B-beta Gene

The gene for the Bβ chain is 8 kilobases in length and consists of 8 exons. It is transcribed in the opposite direction to the γ and Aα gene. Three polyadenylation sites have been identified in the 3' flanking region of the gene (nucleotides 8202, 8268 and 8534) (7). The utilization of these sites does not alter the coding region and does not alter the amino acids of the polypeptide. Among the three fibrinogen genes, the Bβ gene is the only one that contains repetitive elements. Two copies of Alu family repeat, flanked with inverted repeats are present. They are located in intron E (nucleotides 5996-6342) and in the immediate 3' flanking region of the gene (nucleotides 8676-8878). A putative promoter sequence (TATA

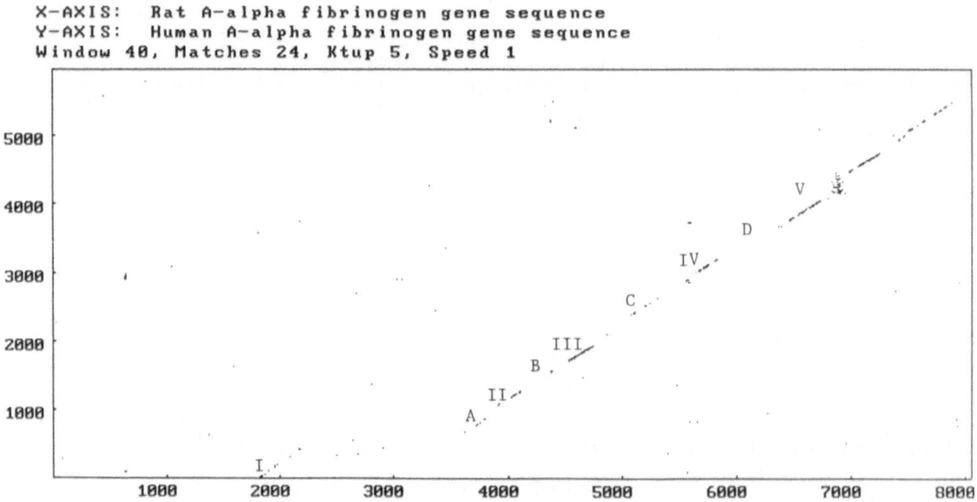

Fig. 3. Matrix comparison of the nucleotide sequences of the genes coding for the Aα chains of human and rat fibrinogen. Exons I to V and introns A to D are identified by the line of homology. Each dot represents 60% or higher sequence identity (24 out of 40 nucleotides).

```
GAATTCATGC CCCTTTTGAA ATAGACTTAT GTCATTGTCA GAAAACATAA GCATTTATGG TATATCATTA ATGAGTCACG ATTTTAGTGG TTGCCTTGTG   100
AGTAGGTCAA ATTTACTAAG CTTAGATTTG TTTTCTCACA TATTCTTTCG GAGCTTGTGT AGTTTCCACA TTAATTTACC AGAAACAAGA TACACACTCT   200
CTTTGAGGAG TGCCCTAACT TCCCATCATT TTGTCCAATT AAATGAATTG AAGAAATTTA ATGTTTCTAA ACTAGACCAA CAAAGAATAA TAGTTGTATG   300
ACAAGTAAAT AAGCTTTGCT GGGAAGATGT TGCTTAAATG ATAAAATGGT TCAGCCAACA AGTGAACCAA AAATTAAATA TTAACTAAGG AAAGGTAACC   400
                                                                                                  -22Phe His
ATTTCTGAAG TCATTCCTAG CAGAGGACTC AGATATATAT AGGATTGAAG ATCTCTCAGT TA AGTCTACA TGAAAAGGAT GGTTTCTTGG AGC TTC CAC   499

Lys Leu Lys Thr Met Lys His Leu Leu Leu Leu Leu Leu Cys Val Phe Leu Val Lys Ser 1Gln Gly Val Asn Asp Asn Glu Glu8
AAA CTT AAA ACC ATG AAA CAT CTA TTA TTG CTA CTA TTG TGT GTT TTT CTA GTT AAG TCC  CAA GGT GTC AAC GAC AAT GAG GAG   583
||[GTGAATTTTT TAAAGCATTA TTATATTATT AGTAGTATTA TTAATATAAG ATGTAACATA ATCATATTAT GTGCTTATTT TAATGAAATT AGCATTGCTT   683
ATAGTTATGA AATGGAATTG TTAACCTCTG ACTTATTGTA TTTAAAGAAT GTTTCATAGT ATTTCTTATA TAAAAACAAA GTAATTTCTT GTTTTCTAGT   783
TTATCACCTT TGTTTTCTTA AGATGAGGAT GGCTTAGCTA ATGTAAGATG TGTTTTTCTC ACTTGCTATT CTGAGTACTG TGATTTTCAT TTACTTCTAG   883
CAATACAGGA TTACAATTAA GAGGACAAGA TCTGAAAATC TCACAAACTA TAAAATAATA AAAGAGCAGA ATTTTAAGAT AAAAGAAACT GGTGGTAGGT   983
AGATTGTTCT TTGGTGAAGG AAGGTAATAT ATATTGTTAC TGAGATTACT ATTTATAAAA ATTATAACTA AGCCTAAAAG CAAAATACAT CAAGTGTAAT  1083
GATAGAAAAT GAAATATTGC TTTTTTCAGA TGAAAAGTTC AAATTAGAGT TAGTGTGTAT TGTTATTATT AATAGTTATG AAACACGGTT CAGTCTAATT  1183
TATTTATTTG TAGAACAGTT TGTCCTCAAC TATTATTTTT GCTGACTTAT TGCTGTTAAT TTGCAGTTAC TAAAAATACA GAAATGCATT TAGGACAATG  1283
GATATTTAAG AAATTTAAAT TTTATCATCA AACGTATCAT GGCCAAATTT CTTACATATA GCATAGTATC ATTAAACTAG AAATAAGAAT ACACAATAAT  1383
ATTTAAATGA AGTGATTCAT TTCGGATCAT TATTGAGTTT CAAGGGAACT TGAGTGTTGT ACTTATCAGA CTCTACATGT AAGAACATAT AGTTAATCTG  1483
GTTGTGTGTG TAAAAACATA TGGTTAATCT GGTTAAGTCT GGTTAATCAT ATTAGGTAAG AAAAATGTAA AGAATGTGTA AGACGAAATT TTTGTAAAGT  1583
ACTCTGCAAA GCACTTTCAC ATTTCTGCTT ATCAACTAAA CCTCACAGAG ATAGTTTAAT AGTTTAGGCT TTAAAATGGA TTTTGATTAT TCAACAAGTG  1683
GCCTTCATAA TTTCTTTAAG TGTTTTTCTT TAAGTATATA CTTTCTTTAA ATATTTTTTA AAATTTCCTT TTCTCTAGTA AAGCCAGACC ATCCATGCTA  1783
CCTCTCTAGT GGCACTCTGA AATAAAAAGA AAATAGTTTT CTCTGTTATA ATTGTATTTG TAATAAGCAG ATGAATCACA TTTCTTAAAA TTTGTTTTAG  1883
AGAGGGTAAG CTCTGACTAG GACCATGACT TCAATGTGAA ATATGTATAT ATCCTCCGAA TCTTTACATA TTAAGAATGT ATATAGTCAA CTGGTTAAAC  1983
AGGAAAATCT GGAACAGCCT GGCTGGGTTT TAATCTTAGC ACCATCCTAC TAAATGTTAA ATAATATTAT AATCTAATGA ATAAATGACA ATGCAATTCC  2083
AAATAGAGTT CATCTGATGA CTTCTAGACT CACAAAATTG CAAGAGAGCT CAGTTGTTGC TCAGTTGTTC CAAATCATGT CGTTTGTTAA TTTGTAATTA  2183
AGCTCCAAAG GATGTATAGC TACTGACAAA AAAAAAAATG AGAATGTAGT TAATCCAAAT CAAAACTTTC CTATTGCAAT GCGTATTTTC TGCTTCATTA  2283
TCCTTTAATA TAATATTTTA AGTTAGCAAG TAATTTTAAT TACAATGCAC AAGCCTTGAG AATTATTTTA AATATAAGAA AATCATAATG TTTGATAAAG  2383
AAATCATGTA AGAAATTTCA AGATAATGGT TTAACAAATA ATTTTGTTGA TAGAAGATAA GACTAAAAGT GAAATTCGAA GTGGAGAGGA CACTTAAACT  2483
GTAGTACTTG TTATGTGTGA TTCCAGTAAA AATAGTAATG AGCACTTATT ATTGCCAAGT ACTGTTCTGA GGGTACCATA TGCAATAAGT TATTTAATCC  2583
TTACAATAAT CTTGTAAGGC AGATTCAAAC TATCATTACA CTTATTTTAC AGATGAGAAA ACTGGGGCAC AGATAAAGCA ACTTGCCCAA GGTCTCATAG  2683
CTGTAAGTCA ACCCTACGGT CAAGACCTAC AAGTAGCCGA GCTCCAGAGT ACATTATGAG GGTCAAAGAT TGTCTTATTA CAAATAAATT CCAAGTAGAA  2783
TCAACCTTTA ATAAGTCTTT AATGTCTCTT AAATATGTTT ATATAGGAGT CTAATCACCA ATTCACAAAA ATGAAAGTAG GGAAATGATT AACAATAATC  2883
ATAGGAATCT AACAATCCAA GTGGCTTGAG AATATTCATT CTTCTTGACA GTATAGATTC TTTACAATTT CGTAAGTTCC AATGTATGTT TTAGGAATAT  2983
GAGGTCATTA CTATTCATAA TCTGATACAG CTTTATCCTA AGGCCTCTCT TTAAAAACTA CACTGCATCA TAGCTTTTTT GTGCAGTTGG TCTTTCTACT  3083
GTTACTGAAC AGTAAGCAAC CTACAGATTC ACTATCACCA ACCAGCCAGT TGATGGATCT TAAGCAAATT ATCAAGCTTG TGATAACCTA AATTATAAAA  3183
                                                                                     9Gly Phe Phe Ser Ala Arg Gly
TGAGGGTGTT GGAATAGTTA CATTCCAAAT CTTCTATAAC ACTCTGTATT ATATTTCTGC CTCATTCCTT GTAG]| GGT TTC TTC AGT GCC CGT GGT  3278

His Arg Pro Leu Asp Lys Lys Arg Glu Glu Ala Pro Ser Leu Arg Pro Ala Pro Pro Pro Ile Ser Gly Gly Gly Tyr Arg Ala
CAT CGA CCC CTT GAC AAG AAG AGA GAA GAG GCT CCC AGC CTG AGG CCT GCC CCA CCG CCC ATC AGT GGA GGT GGC TAT CGG GCT  3362

Arg Pro Ala Lys Ala Ala Ala Thr Gln Lys Lys Val Glu Arg Lys Ala Pro Asp Ala Gly Gly Cys Leu His Ala Asp Pro Asp
CGT CCA GCC AAA GCA GCT GCC ACT CAA AAG AAA GTA GAA AGA AAA GCC CCT GAT GCT GGA GGC TGT CTT CAC GCT GAC CCA GAC  3446
   72
Leu
CTG ||[GTGGGTGCAC TGATGTTTCT TGCAGTGGTG GCTCTCTCAT GCAGAGAAAG CCTGTAGTCA TGGCAGTCTG CTAATGTTTC ACTGACCCAC          3539
ATTACCATCA CTGTTATTTT GTTTGTTTAT TTTGGAAATA AAATTCAAAA CATAAACATA TTGGGCCTTT GGTTTAGGCT TTCTTTCTTG TTTTCTTTGG  3639
TCTGGGCCCA AAATTTCAAA TTAGGATATG TGGGTGCCAC CTTTCCATTT GTATTTTGCC ACTGCCTTTG TTTAGTTGGT AAAATTTTCA TAGCCCAATT  3739
ATATTTTTTC TGGGGTAAGT AATATTTTAA ATCTCTATGA GAGTATGATG ATGACTTTCG AATTTCTGGT CTTACAGAAA ACCAAATAAT AAATTTTTAT  3839
                                                                                                       73Gly
GTTGGCTAAT CGTATCGCTG AATTTTCCTA TGTGCTATTT TAACAAATGT CCATGACCCA AATCCTTCAT CTAATGCCTG CTATTTTCTT TGTTTTTAG]| GGG  3941

Val Leu Cys Pro Thr Gly Cys Gln Leu Gln Glu Ala Leu Leu Gln Gln Glu Arg Pro Ile Arg Asn Ser Val Asp Glu Leu Asn
GTG TTG TGT CCT ACA GGA TGT CAG TTG CAA GAG GCT TTG CTA CAA CAG GAA AGG CCA ATC AGA AAT AGT GTT GAT GAG TTA AAT  4025

Asn Asn Val Glu Ala Val Ser Gln Thr Ser Ser Ser Ser Phe Gln Tyr Met Tyr Leu Leu Lys Asp Leu Trp Gln Lys Arg Gln
AAC AAT GTG GAA GCT GTT TCC CAG ACC TCC TCT TCT TCC TTT CAG TAC ATG TAT TTG CTG AAA GAC CTG TGG CAA AAG AGG CAG  4109
               133
Lys Gln Val Lys
AAG CAA GTA AAA G ||[GTAGATATC CTTGTGCTTT CCATTCGATT TTCAGCTATA AAATTGGAAC CGTTAGACTG CCACGAGAAT GCATGGTTGT     4201
GAGAAGATTA ACATTTCTGG GTTAGTGAAT AGCATTCATA CGCTTTTGGG CACCTTCCCC TGCAACTTGC CAGATAAGCA CTATTCAGCT CTTATTCCCA  4301
GTCTGACATC AGCAAGTGTG ATTTTCTATG AAAAATTCTA CTATGACTCC TTATTTTAAG TATACAAGAA ACTTGTGACT CAGAAGATAA TATTTACAGA  4401
GTGGAAAAAA ACCCCTAGCA TTTATAGTTT TAACATTTGA GGTTTTGAAT GAGAGAGTTA TCCATAATAT ATTCAATTGT GTTGTGGATA ATGACACCTA  4501
ACCTGTGAAT CTTGAGGTCA GAATGTTGAG TGCTGTTGAC TTGGTGGTCA GGAAACAGCT AGTGCGTGAG CCTGGCACAG GCATCTCAGT GAGTAGCATA  4601
CCCACAGTTG GAAATTTTTC AAAGAAATCA AAGGAATCAT GACATCTTAT AAATTTCAAG GTTCTGCTAT ACTTATGTGA AATGGATAAA TAAATCAAGC  4701
ATATCCACTC TGTAAGATTG AACTTCTCAG ATGGAAGACC CCAATACTGC TTTCTCCTCT TTTCCCTCAC CAAAGAAATA AACAACCTAT TTCATTTATT  4801
ACTGGACACA ATCTTTAGCG TATACCTATG GTAAATTACT AGTATGGTGG TTAGGATTTA TGTTAATTTG TATATGTCAT GCGCCAAATC ATTTCCACTA  4901
AATATGACTA TATATCATAA CTGCTTGGTG ATAGCTCAGT GTTTAATAGT TTATTCTCAG AAAATCAAAA TTGTATAGTT AAATACATTA GTTTTATGAG  5001
                                           134Asp Asn Glu Asn Val Val Asn Glu Tyr Ser Ser Glu Leu Glu Lys His Gln
GCAAAAATGC TAACTATTTC TACATAATTT CATTTTTCCA G]| AT AAT GAA AAT GTA GTC AAT GAG TAC TCC TCA GAA CTG GAA AAG CAC CAA  5092

Leu Tyr Ile Asp Glu Thr Val Asn Ser Asn Ile Pro Thr Asn Leu Arg Val Leu Arg Ser Ile Leu Glu Asn Leu Arg Ser Lys
TTA TAT ATA GAT GAG ACT GTG AAT AGC AAT ATC CCA ACT AAC CTT CGT GTG CTT CGT TCA ATC CTG GAA AAC CTG AGA AGC AAA  5176

Ile Gln Lys Leu Glu Ser Asp Val Ser Ala Gln Met Glu Tyr Cys Arg Thr Pro Cys Thr Val Ser Cys Asn Ile Pro Val Val
ATA CAA AAG TTA GAA TCT GAT GTC TCA GCT CAA ATG GAA TAT TGT CGC ACC CCA TGC ACT GTC AGT TGC AAT ATT CCT GTG GTG  5260
```

Fig. 4. Nucleotide sequence of the gene coding for the Bß chain of human fibrinogen.

```
Ser Gly Lys209
TCT GGC AAA G ||[GTAACTGAT TCATAAACAT ATTTTTAGAG AGTTCCAGAA GAACTCACAC ACCAAAAATA AGAGAACAAC AACAACAACA AAAATGCTAA   5359
GTGGATTTTC CCAACAGATC ATAATGACAT TACAGTACAT CATAAAAATA TCCTTAGCCA GTTGTGTTTT GGACTGGCCT GGTGCATTTG CTGGTTTTGA   5459
TGAGCAGGAT GGGGCACAGG TAGTCCCAGG GGTGGCTGAT GTGTGCATCT GCGTACTGGC TTGAACAGAT GGCAGAACCA CAGATAGATG TAGAAGTTTC   5559
TCCATTTTGT GTGTTCTGGG AGCTCATGGA TATTCCAGGA CACAAAAGGT GGAGAAGAGC TTTGTTCATC CTCTTAGCAG ATAAACGTCC TCAAAACTGG   5659
GTTGGACTTA CTAAAGTAAA ATGAAAATCT AATATTTGTT ATATTATTTT CAAAGGTCTA TAATAACACA CTCCTTAGTA ACTTATGTAA TGTTATTTTA   5759

                                                                           210Glu Cys Glu Glu Ile Ile Arg Lys Gly
AAGAATTGGT GACTAAATAC AAAGTAATTA TGTCATAAAC CCCTGAACAT AATGTTGTCT TACATTTGCA G]| AA TGT GAG GAA ATT ATC AGG AAA GGA   5856

Gly Glu Thr Ser Glu Met Tyr Leu Ile Gln Pro Asp Ser Ser Val Lys Pro Tyr Arg Val Tyr Cys Asp Met Asn Thr Glu Asn
GGT GAA ACA TCT GAA ATG TAT CTC ATT CAA CCT GAC AGT TCT GTC AAA CCG TAT AGA GTA TAC TGT GAC ATG AAT ACA GAA AAT   5940

Gly247
GGA G ||[GTAAGCTTTC GACAGTTGTT GACCTGTTGA TCTGTAATTA TTTGGATACC GTAAAATGCC AGGAAACAAG GCCAGGTGTG GTGGCTCATA   6034
CCTGTAATTC CAGCACCTTG GGAGGCCAAA GTGGGCTGAT AGCTTGAGCC TAGGAGTTTG AAACTAGCCT GGGCAACATA ATGAGACCCT AACTCTACAA   6134
AAAAAAAAAA AATACCAAAA AAAAAAAAAA AATCAGCTGT GTTGGTAGTA TGTGCCTGTA GTCCCAGCTA TCCAGGAGGC TGAGATGGGA GATCACCTGA   6234
GCCCACAACC TGGAGTCTTG ATCATGCTAC TGAACTGTAG CCTGGGCAAC AGAGGATAGT GAGATCCTGT CTCAAAAAAA AAAATTAATT AAAAAGCCAG   6334
GAAACAAGAC TTAGCTCTAA CATCTAACAT AGCTGACAAA GGAGTAATTT GATGTGGAAT TCAACCTGAT ATTTAAAAGT TATAAAATAT CTATAATTCA   6434
CAATTTGGGG TAAGATAAAG CACTTGCAGT TTCCAAAGAT TTTACAAGTT TACCTCTCAT ATTTATTTCC TTATTGTGTC TATTTTAGAG CACCAAATAT   6534

                                                                                                       248Gly
ATACTAAATG GAATGGACAG GGGATTCAGA TATTATTTTC AAAGTGACAT TATTTGCTGT TGGTTAATAT ATGCTCTTTT TGTTTCTGTC AACCAAAG]| GA   6634

Trp Thr Val Ile Gln Asn Arg Gln Asp Gly Ser Val Asp Phe Gly Arg Lys Trp Asp Pro Tyr Lys Gln Gly Phe Gly Asn Val
TGG ACA GTG ATT CAG AAC CGT CAA GAC GGT AGT GTT GAC TTT GGC AGG AAA TGG GAT CCA TAT AAA CAG GGA TTT GGA AAT GTT   6718

Ala Thr Asn Thr Asp Gly Lys Asn Tyr Cys Gly Leu Pro289
GCA ACC AAC ACA GAT GGG AAG AAT TAC TGT GGC CTA CCA G ||[GTAACGAACA GGCATGCAAA ATAAAATCAT TCTATTTGAA ATGGGATTTT   6808
TTTTAATTAA AAAACATTCA TTGTTGGAAG CCTGTTTTAG GCAGTTAAGA GGAGTTTCCT GACAAAAATG TGGAAGCTAA AGATAAGGGA AGAAAGGCAG   6908

                                                           290Gly Glu Tyr Trp Leu Gly Asn Asp Lys Ile Ser Gln
TTTTTAGTTT CCCAAAATTT TATTTTTGGT GAGAGATTTT ATTTTGTTTT TCTTTTAG]| GT GAA TAT TGG CTT GGA AAT GAT AAA ATT AGC CAG   7001

Leu Thr Arg Met Gly Pro Thr Glu Leu Leu Ile Glu Met Glu Asp Trp Lys Gly Asp Lys Val Lys Ala His Tyr Gly Gly Phe
CTT ACC AGG ATG GGA CCC ACA GAA CTT TTG ATA GAA ATG GAG GAC TGG AAA GGA GAC AAA GTA AAG GCT CAC TAT GGA GGA TTC   7085

Thr Val Gln Asn Glu Ala Asn Lys Tyr Gln Ile Ser Val Asn Lys Tyr Arg Gly Thr Ala Gly Asn Ala Leu Met Asp Gly
ACT GTA CAG AAT GAA GCC AAC AAA TAC CAG ATC TCA GTG AAC AAA TAC AGA GGA ACA GCC GGT AAT GCC CTC ATG GAT GGA   7166

Ala Ser Gln Leu Met Gly Glu Asn Arg Thr Met Thr Ile His Asn Gly Met Phe Phe Ser Thr Tyr Asp Arg Asp Asn Asp
GCA TCT CAG CTG ATG GGA GAA AAC AGG ACC ATG ACC ATT CAC AAC GGC ATG TTC TTC AGC ACG TAT GAC AGA GAC AAT GAC   7247

Gly Trp385
GGC TG ||[GTATGTGTGG CACTCTTTGC TCCTGCTTTA AAAATCACAC TAATATCATT ACTCAGAATC ATTAACAATA TTTTTAATAG CTACCACTTC   7342
CTGGGCACTT ACTGTCAGCC ACTGTCCTAA GCTCTTTATG CATCACTCGA AAGCATTTCA ACTATAAGGT AGACATTCTT ATTCTCATTT TACAGATGAG   7442
ATTTAGAGAG ATTACGTGAT TTGTCCAATG TCACACAACT ACCCAGAGAT AAAACTAGAA TTTGAGCACA GTTACTTTCT GAATAATGAG CATTTAGATA   7542
AATACCTATA TCTCTATATT CTAAAGTGTG TGTGAAAACT TTCATTTTCA TTTCCAGGGT TCTCTGATAC TAAGGGTTGT AAAAGCTATT ATTCCAGTAT   7642
AAAGTAACAA ACACAGTCCC TAGATGGATT GCCACAAAGG CCCAGTTATC TCTCTTTCTT GCTATAGGGC ACAGGAGGTC TTTGGTGTAT TAGTGTGACT   7742
CTATGTATAG CACCCAAAGG AAAGACTACT GTGCACACGA GTGTAGCAGT CTTTTATGGG TAATCTGCAA AACGTAACTT GACCACCGTA GTTCTGTTTC   7842

                                386Leu Thr Ser Asp Pro Arg Lys Gln Cys Ser Lys Glu Asp Gly Gly Gly Trp Trp Tyr Asn
TAATAACGCC AAACACATTT TCTTTCAG]| G TTA ACA TCA GAT CCC AGA AAA CAG TGT TCT AAA GAA GAC GGT GGT GGA TGG TGG TAT AAT   7931

Arg Cys His Ala Ala Asn Pro Asn Gly Arg Tyr Tyr Trp Gly Gly Gln Tyr Thr Trp Asp Met Ala Lys His Gly Thr Asp Asp
AGA TGT CAT GCA GCC AAT CCA AAC GGC AGA TAC TAC TGG GGT GGA CAG TAC ACC TGG GAC ATG GCA AAG CAT GGC ACA GAT GAT   8015

Gly Val Val Trp Met Asn Trp Lys Gly Ser Trp Tyr Ser Met Arg Lys Met Ser Met Lys Ile Arg Pro Phe Phe Pro Gln Gln461
GGT GTA GTA TGG ATG AAT TGG AAG GGG TCA TGG TAC TCA ATG AGG AAG ATG AGT ATG AAG ATC AGG CCC TTC TTC CCA CAG CAA   8099

stop
TAG TCC CCA ATA CGT AGA TTT TTG CTC TTC TGT ATG TGA CAA CAT TTT TGT ACA TTA TGT TAT TGG AAT TTT CTT TCA TAC ATT   8183
ATA TTC CTC TAA AAC TCT C↓AA GCA GAC GTG AGT GTG ACT TTT TGA AAA AAG TAT AGG ATA AAT TAC ATT AAA ATA GCA CAT GAT   8267
T↓TTC↓TT TTG TTT TCT TCA TTT CTC TTG CTC ACC CAA GAA GTA ACA AAA GTA TAG TTT TGA CAG AGT TGG TGT TCA TAA TTT CAG   8351
TTC TAG TTG ATT GCG AGA ATT TTC AAA TAA GGA AGA GGG GTC TTT TAT CCT TGT CGT AGG AAA ACC ATG ACG GAA AGG AAA AAC   8435
TGA TGT TTA AAA GTC CAC TTT TAA AAC TAT ATT TAT TTA TGT AGG ATC TGT CAA AGA AAA CTT CCA AAA AGA TTT ATT AAT TAA   8519
ACC AGA CTC TGT TGC↓AAT↓AAGTTAA TGTTTTCTTG TTTTGTAATC CACACATTCA ATGAGTTAGG CTTTGCACTT GTAAGGAAGG AGAAGCGTTC
ACAACCTCAA ATAGCTAATA AACCGGTCTT GAATATTTGA AGATTTAAAA TCTGACTCTA GGACGGGCAC GGTGGCTCAC GACTATAATC CCAACACTTT   8714
GGGAGGCTGA GGCGGGCGGT CACAAGGTCA GGAGTTCAAG ACCAGCCTGA CCAATATGGT GAAACCCCAT CTCTACTAAA AATACAAAAA TTAGCCAGGC   8814
GTGGTGGCAG GTGCCTGTAG GTCCCAGCTA GCCTGTGAGG TGGAGATTGC ATTGAGCCAA GATC                                            8878
```

Fig. 4 (Continued).

box) and a CCATT sequence are located 30 and 60 nucleotides from the transcription initiation site. The sequence ATTAAC, which is involved in liver specific expression, is located 83 nucleotides from the initiation site.

Gamma Gene

The gene for the γ chain is 8.5 kilobases in length. Its entire sequence has been published previously from our laboratory (15). The gene consists of 10 exons. Sequences that encode the γ' form are located in intron I (9th intron). An alternative polyadenylation site

```
         CTACACA CTTCTTGAAG GCAAAGGCAA TGCTGAAGTC ACCTTTCATG TTCAAATCAT ATTAAAAAGT TAGCAAGATG TAATTATCAG TGTACTATGT   -1651

AAATCTTTGT GAATGATCAA TAATTACATA TTTTCATTAT ATATATTTTA GTAGATAATA TTTATATACA TTCAACATTC TAAATATAGA AAGTTTACAG AGAAAAATAA   -1541

AGCCTTTTTT TCCAATCCTG TCCTCCACCT CTGCATCCCA TTCTTCTTCA CAGAGGCAAC TGATTCAAGT CATTACATAG TTATTGAGTG TTAACTACAA CTATGTTAAG   -1431

TACAGCTATA TATGTTAGAT GCCGTAGCCA CAGAAATCAG TTTACAATCT AATGCAGTGG ATACAGCATG TATACATATA ATATAAGGTT GCTACAAATG CTATCTGAGG   -1321

TAGAGCTGTT TGAAAGAATA CTAATACTTA AATGTTTAAT TCAACTGACT TGATTGACAA CTGATTAGCT GAGTGGAAAA GATGGATGAG AAAGATTGTG AGACTTAATT   -1211

GGCTGGTGGT ATGGTGATAT GATTGACAAT AACTGCTAAG TCAGAGAGGG ATATATTAAG GAGGAGAAGA AAAGCAACAA ATCTGGTTTT GATGTGTTCA CTTTGTTATA   -1101

ATTATTGATT ATTTACTGAA TATGAATATT TATCTTTGTT TTTGAGTCAA TAAATATACC TTTGTAAAGA CAGAATTAAA GTATTAGTAT TTCTTTCAAA CTGGAGGCAT    -991

TTCTCCCACT AACATATTTC ATCAAAACTT ATAATAAGCT TGGTTCCAGA GGAAGAAATG AGGGATAACC AAAAATAGAG ACATTAATAA TAGTGTAACG CCCAGTGATA    -881

AATCTCAATA GGCAGTGATG ACAGACATGT TTTCCCAAAC ACAAGGATGC TGTAAGGGCC AAACAGAAAT GATGGCCCCT CCCCAGCACC TCATTTTGCC CCTTCCTTCA    -771

GCTATGCCTC TACTCTCCTT TAGATACAAG GGAGGTGGAT TTTTCTCTTC TCTGAGATAG CTTGATGGAA CCACAGGAAC AATGAAGTGG GCTCCTGGCT CTTTTCTCTG    -661

TGGCAGATGG GGTGCCATGC CCACCTTCAG ACAAAGGGAA GATTGAGCTC AAAAGCTCCC TGAGAAGTGA GAGCCTATGA ACATGGTTGA CACAGAGGGA CAGGAATGTA    -551

TTTCCAGGGT CATTCATTCC TGGGAATAGT GAACTGGGAC ATGGGGGAAG TCAGTCTCCT CCTGCCACAG CCACAGATTA AAAATAATAA TGTTAACTGA TCCCTAGGCT    -441
                                                                                           ---------------------------------->
AAAATAATAG TGTTAACTGA TCCCTAAGCT AAGAAAGTTC TTTTGGTAAT TCAGGTGATG GCAGCAGGAC CCATCTTAAG GATAGACTAG GTTTGCTTAG TTCGAGGTCA    -331
-------------------------------->
TATCTGTTTG CTCTCAGCCA TGTACTGGAA GAAGTTGCAT CACACAGCCT CCAGGACTGC CCTCCTCCTC ACAGCAATGG ATAATGCTTC ACTAGCCTTT GCAGATAATT    -221

TTGGATCAGA GAAAAAACCT TGAGCTGGGC CAAAAAGGAG GAGCTTCAAC CTGTGTGCAA AATCTGGGAA CCTGACAGTA TAGGTTGGGG GCCAGGATGA GGAAAAAGGA    -111

ACGGGAAAGA CCTGCCCACC CTTCTGGTAA GGAGGCCCCG TGATCAGCTC CAG[CCAT]TTG CAGTCCTGGC TATCCCAGGA GCTTAC[ATAA A]GGGACAATT GGAGCCTGAG      -1

                                                  -26
                                                  Met Ser Trp Ser Leu His Pro Arg Asn Leu Ile Leu Tyr Phe Tyr Ala
AGGTGACAGT GCTGACACTA CAAGGCTCGG AGCTCCGGGC ACTCAGACATC ATG AGT TGG TCC TTG CAC CCC CGG AAT TTA ATT CTC TAC TTC TAT GCT      99

                                   -1
Leu Leu Phe Leu Ser Ser Thr Cys Val Ala
CTT TTA TTT CTC TCT TCA ACA TGT GTA GCA GTAAGT GTGCTCTTCA CAAAACGTTG TTTAAAATGG AAAGCTGGAA AATAAAACAG ATAATAAACT AGTGA     200

                          +1                                          15
                          Tyr Val Ala Thr Arg Asp Asn Cys Cys Ile Leu Asp Glu Arg Phe
AATTT TCGTATTTTT TCTCTTTTAG TAT GTT GCT ACC AGA GAC AAC TGC TGC ATC TTA GAT GAA AGA TTC GTAAGTAGT TTTTATGTTT CTCCCTTTGT     299

GTGTGAACTG GAGAGGGGCA GAGGAATAGA AATAATTCCC TCATAAATAT CATCTGGCAC TTGTAACTTT TTAAAAACAT AGTCTAGGTT TTACCTATTT TTCTTAATAG     409

                                                  16
                                                  Gly Ser Tyr Cys Pro Thr Thr Cys Gly Ile Ala Asp Phe Leu Ser Thr
ATTTTAAGAG TAGCATCTGT CTACATTTTT AATCACTGTT ATATTTTCAG GGT AGT TAT TGT CCA ACT ACC TGT GGC ATT GCA GAT TTC CTG TCT ACT     507

Tyr Gln Thr Lys Val Asp Lys Asp Leu Gln Ser Leu Glu Asp Ile Leu His Gln Val Glu Asn Lys Thr Ser Glu Val Lys Gln Leu Ile
TAT CAA ACC AAA GTA GAC AAG GAT CTA CAG TCT TTG GAA GAC ATC TTA CAT CAA GTT GAA AAC AAA ACA TCA GAA GTC AAA CAG CTG ATA     597

                                                         76
Lys Ala Ile Gln Leu Thr Tyr Asn Pro Asp Glu Ser Ser Lys Pro
AAA GCA ATC CAA CTC ACT TAT AAT CCT GAT GAA TCA TCA AAA CCA A GTGAGAAAA TAAAGACTAC TGACCAAAAA ATAATAATAA TAATCTGTGA     692

                                                                               77
                                                                               Asn Met Ile Asp Ala Ala Thr Leu Lys Ser Arg
AGTTCTTTTG CTGTTGTTTT AGTTGTTCTA TTTGCTTAAG GATTTTTATG TCTCTGATCC TATATTACAG  AT ATG ATA GAC GCT GCT ACT TTG AAG TCC AGG     794

                                                                   108
Ile Met Leu Glu Glu Ile Met Lys Tyr Glu Ala Ser Ile Leu Thr His Asp Ser Ser Ile Arg
ATA ATG TTA GAA GAA ATT ATG AAA TAT GAA GCA TCG ATT TTA ACA CAT GAC TCA AGT ATT CG  GTAAGGATTT TTGTTTTAAT TTGCTCTGCA     886

AGACTGATTT AGTTTTTATT TAATATTCTA TACTTGAGTG AAAGTAATTT TTAATGTGTT TTCCCCATTT ATAATATCCC AGTGACATTA TGCCTGATTA TGTTGAGCAT     996

AGTAGAGATA GAAGTTTTTA GTGCAATATA AATTATACTG GGTTATAATT GCTTATTAAT AATCACATTG AAGAAAGATG TTCTAGATGT CTTCAAATGC TAGTTTGACC    1106

ATATTTATCA AAAATTTTTT CCCCATCCCC CATTTATCTT ACAACATAAA ATCAATCTCA TAGGAATTTG GGTGTTGAAA ATAAAATCCT CTTTATAAAA ATGCTGACAA    1216

ATTGGTGGTT AAAAAAATTA GCAAGCAGAG GCATAGTAAG GATTTTGGCT CCTAAAGTAA ATTATATTGA ATGTGGAGCA GGAAGAAACA TGTCTTGAGA GACTAAGTGT    1326

GGCAAATATT GCAAAGCTCA TATTGATCAT TGCAGAATGA ACCTGCATAG TCTCTTCCCT TCATTTGGAA GTGAATGTCT CTGTTAAAGC TTCTCAGGGA CTCATAAACT    1436

TTCTGAACAT AAGGTCTCAG ATACAGTTTT AATATTTTTC CCCAATTTTT TTTTCTGAAT TTTTCTCAAA GCAGCTTGAG AAATTGAGAT AAATAGTAGC TAGGGAGAAG    1546

TGGCCCAGGA AAGATTTCTC CTCTTTTTGC TATCAGAGGG CCCTTGTTAT TATTGTTATT ATTATTACTT GCATTATTAT TGTCCATCAT TGAAGTTGAA GGAGGTTATT    1656

GTACAGAAAT TGCCTAAGAC AAGGTAGAGG GAAAACGTGG ACAAATAGTT TGTCTACCCT TTTTTACTTC AAAGAAAGAA CGGTTTATGC ATTGTAGACA GTTTTCTATC    1766

ATTTTTGGAT ATTTGCAAGC CACCCTGTAA GTAACTACAA AAGGAGGGTT TTTACTTCCC CCAGTCCATT CCCAAAGCTA TGTAACCAGA AGCATTAAAG AAGAAAGGGG    1876

AAGTATCTGT TGTTTTATTT TACATACAAT AACGTTCCAG ATCATGTCCC TGTGTAAGTT ATATTTTAGA TTGAAGCTTA TATGTATAGC CTCAGTAGAT CCACAAGTGA    1986

AAGGTATACT CCTTCAGCAC ATGTGAATTA CTGAACTGAG CTTTTCCTGC TTCTAAAGCA TCAGGGGGTG TTCCTATTAA CCAGTCTCGC CACTCTTGCA GGTTGCTATC    2096

TGCTGTCCCT TATGCATAAA GTAAAAAGCA AAATGTCAAT GACATTTGCT TATTGACAAG GACTTTGTTA TTTGTGTTGG GAGTTGAGAC AATATGCCCC ATTCTAAGTA    2206

AAAAGATTCA GGTCCACATT GTATTCCTGT TTTAATTGAT TTTTTGATTT GTTTTTCTTT TTCAAAAAGT TTATAATTTT AATTCATGTT AATTTAGTAA TATAATTTTA    2316

CATTTTCCTC AAGAATGGAA TAATTTATCA GAAAGCACTT CTTAAGAAAA TACTTAGCAG TTTCCAAAGA AAATATAAAA TTACTCTTCT GAAAGGAATA CTTATTTTTG    2426

                                          109
                                          Tyr Leu Gln Glu Ile Tyr Asn Ser Asn Asn Gln Lys Ile Val Asn Leu Lys Glu Lys
TCTTCTTATT TTTGTTATCT TATGTTTCTG TTTGTAG A TAT TTG CAG GAA ATA TAT AAT TCA AAT AAT CAA AAG ATT GTT AAC CTG AAA GAG AAG    2521

                                                                   151
Val Ala Gln Leu Glu Ala Gln Cys Gln Glu Pro Cys Lys Asp Thr Val Gln Ile His Asp Ile Thr Gly Lys
GTA GCC CAG CTT GAA GCA CAG TGC CAG GAA CCT TGC AAA GAC ACG GTG CAA ATC CAT GAT ATC ACT GGG AAA G GTAACTGA TGAAGGTTAT    2612

ATTGGGATTA GGTTCATCAA AGTAAGTAAT GTAAAGGAGA AAGTATGTAC TGGAAAGTAT AGGAATAGTT TAGAAAGTGG CTACCCATTA AGTCTAAGAA TTTCAGTTGT    2722

CTAGACCTTT CTTGAATAGC TAAAAAAAAC AGTTTAAAAG GAATGCTGAT GTGAAAAGTA AGAAAATTAT TCTTGGAAAA TGAATAGTTT ACTACATGTT AAAAGCTATT    2832
```

Fig. 5. Nucleotide sequence of the gene coding for the γ chain of human fibrinogen. Taken from reference (15) and reprinted with permission from the American Chemical Society.

```
                                                                 152
                                                                 Asp Cys Gln Asp Ile Ala Asn Lys Gly Ala Lys Gln
TTTCAAGGCT GGCACAGTCT TACCTGCATT TCAAACCACA GTAAAAGTCG ATTCTCCTTC TCTAG  AT TGT CAA GAC ATT GCC AAT AAG GGA GCT AAA CAG     2932

Ser Gly Leu Tyr Phe Ile Lys Pro Leu Lys Ala Asn Gln Gln Phe Leu Val Tyr Cys Glu Ile Asp Gly Ser Gly Asn Gly Trp Thr Val
AGC GGG CTT TAC TTT ATT AAA CCT CTG AAA GCT AAC CAG CAA TTC TTA GTC TAC TGT GAA ATC GAT GGG TCT GGA AAT GGA TGG ACT GTG     3022
        196
Phe Gln Lys
TTT CAG AAG GTAATTTTT TCCCCACCAT GTGTATTTAA TAAATTCCTA CATTGTTTCT GCCATATGGC AGATACTTTT CTAAGCACCT TGTGAACCGT AGCTCATTTA     3130

ATCCTTGCAA TAGCCCTAAG AGGAAGGTAC TTCTGTTACT CCTATTTACA GAAAAGGAAA CTGAGGCACA CAAGGTTAAA TAACTTGCCC AAGACCACAT AACTAATAAG     3240

CAACAGAGTC AGCATTTGAA CCTAGGCAGT ATAGTTTCAG AGTTTGTGAC TTGACTCTAT ATTGTACTGG CACTGACTTT GTAGATTCAT GGTGGCACAT AATCATAGTA     3350

CCACAGTGAC AAATAAAAAG AAGGAAACTC TTTTGTCAGG TAGGTCAAGA CCTGAGGTTT CCCATCACAA GATGAGGAAG CCCAACACCA CCCCCCACCA CCCCACCACC     3460

ATCACCACCC TTTCACACAC CAGAGGATAC ACTTGGGCTG CTCCAAGACA AGGAACCTGT GTTGCATCTG CCACTTGCTG ATACCCACTA GGAATCTTGG CTCCTTTACT     3570

TTCTGTTTAC CTCCCACCAC TGTTATAACT GTTTCTACAG GGGGCGCTCA GAGGGAATGA ATGGTGGAAG CATTAGTTGC CAGACACCGA TTGAGCAATG GGTTCCATCA     3680

TAAGTGTAAG AATCAGTAAT ATCCAGCTAG AGTTCTGAAG TCGTCTAGGT GTCTTTTTAA TATTACCACT CATTTAGAAT TTATGATGTG CCAGAAACCC TCTTAAGTAT     3790

TTCTCTTATA TTCTCTCTCA TGATCCTTGC AGCAACCCTA AGAAGTAACC ATCATTTTTC CTATTTGATA CATGAGGAAA CTGAGGTAGC TTGGCCAAGA TCACTTAGTT     3900

GGGAGTTGAT AGAACCAGTG CTCTGTATTT TTGACAAAAT GTTGACAGCA TTCTCTTTAC ATGCATTGAT AGTCTATTTT CTCCTTTTGC TCTTGCAAAT GTGTAATTAG     4010
197
Arg Leu Asp Gly Ser Val Asp Phe Lys Lys Asn Trp Ile Gln Tyr Lys Glu Gly Phe Gly His Leu Ser Pro Thr Gly Thr Thr Glu Phe
AGA CTT GAT GGC AGT GTA GAT TTC AAG AAA AAC TGG ATT CAA TAT AAA GAA GGA TTT GGA CAT CTG TCT CCT ACT GGC ACA ACA GAA TTT     4100

Trp Leu Gly Asn Glu Lys Ile His Leu Ile Ser Thr Gln Ser Ala Ile Pro Tyr Ala Leu Arg Val Glu Leu Glu Asp Trp Asn Gly Arg
TGG CTG GGA AAT GAG AAG ATT CAT TTG ATA AGC ACA CAG TCT GCC ATC CCA TAT GCA TTA AGA GTG GAA CTG GAA GAC TGG AAT GGC AGA     4190
    258
Thr Ser
ACC AG  GTACTGTTTT GAAATGACTT CCAACTTTTT ATTGTAAAGA TTGCCTGGAA TGTGCACTTT CCAACTATCA ATAGACAATG GCAAATGCAG CCTGACAAAT     4295

GCAAACAGCA CATCCAGCCA CCATTTTCTC CAGGAGTCTG TTTGGTTCTT GGGCAATCCA AAAAGGTAAA TTCTATTCAG GATGAATCTA AGTGTATTGG TACAATCTAA     4405

TTACCCTGGA ACCATTCAGA GTAATAGCTA ATTACTGAAC TTTTAATCAG TCCCAGGAAT TGAGCATAAA ATTATAATTT TATCTAGTCT AAATTACTAT TTCATGAAGC     4515

AGGTATTATT ATTAATCCCA TTTTATAGAT TAACTTGCTC AAAGTCACAT TGCTGATAAG TGGTAGAGGT AGAATTCAGA CTCAAGTAGT TTAACTTTAG AGCCTGTCCT     4625

CTTAACAACT ATCCTGGTTG AAAAGCAAAT ACAGCCTCTT CAGACTTCTC AGTGCCTTGA TGGCCATTTA TTCTGTCAAA TCATGAGCTA CCCTAAAAGT AAACCAGCTA     4735

GCTCTTTTGA TGATCTAGAG GCTTCTTTTT GCTTGAGATA TTTGAAGGTT TTAAGCATTG TTACCTAATT AAAATGCAGA AAAATATCCA ACCCTCTTGT TATGTTTAAG     4845

GAATAGTGAA ATATATTGTC TTCAAACACA TGGACTTTTT TTTATTGCTT GGTTGGTTTT TAATCCAGAA AGTGCTATAG TCAGTAGACC TTCTTCTAGG AAAGGACCTT     4955

CCATTTCCCA GCCACTGGAG ATTAGAAAAT AAGCTAAATA TTTTCTGGAA ATTTCTGTTC ATTCATTAAG GCCCATCCTT TCCCCCACTC TATAGAAGTG TTGTCCACTT     5065

GCACAATTTT TTCCAGGAAA GAATCTCTCT AACTCCTTCA GCTCACATGC TTTGGACCAC ACAGGGAAGA CTTTGATTGT GTAATGCCCT CAGAAGCTCT CCTTCTTGCC     5175

ACTACCACAC TGATTTGAGG AAGAAAATCC CTTTAGCACC TAACCCTTCA GGTGCTATGA GTGGCTAATG GAACTGTACC TCCTTCAAGT TTTGTGCAAT AATTAAGGGT     5285

CACTCACTGT CAGATACTTT CTGTGATCTA TGATAATGTG TGTGCAACAC ATAACATTTC AATAAAAGTA GAAAATATGA AATTAGAGTC ATCTACACAT CTGGATTTGA     5395

TCTTAGAATG AAACAAGCAA AAAAGCATCC AAGTGAGTGC AATTATTAGT TTTCAGAGAT GCTTCAAAGG CTTCTAGGCC CATCCCGGGA AGTGTTAATG AGCTGTGGAC     5505

TGGTTCACAT ATCTATTGCC TCTTGCCAGA TTTGCAAAAA ACTTCACTCA ATGAGCAAAT TTCAGCCTTA AGAAACAAAG TCAAAAATTC CAAGGAAGCA TCCTACGAAA     5615
                                                                    259
                                                                    Thr Ala Asp Tyr Ala Met Phe Lys Val Gly Pro Glu
GAGGGAACTT CTGAGATCCC TGAGGAGGGT CAGCATGTGA TGGTTGTATT TCCTTCTTCTCAG T ACT GCA GAC TAT GCC ATG TTC AAG GTG GGA CCT GAA     5715

Ala Asp Lys Tyr Arg Leu Thr Tyr Ala Tyr Phe Ala Gly Gly Asp Ala Gly Asp Ala Phe Asp Gly Phe Asp Phe Gly Asp Asp Pro Ser
GCT GAC AAG TAC CGC CTA ACA TAT GCC TAC TTC GCT GGT GGG GAT GCT GGA GAT GCC TTT GAT GGC TTT GAT TTT GGC GAT GAT CCT AGT     5805

Asp Lys Phe Phe Thr Ser His Asn Gly Met Gln Phe Ser Thr Trp Asp Asn Asp Asn Asp Lys Phe Glu Gly Asn Cys Ala Glu Gln Asp
GAC AAG TTT TTC ACA TCC CAT AAT GGC ATG CAG TTC AGT ACC TGG GAC AAT GAC AAT GAT AAG TTT GAA GGC AAC TGT GCT GAA CAG GAT     5895
                                                                           350
Gly Ser Gly Trp Trp Met Asn Lys Cys His Ala Gly His Leu Asn Gly Val Tyr Tyr Gln
GGA TCT GGT TGG TGG ATG AAC AAG TGT CAC GCT GGC CAT CTC AAT GGA GTT TAT TAC CAA G GTATGTTTTC CTTTCTTAGA TTCCAAGTTA     5986

ATGTATAGTG TATACTATTT TCATAAAAAA TAATAAATAG ATATGAAGAA ATGAAGAATA ATTTATAAAG ATAGTAGGGA TTTTATCATG TTCTTTATTT CAACTAAGTT     6096

CTTTGAAACT GGAAGTGGAT AATACCAAGT TCATGCCTAA AATTAGCCCT TCTAAAGAAA TCCACCTGCT GCAAAATATC CAGTAGTTTG GCATTATATG TGAAACTATC     6206

ACCATCATAG CTGGCACTGT GGGTTGTGGG ATCTCCTTTA GACATACAAC ATAAATGATC TGGATGGATT AACATTACTA CATGGATGCT TGTTGACACA TTAACCTGGC     6316

TTCCCATGAG CTTTGTGTCA GATACACGCA GTGAACAGGT GTTTGGAGGA ACAGAATAAA GAGAAGGCAA GCACTGGTAA GGGCAGGGGT TTGTGAAAGC TTGAGAGAAG     6426

AGACCAGTCT GAGGACAGTA GACACTTATT TTAGGATGGG GGTTGGATGA GGAGGCTATA GTTTGCTATA AGCTTGGAAT GGTTTGGAAC ACTGGTTTCA CTCACCTACC     6536

CAGCAGTTAT GTGTGGGGAA GCCTTACCGA TGCTAAAGGA TCCATGTTAC AATAATGGCA TTATTTGGAA ATCCCAGTGG TATTCCATGA ATAAAACCAC TATGAAGATA     6646

ATCCCACTCA ACAGACTCTC CGTTGGAGAA GGACAGCAAC ACCACCCTGG GAAAGCCAAA CAGTCAGACC AGACCTGTTT AGCATCAGTA GGACTTCCCT ACCATATCTG     6756

CTGGGTAGAT GAGTGAAACC AGTGTTCCAA ACCACTCCGG GCTTGTAGCA AACCATAGTC TCCTCATCTA CCAAGATGAG CAACCTTACC TCCTGATGTC CTAGCCAATC     6866

ACCAACTAGG AAACTTTGCA CAGTTTATTT AAAGTAACAG TTTGATTTTC ACAATATTTT TAAATTGGAG AAACATAACT TATCTTTGCA CTCACAAACC ACATAATGAG     6976

AAGAAACTCT AAGGGAAAAT GCTTGATCTG TGTGACCCGG GGCGCCATGC CAGAGCTGTA GTTCATGCCA GTGTTGTGCT CTGACAAGCC TTTTACAGAA TTACATGAGA     7086

TCTGCTTCCC TAGGACAAGG AGAAGGCAAA TCAACAGAGG CTGCACTTTA AAATGGAGAC ATAAAATAAC ATGCCAGAAC CATTTCCTAA AGCTCCTCAA TCAACCAACA     7196

AAATTGTGCT TTCAAATAAC CTGAGTTGAC CTCATCAGGA ATTTTGTGGC TCCTTCTCTT CTAACCTGCC TGAAGAAAGA TGGTCCACAG CAGCTGAGTC CGGGATGGAT     7306

AAGCTTAGGG ACAGAGGCCA ATTAGGGAAC TTTGGGTTTC TAGCCCTACT AGTAGTGAAT AAATTTAAAG TGTGGATGTG ACTATGAGTC ACAGCACAGA TGTTGTTTAA     7416
```

Fig. 5 (Continued).

```
TAATATGTTT ATTTTATAAA TTGATATTTT AGGAATCTTT GGAGATATTT TCAGTTAGCA GATAATACTA TAAATTTTAT GTAACTGGCA ATGCACTTCG TAATAGACAG   7526
                                                                       351
                                                                       Gly Gly Thr Tyr Ser Lys Ala Ser Thr Pro Asn
CTCTTCATAG ACTTGCAGAG GTAAAAAGAT TCCAGAATAA TGATATGTAC ATCTACGACT TGTTTTAG GT GGC ACT TAC TCA AAA GCA TCT ACT CCT AAT   7626

Gly Tyr Asp Asn Gly Ile Ile Trp Ala Thr Trp Lys Thr Arg Trp Tyr Ser Met Lys Lys Thr Thr Met Lys Ile Ile Pro Phe Asn Arg
GGT TAT GAT AAT GGC ATT ATT TGG GCC ACT TGG AAA ACC CGG TGG TAT TCC ATG AAG AAA ACC ACT ATG AAG ATA ATC CCA TTC AAC AGA   7716
                                                            407 408'
Leu Thr Ile Gly Glu Gly Gln Gln His His Leu Gly Gly Ala Lys Gln Val Arg Pro Glu His Pro Ala Glu Thr Glu Tyr Asp Ser Leu
CTC ACA ATT GGA GAA GGA CAG CAA CAC CAC CTG GGG GGA GCC AAA CAG GTC AGA CCA GAG CAC CCT GCG GAA ACA GAA TAT GAC TCA CTT   7806
                    427'
Tyr Pro Glu Asp Asp Leu STOP
TAC CCT GAG GAT GAT TTG TAG AAAATTAACT GCTAACTTCT ATTGACCCAC AAAGTTTCAG AAATTCTCTG AAAGTTTCTT CCTTTTTTCT CTTACTATAT   7907

TTATTGATTT CAAGTCTTCT ATTAAGGACA TTTAGCCTTC AATGGAAATT AAAACTCATT TAGGACTGTA TTTCCAAATT ACTGATATCA GAGTTATTTA AAAATTGTTT   8017
                                                                 ↓γ'
ATTTGAGGAG ATAACATTTC AACTTTGTTC CTAAATATAT AATAATAAAA TGATTGACTT TATTTGCATT TTTATGACCA CTTGTCATTT ATTTTGTCTT CGTAAATTAT   8127

TTTCATTATA TCAAATATTT TAGTATGTAC TTAATAAAAT AGGAGAACAT TTTAGAGTTT CAAATTCCCA GGTATTTTCC TTGTTTATTA CCCCTAAATC ATTCCTATTT   8237
                                                                    408         411
                                                                    Ala Gly Asp Val STOP
AATTCTTCTT TTTAAATGGA GAAAATTATG TCTTTTTAAT ATGGTTTTTG TTTTGTTATA TATTCACAG GCT GGA GAC GTT TAA AAGACCGTTT CAAAAGAGAT   8341

TTACTTTTTT AAAGGACTTT ATCTGAACAG AGAGATATAA TATTTTTCCT ATTGGACAAT GGACTTGCAA AGCTTCACTT CATTTTAAGA GCAAAAGACC CCATGTTGAA   8451
                                                                            ↓γ
AACTCCATAA CAGTTTTATG CTGATGATAA TTTATCTACA TGCATTTCAA TAAACCTTTT GTTTCCTAAG ACTAGATACA TGGTACCTTT ATTGACCATT AAAAAACCAC   8561

CACTTTTTGC CAATTTACCA ATTACAATTG GGCAACCATC AGTAGTAATT GAGTCCTCAT TTTATGCTAA ATGTTATGCC TAACTCTTTG GGAGTTACAA AGGAAATAGC   8671

AATTATGGCT TTTGCCCTCT AGGAGATACA GGACAAATAC AGGAAAATAC AGCAACCCAA ACTGACAATA CTCTATACAA GAACATAATC ACTAAGCAGG AGTCACAGCC   8781

ACACAACCAA GATGCATAGT ATCCAAAGTG CAGCTG                                                                                8817
```

Fig. 5 (Continued).

in intron I has also been identified. Endonucleolytic cleavage and polyadenylation at this alternative site generate a variant form of mRNA in which the 9th intron serves as an extension of the 9th exon and leads to the synthesis of the γ' variant form of the γ chain (16, 17). The gene for the γ chain contains a single-copy repeat of exon IX sequence within the 8th intron. No other repetitive sequences have been identified in the γ chain gene.

The completion of the entire nucleotide sequence for the three fibrinogen genes provides a structural basis for further study of these genes particularly in the analysis of mutations in dysfibrinogenemic patients. These sequences have made it possible to design oligonucleotide primers for use in the polymerase chain reaction to amplify specific regions of these genes directly from total genomic DNA for further study of their sequence or function. This has been particularly useful for determining the nature of mutations in regions of the three polypeptides that are difficult to study by conventional protein sequencing techniques.

The DNA sequences reported in this manuscript will be deposited in the GenBank data base.

This work was supported, in part, by Research Grant HL 16919 from the National Institutes of Health.

REFERENCES

1. D. Collen, G. N. Tygat, H. Claeys, and R. Piessens, Metabolism and distribution of fibrinogen. I. Fibrinogen turnover in physiological conditions in humans, *Br. J. Haematol,* **22**:681 (1972).
2. J. M. Nickerson, and G. M. Fuller, In vitro synthesis of rat fibrinogen: Identification of preAα, preBß, and preγ polypeptides, *Proc. Natl. Acad. Sci. USA,* **78**:303 (1981).
3. G. R. Crabtree, and J. A. Kant, Molecular cloning of cDNA for the alpha, beta, and gamma chains of rat fibrinogen. A family of coordinately regulated genes, *J. Biol. Chem.,* **256**:9718 (1981).
4. J. A. Kant, A. J. Fornace, Jr., D. Saxe, M. L. Simon, O. W. McBride, and G. R. Crabtree, Organization and evolution of the human fibrinogen locus on chromosome four, *Proc. Natl. Acad. Sci. USA,* **82**:2344 (1985).
5. S. L. C. Woo, A sensitive and rapid method for recombinant phage screening, *Methods Enzymol,* **68**:389 (1979).
6. M. W. Rixon, W. Y. Chan, E. W. Davie, and D. W. Chung, Characterization of a complementary deoxynucleic acid coding for the α chain of human fibrinogen, *Biochemistry,* **22**:3237 (1983).

7. D. W. Chung, B. G. Que, M. W. Rixon, M. Mace, and E. W. Davie, Characterization of complementary deoxyribonucleic acid and genomic deoxyribonucleic acid for the ß chain of human fibrinogen, *Biochemistry,* **22:**3244 (1983).
8. D. W. Chung, W. Y. Chan, and E. W. Davie, Characterization of a complementary deoxyribonucleic acid coding for the γ chain of human fibrinogen, *Biochemistry,* **22:**3250 (1983).
9. R. M. Lawn, E. F. Fritsch, R. C. Parker, G. Blake, and T. Maniatis, The isolation and characterization of linked delta- and beta-globin genes from a cloned library of human DNA, *Cell,* **15:**1157 (1978).
10. S. Yoshitake, B. G. Schach, D. C. Foster, E. W. Davie, and K. Kurachi, Nucleotide sequence of the gene for human factor IX (antihemophilic factor B), *Proc. Natl. Acad. Sci. USA,* **24:**3736 (1985).
11. H. R. Rackwitz, G. Zehetner, A. M. Frischauf, and H. Lehrach, Rapid restriction mapping of DNA cloned in lambda phage vectors, *Gene,* **30:**195 (1984).
12. F. Sanger, S. Nicklen, and A. R. Coulson, DNA sequencing with chainterminating inhibitors, *Proc. Natl. Acad. Sci. USA,* **74:**5463 (1977).
13. R. F. Doolittle, The structure of vertebrate fibrinogen, *Ann NY Acad Sci,* **408:**13 (1983).
14. G. R. Crabtree, C. M. Comeau, D. M. Fowlkes, A. J. Fornace, J. D. Malley, and J. A. Kant, Evolution and structure of the fibrinogen genes: random insertion of introns or selective loss? *J. Mol. Biol.,* **185:**1 (1985).
15. M. W. Rixon, D. W. Chung, and E. W. Davie, Nucleotide sequence of the gene for the γ chain of human fibrinogen, *Biochemistry,* **24:**2077 (1985).
16. C. Wolfenstein-Todel, and M. W. Mosesson, Carboxyl-terminal amino acid sequence of a human fibrinogen gamma-chain variant (gamma'), *Biochemistry,* **20:**6146 (1981).
17. D. W. Chung, and E. W. Davie, γ and γ' chains of human fibrinogen are produced by alternative mRNA processing, *Biochemistry,* **23:**4232 (1984).

ON THE IDENTITY OF FIBRIN(OGEN) OLIGOMERS APPEARING DURING FIBRIN POLYMERIZATION

Agnes Henschen

Department of Molecular Biology and Biochemistry
University of California, Irvine
Irvine, CA 92717, USA

INTRODUCTION

Fibrinogen (1) is a dimeric, symmetrical molecule with the overall structure $(A\alpha, B\beta, \gamma)_2$. When fibrinogen is converted into fibrin by the action of thrombin two moles of each fibrinopeptide A and B are finally released. Hereby four N-terminal polymerization sites are revealed in each molecule and these can then interact with four pre-existing C-terminal polymerization sites so that fibrin oligomers and polymers are formed. It is well established that A-peptides are removed from fibrinogen before B-peptides and that A-peptide release alone is sufficient for clotting to occur. The two fibrin types formed have the structures $[(\alpha, \beta, \gamma)_2]_n$ and $[(\alpha, B\beta, \gamma)_2]_n$ respectively.

It is not yet known if already the removal of a single A-peptide per fibrinogen molecule will induce clotting. However, it has been demonstrated that on addition of thrombin to fibrinogen solutions clot formation starts when at an average of only 30 percent of the A-peptides have been cleaved off. During early stages of fibrin polymerization a range of soluble oligomers may be observed by various chromatographic or physicochemical methods. The structures of the oligomers have been suggested to be one of the following:

$$[(A\alpha, B\beta, \gamma)\ (\alpha, B\beta, \gamma)]_n$$

$$[(A\alpha, B\beta, \gamma)\ (\alpha, B\beta, \gamma)]_1\ [(\alpha, B\beta, \gamma)_2]_n\ [(\alpha, B\beta, \gamma)\ (A\alpha, B\beta, \gamma)]_1$$

$$[(A\alpha, B\beta, \gamma)_2]_1\ [(\alpha, B\beta, \gamma)_2]_n\ [(A\alpha, B\beta, \gamma)_2]_1$$

The upper model was proposed by Hnziker et al. (2), the middle by Smith (3), Visser and Payens (4) and Alkjaersig and Fletcher (5) and the lower by Wilf and Minton (6).

In the present investigation the various models were tested using oligomer fractionation by gel sieving chromatography, fibrinopeptide determination by HPLC, fibrin determination by fibrinogen agarose chromatography and identification of central, dimeric fibrin(ogen) (A)α-chain fragments by chemical and enzymatic cleavage followed by HPLC.

RESULTS

Fractionation of Oligomers

Fibrinogen was incubated with a very low concentration of thrombin (0.01 NIH units per ml) at pH 7.4. After various periods of time the action of thrombin was interrupted by the addition of hirudin. The samples were then analyzed for fibrin(ogen) oligomer distribution by gel sieving chromatography on Sepharose CL-4B as shown in Fig. 1 and for fibrinopeptide release by reversed phase HPLC (7), the fibrinopeptide A and B peak areas calculated as

Fibrinogen, Thrombosis, Coagulation, and Fibrinolysis, Edited by
C. Y. Liu and S. Chien, Plenum Press, New York

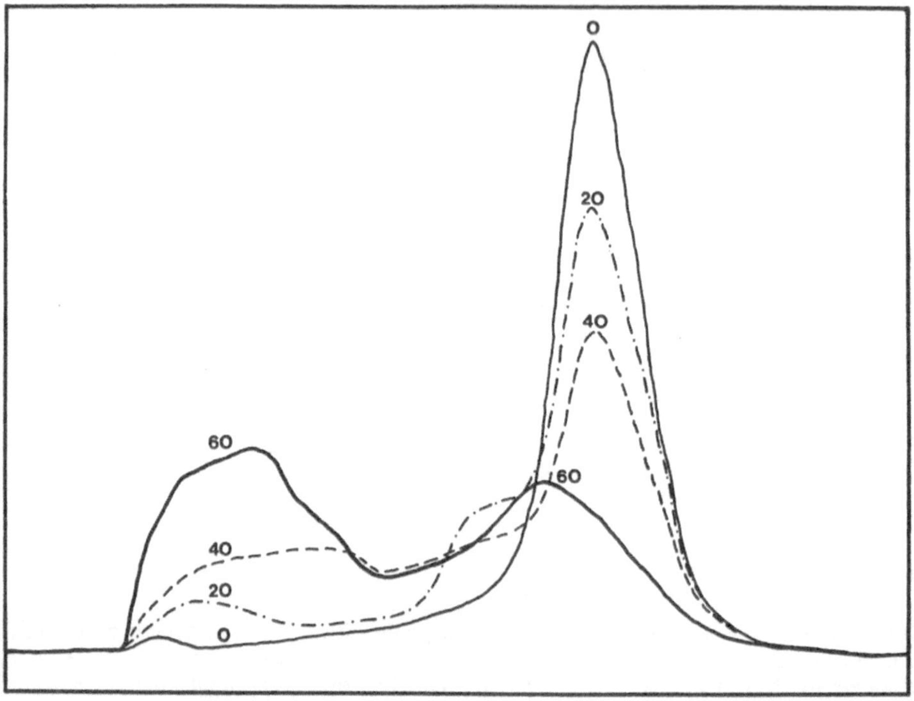

Fig. 1. Gel sieving chromatography on Sepharose CL-4B of fibrinogen incubated for 0, 20, 40 and 60 min with 0.01 NIH units of thrombin per ml; the effluent was monitored at 280 nm.

the percentage of maximally releasable peptide as indicated in Fig. 2. The clotting time in the system selected was approximately 100 min, at which time about 35% of the total A-peptide and 3% of the B-peptide had been cleaved off.

The gel filtration chromatographies demonstrate that with increase in thrombin digestion time from 0 to 60 min there is a gradual shift in the distribution of fibrin(ogen) material, exclusively monomers being present at 0 min and larger and larger oligomers appearing at prolonged incubation times. The shapes of the curves indicate the existence of dimers (mainly at 20 min), trimers, tetramers and larger oligomers (mainly at 40 and 60 min). The time points 20, 40 and 60 min of digestion correspond to 7, 15 and 22% A-peptide release and less than 2% B-peptide release, respectively.

Characterization of Oligomer Fractions

The fractionated material from the 60 min thrombin incubation was characterized more in detail. Protein-containing effluent was combined in the eight consecutive pools indicated in Fig. 3. The pools were analyzed for fibrinopeptide A content by HPLC (7) after extensive thrombin digestion using fibrinopeptide B as a measure for the amount of fibrin(ogen) present in the pool. The upper row of bars in Fig. 3 shows the percentage of A-peptide missing in the material of the various pools. The pools were also analyzed for fibrin content by affinity chromatography on fibrinogen agarose (8) under conditions where material devoid of A-peptide, but not material containing A-peptide, is retained by the column. The lower row of bars in Fig. 3 shows the percentage of the fibrin(ogen) material in the pool which had affinity for the fibrinogen agarose column. The two rows of bars seem to be mirror-images of each other, i.e. in every pool the proportion of molecules with fibrin-like affinity was similar to the proportion which could be predicted to be devoid of both A-peptides and not only one of the A-peptides present in each fibrinogen molecule.

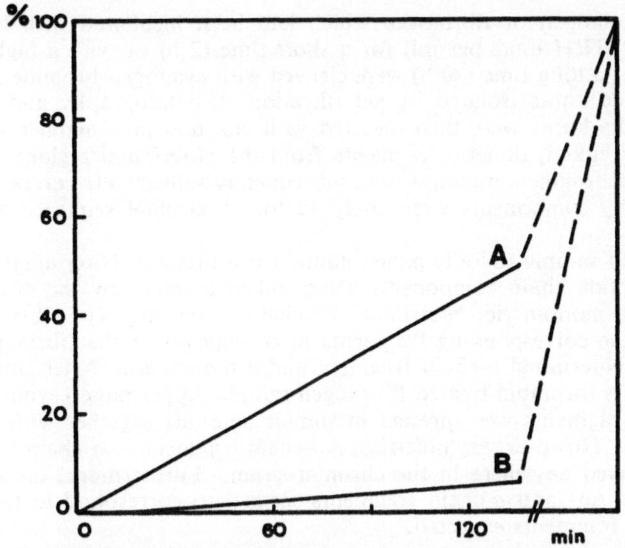

Fig. 2. Fibrinopeptides released during incubation of fibrinogen with 0.01 NIH units of thrombin per ml, expressed as percentage of total amount of fibrinopeptides in sample; the fibrinopeptides were determined by HPLC.

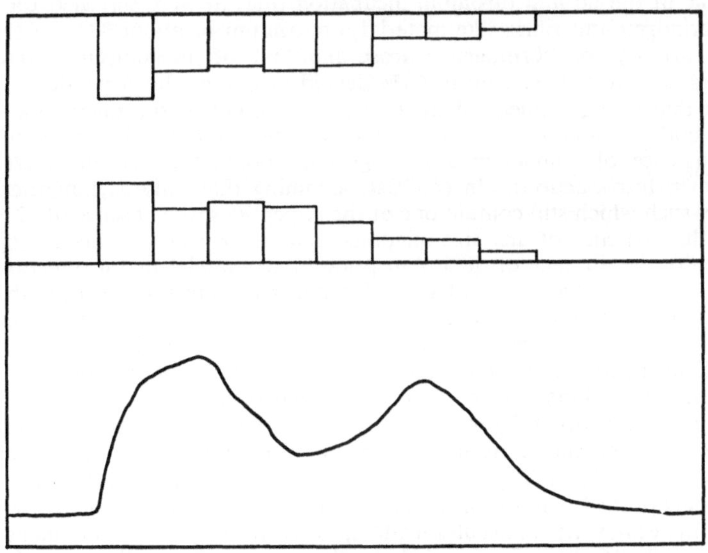

Fig. 3. Gel sieving chromatography on Sepharose CL-4B of fibrinogen incubated for 60 min with 0.01 NIH units of thrombin per ml (lower panel); fibrin-part of material in pool as determined by fibrinogen agarose chromatography (middle panel); fibrinopeptide A missing from material in pool as determined by HPLC after thrombin digestion (upper panel).

Isolation of Dimeric, N-Terminal Fragments

Untreated fibrinogen or fibrinogen which had been incubated with a low thrombin concentration (0.01 NIH units per ml) for a short time (2 h) or with a high concentration (5 units per ml) for a long time (20 h) were cleaved with cyanogen bromide and the various N-terminal disulfide knots isolated by gel filtration chromatography and reversed phase HPLC (9, 10). The knots were then digested with enzymes in a manner which produced central, covalently linked, dimeric fragments from the N-terminal region of all Aα- or α-chains present. The fragment mixtures were subsequently subjected to reversed phase HPLC. All chromatography components were analyzed for N-terminal sequence and amino acid composition.

The fibrinogen sample (middle panel) contained a dimeric, N-terminal Aα-chain fragment, the two peptide chain components being linked together by the disulfide bridge in position 28, and a monomeric, N-terminal Bβ-chain fragment. The fibrin sample (lower panel) contained the corresponding fragments after removal of the fibrinopeptides A and B, i.e. a dimeric, N-terminal α-chain fragment and a monomeric, N-terminal β-chain fragment. In the briefly thrombin-treated fibrinogen sample (upper panel) symmetrical, dimeric Aα- and α-chain fragments were present in similar amounts together with the monomeric Bβ-chain fragment. However, asymmetrical Aα-chain fragment — α-chain fragment hybrids could not be detected anywhere in the chromatogram. Furthermore, the sum of the observed amounts of Aα- and α-chain fragments seemed to correspond to the total amount of Aα- or α-chain fragments expected.

DISCUSSION

The identity of the early, still soluble oligomers formed during thrombin-induced fibrinogen-fibrin conversion was first evaluated for oligomer size distribution by gel sieving chromatography (Fig. 1). The elution patterns indicated that larger oligomers were present in samples incubated for longer periods, as expected, and that dimers, trimers, tetramers, pentamers and higher multimers could be partially separated from each other. When pools from the sample of the 60 min thrombin incubation (Fig. 3) were screened for the relative amounts of fibrinogen and fibrin-like material a maximum of about 32% of the molecules had fibrin-like affinity for fibrinogen agarose and 33% of the molecules were devoid of both A-peptides or, alternatively, up to 66% devoid of one of the A-peptides.

Assuming that only symmetrical molecules are present in the digest, i.e. such which contain either both or none of the A-peptides, the absence of 33% A-peptide could only explain the existence of trimers in a homogeneous population of oligomers (see lower model depicted in Introduction). In contrast, assuming that only asymmetrical molecules are present, i.e. such which still contain one of the A-peptides, the absence of 33% A-peptide could explain the existence of any size oligomer, but never a homogeneous population, as this would require the absence of 50% A-peptide (upper model in Introduction). Finally, assuming the presence of oligomers with central, symmetrical units devoid of both A-peptides, flanked by asymmetrical units devoid of one A-peptide (middle model in Introduction), the population has to be heterogeneous as even a trimer is devoid of 67% of the A-peptides. Thus, it is difficult to interpret the elution patterns in terms of a specific type of oligomer and well-defined, stable dimers, trimers, tetramers and pentamers.

The identity of the early oligomers was also analyzed by isolation and characterization of central, dimeric fragments derived from the N-terminal region of all Aα- and α-chains present in the thrombic digest (Fig. 4). The HPLC elution pattern and the sequence analysis of the components proved that all, or at least virtually all, Aα- and α-chains were contained in symmetrical molecules, i.e. molecules made up of exclusively either Aα-chains or α-chains in combination with Bβ-chains and γ-chains. From these observations can be concluded that all, or virtually all, oligomers have the structure $[(A\alpha, B\beta, \gamma)_2]_1$ $[(\alpha, B\beta, \gamma)_2]_n$ $[(A\alpha, B\beta, \gamma)_2]_1$, i.e. in agreement with the lower model mentioned in the Introduction. It may be suggested that the gel sieving chromatography patterns are caused by the formation and rearrangement of various sizes oligomers and fibrinogen during the progress of the chromatography and that no stable oligomers exist in the early thrombic digest. It can also be concluded that thrombin removes the second A-peptide much faster than the first A-peptide from from the fibrinogen molecule, i.e. the two fibrinopeptides A are released in a conserted fashion by thrombin.

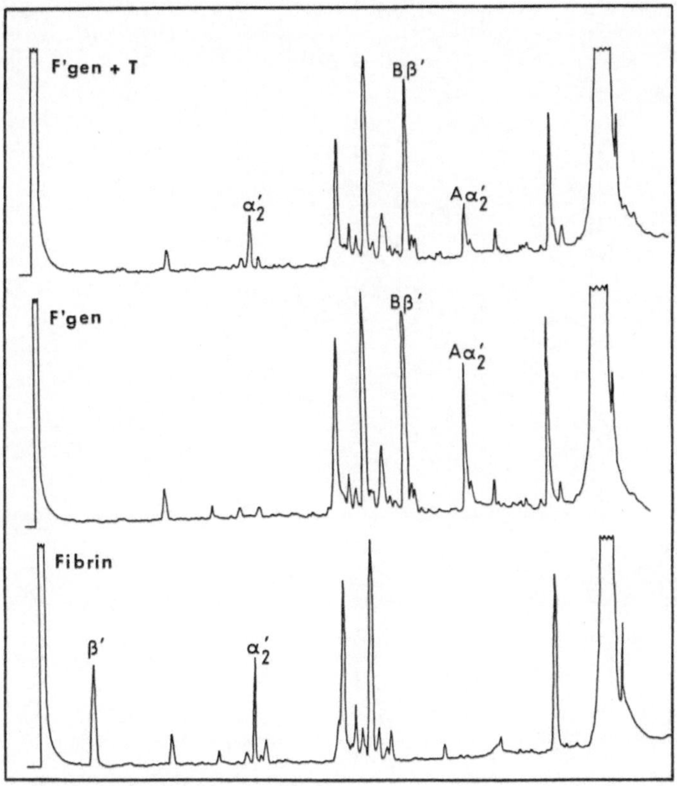

Fig. 4. Reversed phase HPLC of enzymatic fragments of N-terminal disulfide knots derived from fibrin (lower panel), fibrinogen (middle panel) and fibrinogen incubated for 2 h with 0.01 NIH units of thrombin per ml (upper panel); the effluent was monitored at 210 nm; $A\alpha_2'$, α_2', $B\beta'$ and β' indicate N-terminal fragments of the corresponding types of peptide chains.

REFERENCES

1. A. Henschen and J. McDonagh, Fibrinogen, fibrin and factor XIII, *in*: "Blood coagulation", R. F. A. Zwaal and H. C. Hemker, eds., Elsevier Science Publishers, Amsterdam, (1986).
2. E. B. Hunziker, P. W. Straub and A. Haeberli, Molecular morphology of fibrin monomers and early oligomers during fibrin polymerization, *J. Ultrastruct. Molec. Struct. Res.*, **98**:60 (1988).
3. G. F. Smith, Fibrinogen-fibrin conversion, *Biochem. J.*, **185**:1 (1980).
4. A. Visser and T. A. Payens, On the kinetics of the thrombin-controlled polymerization of fibrin, *FEBS Lett.*, **142**:35 (1982).
5. N. Alkjaersig and A. P. Fletcher, Formation of soluble fibrin oligomers in purified systems and in plasma, *Biochem. J.*, **213**:75 (1983).
6. J. Wilf and A. P. Minton, Soluble fibrin-fibrinogen complexes as intermediates in fibrin gel formation, *Biochem.*, **25**:3124 (1986).
7. M. Kehl, F. Lottspeich and A. Henschen, Analysis of human fibrinopeptides by high-performance liquid chromatography, *Hoppe-Seyler's Z. Physiol. Chem.*, **362**:1661 (1981).
8. F. R. Matthias, R. Reinicke and D. L. Heene, Affinity chromatography and quantitation of soluble fibrin from plasma, *Thromb. Res.*, **10**:365 (1977).
9. A. Henschen, Disulfide bridges in the middle part of fibrinogen, *Hoppe-Seyler's Z. Physiol. Chem.*, **359**:1757 (1978).
10. M. Kehl. F. Lottspeich and A. Henschen, High-performance liquid chromatography of proteins as applied to fibrinogen chains, *Hoppe-Seyler's Z. Physiol. Chem.*, **363**:1501 (1982).

IMMUNOCHEMICAL STUDIES OF Aα CHAIN CROSSLINKING

Joan H. Sobel and Robert E. Canfield

College of Physicians and Surgeons of Columbia University
New York, NY 10032, USA

INTRODUCTION

Nearly two decades ago our colleague, the late Dr. Hymie Nossel, considered the possibility that radioimmunoassays might have as much application in the field of coagulation as they had been shown to have in the field of endocrinology. He selected fibrinopeptide A as a likely candidate for measurement (1) and initiated a new era in the quantitation of peptides or proteins that are biochemical markers for various aspects of the coagulation process.

We (JHS and REC) provided the protein chemical components of some of that research and, ultimately, began to pursue other aspects of fibrinogen immunochemistry. We focused our independent efforts on that portion of the fibrinogen Aα chain that serves as a substrate for both Factor XIII_a and plasmin, i.e., the COOH-terminal two-thirds of the molecule whose structure is independent of Fragments E and D. Our goal was to obtain antibodies that recognized defined epitopes within this region of fibrinogen or fibrin (2, 3) for application in biochemical studies of fibrin formation and as potential diagnostic reagents.

Research agenda

Fig. 1 is a cartoon which depicts the assembly of fibrin monomers to form a fibrin polymer. The figure also illustrates the γ chain and α chain crosslinks that are introduced by Factor XIII_a and serve to stabilize the alignments formed during the polymerization process. As already noted, we have directed our attention to studies of the Aα chain crosslinking region and, in particular, are interested in defining the elements that interact to produce the crosslinks that are one of the biochemical hallmarks of thrombus formation. It may be that α chain crosslinks, unlike their γ chain counterpart, provide unique epitopes that will permit an immunologic distinction to be made between the circulating products of pathologic and non-pathologic thrombi. To date, the process of α chain crosslinking is one of the least understood aspects of fibrinogen biochemistry and our studies in this area form the basis of one portion of this presentation.

In developing a repertoire of immunoglobulins that recognize Aα chain crosslinking regions, we have concentrated on the generation of monoclonal antibodies, because they, by definition, react with single epitopes, and can be immortalized in culture. It appears, however, that those portions of the Aα chain that are of greatest interest are highly conserved in evolution. While the sequence of mouse fibrinogen has yet to be determined, we suspect, based on the structural similarity between rat and human Aα chains (4), that many human COOH-terminal Aα chain antigens would not be immunogenic in the mouse. Conversely, when such immunizations are successful and produce murine antibodies that react with human fibrinogen, these same antibodies may also crossreact with circulating murine fibrinogen and thus lead to the animal's death. Attempts to better define this experimental problem form the basis of the second portion of this presentation.

Fibrinogen, Thrombosis, Coagulation, and Fibrinolysis, Edited by
C. Y. Liu and S. Chien, Plenum Press, New York

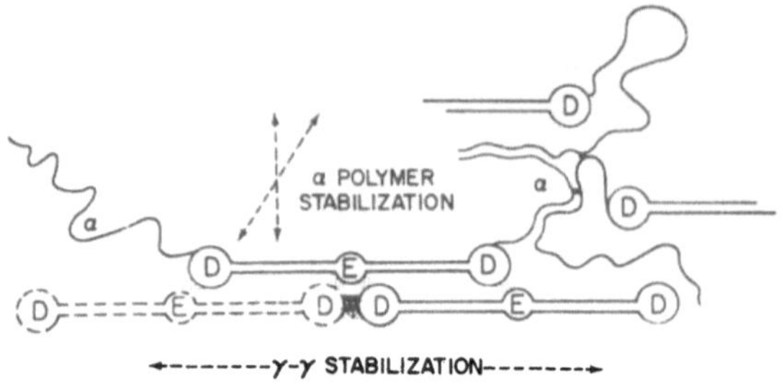

Fig. 1. Fibrin assembly and stabilization: The assembly of trinodular fibrin monomers in a half-staggered overlap is illustrated schematically. The non-covalent D-E and D-D contacts formed between different molecules during polymerization are indicated by stippling. The gamma chain cross-links which serve to stabilize the D-D contacts that form between two fibrin monomers are represented by dark lines. Alpha chains are depicted as "free floating" extensions of the D domain (see figure, left). Two alpha chain crosslinks that serve to create a covelent network involving three different fibrin molecules are shown on the right side of the figure.

The development of monoclonal antibodies for immunochemical studies of α chain crosslinking

The question of how α chains are aligned prior to the introduction of Factor XIII_a-mediated crosslinks has been addressed by several laboratories, including our own (5, 6, 2). Results of these biochemical investigations, which all employed highly crosslinked fibrin as starting material for the isolation of α polymer derivatives, indicate that crosslinking activity is localized, primarily, within the Aα chain regions, Aα #241-476 and Aα #518-584. These two regions are represented by the CNBr fragments referred to as CNI (7) or CNBr VIII (2) and CNIVB (7) or CNBr X (2), respectively. Additional evidence suggests that Aα #477-517 (8), as well as # 208-235 and #585-610 (2), may also contain crosslinking sites. Two glutamine residues, at Aα #328 and 366, selectively incorporate synthetic amines in the presence of activated Factor XIII (9). Based on this finding, CNBr VIII has been implicated as the α chain acceptor crosslinking region. While it has been assumed that CNBr X functions as the α chain donor crosslinking region, the specific lysine residue(s) involved in α polymer formation have not yet been identified. As a result, proposals regarding the alignment of α chains during fibrin stabilization remain speculative.

The biochemical studies of α polymer formation conducted in our laboratory were initiated to characterize a crosslinked α chain derivative that could provide information about the mechanism of fibrin stabilization and could also serve as an immunogen for the generation of monoclonal antibodies that were fibrin-specific because they recognized an epitope that included an α chain crosslink. Immunization with αXLCNBr, a CNBr derivative of α polymer that consisted (for the most part) of multiple units of CNBr VIII crosslinked to CNBr X (2), yielded three monoclonal antibodies of interest (3). The antibody, F-103, binds to CNBr VIII via an epitope that has been localized near Aα #259-276 while the antibody, F-102, binds to an epitope within CNBr X, in the region of Aα #540-554. These features are illustrated schematically in Fig. 2. The antibody, F-101, was serendipitously isolated as a result of the covalent interaction that occurs between fibrin α chains and fibronectin (10). This interaction led to the incorporation, during *in vitro* fibrin formation, of fibronectin into the α polymers that served as the source of the crosslinked immunogen (and screening antigen), αXLCNBr. The antibody, F-101, is an anti-fibronectin immunoglobulin whose epitope appears to be contained within a CNBr (fibronectin) fragment that includes a transglutaminase-sensitive domain (2).

While neither of the anti-Aα chain antibodies, F-103 and F-102, are fibrin specific, both recognize their respective epitopes in a variety of fibrin(ogen) derivatives. These include non-crosslinked and crosslinked, as well as, intact and degraded forms of Aα chains. As a result, these monoclonal antibodies have proven extremely valuable as immunoblotting

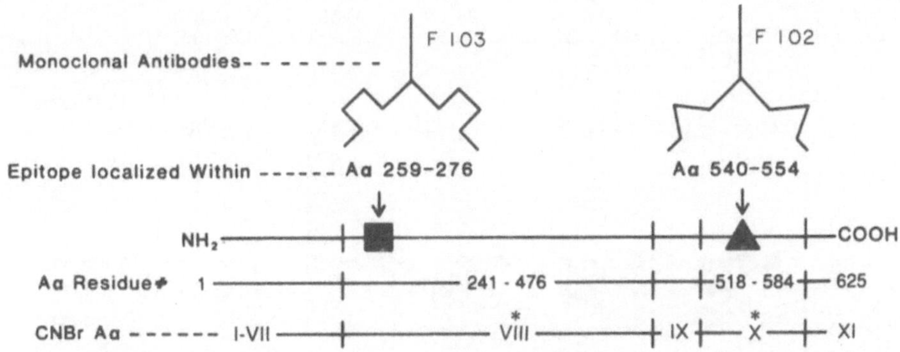

Fig. 2. Monoclonal antibodies that recognize COOH-terminal Aα chain epitopes: The eleven CNBr peptides that comprise the 625 residues of the human Aα chain are referred to by Roman numeral (I-XI) according to their relative position from the NH_2-terminus of the molecule and are schematically depicted (not drawn to scale) at the bottom of the figure. CNBr VIII and CNBr X, regions that contain crosslinking sites, are highlighted by asterisks. The dark square, within CNBr VIII, approximates the position of the epitope recognized by the monoclonal antibody, F-103, while the dark triangle indicates the approximate binding site for the monoclonal antibody, F-102.

reagents and immunosorbents for application in immunochemical studies of α chain crosslinking and fibrino(geno)lysis (see below).

Immunochemical studies of early α chain crosslinking

As an alternate approach to defining the glutamine and lysine residues that partner during α chain crosslinking, we have turned from the use of highly crosslinked fibrin to the characterization of derivatives that originate from lightly crosslinked fibrin preparations. The (crosslinked) α chain component of these preparations consists, primarily, of dimers and trimers. The earliest α chain crosslinks introduced during fibrin stabilization, i.e., those responsible for dimer and trimer formation, may provide the basis for the development of clinical assays for the detection of thrombosis since these crosslinks would be expected to be represented on circulating fragments that are among the earliest products released from particulate fibrin once fibrinolysis is initiated (11, 12). Early α chain crosslinks may also have more physiologic significance than those introduced when fibrin is crosslinked *in vitro* for long periods of time, in view of evidence which suggests that thrombi formed *in vivo* are not extensively crosslinked (6).

Fig. 3 is an immunoblotting study that illustrates the early α chain crosslinking process. Aliquots of purified fibrinogen were clotted under crosslinking conditions for increasing periods of time and the resulting fibrins were subjected to SDS-PAGE, followed by transfer of the electrophoresed components onto nitrocellulose. Duplicate transfers were reacted with F-102 and F-103, respectively, to visualize the incorporation of α chain monomers into higher molecular weight crosslinked species. Immunoreactive bands that migrate slower than α monomer, and are not present at T_0, are visible at the earliest time point examined (see Fig. 3, 10 min. lane, αXLI, II, & III). These persist, presumably as intermediates, throughout the entire incubation. The high molecular weight polymers that characterize later stages of α chain crosslinking are not observed until 2-4 h, under the *in vitro* conditions used in this study. Together with the disappearance of intact α chain monomers (consistent with their incorporation into crosslinked species) there is also a decrease in the intensity of the bands that correspond to α chain remnants, suggesting that degraded α chains can participate in the crosslinking reaction as well (see Fig. 3; compare T_0 & T20h lanes). When purified fibrinogen preparations are depleted of their intact Aα chain population, by F-102 Sepharose immunoaffinity chromatography, the degraded molecules that remain (which are all missing *at least* 70 COOH-terminal Aα chain residues) can still form

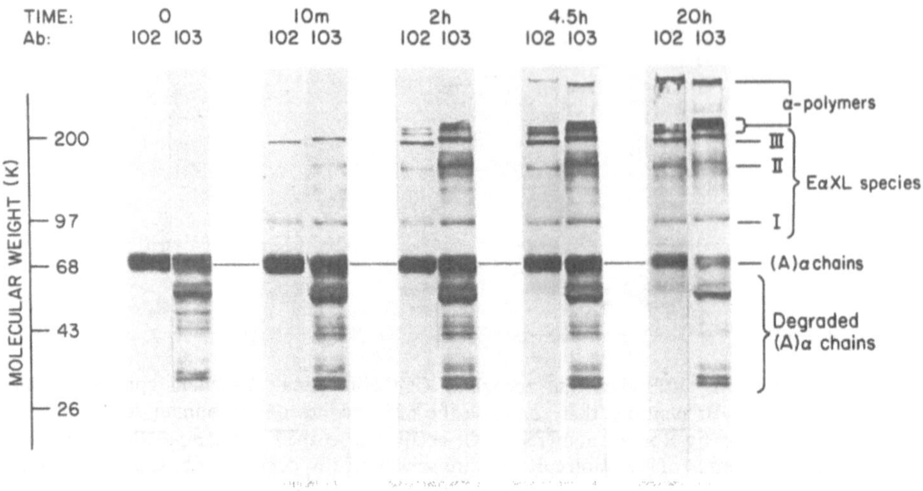

Fig. 3. Immunovisualization of α chain crosslinking by Western blotting with the monoclonal antibodies, F-102 and F-103: Aliquots (800 μl) of purified fibrinogen (2.9 mg/ml) were clotted *in vitro* and the crosslinked fibrins subjected to SDS-PAGE (9% gels, reducing conditions), followed by Western blotting with F-102 and F-103. (Approximately 20 μg of material were applied to duplicate lanes). Horseradish peroxidase-conjugated rabbit anti-mouse Ig was used to detect bound monoclonal antibody. Note that the migration of Bβ and γ chain components is not considered in this study and that all the bands shown represent Aα chain immunoreactivity. (Reprinted with permission of the publisher, from reference (13).

early crosslinked α chain species and, moreover, can support the growth of high molecular weight polymers (13).

In order to extend these collective immunologic observations and biochemically define which α chain regions become crosslinked to one another during the initial stages of fibrin stabilization, α chain dimers and trimers were isolated from lightly crosslinked fibrin preparations and were then digested with cyanogen bromide to release a mixture of non-crosslinked and crosslinked derivatives. These were partially separated by gel filtration and crosslinked fragments selected based on criteria that included size and the presence of F-102 and/or F-103 immunoreactivities. Two types of crosslinked fragments could be distinguished and these were purified by successive immunoaffinity chromatography on F-103, followed by F-102-, Sepharose. Results of NH_2-terminal sequencing indicated that the two different fragments reflected two modes of early α chain crosslinking (14). The first fragment (XL VIII-X, 35-37K), whose formation requires the presence of relatively intact Aα chains, represents crosslinking between CNBr VIII and CNBr X regions of different fibrin molecules. The second fragment (XL VIII-VIII, 53-59K) does not require intact Aα chains for its formation and represents crosslinking between CNBr VIII regions of different fibrin molecules. This finding localizes at least one set of glutamine acceptor *and* donor lysine residues NH_2-terminal to the F-102 epitope, within CNBr VIII (see Fig. 2). It also suggests a mechanism whereby degraded α chains, which circulate as a component of fibrinogen catabolites under a variety of pathophysiologic conditions, can participate in fibrin stabilization despite the loss of considerable COOH-terminal structure.

Immunologic studies of Aα chain proteolysis during fibrinolysis

To what extent can Aα chain degradation proceed before functional activity, ie., crosslinking and fibrin stabilization, is severely impaired or lost? We recently began to address this question by employing the monoclonal antibodies, F-102 and F-103, in immunoblotting studies to visualize the course of Aα chain proteolysis in the plasma of patients undergoing thrombolytic therapy with either streptokinase (SK) or rt-PA (15). Specifically, we wanted to determine whether the patterns of Aα chain degradation differed with the two agents and, if so, whether inferences could be drawn regarding a possible Aα chain contribution to the bleeding and rethrombosis complications that are often associated with current thrombolytic therapies. To date, detailed structural information about the sequence of Aα chain proteolysis *in vivo* has not been forthcoming. The precise bond cleavages responsible for the COOH-terminal Aα chain heterogeneity that characterizes circulating fibrinogen remain, for the most part, undefined, as do the molecular events that lead to Fragment X formation upon induction of a fibrinolytic state.

At the dosages of SK (1.5×10^6 U) and rt-PA (80 mg) employed in the TIMI Phase I Trial (which served as the source of the patient plasmas used in this study), COOH-terminal Aα chain degradation proceeded more rapidly with SK than with rt-PA (data not shown). The entire fibrinogen population appeared to have lost its F-102 epitopes 1/2 h after the start of SK infusion and by 2h after the end of treatment the loss of F-103 epitopes was also complete. In contrast, over the course of treatment with rt-PA, a large proportion of the circulating fibrinogen molecules appeared to retain their F-103 epitope(s), and to a lesser extent, their F-102 epitope(s). Peptides (23 K, 30-33 K) that contained the F-103, but not the F-102, epitope were significant early cleavage products formed during SK infusion. These peptides were not observed to any significant degree either during or after treatment with rt-PA. Rather, large peptides (40 K) that contained *both* the F-102 and F-103 epitopes, and a series of smaller peptides (10-22 K) that contained only the F-102 epitope, were present at the earliest time point examined (15 min).

In view of the findings obtained in our early α chain crosslinking studies (see above), the collective observations described here suggest that, under the treatment regimens employed in the TIMI Phase I Trial, the capacity for effective fibrin stabilization, via α chain crosslinking, would be severely compromised in patients receiving SK since their circulating fibrinogen pool is unable to form either VIII-X or VIII-VIII crosslinks. The patterns of Aα chain proteolysis observed over the course of rt-PA therapy, in contrast, indicate that at least partial crosslinking capacity, ie., VIII-VIII crosslinking, may be retained by the circulating, degraded fibrinogen molecules.

The development of monoclonal antibodies that recognize conserved regions within the COOH-terminal portion of the Aα chain

To date, several attempts to produce additional monoclonal antibodies that recognize defined sequences within the human Aα chain crosslinking region, CNBr VIII, have met with little success. In the course of these investigations, we noted that while two different peptides from this region consistently elicited a strong immune response in most of the immunized mice, the antibodies responsible for this response invariably became "masked" following fusion and the subsequent development of hybridoma cell lines in cultures supplemented with horse serum. This was particularly interesting in light of a second observation which indicated that equine fibrinogen exhibited immunologic crossreactivity with human CNBr VIII, when tested with a rabbit antiserum that was generated against purified preparations of CNBr VIII (16). In order to investigate the possibility that equine Aα chain degradation products (present in horse serum as a result of incomplete defibrination) could interfere with the detection of antibodies of interest by competing with the human screening antigen for antibody binding, we undertook a partial structural characterization of the equine fibrinogen Aα chain, selecting for an CNBr equine fragment that corresponded to the CNBr VIII region of the human Aα chain (17).

The fragment, referred to here as equine CNBr VIII, was isolated from CNBr-treated equine fibrinogen following gel-filtration and hydrophobic chromatography on Phenyl Sepharose. It was identified in the column effluents based on its immunoreactivity with the antiserum to human CNBr VIII (see above). Trypsin-treated equine CNBr VIII was subjected to reversed-phase HPLC and the resolved peptides were characterized by NH_2-terminal sequencing. Fig. 4 illustrates several examples of the homology observed between many of the equine sequences and sequences from the CNBr VIII region of the human Aα chain. These findings provide biochemical evidence to support the potential for cross-reactive interference by equine Aα chain fragments present in horse serum supplements during the development of some anti-human Aα chain hybridoma cell lines. In order to minimize the selection bias created as a result of the presence of these fragments, it may be possible to remove them by pretreating horse serum, prior to its addition as a media supplement, with an immunosorbent constructed from either an antibody that recognizes

#1	HU:	M E L E R P G G
	EQ:	M E L E T A G R
#2	HU:	R N P G S S G T
	EQ:	R H P G S S E P
#3	HU:	R P D S P G S G
	EQ:	R P D S S G H G
#4	HU:	R S C S K T V T
	EQ:	R S C S K T V T

Fig. 4. Homology between equine and human Aα chains in the region corresponding to (human) CNBr VIII: The partial structure of the equine counterpart to human CNBr VIII was determined as described in the text and compared to the sequence of human CNBr VIII reported in reference (18). The four regions illustrated here correspond to the human Aα chain residues: 240-247 (#1); 289-296 (#2); 375-382 (#3); and 440-447 (#4). Amino acids that are shared in the human and equine sequences are underlined.

an epitope common to equine and human Aα chains or one that recognizes an equine Aα chain-specific epitope.

REFERENCES

1. H. L. Nossel, I. Yudelman, R. E. Canfield, V. P. Butler, Jr., K. Spanondis, G. D. Wilner, and G. D. Quereshi, Measurement of fibrinopeptide A in blood. *J. Clin, Invest.,* **54:**43-531 (1974).
2. J. H. Sobel, P. H. Ehrlich, S. B. Birken, A. J. Saffran, and R. E. Canfield, Monoclonal antibody to the region of fibronectin involved in crosslinking to human fibrin. *Biochemistry,* **22:**4175-4183 (1983).
3. P. H. Ehrlich, J. H. Sobel, Z. A. Moustafa, and R. E. Canfield, Monoclonal antibodies to α chain regions of human fibrinogen that participate in polymer formation. *Biochemistry,* **22:**4184-4192 (1983).
4. G. R. Crabtree, C. M. Comeau, D. M. Fowlkes, A. J. Jr. Fornace, J. D. Malley, and J. A. Kant, Evolution and structure of the fibrinogen genes. Random insertion of introns or selective loss? *J. Mol. Biol.,* **185:**1-19 (1985).
5. R. F. Doolittle, K. G. Cassman, B. A. Cottrell, and S. J. Friezner, Amino acid sequence studies on the α chain of human fibrinogen: Isolation and characterization of two linked α-chain cyanogen bromide fragments from fully crosslinked fibrin. *Biochemistry,* **16:**1715-1719 (1977b).
6. L. J. Fretto, and P. A. McKee, Structure of α polymer from *in vitro* highly crosslinked human fibrin. *J. Biol. Chem.,* **253:**6614-6622 (1978).
7. R. F. Doolittle, K. G. Cassman, B. A. Cottrell, S. J. Friezner, J. T. Hucko, and T. Takagi, Amino acid sequence studies on the α chain of human fibrinogen. Characterization of 11 cyanogen bromide fragments. *Biochemistry,* **16:**1703-1709 (1977a).
8. D. H. Corcoran, E. W. Ferguson, L. J. Fretto, and P. A. McKee, Localization of a crosslink donor site in the α chain of human fibrin. *Thrombos. Res.,* **19:**883-888 (1980).
9. B. A. Cottrell, D. D. Strong, K. W. K. Watt, and R. F. Doolittle, Amino acid sequence studies on the α chain of human fibrinogen. Exact location of crosslinking acceptor sites. *Biochemistry,* **18:**5405-5409 (1979).
10. D. Mosher, Crosslinking of cold insoluble globulin by fibrin-stabilizing factor, *J. Biol. Chem.,* **250:**6614-6621 (1975).
11. C. W. Francis, V. J. Marder, and S. E. Martin, Plasmic degradation of crosslinked fibrin I. Structural analysis of the particulate clot and identification of new macromolecular soluble complexes. *Blood,* **56:**456-464 (1980a).
12. C. W. Francis, V. J. Marder, and G. H. Barlow, Plasmic degradation of crosslinked fibrin. Characterization of new macromolecular soluble complexes and a model of their structure. *J. Clin. Invest.,* **66:**1033-1043 (1980b).

13. J. H. Sobel, C. A. Thibodeau, and R. E. Canfield, Early alpha chain crosslinking in human fibrin preparations. *Thromb. Haemostasis,* **60:**153-159 (1988).
14. J. H. Sobel, C. A. Thibodeau, M. A. Gawinowicz Kolks, and R. E. Canfield, Immunochemical characterization of crosslinked derivatives isolated from alpha chain oligomers formed during early stages of fibrin crosslinking. *Thromb. Haemostasis,* **60:**160-169 (1988).
15. J. H. Sobel, H. Backus, and R. E. Canfield, Immunodetection of rt-PA-mediated fibrino(geno)lysis using monoclonal antibodies that recognize (A)α chain crosslinking regions. *Thromb. Haemostasis,* **62:**73 (abst.) (1989).
16. J. H. Sobel. J. A. Koehn, R. Friedman, and R. E. Canfield, Alpha chain crosslinking of human fibrin: Purification and radioimmunoassay development for two Aα chain regions involved in crosslinking. *Thromb. Res.,* **26:**411-424 (1982).
17. J. H. Sobel, C. A. Thibodeau, M. A. Gawinowicz Kolks, and R. E. Canfield, The equine fibrinogen Aα chain: Partial structural comparison with human Aα chain crosslinking regions. in Fibrinogen 3. Biochemistry, biologic functions, gene regulation and expression. M. W. Mosesson, et al. eds. Elsevier Science Publishers, B. V. (1988).
18. D. D. Strong, K. W. K. Watt, B. A. Cottrell, R. F. Doolittle, Amino acid sequence studies on the α chain of fibrinogen. Complete sequence of the largest cyanogen bromide fragment. *Biochemistry,* **24:**5399-5404 (1979).

ABNORMAL FIBRINOGENS WITH TWO STRUCTURAL DEFECTS

Michio Matsuda, Shigeharu Terukina, Nobuhiko Yoshida, Kensuke Yamazumi and Hisato Maekawa

Institute of Hematology, Jichi Medical School
Tochigi-Ken 329-04, Japan

INTRODUCTION

Among the abnormal fibrinogens which we recently analyzed, two types of abnormality were found to have two structural defects: a genetically determined amino acid substitution and an additional secondary modification of the molecule due to the mutation (1, 2). Here we briefly describe these abnormal molecules.

MATERIALS, RESULTS AND DISCUSSION

1) A γ Arg-275 to Cys Substitution with a Free Cysteine Molecule Linked by a Disulfide Bridge — Fibrinogen Osaka II

This abnormal fibrinogen was found in a 48-year-old man and his 15-year-old daughter, both of whom had no bleeding or thrombotic episodes. Since the father had been scheduled to undergo operation, 200 ml of blood was collected from the daughter into a plastic bag containing ACD-solution for this study. Plasma was separated by centrifugation at 3,000 xg for 30 min at 4°C, and fibrinogen was purified as described (3).

Functional Abnormalities of Plasma and Fibrinogen

The thrombin time of plasma was prolonged (107.9 s; normal 14.3 s). The concentration of fibrinogen in plasma determined by the thrombin time method was markedly reduced (< 40 mg/dl; normal 200-400 mg/dl) despite normal levels determined by other methods. The thrombin and ancrod times of purified fibrinogen were both prolonged (> 300 s) as compared with normal fibringoen (15.3 s and 17.6 s, respectively). Fibrin monomer polymerization was severely impaired but release of fibrinopeptides A and B was normal when measured by reverse-phase HPLC (profiles not shown). Cross-linking of the γ chain by factor XIIIa was also normal when analyzed by SDS-PAGE (not shown).

Gross Structure of Fibrinogen and Its Plasmic Digests as Examined by SDS-PAGE

As shown in Fig. 1, the bands corresponding to the fibrinogen γ chain and its remnant in fragment D_1 (γ/D_1) appeared to be somewhat broader than the corresponding normal polypeptides (lanes 2 and 4 as compared with lanes 1 and 3, respectively). The γ remnant in fragment D_2 (γ/D_2) was visualized as a doublet comprising a normal and an additional, apparently higher molecular weight (HMW) γ remnant (lane 6). In fragment D_3, however, the γ remnant (γ/D_3) was noted as a single component (lanes 6 and 8). When the fibrinogen was reduced and S-carboxymethylated, the apparent molecular weight heterogeneity became

Fibrinogen, Thrombosis, Coagulation, and Fibrinolysis, Edited by
C. Y. Liu and S. Chien, Plenum Press, New York

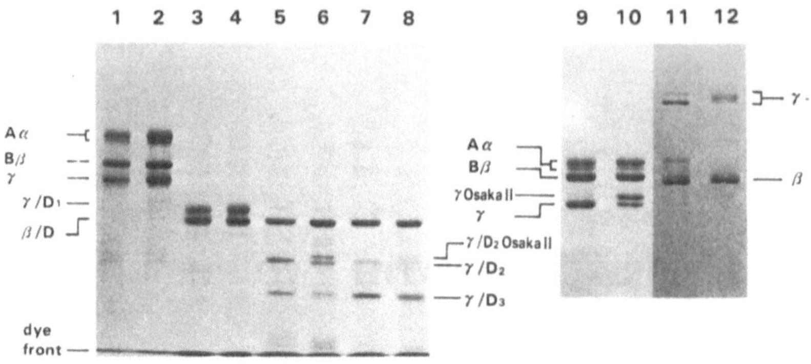

Fig. 1. SDS-PAGE of fibrinogen, its plasmic digests, and cross-linked fibrin. Lanes 1, 3, 5, 7, 9 and 11 are for normal samples and lanes 2, 4, 6, 8, 10, and 12 are for samples derived from the patient's fibrinogen. Lanes 1 and 2, fibrinogen; lanes 3 and 4, plasmic digests in the presence of 5 mM Ca^{2+}; lanes 5-8, plasmic digests in the presence of 10 mM EGTA for 30 min (lanes 5 and 6) and 90 min (lanes 7 and 8); lanes 9 and 10, S-carboxymethylated fibrinogen (Cm-fibrinogen); lanes 11 and 12, Cm-cross-linked fibrin. [Reprinted from Terukina et al. (1)].

more distinct (lane 10). Reduced and S-carboxymethylated cross-linked fibrin gave three γ-γ dimers (lane 12).

Separation of Abnormal Fragment D_1

Since some structural abnormality was suspected to reside in the C-terminal region of the γ remnant of fragment D_1 (γ/D_1), we attempted to separate the abnormal fragment D_1 species by chromatofocusing (4). As shown in Fig. 2, the abnormal fragment D_1 species (peak II) with an apparently higher molecular weight γ-remnant (inset, lane II) was eluted later than the normal D_1 species (peak I). They were separately collected and subjected to structure analysis.

Identification of an Aberrant Peptide in Lysyl Endopeptidase Digests of Abnormal Fragment D_1

Normal and abnormal fragment D_1 species were reduced, S-pyridylethylated and then digested with lysyl endopeptidase, and the digests were fractionated on a Cosmosil 5C18P column. As depicted in Fig. 3, peptide K26 identified in the normal sample (A) was missing in the abnormal one, but instead, peptide K26' was present (B). These peptides were analyzed for total amino acid composition and primary sequence. Table 1 shows the amino acid compositions of K26 and K26'. An Arg residue that is present in peptide K26 is missing, but instead, an S-pyridylethylated Cys (Pe-Cys) is present in K26'. The data for K26 coincided with the theoretical values for the normal peptide segment composing γ 274-302 residues. Since the γ 274-302 peptide contains a single Arg residue at position 275, this Arg residue was assumed to be replaced by Cys in the aberrant peptide. This assumption was verified by sequence analysis as shown in Table 2.

Characterization of the Cys Substitute at Position 275 of the Mutant γ Chain

We then characterized the status of the Cys-275 substitute as to whether it was present as a free cysteine or it was linked by a disulfide bridge with some SH-containing substance. For this purpose, we tried to isolate native K26' from abnormal fragment D_1 without prior reduction or S-alkylation. A small peak, which was not present in the normal sample, was siolated and confirmed to be the peptide comprising γ 274-302 residues. This peptide was analyzed for its mass value by fast atom bombardment (FAB) mass spectrometry (7). We obtained a mass value of 3153.3 for normal peptide K26, which was consistent with the

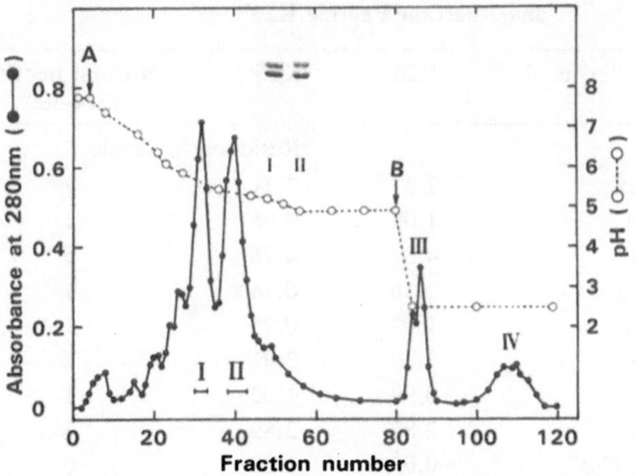

Fig. 2. Chromatofocusing chromatography of plasmic digests of d-fibrinogen. The plasmic digests were dialyzed against 0.025 M Tris-HCl, pH 7.8, loaded onto a column of polybuffer exchanger 94 gel (1.2 × 18 cm) at 4°C, and then eluted with 400 ml of degassed polybuffer 74, pH 5.0 (A). The elution with polybuffer was followed by 400 ml of 0.025 M glycine-HCl, pH 2.5, containing 0.1 M NaCl (B). The fragment D_1 species were eluted in two peaks, one with a normal γ chain remnant (peak I) and the other with a larger molecular weight γ chain remnant (peak II) as evidenced by SDS-PAGE after reduction (inset). Peak III corresponds to fragment E and peak IV to low molecular weight fragments. [Reprinted from Terukina et al. (1)]

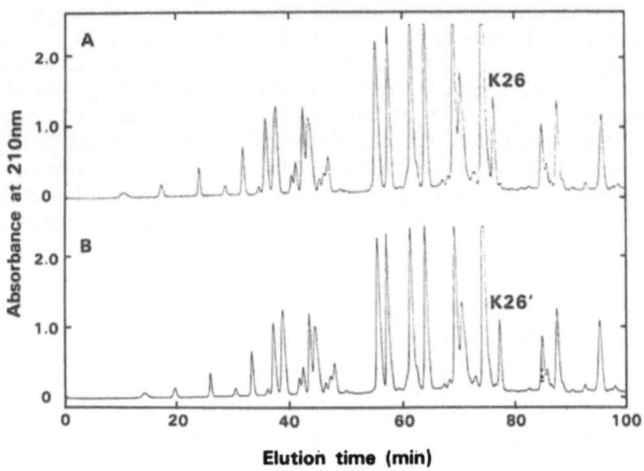

Fig. 3. HPLC elution profile of lysyl endopeptidase digests of reduced and S-pyridylethylated fragment D_1. Normal (panel A) and abnormal (panel B) fragment D_1 species derived from the patient's fibrinogen were digested with lysyl endopeptidase after reduction and S-pyridylethylation and analyzed by HPLC using a Cosmosil 5C18P column (4.6 × 150 mm). Peptides were eluted by a linear gradient from 0 to 60% acetonitrile in 120 min. [Reprinted from Terukina et al. (1)]

Table 1. Amino Acid Compositions of Normal Peptide K26 and Aberrant Peptide K26'

Amino acid	K26	K26'	Normal peptide γ274-302[a]
		Residues/molecule	
Asx	7.37	7.31	7
Ser	1.00	0.95	1
Gly	4.67	4.76	5
Arg	1.16	0.14	1
Thr	1.02	0.95	1
Ala	3.80	3.91	4
Pro	0.99	1.02	1
Tyr	2.87	2.89	3
Pe-Cys[b]	0.09	1.00	0
Leu	1.00	1.00	1
Phe	3.98	3.88	4
Lys	1.00	1.00	1

[a]The amino acid composition of γ274-302 peptide segment was calculated on the basis of the data for amino acid sequence (5, 6).
[b]Pyridylethylcysteine.
[Reprinted from Terukina et al. (1)]

theoretical value of 3153.3 for the peptide with a normal γ 274-302 sequence. For the aberrant peptide, we obtained two mass values, 3219.0 and 3099.9 as shown in Fig. 4. The larger value was in good agreement with the value 3219.2 for a putative peptide with a single Cys linked by a disulfide bridge to the Cys substitution at position 2. The small value was consistent with the theoretical value 3100.2 for a reduced peptide produced during FAB mass measurement. It thus appeared that the Cys substitution at position 2 of K26' was linked to a single free Cys molecule by a disulfide bridge. This was confirmed by identification of a single Pe-Cys recovered from the abnormal fragment D_1 and also from the patient's fibrinogen after reduction and S-pyridylethylation. The replacement of an Arg by a Cys has been identified in various abnormal plasma proteins (8-12). In fibrinogen Kawaguchi and Osaka I, we showed that the Cys substitution at position 16 of one Aα chain formed an extra disulfide bridge with the Cys substitution of the other mutant Aα chain in the same molecule (9). The solitary cystine structure identified at the mutation

Table 2. Amino Acid Sequences of Peptide K26 and Aberrant Peptides, K26'

Cycle	K26		K26'	
	Amino acid	pmol	Amino acid	pmol
1	Tyr	163	Tyr	72
2	Arg	105	Pe-Cys	NQ[#]
3	Leu	130	Leu	58
4	Thr	99	Thr	46
5	Tyr	121	Tyr	58

[#]Not quantitated
[Adapted from Terukina et al. (1)]

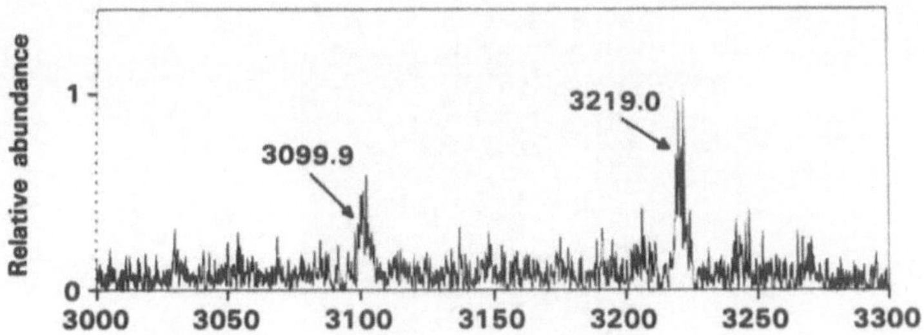

Fig. 4. Positive FAB mass spectrum of non-reduced peptide K26'. [Reprinted from Terukina et al. (1)]

site in fibrinogen Osaka II seems to be unique for abnormal fibrinogens and may have contributed to the perturbation of the tertiary structure required for the expression of the polymerization site assigned to the D domain.

2) A γMet-310 to Thr Substitution with an Additional Oligosaccharide Attached to γ Asn-308

This unique mutation was identified in an abnormal fibrinogen found in a 33-year-old man with posttraumatic bleeding and delayed wound healing. Routine hemostasis and coagulation studies revealed that the patient had markedly prolonged thrombin time (> 300 s, control 15.8 s) and ancord time (> 300 s, control 9.8 s), and an apparently reduced fibrinogen level in plasma (< 50 mg/dl, normal 200-400 mg/dl) as determined by the thrombin time method. When plasma fibrinogen was measured by other methods, rather high values were obtained: 525 mg/dl by Laurell's method and 565 mg/dl by the turbidimetric method. This fibrinogen was designated as fibrinogen Asahi.

Studies on Purified Fibrinogen

Fibrinogen was purified from the patient-derived ACD plasma and clotting tests were performed. Here again, both thrombin and ancrod times were markedly prolonged even after addition of $CaCl_2$ (> 180 s; normal, 12.8 s for thrombin and 5.5 s for ancord times). By SDS-PAGE after reduction, we noticed an additional HMW γ chain as indicated by γ Asahi in Fig. 5. We then analyzed cross-linking profiles of fibrin mediated by factor XIIIa by SDS-PAGE after reduction. The abnormal γ chain species was converted to the γ-γ dimer very slowly as compared with the normal one (Fig. 6). This extremely retarded cross-linking between the two abnormal γ chains may be accounted for by severely impaired

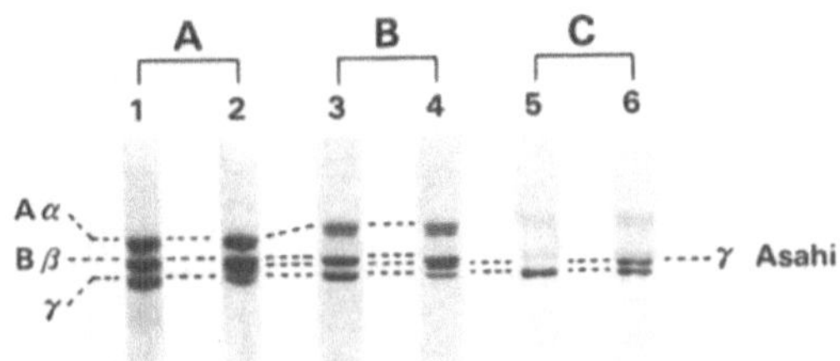

Fig. 5. Subunit polypeptides of fibrinogen examined by SDS-PAGE and Western blotting. (A) Weber and Osborn gel, 5%; (B) Laemmli gel, 10%; (C) Western blot analysis with an anti-γ core (the γ 85-302 residue segment) antibody. Lanes 1, 3, and 5 are normal fibrinogen and lanes 2, 4 and 6 are the patient's fibrinogen. To each lane, 1.5 μg of protein was loaded. [Reprinted from Yamazumi et al. (2)]

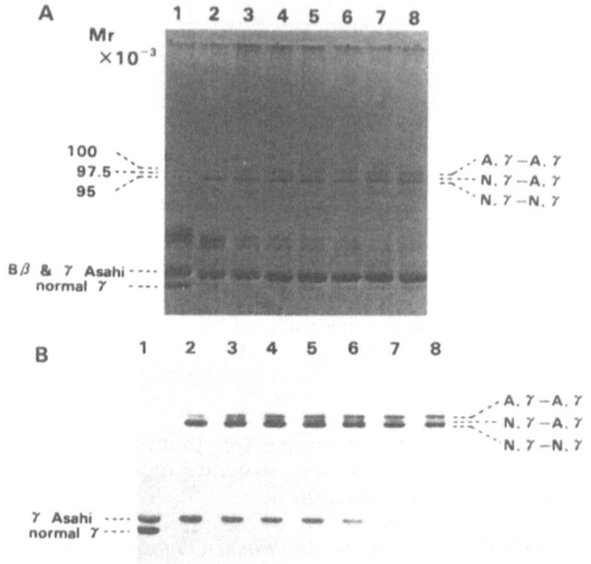

Fig. 6. Factor XIIIa-mediated cross-linking of fibrin analyzed by SDS-PAGE and Western blotting after reduction. One gel was stained with CBB for proteins (A), and the other gel was subjected to Western blot analysis using the anti-γ core MAb (B). Lane 1, 0 min; lane 2, 2 min; lane 3, 5 min; lane 4, 10 min; lane 5, 15 min; lane 6, 30 min; lane 7, 2 h and lane 8, 24 h. [Reprinted from Yamazumi et al. (2)]

fibrin monomer polymerization. Therefore, we attempted to see the cross-linking profile of fibrinogen molecules, but not fibrin monomers, to minimize the effect of polymerization. Here again, the mutant γ chain spcies was not cross-linked even after 24 h of incubation (lane 4, Fig. 7A). To see whether the cross-linking sites of the γ Asahi chain function normally, we examined the amine acceptor γ Gln-398 by monitoring the incorporation of fluorescent monodansylcadaverine. As shown in lane 4, Fig. 7B, fluorescence was equaly incorporated into both the normal and Asahi γ chains, indicating that the amine acceptor of the mutant γ chain functioned normally.

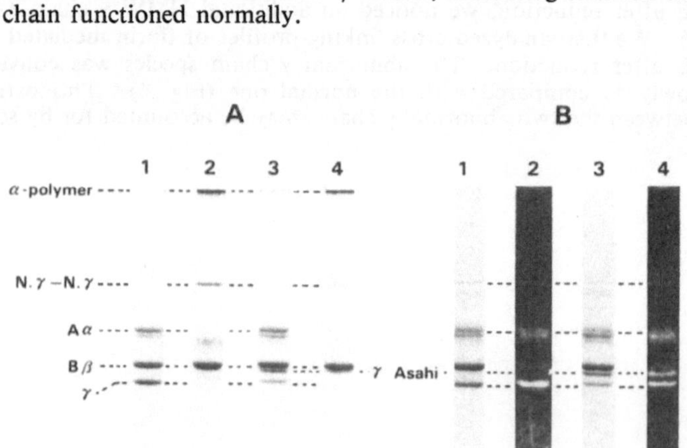

Fig. 7. Factor XIIIa-mediated cross-linking of fibrinogen (A) and incorporation of monodansylcadaverine (B), analyzed by SDS-PAGE. N: normal fibrinogen; P: the patient's fibrinogen.

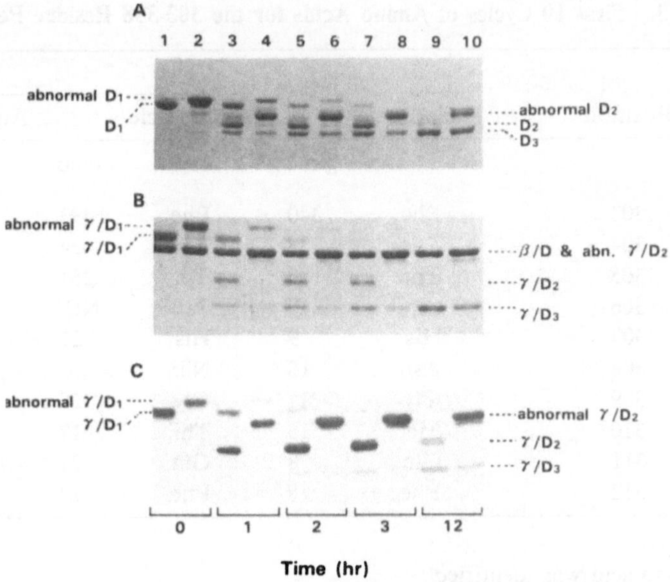

Fig. 8. Sequential digestion of fragment D_1 by plasmin. A. SDS-PAGE. B. SDS-PAGE. C. Western-Blot.

Sequential Digestion of the Abnormal Fragment D_1 by Plasmin

The patient-derived fibrinogen was digested with plasmin in the presence of 5 mM $CaCl_2$. Fragment D_1 was separated into normal and abnormal species by chromatofocusing, in which the abnormal fragment D_1 species was eluted after the normal one (profile not shown). These two fragment D_1 species were separately collected and further digested with plasmin in the presence of 10 mM EGTA, and the digests were analyzed by SDS-PAGE and Western blotting. As shown in Fig. 8A, abnormal fragments D_1 and D_2 were apparently larger than their normal counterparts, whereas both normal and abnormal fragment D_3 species were identical in size with each other. By SDS-PAGE after reduction and Western blot analysis, we found that γ/D_1 and γ/D_2 were larger than the normal counterparts but

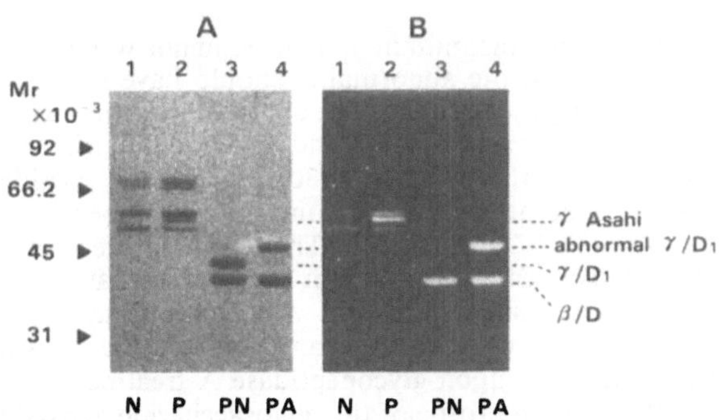

Fig. 9. Analysis of the patient's fibrinogen and its two fragment D_1 species by carbohydrate staining. SDS-PAGE was performed by Laemmli's method using 10% gels. (A) CBB staining for proteins; (B) carbohydrate staining by dansylhydrazine. Lane 1, normal fibrinogen; lane 2, patient's fibrinogen; lane 3, normal fragment D_1 species derived from the patient's fibrinogen; and lane 4, abnormal fragment D_1 species derived from the patient's fibrinogen. [Reprinted from Yamazumi et al. (2)]

Table 3. First 10 Cycles of Amino Acids for the 303-338 Residue Peptides

Cycle	Position	P Amino acid	pmol	P' Amino acid	pmol	P'-G Amino acid	pmol
1	303	Phe	340	Phe	493	Phe	495
2	304	Phe	283	Phe	424	Phe	334
3	305	Thr	139	Thr	251	Thr	140
4	306	Ser	NQ*	Ser	NQ*	Ser	NQ*
5	307	His	9	His	21	His	6
6	308	Asn	15	Nil‡		Asp	10
7	309	Gly	11	Gly	29	Gly	7
8	310	Met	15	Thr	17	ND§	
9	311	Gln	8	Gln	21	ND§	
10	312	Phe	8	Phe	24	ND§	

* Not quantitated
‡ No PTH amino acid was identified
§ Not determined
P, normal and P', aberrant 303-338 residue peptides.
P'-G, glycopeptidase-treated P'.
[Reprinted from Yamazumi et al. (2)]

γ/D_3 was not so. These findings indicated that the C-terminal peptide segment of fragment D_2 that should be cleaved by plasmin upon conversion of fragemnt D_2 to fragment D_3 must have been elongated or attached with some additives. Therefore, we first examined possible extra glycosylation to the mutant γ chain by staining the polypeptide remnants of fragment D_1 with dansylhydrazine. As shown in Fig. 9B, the γ Asahi remnant in fragment D_1 was clearly stained for carbohydrate (lane 4, B) whereas the normal one that lacks carbohydrate was not stained (lane 3, B). Furthermore, γ Asahi chain of the patient's fibrinogen showed strongly positive staining for carbohydrate (lane 2, B) as compared with the normal Bβ and γ chains (lanes 1 and 2, B).

Identification of the Point Mutation and Extra Glycosylation in the Mutant γ Chain

Since the molecular weight for the mutant γ chain remnant was higher in fragments D_1 and D_2, but not so in fragment D_3, the abnormality should have resided in the C-terminal peptide released from fragment D_2 by plasmin. The aberrant peptide was isolated by reverse-phase HPLC and analyzed for the carbohydrate and also for the primary sequence. Table 3 illustrates the first 10 cycles of the aberrant γ 303-338 residue peptide (separated after reduction of the γ 303-356 residue peptide) before and after glycopeptidase A-treatment (P' and P'-G, respectively). In both P' and P'-G, a Thr was identified for a Met at cycle 10 corresponding to position 310 of the γ chain. An Asn present at cycle 6 in the normal peptide was not identified in P', suggesting that Asn-308 was most likely linked to an oligosaccharide. Indeed, in P'-G, we identified an Asp at cycle 6, which was certainly derived from an N-glycosylated Asn upon glycopeptidase A treatment.

Fig. 10 shows the HPLC-elution profile of the oligosaccharide derived from the aberrant 303-356 peptide (A) as compared with those derived from the normal γ chain (B), the Bβ chain of the patient's fibrinogen (C) and the normal Bβ chain (D). They were all quite identical consisting of essentially three types of oligosaccharides, shown at the bottom in Fig. 10. Attachment of such a bulky oligosaccharide is certainly due to a newly constructed tripeptide sequence: Asn-308•Gly-309•Thr-310, a consensus sequence for N-glycosylation, where a Thr was a substitution for Met-310 in the mutant γ chain. Thus in fibrinogen Asahi with severely altered fibrin monomer polymerization and impaired γ-γ formation, the structure in the D main seems to be critically perturbed by a genetically determined

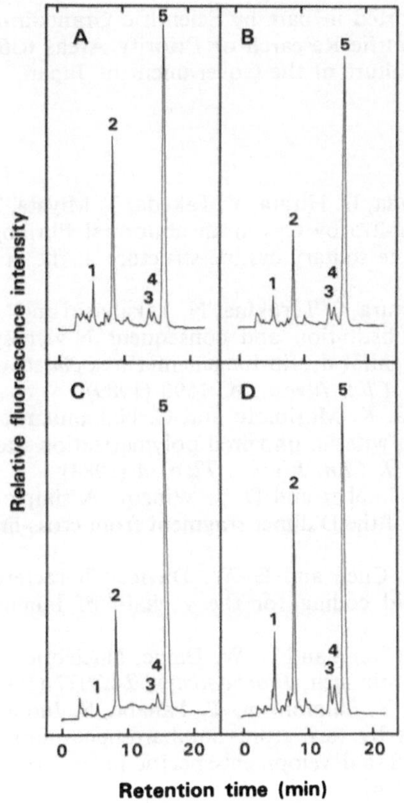

Peaks 1, 2 and 5 :

Galβ(1,4)GlcNAcβ(1,2)Manα(1,3) ↘
Manβ(1,4)GlcNAcβ(1,4)GlcNAcβ-Asn
Galβ(1,4)GlcNAcβ(1,2)Manα(1,6) ↗

Peak 3 :

GlcNacβ(1,2)Manα(1,3) ↘
Manβ(1,4)GlcNAcβ(1,4)GlcNAcβ-Asn
Galβ(1,4)GlcNacβ(1,2)Manα(1,6) ↗

Peak 4 :

Galβ(1,4)GlcNAcβ(1,2)Manα(1,3) ↘
Manβ(1,4)GlcNAcβ(1,4)GlcNAcβ-Asn
GlcNAcβ(1,2)Manα(1,6) ↗

Fig. 10 Analysis of the extra oligosaccharide attached to the mutant γ chain by reverse-phase HPLC. HPLC elution profiles of oligosaccharides derived from A, the aberrant γ 303-356 residue peptide; B, the normal γ chain, C, the Bβ chain of the patient's fibrinogen; and D, the normal Bβ chain. [Adapted from Yamazumi et al. (2))

amino acid substitution together with an additional bulky oligosaccharide attached to an Asn located at two residues N-terminal side of the mutation site.

ACKNOWLEDGMENTS

The authors are indebted to Dr. K. Shimura, Asahi General Hospital and Mr. Y. Takeda, Osaka Kosei Nenkin Hospital for referring their patients to us.

This work was supported in part by Scientific Grants-in-Aid 61480272 and 63480293 and Grant-in-aid for Scientific Research or Priority Areas 63616512 from the Ministry of Education, Science and Culture of the Government of Japan.

REFEENCES

1. S. Terukina, M. Matsuda, H. Hirata, Y. Takeda, T. Miyata, T. Takao and Y. Shimonishi, Substitution of γArg-275 by Cys in an abnormal fibrinogen "fibrinogen Osaka II." Evidence for a unique solitary cystine structure at the mutation site, *J. Biol. Chem.*, **263:**13579 (1988).
2. K. Yamazumi, K. Shimura, S. Terukina, N. Takahashi and M. Matsuda, A γ-methionine-310 to threonine substitution and consequent N-glycosylation at γ asparagine-308 identified in a congenital dysfibrinogenemia associated with posttraumatic bleeding, fibrinogen Asahi, *J. Clin. Invest.*, **83:**1590 (1989).
3. M. Matsuda, M. Baba, K. Morimoto and C. Nakamikawa, Fibrinogen Tokyo II. An abnormal fibrinogen with an impaired polymerization site on the aligned DD domain of fibrin molecules, *J. Clin. Invest.*, **72:**1034 (1983).
4. P. P. Masci, A. N. Whitaker and D. J. Winzor, A simple chromatographic procedure for the purification of the D dimer fragment from cross-linked fibrin, *Anal. Biochem.*, **147:**128 (1985).
5. D. W. Chung, W.-Y. Chen and E. W. Davie, Characterization of a complementary deoxyribonucleic acid coding for the γ chain of human fibrinogen, *Biochemistry,* **22:**3250 (1983).
6. M. W. Rixon, D. W. Chung and E. W. Davie, Nucleotide sequence of the gene for the γ chain of human fibrinogen, *Biochemistry,* **24:**2077 (1985).
7. T. Takao, T. Hitouji, Y. Shimonishi, T. Tanabe, S. Inoue and M. Inoue, Verification of protein sequence by fast atom bombardment mass spectrometry. Amino acid sequence of protein S, a development-specific protein of myxococcus xanthus, *J. Biol. Chem.,* **259:**6105 (1984).
8. C. Southan, A. Henschen and F. Lottspeich, The search for molecular defects in abnormal fibrinogens, *in*: "Fibrinogen, Recent Biochemical and Medical Aspects," A. Henschen, H. Graeff and F. Lottspeich, eds., Walter de Gruyter, Berlin (1982).
9. T. Miyata, S. Terukina, M. Matsuda, A. Kasamatsu, Y. Takeda, T. Murakami and S. Iwanaga, Fibrinogens Kawaguchi and Osaka: An amino acid substitution of Aα arginine-16 to cysteine which forms an extra interchain disulfide bridge between the two Aα chains, *J. Biochem.,* **102:**93 (1987).
10. M.-J. Rabiet, B. C. Furie and B. Furie, Molecular defect of prothrombin Barcelona. Substitution of cysteine for arginine at residue 273, *J. Biol. Chem.,* **261:**15045 (1986).
11. T. Koide, S. Odani, K. Takahashi, T. Ono and N. Sakuragawa, Antithrombin III Toyama: Replacement of arginine-47 by cysteine in hereditary abnormal antithrombin III that lacks heparin-binding ability, *Proc. Natl. Acad. Sci. USA,* **81:**289 (1984).
12. N. Yoshida, K. Ota, M. Moroi and M. Matsuda, An apparently higher molecular weight γ-chain variant in a new congenital abnormal fibrinogen Tochigi characterized by the replacement of γ arginine-275 by cysteine, *Blood,* **71:**480 (1988).

ELECTROPHORETIC CHARACTERIZATIONS OF CROSS-LINKED FIBRINOGEN DERIVATIVES IN BLOOD AND VASCULAR TISSUE BY ZONAL IMMOBILIZATION ON GLYOXYL AGAROSE

John R. Shainoff, Rafael Valenzuela, Robert Graor, David A. Urbanic, and Patricia M. DiBello

Research Institute of the Cleveland Clinic Foundation
Cleveland, OH 44195, USA

INTRODUCTION

Current definitions of fibrinogen derivatives produced in course of clot formation and lysis derive from electrophoretic and immunochemical analyses of the alterations that the purified protein undergoes during these reactions (1, 2). Methods for profiling the distribution of fibrinogen-derivatives in blood continue to improve as new methods arise for immunoprobing electrophoregrams. Polyacrylamide gel electrophoregrams (PAGE) are too impermeable to be probed directly with antibody. The earliest profiles of molecular weight distributions of the fibrinogen-related antigens in blood used immunosorbents for specific absorption of the protein from blood, followed by electrophoretic analysis of desorbed protein (3). Western-blotting (4) in which proteins are electro-transferred out of the electrophoregram to produce a replica on cellulose nitrate or other adsorptive membrane is now commonly used for immunoprobing. However, as might be anticipated (5), this technique does not work well with fibrinogen; it and its polymers do not transfer well due to their precipitation when separated from SDS. Because of relatively poor transfer of the intact protein, blots made of fibrinogen-related antigens usually show the degraded and readily transferred forms at levels far out of proportion to their actual concentrations in the samples. To eliminate the uncertainties that arise from transfer procedures, we have devised methods for directly immunoprobing the electrophoregrams. The present communication outlines underlying principles of the methodology, and surveys some of the findings regarding the composition of fibrinogen which were discovered with it.

The methodology, one we call zonal immobilization, is based in part on development of a novel electrophoretic medium, glyoxyl agarose (6), that can function not only as a supporting matrix for electrophoresis, but can be activated with an additive to link the separated protein covalently in place on the matrix for immunoprobing (7). The covalent fixation eliminates risk of washout of the antigens during immunoprobing, and even enables fibrinopeptide A itself to be fixed in the electrophoregram. We have just recently developed methods for compositing that medium with a removable polyacrylamide filler to provide resolving power equivalent to conventional polyacrylamide-gel electrophoresis. With the electrophoretically arrayed protein linked only to the high porosity agarose-matrix, the polyacrylamide filler can be washed away, thus rendering the electrophoregram back to one capable of being probed with antibody. We describe here the preparation of composites with a continuous buffer system and photopolymerized non-cross-linked filler, the simplest mode of application of the system. Other modes of preparation and application of the composites, including use of persulfate to catalyze polymerization, degradable cross-linked fillers for maximal sieving, and discontinuous buffers for band sharpening are described in a manual submitted for publication (8).

Glyoxyl agarose is an aldehydic medium produced by oxidation of glyceryl agarose, and contains substituent groups equivalent to acetaldehyde. These groups form only weak Schiff base linkages with protein amino groups at neutral pH, so weak that binding of plasma proteins to glyoxyl agarose gel is not perceptible over a 24 hour period. However, the weak Schiff-base linkages which form only transiently in neutral solution can be rapidly converted to stable alkyl-amine linkages by the selective reducing power of $NaCNBH_3$ (9). The only requirement for fixation (immobilization) is that the protein or peptide must contain at least one amino group (eqn. 1).

$$1)\ \text{Agarose-O-}CH_2CH{=}O + H_2N\text{-}R \rightleftharpoons \text{Agarose-O-}CH_2CH{=}\overset{+}{N}H\text{-}R \xrightarrow{BH_3CN^-} \text{Agarose-O-}CH_2CH_2\text{-}NH\text{-}R$$

Because of high porosity of agarose gels their utility for separating proteins according to size <100 KDa is poor, but resolving power equivalent to polyacrylamide can be imparted by polymerizing acrylamide within the agarose matrix, and the resultant composites have superior strength and resiliency (10). However, the polymers produced by acrylamide itself are too long and stringy to be removed from the agarose matrix. To produce short polymers, we copolymerize acrylamide with 1-allyloxy-2,3-propanediol (APD) which acts to inhibit growth of the polymer-chains by polymerizing with the acrylamide much more slowly than the acrylamide polymerizes with itself. Copolymers produced with acrylamide and APD in 5:1 proportion can be removed adequately to allow immunoprobing of the gels, and impart approximately 70% of the sieving capability of a regular polyacrylamide gel of the same concentration. Sieving to a degree fully equivalent to a regular polyacrylamide gel can be obtained by adding a degradable cross-linker (11) N,N'-(1,2-dihydroxyethylene)bis-acrylamide (DHEBA) to extend the chain length of the polymers, and these polymers can be removed by hydrolytically cleaving the cross-links with 0.2 M ethanolamine.

MATERIALS AND METHODS

Chemicals used were obtained from sources indicated in parentheses: NuFix® glyoxyl agarose and GelBond® (FMC Corp., Rockland, ME), electrophoresis grade acrylamide (Eastman Kodak, Rochester, NY), 1-allyloxy-2,3-propanediol, butylated hydroxytoluene, and sodium cyanoborohydride (Aldrich Chemical Co., Milwaukee, WI), Brij®, Tween20®, Thimerosal® and horseradish peroxidase (Sigma Chemical Co., St. Louis, MO). Since cyanoborohydride can release cyanide and hydrogen, fixative solutions containing it should be kept in sealed refrigerator dishes, and containers should be opened in a fume-hood.

Anti-fibrinogen antibodies were obtained from the IgG fraction of antisera from rabbits immunized with chromatographic fraction I fibrinogen (12), and were affinity purified by 1) adsorption onto fibrinogen immobilized on beaded NuFix® (prepared by admixing 4 mg of fibrinogen per ml of gel in 0.3 M phosphate-buffered saline and 0.05 M $NaCNBH_3$ at pH 6.4 overnight, and then washing with PBS), and 2) desorbing the immunosorbed antibody with neutral 4M guanidine-HCl followed by immediate dialysis into PBS containing 0.05% Thimerosal®. Antibodies to α-, β-, and γ-chains were prepared from antisera of rabbits that were immunized with purified (13) carboxymethylated chains. These polyclonal antibodies were affinity purified from the IgG fraction of rabbit-antisera by 1) absorbing them with immobilized reduced and carboxymethylated fibrinogen (α, β, and γ chains combined for economy of purified chains), and 2) eluting (4 M guanidine-HCl) them, and then 3) removing cross-reacting antibodies by absorption with NuFix® containing the unwanted chains; e.g. anti-α antibodies were passed through adsorbents containing β-and γ-chains to remove cross-reactants. Those affinity adsorbents were prepared by dissolving the chains in 4 M guanidine-HCl and admixing overnight with beaded NuFix® in the solvent containing 0.05 M $NaCNBH_3$ at pH 6.4, and subsequently washing with guanidine-HCl solution followed by PBS. Murine anti-fibrinopeptide-A-antibodies (14) were affinity-purifed by adsorption with peptide-A immobilized on NuFix®, and were eluted with 0.32 M citrate at pH 3.2, elution with guanidine-HCl being contraindicated with murine antibodies. The fibrinopeptides were 1) prepared by coagulating purified fibrinogen (10 mg/ml) with thrombin in 0.15 M ammonium acetate, freeze-drying and redissolving in 1/20th vol water, removing all protein with 10% TCA, and then removing the TCA by ether-extraction, and 2) immobilized by admixing neutralized concentrate (0.6 mM) overnight with an equal volume of NuFix® in 0.1 M $NaCNBH_3$ and 0.1 M phosphate at pH 6.4, followed by washing with PBS to remove fibrinopeptide B which is not retained because it lacks amino groups. Antibodies were labeled with peroxidase by the two-step glutaraldehyde method (15). All peroxidase-labeled antibodies were stored and diluted in solutions containing 1 μM hemin (16).

Glyoxyl Agarose Electrophoresis

For analysis of non-reduced plasma protein, the electrophoresis is carried out essentially the same way as for simple agarose electrophoresis, but using 3% NuFix®. The gel is prepared by dissolving the NuFix® in 0.1 mM HCl and adding one-tenth volume of 10 X (0.3 M) neutral phosphate buffer, and subsequently admixing 1/100th volume of 10% SDS. After electrophoresis, the gel is fixed by agitation (1 hour minimum, no max.) in 8-volumes of pH 10 buffer (0.2 M NaOH and 0.22 M boric acid) containing 0.02 M $NaCNBH_3$ and either 1% Lubrol® or 3%Brij®. A sheet of plastic screening should be placed beneath the gel in the dish. The $NaCNBH_3$ is not added to the buffer directly, but by dilution of a freshly boild 2 M solution. When electrophoregrams are to be stained with diaminobenzidine, the aldehyde groups in the gel should be reduced before probing with antibody, and this can be done by adjusting the fixative solution to pH 4-4.5 for an hour and then washing. The aldehyde-groups do not need to be inactivated for other modes of immunochemical staining, but if they are not reduced the cyanoborohydride should be thoroughly washed out before probing the gel with antibody.

Composite Polyacrylamide/Glyoxyl Agarose Gel Electrophoresis

The steps of procedure for preparation, runniing, and fixation of the composite gels all differ from those used (17) with regular polyacrylamide gels, as outlined in a guide to the steps of procedure (Table 1) for gels prepared by photopolymerization. The composites are best made by casting the stacking/sample application gel first, and then trimming it and pouring the resolving gel over it. The gels are cast in a cassette consisting of two glass plates sandwiching a 1.5 mm thickness U-frame.

Stacking/sample application gel. This gel consists simply of 2% NuFix® agarose in buffer containing riboflavin and TEMED. For 20 ml, 1) dissolve 0.4 g NuFix® by intermittent heating in 17 ml of 0.1 mM HCl in a 50 ml round-bottom flask until bubbles in the suspension disappear, 2) add water to readjust to tared weight, 3) admix 2 ml of 10 X buffer (0.15 M Na_2HPO_4 and 0.15 M NaH_2PO_4), 4) reheat briefly, 5) add catalysts (1 ml 0.015% riboflavin, 13 μL TEMED) and SDS (0.2 ml of 10%). The molten solution is poured into a heated (50-60°) gel-casting cassette using a folded slip of 0.1 mil GelBond® (FMC Corp.) as a funnel. The cassette is then chilled to promote gelation of the agarose, and is opened so that a straight edge can be cut into the gel and excess gel scraped away with a blade. The cassette is then reassembled, placed in a sandwich-bag, and warmed in a bath of hot tap-water prior to pouring the resolving gel.

Resolving gel. The formulation provided here is for 30 ml of gel with 16% non-cross-linked filler. Lower concentrations of filler are obtained by diluting the monomer stock-solution. Cross-linked filler is obtained by adding 0.3 g of dihydroxy-ethylene bis-acrylamide per 10 ml of monomer stock-solution, and its removal from the gel is effected by overnight digestion with 0.2 M ethanolamine following fixation of the electrophoregram and washout of the cyanoborohydride.

Monomer stock-solution (60%) is prepared with freshly dissolved components (4.8 g acrylamide, 1.2 ml 1-allyloxy-2,3-propanediol, and water to 10 ml). A buffered glyoxyl agarose solution is prepared separately in a 50 ml round-bottom flask by heating 0.7 g NuFix® in 17 ml of 0.1 mM HCl and admixing 2 ml of 0.3 M neutral phosphate buffer. The gel casting cassette with stacking gel trimmed is warmed by placing it in a bath of hot tap water (~60°) before pouring the resolving gel. While the cassette is warming, the agarose solution is heated intermittently for about 5 min until small bubbles disappear, and it is then adjusted to its original weight by adding water. The monomer stock-solution at ambient temperature is poured into the hot agarose, mixed briefly by swirling, and then catalysts (0.02 ml TEMED and 1 ml of 0.015% riboflavin) and SDS (0.3 ml of a 10%

Table 1. Comparison of steps of procedure for SDS-PAGE

	Conventional Gels	*Composite Gels*
Gel Components	Monomer concentrate Buffer concentrate SDS Riboflavin + Add water & Deaerate	Monomer concentrate Buffer concentrate SDS Riboflavin + Add hot agarose
⇓		
Gel Preparation	1. Pour *resolving* gel 2. Overlay buffer for sharp junction 3. Photopolymerize 4. Pour *stacking* gel 5. Photopolymerize	1. Pour *stacking* gel 2. Cut gel for sharp junction 0 3. Pour *resolving* gel 4. Photopolymerize
⇓		
Gel Handling	0	Cassette removed
⇓		
Sample Application	Into wells/slots	Through foil
⇓		
Electrophoresis	Vertical No wicks	Horizontal Wicks
⇓		
Disassembly	Fragile	0
⇓		
Fixation	Blot transfer (adsorptive)	Direct (covalent)

solution) are added and admixed by briefly swirling (~8 sec), and the mixture is poured immediately into the prewarmed cassette using a folded slip of GelBond® as a funnel. The filled cassette is transferred to a cold-bath for about 5 min to promote gelation of the agarose, and is then allowed to stand for about an hour at room temperature in a dark cabinet before photopolymerizing the filler. The photopolymerization is carried out for an hour by backing the cassette with a sheet of aluminum foil and stationing the assembly 8 cm in front of a desk-lamp equipped with two 15 W daylight-type fluorescent tubes. The cassette can then be stored tightly sealed in food-wrap or opened immediately for use.

Running. Samples are applied to the stacking-gel through an application foil (FMC #56803). Electrophoresis is carried out on a flatbed apparatus with 0.03 M neutral phosphate buffer in the electrode chambers. SDS is not added to the electrode buffer because it dissolves contaminants in the wicks; instead, SDS is supplied by interposing a thick ($\sim 1 \times 2$ cm^2) slab of contact-gel (1% agarose gel in 0.015 M buffer containing 0.1% SDS) between the cathodal wick and the edge of the specimen-gel.

Fixation. After electrophoresis, the composite gel is rinsed briefly and immersed and agitated (1 hour minimum) in a covered dish with 8-volumes of fixative solution containing 0.1 M $NaCNBH_3$ to promote covalent linkage of the protein to the agarose, 6% non-ionic detergent Brij® to quench SDS and also provide for dissolution of a free-radical scavenger butylated hydroxytoluene (BHT), and 0.15 M citrate at pH 6.4. A sheet of plastic-screening should be placed beneath the gel in the dish. The fixative solution is prepared by dissolving 0.22 g BHT in 40 ml of 30% Brij®, then diluting to 100 ml with water, and admixing 100 ml of 0.3 M citrate buffer and 10 ml of freshly boiled 2 M $NaCNBH_3$. The fixation is best started with a warm solution (40°), but need not be kept warm. If diaminobenzidine is to be used for immunostaining, the aldehyde groups in the gel should be reduced by adjusting to pH 4 for an hour with the $NaCNBH_3$ present. The gel should also be agitated overnight with several changes of PBS containing 0.05% Brij® to wash out the polyacrylamide filler.

Immunoprobing. The gel with electrophoretically arrayed protein fixed in-place is rinsed in PBS supplemented with 1% albumin and 0.05% Tween®-20, and then exposed overnight

(>4 h) to peroxidase-labeled antibody (10 μg/ml) in the supplemented PBS. The antibody is then washed out by actively suctioning 1.5 gel-volumes of PBS-Tween through the gel on a gel-dryer as described (14). To use the dryer for washing, a sheet of food-wrap is spread over the porous polyethylene mat and an overlaid sheet of glass-fiber paper, and is held in place by vacuum so that a windown can be razor-cut to size that would overlap the gel by a few mm on each side. The gel is then placed over the window made in the food-wrap. The wash fluid is applied in four aliquots over the surface of the gel, and is suctioned through the gel with a vacuum of 30 cm Hg. At this vacuum, the gel undergoes only slight compression during washing, such that its thickness is reduced from 1.5 to no less than 1 mm. The washing takes about 40 minutes. The gel is then rinsed briefly in buffer used for peroxidase-reaction, and then immersed in substrate for production of peroxidase reaction-product from retained antibody. Substrates used for peroxidase localization were either diaminobenzidine (DAB) or 4-chloronaphthol. The DAB is dissolved in PBS (25 mg/50 ml) and filtered, and then admixed with 8 μL of 30% H_2O_2, and reaction halted at about 8 min with 1% sodium azide. The 4-chloronaphthol is dissolved in ethanol (34 mg/5 ml), and then diluted with 45 ml of 0.05 M sodium acetate at pH 5, and admixed with 8 μL of 30% H_2O_2; and, reactions are terminated after 30 min by washing with water.

Specimens. Blood was taken into EDTA, Trasylol®, and PPACK for anticoagulation, and plasma was immediately diluted with 5 volumes of 4% SDS. Specimens of aortic intima were extracted and prepared for electrophoresis as previously described to yield protein fractions that were either 1) freely soluble in saline-inhibitor cocktail, 2) chaotrope-soluble in SDS/urea, or 3) soluble only after reduction and alkylation (18). All SDS-treated specimens were stored at ambient rather than low temperature. After the specimens were electrophoresed and fixed on glyoxyl agarose-gels, the electrophoregrams could be stored in saline at 4° for periods of several months before immunoprobing.

For reduction and alkylation of plasma proteins, 0.2 ml of SDS-treated plasma (plasma diluted 1:5 with 4% SDS) was admixed with 0.1 ml of 0.3 M Tris-HCl (0.3 M chloride) at pH 8.3 and 0.06 ml of 10% dithiothreitol, and then heated to 100°C for 10 min. After heating, the samples were chilled and admixed with 0.036 ml of iodoacetic acid (122 mg/ml) and 0.025 ml 1N NaOH, followed by 0.025 ml of the 0.3 M Tris-HCl after 5 min. The alkylation was allowed to proceed for 15 min, and then a second reduction of the protein was performed by adding 0.03 ml more of dithiothreitol and heating to 100° again.

RESULTS AND DISCUSSION

By eliminating the need for blot-transfer to immunoprobe SDS-PAGE electrophoregrams, glyoxyl agarose and composite electrophoregrams provide direct profiles of the molecular weight distribution of fibrinogen and its component polypeptide-chains. Several new findings have come from this method of analysis in applications directed to study the cross-linking of fibrinogen in blood, in tissue, and in course of clot stabilization-reactions; 1) fibrinogen dimers in blood frequently contain cross-linked Aα-chains more so than γ-chain dimers, 2) the intima of uncomplicated atherosclerotic aortas contain large quantities Aα-chain polymers, 3) the fibrinogen within those cadaveric tissues is largely undegraded, unlike the fibrinogen in cadaver blood, and 4) hybrid cross-linking between the α- and γ-chains proceeds much more rapidly in course of clot stabilization than anticipated from previous studies.

Plasma Fibrinogen-Dimers

Plasma from subjects with arterial peripheral vascular disease regularly contain elevated levels of dimeric derivatives of fibrinogen, at levels ranging between 1-6% of the total fibrinogen, as analyzed by SDS-glyoxyl agarose electrophoresis of non-reduced plasma protein. The illustrated electrophoregram (Fig. 1) is representative of over 40 specimens analyzed. Immunoprobing of reduced and alkylated plasma samples (Fig. 2) to profile the α- and γ-chains usually show higher levels of cross-linked α-chains than γ-chains. Also, immunoprobing with anti-fibrinopeptide-A monoclonal antibody (Fig. 3) indicates that fibrinopeptide A is essentially fully intact in the cross-linked chains, the relative intensity of staining of cross-linked/non-cross-linked α-chains being essentially the same whether stained with polyclonal rabbit anti-α-chain antibodies or monoclonal murine anti-fibrinopeptide-A antibodies. Two species of the cross-linked Aα-chains are usually present, and their mobilities are somewhat

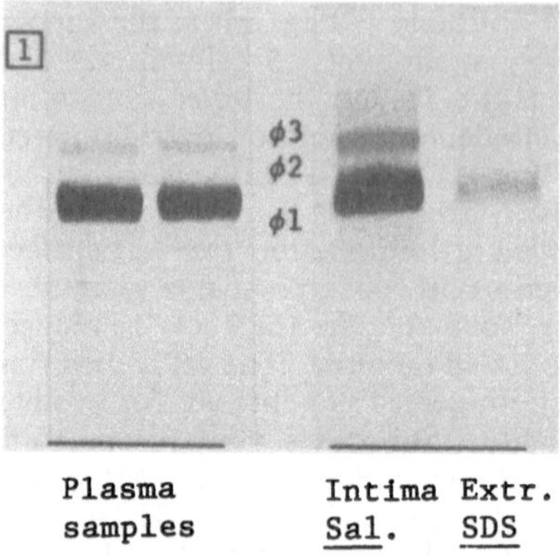

Fig. 1. Non-reduced plasma-specimens and intima-extracts on 3% glyoxyl agarose, stained with peroxidase-labeled rabbit anti-human fibrinogen antibodies. The two plasma-specimens portray the usual range of distributions of dimers and degradation products that we observe in subjects with arterial vascular disease. The intimal extracts were obtained at autopsy, and were made after washing and stripping away the aortic media; the "Sal" extract was made by grinding the intima in saline/PMSF, and the "SDS" extract was made by grinding in SDS-urea as described (18). Electrophoresis: 3% NuFix®, no filler.

faster than that of the principal form of α-α dimer that is produced in course of cross-linking of fibrin by factor XIIIa, mobilities that correspond to 112 and 102 *vs* 128 KDa of fibrin α-α dimers. We designate the cross-linked chains as Aα-X_1 (112 kDa) and Aα-X_2 (102 kDa), becuase their origins are uncertain. Chains with the same mobilities occur at elevated levels compared to levels of monomer in extracts of atherosclerotic aortas (18), as illustrated (Fig. 3). Similar components with mobilities in between γ-γ and α-α dimers are not produced in course of cross-linking of purified fibrin by factor XIIIa, but are evident in published electrophoregrams of cross-linking by tissue-transglutaminasė (19). The chain-composition of the products of cross-linking by tissue-transglutaminase is currently under investigation.

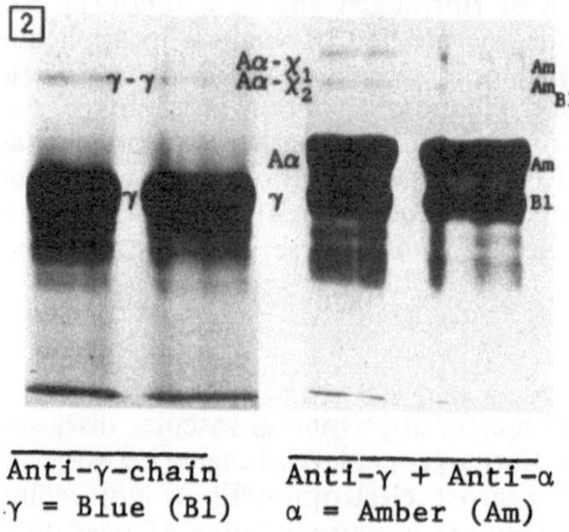

Fig. 2. Reduced and alkylated plasma on composite gels with 8% cross-linked filler. Specimens marked as anti-γ were probed with peroxidase-labeled rabbit anti-human-γ-chain antibody and stained blue with 4-chloronaphthol, and those marked as anti-α + anti-γ are the same specimens re-probed with anti-α-chain antibody which was stained amber with DAB. The lower density of the chloronaphthol-bands in the compound pattern is due to bleaching during secondary probing. Electrophoresis: 8% cross-linked filler.

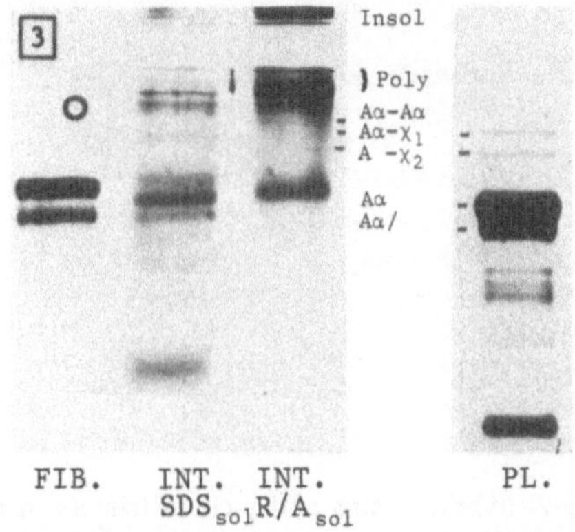

Fig. 3. Reduced and alkylated intimal extracts and a plasma-specimen stained with anti-fibrinopeptide-A antibody. The intimal extracts (INT) were made by grinding saline-insoluble protein in SDS/urea (labeled SDS_{sol}) and subsequently extracting again after reduction and alkylation (R/A_{sol}) as described (18). In addition to the high molecular weight Aα-bands, insolubles containing the peptide are evident in the R/A extract.

Intimal Fibrinogen

As characterized by Western blotting, freely extractable forms of fibrinogen in atherosclerotic aortic intimas obtained from cadavers is only 40-60% intact (20, 21). By contrast, we find by direct immunoprobing that soluble forms of fibrinogen consist almost entirely of molecules with mobilities essentially the same as plasma-fibrinogen monomers and oligomers; degraded forms detectable as fast migrating components comprise only a small percentage of the total (Fig. 1, and ref. 17). The presence of undegraded fibrinogen with fibrinopeptides intact contrasts with the virtual absence of undegraded forms of soluble fibrinogen in cadaveric clots and serum. The observation that cadaveric blood contains only degradation products and clots devoid of fibrinopeptide, while levels of soluble and undegraded fibrinogen remain high in the vessel wall may be taken as an indication that the intimal tissue is quite rich in enzyme inhibitors.

As known (22), atherosclerotic aortic intimas are very rich in insoluble forms of fibrinogen, and there have been indications (23, 24) from thrombin-digests of reduced and alkylated tissue that the fibrinopeptides are largely (~60%) intact, even in the insoluble forms. We find (Fig. 3) that most of the fibrinopeptide A is contained in high molecular Aα-chain polymers. As illustrated previously (18), levels of non-cross-linked α- and Aα-chains are lower than the levels of non-cross-linked γ-chains, a pattern contrary to that which would be expected if the cross-linking was due to factor XIIIa. The high degree of cross-linking of Aα-/α-chains raises a prospect that much of the cross-linking may be due to tissue-transglutaminase which, unlike factor XIIIa, cross-links α-chains faster than γ-chains (25). The prevalence of chains with mobilities equal to those of α-X_1 and α-X_2 (Fig. 3) raises a prospect that the corresponding chains occurring in plasma of subjects with vascular disease are leaching from atherosclerotic aortic intimas.

There are high concentrations of polymeric γ-chains in the intimal fibrinogen deposits (18). The polymers are probably copolymers of γ- and α-chains. We find (Fig. 4) that factor XIIIa produces hybrid cross-links between γ- and α-chains of fibrin at a rate second only to the rate of γ-chain dimerization (app. one-fourth the rate). However, we suspect that the Aα/γ-chain polymers in the intima are being formed by tissue-transglutaminase rather than factor XIIIa, because a preceding study indicated that only the γ-chains underwent rapid cross-linking in course of polymerization of fibrinogen by factor XIIIa (26).

Cross-linking of Fibrin Clots

Study of the course of cross-linking of fibrin by factor XIIIa using direct immunoprobing of electrophoregrams for positive identification of the chains shows that the

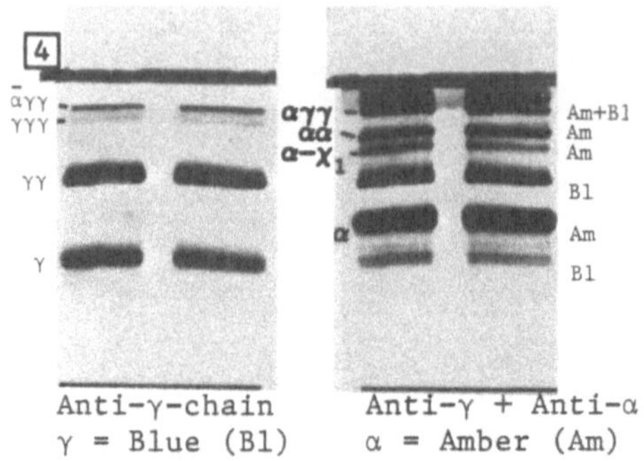

Fig. 4. Formation of α-γ-γ- hybrid chains and γ-chain trimers in a partially cross-linked fibrin clot. The electrophoregrams are of clots formed by fibrinogen fraction I-0 at 8 and 16 min after coagulation with thrombin in presence of 5 mM calcium. They were probed initially with peroxidase-labeled anti-γ-chain antibodies and stained to produce blue bands with 4-chloronaphthol, and after photographing were washed with SDS and probed with anti-α-antibodies which were stained amber with DAB to produce the compound staining pattern (labeled: anti-γ + anti-α).

products usually designated as α-polymers are mixtures of α- and γ-polymers, and products of co-cross-linking of α- and γ-chains. The formation of γ-γ-dimers proceeds faster than all other modes of cross-linking at the outset, as anticipated; (1) however, the second fastest mode is the formation of α-γ-γ-hybrids, and this is accompanied at somewhat slower rate by formation of α-α-dimers. As illustrated (Fig. 4) the principal component with KDa greater than γ-γ-dimers are α-γ-γ-hybrids. The fact that hybrid cross-linking of α- and γ-chains accompanies the formation of γ-γ-dimers had not been known before. Interestingly, the rapid formation of the hybrid α-γ-γ-trimers without accompanying formation of α-γ-dimers suggests that the hybrid cross-linking is promoted by the γ-chain dimerization. Anticipated formation of γ-γ-γ-trimers is also observed (27), but only as minor products. The hybrid cross-linking progresses further with the formation of large polymers of the hybrid chains on prolonged incubation. Such polymers are found in extracts of atherosclerotic aortas, but unlike fibrin clots those polymers contain fibrinopeptide A (18).

SUMMARY

Direct immunoprobing of electrophoregrams of plasma and intimal protein on glyoxyl agarose and composite-gels with polyacrylamide have uncovered novel modes of cross-linking of fibrinogen that differ from those previously characterized. These modes of cross-linking involve the Aα-chains of fibrinogen and hybrid cross-linking of α- and γ-chains.

REFERENCES

1. M. L. Schwartz S. V. Pizzo, R. L. Hill, and P. A. McKee, Human factor XIII from plasma and platelets. Molecular weights, subunit structures, proteolytic activation, and cross-linking of fibrinogen and fibrin, *J. Biol. Chem.*, **248**:1395 (1973).
2. V. J. Marder, R. N. Shulman, and W. R. Carroll, High molecular weight derivatives of human fibrinogen produced by plasmin, *J. Biol. Chem.*, **244**:2111 (1969).
3. H. Graeff and R. Hafter, Detection and relevance of cross-linked fibrin derivatives in blood, *Sem. in Thromb. & Hemost.*, **8**:57 (1982).
4. H. Towbin, T. Taehelin, and J. Gordon, Electrotransfer of proteins from polyacrylamide gels to nitrocellulose sheets, *Proc. Natl. Acad. Sci. USA*, **76**:4350 (1979).
5. W. Lin and H. Kasamatsu, On the electrotransfer of polypeptides from gels to nitrocellulose membranes, *Anal. Biochem.*, **128**:302 (1983).

6. J. R. Shainoff, Zonal immobilization of proteins, *Biochem. Biophys. Res. Commun.*, **95**:690 (1980).
7. J. R. Shainoff and B. N. Dardik, Cascade immunoelectrophoresis: combined electrophoretic and solid-phase processing of immunoreactive protein by zonal immobilization, *J. Immunol. Methods* **42**:229 (1981).
8. J. R. Shainoff, Zonal Immobilization of Proteins and Peptides, Guidelines for Electrophoretic, Chromatographic, and Dot-Blot Analyses on Glyoxyl Agarose and Composite Gels. (Manuscript submitted) (1989).
9. R. F. Borch, M. D. Berstein, and H. D. Durst, The cyanohydridoborate anion as a selective reducing agent, *J. Am. Chem. Soc.*, **93**:2897 (1971).
10. A. C. Peacock and C. W. Dingman, Molecular weight estimation and separation of ribonucleic acid by electrophoresis in agarose-acrylamide composite gels, Biochemistry, **7**:668 (1968).
11. P. B. H. O'Connell and C. J. Brady, Polyacrylamide gels with modified cross-linkages, *Anal. Biochem.*, **76**:63 (1976).
12. J. S. Finlayson and M. W. Mosesson, Heterogeneity of human fibrinogen, *Biochemistry,* **2**:42 (1963).
13. A. Henschen and P. Edman, Large scale preparation of S-carboxymethylated chains of human fibrin and fibrinogen and the occurrence of γ-chain variants, *Biochim. Biophys. Acta,* **263**:351 (1972).
14. J. R. Shainoff, R. Valenzuela, S. R. Gonda, and M. Caulfield, Blot-screening of culture fluid immunoreactivities on glyoxyl agarose films, *Biotechniques,* **4**:120 (1986).
15. D. M. Boorsma, Preparation of HRP-labelled antibodies, *In:* A. C. Cuello, ed. Immunohistochemistry, NY: John Wiley & Sons, 87, (1983).
16. J. M. Coll, Heme increases peroxidase-antibody activity in aged conjugates, *J. Immunol. Methods,* **104**:259 (1987).
17. B. D. Hames and D. Rickwood, Gel Electrophoresis of Proteins, Oxford: IRL Press (1981).
18. R. Valenzuela R, J. R. Shainoff, F. V. Lucas, B. J. Kudryk, and J. M. Anderson, Polypeptide-chain composition of fibrinogen oligomers in the human aortic intima and blood plasma. *In:* M. W. Mosesson et al. eds. Fibrinogen 3. Biochemistry, Biological Functions, Gene Regulation and Expression. Amsterdam: Excerpta Medica, 313 (1988).
19. P. A. McKee, M. L. Schwartz, S. V. Pizzo, and R. L. Hill, Cross-linking of fibrin by fibrin-stabilizing factor, *Ann. NY Acad. Sci.,* **202**:128 (1972).
20. E. B. Smith, G. A. Keen, and A. Grant, Origin of fibrin/fibrinogen degradation products in atherosclerotic lesions, *in:* M. W. Mosesson, D. L. Amrani, K. R. Siebenblist, J. P. Diorio, eds. Fibrinogen 3: Biochemistry, Biological Functions, Gene Regulation and Expression. Amsterdam: Excerpta Medica, 317 (1988).
21. E. B. Smith, Fibrinogen, fibrin and fibrin degradation products in relation to atherosclerosis, *Clinics in Haematology,* **15**:355 (1986).
22. V. Kao and R. W. Wissler, A study of the immunohistochemical localization of serum lipoproteins and other plasma proteins in human atherosclerotic lesions, *Exper. Molec. Path.,* **4**:465 (1965).
23. J. R. Shainoff and I. H. Page, Deposition of modified fibrinogen within the aortic intima, *Atherosclerosis,* **16**:287 (1972).
24. A. Bini, J. Jr. Fenoglio, J. Sobel, J. Owen, M. Fejgl, and K. L. Kaplan, Immunochemical characterization of fibrinogen, fibrin I, and fibrin II in human thrombi and atherosclerotic lesions, *Blood,* **69**:1038 (1987).
25. S. I. Chung, Comparative studies on tissue transglutaminase and factor XIII, *Ann. NY Acad. Sci.,* **202**:240 (1972).
26. H. Kanaide and J. R. Shainoff, Cross-linking of fibrinogen and fibrin by fibrin-stabilizing factor, *J. Lab. Clin. Med.,* **86**:574 (1975).
27. M. W. Mosesson, K. R. Siebenblist, D. L. Amrani, and J. P. Diorio, Identification of covalently linked trimeric and tetrameric D domains in crosslinked fibrin, *Proc. Natl. Acad. Sci. USA,* **86**:1113 (1989).

STUDIES ON THE LOCALIZATION AND ACCESSIBILITY OF SITES IN FIBRIN WHICH ARE INVOLVED IN THE ACCELERATION OF THE ACTIVATION OF PLASMINOGEN BY TISSUE-TYPE PLASMINOGEN ACTIVATOR

W. Nieuwenhuizen*, W.J.G. Schielen**, O. Yonekawa*, G. I. Tesser**, and M. Voskuilen*

*Gaubius Institute TNO, P.O. Box 612, 2300 AP Leiden
**The Laboratory of Organic Chemistry of the Catholic University Nijmegen, the Netherlands

INTRODUCTION

Under normal physiological conditions there is a dynamic equilibrium between the two opposing systems of coagulation and fibrinolysis. Activation of the coagulation system results in the eventual formation of fibrin. Fibrin plays an important role as the insoluble protein matrix of a blood clot, and in tissue repair. Activation of the fibrinolytic system results in the activation of plasminogen to plasmin which converts fibrin to fibrin degradation products. The activation of plasminogen is mediated by plasminogen activators such as tissue-type plasminogen activator (t-PA).

Fibrin, however, is not merely a passive substrate of the fibrinolytic system, but it also acts as a cofactor in that system, since it enhances the rate of activation of plasminogen by t-PA by at least two orders of magnitude. Fibrinogen, the soluble precursor of fibrin, does not exhibit this cofactor activity (1).

We have shown previously that the activation rate-enhancing properties of fibrin can be partly recovered in the plasmin-generated fragments of fibrin and fibrinogen, D-dimer and D_{EGTA}, respectively (2), and also in a fibrin (ogen) fragment obtained by CNBr cleavage i.e. FCB-2 (3) also known as Hol DSK (4). Both D_{EGTA} and FCB-2 are composed of remnants of the Aα-, Bß- and γ-chains of the parent fibrinogen molecule.

In later studies (5,6) we observed that the rate-enhancing capacity of both D_{EGTA} and FCB-2 resides in their Aα-chain remnants, i.e. in Aα-[111-197] and Aα-[148-207], respectively. We concluded that the overlapping sequence Aα-[148-197] harbours a rate-enhancing site. A more detailed study (6) showed that the stretch Aα-[148-160] contains essential structural elements, which are required for enhancement of the t-PA catalysed plasminogen activation.

Obviously, the sequence Aα-[148-160] is contained in fibrinogen, and yet fibrinogen, in contrast to fibrin, is not capable of enhancing the rate of plasminogen activation. We formulated the hypothesis, that Aα-[148-160] is buried in fibrinogen, and becomes accessible to plasminogen and/or t-PA upon conversion of fibrinogen to fibrin, or when fibrinogen is fragmented with plasmin or CNBr (5).

In a recent study we tested this hypothesis. The approach was to raise monoclonal antibodies (MoAb) against Aα-[148-160]. If our hypothesis on the accessibility is correct, the anti-Aα-[148-160] MoAb's will react with fibrin and with the rate-enhancing Aα-[148-160] containing fibrinogen fragments, but not with fibrinogen. Furthermore, it would be possible with such fibrin-specific MoAb's to develop an enzyme-immunoassay (EIA) for the

Fibrinogen, Thrombosis, Coagulation, and Fibrinolysis, Edited by C. Y. Liu and S. Chien, Plenum Press, New York measurement of plasma levels of soluble fibrin. Such an EIA is clinically relevant, since elevated levels of soluble fibrin in the circulation are considered as a molecular marker of a prethrombotic state.

Several observations indicate that more sites in fibrin that Aα-[148-160] are involved in the fibrin-induced enhancement of the t-PA catalysed plasminogen activation. The most important indication is that fibrin can enhance the activation of both (native) glu-plasminogen and mini-plasminogen (Val 442-plasminogen), whereas FCB-2 enhances only glu-plasminogen activation (7).

In another recent study we investigated the possible presence of additional rate-enhancing fibrinogen fragments in a CNBr digest of fibrinogen.

In this paper we describe the results obtained in both these recent investigations.

MATERIALS AND METHODS

Proteins

Fibrinogen (human, grade L, 90% clottable) was obtained from Kabi (Stockholm, Sweden) or prepared from fresh plasma as described before (8). Fibrinogen Aα-, Bß- and γ-chains were separated and purified as described by Doolittle et al. (9).

Fibrin monomers were prepared as described by Haverkate and Timan (10) and stored as a 13.4 mg/ml solution in 20 mM acetic acid at −20°C. Prior to use they were diluted with 20 mM acetic acid to 0.45 mg/ml. Plasmin generated fibrin(ogen) fragments D-dimer, D_{EGTA}, D_{cate}, E, X and Y were prepared and purified as in (11).

A CNBr digest of fibrinogen was prepared according to Blombäck et al. (12). CNBr fragment FCB-2 was isolated from the digest as previously described by Nieuwenhuizen et al. (3) using a method adapted partly from the procedure of Olexa et al. (13).

Two-chain t-PA was purified from melanoma cell culture medium according to Rijken et al. (14) as modified by Kluft et al. (15). Glu-plasminogen was prepared from fresh human plasma according to Deutsch and Mertz (16) by affinity chromatography on lysine-Sepharose (Pharmacia, Uppsala, Sweden).

Mini-plasminogen (Val 442-plasminogen) was prepared by limited elastase digestion of glu-plasminogen, and purified as described by Sottrup-Jensen et al. (17).

Methods

Synthesis of the immunogen used to elicit MoAb's to fibrinogen Aα-[148-160]. Fibrinogen Aα-[148-160]-tridecapeptide (Lys-Arg-Leu-Glu-Val-Asp-Ile-Asp-Ile-Lys-Ile-Arg-Ser) was prepared according to methods published in detail elsewhere (18). In brief, the synthesis of fibrinogen Aα-[148-160] was performed on a p-alkoxy-benzyl alcohol resin (19), using the Nα-(9-fluorenyl) methyloxycarbonyl (Fmoc) chemistry (20) and *in situ* carboxyl activation with dicyclohexylcarbodiimide and 1-hydroxybenzotriazole (21).

Fibrinogen Aα-[148-160]-tridecapeptide was extended at its aminoterminus with acetylthioacetyl-D, L-norleucine (Ata-Nle). Thus, methionine present in fibrinogen at Aα 147 is replaced by D, L-Nle. This procedure allows both the coupling to a carrier protein and the subsequent determination of the number of peptides coupled per carrier protein molecule.

Bovine serum albumin (BSA) (Roth, Karlsruhe, FRG, 99% pure) was treated with 6-(1-maleimido) hexanoic acid 1-succinimidyl ester (22) to introduce maleimido functions. The acetyl group was removed from the Ata-Nle peptide derivative by dissolution in a mixture of 4N NaOH/methanol/dioxane (1/5/14) for exactly 15 seconds, and subsequent neutralisation with acetic acid. The peptide, with now a free sulphydryl group at its aminoterminus, was added to the maleimidated BSA, and allowed to react (23). The mixture was dialysed against distilled water and freeze-dried. Amino acid analysis showed that between 4 and 5 peptides had been coupled per BSA molecule.

Production of MoAb's. Immunisation with the immunogen, cell fusions (24) selection and limiting dilutions (25) were done according to standard procedures and will be published in detail elsewhere (18). Screening was done by ELISA, using purified Aα-chain and FCB-2 as antigens.

Assessment of rate-enhancing properties of fibrin(ogen) fragments. Rate-enhancement of the t-PA catalysed plasminogen activation was assessed as described by Verheijen et al. (26).

Sandwich enzyme immuno assay (EIA) to assess the reactivity of the MoAb's with different fibrin(ogen) derivatives. MoAb was absorbed to polystyrene microtitre plates by overnight incubation at 4°C of a 10 μg/ml solution of the MoAb in 0.05 M phosphate buffer pH 6, containing 1.6 M NaCl. The plates were then washed with PBS containing 0.05% Tween 20 (PBST) and incubated with solutions of the different fibrin(ogen) fragments in PBST for 2 hours at ambient temperatures. After washing with PBST, the wells were incubated with a horse radish peroxidase conjugate of a rabbit polyclonal antibody against all fibrin(ogen) derived materials. Reactivity was visualized using 3,3', 5,5'-tetramethyl benzidine (Aldrich Chemical Co, Milwaukee, USA) and H_2O_2 as the substrate mixture. The reaction was arrested by addition of 1 N H_2SO_4.

Isolation of a rate-enhancing fibrinogen fragment, other than FCB-2 from a fibrinogen CNBr digest. A CNBr digest of fibrinogen was dialysed against distilled water and lyophilized. The dry protein was suspended in a minimum amount of distilled water, and acidified with formic acid until a clear solution was obtained. This was applied to a 2.5 × 80 cm Biogel P-2 (Biorad, Richmond, USA) column, run in water acidified to pH = 4 with acetic acid.

Miscellaneous methods. SDS-PAGE was carried out according to Schägger and von Jagow (27); amino acid sequences were determined with a gas-phase sequenator (Applied Biosystems, model 490A protein sequencer, on-line equipped with a model 120A PTH (phenyl-thiohydantoin) analyser by Dr. Amons, State University, Leiden.

RESULTS

Preparation and Specificity of MoAb's Against Fibrinogen Aα-[148-160]

As will be described elsewhere in more detail (18), four clones were obtained, all producing MoAb's of the IgM subclass with kappa light chains. One was studied in detail, and will be referred to as anti-Fb-1/2. Anti-Fb-1/2 can obviously not be purified by affinity chromatography on protein A-Sepharose. It precipitates, however, during dialysis against a low ionic strength buffer (0.005 M sodium phosphate pH 6), and can easily be redissolved in a high ionic strength buffer (1.6 M NaCl, 0.05 M sodium phosphate, pH 6) and remains in solution upon dilution to physiological ionic strengths.

Specificity of Anti-Fb-1/2

Table 1 shows the reactivity of anti-Fb-1/2 with different fibrin(ogen) derivatives in the sandwich EIA. For comparison the rate-enhancing properties of the derivatives in the t-PA catalysed plasminogen activation are also given. The correlation between rate-enhancement and immunoreactivity is perfect. Fig. 1A and 1B illustrate, that anti-Fb-1/2 is not reactive with fibrinogen i.e. normal pooled plasma (2.4 mg fibrinogen/ml) virtually does not react in the EIA, whereas the same plasma spiked with 15 μg fibrin monomers/ml shows a strong reactivity (Fig. 1A). Virtually identical dose-response curves were obtained when the fibrin monomer concentration was varied and the plasma dilution kept constant. Plasma treated with thrombin shows time-dependent progressive increase of response (Fig. 1B). Threatment with Arvin (which releases only fibrinopeptide A) instead of thrombin yields comparable results. This and the above strongly supports our theory that Aα-[148-160] which contains the epitope for anti-Fb-1/2 is accessible in fibrin and in the rate-enhancing fibrin(ogen) derivatives, but not in fibrinogen itself.

Identification of a New Site in Fibrin(ogen) which Enhances the t-PA Catalysed Plasminogen Activation

Fig. 2 shows the elution pattern of a CNBr digest of fibrinogen on a Biogel P-2 column (see Materials and methods section). Two peaks with rate-enhancing capacity were observed. One eluted at the void volume of the column, and had all the properties (results not shown) of the previously described (3) FCB-2 fragment; the other peak eluted much later, which suggests that the material in this peak has a rather low molecular weight. The material in this peak was pure upon precipitation at pH 8.3 and rechromatography on Biogel P-2. Upon SDS-PAGE without reduction according to Schägger and von Jagow (27), it showed only one band with an apparent molecular weight of 6500; after reduction,

Table 1. Immunoreactivity in a Sandwich-type EIA (see Materials and Methods), and Rate-Enhancing Properties in the Activation of Plasminogen by t-PA of Different Fibrin(ogen) Derivatives

Fibrin(ogen) Derivative	Immunoreactivity	Rate-enhancement (2)
Fibrin Monomers	+ + +	+ + +
D_{cate}	−	−
D_{EGTA}	+ +	+ +
D-dimer	+ +	+ +
E	−	−
X	−	−
Y	−	±

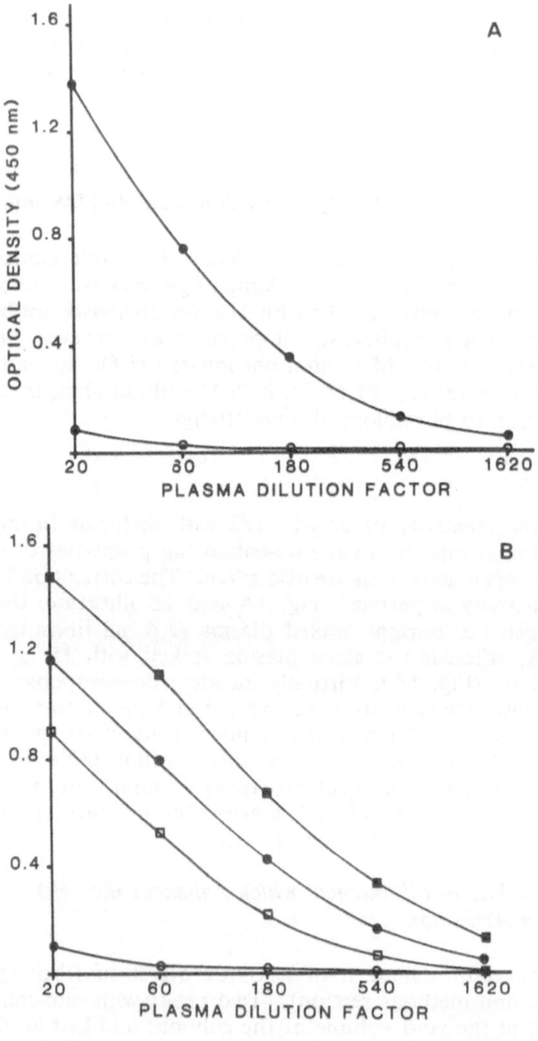

Fig. 1. Dose-response curves of normal plasma with a fibrinogen concentration of 2.4 mg/ml (○), and of the same plasma spiked with fibrin monomers to a concentration of 15 μg/ml (Fig. 1A, ●) or treated with a low concentration of thrombin for different periods of time (Fig. 1B; □ 25 seconds, ● 50 seconds and ■ 75 seconds).

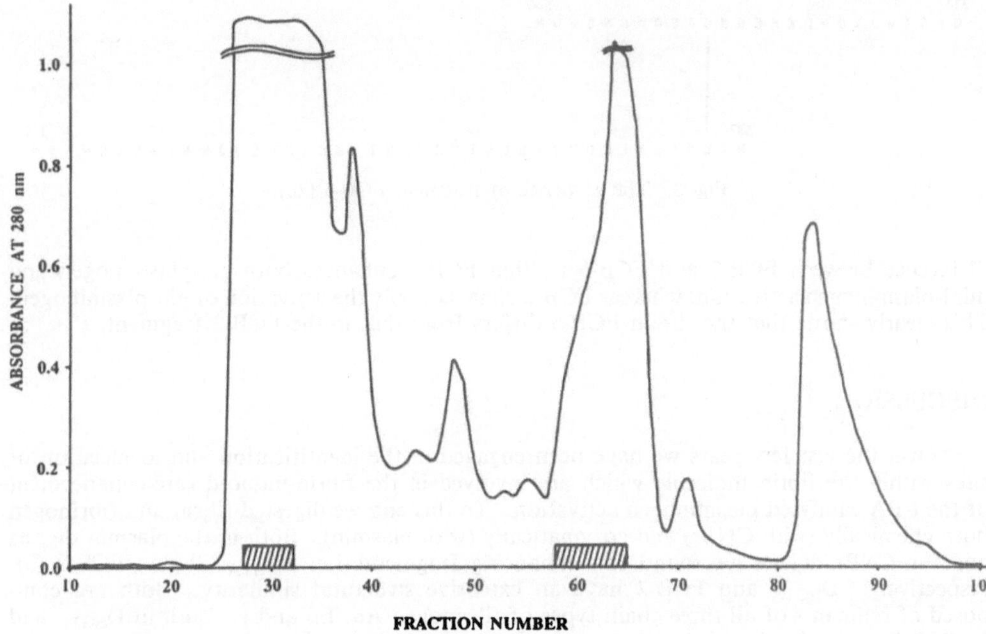

Fig. 2. Elution pattern of a CNBr digest of fibrinogen on a Biogel P-2 column (2.5 × 80cm) run in distilled water, pH 4. Hatched bars indicate the positions of fractions with rate-enhancing capacity in the t-PA catalysed plasminogen activation.

one strongly-stained band was observed with an apparent molecular weight of 2700, plus a weakly-stained band with a molecular weight of 4000 (gel not shown). This indicates, that the rate-enhancing material with molecular weight 6500 consists of two polypeptide chains. This was supported by the results of the amino acid sequence determinations which are shown in Table 2.

These properties unequivocally identify the 6.5 kD fragment as a fragment of the γ-chain which is known as FCB-5 (28), which has the sequence shown in Fig. 3.

Fig. 4 shows the effects of FCB-2 and the newly identified fragment FCB-5 on the activation rate of glu-plasminogen and mini-plasminogen by t-PA. The most remarkable

Table 2. Amino Acid Sequence of the 6.5 kD Rate-Enhancing Fragment

Cycle	Amino Acid	Picomoles
1	gln	19
	asn	37
2	phe	22
	lys	84
3	ser	22
	–	–
4	thr	10
	his	15
5	trp	11
	ala	41
6	asp	22
	gly	32
7	asn	7
	his	7

Fig. 3. The sequence of fragment FCB-5 (28).

difference between FCB-2 and FCB-5 is, that FCB-5 enhances both glu-plasminogen and mini-plasminogen activation, whereas FCB-2 enhances only the activation of glu-plasminogen. This clearly shows that the site in FCB-5 differs from that in the FCB-2 fragment.

DISCUSSION

Over the last few years we have been engaged in the identification and localisation of sites within the fibrin molecule which are involved in the fibrin-induced rate-enhancement of the t-PA catalysed plasminogen activation. To this end we digested fibrin and fibrinogen both chemically (with CNBr) and enzymatically (with plasmin). Both in the plasmin digests and the CNBr digests we found rate-enhancing fragments i.e. D_{EGTA} (2) and FCB-2 (3), respectively. D_{EGTA} and FCB-2 have an extensive structural similarity. Both are composed of remnants of all three chain types of fibrinogen Aα, Bß and γ. Both in D_{EGTA} and in FCB-2, only the Aα chain remnants are rate-enhancing. Since the Aα-chain remnants in D_{EGTA} and FCB-2 are Aα-[111-197] and Aα-[148-207], respectively, we concluded that a rate-enhancing site must be localised in the overlapping Aα-chain stretch i.e. in Aα-[148-197] (3). In a more detailed study (6) we could show that Aα-[148-160] contains the structural elements required for rate-enhancement. Obviously, this stretch Aα-[148-160] is present in

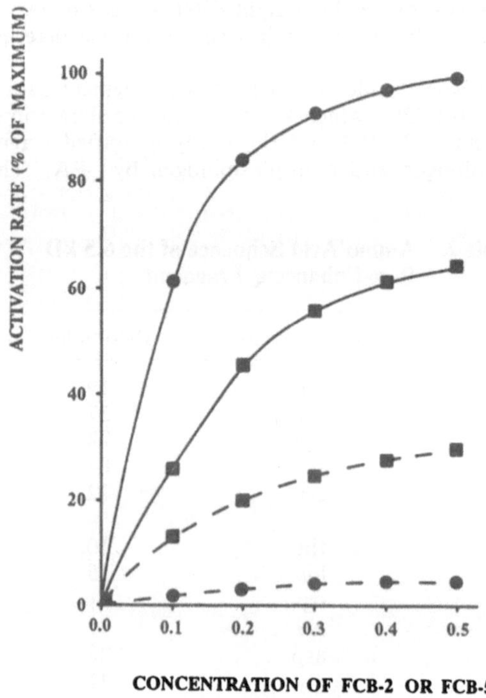

Fig. 4. Rate-enhancement of the t-PA catalysed activation of glu-plasminogen (———) and mini-plasminogen (---) by FCB-2 (●) and FCB-5 (■).

fibrinogen and yet fibrinogen (in contrast to fibrin) does not enhance the rate of plasminogen activation by t-PA. We forwarded the working hypothesis that Aα-[148-160] in fibrin, but not in fibrinogen, is accessible to proteins such as t-PA and plasminogen. To find support for this hypothesis we raised MoAb's against synthetic Aα-[148-160] coupled exclusively via its amino-terminal end to BSA. Several clones were obtained. One of these (anti-Fb-1/2) was characterized and was found to react with fibrin and the rate-enhancing fibrin(ogen) fragments FCB-2, D_{EGTA} and D-dimer, which all comprise Aα-[148-160]. D_{cate}, which also comprises Aα-[148-160], does not react with anti-Fb-1/2. This is in agreement with the fact that D_{cate} is not rate-enhancing and does apparently not expose Aα-[148-160].

Anti-Fb-1/2 reacts very weakly with normal pooled plasma with a fibrinogen concentration of 2.4 mg/ml. This low reactivity is probably due to low concentrations of soluble fibrin in the plasma, since it has been reported (29) that the fibrin concentration in plasma of normals in between 0.5 and 13.5 μg/ml.

Fibrin monomers added to plasma in the μg/ml range, were detected in a sandwich EIA based on anti-Fb-1/2. This indicates that fibrinogen when present in a more than thousand-fold excess does not interfere with the fibrin-anti-Fb-1/2 interaction.

In conclusion, the data strongly support our hypothesis that Aα-[148-160] in fibrin, but not in fibrinogen, is accessible to proteins. Anti-Fb-1/2 appears to be a valuable tool to assess levels of soluble fibrin in plasma, which is considered as a molecular marker of a prethrombotic state.

Several observations indicate that more rate-enhancing sites that Aα-[148-160] exist in the fibrin molecule. The strongest indication is that fibrin enhances the activation rate of glu-plasminogen and mini-plasminogen. The site Aα-[148-160], however, accelerates only the glu-plasminogen activation, and has no effect on the mini-plasminogen activation. When we fractionated a whole CNBr digest of fibrinogen on Biogel P-2 run in water, we observed two peaks with rate-enhancing capacity. One was identical to the previously described FCB-2. The other was found to be identical to FCB-5. This fragment is derived from the γ-chain i.e. γ-[311-336] and γ-[337-379] connected by a single disulphide bond between cysteines γ 326 and 339.

In native fibrinogen, the residues γ-[326-339] constitute a loop. Apparently, this loop is not essential for the rate-enhancing capacity, since the loop has been broken in FCB-5 as a result of the CNBr cleavage. Furthermore, preliminary data indicate that only one of the two FCB-5 polypeptide chains, i.e. γ-[311-336] is required for the rate-enhancing properties. The most remarkable property of FCB-5 is that it enhances the rate of mini-plasminogen activation (and also that of glu-plasminogen).

Since fibrinogen has no rate-enhancing capacity, the site within FCB-5 is probably not accessible in fibrinogen, but is exposed upon the conversion of fibrinogen to fibrin. This situation, if proved, would be similar to that found for the Aα-[148-160] site. At present we are engaged in the production of MoAb's against FCB-5 to test this assumption.

REFERENCES

1. P. Wallén, Activation of plasminogen with urokinase and tissue activator, *in:* Thrombosis and Urokinase, R. Paoletti and S. Sherry, eds., Academic Press, New York, Vol. 9, pp 91-102 (1977).
2. J. H. Verheijen, W. Nieuwenhuizen, and G. Wijngaards, Activation of plasminogen by tissue activator is increased specifically in the presence of certain soluble fibrin(ogen) fragments, *Throm. Res.,* **27**:337-385 (1982).
3. W. Nieuwenhuizen, J. H. Verheijen, A. Vermond A, and G. T. G. Chang, Plasminogen activator by tissue activator is accelerated in the presence of fibrin(ogen) cyanogen bromide fragment FCB-2, *Biochim. Biophys. Acta,* **755**:531-533 (1983).
4. B. Gårdlund, B. Hessel, G. Marguerie, G. Murano, and B. Blombäck, Primary structure of human fibrinogen. Characterization of disulphide containing cyanogen-bromide fragments, *Eur. J. Biochem.,* **77**:595-610 (1977).
5. W. Nieuwenhuizen, A. Vermond, M. Voskuilen, D. W. Traas, and J. H. Verheijen, Identification of a site in fibrin(ogen) which is involved in the acceleration of plasminogen activation by tissue-type plasminogen activator, *Biochim. Biophys. Acta,* **748**:86-92 (1983).
6. M. Voskuilen, A. Vermond, G. H. Veeneman, J. H. Van Boom, E. A. Klasen, N. D. Zegers, and W. Nieuwenhuizen, Fibrinogen lysine residue Aα 157 plays a crucial role

in the fibrin-induced acceleration of plasminogen activation, catalyzed by tissue-type plasminogen activator, *J. Biol. Chem.,* **262:**5944-5946 (1987).
7. W. Nieuwenhuizen, M. Voskuilen, A. Vermond, B. Hoegee-De Nobel, and D. W. Traas, The influence of fibrin(ogen) fragments on the kinetic parameters of the tissue-type plasminogen-activator-mediated activation of different forms of plasminogen, *Eur. J. Biochem.,* **174:**163-169 (1988).
8. I. A. M. Van Ruijven-Vermeer, and W. Nieuwenhuizen, Purification of rat fibrinogen and its consitituent chains, *Biochem. J.,* **169:**653-658 (1978).
9. R. F. Doolittle, K. G. Cassman, B. A. Cottrell, S. J. Friezner, J. T. Hucko, and T. Takagi, Amino acid sequence studies on the α chain of human fibrinogen. Characterization of 11 cyanogen bromide fragments, *Biochemistry* **16:**1703-1709 (1977).
10. F. Haverkate, and G. Timan, Preparation of plasminogen-free fibrinogen for fibrin plates, *in:* "Progress in Chemical Fibrinolysis and Thrombolysis," J. F. Davidson, M. M. Samama, and P. C. Desnoyers, eds., Raven Press, New York, Vol. 2, pp 67-71 (1976).
11. I. A. M. Van Ruijven-Vermeer, W. Nieuwenhuizen, F. Haverkate, and T. Timan, A novel method for the rapid purification of human and rat fibrin(ogen) degradation products in high yields, *Hoppe Seyler's Z. Physiol. Chem.,* **360:**633-637 (1979).
12. B. Blombäck, M. Blombäck, A. Henschen, B. Hessel, S. Iwanaga, and K. R. Woods, N-terminal disulphide knot of human fibrinogen, *Nature,* **218:**130-134 (1968).
13. S. A. Olexa, and A. Z. Budzynski, Binding phenomena of isolated unique plasmic degradation products of human cross-linked fibrin, *J. Biol. Chem.,* **254:**4925-4932 (1979).
14. D. C. Rijken, G. Wijngaards, M. Zaal-De Jong, and J. Welbergen, Purification and partial characterization of plasminogen activator from human uterine tissue, *Biochim. Biophys. Acta,* **580:**140-153 (1979).
15. C. Kluft, A. L. Van Wezel, G. A. M. Van der Velden, J. J. Emeis, J. H. Verheijen, and G. Wijngaards, Large-scale production of extrinsic (tissue-type) plasminogen activator from human melanoma cells, *in:* "Advances in Biotechnological Processes," A. Mizrahi, and A. L. Van Wezel, eds., Alan R. Liss, New York, Vol. 2, pp 97-110 (1983).
16. D. G. Deutsch, and E. T. Mertz, Plasminogen: purification from human plasma by affinity chromatography, *Science,* **170:**1095-1096 (1970).
17. L. Sottrup-Jensen, H. Claeys, M. Zajdel, T. E. Petersen, and S. Magnusson, The primary structure of human plasminogen: isolation of two lysine-binding fragments and one "mini-" plasminogen (MW 38.000) by elastase-catalyzed-specific limited proteolysis, *in:* "Progress in Chemical Fibrinolysis and Thrombolysis," J. F. Davidson, M. M. Samama, and P. C. Desnoyers, eds., Raven Press, New York, Vol. 3, pp. 191-209 (1978).
18. W. J. G. Schielen, M. Voskuilen, G. I. Tesser, and W. Nieuwenhuizen, The sequence Aα-(148-160) in fibrin, but not in fibrinogen, is accessible to monoclonal antibodies, *Proc. Natl. Acad. Sci.,* **86:**8951-8954 (1989).
19. S. S. Wang, *p*-alkoxybenzyl alcohol resin and *p*-alkoxybenzyloxycarbonyl-hydrazide resin for solid phase synthesis of protected peptide fragments, *J. Am. Chem. Soc.,* **95:**1328-1333 (1973).
20. P. B. W. Ten Kortenaar, B. G. Van Dijk, J. M. Peeters, B. J. Raaben, P. J. H. M. Adams, and G. I. Tesser, Rapid and efficient method for the preparation of Fmoc-amino acids starting from 9-fluorenylmethanol, *Int. J. Pept. Protein Res.,* **27:**398-400 (1986).
21. J. W. Van Nispen, J. P. Polderdijk, and H. M. Greven, Suppression of side-reactions during the attachment of Fmoc-amino acids to hydroxymethyl polymers, *Recl. Trav. Chim. Pays-Bas,* **104:**99-100 (1985).
22. J. M. Peeters, T. Hazendonk, E. C. Beuvery, and G. I. Tesser, Comparison of four bifunctional reagents for coupling peptides to proteins and the effect of the three moieties on the immunogenicity of the conjugates, *J. Immunol. Methods,* **120:**133-143 (1989).
23. O. Keller, and J. Rudinger, Preparation and some properties of maleimido acids and maleoyl derivatives of peptides, *Helv. Chim. Acta,* **58:**531-547 (1975).
24. J. Köhler, and C. Milstein, Continuous cultures of fused cells secreting antibody of predefined specificity, *Nature,* **256:**495-497 (1975).

25. V. T. Oi, and L. A. Herzenberg, Immunoglobulin-producing hybrid cell lines, *in:* "Selected Methods in Cellular Immunology," B. B. Mishell, and S. M. Shiigi, eds., Freeman and Company, New York, pp 351-370 (1980).
26. J. H. Verheijen, E. Mullaart, G. T. G. Chang, C. Kluft, and G. Wijngaards, A simple, sensitive spectrophotometric assay for extrinsic (tissue-type) plasminogen activator applicable to measurements in plasma, *Thromb. Haemostas.,* **48**:266-269 (1982).
27. H. Schägger, and G. Von Jagow, Tricine-sodium dodecyl sulfate-poly-acrylamide gel electrophoresis for the separation of proteins in the range from 1 to 100 kDa, *Anal. Biochem.,* **166**:368-379 (1987).
28. A. Henschen, Fibrinogen — Blutgerinnungsfaktor I. Biochemische Aspekte, *Haemostaseologie,* **1**:49-61 (1981).
29. W. Nieuwenhuizen, L. C. Creighton, P. J. Gaffney, H. Graeff, R. Hafter, B. Hoegee-De Nobel, G. Müller-Berghaus, U. Scheefers-Borchel, R. Thurmayer, J. A. Davies, F. Duckert, J. Jespersen, F. R. Matthias, C. R. M. Prentice, F. E. Preston, and M. M. Samama, A double blind comparative study of six monoclonal antibody-based plasma assays for fibrinogen derivatives, *in:* Fibrinogen: Biochemistry, Physiology and Clinical Relevance, G. D. O. Lowe, J. T. Douglas, C. D. Forbes, and A. Henschen, eds., Excerpta Medica, Amsterdam, Vol. 2, pp 181-186 (1987).

PROTHROMBINASE: RECOGNITION AND DEVELOPMENTS

Walter H. Seegers

Department of Physiology
Wayne State University, School of Medicine
Detroit, MI, USA

PERSPECTIVE

At the beginning of this century the crux of the blood coagulation problem centered on the question: what activates prothrombin? Answers commonly turned out to be found inside the diciplines of basic sciences. As these and clinical sciences developed, their language accomodated the phenomenon designated prothrombin activation.

Foundations for current concepts were primarily laid by P. Nolf of Belgium and K. Lenggenhager in Switzerland. The latter advanced the idea that a *thrombokinin* is an activator of prothrombin. Similarily Nolf used the word *thrombozym* with the intended meaning that there is an activator of prothrombin. Both Nolf and Lengenhagger were rediscovered when it was stated that Factor Xa is the activator of prothrombin.

In 1952 H. Milstone published a review entitled "On the evolution of blood clotting theory" and wrote: "Therefore, the capacity to activate prothrombin is taken as the defining property of thrombokinase. Accumulating evidence favors the view that thrombokinase is an enzyme".

In 1962, in several papers from my laboratory, we outlined the nature of a 5 component system in which prothrombin produces alpha-thrombin. Evidence for the system was based on experiments with purified components, and projected as follows:

$$\text{Prothrombin} \xrightarrow[\text{Autoprothrombin C}]{\text{Ca}^{2+}\text{, Ac-globulin, Platelet factor 3}} \text{Thrombin}$$

Converting this to use other names for the same factor one states as follows:

$$\text{Prothrombin} \xrightarrow[\text{Factor Xa}]{\text{Ca}^{2+}\text{, Factor V/Va, Phospholipid}} \text{Thrombin}$$

The discovery of this complex, called prothrombinase, needed elaboration to take into account experimental facts. The foremost one was that prothrombin and thrombin were purified and the molecular weight of thrombin was found to be approximately half of that for prothrombin. What could be the role of the nonthrombin "half" of prothrombin? The question first arose in my laboratory in 1950. The answer was soon found and offered many ramifications. The nonthrombin portion of the molecule has an essential role in the prothrombinase complex for generating thrombin. Modified Xa is the activator of prothrombin.

Factor V was recognized during the midcentury period (circa 1945-1947) and recognized as an accelerator of thrombin formation. Its designation as a factor in the five component system ended about 15 years of dormancy with regard to its role in answering the question: what activates prothrombin? It becomes more active with small amounts of thrombin.

PROTHROMBIN STRUCTURE AND MODIFICATION

Bovine prothrombin is a single chain glycoprotein with 582 amino acids. Only 53% of these are needed for the thrombin structure. In the activation of purified bovine prothrombin clevage of the chain occurs first between A and B chain segments. Then fragments 1 and 2 are separated from the A chain segment. Activation occurs in the given order. Disulfide bridges are involved (Fig. 1). On a w/w basis Factor Xa is approximately as effective as trypsin in certain thrombin generating experiments. In the complex it is thousands of times more active than the free standing enzyme. Except for the gamma-carboxy-glutamic acid and kringle structures the non-thrombin proteins are simply orderly stretches of polypeptide chains (Fig. 2).

To demonstrate the essential role of prothrombin fragment 1 and fragment 2 it was first necessary to discover them and have them in purified form. The same is true for Factors X and Xa, Factors V and Va, prethrombin 1, peptide 1-44 from prothrombin, meizothrombin 1, prethrombin 2 and other proteins (Fig. 1 and 2).

PRETHROMBIN 2

One can say that prethrombin 2 is the real prothrombin. It is a strategic location for beginning an explanation of the molecular nature of prothrombinase. We obtained it as a purified protein. The method developed consists of removing prothrombin fragment 2 from purified prethrombin 1. This was done with use of purified acutin. This thrombin-like enzyme was isolated from the venom of *Agkistrodon acutus* snakes.

Prethrombin 2 has the same structure as thrombin except for one difference; namely, in thrombin the bond between the A and B chains is broken. This bond is specifically cleaved by ecarin, an enzyme isolated from *Echis carinatus* snake venom. The enzyme hardly recognizes this bond, but functions effectively when supported by prothrombin fragment 2. The latter is a component of the prothrombinase complex. The molecular basis for modifying the specificity of ecarin is not known. Perhaps other enzymes will be found that can have their function modified by prothrombin fragment 2 (Fig. 3).

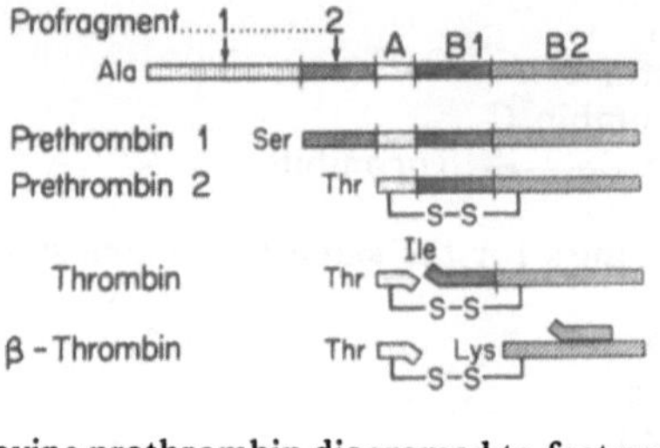

Fig. 1. Prothrombin. Bovine prothrombin diagramed to feature and illustrate the function of the prothrombin molecule in the formation of thrombin and beta-thrombin. In the prothrombin activating system called prothrombinase complex prothrombin gives the maximum yield of thrombin. Less thrombin forms if prethrombin 1 is the substrate zymogen, but by also adding purified profragment 1 results in making the thrombin yield equal to that obtained from prothrombin. If purified prethrombin 2 is the zymogen the thrombin yield is very low, but can be made equal to prothrombin if both purified profragment one and profragment 2 are also added. Alpha-thrombin slowly degrades by autolysis to produce a thrombin to which fibrinogen is refractory, but hydrolysis of synthetic substrates is not impared. The autolysis fragment remains bound to the thrombin by binding forces that are not covalent.

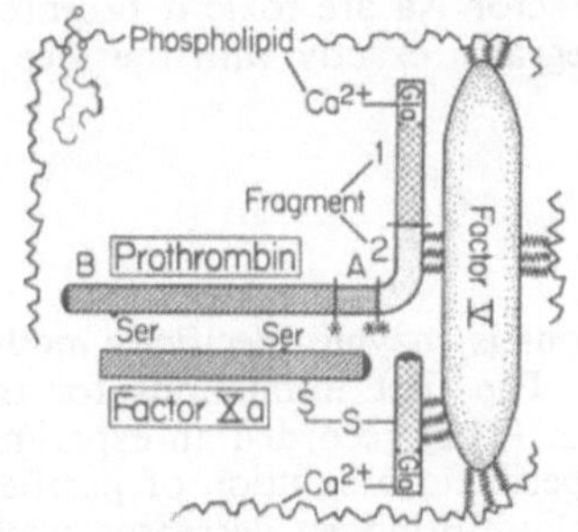

Fig. 2. Imagined nature to the prothrombinase complex as a visual perception. The 5 component complex consists of prothrombin substrate, Factor Xa as enzyme that becomes modified in specificity, and Factor V/Va as cofactor. Fragments 1 and 2 are essential, but need not be covalently restricted. A and B chains represent the thrombin portion of prothrombin. Physiologically, platelets are the source of lipoproteins. Lipids from thromboplastin function in case of injury. The gamma-carboxy-glutamic acids of prothrombin and Factor Xa are well suited for binding with calcium ions. Profragment 2 binds Factor V/Va. In prothrombinase Factor Xa functions as modified Factor Xa. Ser = serine protease, Gla = gamma-carboxy-glutamic acid, S-S = arbitrary positioning of disulfide bonds.

PROTHROMBINASE ASSAY METHODS

In principle thrombin yield in prothrombinase is an expression of the activity of ever single component. In practice all components are provided in optimum concentration. If any one of the components is reduced in concentration the thrombin yield is reduced. Several variables need to be controlled. At a concentration below optimum for a specific component, thrombin production is limited. It is not a case where thrombin continues to be produced if some substrate is left in the reaction mixture.

Applying the assay procedure to phospholipids yields data from which it can be concluded that platelets are the source of phospholipids for the generation of thrombin under physiological conditions. Platelets contain the necessary phospholipids while lipid low plasma does not. Red cells can contribute if disrupted, and if their membranes are tested as inside-out vesicles.

Another fact that favors viewing platelets as the main source of active lipids is the conditions needed for the intravenous toxicity of purified Factor Xa. If infused in "high" concentrations disseminated intravascular coagulation is produced. In lower concentrations the enzyme is tolerated and inhibitors neutralize the enzyme's procoagulant properties.

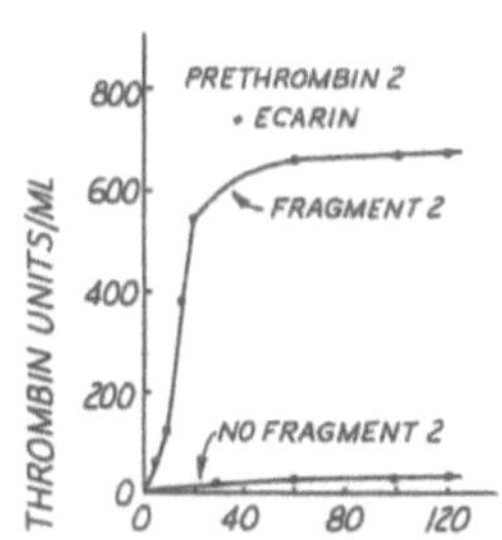

Fig. 3. Conversion of purified prethrombin 2 to beta-thrombin with purified ecarin in the ratio (w/w) of 500:1). Small amounts of thrombin generated. With increments of added purified prothrombin fragment 2 the yield reached a maximum. At that time course the molar ratio of prethrombin 2 and prothrombin fragment 2 was 1:1.

These lower concentrations of Factor Xa are toxic if prothrombinase active phospholipids are first adminstered. This integrates exactly with the idea that platelet anti-aggregating agents retard thrombosis.

CLOSING PERSPECTIVE

The crux of the considerations is enzyme specificity modifications that are compatable with physiological functioning. The first intimations for this perspective came in 1954 paper by Landaburu and Seegers. They recorded an experiment that showed modification of thrombin activity. In the experiment a solution of purified enzyme was stored at 4°C. In a few days the capacity to clot fibrinogen decreased while for the esterolytic titre remained with very little loss.

THE INITIATION OF THE TISSUE FACTOR DEPENDENT PATHWAY OF BLOOD COAGULATION

Samuel I. Rapaport

Departments of Medicine and Pathology
University of California, San Diego
La Jolla, CA 92037, USA

INTRODUCTION

A schematic representation of a present concept of the blood coagulation reactions is shown in Figure 1. In my lecture I will discuss two questions related to the reactions shown in the upper left hand portion of this figure, ie., to the tissue factor dependent pathway of blood coagulation.

The *first question* is this. Factor VIIa/TF complexes catalyze the activation of two substrates, factor IX and factor X. Since factor VIIa/TF can directly activate factor X and the role of factor IXa is also to activate factor X (see Fig. 1) — is the factor VIIa/TF activation of factor IX a kinetically significant reaction in initiating hemostasis by way of the tissue factor pathway?

The *second question* is concerned with how the reactions of the TF pathway really get started. Exposure of blood to TF on the surface of cells in the outer lining of a damaged vessel wall will lead to the formation of factor VII/TF complexes. But why should the formation of such *zymogen* factor VII/TF complexes suffice to start the reactions off?

THE SIGNIFICANCE OF THE FACTOR VIIa/TF ACTIVATION OF FACTOR IX

After Østerud and I described the factor VIIa/TF activation of factor IX in 1977 (1), initial skepticism was expressed as to the physiologic significance of the reaction. This skepticism arose from kinetic studies in bovine purified and plasma systems in which factor VIIa/TF was shown to activate factor X many fold faster than it activated factor IX (2, 3). Subsequently, factor VIIa/TF was also shown to activate factor X more efficiently than factor IX in human systems (4).

However, a substantial body of data has now accumulated which provide substantial evidence that factor IXa molecules generated through the factor VIIa/TF activation of factor IX can—in the presence of factor VIIIa— *significantly amplify* the direct factor VIIa/TF activation of factor X. Østerud and I provided the first such data from semi-purified systems in our 1977 paper (1). Subsequently, Marlar and colleagues (5) reported that adding a diluted tissue factor preparation to plasma deficient in factor VIII or factor IX resulted in a much slower initial rate of activation of factor X than when the preparation was added to normal plasma or plasma deficient in factor XI. Finally, Stern and his collaborators reported that when coagulation was initiated by factor VIIa on vascular endothelium perturbed to express TF activity, the formation of factor Xa was 10 times greater when factors IX and VIII were also present than when only factor X was present (6).

Moreover, one should be aware of an important abstract (7) reporting that the addition of heparin to reaction mixtures was without effect upon factor VIIa/TF activation of factor

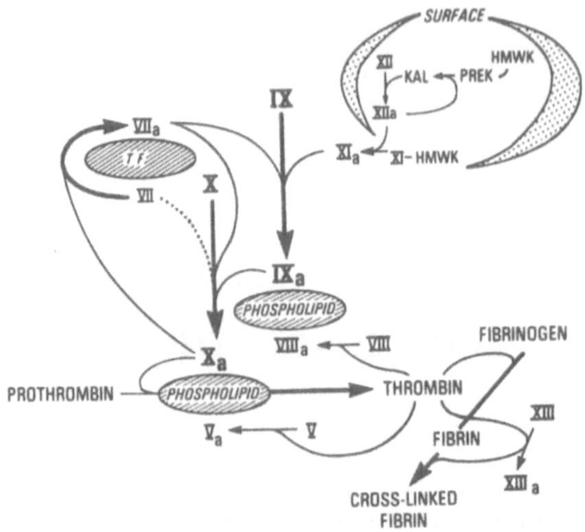

Fig. 1. An overall diagram of the blood coagulation reactions. Abbreviations are: TF, tissue factor; HMWK, high molecular weight kininogen; KAL, kallikrein; PREK, prekallikrein. (Modified from Rapaport SI in West JB (ed): Best and Taylor's Physiologic Basis of Medical Practice, 11th ed., Baltimore, Williams & Wilkins, 1984).

X but enhanced the rate of factor VIIa/TF activation of factor IX almost 3-fold. Recently, Drs. Almus, Rao and I have obtained data in an umbilical vein model system (8) pertinent to the issue of whether physiologic materials such as glycosaminoglycans in extracellular matrix may modulate the substrate specifity of factor VIIa/TF in favor of factor IX. In this model the umbilical vein of a fresh umbilical cord was canulated, washed free of blood, and then paired vein segments were filled with a reaction mixture containing factor VIIa, calcium ions, and either tritiated factor X or tritiated factor IX to measure activation of either factor X or factor IX by activation peptide release assays.

The model system was originally designed with the intent of studying whether endothelial cells in situ would develop TF activity after perturbing stimuli such as endotoxin. However, to our initial surprise, the vein wall was found to provide TF activity supporting factor VIIa catalyzed activation of factors X and IX without having first to expose the vein to a known perturbing stimulus. Histologic studies demonstrated that our processing technique had disrupted the umbilical vein endothelium. Immunostaining with anti-TF antibodies revealed no staining of endothelium; in contrast, bright staining was seen in extensions of Wharton's jelly penetrating fenestrations of the muscularis media of the vein. Using this model system, which thus permitted us to evaluate the catalytic activity of factor VIIa/TF complexes formed within the mucoid connective tissue of Wharton's jelly, we found that factors IX and X were activated at equivalent rates (mean activation rate for F IX, 18.8 nMh^{-1}, mean activation rate for F. X, 17.8 nM^{-1}, n = nine paired vein segments). These data provide convincing confirmatory evidence that the factor VIIa/TF activation of factor IX plays an important role in the initiation of tissue factor dependent coagulation. They also provide clear-cut evidence that materials in extracellular matrix can modify the functional activity of factor VIIa/TF complexes.

CONSEQUENCES OF FORMING ZYMOGEN FACTOR VII/TF COMPLEXES

The blood as it circulates normally does not come into contact with cells expressing surface membrane TF activity. Vascular endothelium does not stain in situ for TF antigen. Where TF antigen can be demonstrated is on the surface membrane of fibroblasts and pericytes in the outer walls of the blood vessels. When a blood vessel is damaged, blood can come into contact with TF on such cells with presumed resultant formation of factor VII/TF complexes. An unanswered question has been whether such zymogen factor VII/TF

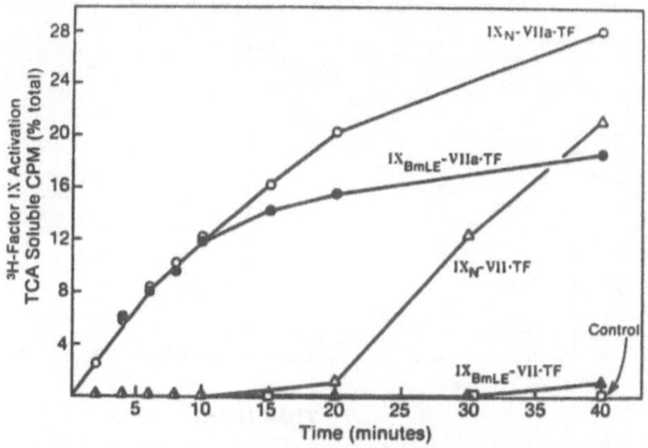

Fig. 2. Activation of factor IX (IX_N) and factor IX Bm Lake Elsinore (IX BmLE) by factor VII/TF and by factor VIIa/TF in a purified system. Final concentrations: ^{3}H-factor IX_N or ^{3}H-factor IX_{BmLE}, 88 nmol/L; factor VII or factor VIIa, 1.0 nmol/L, tissue factor, 10% v/v; Ca^{2+}, 5 mmol/L and anti-factor X antibodies, 2% v/v. Neither factor VII nor tissue factor were added to ^{3}H-factor IX_N in the control reaction mixture (□). Other symbols are: (○) IX_N, VIIa/TF; (△) IX_N, VII/TF; (●) IX_{BmLE}, VIIa/TF; (▲) IX_{BmLE}, VII/TF. (Reproduced by permission from Rao LVM, Rapaport SI, Bajaj SP. Blood 68: 685, 1986.)

complexes possess a minimal enzymatic activity capable in itself of triggering the TF-dependent pathway of coagulation. In bovine systems, zymogen factor VII has been reported to possess a minimal enzymatic activity thought capable of initiating tissue factor-dependent coagulation (9). However, the question has been very difficult to study because both of the physiologic substrates of factor VII(a)/TF, factor X and factor IX, back-activate factor VII once they are activated.

Drs. Rao, Bajaj and I turned our attention to this question several years ago (4) when we had the opportunity to purify a variant factor IX molecule, factor IX Bm Lake Elsinore. This molecule can be activated normally but possesses essentially no enzymatic activity after activation, and so can not back-activate factor VII. What we found is shown in Fig. 2. In this figure the ordinate is activation of factor IX as measured by activation peptide release from tritiated normal factor IX or factor IX Bm Lake Elsinore. The abscissa is incubation time in minutes. Reaction mixtures contained a crude tissue factor preparation, purified factor VII or factor VIIa, calcium ions, antibodies to factor X to neutralize any traces of factor X or factor Xa that could have contaminated reagents, and as substrate either tritiated purified normal factor IX or tritiated purified factor IX Bm Lake Elsinore.

As can be seen in Fig. 2, in reaction mixtures made with factor VIIa, both normal factor IX and factor IX Bm Lake Elsinore were activated without delay at equivalent initial rates. In a reaction mixture containing factor VII and normal factor IX, about 15 minutes of incubation time were required before beginning evidence of activation of normal factor IX was noted. Shortly thereafter, the activation rate rose rapidly secondary to factor IXa back-activating factor VII. In a reaction mixture containing factor VII and factor IX Bm Lake Elsinore as the substrate, activation was not measurable after 30 minutes of incubation time and only trace activation was evident after 40 minutes.

These data do not rule out the possibility that zymogen factor VII/TF complexes possessed a minimal enzymatic activity responsible for the activation of trace amounts of factor IX Bm Lake Elsinore after a 40 minute incubation period. However, I began this lecture by summarizing evidence for believing that activation of factor IX in the tissue factor pathway is a physiologically important step in the initiation of blood coagulation during hemostasis. Clearly for this to be so, some factor VIIa/TF complexes have to be formed very early, for these, rather than zymogen factor VII/TF complexes, are the complexes that can trigger the generation of meaningful concentrations of factor IXa by way of the tissue factor pathway.

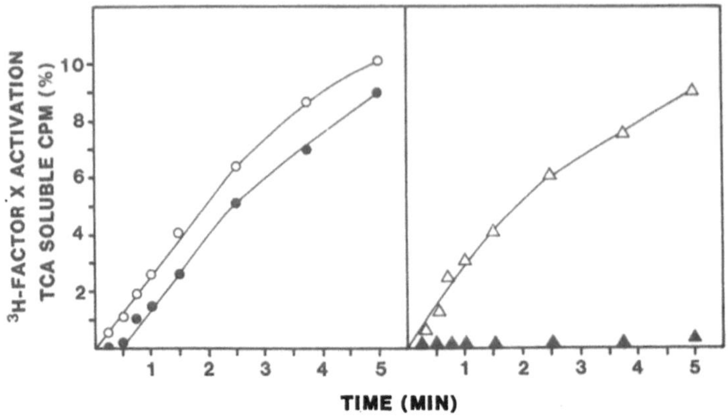

Fig. 3. Activation of factor X by factor VIIa/TF and by factor VII/TF in the absence (A) and in the presence (B) of antithrombin III and heparin. Composition of reaction mixtures and concentrations of reactants are given in the text. In A, ○, factor VIIa/TF; and ●, factor VII/TF; in B, △, factor VIIa/TF and ▲, factor VII/TF. Factor VII or factor VIIa/TF and ^{3}H-factor X were incubated with antithrombin III and heparin for 3 min at 37°C before the calcium was added. (Reproduced from Rao, LVM and Rapaport SI. Proc. Natl. Acad. Sci. USA, 85:6687, 1988.)

More recently, Dr. Rao and I (10) have examined the question of whether zymogen factor VII/TF complexes can activate factor X. An example of the data we obtained is shown in Fig. 3. In this figure the ordinate is per cent activation of tritiated factor X as measured by activation peptide release and the abscissa is incubation time in minutes. Reaction mixtures contained reconstituted, purified human brain TF at a 2 ng/ml concentration, factor VII or factor VIIa, at a plasma concentration of 500 ng/ml, and tritiated factor X at a concentration of 5 ug/ml. In Panel A of Fig. 3 one notes a rapid rate of activation of factor X in reaction mixtures made with either factor VIIa or factor VII. The only difference is a 30 second delay in the reaction mixture made with factor VII. In Panel B, reaction conditions were the same as in Panel A except that antithrombin III, 150 ug/ml final concentration, and heparin 5 U/ml, were added to the reaction mixtures. This was without effect upon a reaction mixture made with factor VIIa, but in a reaction mixture made with native factor VII the activation of factor X was markedly suppressed. We have concluded from this experiment that zymogen factor VII/TF complexes can not activate physiologically significant amounts of factor X. The data suggest further that in the reaction mixture of panel A made with zymogen factor VII all of the factor VII in initial factor VII/TF complexes had been very rapidly converted to factor VIIa. This presumably reflected activation by otherwise unmeasurable traces of factor Xa, presumably generated by a trace of factor VIIa contaminating the factor VII preparation.

Further support for this interpretation was obtained from experiments in which the order of addition of antithrombin III and heparin to a reaction mixture containing zymogen factor VII, calcium ions, TF, and factor X was varied. When antithrombin III and heparin were added 1 minute before the tritiated factor X, activation of factor X was prevented. However, if the antithrombin III and heparin were added 1 minute after adding the factor X, then activation proceeded as if the reaction mixture contained no antithrombin III and heparin. The factor Xa that had been generated in the reaction mixture before adding the antithrombin III and heparin had converted, within 1 minute, all of the factor VII/TF complexes in the reaction mixture to factor VIIa/TF complexes.

Factor VII and factor VIIa have very similar binding affinities for TF (11). Consequently, if within 1 minute in a reaction mixture starting with zymogen factor VII all of the factor VII bound to the TF is present as factor VIIa/TF complexes, one of two events must have occurred. One possible event is that all factor VII in the reaction mixture has been converted to factor VIIa. The second possible event is that the factor VII bound to

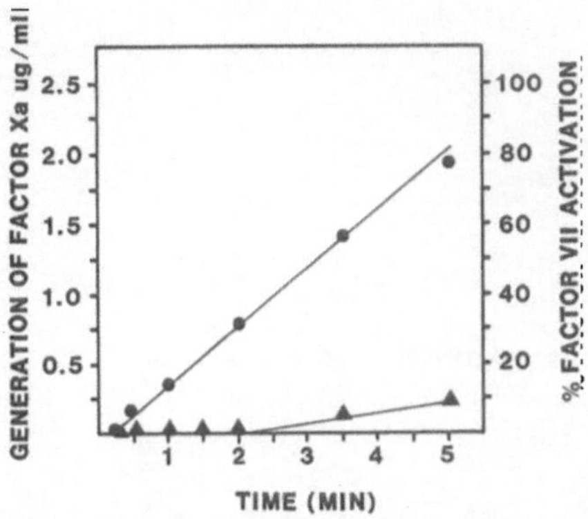

Fig. 4. Relationship between activation of factor VII and activation of factor X in a reaction mixture made with native factor VII. Final concentrations are unlabeled factor X, 10 ug/ml: ^{125}I-factor VII, 0.5 ug/ml; TF, 2.0 ng/ml; and $CaCl_2$, 5 mM. The progress curve of activation of factor VII (▲) was determined from radioactivity profiles; the progress curve of activation of factor X (●) was determined by assay of amidolytic activity. (Reproduced from Rao, LVM and Rapaport SI. Proc. Natl. Acad. Sci. USA, 85:6687, 1988.)

TF has been specifically and preferentially activated to factor VIIa while the free factor VII in the reaction mixture is still primarily in the native form.

There was reason to suspect that the second possibility would prove correct because Nemerson and Repke (12) had already shown that factor Xa catalyzes activation of bovine factor VII much more rapidly in reaction mixtures containing relipidated bovine TF than in reaction mixtures containing only the phospholipid used for the relipidation. Therefore, Dr. Rao and I were not too surprised when we obtained the data depicted in Fig. 4.

In the experiment of Fig. 4, a reaction mixture was made containing zymogen factor VII, TF, calcium ions and factor X. The factor X was unlabeled and the generation of factor Xa was measured by assaying serial subsamples for their ability to cleave a chromogenic substrate (S 2222). The factor VII was labelled with ^{125}I and, therefore, activation of total factor VII in the reaction mixture could be determined by removing serial subsamples and analyzing their radioactivity profiles after reduced SDS polyacrylamide gel electrophoresis. As you can see from Fig. 4 at a time when a linear rate of activation of factor X had been established, the activation of total factor VII in the reaction mixture was unmeasurable. This is illustrated by the radioactivity profile of panel B of Fig. 5.

Recognition of the key importance of the rapid, preferential activation of factor VII bound to TF for the initiation of the tissue factor pathway, in my mind makes physiologically insignificant the question of whether or not native factor VII/TF complexes possess a minimal catalytic activity. It is now clear that the binding of factor VII to TF has two consequences for the initiation of the tissue factor dependent pathway of coagulation. The *first* is a consequence that has been recognized since 1977 (1), namely, that the TF serves as a cofactor necessary for factor VIIa to function efficiently in catalyzing activation of its two physiologic substrates—factor X and factor IX. The *second* is the equally important, newly recognized consequence that binding of factor VII to TF markedly enhances the ability of trace concentrations of factor Xa, and also of factor IXa, to catalyze the activation of the factor VII bound to the TF. Indeed, it is reasonable to suspect that binding to TF may allow the preferential activation of factor VII by other proteolytic enzymes released from cells at a site of tissue injury. I believe now that this reaction—the rapid, preferential activation to factor VIIa of factor VII bound to TF—should be viewed as the key trigger reaction initiating the tissue factor pathway of blood coagulation.

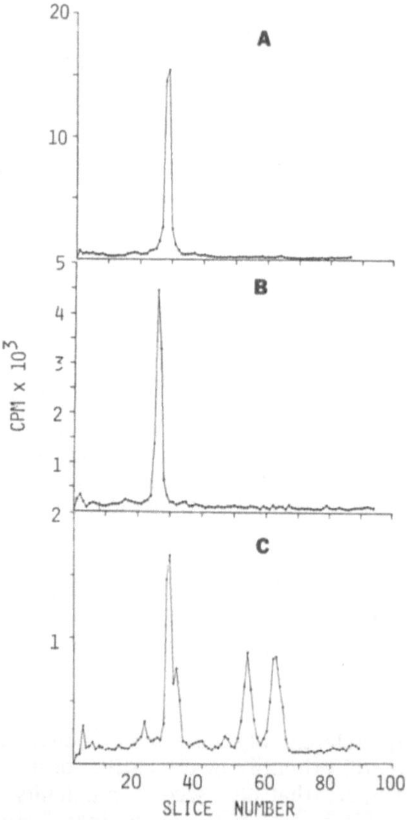

Fig. 5. Radioactivity profiles obtained on SDS polyacrylamide gel electrophoresis under reducing conditions of ^{125}I-factor VII and unlabeled factor X. The reaction mixture is that described in the legend to Fig. 4. (A) Profile of a zero-time subsample removed just before addition of calcium. (B) Profile of a subsample removed 1 min after the addition of the calcium, at which time a linear activation of factor X was evident. (C) Profile of a positive control sample containing ^{125}I-factor VIIa prepared by incubating ^{125}I-factor VII with factor Xa and phosphatidylcholine/phosphatidylserine vesicles in the presence of calcium ions for 20 min at 37°C. (Reproduced from Rao LVM and Rapaport SI. Proc. Natl. Acad. Sci. USA. 85:6687, 1988.)

REFERENCES

1. B. Østerud and S. I. Rapaport, Activation of factor IX by the reaction product of tissue factor and factor VII: additional pathway for initiating blood coagulation, *Proc. Natl. Acad. Sci. USA,* **74:**5260-5264 (1977).
2. J. Jesty and S. A. Silverberg, Kinetics of the tissue factor-dependent activation of coagulation factors IX and X in a bovine plasma system, *J. Biol. Chem.,* **254:**12337-12345 (1979).
3. M. Zur and Y. Nemerson, Kinetics of factor IX activation via the extrinsic pathway, *J. Biol. Chem.,* **255:**5703-5707 (1980).
4. L. V. M. Rao, S. I. Rapaport, and S. P. Bajaj, Activation of human factor VII in the initiation of tissue factor-dependent coagulation, *Blood,* **68:**685-691 (1986).
5. R. A. Marlar, A. J. Kleiss, and J. H. Griffin, An alternative extrinsic pathway of human blood coagulation, *Blood,* **60:**1353-1358 (1982).
6. D. Stern, P. Nawroth, D. Handley, W. Kisiel, An endothelial cell-dependent pathway of coagulation, *Proc. Natl. Acad. Sci. USA,* **82:**2523-2527 (1985).
7. D. I. Repke, D. MacLean, Y. Nemerson, Heparin affects the substrate specificity of the tissue factor pathway of coagulation, *Fed. Proc.,* **40:**274 (Abstr.) (1981).

8. F. E. Almus, L. V. M. Rao, and S. I. Rapaport, Tissue factor-VIIa activation of factors IX and X in umbilical veins, *Circulation,* **78:**no. 4, Part II. Abstract 1251 (1988).
9. M. Zur, R. D. Radcliffe, J. Oberdick, and Y. Nemerson, The dual role of factor VII in blood coagulation. Initiation and inhibition of a proteolytic system by a zymogen, *J. Biol. Chem.,* **257:**5623-5631 (1982).
10. L. V. M. Rao and S. I. Rapaport, Activation of factor VII bound to tissue factor: A key early step in the tissue factor pathway of blood coagulation, *Proc. Natl. Acad. Sci. USA,* **85:**6687-6691 (1988).
11. R. Bach, R. Gentry, and Y. Nemerson, Factor VII binding to tissue factor in reconstituted phospholipid vesicles: Induction of cooperativity by phosphatidylserine, *Biochemistry,* **25:**4007-4020 (1986).
12. Y. Nemerson and D. Repke, Tissue factor accelerates the activation of coagulation factor VII: The role of a bifunctional coagulation cofactor, *Thromb. Res.,* **46:**325-335 (1985).

INTERACTIONS BETWEEN THE CONTACT SYSTEM, NEUTROPHILS AND FIBRINOGEN

Robert W. Colman

Thrombosis Research Center
Temple University School of Medicine
Philadelphia, PA 19140, USA

SUMMARY

Since plasma kallikrein activates human neutrophils, and in plasma prekallikrein (PK) circulates complexed with high molecular weight kininogen (HK), we determined whether HK could mediate kallikrein's association with neutrophils. Human neutrophils were found to possess surface-membrane binding sites for HK but no internalization was detected at 37°C. ^{125}I-HK binding to neutrophils was dependent upon Zn^{++}, specific, saturable and reversible with a Kd of 9-18 nM and 40,000-70,000 sites per cell. Furthermore, HK found in neutrophils (240 ng/10^7 neutrophils) also served as a cofactor for HNE secretion since neutrophils deficient in HK have reduced HNE secretion when stimulated in plasma deficient in HK or with purified kallikrein. Thus, neutrophil surface HK may serve as a receptor for kallikrein. Fibrinogen inhibited ^{125}I-fibrinogen bound specifically and reversibly to human neutrophils. Zn^{++} (50 μM) was required for binding of ^{125}I-fibrinogen to neutrophils and the addition of Ca^{++} (2 mM) increased the binding 2-fold. Excess HK completely inhibited binding of and displaced labeled fibringogen as well as unlabeled fibrinogen. Binding of ^{125}I-fibrinogen was saturable with an apparent Kd of 170 nM and 140,000 sites/neutrophil. The binding of ^{125}I-fibrinogen to neutrophils was not inhibited by the peptide RGDS derived from the α-chain of fibrinogen, nor by the monoclonal antibodies (MAB) 10E5 to the platelet glycoprotein IIb/IIIa heterodimer. Fibrinogen binding was inhibited by a γ-chain peptide CYGHHLGGAKQAGDV and by MAB OKM1 but was not inhibited by OKM10, a MAB to a different domain of the adhesion glycoprotein Mac-1 (CR3). HK binding to neutrophils was not inhibited by OKM1. These observations were consistent with a further finding that fibrinogen is a noncompetitive inhibitor of ^{125}I-HK binding to neutrophils. These studies indicate that fibrinogen specifically binds to an integrin receptor (Mac-1) on the neutrophil surface through the carboxy terminal of the γ-chain and that HK inhibits this interaction.

Abbreviations used in this paper:
HK, high molecular weight kininogen
HNE, human neutrophil elastase
FNDP, fibronectin degradation products
MAB, monoclonal antibody
GP, glycoprotein
CR_3, complement receptor type 3
PK, prekallikrein
ELISA, enzyme-linked immunosorbent assay

Fibrinogen, Thrombosis, Coagulation, and Fibrinolysis, Edited by
C. Y. Liu and S. Chien, Plenum Press, New York

INTRODUCTION

The binding of HK to negatively charged surfaces enables PK to attach to negatively charged surfaces where the zymogen can be optimally activated by factor XIIa (1, 2). Kallikrein stimulates neutrophil chemotaxis (3), aggregation (4), and degranulation (5). The receptor(s) for plasma kallikrein on neutrophils is not known. We postulated that kallikrein might bind to neutrophils through its cofactor, HK, with which it complexes.

Vroman et al. (6) noted that HK can displace fibrinogen from artificial hydrophilic surfaces such as glass. This effect is specifically due to the light chain of HK since purified low molecular weight kininogen did not alter the surface expression of fibrinogen (7). To date, no evidence exists to indicate whether such a phenomenon exists on biological surfaces. Studies were initiated to determine if neutrophils also contain a form of HK and whether the external membrane of human neutrophils possesses a specific binding site for HK. We also sought to determine whether ^{125}I-fibrinogen binds to the neutrophil surface and, if so, to which receptor. Further, we investigated whether HK can modify the interaction of fibrinogen with neutrophils. The results of this investigation demonstrate that neutrophils contain HK and possess a specific, reversible, and saturable binding site for HK, and that HK from either plasma or neutrophils is a cofactor for HNE secretion. Furthermore, ^{125}I-fibrinogen binds to CR_3 on human neutrophils in a specific, reversible and saturable manner and HK and fibrinogen reciprocally inhibit binding of the other protein to the neutrophil surface.

METHODS

Plasma and Neutrophils

Pooled normal plasma (lot 6130) was purchased from George King Biomedical Inc., Overland Park, KS. Total kininogen-deficient plasma (plasma deficient in both HK and low molecular weight kininogen), and neutrophils deficient in HK were donated by Mrs. Williams. Normal donors were young males and females (age 21-45 years) receiving no medication, who had given their written, informed consent.

Neutrophil Isolation

Human neutrophils were isolated from whole blood anticoagulated with 1/10 volume acid-citrate-dextrose by sedimentation at 1 g in dextran (1.5%) followed by sedimentation of the upper leukocyte-enriched plasma through Ficoll-Paque (5). After the final saline wash, the cells were resuspended in HBSS without magnesium of calcium (1×10^7-10^9 cells/ml). This procedure yielded approximaely 8×10^8 neutrophils/unit whole blood and the cells were isolated to 96% purity. The cells were either solubilized with 0.5% Triton X-100 (final concentration) and stored at −70°C for antigen determination or used immediately in binding experiments.

Proteins

HK was purified using a modified method (8) of Kerbiriou and Griffin (9). Purified HK was radiolabeled with ^{125}I-Na using Iodogen by the method of Fraker and Speck (10) under conditions previously described (16). The radiolabeled protein retained > 95% of its procoagulant activity as well as its antigenic properties as previously reported (11).

Purified factor XII (70 μg/ml) and PK (1 mg/ml) were provided by Dr. Robin Pixley; fibronectin and degradation products (FNDPs) (1.3 mg/ml) was donated by Dr. Andrei Budzynski, and factor V (0.93 mg/ml) by Dr. Anjanayaki Annamalai, all of Temple University School of Medicine, Philadelphia, PA. HNE was purified to homogeneity by the procedure of Baugh and Travis (12). Human fibrinogen (Kabi, Stockholm, Sweden) was further purified by ammonium sulfate precipitation (34), radiolabeled with ^{125}I-Na with the aid of Enzymo beads (Pierce Chemical Company, Rockford, IL) or BioRad-beads (BioRad, Inc., Richmond, CA), and was separated from free iodine by gel filtration using a Sephadex G25 column. The radiolabeled fibrinogen demonstrated 95% clottability. A septadecapeptide, CYGHHLGGAKQAGDV, modeled from a portion of the γ-chain of fibrinogen, was purchased from Penninsula Laboratories (Belmont, CA).

Antibodies

The MAB, 10E5 was a generous gift of Dr. Barry Coller, State University of New York at Stony Brook, Stony Brook, NY (13). The MAB, IMY8 was purchased from Ortho Diagnostics, Raritan, NJ (14). The MAB, OKM1 was purchased from Coulter Diagnostics, Hialeah, FL. The MAB, OKM10 was obtained from Ortho Diagnostics, Raritan, NJ (courtesy of Dr. Pat Rao) in the form of IgG purified using protein A.

Assays

HK procoagulant activity was measured by a one-stage kaolin activation assay (15) using a total kininogen-deficient plasma as substrate. HK antigen was measured by a competitive-enzyme-linked immunosorbent assay (ELISA), performed as previously reported (8) using a monospecific polyclonal antibody directed to the 56 kDa light chain of HK (8). HNE activity was measured by a chromogenic assay using the substrate, methoxysuccinyl-Ala-Ala-Pro-Val-pNA (16) and antigenic levels were measured using a competitive ELISA specific for HNE (5).

Neutrophil Aggregation

Aggregation was performed with modifications of the previously described method (4). Neutrophils (1×10^7/ml) were resuspended in a total volume of 1 ml HBSS containing a final concentration of $CaCl_2$ (2 mM), $MgCl_2$ (1.5 mM), and cytochlasin B (5 μg) in the presence of the agonist. Aggregation was measured using a Monitor IV Plus platelet aggregometer (Helena Laboratories, Beaumont, TX) stirred at 900 rpm.

Neutrophil Degranulation in an In Vitro Plasma-Free System

Supernatants were harvested for subsequent HNE activity determinations by centrifuging the contents of the aggregation cuvette at 13,000 g for 5 min at 23°C in a Micro-Centrifuge. The supernatants were assayed using the chromogenic substrate methoxysuccinyl-Ala-Ala-Pro-Val-PNA (5).

Neutrophil Degranulation in an In Vitro Plasma System

Neutrophils (1×10^7/ml) were resuspended in polypropylene tubes containing plasma whereupon Ca^{++} (final concentration 30 mM) was added. Incubations were performed for various times at 23°C, and aliquots were then removed and centrifuged at 13,000 g for 5 min at 23°C in a Micro-Centrifuge (Model 235A, Fisher Scientific Co., Pittsburgh, PA). The plasma supernatants were stored at −70°C until immunochemical determinations of elastase were made, using an indirect competitive ELISA (5, 17).

Binding Experiments

In all binding experiments, neutrophils were at a final concentration of 1×10^7/ml. In a typical binding experiment, 300-400 μl of washed neutrophils in HBSS without added calcium or magnesium, pH 7.4, were incubated at 23°C with the radiolabeled protein without stirring in a 1.5 ml conical polypropylene centrifuge tube (Sarstedt, Inc., Princeton, NJ) with ^{125}I-HK and additions to a total volume of 350-450 μl made. After incubation of appropriately expanded volumes, 50 μl aliquots were removed (in triplicate) for each experimental point and centrifuged at 9,650 g at 23°C in a microfuge (Model B, Beckman Instruments, Inc., Fullerton, CA) through a 200 μl mixture of silicon oils (one part Apiezion/nine parts N-butyl-phthalate) in polypropylene microsedimentation tubes with narrow bore extended tips (Sarstedt, Inc., Princeton, NJ) for 2 min. After the supernatant was removed, the tips containing the pellet were amputated and counted in a Rack gamma counter (LKB Instruments, Inc., Gaithersburg, MD).

Statistical Analysis

Calculation of bound HK was based on the specific activities of the radiolabeled ligand, and the results were expressed as nanograms of HK per 10^7 neutrophils. In a typical experiment, total binding was the amount of ^{125}I-HK bound in the absence of unlabeled ligand, while nonspecific binding was the amount of ^{125}I-HK bound in the presence of a 50-fold molar excess of unlabeled ligand. Specific binding was obtained by subtracting the nonspecific binding from the total binding.

In competition, inhibition experiments with unlabeled HK, the affinity of binding of ^{125}I-HK was calculated from the IC_{50} (18) as previously reported (11, 19) using a computer program to determine the 50% inhibition point (20). In concentration-dependent experiments, binding ^{125}I-HK to PMNs was also analyzed by the graphical methods of Scatchard (30), as well as the computer programs of Munson and Rodbard (31) and Brass and Shattil (32) using an Apple IIe computer (Apple Computer Corp., Cupertino, CA). Experimental results at each concentration of the total dose of ligand (both labeled and unlabeled) and bound radioligand were fed into a preanalysis program (21). After calculating the amount of specifically bound ligand, the data were also fit by the program developed by Brass (22). Comparison of the differences of experimental groups in the binding studies were performed by the Students' test (paired). Significant differences are defined as $p < 0.05$. All values are expressed as mean ± SEM.

RESULTS

Identification of Neutrophil HK

Washed solubilized neutrophils from normal blood donors were found to have 237 ± 61 ng HK antigen/10^8 cells (mean ± SEM, n = 8). This result was specific for the identification of HK antigen because a lysate of neutrophils from a patient with a known total deficiency of plasma and platelet HK (8, 15) had a neutrophil HK value of ≤ 2.7 ng/10^8 cells. Neutrophil HK was immunochemically identical to plasma HK.

Binding of ^{125}I-labeled HK to Isolated Neutrophils

Binding experiments of ^{125}I-HK to washed neutrophils were performed at 4°C, 23°C, and 37°C. Neutrophils bound ^{125}I-HK at all temperatures and the amount of ^{125}I-HK bound was equivalent at all temperatures (Fig. 1) indicating that internalization had not occurred. Nonspecific binding which was determined by binding of ^{125}I-HK in the presence of a 50-fold molar excess of unlabeled HK, was approximately 15% of total binding.

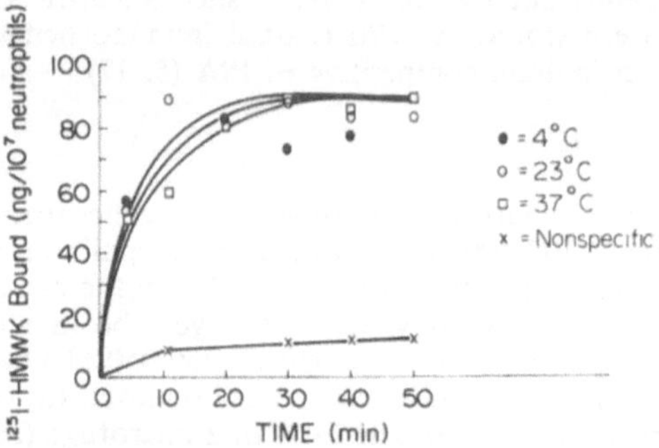

Fig. 1. Binding of ^{125}I-HK to neutrophils at various temperatures. Isolated PMNs (1×10^7/ml) in HBSS without Ca^{++} or Mg^{++} but with 50 nM Zn^{++} pH 7.4, were incubated at 4°C (•), 23°C (o) or 37°C (□) with ^{125}I-HK (1 μg/ml) in the presence of Zn^{++} (50 μM). Nonspecific binding (nonspecific) of ^{125}I-HK was measured in the presence of 50 μM Zn^{++} and a 50 fold molar excess of unlabeled HK at 23°C (X). At the designated time points, samples were removed and the amount of ^{125}I-HK bound to neutrophils was determined as indicated in Methods. The results are the mean of four separate experiments and the plotted data represent total binding.

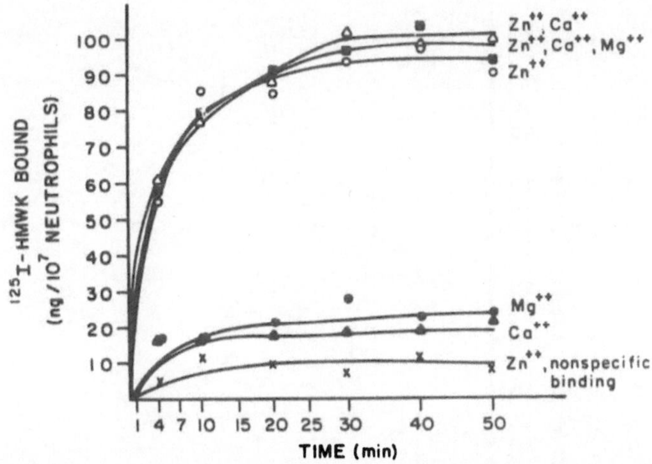

Fig. 2. Effect of divalent cations on binding of ^{125}I-HK to neutrophils. Isolated neutrophils (1×10^7/ml) in HBSS without Ca^{++} or Mg^{++} were incubated at 37°C for 4 to 50 mins with ^{125}I-HK (1 μg/ml) and various cations. In each experiment where an addition was made, the concentration of Zn^{++} was 50 μM and Ca^{++} 2 mM. At the designated time points, binding of ^{125}I-HK to neutrophils was determined as indicated in Methods. Binding was determined in the presence of Zn^{++} alone (o), Zn^{++} and Ca^{++} (△), Zn^{++}, Ca^{++}, and Mg^{++} (■), Ca^{++} alone (▲), and Mg^{++} alone (•). Nonspecific binding as determined by radioligand binding in the presence of Zn^{++} (50 μM) and a 50-fold molar excess of unlabeled HK (X). The plotted data are the mean of four experiments and represent total binding. For clarity, the SEMs are omitted but they represent about 10% of the mean.

Subsequent investigations determined the divalent cation requirements for binding of ^{125}I-HK to neutrophils (Fig. 2). In the presence of 50 μM Zn^{++}, binding of ^{125}I-HK to neutrophils increased over time to reach a plateau by 15 min. The amount of ^{125}I-HK in the presence of 2 mM Ca^{++} or 2 mM Mg^{++} was not significantly different from the level of nonspecific binding. Moreover, binding of ^{125}I-HK in the presence of Zn^{++} and Ca^{++} or Zn^{++}, Ca^{++} and Mg^{++} resulted in a level of binding that was not significantly different ($p \geq 0.45$) from the level of binding in the presence of Zn^{++} alone, thus demonstrating that either Ca^{++} or Mg^{++} contributed nothing to the binding of ^{125}I-HK to neutrophils. Therefore, Zn^{++} was the only divalent cation tested required for binding of ^{125}I-HK to neutrophils. ^{125}I-HK binding to neutrophils was not inhibited by a 50- to 100-fold molar excess of factor V, factor XII, PK, or FNDPs.

The ability of HK to inhibit binding of ^{125}I-HK to neutrophils was also found to be concentration dependent (Fig. 3). Unlabeled HK inhibited the binding of ^{125}I-HK (6.5 nM) to neutrophils with a 50% inhibitory concentration of 32 ± 16 nM. Using this mean 50% inhibitory concentration from these experiments, the apparent Ki was calculated to be about 9.6 nM (18).

Reversibility and Saturability of the Binding of ^{125}I-HK to Neutrophils

Binding of ^{125}I-HK to neutrophils was reversible (Fig. 4). When a 50-fold molar excess of unlabeled HK was added to the binding reaction at 8 and 18 min, rapid dissociation of the bound ligand occurred. At 8 min, 96% of the bound ligand was displaced within 1 min, and at 18 min, 93% of the bound ligand was displaced. Increasing concentrations of ^{125}I-HK were added to neutrophils in the absence of presence of a 50-fold molar excess of unlabeled ligand (Fig. 5). As the concentration of added ^{125}I-HK increased, the level of binding increased until it leveled off at approximately 3.5 μg/ml (29 nM) of the added ^{125}I-HK (Fig. 5, on left). Using the graphical method of Scatchard (30), a saturable binding site was characterized to be one of the high affinity with an apparent Kd of 9 nM and

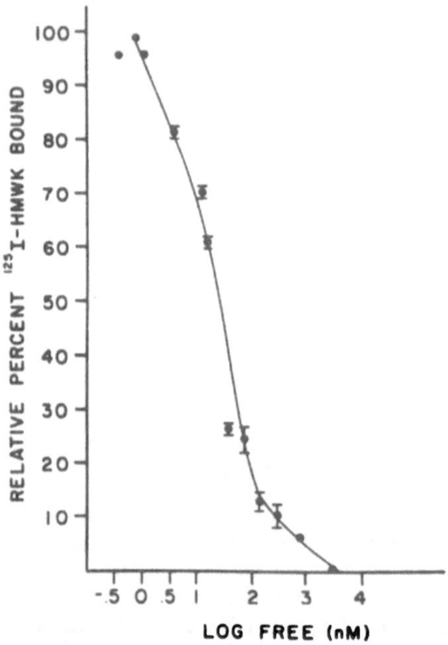

Fig. 3. Ability of HK to inhibit ^{125}I-HK binding to neutrophils. ^{125}I-HK (0.8 μg/ml or 6.55 nM) was incubated with isolated neutrophils (10^7/ml) in HBSS with Ca^{++} or Mg^{++} containing 50 nM Zn^{++} for 10 min at 23°C in the presence of the indicated log concentration of HK given on the abscissa. The mean ± SEM of each individual data point from three individual experiments are plotted and the line drawn is calculated from the mean of the three experiments using a four-parameter logistic function which fits the values of the ordinate into relative values between 0 and 100% (20).

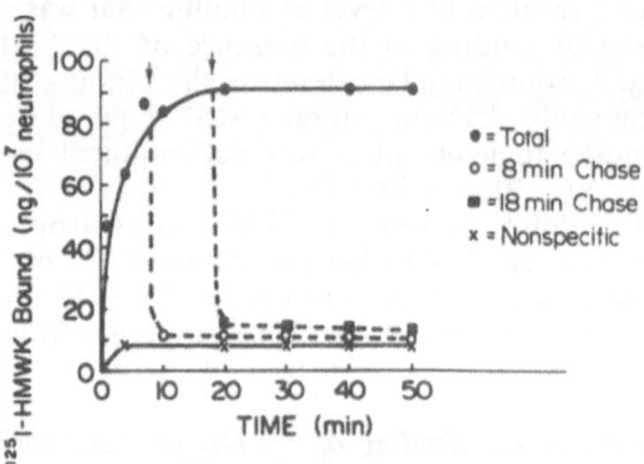

Fig. 4. Reversibility of ^{125}I-HK binding to neutrophils. Isolated neutrophils (10^7/ml) in HBSS without Ca^{++} or Mg^{++} were incubated at 23°C in the presence of Zn^{++} (50 μM) and ^{125}I-HK (1 μg/ml). (•) represents total binding. At 8 min (o) and 18 min (■), a 50-fold molar excess of unlabled HK was added. The binding was determined at the indicated time points as described in Methods. The plotted data are the mean of two experiments and represent total binding.

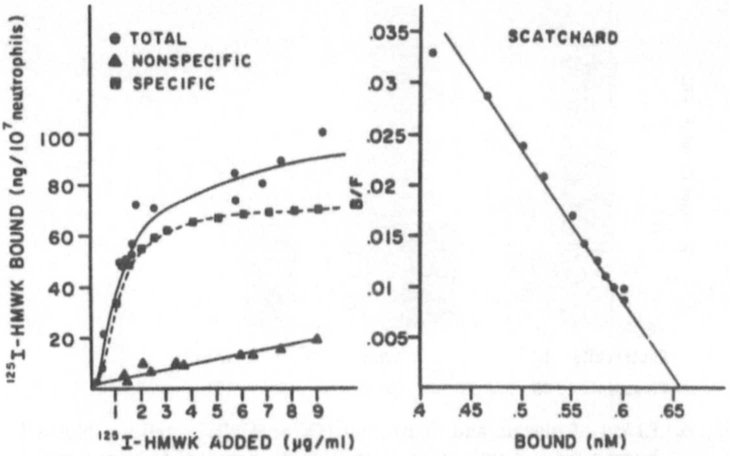

Fig. 5. Concentration-dependent binding of ^{125}I-HK to neutrophils. Neutrophils (1×10^7/ml) in HBSS without Ca^{++} or Mg^{++} were incubated with increasing concentrations of ^{125}I-HK in the presence of Zn^{++} (50 μM) at 23°C for 15 min. Binding of ^{125}I-HK to neutrophils was determined as indicated in Methods. In the panel on the left, total binding (-•-) and nonspecific binding (-▲-) were directly determined and specific binding (-■-) was calculated by subtracting nonspecific binding from total binding. This figure is a representative experiment of three, performed identically. In the panel on the right, entitled "Scatchard", the "bestfit" data of the ratio of bound/free versus bound (nM), as determined by linear regression analysis of the data from the panel on the left, is plotted.

40,000 sites per cell (Fig. 5, on right). Using a one site model for calculating the data for three experiments, a high affinity binding site with an apparent Kd of 18.0 ± 2.6 nM (mean ± SD) and 70,000 ± 25,000 molecules per cell (mean ± SD) was described.

Role of HK in HNE Secretion in Plasma

Ten million normal human neutrophils were resuspended and preincubated for either 30 sec or 30 min in normal plasma followed by recalcification for 60 min. Normal neutrophils released 1.08 μg/ml HNE after 30 sec and 0.7 μg/ml HNE, respectively, after 30 min preincubation (Fig. 6, bar A). In contrast, the extent of HNE secretion from HK-deficient neutrophils suspended in total kininogen-deficient plasma was 0.4 μg/ml HNE when preincubated for 30 sec and 0.43 μg/ml HNE for 30 min (Fig. 6, bar B). The release of HNE from HK-deficient neutrophils suspended in normal plasma was 0.42 μg/ml HNE when preincubated for 30 sec but became normal, 0.98 μg/ml HNE, when preincubated for 30 min (Fig. 6, bar C). These results suggest that HK bound to the neutrophil surface from plasma could contribute to HNE secretion and that neutrophil HK could also serve the same function.

Role of HK in Neutrophil Activation in a Plasma-free System

Normal neutrophils stimulated with purified human plasma kallikrein (0.2 U/ml) released 16.6 ± 8% of the total HNE and induced an initial rate of aggregation that was 36% of that which was induced by N-formyl-Methionyl-Leucyl-Phenylalanine (0.2 μM). Alternatively, HK-deficient neutrophils stimulated with human plasma kallikrein (0.2 U/ml) released only 2.5% of the total HNE and induced an initial rate of aggregation that was 8% of that which was induced by N-formyl-Methionyl-Leucyl-oPhenylalanine (0.2 μM). These results confirm the importance of neutrophil HK in the absence of plasma HK.

Displacement of Bound ^{125}I-HK by Fibrinogen from the Neutrophil Surface

While investigating the binding of ^{125}I-HK to neutrophils, it was found that a 50-fold

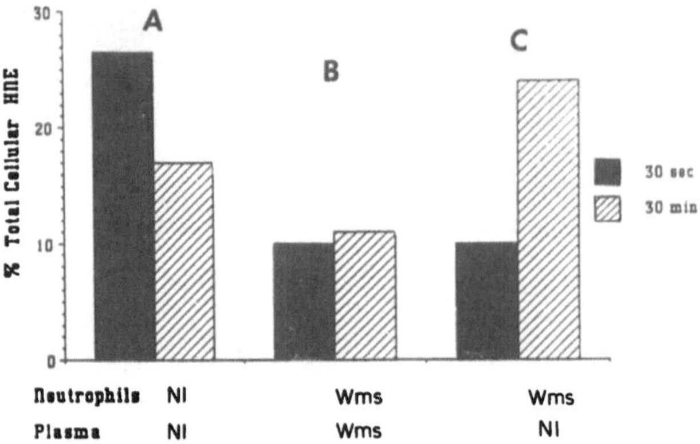

Fig. 6. Effect of plasma and neutrophil HK on HNE secretion. Normal human neutrophils (1×10^7/ml) were resuspended in normal human plasma (A) and HK-deficient neutrophils (1×10^7/ml) were resuspended in normal human (C) or total kininogen-deficient plasma (B). After 30 sec or 30 min preincubations, the cell suspension in plasma in polypropylene tubes were recalcified at 23°C by the addition of 30 mM $CaCl_2$. At 60 min aliquots from each cell suspension were removed and centrifuged at 13,000 g for 5 min. HNE antigen was assayed by competitive ELISA (17). Each HNE measurement was performed in duplicate and the means are graphed. The coefficient of variation of the ELISA for HNE is 10% (43).

molar excess of fibrinogen could inhibit HK binding. Studies were conducted to determine if fibrinogen could displace ^{125}I-HK already bond to the surface of the neutrophil (Fig. 7). ^{125}I-HK binding to neutrophils increased over time and reached a plateau by 20 min. A 50-fold molar excess of fibrinogen added at either 8 or 12 min was able to displace at least 86% of ^{125}I-HK bound to the neutrophil surface.

Role of Divalent Cations in Fibrinogen Binding

Binding studies were performed in the presence of these divalent cations (Fig. 8). Binding of ^{125}I-fibrinogen to neutrophils was maximal in the presence of plasma concentrations of both Ca^{++} (2 mM) and Zn^{++} (50 μM). Ca^{++} alone could not support ^{125}I-fibrinogen binding to neutrophils. In the presence of Zn^{++} alone, binding was one-half that of the maximal level obtained when both Zn^{++} and Ca^{++} were present. These studies indicated that both Zn^{++} and Ca^{++} were required for optimal binding of ^{125}I-fibrinogen to neutrophils. Nonspecific binding was the same regardless of the absence or presence of any single or combination of divalent cations.

Specificity of Binding of ^{125}I-Fibrinogen to Neutrophils

The binding of ^{125}I-fibrinogen to neutrophils was not inhibited by a 50-fold molar excess of factor XII or prekallikrein. FNDPs at a 50-fold molar excess inhibited fibrinogen binding by 26%, while a 50-fold molar excess of HK was able to inhibit the binding by 94%. The ability of fibrinogen to inhibit the binding of ^{125}I-fibrinogen to neutrophils was concentration-dependent. Using the mean $\pm$ SEM for each point from four experiments, unlabeled fibrinogen inhibited the binding of ^{125}I-fibrinogen to neutrophils 50% at a concentration of 2.8 ± 1.3 μM, which gave a calculated apparent Ki of 0.49 ± 0.30 μM. This value was not significantly different from the calculated apparent Ki obtained from the IC_{50} for each individual experiment.

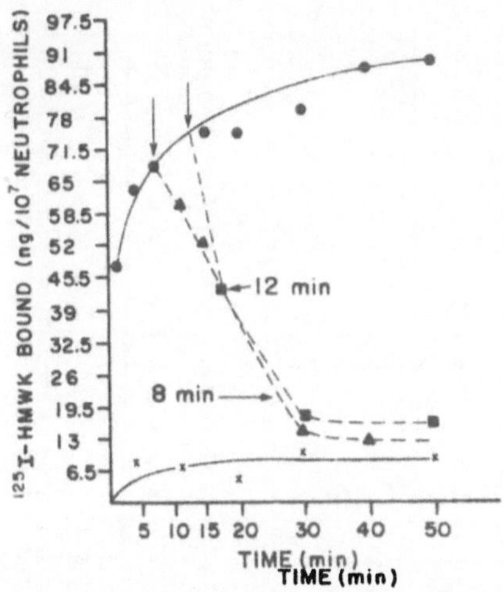

Fig. 7. Displacement of bound ^{125}I-HK by fibrinogen from the neutrophil surface. Isolated neutrophils (1×10^7/ml) in HBSS without Ca^{++} or Mg^{++}, pH 7.4, were incubated with Zn^{++} (50 μM), Ca^{++} (2 mM) and ^{125}I-HK (1 μg/ml) at 23°C (•). At 8 (▲) and 12 min (■), a 50-fold molar excess of unlabeled fibrinogen was added. Nonspecific binding (X) was measured by ^{125}I-HK binding in the presence of above divalent cations and a 50-fold molar excess of unlabeled HK. The binding was determined at the indicated time points as described in Methods. The plotted date are the mean of four experiments.

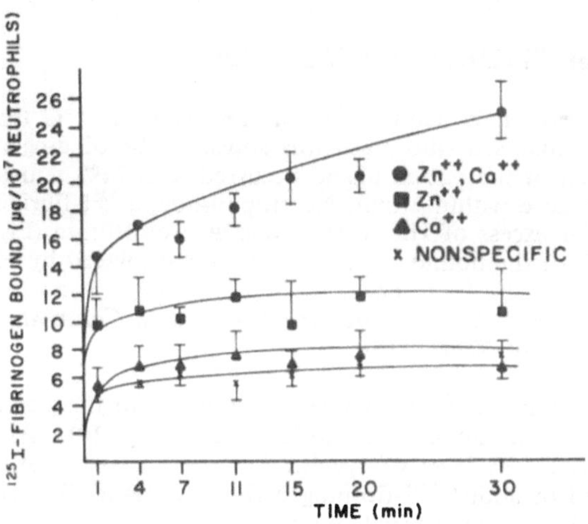

Fig. 8. Effect of divalent cations on binding of ^{125}I-fibrinogen to neutrophils. Isolated neutrophils (1×10^7/ml) in HBSS without Ca^{++} or Mg^{++}, pH 7.4, were incubated at 4°C with ^{125}I-fibrinogen (400 μg/ml) in buffer containing various cations. In each experiment where an addition was made, the concentration of Zn^{++} was 50 μM and Ca^{++}, 2 mM. At the designated time points, binding of ^{125}I-fibrinogen was determined as indicated in Methods. Binding was determined in the presence of Zn^{++} alone (■), Zn^{++} and Ca^{++} (•) and Ca^{++} alone (▲). Nonspecific binding was measured in the presence of all divalent cations, ^{125}I-fibrinogen and a 50-fold molar excess of unlabeled fibrinogen (X). The plotted data are the mean ± SEM Of four experiments.

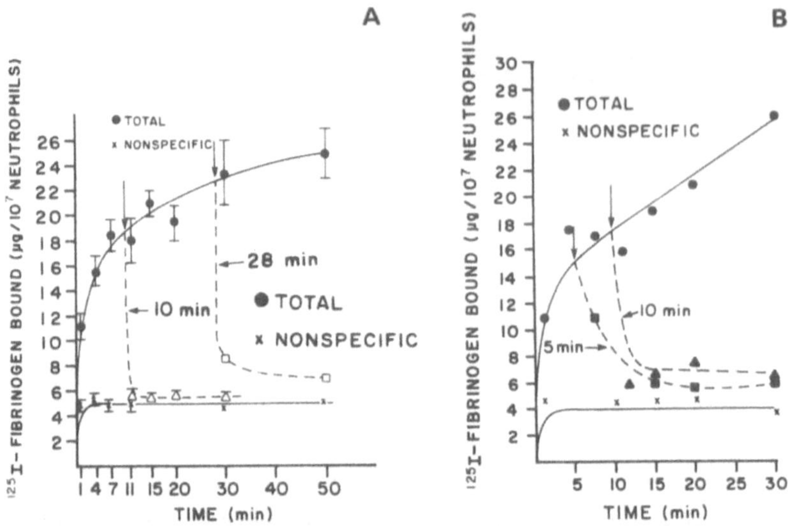

Fig. 9A. Displacement of bound ^{125}I-fibrinogen by HK and unlabeled fibrinogen from the neutrophil surface. Isolated neutrophils (1×10^7/ml) in HBSS without Ca^{++} or Mg^{++}, pH 7.4, were incubated at 4°C in the presence of Zn^{++} (50 μM), Ca^{++} (2 mM) and ^{125}I-fibrinogen (400 μg/ml). In A) at 10 (△) and 28 (□) min, a 50-fold molar excess of unlabeled fibrinogen was added.

Fig. 9B. Displacement of bound ^{125}I-fibrinogen by HK and unlabeled fibrinogen from the neutrophil surface. In B) at 5 (▲) and 10 (■) min, a 50-fold molar excess of unlabeled HK was added. Nonspecific binding was binding measured in the presence of the above additives and a 50-fold molar excess of unlabeled fibrinogen (X). Binding was determined at the indicated time points as described in Methods. The plotted data are the mean of three experiments.

Reversibility of Binding of ^{125}I-Fibrinogen to Neutrophils

Binding of ^{125}I-fibrinogen to neutrophils was reversible at 4°C (Fig. 9A). When a 50-fold molar excess of unlabeled fibrinogen was added to the binding reaction at 10 and 28 min, rapid dissociation of the bound ligand occurred with 94% and 88% of the bound ligand, respectively, displaced within 1 min. Neutrophil-bound ^{125}I-fibrinogen also was displaced by a 50-fold molar excess of HK when added at 5 or 10 min (Fig. 9B). At 5 min, 82% and at 10 min 77% of the bound ^{125}I-fibrinogen was displaced by HK.

Determination of the Number of Binding Sites and Dissociation Constant of Binding of ^{125}I-Fibrinogen to Neutrophils

Increasing concentrations of ^{125}I-fibrinogen were added to neutrophils in the absence or presene of a 50-fold molar excess of unlabeled ligand (Fig. 10). As the concentration of ^{125}I-fibrinogen increased, the level of specific binding increased until it leveled off at approximately 120 μg/ml of added ^{125}I-fibrinogen (Fig. 10, panel B). Using the graphical method of Scatchard (45), a single saturable binding site was characterized with an apparent Kd of 0.17 μM and 140,000 sites/cell (Fig. 10, panel A). A plot of the computer fitted points from the three individual experiments showed (Fig. 10, panel C) a sigmoid curve with a plateau at about 0.2 μM added fibrinogen. This result characterizes one saturable binding site with an apparent Kd of 0.15 μM.

Characterization of Fibrinogen/HK Interaction on the Neutrophil Surface

Since the binding of fibrinogen to platelets is inhibited by certain MABs to the GP IIb/IIIa complex (13), by the tetrapeptide RGDS (23) and by a dodecapeptide from the γ-chain of fibrinogen (24), the effect of these agents on fibrinogen binding to neutrophils was investi-

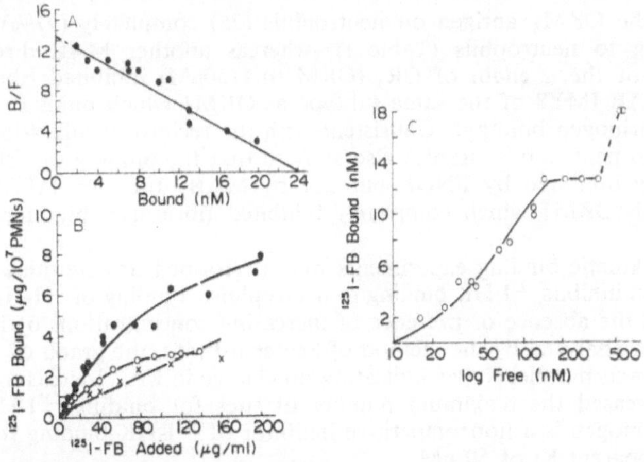

Fig. 10. Concentration-dependence of binding of ^{125}I-fibrinogen to neutrophils. Isolated neutrophils (PMNs) (1×10^7/ml) in HBSS without Ca^{++} or Mg^{++}. pH 7.4, were incubated with increasing concentrations of ^{125}I-fibrino-gen (^{125}I-FB) in the presence or absence of 50-fold molar excess of unlabeled fibrinogen. Panel B shows the total nonspecific and specific binding. The figure is a represnetative of three performed identically. Panel A represents a Scatchard plot of the data in Panel B. Panel C represents a plot of bound ^{125}I-FB (µM) on the ordinate versus log free (nM) FB on the abscissa. The line running through the points represents a manual graph of the computer-fitted data (8) from three experiments performed identically.

gated. The binding of ^{125}I-fibrinogen to neutrophils was not inhibited by 10E5 (a MAB to GPIIb/IIIa complex) (Table I). RGDS at concentrations up to 1 mM showed no inhibition.

A second site on fibrinogen which is important for its binding to platelets is located in the carboxy terminal section of the γ-chain. Therefore, we tested the effect of the septadecapeptide CYGHHLGGAKQAGDV on fibrinogen binding to neutrophils. Although minimal inhibition was noted at 150 µM, at concentrations of 250-1000 µM inhibition ranged from 62-79%.

Table 1. Effect of RGDS and MABs on ^{125}I-Fibrinogen and ^{125}I-HK Binding to Neutrophils

Competitor	^{125}I-Fibrinogen Binding % Inhibition	^{125}I-HK Binding % Inhibition
None	0	0 ± 3
Fibrinogen	100	86 ± 2.6
HK	94 ± 5.0	100
FNDPs	26 ± 1.0	22 ± 4
10E5	7.0 ± 1.2	0 ± 3
OKM1	97 ± 1	9.8 ± 5
OKM10	17.0	ND
IMY8	0	0 ± 4

Human neutrophils (1×10^7/ml) in HBSS without Ca^{++} or Mg^{++} pH 7.4 were preincubated with each competitor in a 50-fold molar excess in the presence of 50 µM Zn^{++} and 2mM Ca^{++} at 4°C for 60 min. ^{125}I-fibrinogen (1.18 µM) was then added and binding measured at 4°C after 30 min. Values present are mean ± SEM of three experiments. ND = not done.

A MAB to the OKM1 antigen on neutrophils (25) completely (97%) prevented ^{125}I-fibrinogen binding to neutrophils (Table I), whereas another MAB directed against a different domain of the α chain of CR_3 (OKM10) (150μM) inhibited fibrinogen binding by only 17%. MAB IMY8 of the same subtype as OKM1 which binds to the neutrophil, did not inhibit fibrinogen binding. Consistent with the reciprocal inhibition of fibrinogen and HK binding to neutrophils, studies also showed that the binding of ^{125}I-HK to neutrophils was partially inhibited by FNDP but not by MABs 10E5 or IMY8 (Table I). In addition, the MAB OKM1 which completely inhibited fibrinogen binding did not inhibit HK binding.

Competition kinetic binding experiments were performed to determine the mechanism by which fibrinogen inhibits ^{125}I-HK binding to neutrophils. Binding of ^{125}I-HK to neutrophils was determined in the absence or presence of increasing concentrations of fibrinogen (data not shown). When analyzed by the method of Scatchard (45), the graph of ^{125}I-HK binding to neutrophils showed parallel slopes indicating no change in Kd. Increasing the fibrinogen concentration decreased the maximum number of sites for binding ^{125}I-HK. This result indicated that fibrinogen is a noncompetitive inhibitor of ^{125}I-HK binding to the neutrophil surface with an apparent Ki of 50 nM.

DISCUSSION

Since the amount of ^{125}I-HK that binds to neutrophils is equivalent at 0°C, 23°C, and 37°C remains constant over a 50 min incubation period (Fig. 1), and is completely reversible (Fig. 4), HK associated with neutrophil does not result from internalization of plasma HK. Demonstration that ^{125}I-HK binds to the surface of neutrophils indicates a means whereby HK from either plasma or neutrophils might be placed in a position such that it could influence the interaction of kallikrein with neutrophils. The specificity of binding of ^{125}I-HK to neutrophils is shown since excess unlabeled HK completely inhibits binding of ^{125}I-HK to neutrophils (Fig. 3) and only unlabeled HK, but not factor V, factor XII, PK or FNDPs is able to prevent the binding of ^{125}I-HK to the surface of the neutrophil.

Binding of ^{125}I-HK to neutrophils requires the presence of Zn^{++}. Ca^{++} alone is not able to support binding of ^{125}I-HK to neutrophils, nor does it potentiate binding when Zn^{++} is present (Fig. 2). The same findings were observed with Mg^{++}. This observation as well as the fact that Zn^{++} has not been reported to be essential for binding of other proteins to the platelet, endothelial cell or neutrophil surface supports the postulate that the effect of Zn^{++} is on the HK molecule itself. Human and rat HK or HK light chains (26) but not their heavy chains, specifically bind to Zn^{++} affinity columns. The observation has also been made that 50 μM Zn^{++} induces a conformational, change of bovine HK and its light chain as measured by scanning absorption sectrophotometry (27).

Binding of ^{125}I-HK to neutrophils is greater than 90% reversible at 8 and 18 min (Fig. 4). The binding is characterized by one high affinity, saturable site with an apparent Kd of 18.0 $\pm$ 1.5 nM by computer fitting of the experimental points from three experiments (Fig. 5). The estimated Ki computed from the midpoints of the competition inhibition curves from three experiments (Fig. 3) (18) is found to be similar at 10.4 $\pm$ 4.0 nM.

The functional significance of binding of HK to neutrophils is not yet fully delineated. Kallikrein has been shown to result in neutrophil chemotaxis (3), aggregation (4) and degranulation (5). Since prekallikrein circulates in plasma complexed with HK (28, 29), the ability of kallikrein to interact with neutrophils may result from the binding of the enzyme to the neutrophils surface in the presence of HK. The inability of β-kallikrein, which contains an additional cleavage in the heavy chain, to cause neutrophil activation (30) and the impaired cleavage of HK by β-kallikrein are consistent with this possibility since the heavy chain of kallikrein contains the binding region for HK (31).

HK from either plasma or neutrophils appears to be a cofactor for HNE secretion from neutrophils (Fig. 6). The finding that HK-deficient neutrophils suspended in total kininogen-deficient plasma had a reduced level of HNE secretion at 30 sec and 30 min indicates a specific function of neutrophil HK in the mechanism of HNE secretion. Our previous studies (8) show that plasma kallikrein plays a role in the secretion of HNE. Since the release of HNE by purified kallikrein from HK-deficient neutrophils is only 15% compared to normal neutrophils, we suggest that HK may serve as an acquired receptor for plasma kallikrein.

This study also extends observations of the interaction of fibrinogen and HK on artificial surfaces to biological surfaces and demonstrates that fibrinogen binds to neutrophils in a specific, reversible (Fig. 9) and saturable manner (Fig. 10). In addition, fibrinogen and HK are able to act as reciprocal inhibitors on the surface of neutrophils (Fig. 7, 9). The binding of fibrinogen to platelets requires the presence of Ca^{++} as this divalent cation is necessary for the association of GPIIb and IIIa, which when in complex, function as the fibrinogen receptor on the platelet surface (32). The amount of ^{125}I-fibrinogen bound to the neutrophils is increased if physiologic concentrations of Zn^{++} are present in addition to the Ca^{++} (Fig. 8). An increase of fibrinogen binding sites on the surface of neutrophils by Zn^{++} may result from an effect on fibrinogen (33) or from the direct action of this cation on the cell surface receptors. In support of the former explanation, it is known that fibrinogen will bind to a zinc affinity column (52). However, Zn^{++} at the concentrations used in this study did not cause fibrinogen aggreagates. Therefore, we favor the latter explanation. The mechanism of the Zn^{++} action on the cell surface is unknown; it has been recently reported that Zn^{++} stabilizes platelet cytoskeleton by preventing proteolysis of structural elements (35).

The sequence RGD which is a recognition site for certain integrins (36) is present in the fibrinogen molecule at two separate sites (37) and appears to be important in cell binding (38). A peptide from the C terminal of the γ-chain is also important (24). ^{125}I-fibrinogen binding to neutrophils, however, is not inhibited by the MABs 10E5 nor by the adhesive peptide RGDS (Table I). Moreover, RGD is also present in fibronection but FNDP which are known to stimulate neutrophils (39) and recognize macrophages (37, 40), only weakly inhibits fibrinogen binding to neutrophils (Table I). In contrast, the septadecapeptide derived from the fibrinogen γ-chain inhibited fibrinogen binding to neutrophils at concentrations of 250 μM and above. This finding suggests that the γ-chain plays a role in the interaction with the CR_3 receptor on the neutrophil surface similar to its interaction with the integrin GPIIb on the platelet surface.

We tested two MABs (OKM1 and OKM10) directed toward different epitopes (41) on the α chain of CR_3 for their effect on ^{125}I-fibrinogen binding to neutrophils. The heterodimer complex recognized by OKM1 and OKM10 is one of a family of human leukocyte differentiation antigens with distinct α subunits and a common ß subunit which include the lymphocyte-function associated antigen (LAF-1), complement receptor type (CR_3 or Mac-1), and the P150,95 molecule. Recent epitope mapping studies have shown that CR_3 is a multivalent receptor with at least two independent adhesion related functions; one is identified by OKM1 and the other by OKM10 (41). OKM1 has a functional domain which is involved with such neutrophil functions as aggregation (41), spreading on plastic surfaces (42) and chemotaxis (41). Most recently, Weitz et al (43) studied the adhesion of neutrophils to fibrinogen coated surfaces. In agreement with our data, this reaction was not affected by MABs directed against GPIIb/IIIa complex or by RGD derived peptides and was inhibited by the C-terminal peptide of the fibrinogen γ-chain.

Our results agree with those of Altieri et al (44) who demonstrated that neither OKM10 nor RGD-containing peptide inhibited binding of fibrinogen to monocytes. In addition, Altieri et al have previously shown that binding of fibrinogen to monocytes is inhibited by OKM1 (45). These leukocyte differentiation antigens are part of the integrin receptor superfamily (46, 47, 48). We found that there are 140,000 sites for fibrinogen on the neutrophil in agreement with Springer et al (49) who found 140,000 binding sites on Mac-1 consistent with our hypothesis that complement receptor type three (C11b/CD18) of human neutrophils functions as a fibrinogen receptor through a RGD independent mechanism. The data by Altieri et al (44) suggest that the same mechanism operates during the interaction of monocytes with fibrinogen. Since the inhibition of ^{125}I-HK binding to neutrophils by fibrinogen is noncompetitive, fibrinogen and HK probably do not share the same receptor on the neutrophil surface. This interpretation is reinforced by the results obtained from the experiments with various MABs. The MAB, OKM1 inhibits binding of fibrinogen to the neutrophil surface but does not inhibit binding of HK (Table I). The inhibitory effect of HK on the binding of fibrinogen to neutrophils may result from steric hindrance since both HK and fibrinogen are large asymmetric proteins. It is likely that the fibrinogen and HK binding sites, while distinct, are located in close proximity on the neutrophil membrane.

Although the functional significance of binding of ^{125}I-fibrinogen to neutrophils is not completely elucidated, our demonstration that binding of fibrinogen is inhibited by a MAB to CR_3 suggests that fibrinogen may play a role in such neutrophil functions such as aggregation (45), spreading on surfaces, adhesion and chemotaxis.

ACKNOWLEDGMENTS

This work was supported in part by grants HL24365 and HL19055 from the National Institutes of Health. Additional support was supplied by grants from the Ben Franklin Partnership of the Commonwealth of Pennsylvania, the American Lung Association of Pennsylvania, and a grant from the Cystic Fibrosis Foundation. We thank Cathy Spiotta for typing the manuscripe.

Figs. 1-6 were reprinted from the *Journal of Clinical Investigation*. Figs. 7-10 and Table I were reprinted from the *Journal of Cellular Biology*. These items were used with the permission of the Rockefeller University Press.

REFERENCES

1. M. Silverberg, J. E. Nicoll, and A. P. Kaplan. The mechanism by which the light chain of cleaved HMW-kininogen augments the activation of prekallikrein, factor XI and Hageman factor. *Thromb. Res.* **20**:173-189, (1980).
2. R. E. Thompson, Jr. R. Mandle, and A. P. Kaplan. Studies of binding of prekallikrein and factor XI to high molecular weight kininogen and its light chain. *Proc. Natl. Acad. Sci. USA* **76**:4862-4866, (1979).
3. A. P. Kaplan, A. B. Kay, and K. F. Austen. A prealbumin activator of prekallikrein. III. Appearance of chemotactic activity for human neutrophils. J. Exp. Med. **135**:81-97, (1972).
4. M. Schapira, E. Despland, C. F. Scott, L. A. Boxer, and R. W. Colman. Purified human plasma kallikrein aggregates human blood neutrophils. *J. Clin. Invest.* **69**:1199-1201, (1972).
5. Y. T. Wachtfogel, U. Kucich, H. L. James, C. F. Scott, M. Schapira, M. Zimmerman, A. B. Cohen, and R. W. Colman. Human plasma kallikrein releases neutrophil elastase during blood coagulation. *J. Clin. Invest.* **72**:1672-1677, (1983).
6. L. Vroman, A. L. Adams, G. C. Fischer, and P. C. Munoz. Interaction of high molecular weight kininogen, factor XII and fibrinogen in plasma at interfaces. *Blood* **55**:156-159, (1980).
7. A. H. Schmaier, L. Silver, A. L. Adams, G. C. Fischer, P. C. Mounoz, L. Vroman, and R. W. Colman. The effect of high molecular weight kininogen on surface-adsorbed fibrinogen. *Thromb. Res.* **33**:51-67, (1984).
8. A. H. Schmaier, A. Zuckerberg, C. Silverman, J. Kuchibhotla, G. P. Tuszynski, and R. W. Colman. High molecular weight kininogen. A secreted platelet protein. *J. Clin. Invest.* **71**:1477-1489, (1983).
9. D. M. Kerbiriou-Nabias, F. O. Garcia, and M.-J. Larrieu. Radioimmunoassays of human high and low molecular weight kininogens in plasma and platelets. *Brit. J. Haem.* **56**:273-286, (1984).
10. P. J. Fraker and Jr. S. C. Speck. Protein and cell membrane iodinations with a sparingly soluble chloroamide 1,3,4,6-tetrachloro-3 alpha, 6 alpha-diphenylglycoluril. *Biochem. Biophys. Res. Commun.* **80**:849-857, (1978).
11. E. J. Gustafson, D. Schutsky, L. C. Knight, and A. H. Schmaier. High molecular weight kininogen binds to unstimulated platelets. *J. Clin. Invest.* **78**:310-318, (1986).
12. R. J. Baugh and J. Travis. Human leukocyte granule elastase: rapid isolation and characterization. *Biochemistry* **15**:836-841, (1976).
13. B. S. Coller, E. I. Peerschke, L. E. Scudder, and C. A. Sullivan. A murine monoclonal antibody that completely blocks the binding of fibrinogen to platelets produces a thrombasthenic-like state in normal platelets and binds to glycoproteins IIb and/or IIIa. *J. Clin. Invest.* **72**:325-338, (1983).
14. R. F. Todd III, J. A. Roach, and M. A. Arnaout. The modulated expression of MO5, a human myelomonocytic plasma membrane antigen. *Blood* **65**:964-973, (1985).
15. R. W. Colman, A. Bagdasarian, R. C. Talamo, C. F. Scott, M. Sevy, J. A. Guimares, J. V. Pierce, and A. P. Kaplan. Williams trait. Human kininogen deficiency with diminished levels of plasminogen proactivator and prekallikrein associated with abnormalities of the Hageman-dependent pathways. J. Clin. Invest. **56**:1650-1662, (1975).
16. M. J. Castillo, K. Nakajima, M. Zimmerman, and J. C. Powers. Sensitive substrates for human leukocyte and porcine pancreatic elastase. A study of the merits of various

chromophoric and fluorogenic leaving groups in assays for serine proteases. *Anal. Biochem.* **99**:53-64, (1979).
17. Y. T. Wachtfogel, R. A. Pixley, U. Kucich, W. Abrams, G. Weinbaum, M. Schapira, and R. W. Colman. Purified plasma factor XIIa aggregates human neutrophils and causes degranulation. *Blood* **67**:1731-1737, (1986).
18. R. Mueller. Determination of affinity and specificity of anti-Hapten antibodies by competitive radioimmunoassay. *Methods Enzymol.* **92**:589-601, (1983).
19. A. J. Schmaier, D. Schutsky, A. Farber, L. D. Silver, H. N. Bradford, and R. W. Colman. Determination of the bifunctional properties of high molecular weight kininogen by studies with monoclonal antibodies directed to each of its chains. *J. Biol. Chem.*, **262**:1405-1411, (1987).
20. P. F. Canellas and A. E. Karu. Statistical package for analysis of competitive ELISA results. *J. Immunol. Methods* **47**:375-385, (1981).
21. P. J. Munson and D. Rodbard. LIGAND: A versatile computerized approach for characterization of ligand-binding systems. *Anal. Biochem.* **107**:220-239, (1980).
22. L. F. Brass and S. J. Shattil. Changes in surface-bound and exchangable calcium during platelet activation. *J. Biol. Chem.* **257**:14000-14005, (1982).
23. R. Pytela, M. D. Pierschbacher, M. H. Ginsberg, E. F. Plow, and E. Ruoslahti. Platelet membrane glycoprotein IIb/IIIa: member of a family of Arg-Gly-Asp-Asp-specific adhesion receptors. *Science* **231**:1559-1562, (1986).
24. M. Kloczewiak, S. Timmons, T. J. Lukas, and J. Hawiger. Platelet receptor recognition site of human fibrinogen synthesis and structure-function relationship of peptides corresponding to the carboxyterminal segment of gamma chain. *Biochemistry* **23**:1767-1774, (1984).
25. R. M. Senior, W. F. Skogen, G. L. Griffin, and G. D. Wilner. Effects of fibrinogen derivatives upon the inflammatory response. *J. Clin. Invest.* **77**:1014-1019, (1986).
26. I. Hayashi, H. Kato, S. Iwanaga, and S. Oh-ishi. Rat plasma high molecular weight kininogen. *J. Biol. Chem.* **260**:6115-6123, (1985).
27. T. Shimada, H. Kato, and S. Iwanaga. Effect of metal ions on the surface mediated activation of factor XII and prekallikrein. *In*: Kinins '84. L. Greenberg, ed. Conference at Savannah, GA (abstract), (1984).
28. C. F. Scott, L. D. Silver, M. Schapiro, and R. W. Colman. Cleavage of human high molecular weight kininogen markedly enhances its coagulant activity. Evidence that this molecule exists as a procofactor. *J. Clin. Invest.* **73**:954-962, (1984).
29. Jr. R. Mandle, R. W. Colman, and A. P. Kaplan. Identification of prekallikrein and high molecular weight kininogen as a complex in human plasma. *Proc. Natl. Acad. Sci. USA* **73**:4179-4183, (1976).
30. R. W. Colman, Y. Wachtfogel, U. Kucich, G. Weinbaum, S. Hahn, R. A. Pixley, C. F. Scott, de A. Agostini, D. Burger, and M. Schapira. Effect of cleavage of the heavy chain of human plasma kallikrein on its functional properties. *Blood* **65**:311-318, (1985).
31. F. Van der Graaf, G. Tans, B. N. Bouma, and J. H. Griffin. Isolation and functional properties of the heavy and light chains of human plasma kallikrein. *J. Biol. Chem.* **257**:14300-14305, (1982).
32. G. A. Marguerie, E. F. Plow, and T. S. Edgington. Human platelets possess an inducible and saturable receptor specific for fibrinogen. *J. Biol. Chem.* **254**:5357-5363, (1979).
33. J. N. Lindon, G. McManama, L. Kushner, E. W. Merrill, and E. W. Salzman. Does the confirmation of adsorbed fibrinogen dictate platelet interactions with artificial surfaces? *Blood* **68**:355-362, (1986).
34. M. F. Scully and V. V. Kakkar. Structural features of fibrinogen associated with binding to chelated c zinc. *Biochem. Biophys. Acta* **700**:130-133, (1982).
35. C. R. Zobel. The platelet cytoskeleton: evidence for its structure from interactions with $ZnCl_2$. *J. Submicrosc. Cytol. Pathol.* **20**:268-275, (1988).
36. E. Ruoslahti and M. D. Pierschbacher. New perspectives in cell adhesion: RGD and integrins. *Science* **238**:491-497, (1987).
37. R. W. Watt, B. A. Cottrell, D. D. Strong, and R. F. Doolittle. Amino acid sequence studies on the α chain of human fibrinogen. Overlapping sequences providing the complete sequence. *Biochemistry* **18**:5410-5422, (1979).
38. T. K. Gartner and J. S. Bennett. The tetrapeptide analogue of the cell attachment site of fibronectin inhibits platelet agga and fibrinogen binding to activated platelet. *J. Biol. Chem.* **260**:11891-11896, (1985).

39. Y. T. Wachtfogel, W. Abrams, U. Kucich, G. Weinbaum, M. Schapira, and R. W. Colman. Fibronectin degradation products containing the cytoadhesive tetrapeptide stimulate human neutrophil degranulation. *J. Clin. Invest.* **81**:1310-1316, (1988).
40. J. I. Weitz, S. L. Landman, K. A. Crowly, S. Birken, and F. J. Morgan. Development of an assay for *in vivo* human neutrophil elastase activity. *J. Clin. Invest.* **78**:155-162, (1986).
41. N. Dana, B. Styrt, J. Griffin, R. F. Todd III, M. Klempher, and M. A. Arnaout. Two functional domains in the phagocyte membrane. Glycoprotein MoI identified by monoclonal antibodies. *J. Immunology* **137**:3259-3263, (1986).
42. M. A. Arnaout, N. Dana, J. Pitt, and R. F. Todd III. Deficiency of two human leukocyte surface membrane glycoproteins (MoI and LFA-1). Fed. Proc. **44**:2664-2670, (1985).
44. D. C. Altieri, T. Edgington, and P. M. Mannucci. Oligospecificity of cellular adhesion receptor Mac-1 encompasses an inducable recognition specificity for fibrinogen. *J. Cell Biol.* **107**:1893-1900, (1988).
45. D. C. Altieri, R. Bader, and P. M. Mannucci. Structural diversity among cellular adhesion receptors: fibrinogen binding is a novel biological property of the monocyte differentiation antigen OKM1. *Thromb. Hemost.* **58**:1052, (1987).
46. R. O. Hynes. Integrins: A family of cell surface receptors. *Cell* **48**:549-554, (1987).
47. E. F. Plow, J. C. Loftus, E. G. Levin, D. S. Lou, D. Dixon, J. Forsyth, and M. H. Ginsberg. Immunologic relationship between platelet membrane glycoprotein IIb/IIIa and cell surface molecules expressed by a variety of cells. *Proc. Natl. Acad. Sci. USA* **83**:6002-6006, (1986).
48. F. Sanchez-Madrid, J. A. Nagy, E. Robbins, P. Simm, and T. A. Springer. A human leukocyte differentiation antigen family with distinct alpha subunits and a common beta subunit: the lymphocyte-function associated antigen (LFA-1), the C3bi complement receptor (OKM1/Mac-1), and the p150,95 molecule. *J. Exp. Med.* **158**:1785-1803, (1983).
49. T. A. Springer, L. J. Miller, and D. C. Anderson. p150,95,the third member of the Mac-1, LFA-1 human leukocyte adhesion glycoprotein family. *J. Immunol.* **136**: 240, (1986).

A NEW TRISACCHARIDE SUGAR CHAIN LINKED TO A SERINE RESIDUE IN THE FIRST EGF-LIKE DOMAIN OF CLOTTING FACTORS VII AND IX AND PROTEIN Z

Sadaaki Iwanaga, Hitoshi Nishimura, Shun-ichiro Kawabata, Walter Kisiel,* Sumihiro Hase,** and Tokuji Ikenaka**

Department of Biology, Faculty of Science
Kyushu University 33, Fukuoka 812, Japan
*Blood System Foundation Laboratory
Department of Pathology, University of New Mexico
School of Medicine, Albuquerque, NM 87131, USA
**Department of Chemistry, Osaka University College of Science
Osaka 565, Japan

ABSTRACT

Recently, we determined the complete amino acid sequence of bovine factor VII (Takeya, H. et al. (1988) J. Biol. Chem. *263*, 14868-14877). In the course of the studies, we found an unknown serine derivative at position 52 in the first epidermal growth factor-like domain of factor VII. A pentapeptide isolated from the *S*-aminoethylated factor VII contained Ser-52, which could not be identified with a gas-phase sequencer. The same results were also obtained for a pentapeptide containing Ser-53 of factor IX and protein Z. Component sugar analysis revealed that the peptide contained 1 mol of glucose and 2 mol of xylose. This sugar component was also confirmed by high-resolution fast atom bombardment mass spectrometric analysis of the pentapeptide. The trisaccharide was released from the peptides by means of ß-elimination reaction and its reducing end was identified as pyridylamino-glucose. These results indicate the existence of a (Xyl_2)Glc-Ser structure in factors VII, IX and protein Z. Similar results were obtained for human factors VII, IX and protein Z. This is the first report of a (Xyl_2)-Glc-Ser structure in glycoproteins to our knowledge. The presence of the unique trisaccharide structure in factors VII, IX and protein Z leads us to anticipate its biological role in the tissue factor pathway.

INTRODUCTION

The serine protease zymogen factor VII is a vitamin K-dependent plasma protein that plays a key role in the initiation of the extrinsic pathway of blood clotting (1). Factor VII has no coagulant activity in the absence of tissue factor, phospholipids, and calcium ions. However, in the presence of these cofactors, factor VIIa rapidly activates both factors X and IX. Fig. 1 shows the entire amino acid sequence of bovine factor VII, which was recently established in our laboratory (2). Bovine factor VII consists of 407 residues and has 71% sequence identity with the human molecule (406 residues) based on the cDNA sequence (3). Factor VII is cleaved to a light chain and a heavy chain by factors Xa and XIIa, and both chains are located in the NH_2-terminal and COOH-terminal portions of the zymogen and held together by a disulfide linkage. The light chain consists of 152 residues with one carbohydrate chain at Asn^{145}, and 11 γ-carboxyglutamic acid residues are found within the

Fibrinogen, Thrombosis, Coagulation, and Fibrinolysis, Edited by
C. Y. Liu and S. Chien, Plenum Press, New York

NH_2-terminal 35 residues. The light chain contains 0.2 to 0.3 mol of ß-hydroxyaspartic acid/mol of protein, indicating that an aspartic acid residue in bovine factor VII is incompletely hydroxylated. On the other hand, the heavy chain is composed of 255 residues and one Asn-linked carbohydrate chain at position 203, and constitutes a serine protease domain.

In the course of these studies, we discovered an unknown serine derivative at position 52 in the first epidermal growth factor (EGF)-like domain of bovine factor VII (Fig. 1). When we compared the sequence of the first EGF-like domain in factor VII with other vitamin K-dependent plasma proteins, the sequence prior to Ser-52 is highly conserved in factor IX and protein Z, and these factors contain a serine residue at the same position as that of factor VII, as shown in Fig. 2. On the chemical analysis, we also detected the same derivative at Ser-53 in bovine and human factors IX, protein Z and it was identified to be a new trisaccharide sugar chain (4).

In the present study, we would like to focus on the chemical structure of this carbohydrate chain isolated from human clotting factors.

RESULTS

Isolation and sequence analysis of sugar containing peptides

Our strategy to identify the carbohydrate moiety is as follows (4, 5). First of all, the isolation of a pentapeptide from their *S*-aminoethylated proteins by lysylendopeptidase

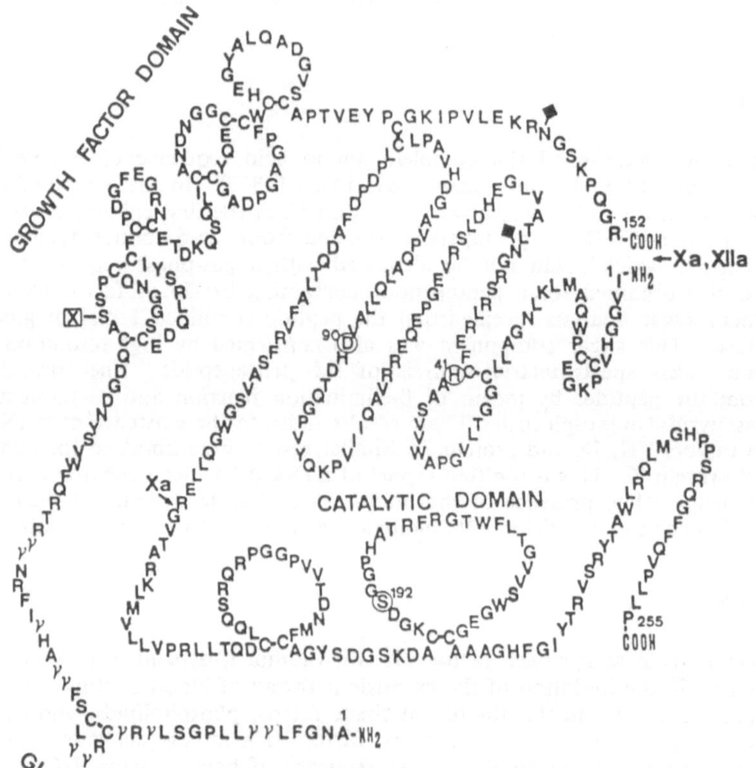

Fig. 1. Whole amino acid sequence of bovine factor VII (2). The site of cleavage in factor VII by factors Xa and XIIa is shown by an arrow. The carbohydrate binding sites are indicated by triangles and X(Ser-52). The amino acid residues which form an active site triads are circled.

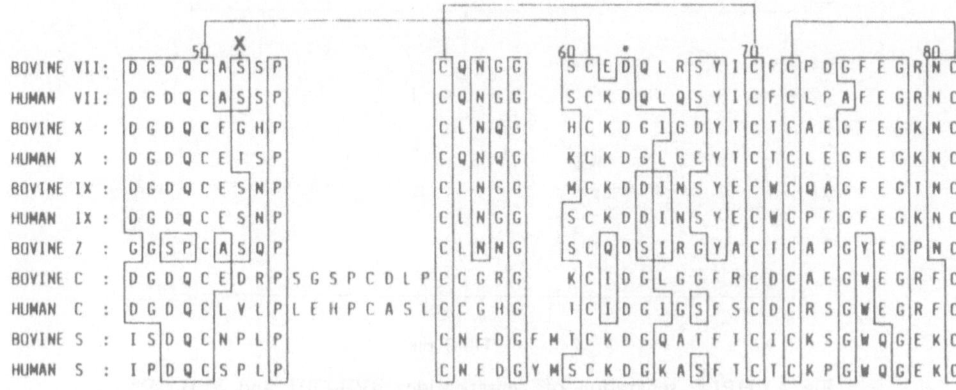

Fig. 2. Comparison of the first epidermal growth factor (EGF)-like domain of bovine factor VII with the corresponding regions in the vitamin K-dependent clotting factors. The numbers refer to bovine factor VII. Identical residues shared by at least five proteins are boxed. Star indicates the position of a putative ß-hydroxyaspartic acid residue in bovine factor VII.

digestions, secondly, the amino acid sequence analyses of the isolated pentapeptides using a gas-phase sequencer, thirdly, the component sugar analyses of these pentapeptides by the fluorescent pyridylamino derivative method, and finally, the confirmation of the deduced structures by FAB mass spectrometry.

One of the examples of lysylendopeptidase peptide mapping for the *S*-aminoethylated human factor IX preparation is shown in Fig. 3. Based on the amino acid analysis data, a peak eluted at 47.6 min appeared to be a pentapeptide, corresponding to residues 52-56 in the first EGF-like domain. This peptide, designated hIX-GP, was further purified by gel filtration on a column of Bio-Beads SM-2, and the purified peptides were subjected to further analyses. The same procedures as used for the isolation of the human factor IX pentapeptide were also employed to isolate the corresponding pentapeptides from the lysylendopeptidase digests of human factor VII, human protein Z, and bovine protein Z.

In the case of human factor VII, the digest yielded two pentapeptides, designated hVII-GP1 and hVII-GP2, as shown in Fig. 4, suggesting microheterogeneity of the carbohydrate moiety. Similar results to that of the factor VII-digest were obtained with the human protein Z digest (data not shown), and these pentapeptides were designated hZ-GP1 and hZ-GP2. In contrast, the bovine protein Z digest yielded a single pentapeptide.

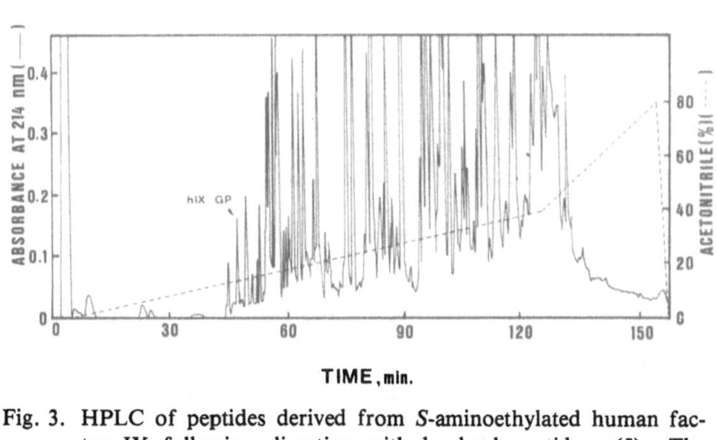

Fig. 3. HPLC of peptides derived from *S*-aminoethylated human factor IX following digestion with lysylendopeptidase (5). The lysylendopeptidase digest was subjected to HPLC on a 0.46 × 10 cm Cosmosil $3C_{18}$ column at a flow rate of 0.5 ml/min.

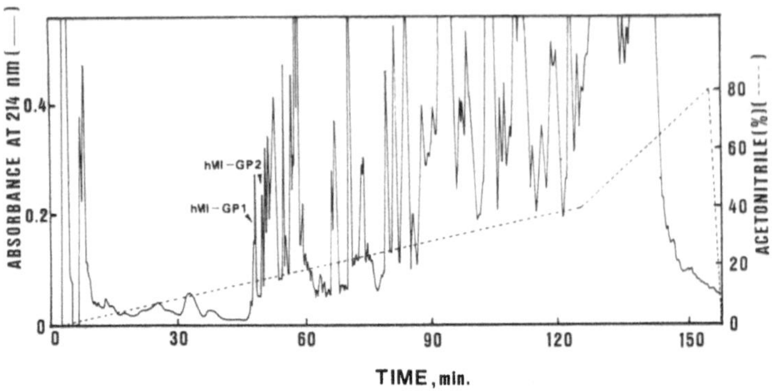

Fig. 4. HPLC separation of pentapeptides hVII-GP1 and hVII-GP2 from the lysylendopeptidase digest of *S*-aminoethylated human factor VII (5). The digest was applied to a Cosmosil $3C_{18}$ column (0.46 × 10 cm) at a flow rate of 0.5 ml/min. Peptides were eluted at room temperature with a linear gradient of acetonitrile in 0.1% (v/v) trifluoroacetic acid.

Table 1. Amino Acid Compositions of Pentapeptides Derived from Human Factors IX, VII and Protein Z and Bovine Protein Z

Amino acid	hIX-GP	hVII-GP1[c]	hVII-GP2	hZ-GP1[c]	bZ-GP
	Residues/molecule				
Asp	1.1 (1)[a]	0.7		0.5	0.4
Glu	1.0 (1)	0.4		2.0 (1)	1.4 (1)
Ser	0.9 (1)	2.2 (2)	2.2 (2)	1.3 (1)	1.3 (1)
Gly		0.6	0.4	0.7	0.5
Ala		1.0 (1)	1.0 (1)		1.0 (1)
Pro	1.0 (1)	0.8 (1)	1.1 (1)	1.3 (1)	1.0 (1)
Tyr					
Val				0.4	
Aec[b]	0.7 (1)	0.9 (1)	1.3 (1)	1.5 (1)	0.9 (1)
Ile				1.0 (1)	
Lys		0.8			
Yield (%)	3.4	9.1	6.1	16.2	14.9

[a]Values in parentheses are taken from the sequence data.
[b]*S*-aminoethylcysteine.
[c]The pentapeptides, hVII-GP1 and hZ-GP1, contained a trace amount of contaminant peptides which were not separable on HPLC used. The amino acid analysis of hZ-GP2 was not performed, since its yield was low.

The amino acid compositions of the isolated pentapeptides are shown in Table 1. Based on their compositions, hZ-GP1, hVII-GP1 and hVII-GP2 corresponded to residues 52-56 and 51-55 in the first EGF-like domains of factors IX and VII, respectively. Furthermore, the amino acid composition of the isolated bZ-GP was identical to that calculated for residues 52-56 located in the first EGF-like domain of bovine protein Z. Pentapeptides hZ-GP1 and hZ-GP2 also gave a similar composition to that of bZ-GP.

Table 2 shows the amino acid sequence analyses on the isolated pentapeptides. As shown in the table, Ser-53 (hIX-GP), Ser-52 (hVII-GP1 and hVII-GP2) and Ser-53 (bZ-GP)

Table 2. NH_2-Terminal Sequence of Pentapeptides Derived from Human Factors IX, VII and Protein Z and Bovine Protein Z

	hIX-GP		hVII-GP1		hVII-GP2		hZ-GP1		bZ-GP	
Cycle No.	PTH amino acid	Yield (pmol)	PTH amino acid	Yield (pmol)	PTH amino acid	Yield (pmol)	PTH amino acid	Yield (pmol)	PTH amino acid	Yield (pmol)
1	Glu	290	Ala	250	Ala	560	Ile	310	Ala	610
2	—		—		—		—		—	
3	Asn	270	Ser	58	Ser	310	Gln	260	Gln	340
4	Pro	290	Pro	100	Pro	390	Pro	280	Pro	340
5	Aec	ND	Aec	ND	Aec	ND	Aec	ND	Aec	ND
Position	52-56		51-55		51-55				52-56	
Amount of peptide used (pmol)	530		500		650		510		610	

—: not identified. ND: not quantitatively determined. The sequences of these pentapeptides, hIX-GP, hVII-GP1, hVII-GP2 and bZ-GP, except for hZ-GP1 and hZ-GP2, agreed well with those predicted from their cDNA sequences of human factors VII, IX and bovine protein Z. The amino acid sequence of human protein Z has not yet been determined.

could not be identified. These serine residues were, however, detected by amino acid analyses after acid hydrolysis. Moreover, the sequence analyses of pentapeptides hZ-GP1 and hZ-GP2 yielded the same sequence of Ile-X-Gln-Pro-Aec demonstrating homology with bZ-GP. Again, no PTH-Ser at the second cycles of these peptides was detected, suggesting the existence of a serine derivative.

Component sugar analysis of isolated pentapeptides

For sugar analyses, the pentapeptides thus isolated were initially hydrolyzed in 4 M trifluoroacetic acid at 100°C for 3 h, and then the hydrolysates were pyridylaminated, and subsequently analyzed by HPLC. One of the chromatograms obtained from the hydrolysate of hIX-GP is shown in Fig. 5. A standard sugar mixture containing Gal. Glc, Man, Xyl, Fuc, Rha (internal standard) and other sugars were also hydrolyzed, pyridylaminated, and chromatographed on the same column as a control sample. The result is shown in panel A. In panel B, the PA-sugar components released from hIX-GP were identified as Xyl and Glc, and quantified by comparing the ratios of their peak heights to that of PA-Rha. The same method was also employed for component sugar analyses of the isolated pentapeptides, hVII-GP1, hVII-GP2, hZ-GP1, hZ-GP2 and bZ-GP, and the results are summarized in Table 3.

Although these pentapeptides contained Xyl and Glc in common, the molar ratios of these sugars in GP1 and GP2 were different. Pentapeptides hVII-GP1 and hZ-GP1 were found to contain Xyl:Glc at a molar ratio of 1 to 1, whereas hVII-GP2 and hZ-GP2 contained Xyl:Glc at a molar ratio of 2 to 1, indicating the existence of either an *O*-linked disaccharide or trisaccharide chain in these pentapeptides. In contrast, the molar ratio of Xyl:Glc in bZ-GP was 2 to 1, similar to that observed for bovine factors VII and IX (4).

We also performed component sugar analyses on intact human factor VII, human factor IX, human protein Z and bovine protein Z, using the method of fluorescent PA-derivatives. The results are also shown in Table 3. While the presence of Xyl and Glc was confirmed using the intact proteins, their molar ratios were far from the expected integral values with the exception of bovine protein Z, which was in good agreement with the molar ratio determined for the pentapeptide, bZ-GP.

Moreover, no microheterogeneity was so far observed for the sugar chains corresponding to the pentapeptides derived from bovine factors VII and IX, in addition to bovine protein

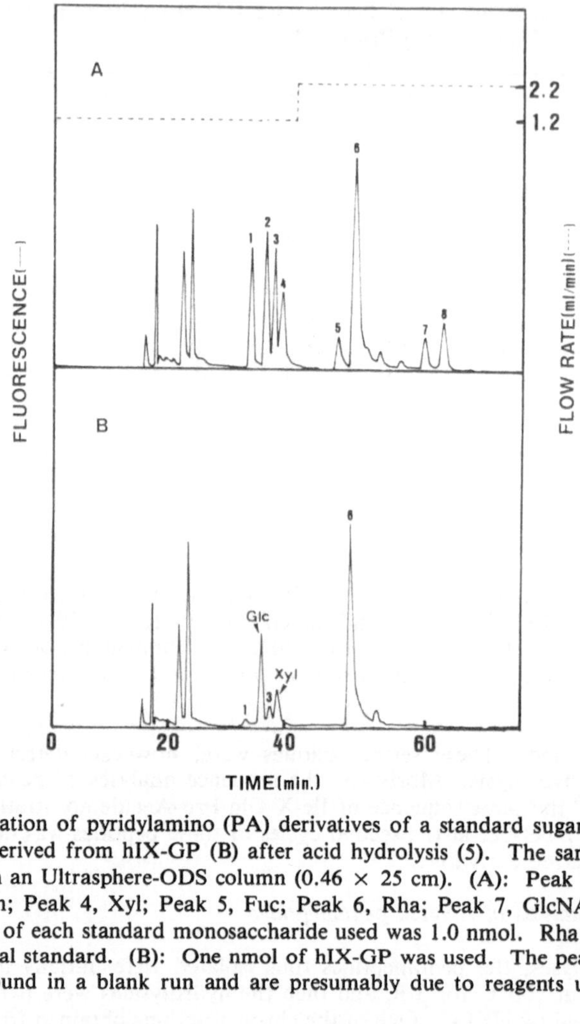

Fig. 5. HPLC-separation of pyridylamino (PA) derivatives of a standard sugar mixture (A) and of PA-sugars derived from hIX-GP (B) after acid hydrolysis (5). The samples were subjected to HPLC on an Ultrasphere-ODS column (0.46 × 25 cm). (A): Peak 1, Gal; Peak 2, Glc; Peak 3, Man; Peak 4, Xyl; Peak 5, Fuc; Peak 6, Rha; Peak 7, GlcNAc; Peak 8, GalNac. The amount of each standard monosaccharide used was 1.0 nmol. Rha (4.5 nmol) was used as the internal standard. (B): One nmol of hIX-GP was used. The peaks without numbers were also found in a blank run and are presumably due to reagents used in the pyridylamination.

Table 3. Glc and Xyl Contents (in Average) of Pentapeptides Derived from Clotting Factors and Protein Z and of These Whole Proteins

	Human factor IX		Human factor VII			Human protein Z			Bovine protein Z	
	hIX-GP	Whole[a]	hVII-GP1	hVII-GP2	Whole	hZ-GP1	hZ-GP2	Whole	bZ-GP	Whole
	n[b] = 3	n = 3	n = 2	n = 2	n = 2	n = 2	n = 1	n = 3	n = 3	n = 2
	Residues/molecule									
Glc	1.0	1.1	1.3	1.2	0.9	0.8	1.1	1.0	1.0	0.9
Xyl	1.1	1.6	1.2	2.4	1.2	0.9	2.0	1.4	1.9	2.0
Xyl/Glc (mol/mol)	1.1	1.5	0.9	2.0	1.3	1.1	1.8	1.4	1.9	2.2

[a]Whole: Whole protein. [b]n: The number of component sugar analyses.

Z presented here. As yet, a pentapeptide linked with a disaccharide chain has not been found in those vitamin K-dependent proteins of bovine origin. These results may be suggestive that a Xyl residue in the trisaccharide chain is cleaved by some endo/exoglycosidase in circulating human blood or during secretion and transport of these proteins in human hepatocytes.

FAB mass spectrometric analysis of pentapeptides

To determine the molecular weight of the glycopeptides, these were submitted to FAB mass measurement. Fig. 6 shows a low-resolution mass spectrum of hIX-GP, giving a molecular ion $(M + H)^+$ as a single peak at m/z = 886.2. The accurate mass 295.101 of the residue linked to Ser-53 of hIX-GP, which was obtained by subtraction of theoretical mass value of the peptide portion from the observed mass 886.333, in a high-resolution mass spectrum, as shown in the inset of this figure, agreed well with that calculated from the elemental composition corresponding to the sum of 1 mole each of pentose (Xyl) and hexose (Glc). This result supports the chemical analysis.

In the case of hVII-GP1 and hVII-GP2, their molecular ions were observed at 801.1 and at 933.2, indicating the existence of di- and trisaccharide moieties with the same sugar as that of hIX-GP (Fig. 7). FAB mass spectra of the glycopeptides hZ-GP1, hZ-GP2 and bZ-GP provided the similar results. Therefore, the amino acid and sugar compositions of these glycopeptides obtained by chemical analyses were confirmed again by FAB mass spectrometry.

Identification of Glc linked to serine residue

To obtain further information of the sugar residue which links to Ser-53 in hIX-GP, the disaccharide was released from the glycopeptide by hydrazinolysis followed by *N*-acetylation. The resulting disaccharide was pyridylaminated and a PA-disaccharide was purified on a TSK-GEL HW-40F column. The total yield of the PA-disaccharide was 15%. The PA-disaccharide fraction was further purified by reversed-phase HPLC on a Cosmosil $5C_{18}$-P column. As shown in Fig. 8, only PA-Glc was found after acid hydrolysis of the PA-disaccharide, and no PA-Xyl peak was observed by reversed-phase HPLC on an Ultrasphere-ODS column. This result clearly indicates that the Glc residue is linked to Ser-53 in human factor IX, which is in agreement with that obtained previously for bovine factor IX (4).

DISCUSSION

Fig. 9 shows the chemical structure of new trisaccharide sugar chain *O*-glycosidically linked to Ser-53 of bovine factor IX. For structural analysis, the methylation analysis, the periodide oxidation, and ^{1}H-NMR spectroscopy were performed to elucidate the glycosidic

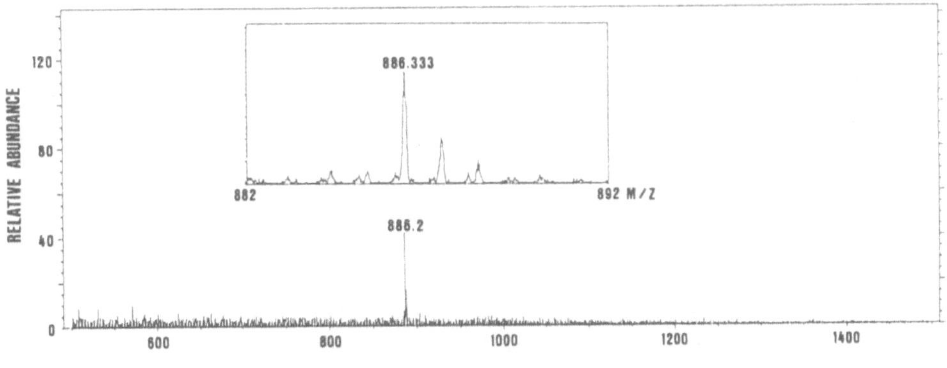

Fig. 6. Positive low and high (inset)-resolution FAB mass spectra of hIX-GP (5). The accurate mass (886.333) was the mean value of the masses obtained from seven scan data.

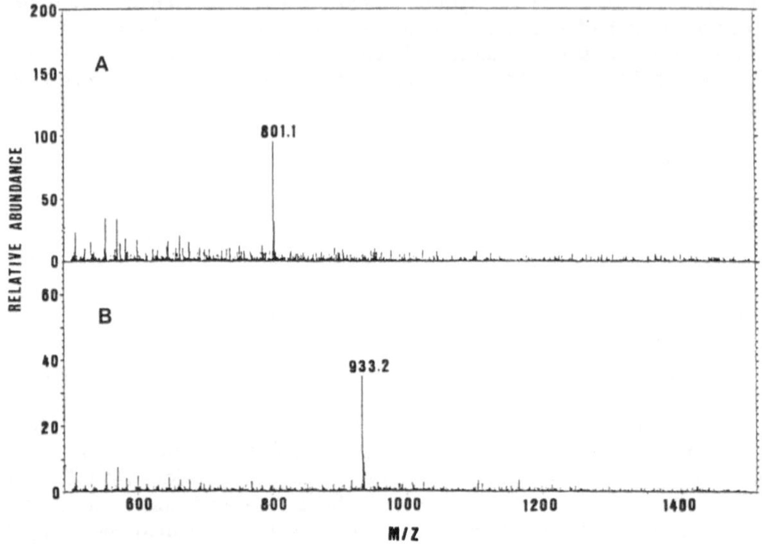

Fig. 7. Positive FAB mass spectrum of hVII-GP1 (A) and hVII-GP2 (B).

linkage and anomeric configuration of the trisaccharide. D-Glucose oxidase was also used to determine the optical isomers. Although the detailed data are not shown here, these results indicate that it consists of Xyl(pyranose) α1-3Xylα1-3-D-Glc(pyranose) β1-O-Ser (6).

Fig. 10 shows an alignment of regions corresponding to the lysylendopeptidase-released pentapeptides from the first EGF-like domains of several vitamin K-dependent proteins and human thrombospondin. All the amino acid sequences surrounding Ser-52 and Ser-53 are highly conserved in these proteins, and factors VII, IX and protein Z from both human

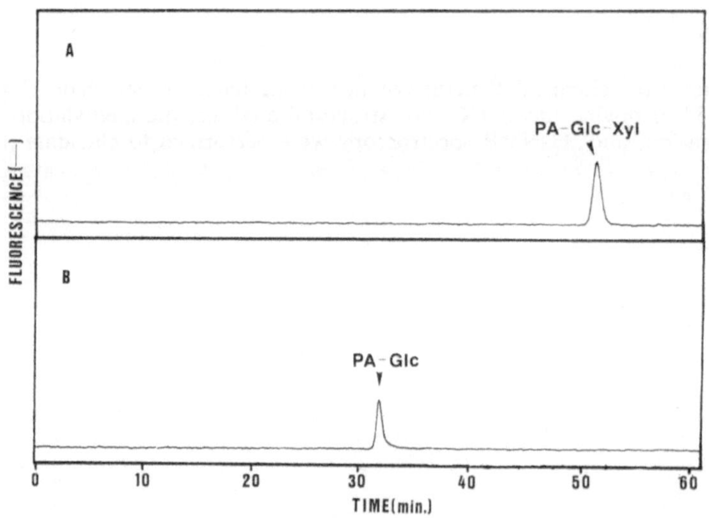

Fig. 8. HPLC of PA-sugar derived from hIX-GP before and after acid hydrolysis (5). (A): HPLC of the PA-disaccharide on an Ultrasphere-ODS column under the conditions reported previously (4). (B): HPLC of an acid hydrolysate of the PA-disaccharide under the same conditions as (A). The elution position of this single peak corresponded to that of standard PA-Glc.

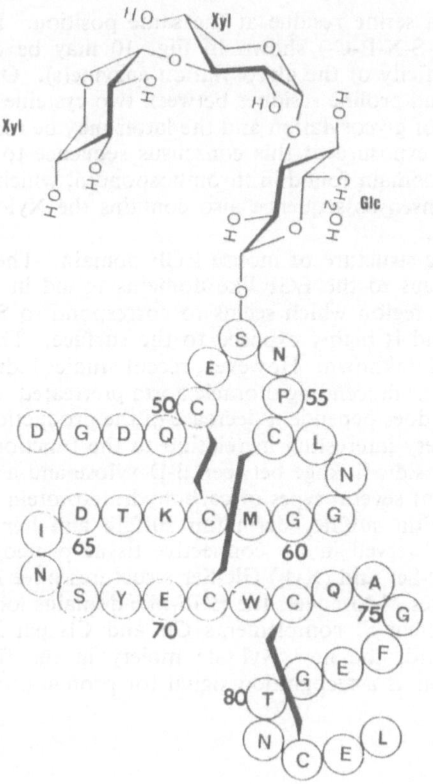

Fig. 9. Structure of new trisaccharide found in the first EGF-like domain of bovine factor IX (6). The formula is Xylpα1-3Xylpα1-3-D-Glcpß1-O-Ser-53.

				X			
Bovine Factor VII	(50-55):	C	A	S	S	P	C
Human Factor VII	(50-55):	C	A	S	S	P	C
Bovine Factor IX	(51-56):	C	E	S	N	P	C
Human Factor IX	(51-56):	C	E	S	N	P	C
Bovine Protein Z	(51-56):	C	A	S	Q	P	C
Human Protein Z	:	C	I	S	Q	P	C
Human Thrombospondin	(533-538):	C	L	S	N	P	C

Fig. 10. An alignment of amino acid sequences from EGF-like domains with *O*-linked sugar chain of factors VII, IX, protein Z and human thrombospondin. Identical residues shared by all proteins are boxed and shadowed. X indicates *O*-linked sugar chain moieties in factors VII, IX and protein Z. The NH_2-terminal cysteine residue of human protein Z is a tentative assignment, as the sequence of human protein Z is highly homologous to that of bovine protein Z.

and bovine sources have a serine residue at the same position. Based on these data, the consensus sequence (-C-X-S-X-P-C-) shown in Fig. 10 may be considered as a minimal representation of the specificity of the glycosylation enzyme(s). Of these four residues, the importance of the serine and proline residues between two cysteines is emphasized, since the former is the putative site of glycosylation and the latter may be required to form the ß-turn structure, resulting in the exposure of this consensus sequence to the surface. Thus, it is possible that an EGF-like domain found in thrombospondin, which has a sequence (residues 533-538) similar to the consensus sequence also contains the Xyl-Glc carbohydrate moiety at Ser-535.

Fig. 11 is a schematic structure of mouse EGF domain. The recent NMR studies on mouse EGF (7), homologous to the EGF-like domains found in the vitamin K-dependent proteins, indicate that the region which seems to correspond to Ser-52 of factor VII contains a ß-turn structure and is highly exposed to the surface. The functional role of this carbohydrate moiety is still unknown. However, recent studies indicate that pig aortic endothelial cells, isolated from the descending thoracic aorta pretreated with 4-methylumbelliferyl-ß-D-xyloside, experience a does-dependent decrease in the production of cell surface heparan sulfate (8). This fact is very interesting in relation to the function of such *O*-linked sugar chains. So far, an *O*-glycosidic linkage between ß-D-xylose and a hydroxyl group of serine has been identified as one of several types of carbohydrate-protein linkages in proteoglycans such as heparin, chondroitin sulfate, dermatan sulfate and heparan sulfate. Thus, the Xyl-Ser linkage is often observed in the connective tissue proteoglycan group. However, to our knowledge, Xyl-Glc-Ser and (Xyl_2)-Glc-Ser structures have not as yet been identified in any other glycoconjugates. Moreover, the EGF-like domains located in the NH_2-terminal portion of protein C, protein S, complements Clr and Cls participate in protein-protein interactions. In this regard, the new Xyl-Glc moiety in the first EGF-like domain in these proteins may function as a recognition signal for protein-protein or protein-cell interactions.

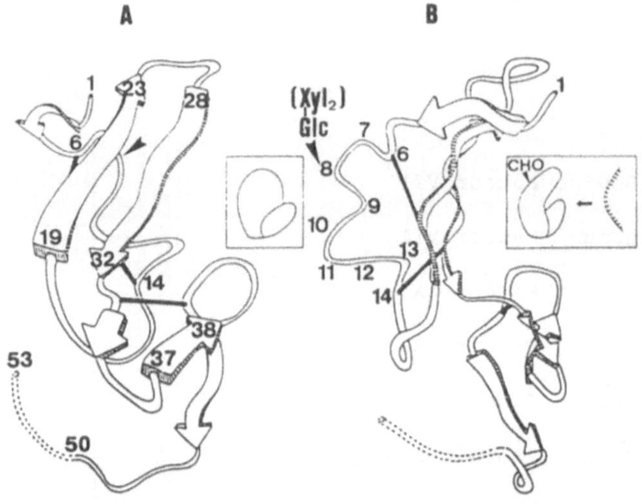

Fig. 11. Schematic drawings of the backborn topology of mouse EGF (7). (A) Corresponding to the view shown Montelione et al. (9). (B) A different view from (A), being rotated by 90° around the vertical axis on the paper. Since the sequence homology between mouse EGF and the EGF-like domain found in bovine factor VII is approximately 40%, the sugar-linked Ser-52 in factor VII seems to be located in a *β*-turn structure corresponding to Ser-8 in mouse EGF. Based on these data, the sugar moiety (Xyl-Xyl-Glc) is highly exposed to the surface. Inset shows a mitten model which represents the overall structure of mouse EGF (left) and its receptor binding (right).

CONCLUSION

A new trisaccharide sugar chain was first identified in bovine factors VII and IX. A pentapeptide isolated from factor VII contained Ser-52, which could not be identified with a gas phase sequencer. The same results were obtained for each pentapeptide containing Ser-52 of human factor VII, and Ser-53 of human factor IX and bovine protein Z. The component sugar analysis revealed that all the bovine peptides contained 1 mol of Glc and 2 mol of Xyl. In contrast, the human peptides from factor VII and protein Z contained Glc and Xyl at molar ratios of 1:1 and 1:2, suggesting microheterogeneity of these *O*-linked sugar chains. Based on the hydrazinolysis, methylation, periodide oxidation methods, in addition to ^{3}H-NMR spectroscopic analysis, the chemical structure of the trisaccharide is proposed as follows:

Xylα1-3Xylα1-3Glcß1-O-Ser-53 (factor IX)

The consensus sequence, C-X-S-X-P-C, found in the first EGF-like domain of factors VII, IX and protein Z may be considered as a minimal representation of the specificity of the glycosylation enzyme(s).

ACKNOWLEDGMENTS

This work was supported by a Grant-in-aid for Scientific Research from the Ministry of Education, Japan, and by research grants from the National Institutes of Health (HL35246) and Blood Systems, Inc. (to W. K.). We wish to express our thanks to Drs T. Takao and Y. Shimonishi in Institute for Protein Research, Osaka University, for FAB mass spectrometric analysis. We also thank S. Kajiyama and Y. Nishina for performing amino acid analysis and N. Ueno for her expert secretarial assistance.

FOOTNOTE

A part of this work has been published in *J. Biol. Chem.*, **264,** 20320-30325, 1989, and *J. Biol. Chem.*, **265,** 1858-1861, 1990.

REFERENCES

1. C. M. Jackson and Y. Nemerson, *Annu. Rev. Biochem.*, **49:**765 (1980).
2. H. Takeya, S. Kawabata, K. Nakagawa, Y. Yamamichi, T. Miyata, S. Iwanaga, T. Takao, and Y. Shimonishi, *J. Biol. Chem.*, **263:**14868 (1988).
3. F. S. Hagen, C. L. Gray, P. O'Hara, F. J. Grant, G. C. Saari, R. G. Woodbury, C. E. Hart, M. Insley, W. Kisiel, K. Kurachi, and E. Q. Davie, *Proc. Natl. Acad. Sci. U.S.A.*, **83:**2412 (1986).
4. S. Hase, S. Kawabata, H. Nishimura, H. Takeya, T. Sueyoshi, T. Miyata, S. Iwanaga, T. Takao, Y. Shimonishi, and T. Ikenaka, *J. Biochem.*, (Tokyo) **104:**867 (1988).
5. H. Nishimura, S. Kawabata, W. Kisiel, S. Hase, T. Ikenaka, T. Takao, Y. Shimonishi, and S. Iwanaga, *J. Biol. Chem.*, **264:**20320 (1989).
6. S. Hase, H. Nishimura, S. Kawabata, S. Iwanaga, and T. Ikenaka, *J. Biol. Chem.*, **265:**1858 (1990).
7. D. Kohda, N. Go, K. Hayashi, and F. Inagaki, *J. Biochem.* (Tokyo), **103:**741 (1988).
8. K. Shimada and T. Ozawa, *Arteriosclerosis* **7:**627 (1987).
9. G. T. Montelione, K. Wüthrich, E. C. Nice, A. W. Burgess, and H. A. Scherage, *Proc. Natl. Acad. Sci. U.S.A.*, **84:**5226 (1987).

MULTIPLE EPITOPE SPECIFICITY OF MONOCLONAL ANTIBODIES TO A SINGLE SYNTHETIC PEPTIDE: USE IN THE CHARACTERIZATION OF THE GP IIb-IIIa BINDING DOMAIN OF VON WILLEBRAND FACTOR

Shlomo A. Berliner, Richard A. Houghten, James R. Roberts, and Zaverio M. Ruggeri*

Division of Experimental Hemostasis and Thrombosis
Roon Research Center for Arteriosclerosis and Thrombosis
Department of Molecular and Experimental Medicine
Committee on Vascular Biology and Department of Molecular Biology
Research Institute of Scripps Clinic, La Jolla, CA 92037, USA

ABSTRACT

Monoclonal antibodies have been induced against the synthetic peptide with sequence Tyr-Glu-Val-Val-Thr-Gly-Ser-Pro-Arg-Gly-Asp-Ser-Gln-Ser-Ser. This peptide represents residues Glu^{1737}-Ser^{1750} of the mature von Willebrand factor (vWF) subunit and contains the sequence Arg-Gly-Asp, thought to be important in mediating binding to the platelet receptor glycoprotein (GP) IIb-IIIa complex. Twelve antibodies were obtained, eight of which bound to native vWF as well as to the peptide immunogen insolubilized onto agarose beads. These antibodies defined at least three distinct epitopes, as demonstrated by antibody interaction with peptides having a single phenylalanine substitution at each position in the sequence. In particular, two antibodies bound to epitopes on vWF that included one or more of the three residues (arginine, glycine, aspartic acid) thought to be involved in binding to GP IIb-IIIa, whereas one antibody bound to an epitope that did not include any of those residues. Nevertheless, the three antibodies cross-reacted with each other, a finding explained by the fact that the corresponding epitopes had at least two residues in common, namely Gly^{1741} and Ser^{1742}. In spite of the cross-reactivity for binding to vWF, only the two antibodies whose epitopes included residues in the Arg-Gly-Asp sequence inhibited vWF interaction with GP IIb-IIIa. The third antibody had no inhibitory effect even though it was bound to an epitope located at a distance of only few residues on the amino terminal side of Arg-Gly-Asp. These results demonstrate that monoclonal antibodies raised against a single synthetic peptide with sequence limited to fifteen residues may exhibit distinct epitope specificity and may be used to define functional domains in macromolecules with a high degree of resolution.

INTRODUCTION

Antibodies have been used extensively to study the structure and function of target antigens. Their usefulness has been greatly increased by the possibility of predetermining

*To whom reprint requests should be addressed at: Scripps Clinic and Research Foundation; BCR-8; 10666 N. Torry Pines Road; La Jolla, CA 92037, USA

their specificity with synthetic peptides reproducing immunogenic epitopes (1-5). Peptide-specific antibodies which react with cognate sequences in native macromolecules have been shown to be unique tools for studying the nature of the antigen-antibody interaction and for characterizing structural correlates of defined molecular activities (4-15). Utilizing this approach, we have obtained initial results in an ongoing effort aimed at characterizing the role of adhesive proteins, and in particular von Willebrand factor (vWF)[1], in platelet function (16).

Platelets, and cells in general, express the ability to adhere to substrata or to each other through the activity of specific membrane receptors interacting with adhesive molecules either in solution or bound to surfaces. Most adhesion receptors, like the platelet glycoprotein (GP) IIb-IIIa complex, have related structures and are presently grouped in the integrin superfamily (17). Functionally, they are thought to interact with the Arg-Gly-Asp (RGD with one letter notation) sequence present in adhesive molecules (18). In spite of the putative common recognition sequence, individual integrins express restricted ligand specificity, presumably based on unique three-dimensional conformations pertaining to the binding domain of each individual adhesive molecule. Platelet GP IIb-IIIa, however, interacts with four different ligands, namely vWF (19, 20), fibrinogen (21-23), fibronectin (24), and vitronectin (25). The molecular bases for this multiple binding specificity are presently poorly understood. Moreover, the respective role of the four adhesive proteins in platelet hemostatic functions and in the development of pathological thrombosis remains to be elucidated.

In order to address these problems, we have considered the possibility of generating antibodies specific for the putative GP IIb-IIIa binding domains of adhesive molecules using synthetic peptides reproducing the Arg-Gly-Asp containing regions of each ligand. These antibodies could provide information on the structural organization and possible homologies in the functional domains of the four proteins, and could be used to obtain selective inhibition of individual ligands and help clarifying the consequences on platelet function. In this report we present the results obtained to date with monoclonal anti-peptide antibodies directed at the GP IIb-IIIa binding domain of vWF. Antibodies were selected by immunization with a fifteen residue synthetic peptide representing residues 1737-1750 of the mature vWF subunit, including Arg^{1744}, Gly^{1745}, and Asp^{1746}. We have identified three antibodies defining distinct epitopes, one of which did not include any of the Arg-Gly-Asp residues constituting the putative receptor recognition sequence. The antibody reacting with this epitope failed to inhibit vWF binding to platelets, even though it was bound at a distance of only few amino acids from the active sequence. The present findings demonstrate the potential usefulness and high degree of specificity of peptide-specific monoclonal antibodies in the characterization of functional domains of macromolecules.

EXPERIMENTAL PROCEDURES

Synthesis and purification of synthetic peptides

The peptides used in these studies were based on the sequence Tyr-Glu-Val—Val-Thr-Gly-Ser-Pro-Arg-Gly-Asp-Ser-Gln-Ser-Ser, corresponding to residues Glu^{1737} to Ser^{1750} of the vWF subunit (26). The Tyr residue at the amino terminus was added to the native sequence of vWF for the purpose of labeling with ^{125}I. In addition to the native one, fifteen modified peptides were also prepared, each with a single phenylalanine substitution for one of the residues in the original sequence. All peptides were synthesized using the method of simultaneous multiple peptide synthesis which has been described in detail (5). After synthesis, they were analyzed by reversed phase HPLC (Perkin Elmer) using a 1×25 cm Vydac C-18 column (TP Silica; pore diameter 300 Angstrom) with a 0-60% acetonitrile linear gradient in 0.1% trifluoroacetic acid. The peptides were then purified using the best conditions suggested by the analytical chromatography. Amino acid compositions were determined with an automated amino acid analyzer (LKB) after 24-hour hydrolysis in 6 M HCl at 110°C.

[1]The abbreviations used are: vWF, von Willebrand factor; GP, membrane glycoprotein; HPLC, high performance liquid chromatography; ELISA, enzyme linked immunoassay; IC_{50}, concentration of a peptide necessary to inhibit 50% of antibody binding to vWF; K_a, affinity constant.

Preparation of monoclonal anti-peptide antibodies

The synthetic peptide representing the native sequence of vWF was used as immunogen after coupling to the carrier protein, keyhole limpet hemocyanin, as described (16). The amount of peptide coupled varied between 30-80% of the weight of the complex, and was calculated approximately by weighing the lyophilized complex and assuming 90% recovery of the carrier protein. Monoclonal anti-peptide antibodies were produced by immunizing BALB/c mice with intraperitoneal injections. The first one consisted of 0.8 mg of peptide-carrier complex in complete Freund's adjuvant. Two booster injections of the same amount of antigen in incomplete Freund's adjuvant were given at weekly intervals. A final booster was given four days before fusion. Mouse spleen cells were fused with mouse plasmacytoma cells of the line P3X63-Ag8.653 at a ratio of 6:1 using hybridoma technology that has been described in detail by Liu and coworkers (27). Culture supernatant from hybridomas were tested for reactivity against vWF using an enzyme-linked immunoassay (ELISA) (28) as described previously (16). Positive hybridomas detected in this manner were subcloned twice by limiting cell dilution according to Coller and Coller (29). Monoclonal immunoglobulins were then produced in mouse ascites fluid as described previously (27).

Purified immunoglobulin from serum or ascitic fluid was obtained by affinity chromatography on Protein A-Sepharose (Sigma) (30) or Affi Gel Blue (Bio Rad) (31) or by anion exchange chromatography. Protein concentration was calculated from the optical density of the purified solutions at 280 nm using an extinction coefficient of 1.4. All antibodies used in these studies were purified by passing them through a column of native peptide linked to cyanogen bromide-activated Sepharose 4 B (Sigma). Coupling was obtained with a peptide concentration of 5 mg/ml. The immunoglobulin molecules bound to the peptide were eluted with 0.1 M glycine at pH 2.4, then dialyzed against a buffer composed of 0.02 M HEPES and 0.15 NaCl adjusted to pH 7.35 with 1N HC1 (Hepes buffer), and concentrated before storage. All purified proteins were stored at -70°C. Radiolabeling of immunoglobulins with ^{125}I was performed with Iodogen (Pierce) as described (32).

Definition of the epitopes recognized by the anti-peptide antibodies

The approach used for defining the amino acid residues involved in antibody binding was that of measuring the ability of the native peptide and of analogous peptides containing single amino acid substitutions to block binding of the antibodies to native vWF, as described previously (16). Anti-peptide antibodies were used at a final concentration that gave an optical density reading of between 1.0 and 2.0 in the ELISA assay used when incubated with Hepes buffer instead of test peptides. Two sets of experiments were performed. In one, the concentration of each peptide used corresponded to the concentration of native peptide that inhibited binding of the antibodies to vWF by 75% or more. In the other, peptides were used in varying concentrations to determine the one that inhibited 50% of antibody binding to vWF (IC_{50}). All the dilutions of antibodies and peptides were prepared in Hepes buffer.

Cross-reactivity studies were performed using ^{125}I-labeled, affinity purified monoclonal immunoglobulins. Each labeled antibody was added to vWF coated wells in the presence of increasing concentrations of unlabeled immunoglobulin, either the same or the other two antibodies. Control wells contained Hepes buffer instead of unlabeled immunoglobulin. After incubation for one h at 22-25°C, the wells were washed, bound radioactivity was counted, and the corresponding values were used to calculate the amount of immunoglobulin associated with vWF in each well.

Measurement of platelet adhesion to vWF and of vWF binding to GP IIb-IIIa

Polystyrene plates with removable flat bottom wells (Immulon 1 Removawell, Dynatech Laboratories) were coated with purified vWF at the concentration of 5 μg/ml, as described previously for the ELISA assay (16). Platelets were isolated free of plasma constituents from blood collected in acid/citrate/dextrose anticoagulant. Platelet-rich plasma obtained from the blood was incubated with 1 μM prostaglandin E_1 (Sigma) for 10 min at 22-25°C. Platelets were pelleted by centrifugation at 700 g for 14 min at 22-25°C and then resuspended in a buffer composed of 0.02 M Tris, 0.15 M NaCl, pH 7.35 (Tris buffer). This step was repeated once, the platelets resuspended in 1 ml of Tris buffer at approximately 2×10^9 ml, and incubated with 40-60 μCi of ^{111}In oxine (Amersham) for 20 min at 37°C. Platelets were then pelleted as described above, the supernatant was discarded, and the platelets resuspended in a buffer composed of 0.02 M HEPES, 0.15 M NaCl, 0.005 M Dextrose, 1 mg/ml bovine serum albumin, pH 7.35 (adhesion buffer). Pelleting and resuspension in adhesion buffer were repeated once, and the platelets were finally resuspended at a count of 2×10^8 ml. The specific activity corresponded approximately to 0.002 cpm/platelet. Platelets were stimulated with 0.5 NIH units/ml of α-thrombin (a gift of Dr. John W. Fenton, II, Griffin Laboratories, New York State Department of Health, Albany, NY) for 10 min at 22-25°C. Hirudin (Sigma) was then added at a 25-fold excess on a U/U basis, followed by 1 mM final concentration each of $CaCl_2$ and $MgCl_2$. One hundred μl of the platelet suspension was added to each microtiter well and incubated for 30 min at 22-25°C. The wells were then washed three times with adhesion buffer to remove non adhering platelets, and bound radioactivity was measured in a γ scintillation spectrometer (Packard Instruments) and taken as an estimate of adhering platelets. In experiments where the effect of anti-vWF antibodies was tested, these were added to the wells coated with vWF for 1 h at 22-25°C, and the wells were then washed three times with Hepes buffer before addition of platelets. Concurrent measurement of ^{125}I-labeled immunoglobulin binding and ^{111}In-labeled platelet adhesion to vWF was performed by counting the ^{125}I-associated radioactivity three weeks after counting the ^{111}In-associated radioactivity, at a time when interference from the latter was negligible due to the short half-life of this isotope.

Binding of purified radiolabeled vWF to thrombin-stimulated platelets (therefore to the GP IIb-IIIa receptor) was tested as reported previously (19, 20). Purification and iodination of vWF with ^{125}I have already been described in detail (19, 20).

RESULTS

Epitope specificity of three monoclonal anti-peptide antibodies

The three monoclonal anti-vWF peptide antibodies described in the present study were selected among eight on the basis of their higher affinity for native vWF in the ELISA assay. Each of the three antibodies recognized a different epitope, as judged by the different effects of single amino acid substitutions on the ability of the modified peptides to react with the antibodies (Fig. 1). The epitope of one antibody, 152B-18, appeared to include three residues in linear sequence, while the epitopes of 152B-3 and 152B-20 appeared to be constituted by residues not in linear sequence (Fig. 1). Only the epitopes of 152B-3 and 152B-20 included residues thought to be important for vWF binding to GP IIb-IIIa (namely, GLy^{1745} and Asp^{1746}, but not Arg^{1744}); none of these residues was part of the epitope of 152B-18 (Fig. 1).

Confirmation that the anti-peptide antibodies tested recognized distinct epitopes was obtained by measuring the amount of each substituted peptide necessary to inhibit 50% of antibody binding to vWF (Fig. 2). Thus, the epitope of 152B-18 was formed mainly by Thr^{1740}, Gly^{1741}, and Ser^{1742}, with minor contribution by Val^{1738}, Val^{1739}, and Pro^{1743}; that of 152B-3 mainly by Gly^{1741} and Asp^{1746}, with contribution by all the residues between Glu^{1737} and Asp^{1746}; and that of 152B-20 included GLy^{1741}, Ser^{1742}, Gly^{1745} and Asp^{1746} (Fig. 2). In no case was any of the residues on the carboxyl terminal side of Asp^{1746} essential for antibody binding (Fig. 1 and 2).

Cross-reactivity of the anti-vWF peptide antibodies

The three epitopes recognized by the corresponding antibodies were distinct, but they had at least two residues in common, namely Gly^{1741} and Ser^{1742}. Thus, it was not unexpected that the three antibodies competed with each other for binding to vWF (Fig. 3). Binding of radiolabeled antibody 152B-3 was inhibited equally well by unlabeled 152B-3, 152B-18 and 142B-20. In contrast, relatively large amounts of 152B-3 and 152B-18 were necessary to inhibit binding of 152B-20. In particular, antibody 152B-3 was the least effective in inhibiting binding of the other two antibodies (Fig. 3). As shown below, this may be due, in part, to the fact that less antibody 152B-3 bound to vWF than any of the other two.

Antibody effect on the vWF-platelet interaction

The effect of the three antibodies on the interaction between vWF and GP IIb-IIIa was of considerable interest. Antibodies 152B-3 and 152B-20 both inhibited binding of

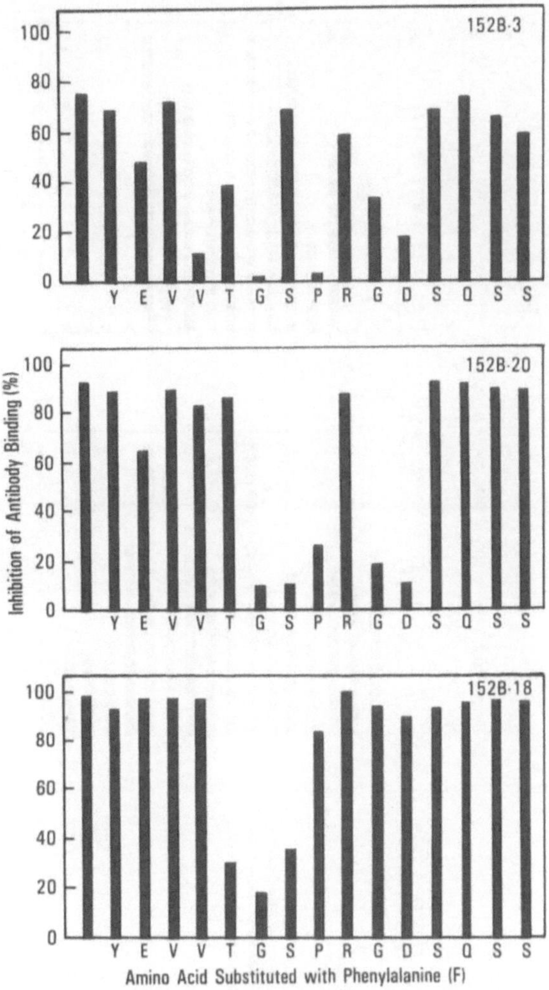

Fig. 1. Inhibition of peptide-specific antibody binding to vWF by analogous synthetic peptides with single amino acid substitutions. Polystyrene microtiter wells were coated with a solution of purified vWF at 5 μg/ml. The concentration of each anti-peptide antibody used (indicated with the corresponding designation in each panel) was such to give an optical density of approximately 1.0-1.5 in the ELISA assay (between 3-8 μg/ml of affinity purified immunoglobulin). In the control experiment (column to the left; unmarked) each antibody was incubated for 1 hour at 22-25°C with 100 μmoles/liter of unmodified peptide (the same used as immunogen for antibody production) before adding to the wells coated with vWF; this resulted in almost complete inhibition of antibody binding to vWF (decreased optical density) demonstrating that the antibody interacted with the peptide in solution. The inhibition of antibody binding was arbitrarily expressed as percentage (ordinate) by subtracting from 100 the percent value of optical density in the well containing competing peptide as compared to a well containing Hepes buffer instead of peptide. In all the other mixtures, each antibody was incubated with a modified synthetic peptide having a single substitution of phenylalanine (F) for one of the amino acids in the sequence, as indicated in the figure with the single letter abbreviation for the substituted residue. Lack of inhibition of antibody binding indicates those residues whose substitution with phenylalanine results in a peptide with markedly decreased affinity for a given antibody. Note that the three antibodies tested exhibit a different pattern of reactivity with the modified peptides.

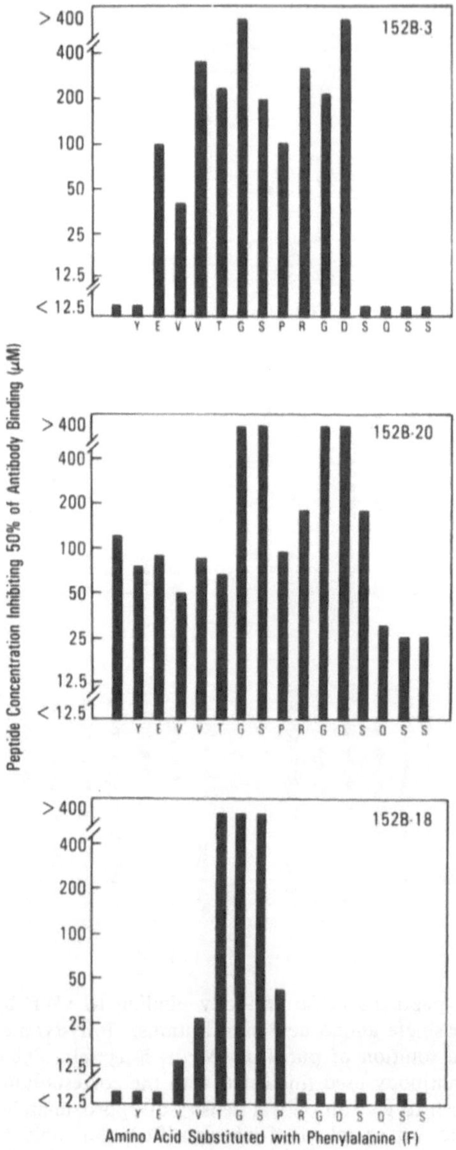

Fig. 2. Characterization of the interaction of modified peptides with the monoclonal anti-vWF peptide antibodies. These experiments were conducted in a manner similar to that described in Fig. 1 except that each peptide, either unmodified (left column, unmarked) or with a single phenylalanine substitution at the position indicated, was tested at decreasing concentrations (from 400 μM to 12.5 μM) in order to determine the amount necessary to inhibit antibody binding to vWF by 50% (IC_{50}). Inhibition of antibody binding was calculated as described in Fig. 1. Those substitutions resulting in peptides with decreased antibody binding capacity are evidenced by an increase in IC_{50}. The results depicted here are in excellent agreement with those of Fig. 1 in indicating amino acid residues important for vWF interaction with each antibody (identified in each panel with the corresponding designation).

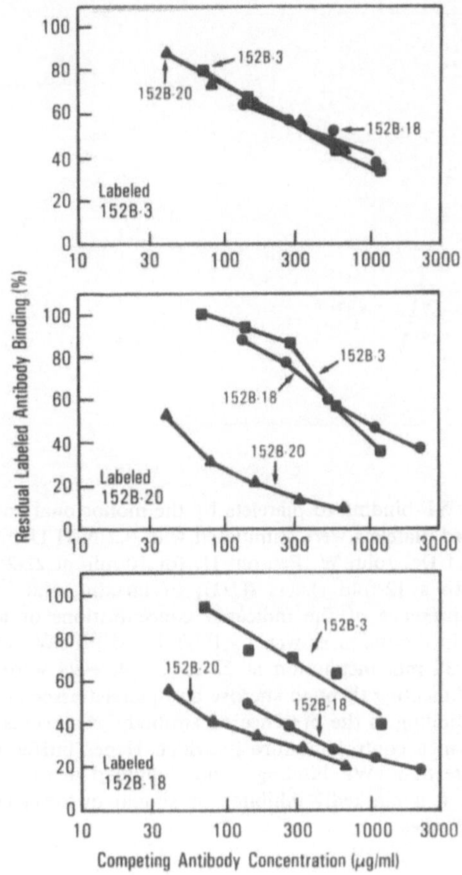

Fig. 3. Cross-inhibition studies with the three monoclonal anti-vWF peptide antibodies. Binding of ^{125}I-labeled monoclonal immunoglobulins (identified with the corresponding designation in each panel) to vWF insolubilized onto polystyrene microtiter wells was measured in the presence of increasing concentrations of each of the three monoclonal anti-vWF peptide immunoglobulins, as indicated for the corresponding line in the three panels. Note that in all cases binding of the labeled immunoglobulin is inhibited in a dose-dependent manner by all the unlabeled immunoglobulins tested, demonstrating that the three antibodies competed with each other for binding to vWF. The concentration of labeled immunoglobulin used corresponded approximately to that necessary to give half maximal binding to the vWF coated wells under the experimental conditions used; it was 20 μg/ml for antibody 152B-3 (top panel), 6 μg/ml for antibody 152B-20 (middle panel), and 30 μg/ml for antibody 152B-18 (lower panel).

soluble vWF to thrombin-stimulated platelets, i.e. to GP IIb-IIIa, but the latter was over 100-fold more effective on a concentration basis; in contrast, 152B-18 was without effect even at the highest concentration tested (Fig. 4). In order to demonstrate more directly the correlation between antibody binding to vWF and effect on the interaction with GP IIb-IIIa, adhesion of thrombin-stimulated platelets to surface-bound vWF was measured. In this manner, binding of ^{125}I-labeled antibody and adhesion of ^{111}In-labeled platelets to vWF could be measured concurrently (Fig. 5). Thus, increasing binding of antibodies 152B-3 and 152B-20 correlated with progressive decrease in platelet adhesion, whereas maximal binding of antibody 152B-18 did not perturb platelet adhesion. These experiments did, in fact, suggest that antibody 152B-18 remained bound to vWF at a time when this was interacting with platelet GP IIb-IIIa (Fig. 5). Similar amounts of antibodies 152B-18 and

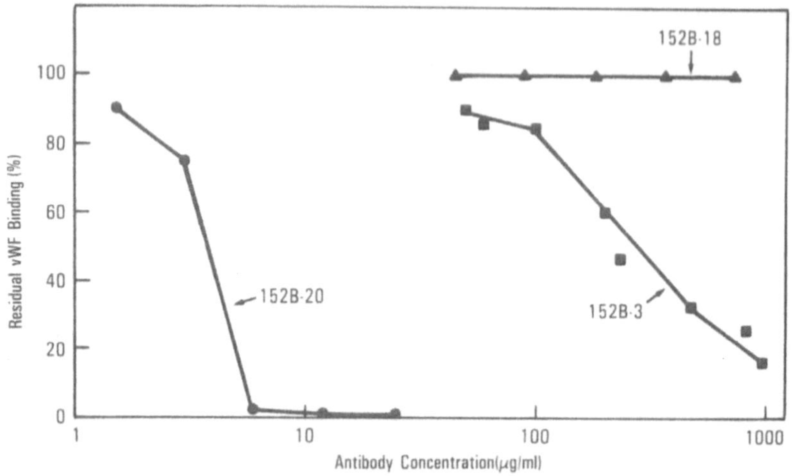

Fig. 4. Inhibition of vWF binding to platelets by the monoclonal anti-vWF peptide antibodies. Washed platelets were stimulated with 0.2 NIH U/ml of α-thrombin (the generous gift of Dr. John W. Fenton, II) for 10 min at 22-25°C. Thrombin was neutralized with a 12-fold excess (U/U) of hirudin, and ^{125}I-labeled vWF was added in the presence of the indicated concentrations of antibody. The final platelet count in the mixtures was 1×10^8/ml and ^{125}I-vWF concentration was 10 μg/ml. After 30 min incubation at 22-25°C, platelets were separated from the mixture by sedimenting through sucrose and platelet-associated radio-activity was determined. Binding in the presence of antibody was expressed as percentage of that observed in a control mixture in which Hepes buffer was used instead of antibody. Note that vWF binding is not inhibited in the presence of antibody 152B-18, while it is markedly inhibited by similar or lower concentrations of the other two antibodies.

152B-20 bound to vWF as a function of concentration, whereas binding of antibody 152B-3 was considerably lower (Fig. 5). This suggested the possibility that not all surface-bound vWF molecules expressed the three corresponding epitopes in equivalent manner.

In order to prove that antibody 152B-18 was actually binding to vWF molecules expressing GP IIb-IIIa binding sites, an experiment was performed to demonstrate that this antibody could compete with the inhibitory effect of antibody 152B-20. Increasing concentrations of antibody 152B-18 were added to six different inhibitory concentrations of antibody 152B-20, and the effects of these mixtures on platelet adhesion to vWF were recorded. In accordance with the fact that the two antibodies competed for binding to vWF (Fig. 3), platelet adherence progressively increased, to the point of exhibiting no inhibition, when increasing concentrations of antibody 152B-18 were present (Fig. 6). These findings suggested that antibody 152B-18 was binding to the same vWF molecules expressing the epitope for antibody 152B-20, and actually to an epitope overlapping partially with that recognized by the latter antibody, yet did not interfere with platelet adhesion.

DISCUSSION

The results presented in this report demonstrate the applicability of monoclonal anti-peptide antibodies for analyzing the function of cell binding domains in adhesive molecules containing the sequence Arg-Gly-Asp. In a previous publication (16) we have shown that antibodies reacting with a synthetic peptide whose sequence corresponded to residues 1737-1750 of the mature vWF subunit interfered specifically with vWF binding to platelet GP IIb-IIIa without reacting with other adhesive molecules known to bind to the same receptor. Now we demonstrate that the inhibitory effect of these antibodies is the consequence of their interacting with residues in the Arg-Gly-Asp sequence, and that only antibodies directly interfering with this putative receptor recognition site for integrins block vWF binding to

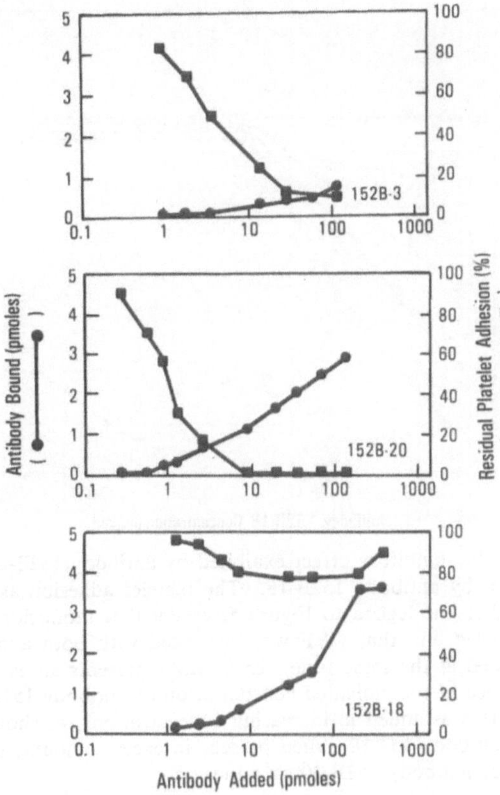

Fig. 5. Inhibition of platelet adhesion to vWF by the monoclonal anti-vWF peptide antibodies. Adhesion of thrombin-stimulated, ^{111}In-labeled washed platelets to vWF insolubilized onto a polystyrene surface was measured as described in Experimental Procedures, after incubating vWF with ^{125}I-labeled monoclonal immunoglobulin added to the wells at the concentrations indicated on the abscissa. Platelet adhesion was expressed as percentage of that observed in control wells in which vWF was incubated with Hepes buffer instead of antibody. Immunoglobulin binding, as indicated on the ordinate, was calculated from the specific activity of the labeled immunoglobulin after decay of greater than 90% of the ^{111}In activity associated with the adhering platelets. Note that platelet adhesion is not inhibited in the presence of antibody 152B-18 (lower panel), whereas it is completely inhibited by the other two antibodies. The amount of immunoglobulin bound was similar for antibody 152B-18 (lower panel) and for antibody 152B-20 (middle panel). Binding of antibody 152B-3 (upper panel) was lower than for the other two, but the inhibitory effect on platelet adhesion was, nevertheless, very pronounced. The affinity of interaction with vWF, K_a, calculated from these binding isotherms using computer assisted Scatchard-type analysis (35), was similar for all three antibodies. It was (liter/mole): 5.2×10^6 for 152B-3; 7.8×10^6 for 152B-20; and 6.5×10^6 for 152B-18. Thus, the difference in inhibitory effect could not be explained by differences in affinity for the native vWF protein.

GP IIb-IIIa. Thus, the present results provide one example in which the biological function of a domain in a complex protein cannot be impaired as a consequence of nonspecific steric hindrance by immunoglobulin molecules bound to a contiguous epitope, and reaffirm with immunochemical evidence the direct role of the Arg-Gly-Asp sequence as part of the sole GP IIb-IIIa binding site of vWF.

It is apparent from the results presented here that an immunoglobulin molecule (antibody 152B-18) bound to residues Thr^{1740}-Gly^{1741}-Ser^{1742} did not interfere with the cell binding

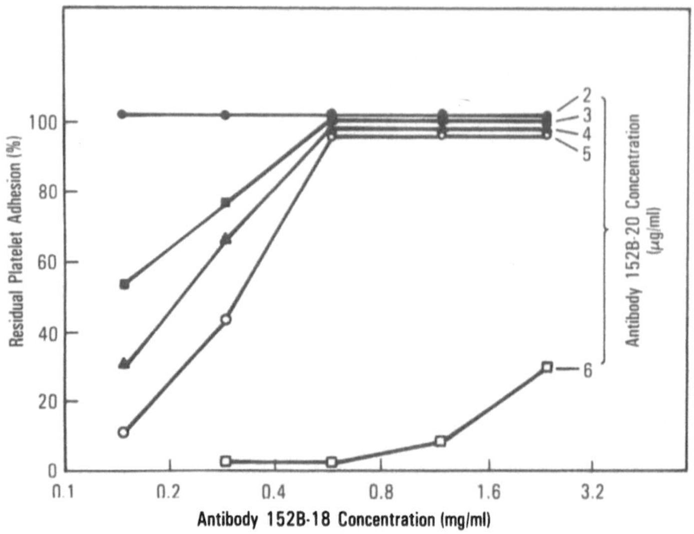

Fig. 6. Reversal of the inhibitory effect exhibited by antibody 152B-20 on platelet adhesion to vWF by antibody 152B-18. The platelet adhesion assay was performed as described in the legend to Figure 5, except that monoclonal immunoglobulin was not labeled and that vWF was incubated with both antibody 152B-20 and 152B-18 added at the same time. Each line represents an experiment performed in the presence of the indicated concentration of antibody 152B-20 to which antibody 152B-18 was added in increasing concentrations, as shown on the abscissa. Note that antibody 152-18, when present in excess amount, prevents the inhibitory effect of antibody 152B-20 on adhesion.

function of the nearby Arg-Gly-Asp sequence located at positions 1744-1746 in the mature vWF subunit. Since the cell binding site and the epitope recognized by this noninhibitory antibody appear to be in close spatial proximity, we considered it essential to obtain convincing evidence that the antibody was indeed bound to the same vWF molecules that interacted with platelet GP IIb-IIIa. The demonstration that binding of antibody 152B-18 and platelet adhesion to vWF occurred concurrently, as shown in Figure 5, suggests that this is likely to be the case. It is not, however, a conclusive finding since it could be explained, at least theoretically, by postulating the existence of two distinct states for surface-bound vWF molecules, one permitting binding of antibody 152B-18 but not expressing the GP IIb-IIIa binding site, and the other capable of interacting with GP IIb-IIIa but not expressing the epitope for antibody 152B-18. A more convincing demonstration that binding of antibody 152B-18 and interaction with GP IIb-IIIa can occur in the same vWF molecules rests on the finding that 152B-18 in excess amount prevented inhibition of platelet adhesion to vWF by 152B-20, as shown in Figure 6. This result can be explained on the basis of the cross reactivity between the two antibodies, which cannot bind to vWF at the same time since the corresponding epitopes share the two essential amino acid residues Gly^{1741} and Ser^{1742}. The fact that platelets could adhere to vWF when excess antibody 152B-18 prevented binding of antibody 152B-20 directly demonstrates that platelets interact with the Arg-Gly-Asp sequence in the same vWF molecules to which antibody 152B-18 is attached. Under the experimental conditions used, in the presence of inhibitory concentrations of antibody 152B-20, only binding of the noninhibitory immunoglobulin 152B-18 to a vWF molecule can prevent interaction with the other antibody and permit expression of the platelet binding function of the adjacent Arg-Gly-Asp sequence.

Some concepts of general interest can be deduced from the findings reported here and pertaining specifically to the interaction between vWF and GP IIb-IIIa. Monoclonal antibodies raised against a synthetic peptide composed of fifteen residues may differ significantly in epitope specificity, suggesting that their diversity depends on the recognition of different conformations in the peptide immunogen and that these can be partly reproduced in the native protein containing the cognate sequence. It is also apparent that monoclonal antibodies to a single peptide may exhibit different effects on molecular functions. Both these properties may be the consequence of the limited extension of the corresponding epitopes in the native molecule. For example, the three antibodies used for the present studies appear to recognize three to ten residues crucial for their interaction with the antigen. This is particularly evident for the noninhibitory antibody 152B-18, whose continuous epitope consists mainly of three residues and is thus considerably smaller than the size of the combining site of anti-protein antibodies. Crystallographic data of lysozyme-antibody complexes (33) suggest that, for antibodies with affinity constants between 10^4–10^8 liter/mol, the area representing the combining site is between 300-600 A^2 and involves 12-17 residues (33, 34). Since it has been calculated that no area of a protein surface constituting an epitope of this size can be continuous (34), our findings are in agreement with the contention that anti-peptide antibodies that cross react with cognate sequences in a protein recognize only parts of immunogenic epitopes present in the native molecule (34). The limited extension of the epitopes recognized by anti-peptide antibodies in native molecules may be an important useful characteristic of these reagents. For example, it is clear from the present findings that spatial proximity of a functional sequence and an immunologic epitope does not necessarily result in inhibition of function by steric hindrance when the antibody is bound to the antigen. Rather, the inhibitory effect of an antibody is likely to indicate that a functional site and an immunologic epitope are partially overlapping, thus providing a powerful tool for the identification of amino acid residues involved in specific activities. Our studies, therefore, lend further experimental support to the suitability of synthetic peptides and anti-peptide antibodies for evaluating the structure of functional domains in macromolecules (4-15).

The use of a single amino acid substitution with phenylalanine for identifying residues contributing to an antibody epitope represents a limit to the conclusions that may be reached on the structure of the corresponding combining site in the native protein, since the nature of the substituting amino acid may affect the results observed (4-5). The choice of phenylalanine was based on the rationale of introducing a significant change in the peptide structure with a nonconservative substitution, and our present findings clearly indicate which residues in the vWF peptide sequence cannot be substituted with phenylalanine without a drastic decrease in affinity for the corresponding antibody. It is obvious, however, that substitutions with other residues at the same positions might have resulted in peptides retaining reactivity or, alternatively, that substitutions with other residues in different positions might have resulted in peptides with less reactivity. This may affect interpretation of the role that single residues in the corresponding native molecule play in determining the properties of the antibody combining site. Nevertheless, when related antibodies with different inhibitory effects on a functional activity can be shown to recognize distinct epitopes, as in the case described here, the results may be taken to indicate which residues are likely to be involved in performing a molecular function even when based on a single amino acid substitution for each position in the peptide sequence.

In conclusion, the characterization of monoclonal antibodies directed against a single synthetic peptide has provided clear immunochemical and functional evidence of the varied epitope specificity that can be obtained with these reagents. Antibodies against a peptide as small as fifteen residues may express significant differences in their inhibitory effects on a given function in a macromolecule. Thus, evaluation of the epitope corresponding to each distinct antibody is essential for a correct interpretation of their functional effects. When appropriately characterized, these reagents represent a potentially useful tool for the definition of the structure of functional domains in macromolecules.

ACKNOWLEDGMENTS

This work was supported in part by Grants HL 31950 and HL 37522 from the National Institute of Health.

This is publication number 5725-MEM from the Research Institute of Scripps Clinic.

REFERENCES

1. J. L. Bittle, R. A. Houghten, A. Hannah, T. M. Shinnick, J. G. Sutcliffe, R. A. Lerner, D. J. Rowlands, and F. Brown, *Nature,* **298**:30-33 (1982).

2. H. M. Geysen, S. J. Barteling, and R. H. Meloen, *Proc. Natl. Acad. Sci. USA,* **82:**178-182 (1985).
3. J. Bouhnik, F. X. Galen, J. Menard, P. Corvol, R. Seyer, J. A. Fehrentz, D. L. Nguyen, P. Fulcrand, and B. Castro, *J. Biol. Chem.,* **262:**2913-2918 (1987).
4. H. M. Geysen, R. H. Meloen, and S. J. Barteling, *Proc. Natl. Acad. Sci. USA,* **81:**3998-4002 (1984).
5. R. A. Houghten, *Proc. Natl. Acad. Sci. USA,* **82:**5131-5135 (1985).
6. J.-M. Bidart, F. Troalen, C. J. Bohuon, G. Hennen, and D. H. Bellet, *J. Biol. Chem.,* **262:**15483-15489 (1987).
7. S. H. Socher, M. W. Riemen, D. Martinez, A. Friedman, J. Tai, J. C. Quintero, V. Garsky, and A. Oliff, *Proc. Natl. Acad. Sci. USA,* **84:**8829-8833 (1987).
8. F. Troalen, D. H. Bellet, P. Ghillani, A. Puisieux, C. J. Bohuon, and J.-M. Bidart, *J. Biol. Chem.,* **263:**10370-10376 (1988).
9. G. Noe, J. Hofsteenge, G. Rovelli, and S. R. Stone, *J. Bio. Chem.,* **263:**11729-11735 (1988).
10. R. S. Hodges, R. J. Heaton, J. M. Robert Parker, L. Molday, and R. S. Molday, *J. Biol. Chem.,* **263:**11768-11775 (1988).
11. J. C. Vera, C. I. Rivas, and R. B. Maccioni, *Proc. Natl. Acad. Sci. USA,* **85:**6763-6767 (1988).
12. L. K. Curtiss, and R. S. Smith, *J. Biol. Chem.,* **263:**13779-13785 (1988).
13. S. Bishayee, S. Majumdar, C. D. Scher, and S. Khan, *Molec. Cell Biol.,* **8:**3696-3702 (1988).
14. B. H. Devens, G. Semenuk, and D. R. Webb, *J. Immunol.,* **141:**3148-3155 (1988).
15. P. Ghillani, P. Motte, C. Bohuon, and D. Bellet, *J. Immunol.,* **141:**3156-3163 (1988).
16. S. Berliner, K. Niiya, J. R. Roberts, R. A. Houghten, and Z. M. Ruggeri, *J. Biol. Chem.,* **263:**7500-7505 (1988).
17. R. O. Hynes, *Cell,* **48:**549-554 (1987).
18. E. Rusolahti and M. D. Pierschbacher, *Science,* **238:**491-497 (1987).
19. Z. M. Ruggeri, R. Bader, and L. De Marco, *Proc. Natl. Acad. Sci. USA,* **79:**6038-6041 (1982).
20. Z. M. Ruggeri, L. De Marco, L. Gatti, R. Bader, and R. R. Montgomery, *J. Clin. Invest.,* **72:**1-12 (1983).
21. R. L. Nachman and L. L. K. Leung, *J. Clin. Invest.,* **69:**263-269 (1982).
22. J. S. Bennett, G. Vilaire, and D. B. Cines, *J. Biol. Chem.,* **257:**8049-8064 (1982).
23. G. O. Gogstad, F. Brosstad, M.-B. Krutnes, I. Hagen, and N. O. Solum, *Blood,* **60:**663-671 (1982).
24. E. F. Plow, R. P. McEver, B. S. Coller, V. L. Woods, G. A. Jr. Marguerie, and M. H. Ginsberg, *Blood,* **66:**724-727 (1985).
25. P. Thiagarajan, K. L. Kelly, *J. Biol. Chem.,* **263:**3035-3038 (1988).
26. K. Titani, S. Kumar, K. Takio, L. H. Ericsson, R. D. Wade, K. Ashida, K. A. Walsh, M. W. Chopek, J. E. Sadler, and K. Fujikawa, *Biochemistry,* **25:**3171-3184 (1986).
27. F. T. Liu, J. W. Bohn, E. L. Terry, H. Yamamoto, C. A. Molinaro, L. A. Sherman, N. R. Klinman, and D. H. Katz, *J. Immunol.,* **124:**2728-2737 (1980).
28. J. R. Clamp, K. W. Hughes, and J. C. McPherson, *Immunochemistry,* **8:**871-879 (1971).
29. H. A. Coller and B. S. Coller, *Hybridoma,* **2:**91-96 (1983).
30. P. L. Ey, S. J. Prowse, and C. R. Jenkin, *Immunochemistry,* **15:**429-436 (1978).
31. C. Bruck, D. Portetelle, C. Glineur, and A. Bollen, *J. Immunol. Methods,* **53:**313-319 (1982).
32. D. J. Fraker and J. C. Speck, *Biochem. Biophys. Res. Commun.,* **80:**849-857 (1978).
33. A. G. Amit, R. A. Mariuzza, and R. J. Poljac, *Science,* **233:**747-753 (1986).
34. D. J. Barlow, M. S. Edwards, and J. M. Thornton, *Nature* **322:**747-748 (1985).
35. P. J. Munson and D. Rodbard, *Anal. Biochem.,* **107:**220-239 (1980).

FACTOR VII AND DIETARY FAT INTAKE

G. J. Miller, J. C. Martin, K. A. Mitropoulos, and J. K. Cruickshank

Medical Research Council, MRC Epidemiology and Medical Care Unit
Northwick Park Hospital, Harrow
Middlesex, HA1 3UJ, UK

INTRODUCTION

Evidence for an increased factor VII coagulant activity (VII_c) in acute myocardial infarction was first provided by Poller in 1957 (1). In this study a group of 20 patients examined within 48 hours of onset of an acute thrombotic episode (18 were cases of coronary thrombosis) had a significantly higher mean VII_c than control subjects. This finding has since been confirmed by some (2, 3) but not all (4) more recent studies.

The Northwick Park Heart Study (5) was established to determine whether changes in the haemostatic pathway, including factor VII, actually preceded the onset of acute coronary events (myocardial infarction and sudden coronary death). Among 1511 men aged 40 to 64 years at entry, 109 subsequently developed an acute coronary episode. A high VII_c was associated with increased risk, especially for events within 5 years of recruitment. When allowance was made for the within-subject variability of VII_c from day-to-day, a one standard deviation increase in VII activity appeared to increase the risk of a fatal coronary event within 5 years by about 2.4 fold. Furthermore, the association was independent of serum cholesterol and plasma fibrinogen concentration. These observations suggested that a high VII_c acted as a marker of a hypercoagulable or pre-coronary thrombotic state.

Even though the predictive power of VII_c for coronary heart disease was independent of serum cholesterol, a positive association was observed between the baseline levels of these two risk factors in the Northwick Park Heart Study. The relation was:

$$VII_c\ (\%\ \text{standard}) = 76.6 + 5.6\ \text{cholesterol (mmol/l)};\ p < 0.001$$

These findings led to the suspicion that both factors were themselves influenced by a third underlying factor itself of importance for coronary heart disease. Dietary fat being recognised as a major determinant of serum cholesterol concentration (6, 7), the hypothesis was proposed that a high fat consumption would also be associated with a high VII_c, thereby increasing the risk of coronary thrombogenesis.

EXPERIMENTAL STUDIES

In order to test this idea, 6 healthy individuals recorded their usual diet for 7 days while maintaining their daily routine. Food intake was measured by precise weighing (8) on an electronic scale and converted to energy and nutrient intake with published tables of food composition (9). This record was used to formulate guidelines for two experimental diets of similar energy value, one providing about 70% of energy as fat and the other 10% energy as fat. These dietary targets were introduced for 3 and 2 weeks respectively, in randomised alternate order. Throughout the study, subjects recorded their dietary intakes by weighed inventory, and a 25 ml blood sample was drawn before breakfast at 2-3 day intervals. Citrated plasma and serum were stored in plastic tubes at -150°C within 2 hours

Fibrinogen, Thrombosis, Coagulation, and Fibrinolysis, Edited by
C. Y. Liu and S. Chien, Plenum Press, New York

Table 1. Dietary Intake, Factor VII and Serum Lipids on Low-Fat and High-Fat Diets. Values are Means of Individual Averages (N=6)

	Low-fat	High-fat	Mean Difference (SE)	p*
Energy intake kJ/day	9,339	11,958		
Fat intake kJ/day	1,197	7,422		
Plasma VII_c % standard	81.7	97.5	15.8 (4.8)	0.02
Serum VII_t % standard	102.7	110.7	8.0 (4.7)	>0.05
Serum cholesterol mmol/l	4.24	4.70	0.46 (0.18)	0.05
Serum triglyceride mmol/l	1.21	0.83	0.38 (0.12)	0.03

*paired t test

of collection. At the end of the study the stored samples were assayed in duplicate and in random order as a single batch for VII_c and serum lipids as described elsewhere (10).

Table 1 shows that, as expected, serum cholesterol concentration was higher on the high-fat than on the low-fat regimen, and that serum triglyceride concentration was higher on the low-fat (high-carbohydrate) diet. Plasma VII_c was on average about 16% of standard higher on the high-fat diet.

This result prompted the question as to whether the increase in VII_c on the high-fat regimen was due to an increase in factor VII concentration or to an increase in factor VII reactivity (ie. conversion of the low-activity single-chain native protein to the fully-activated two-chain species VII_a). To assess the relative contributions of these two possibilities, a second assay was performed in which VII activity in serum is estimated from its rate of activation of a tritiated preparation of bovine factor X (VII_t). The preliminary conversion of test plasma to test serum in the presence of a high concentration of tissue factor so activates factor VII that its behaviour in the VII_t assay is the same whether it existed in vivo as the single-chain or two-chain protein (11). Table 1 shows that VII_t was on average 8% of standard higher on the high-fat diet, but that this difference was not statistically significant.

A far better indication of the effect of dietary fat on VII_c was given by examination of the data within individuals across the duration of the study. When the day-to-day changes in fat intake and VII_c were expressed as deviations from the individual means, a positive correlation was observed between the deviation in fat intake and the deviation in VII_c measured the following morning ($r = 0.56$; $p < 0.001$).

There was, however, a potential flaw in the design of the preliminary study which arose because the participants consumed the diets unsupervised and the high-fat regimen was exceptionally rich in fat. On this account, energy intake was considerably higher on the high-fat regimen than on the low-fat diet (Table 1). The statistical analysis indicated that only the fat component of energy was of importance for VII_c, but nevertheless a second study was undertaken in which energy intake was standardised. Ten healthy adults recorded prescribed diets by precise weighing over 24 hours on two days, one week apart. The diets were isocaloric, fat providing 54% of energy in one and 23% of energy in the other. As shown in Table 2, the results conformed entirely with the first study, there being increases in both VII_c and VII_t after the high-fat regimen, but only that in VII_c being statistically significant.

COMMUNITY STUDIES

The dietary manipulations used in these experimental studies were fairly extreme. In order to determine whether similar relations existed between VII_c and fat intake in the general community, 275 men aged 40-59 years were recruited from the register of a general medical practice (12).

Table 2. Effect of Increased Fat Consumption on Factor VII; Dietary Energy Constant

	Low-fat	High-fat	Mean difference	p*
Energy intake kJ/day	10,925	10,824	71	
Fat intake % energy	23	54	31	
Plasma VII_c % standard	99	107	8.4 (3.4)	0.02
Serum VII_t % standard	114	119	4.6 (4.0)	>0.05

*paired t test.

Of these, 203 participated in the study and 170 (62%) produced a satisfactory weighed inventory of what they ate and drank over 5 days (Wednesday to Sunday). A non-fasting blood sample was taken for VII_c after completion of the dietary record. Six samples were technically unsatisfactory, and of the remainder, 137 (84%) were obtained within 48 hours of the last recorded meal. As is standard practice in this laboratory, VII_c was determined on each sample in duplicate by semi-automated one-stage assay (13) and the mean expressed as a percentage of a standard plasma.

The first finding of note was a positive correlation between daily average fat intake and body height, indicating that in this largely sedentary middle-aged population tall men tended to eat more than small men ($r = 0.29$; $p < 0.001$). This body-size effect was allowed for by expressing fat intake over $height_a$, an index that was unrelated to body size. There was a clear tendency for VII_c to increase with the body-size adjusted fat intake. The difference of 12% of standard between men in the highest quarter and the lowest quarter of the distribution of fat intake was similar to that found in the Northwick Park Heart Study between men who developed an acute coronary event and those who remained free during following-up (5).

In multiple linear regression analysis of these data, not only fat intake ($p = 0.013$) but also serum triglyceride concentration ($p < 0.001$) was found to have made an independent contribution to VII_c in these men. The relationship was:

$$\log_{10} VII_c \text{ (\% standard)} = 1.93 + 0.0019 \text{ fat (g day}^{-1} \text{ height}^{-2}) + 0.11 \log_{10} \text{triglyceride (mmol l}^{-1})$$

Dietary fat is the source of the circulating chylomicron triglyceride and a varying proportion of plasma triglyceride also resides in very-low density lipoproteins of hepatic origin. The regression relation therefore suggested that large triglyceride-rich lipoproteins of splanchnic or hepatic origin, or their lipolytic products, induce an increase in VII_c. Overall, the results of the community-based study accorded well with the earlier experimental investigations.

ANIMAL STUDIES

Further evidence for an activating effect of large lipoprotein particles on factor VII has been provided by studies of rabbits fed a 1% cholesterol-supplemented diet (14). Male New Zealand white rabbits were fed a standard diet for one week before random separation into two groups, each of 6 animals. One group continued with the standard feed while the other received the supplemented diet. Blood samples were drawn for VII_c, VII_t and blood lipid concentrations at 2-7 day intervals for 100 days. Plasma cholesterol concentration increased steadily in the experimental group to reach its maximum value at about 30 days, thereafter declining progressively but remaining significantly elevated in comparison with the control group at 100 days (mean values, 11864 and 609 μg/ml respectively). About 60% of plasma cholesterol resided in large lipoproteins of very-low density or chylomicron particle size in the cholesterol-fed group, as compared with 7% in the control animals.

From about the fourth study-day onwards the experimental group showed a higher VII_c than the control group. By about the mid-point of the study, VII_c was in the region of 500% of standard in the cholesterol-fed group, as compared with 130% of standard in the animals on the non-supplemented diet. Even more impressively (and reminiscent of the first experimental study described on human volunteers) peaks and troughs of VII activity across the 100 days tended to coincide with those in plasma cholesterol in both groups. These findings were confirmed when the experiment was repeated in another set of rabbits.

At day 10 and on all but one of 15 subsequent test days VII_t was higher in the experimental group than in the control group, although in most comparisons the difference was not statistically significant. Overall, however, the mean average VII_t in the cholesterol-fed rabbits (128% of standard) was significantly higher ($p < 0.001$) than in the control rabbits (105% of standard). This increase in VII_t, although very likely to have been real, was far less impressive than the elevation in VII_t in the experimental group.

The major component of the core of the large lipoprotein particles that were associated with VII_c in the human studies would have been triacylglycerol. By contrast in the rabbit experiments, the core will have been predominantly cholesterol ester. This suggests that the mechanism by which these particles influence VII_c probably involved their surface rather than the core. A negatively-charged surface will activate the intrinsic coagulation pathway, and the enzyme activity generated is able to convert single-chain VII to the two-chain species by a single proteolytic cleavage (15, 16). Although there is some indirect evidence to support the proposition that the negatively-charged surface of large lipoproteins has a role in VII activation (17), direct evidence is still awaited. Whatever the eventual mechanism, the evidence to date suggests that the high-fat diet of Western society is responsible not only for hypercholesterolaemia and atheroma, but also for changes in VII_c which precede the onset of coronary thrombosis. Further research is required to determine whether the composition of dietary fat is of any relevance for the reactivity of factor VII. Preliminary studies in this laboratory, in which healthy volunteers have consumed saturated-fat-rich and polyunsaturated-fat-rich diets of standard energy and total fat content, suggest that the quality of fat consumed is of less importance in this respect than the quantity.

REFERENCES

1. L. Poller, Thrombosis and factor VII activity, *J. Clin. Path.*, **10**:348-350 (1957).
2. J. Carvalho de Sousa, J. Azevedo, C. Soria, F. Barros, C. Ribeiro, F. Parreira, and J. P. Caen, Factor VII hyperactivity in acute myocardial thrombosis. A relation to the coagulation activation, *Thromb. Res.*, **51**:165-173 (1988).
3. C. J. Hoffman, R. H. Miller, W. E. Lawson, and M. B. Hultin, Elevation of factor VII activity and mass in ischemic heart disease and in young subjects at high risk of ischemic heart disease, *Circulation,* **78 (Suppl. II)**:316 (1988).
4. P. M. Sandset, P. A. Sirnes, and U. Abildgaard, Factor VII and extrinsic pathway inhibitor in acute coronary disease, *Br. J. Haematol.*, **72**:391-396 (1989).
5. T. W. Meade, S. Mellows, M. Brozovic, G. J. Miller, R. R. Chakrabarti, W. R. S. North, A. P. Haines, Y. Stirling, J. D. Imeson, and S. G. Thompson, Haemostatic function and ischaemic heart disease: principal results of the Northwick Park Heart Study, *Lancet,* **2**:533-537 (1986).
6. A. Keys, J. T. Anderson, and F. Grande, Serum cholesterol response to changes in the diet, *Metabolism,* **14**:747-787 (1965).
7. D. M. Hegsted, R. B. McGandy, M. L. Myers, and F. J. Stare, Quantitative effects of dietary fat on serum cholesterol in men, *Am. J. Clin. Nutr.*, **17**:281-295 (1965).
8. J. W. Marr, Individual dietary surveys — purposes and methods, *World Rev. Nutr. Diet.*, **13**:105-164 (1971).
9. A. A. Paul, and D. A. T. Southgate, McCance and Widdowson's The Composition of Foods, 4th revised and extended edition of MRC special report No. 297, HMSO, London (1978).
10. G. J. Miller, J. C. Martin, J. Webster, H. C. Wilkes, N. E. Miller, W. H. Wilkinson, and T. W. Meade, Association between dietary fat intake and plasma factor VII coagulant activity — a predictor of cardiovascular mortality, *Atherosclerosis,* **60**:269-277 (1986).
11. G. J. Miller, M. J. Seghatchian, S. J. Walter, D. J. Howarth, S. G. Thompson, M. P. Esnouf, and T. W. Meade, An association between the factor VII coagulant activity

and thrombin activity induced by surface/cold exposure of normal human plasma, *Br. J. Haematol.*, **62**:379-384 (1986).
12. G. J. Miller, J. K. Cruickshank, L. V. Ellis, R. L. Thompson, H. C. Wilkes, Y. Stirling, K. A. Mitropoulos, J. V. Allison, T. E. Fox, and A. O. Walker, Fat consumption and factor VII coagulant activity in middle-aged men. An association between a dietary and thrombogenic coronary risk factor, *Atherosclerosis*, **78**:19-24 (1989).
13. M. Brozovic, Y. Stirling, C. Harricks, W. R. S. North, and T. W. Meade, Factor VII in an industrial population, *Br. J. Haematol.*, **28**:381-391 (1974).
14. K. A. Mitropoulos, M. P. Esnouf, and T. W. Meade, Increased factor VII coagulant activity in the rabbit following diet-induced hypercholesterolaemia. Evidence for increased conversion of VII to αVII_a and higher flux within the coagulation pathway, *Atherosclerosis*, **63**:43-52 (1987).
15. R. D. Radcliffe, A. Bagdasarian, R. W. Colman, and Y. Nemerson, Activation of bovine factor VII by Hageman factor fragments, *Blood*, **50**:611-617 (1977).
16. U. Seligsohn, B. Osterud, S. F. Brown, J. H. Griffin, and S. I. Rapaport, Activation of human factor VII in plasma and in purified systems. Roles of activated factor IX, kallikrein and activated factor XII, *J. Clin. Invest.*, **64**:1056-1065 (1979).
17. K. A. Mitropoulos, J. C. Martin, B. E. A. Reeves, and M. P. Esnouf, The activation of the contact phase of coagulation by physiologic surfaces in plasma: the effect of large negatively charged liposomal vesicles, *Blood*, **73**:1525-1533 (1989).

CHARACTERIZATION OF SNAKE VENOM PRINCIPLES AFFECTING BLOOD COAGULATION AND PLATELET AGGREGATION

Chaoho Ouyang, Che-Ming Teng, and Tur-Fu Huang

Pharmacological Institute, College of Medicine
National Taiwan University
Taipei 10018, Taiwan, ROC

INTRODUCTION

Snake venoms can affect blood coagulation and platelet function in various ways (1-6). The literature is extensive on the use of snake venoms for elucidating the nature of the mechanisms of blood coagulation and platelet aggregation as well as for the need to understand snake venoms more fully. One of the main attractions for investigators in the field of hemostasis and thrombosis has been the use of venoms and venom components in practical laboratory tests, in clinical use, and more recently, for application in theoretical studies. In this article, the characteristics of snake venom principles affecting blood coagulation and platelet aggregation are reviewed. The mechanisms of action of snake venom principles affecting blood coagulation and platelet aggregation are summarized in Fig. 1.

SNAKE VENOMS AND FIBRINOGEN

Thrombin-like Enzymes (TLE)

The clotting of fibrinogen by venoms was recognized at the beginning of this century and probably earlier (7, 8). The thrombin-like activity was found in most of the *Crotalidae* venoms. None was found in *Elapidae* and *Hydrophidae* venoms. Only one Viperid venom (*Vipera ammodytes*) was at this time recorded as possessing this activity.

Thrombin-like enzymes (TLE) were first separated by Habermann in 1958 from the venom of Bothrops jararaca (105). TEL's have been purified and characterized from the venoms of *Calloselasma* (formerly *Ancistrodon, Agkistrodon*) *rhodostoma* (10), *Agkistrondon contortrix contortrix* (15), *Agkistrodon acutus* (AAV) (11-13), *Crotalus adamanteus* (21), *Trimeresurus okinavensis* (111), *Trimeresurus gramineus* (TGV) (14), *Bitis gabonica* (107), *Crotalus h. horridus* (109) and *Bothrops atrox* (108). Their molecular weights are in the region of 30,000, with the exception of *Crotalus horridus horridus* venom (19,400) (110). They are acidic glycoproteins, rich in aspartic acid and glutamic acid (10, 12, 14, 21, 109, 110).

Like bovine thrombin, TLE's are esterases and are inhibited by hydroxy group (serine) reagent (diisopropyl fluorophosphate). However, unlike thrombin, TLE's are not affected by heparin-antithrombin III, with the exception of the TLE from *Agkistrodon c. contortrix* venom. Neither clot retraction nor Factor XIII (fibrin stabilizing factor) activation is found with TLE's from the AAV (12) and TGV (14), while TLE's from *Bitis gabonica* (20), *Bothrops moojeni, B. asper* (111) and *Trimeresurus* species (112) do activate Factor XIII.

The clotting effects of bovine thrombin on rabbit, dog and human plasmas are similar; guinea-pig plasma is less sensitive. The sensitivity of different plasmas to AAV-TLE is

Fibrinogen, Thrombosis, Coagulation, and Fibrinolysis, Edited by
C. Y. Liu and S. Chien, Plenum Press, New York

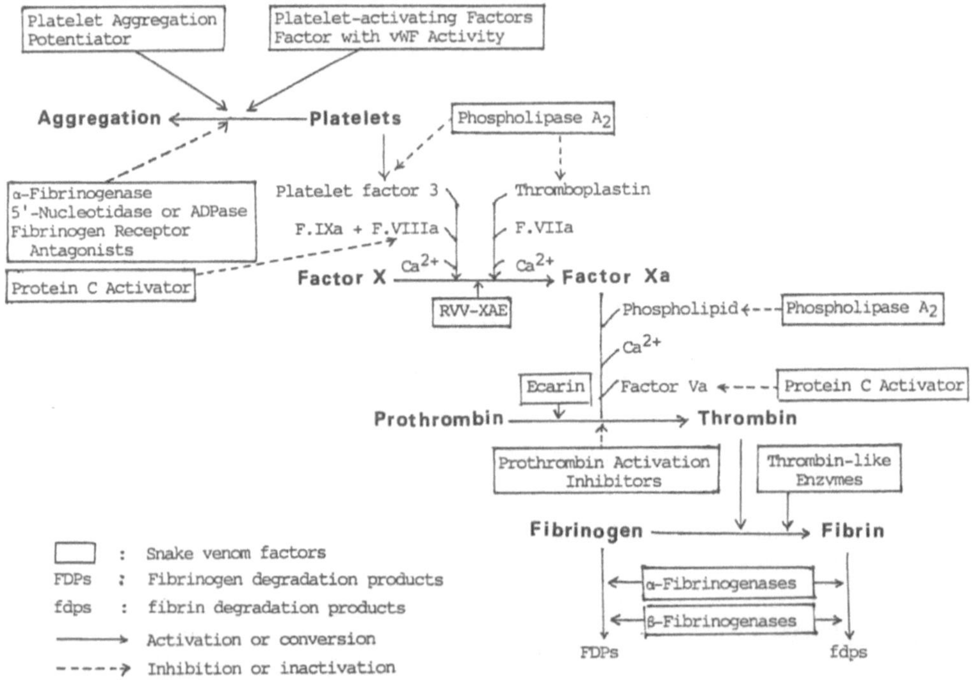

Fig. 1. Mechanisms of action of snake venom factors affecting blood coagulation and platelet aggregation.

human > dog > guinea-pig > rabbit. The susceptibilities of human, bovine and rabbit fibrinogens to thrombin are almost the same. However, the clotting effects of AAV-TLE on fibrinogens are human > bovine > > rabbit. The differences in the effects of plasmas on AAV-TLE are determined by the antithrombic effect of different plasma and the reactivity of fibrinogen to this venom enzyme (16). The TLE's from the venoms of *Calloselasma* (formerly Ancistrodon, Agkistrodon) *rhodostoma, Bothrops atrox, Crotalus adamanteus* and *Trimeresurus okinavensis* (111, 113, 114) release fibrinopeptide A only, while the TLE's from the venoms of *A. c. contortrix* (15) and *A. halys* pallas (115) preferentially release fibrinopeptide B, and like thrombin, TLE from *Bitis gabonica* venom cleaves both fibrinopeptides A and B (20).

AAV-TLE degrades purified bovine prothrombin in several steps. A bond at Lys_{44}-Tyr_{45} is broken with the formation of prothrombin intermediate D. Prethrombin 1 is formed and the digestion stops at the level of prethrombin 2. The latter is then easily isolated (22-24).

By using AAV-TLE, β-Factor X has been digested to obtain γ-Factor X and δ-Factor X by shortening the heavy chain from the N-terminal end. β,γ and δ-Factor X have been purified and each one is converted to the same active β-Factor Xa by the purified activator of Russell's viper venom (25).

Fibrin formed by TLE is not cross-linked. AAV-TLE can further digest the α- chain of fibrin, and the fibrin formed by this enzyme is more susceptible to plasmin degradation than the fibrin formed by thrombin. TLE does not induce platelet aggregation (26).

In vivo effects of AAV-TLE in rabbits have been studied (26, 27). AAV-TLE causes a marked prolongation of whole blood coagulation and one-stage prothrombin times and a marked decrease in the fibrinogen level. However, no significant change in the two-stage plasma prothrombin level is detected. The retardation of blood clotting by AVV-TLE *in vivo* is chiefly due to a decrease of plasma fibrinogen levels.

AAV-TLE does not cause significant changes in blood pressure, heart rate, E.K.G. and respiration at a defibrinating dose (0.1 mg/kg, i.v.) in rabbits. TGV-TLE causes similar but less profound effects on blood coagulation (28).

A coagulant fraction of *Bothrops atrox* venom (Reptilase®) containing a TLE as well as a Factor X activator which requires phospholipid has been used as a hemostatic agent.

Two thrombin-like enzymes purified from the venoms of snakes, *Agkistrodon (Calloselasma rhodostoma* (Ancrod, Arvin® or Venacil®) and *B. atrox moojeni* (Defibrase®) have been used as defibrinogenating agents in the treatment of venous thrombosis (29, 30). Ancord has been used in hemodialysis as an anticoagulant (31). Reptilase-Reagent®, a purified thrombin-like enzyme from *B. atrox* (proposed subspecies *B. moojeni*) free from Factor X activator, has been widely used in the determination of clotting time of citrated plasma (Reptilase time) (19). It is especially valuable in screening for dysfibrinogenemia, since the Reptilase time may be greatly prolonged, while the thrombin time may be only slightly prolonged. The Reptilase time is not affected by the presence of heparin-antithrombin III and only slightly prolonged by fibrin(ogen) degradation products. Hence the fibrinogen levels in both heparinized plasma and plasma from patients with DIC can be determined by the Reptilase time (128).

SNAKE VENOMS AND DIRECT FIBRINO(GENO)LYSIS

Snake venom principles can be classified into the following three groups, depending on which chain of the fibrinogen molecule is affected.

A. Digestion of α(A) Chain of Fibrinogen

Venom principles, which digest α(A) chain of monomeric fibrinogen specifically, have been purified from the venoms of *T. mucrosquamatus, A. acutus, T. gramineus, C. rhodostoma* (α-fibrinogenases), and *C. atrox* (Fibrinogenase). Their estimated molecular weights range from 21,500 to 31,000 (32-39).

B. Digestion of β(B) Chain of Fibrinogen

Venom principles, which preferentially attack the β(B) chain of fibrinogen, have been purified from the venoms of *T. mucrosquamatus, T. gramineus* (β-fibrinogenases), and *C. atrox* (fibrinogenase). Their molecular weights range from 22,900 to 26,000 (32, 34-40).

C. Digestion of γ Chain of Fibrinogen

Only one venom principle, which digests the γ chain of fibrinogen, has been purified from *C. atrox* venom (hemorrhagic toxin). The estimated molecular weight is 64,000 (41).

α-Fibrinogenases possess hemorrhagic activity without TAMe esterase activity, while β-fibrinogenases have TAMe esterase activity without hemorrhagic activity (32-38).

The fibrinogenolytic activities of α-fibrinogenases are markedly inhibited by EDTA and cysteine, while those of β-fibrinogenases are inhibited markedly by phenylmethane sulfonylfluoride and slightly by tosyl-L-lysine chloromethyl ketone and cysteine (32-38).

α-Fibrinogenase from *T. mucrosquamatus* venom digests bovine fibrinogen better than human fibrinogen, while β-fibrinogenase from the same venom digests human fibrinogen more extensively than bovine fibrinogen; Human fibrin is digested by both enzymes. Plasma fibrinogens of 4 animal species are digested by α-fibrinogenase to almost the same degree, while by β-fibrinogenase in the following order: human > dog > guineapig > rabbit. The fibrinogenolytic effects of α-fibrinogenase on human fibrinogen are strongly inhibited by sera of the 4 animal species, while those of β-fibrinogenase are inhibited in the following order: rabbit > guinea-pig > dog > human. It is concluded that the different activities of the protease inhibitors in the plasma of different animal species are mainly responsible for the sensitivity differences (42).

The venoms from *C. atrox, C. adamanteus, A. p.piscivorus* and *A. c. contortrix* have fibrinolytic activity with molecular weights of 21,500-34,000 (116-118).

SNAKE VENOMS AND THROMBIN FORMATION

A. Prothrombin Activators

Without further purification, the venom of *Oxyuranus s. scutellatus* has been used for the assay of prothrombin (119-121). The venom is convenient for it neither requires nor is influenced by other coagulation factors. Consequently, a one-stage assay can be used to measure prothrombin concentration in many deficient plasmas. Prothrombin activation by the venom of *Notechis scutatus* is dependent upon Factor V (119-121).

A prothrombin activator (Ecarin, MW 86,000) has been purified from the venom of *Echis carinatus*. It is an acidic protein, DFP-resistant, and converts prothrombin to thrombin (43, 44). It also converts prethrombin 1 to meizothrombin 1 (45).

Another prothrombin activator has been purified from the venom of *Bothrops atrox*.

It is a single chain polypeptide (molecular weight 70,000), DFP-resistant, Ca^{2+}-independent, and inhibited by EDTA. It converts prothrombin to meizothrombin (46).

B. Prothrombin Activation Inhibitors (PAI)

Like antithrombin III-heparin, all prothrombin activation inhibitors do not destroy fibrinogen and prothrombin, or interfere with the interaction between thrombin and fibrinogen. However, unlike antithrombin III-heparin, they do not inactivate thrombin (47-53).

1. Inhibitors without Recognizable Enzymatic Activity. The molecular weights of PAI from AAV and TGV are approximately 20,000. They are acidic glycoproteins without recognizable enzymatic activity (47-49). They are capable of prolonging whole blood coagulation time, calcium clotting time and prothrombin time of rabbit plasma *in vitro*.

The inhibition of prothrombin activation by the PAI's of AVV and TGV is due to interference in the interaction between prothrombin and its activation factors because of the reversible binding of these factors with PAI of AAV or TGV (47-49). It has been shown that AAV-PAI binds to purified bovine F Xa in the presence of Ca^{2+} (52). *In vivo* effects of AAV-PAI have been studied in rabbits (26, 27); AAV-PAI produces a marked, but transient prolongation of whole blood coagulation time and one-stage plasma prothrombin time, with no significant change in the two-stage plasma prothrombin level or plasma fibrinogen level. AAV-PAI does not affect platelet aggregation induced by ADP; nor does it cause significant changes in blood pressure, heart rate, EKG or respiration at an i.v. dose of 1-10 mg/kg in rabbits (26, 27). Similar effects of TGV-PAI on blood coagulation have been observed (28).

2. Phospholipase A_2 (PLA_2). The PAI from *Trimeresurus mucrosquamatus* venom (TMV) is a basic phospholipase A_2 (mol. wt. 11,700) (50, 51). The anticoagulant activity of this PLA_2 is due to the inhibition of Factor X and prothrombin activation through the inactivation of the procoagulant activity of phospholipids, mediated partly by the phospholipid-binding activity and partly by its enzymatic hydrolysis of phospholipids (50, 51). The anticoagulant action of basic PLA_2 from snake venoms is stronger than that of the neutral or acidic PLA_2, and their enzymatic activities are not parallel with their anticoagulant activities. When the enzymatic activity of TMV-PLA, is almost completely (> 99%) inhibited with p-bromophenacyl bromide, the anticoagulant activity decreases markedly. The procoagulant activity of lysocephalin, derived from cephalin by the action of PLA_2, is only about one hundredth of that of cephalin. When p-bromophenacyl bromide-modified TMV-PLA_2 is added to the mixture of cephalin and lysocephalin, it still can prolong the Stypven clotting time. The anticoagulant activity may be due to both enzymatic and binding activities. The anticoagulant activity of PLA_2 from *Naja naja atra* venom (NNAV) is very weak in the presence of plasma and cephalin. However, when NNAV-PLA_2 has been pre-incubated with cephalin at 37°C for 2 min. and plasma is subsequently added to the pre-incubation mixture, the anticoagulant effect of NNAV-PLA_2 is more pronounced. This indicates that the weak anticoagulant activity of NNAV-PLA_2 is due to interference of phospholipid binding to NNAV-PLA_2 by plasma binding proteins. The phospholipid-binding activity of PLA_2 is thought to be a pre-requisite for its anticoagulant activity, with the enzymatic activity potentiating its anticoagulant activity. The anticoagulant activity of basic PLA_2 is usually stronger than that of acidic PLA_2, possibly due to a stronger binding activity (50, 51). The anticoagulant action of NNAV has been ascribed to the synergistic effect of both NNAV-PLA_2 and cardiotoxin (104). Boffa *et al.* (53) reported that the anticoagulant effect of phospholipase A_2 was related, besides the catalytic effect of the enzyme, to an additional interaction between the phospholipid/phospholipase complex and an undetermined clotting factor.

SNAKE VENOMS AND FACTOR X ACTIVATION

Arthus (54) was the first to report that *Vipera russellii* venom possessed a strong procoagulant activity. Macfarlane and Barnett reported the hemostatic possibilities of snake venom (122). In a series of toxicological studies on the venom of *Vipera russellii formosensis*, Maki, Lee (formerly Ri) demonstrated that the venom had a potent coagulant action both *in vivo* and *in vitro* and that its action resembled that of thromboplastin but not that of thrombin (55, 56). Lee *et al.* (57) suggested that *V. russellii* venom might function in conjunction with platelet material in the activation of purified prothrombin. Esnouf and Williams (58) have reported that *V. russellii* venom contains a Factor-X activator. A use for the venom is to differentiate Factor VII from Factor X deficiency (124). It has been partially purified from *V. russellii* venom, has a molecular weight of 145,000 and is calcium-dependent (59). Factor X activator has been reported to be a two-chain (60), rather than a single-chain protein (67). It is also considered to be an acidic protein (molecular weight 79,000) (61). The venom of *Bothrops atrox* has also been found to be Factor X activator (119). Recently, two Factor X activators differing only in their electric charges have been purified from *Bothrops atrox* venom. The molecular weights of both activators are estimated to be 77,000 daltons, consisting of one heavy chain (59,000 daltons) and one or two light chains (15,000-16,000 daltons) linked by disulfide bridges. They are Ca^{2+}-dependent and DFP-resistant (125). The venom of *Crotalus viridis helleri* has also been found to be Factor X activator (123).

SNAKE VENOMS AND FACTOR XI ACTIVATION

Phillips *et al.* have reported that the venom of puff adder (*Bitis arietans*) activates Factor XI in studies of the *in vitro* effects of the venom on the coagulation system of human plasma (129).

SNAKE VENOMS AND PROTEIN C ACTIVATION

A specific protein C activator, a single-chain polypeptide (MW 37,000, pI. 3.0), has been detected in *Agkisrodon contortrix* venom. (62-66, 130). Activated protein C catalyzes the proteolytic inactivation of the cofactor proteins, factor Va and VIIIa, thereby exerting a potent anticoagulant effect. The venoms of various *A. contortrix, A. piscivorus* and A. bilineatus subspecies contain similar protein C activators (130).

SNAKE VENOMS AND PLATELET AGGREGATION

A. Platelet Aggregation Inducers

1. Non-Clotting Platelet Aggregation Inducers. Only more recently the platelet-activating effects of snake venoms have been recognized for crude venom (68). The non-clotting platelet aggregation inducers have been purified from the venoms of *T. okinavensis* (69), *T. mucrosquamatus (aggregoserpentin)* (70-73), *T. gramineus* (74), *B. atrox (thrombocytin)* (75, 76), *C. horridus horridus* (crotalocytin) (77, 78), *C. durissus terrificus, C. durissus cascavella (convulxin)* (79, 84) and *T. wagleri* (134). They are glycoproteins without recognized enzymatic activity except that an amidolytic activity is found in crotalocytin and thrombocytin, and a weak amidolytic activity in the component isolated from *T. gramineus* venom. The component from *T. okinavensis* venom and convulxin have been reported to be DFP-resistant, while thrombocytin and crotalocytin are DFP-sensitive. The platelet-activating activity of convulxin, aggregoserpentin, the components from the venoms of *T. gramineus* and *T. wagleri* are claimed to be ADP-, thromboxane A_2- and PAF-independent (134, 135). The inducer from *T. gramineus* venom has also been found to be able to lower cAMP levels of platelets. A platelet-activating glycoprotein (molecular weight 78,930) from *C. d. cascavella* venom has been reported to have a hexameric structure involving two distinct subunits (12,540 and 13,770 daltons) (85). The *in vivo* effects of the platelet aggregation inducer from *T. mucrosquamatus* venom in rabbits have been studied (71); an intravenous dose of 35 $\mu g/kg$ of the venom platelet aggregation inducer injected into the rabbit marginal ear vein caused a marked thrombocytopenia with the platelet number decreasing to about 10% of the control value within 30 min. In contrast to the long duration of action of the crude venom (70), the thrombocytopeina caused by the venom platelet aggregation inducer is transient. It is possible that other factor(s) in the crude venom may potentiate the aggregating effect of the venom inducer and cause more irreversible aggregation.

2. Clotting Platelet Aggregation Inducer. The prothrombin activator from *Echis carinatus* venom, Ecarin, induces platelet aggregation in the presence of a trace amount of prothrombin; this effect is inhibited by heparin or hirudin (86, 87).

B. Platelet Aggregation Inhibitors

Platelet aggregation inhibitors have been isolated from the venoms of *T. gramineus* (88, 89), *A. halys* (90, 91), *T. mucrosquamatus* (36), *C. rhodostoma* (92). They can be classified into three groups, α-fibrinogenase, 5'-nucleotidase and fibrinogen receptor antagonists.

1. α-Fibrinogenase. Two α-fibrinogenases having an inhibitory effect on platelet aggregation have been purified from the venoms of *T. mucrosquamatus* (36) and *C. rhodostoma* (92). These are basic proteins (molecular weights 22,400 and 25,360, respectively). Their inhibitory effects on platelet aggregation depend on their fibrinogenolytic activities.

2. 5'-Nucleotidase or ADPase. A 5'-nucleotidase, a single peptide chain glycoprotein (molecular weight 74,000), was purified from *T. gramineus* venom (93). The removal of ADP, which is released by the platelet aggregation inducers and the subsequent accumulation of adenosine are responsible for the inhibitory effect on platelet aggregation.

An ADPase, a single-chain polypeptide (molecular weight 94,000), has been purified from *A. acutus* venom (94). It has a marked inhibitory action on ADP-, collagen- and sodium arachidonate-induced platelet aggregations of rabbit platelet-rich plasma.

3. Fibrinogen receptor antagonists. Fibrinogen receptor antagonists, proteins without recognizable enzymatic activity, have been isolated from the venoms of *T. gramineus* (Trigramin, 7,500 daltons, 72 aa residues) (88, 89, 131). *A. halys* (Halysin, 7,500 daltons, 71 aa residues) (90, 91, 132, 133), *Echis carinatus* (6,800 daltons) (86, 87), *Calloselasma rhodostoma* (Rhodostomin, 7,500 daltons, 71 aa residues) (95, 133) and *Bitis arietans* (Arietin, 7,500 daltons, 70 aa residues) (133). Trigramin, halysin, rhodostomin and arietin have been claimed to be specific antagonists of the specific receptor associated with glycoprotein IIb-IIIa complex on platelet membrane (131, 132, 133).

C. Platelet Aggregation Potentiator

Cardiotoxin, isolated from *Naja naja atra* snake venom (126), potentiates platelet aggregation induced by ADP, thrombin, collagen and venom PLA_2. The production of malondialdehyde caused by ADP, thrombin and venom PLA_2 is also increased in the presence of cardiotoxin. Both the potentiation of aggregation and the increase in malondialdehyde are blocked by indomethacin or Ca^{2+} (5 mM or 0.05 mM, respectively). Cardiotoxin does not potentiate thrombin-induced aggregation of p-bromophenacyl bromide-modified platelets. Thromboxane B_2 formation induced by thrombin or collagen is also increased by cardiotoxin, while that induced by arachidonate is not affected. As a membrane-active polypeptide, cardiotoxin might augment the Ca^{2+}-flux during the activation of the platelet membrane by aggregation inducers and subsequently increase the activation of endogenous PLA_2 (96).

D. Biphasic Effect on Platelet Aggregation

A basic PLA_2 has been isolated from *Vipera russellii* snake venom (127). It induces a biphasic effect on washed rabbit platelets suspended in Tyrodes' solution. The first phase is a reversible aggregation which is dependent on stirring and extracellular calcium. The second phase is an inhibition of platelet aggregation; this occurs 5 min. after the addition of the venom PLA_2 without stirring or following recovery from the reversible aggregation. The aggregating phase can be inhibited by indomethacin, tetracaine, papaverine, creatine phosphate/creatine phosphokinase, mepacrine, verapamil, sodium nitroprusside, PGE_1 or bovine serum albumin. The venom PLA_2 releases free fatty acids from synthetic phosphatidylcholine and intact platelets. p-Bromophenacyl bromide-modified venom PLA_2 loses its PLA_2 enzymatic and platelet-aggregating activities, but protects platelets from the aggregation induced by the native venom PLA_2. The second phase of the venom PLA_2 action demonstrates a different degree of inhibition of platelet aggregation induced by some activators in the following order: arachidonic acid > collagen > thrombin > ionophore A-23187. The longer the incubation time or the higher the concentration of the venom PLA_2, the more pronounced is the inhibitory effect. The venom PLA_2 does not affect the thrombin-induced release reaction which is caused by intracellular Ca^{2+} mobilization in the presence of EDTA, but inhibits collagen-induced release reaction which is caused by Ca^{2+} influx from extracellular medium. The inhibitory effect of the venom PLA_2 and also lysophosphatidylcholine or arachidonic acid can be antagonized or reversed by bovine serum albumin. It is concluded that the first stimulatory effect of the venom PLA_2 action may be due to arachidonate liberation from the platelet membrane. The second phase of inhibition of platelet aggregation and the release of ATP may be due to the inhibitory action of the split products produced by this venom PLA_2 (97).

Effects of seven purified PLA_2s from the venoms of both snakes (*N. n. atra, T. mucrosquamatus* and *T. gramineus*) and honey bee (*Apis mellifera*) on rabbit washed platelet suspensions in the absence of bovine serum albumin have been studied (98, 103). Only phospholipases A_2 from *N. n. atra, T. mucrosquamatus* and *A. mellifera* venoms induces platelet aggregation with small amounts of ^{14}C-serotonin release. They show tachyphylaxis and also cross-tachyphylaxis in inducing platelet aggregation. The former two phospholipases A_2 exhibit biphasic responses in which irreversible aggregation occurs at concentrations of 1-10 μg/ml. At higher concentrations, they elicit reversible aggregation. Exogenous Ca^{2+} is essential to their activity. Indomethacin and EDTA completely abolish both PLA_2-induced platelet shape change and aggregation, while mepacrine, prostaglandin E_1, verapamil and nitroprusside inhibit only the aggregation response. p-Bromophenacyl bromide-modified phospholipases A_2, which has almost completely lost enzymatic activity, fails to induce platelet aggregation. Phosphatidylcholine, phosphatidylethanolamine and phosphatidylinositol inhibit the phospholipase A_2-induced platelet aggregation. These PLA_2's induce thromboxane B_2 formation which is inhibited by EDTA and indomethacin, but not by prostaglandin E_1. Pretreatment of platelet suspension with phospholipase A_2 from *N. n. atra* or *A. mellifera* venom (50 μg/ml) inhibits platelet aggregation induced by sodium arachildonate or collagen, but not that caused by thrombin or ionophore A-23187. Exogenous sodium arachidonate or lysophosphatidylcholine also shows unaltered inhibitory spectrum on platelet aggregation. It is concluded that PLA_2's induce platelet aggregation by virtue of their enzymatic activity, cleaving the membrane phospholipids to result in arachidonic acid release and formation of thromboxane A_2. On the other hand, the cleaved products, lysophosphatidylcholine, arachidonic acid or arachidonate metabolites (via lipoxygenase pathway), may be responsible for anti-platelet activity (103).

SNAKE VENOM AND VON WILLEBRAND FACTOR ACTIVITY

Botrocetin (venom coagglutinin), purified from *Bothrops atrox* venom, elicits a von Willebrand factor-dependent platelet agglutination in several animal plasmas. It may be useful for the detection of von Willebrand-factor activity in animal plasmas (99, 100).

ACKNOWLEDGMENTS

We wish to express our sincere appreciation to Prof. Chen-Yuan Lee, former Dean of the College of Medicine, National Taiwan University for his invaluable advice and encouragement. We also thank Prof. Walter H. Seegers, former Chairman of Department of Physiology and Prof. Lowell E. McCoy, Department of Physiology, Wayne State University for their cooperation in the sedimentation studies and amino acid analyses on these proteins.

This work was supported by the National Science Council Research Grants of the Republic of China, the Chung-Cheng Science Foundation of the Republic of China, and China Medical Board of New York, Inc.

The authors wish to express their thanks to Mr. T.P. Hsu and Mr. I.S. Peng for drawing the figures and typing this manuscript.

REFERENCES

1. P. Boquet, Venins de serpents. Physiopthologie de I'envenimation et proprieties biologigues des venins, *Toxicon,* **2:**5-44 (1954).
2. J. M. Jimenez-Porras, Pharmacology of peptides and proteins in snake venoms, *Annual Review of Pharmacology,* **8:**299-318 (1968).
3. J. Meaume, Les venins de serpents agents modificateurs de la coagulation sanguine, *Toxicon,* **4:**25-28 (1966).
4. C. Ouyang, The effects of Formosan snake venoms on blood coagulation in vitro, *Journal of the Formosan Medical Association,* **56:**435-448 (1957).
5. W. H. Seegers, and C. Ouyang, Snake venoms and blood coagulation, *in:* C. Y. Lee, ed., Snake venoms, Handbook of Experimental Pharmacology, Berlin-Heidelberg-New York: Springer-Verlag, **52:**684-750 (1979).
6. C. Ouyang, C. M. Teng, and T. F. Huang, Characterization of the purified principles of Formosan snake venoms which affect blood coagulation and platelet aggregation, *Journal of Formosan Medical Association,* **81:**781-790 (1982).
7. J. Mellanby, The coagulation of blood, Part II. The action of snake venoms, peptone and leech extract, *Journal of Physiology,* **38:**441-503 (1909).
8. B. A. Houssay, and A. Sordelia, Action des venins de serpents sur la coagulation sanguine, *J. Physiol. Path. Gen.,* **18:**731 (1919).

9. A. L. Copley, S. Banerjee, and A. Devi, Studies of snake venoms on blood coagulation. I. The thromboserpentin (thrombin-like) enzyme in the venoms, Thrombosis Research, **2**:487-508 (1973).
10. M. P. Esnouf, and G. W. Tunnah, The isolation and properties of the thrombin-like activity from *Ancistrodon rhodostoma* venom, *British Journal of Haematology,* **13**:581-590 (1967).
11. C. Ouyang, and J. S. Hong. Inhibition of the thrombin-like principle of *Agkistrodon acutus* venom by group-specific enzyme inhibitors, *Toxicon,* **12**:449-453 (1974).
12. C. Ouyang, J. S. Hong, and C. M. Teng, Purification and properties of the thrombin-like principle of *Agkistrodon acutus* venom and its comparison with bovine thrombin, *Thrombosis et Diathesis haemorrhagica,* **26**:224-234 (1971).
13. C. Ouyang, C. M. Teng, J. S. Hong, Purification and properties of the coagulant and anticoagulant principles of *Agkistrodon acutus* venom, *Journal of Formosan Medical Association,* **71**:401-407 (1972).
14. C. Ouyang, and F. Y. Yang, Purification and properties of the thrombin-like enzyme from *Trimeresurus gramineus* venom, *Biochimica et Biophysica Acta,* **351**:354-363 (1974).
15. R. H. Herzig, O. D. Ratnoff, and J. R. Shainoff, Studies on a procoagulant fraction of southern copperthead snake venom: the preferential release of fibrinopeptide B, *Journal of Laboratory and Clinical Medicine,* **76**:451 (1970).
16. C. Ouyang, Y. C. Chen, and C. M. Teng, The clotting activity of the thrombin-like enzyme of *Agkistrodon acutus* (Hundred pace snake) venom, *Toxicon,* **17**:313-316 (1979).
17. W. R. Bell, Defibrinogenation with arvin in thrombic disorders. In Sherry S, Scriabine A, eds., Platelets and Thrombosis, Munich: *Berlin, Vienna, Urban-Schwarzenberg,* 274-298 (1974).
18. N. Egberg, Experimental and clinical studies on the thrombin-like enzymes from the venom of *Bothrops atrox.* On the primary structure of fragment E. Acta Physiologica Scandinavica supplement 400 (1973).
19. C. Funk, J. Gmur, R. Herold, and P. W. Straub, Reptilase -R- A new reagent in blood coagulation, *British Journal of Haematology,* **21**:43-52 (1971).
20. P. J. Gaffney, N. A. Marsh, and B. C. Whaler, a coagulant enzyme from gaboon viper venom: some aspects of its mode of action, *Biochemical Society Transactions,* **1**:1208-1209 (1973).
21. F. S. Markland, and P. S. Damus, Purification and properties of a thrombin-like enzyme from the venom of *Crotalus adamanteus, Journal of Biological Chemistry,* **246**:6460-6473 (1971).
22. W. H. Seegers, C. M. Teng, and E. Novoa, Preparation of bovine prethrombin 2: use of acutin and activation with prothrombinase or ecarin, *Thrombosis Research,* **19**:11-20 (1980).
23. C. M. Teng, and W. H. Seegers, Production of prothrombin fragment 1-44 with acutin and some effects on thrombin generation, *Thrombosis Research,* **20**:217-279 (1980).
24. W. H. Seegers, C. M. Teng, A. Ghosh, and E. Novoa, Three aspects of prothrombin activation related to protein M, ecarin, acutin, meizothrombin 1 and prethrombin 2, *Annals of the New York Academy of Sciences,* **370**:453-467 (1981).
25. C. M. Teng, and W. H. Seegers, Production of Factor X and Factor Xa variants with thrombin, acutin and by autolysis, *Thrombosis Research,* **22**:213-220 (1981).
26. C. Ouyang, and C. M. Teng, *In vivo* effects of the purified thrombin-like and anti-coagulant principles of *Agkistrodon acutus* (Hundred pace snake) venom, *Toxicon,* **16**:583-593 (1978).
27. C. Ouyang, and C. M. Teng, The effect of the purified thrombin-like and anticoagulant principles of *Agkistrodon acutus* venom on blood coagulation *in vivo, Toxicon,* **14**:49-54 (1976).
28. C. Ouyang, and F. Y. Yang, The effects of the purified thrombin-like enzyme and anticoagulant principle of *Trimeresurus gramineus* venom on blood coagulation *in vivo, Toxicon,* **14**:197-201 (1976).
29. H. A. Reid, and K. E. Chan, The paradox in therapeutic defibrination, *Lancet,* **1**:485-486 (1968).
30. Z. S. Latallo, Report of the task force on clinical use of snake venom enzymes, *Thrombosis and Haemostasis,* **39**:768-774 (1978).
31. G. H. Hall, H. M. Holman, and A. D. B. Webster, Anticoagulation by ancrod for haemodialysis, *British Medical Journal,* **4**:591-593 (1970).

32. C. Ouyang, and C. M. Teng, Fibrinogenolytic enzymes of *Trimeresurus muscrosquamatus* venom, *Biochimica et Biophysica Acta,* **420:**298-308 (1976).
33. C. Ouyang, and T. F. Huang, Purification and characterization of the fibrinolytic principle of *Agkistrodon acutus* venom, *Biochimica et Biophysica Acta,* **439:**146-153 (1976).
34. C. Ouyang, C. M. Teng, and Y. C. Chen, Physicochemical properties of α- and β-fibrinogenases of *Trimeresurus mucrosquamatus* venom, *Biochimica et Biophysica Acta,* **481:**622-630 (1977).
35. C. Ouyang, and T. F. Huang, The properties of the purified fibrinolytic principle from *Agkistrodon acutus* snake venom, *Toxicon,* **15:**161-167 (1977).
36. C. Ouyang, and T. F. Huang, α- and β-fibrinogenases from *Trimeresurus gramineus* snake venom, *Biochimica et Biophysica Acta,* **571:**270-283 (1979).
37. C. Ouyang, C. M. Teng, Y. C. Cheng, Properties of fibrinogen degradation products produced by α- and β-fibrinogenase of *Trimeresurus mucrosquamatus* snake venom, *Toxicon,* **17:**121-126 (1979).
38. C. Ouyang, L. J. Hwang, and T. F. Huang, α-Fibriongenase from *Agkistrodon rhodostoma* (Malayan pit viper) snake venom, *Toxicon,* **21:**25-33 (1983).
39. T. Nikai, R. Kito, N. Mori, H. Sugihara, and A. T. Tu, Comparative Biochemical Physiology, **76B:**679-686 (1983).
40. Z. Z. Sapru, A. T. Tu, and G. S. Bailey, Purification and characterization of a fibrinogenase from the venom of western diamondback rattlesnake (*Crotalus atrox*), *Biochimica et Biophysca Acta,* **747:**225-231 (1983).
41. T. Nikai, N. Mori, M. Kishida, H. Sugihara, and A. T. Tu, Archieves of Biochemistry and Biophysics, **231:**309-319 (1984).
42. C. M. Teng, C. Ouyang, and S. C. Lin, Species difference in the fibrinogenolytic effects of α- and β-fibrinogenases from *Trimeresurus mucrosquamatus* snake venom, *Toxicon (Oxford),* **23:**777-782 (1985).
43. A. Schieck, F. Kornalik, and E. Habermann, The prothrombin activating principle from *Echis carinatus* venom, I. Preparation and biochemical properties, *Naunyn-Schiedeberg's Archives of Pharmacology,* **272:**402-416 (1972).
44. A. Schieck, E. Habermann, and F. Kornalik, The prothrombin activating principle from *Echis carinatus* venom. II. Coagulation studies *in vitro* and *in vivo.* Naunyn-Schmiedeberg's Archives of Pharmacology, **274:**7-17 (1972).
45. E. Novoa, and W. H. Seegers, Mechanism of α-thrombin and β-thrombin-E formation: Use of Ecarin for isolation of meizo-thrombin I, *Thrombosis Research,* **18:**657-668 (1980).
46. C. Bon, and H. Hofmann, Prothrombin and Factor X activators from *Bothrops atrox* venom, *Toxicon,* **23:**553 (1985).
47. C. Ouyang, and C. M. Teng, Purification and properties of the anticoagulant principle of *Agkistrodon acutus* venom, *Biochimica et Biophysica Acta,* **278:**155-162 (1972).
48. C. Ouyang, and C. M. Teng, The effect of the purified anticoagulant principle of *Agkistrodon acutus* venom on blood coagulation, *Toxicon,* **11:**287-292 (1973).
49. C. Ouyang, and F. Y. Yang, Purification and properties of the anticoagulant principle of *Trimeresurus gramineus* venom, *Biochimica et Biophysica Acta,* **386:**479-492 (1975).
50. C. Ouyang, C. M. Teng, Y. C. Chen, and S. C. Lin, Purification and characterization of the anticoagulant principle of *Trimeresurus mucrosquamatus* venom, *Biochimica et Biophysica Acta,* **541:**394-407 (1978).
51. C. Ouyang, W. Jy, Y. P. Zan, and C. M. Teng, Mechanism of the anticoagulant action of phospholipase A purified from *Trimeresurus mucrosquamatus* (Formosan Habu) snake venom, *Toxicon,* **19:**113-120 (1981).
52. C. M. Teng, and W. H. Seegers, *Agkistrodon acutus* snake venom inhibits prothrombinase complex formation, *Thrombosis Research,* **23:**255-263 (1981).
53. M. C. Boffa, C. Rothen, B. Verhelj, R. Verger, and G. De Haas, Enzymatic and anticoagulant activity of phospholipase A_2, *Toxicon,* **17:**supplement No. 1, p. 11 (1979).
54. M. Arthus, Actions coagulants et anticoagulants des venins, *Archives Internationales de Physiologie,* **15:**203 (1919).
55. T. Ri, Folia Pharmacologica Japonica, **27:**13 (1939).
56. C. Y. Lee, Toxicological studies on the venom of *Vipera russellii formosensis*, Maki. Part 1. Toxicity and Pharmacological Properties, *Journal of the Formosan Medical Association,* **47:**65-84 (1948).

57. C. Y. Lee, S. A. Johnson, and W. H. Seegers, Clotting of blood with Russell's viper venom, *Journal of Michigan State Society of Medicine,* **54:**801-804 (1955).
58. M. P. Esnouf, and W. J. Williams, The isolation and purification of a bovine plasma protein which is a substrate for the coagulation fraction of Russell's viper venom, *Biochemical Journal,* **84:**62-71 (1962).
59. S. Schiffman, I. Theodor, and S. I. Rapaport, Separation from Russell's viper venom of one fraction reacting with Factor X and another reacting with Factor V. *Biochemistry,* **8:**1397-1405 (1969).
60. W. Kisiel, M. A. Hermondson, and E. W. Davie, Interaction of Lanthanide ions with bovine factor X and their use in the affinity chromatography of the venom coagulant protein of *Vipera russellii, Biochemistry,* **15:**4901-4906 (1976).
61. G. W. Amphlet, R. Byrne, and F. Castellino, Cation binding properties of the multiple subforms of Russell's viper venom Factor X activating enzyme, the coagulant protein from *Vipera russellii* venom, *Biochemistry,* **21:**125-132 (1982).
62. K. Stocker, and J. Meier, Thrombin-like snake venom enzymes, Proc. Symp. on Animal Venoms and Haemostasis, San Diego CA, July 20-21 (1985).
63. K. Stocker, H. Fischer, J. Meier, M. Brogil, and L. Svendsen, Protein C activators in snake venoms, *Behring Inst. Mitt.,* **79:**37 (1986).
64. J. L. Martionli, and K. Stocker, Fast functional protein C assay using Protac®, a novel protein C activator, *Thrombosis Research,* **43:**253-264 (1986).
65. H. Loebermann, H. J. Kolde, R. Denbel, R. Peter, E. Tourte, and U. Becker, Determination of protein C in plasma, *Behring Inst. Mitt.,* **79:**112 (1986).
66. T. Exner, B. Cotton, and M. Howden, Detection of specific proenzyme activators in snake venoms by a new immunoabsorbant chromogenic substrate method, *Biochimica et Biophysica Acta,* **832:**351 (1985).
67. B. C. Furie, and B. Furie, Factor X activating enzyme from Russell's viper venom: isolation and characterization, *Journal of Biological Chemistry,* **250:**601-608 (1975).
68. M. G. Davey, and E. F. Lüscher, Action of some coagulant snake venoms upon blood platelets, *Nature,* **207:**1037-1039 (1965).
69. M. G. Davey, and M. P. Esnouf, The isolation of a component of the venom of *Trimeresurus okinavenis* that causes the aggregation of blood platelet, *Biochemistry,* **111:**733-743 (1969).
70. C. Ouyang, and C. M. Teng, The effect of *Trimeresurus mucrosquamatus* snake venom on platelet aggregation, *Toxicon,* **16:**575-582 (1978).
71. C. Ouyang, J. P. Wang, and C. M. Teng, A potent platelet aggregation inducer purified from *Trimeresurus mucrosquamatus* snake venom, *Biochimica et Biophysica Acta,* **630:**246-253 (1980).
72. C. Ouyang, and C. M. Teng, The action mechanism of the purified platelet aggregation principle of *Trimeresurus mucrosquamatus* venom, *Thrombosis and Haemostasis,* **41:**475-490 (1979).
73. C. M. Teng, K. K. Liao, J. P. Wang, H. S. Lin, and C. Ouyang, Ultrastructural changes and release reaction of platelets induced by an aggregation inducer purified from *Trimeresurus mucrosquamatus* (Formosan Habu) snake venom, *Toxicon,* **19:**121-130 (1981).
74. C. Ouyang, and T. F. Huang, A potent platelet aggregation inducer from *Trimeresurus gramineus* snake venom, *Biochimica et Biophysica Acta,* **761:**126-134 (1983).
75. E. P. Kirby, S. Niewiarowski, K. Stocker, E. Kettner, E. Shaw, and T. M. Brudzynski, *Biochemistry,* **18:**3564-3570 (1979).
76. S. Niewiarowski, E. P. Kirby, T. M. Brudzynski, and K. Stocker, Thrombocytin, a serine protease from *Bothrops atrox* venom. 2. Interaction with platelets and plasma-clotting factors, *Biochemistry,* **18:**3570-3577 (1979).
77. A. H. Schmaier, W. Claypool, and R. W. Colman, Crotalocytin: Recognition and purification of a timber rattlesnake platelet aggregating protein, *Blood,* **56:**1013-1019 (1980).
78. A. H. Schmaier, and R. W. Colman, Crotalocytin: Characterization of the timber rattlesnake platelet activating protein, *Blood,* **56:**1020-1028 (1980).
79. F. Markwardt, W. Barthel, E. Glusa, and M. Hoffman, Über die Freisatzung biogener Amine aus Blutpättchen durch tierische Gifte. N-S Archiv für experimentelle Pathologie und Pharmakologie, **252:**297 (1966).
80. J. Prado-Francesci, Thesis, University of Campinas, Brazil (1970).
81. B. B. Vargaftig, J. Prado-Franceschi, M. Chignard, J. Lefort, and G. Marlas, European Journal of Pharmacology, **68:**451-464 (1980).

82. B. B. Vargaftig, M. Chignarad, J. Benveniste, J. Lefort, and F. Wal, Annals of New York Academy of Sciences, **370:**119-137 (1981).
83. G. Marlas, D. Joseph, and C. Huet, *Biochimie,* **65:**619-628 (1983).
84. J. Prado-Franceschi, D. Q. Tavares, R. Heritel, and Lobo de Araujo, Effects of convulxin, a toxin from rattlesnake venom, on platelets and leukocytes of anesthetized rabbits, *Toxicon,* **19:**661-666 (1981).
85. G. Marlas, The potent platelet-activating glycoprotein from the venom of *Crotalus durissus cascavella:* Separation and characterization of its α and ß subunits, *Toxicon,* **23:**592 (1985).
86. C. Ouyang, Y. H. Ma, H. C. Jih, and C. M. Teng, Characterization of the platelet aggregation inducer and inhibitor from *Echis carinatus* snake venom, *Biochimica et Biophysica Acta,* **841:**1-7 (1985).
87. C. M. Teng, Y. H. Ma, and C. Ouyang, Action mechanism of the platelet aggregation inducer and inhibitor from *Echis carinatus* snake venom, *Biochimica et Biophysica Acta,* **841:**8-14 (1985).
88. C. Ouyang, and T. F. Huang, Platelet aggregation inhibitor from *Trimeresurus grmineus* snake venom, *Biochimica et Biophysica Acta,* **757:**332-341 (1983).
89. T. F. Huang, and C. Ouyang, Action mechanism of the potent platelet aggregation inhibitor from *Trimeresurus gramineus* snake venom, *Thrombosis Research,* **33:**124-138 (1984).
90. C. Ouyang, H. I. Yeh, and T. F. Huang, A potent platelet aggregation inhibitor purified from *Agkistrodon halys* (Mamushi) snake venom, *Toxicon,* **21:**797-804 (1983).
91. T. F. Huang, H. I. Yeh, and C. Ouyang, Mechanism of action of the platelet aggregation inhibitor purified from *Agkistrodon halys* (Mamushi) snake venom, *Toxicon,* **22:**243-251 (1984).
92. C. Ouyang, L. J. Hwang, and T. F. Huang, Inhibition of rabbit platelet aggregation by α-fibrinogenase purified from *Agkistrodon rhodostoma* (Malyan pit viper) snake venom, *Journal of the Formosan Medical Association,* **84:**1197-1206 (1985).
93. C. Ouyang, and T. F. Huang, Inhibition of platelet aggregation by 5'-nucleotidase purified from *Trimeresurus gramineus* snake venom, *Toxicon,* **21:**491-501 (1983).
94. C. Ouyang and T. F. Huang, Platelet aggregation inhibitors from *Agkistrodon acutus* snake venom, *Toxicon,* **24:**1099-1106 (1986).
95. T. F. Huang, Y. J. Wu, and C. Ouyang, Characterization of platelet aggregation inhibitor from *Agkistrodon rhodostoma* venom, *Biochimica et Biophysica Acta,* **925:**248-257 (1987).
96. C. M. Teng, W. Jy, and C. Ouyang, Cardiotoxin from *Naja naja atra* snake venom: a potentiator of platelet aggregation, *Toxicon,* **22:**463-470 (1984).
97. C. M. Teng, Y. H. Chen, and C. Ouyang, Biphasic effect on platelet aggregation by phospholipase A purified from *Vipera russellii* snake venom, *Biochimica et Biophysica Acta,* **772:**393-402 (1984).
98. C. Ouyang, and T. F. Huang, Effect of the purified phospholipases A_2 from snake and bee venoms on rabbit platelet function, *Toxicon,* **22:**705-718 (1984).
99. M. S. Read, R. W. Shermer, and K. M. Brinkhous, Venom coagglutinin: An activator of platelet aggregation dependent on von Willebrand factor. Proceedings of the National Academy of Sciences, U.S.A. **75:**4514-4518 (1978).
100. M. S. Read, J. Y. Potter, and K. M. Brinkhous, Venom coagglutinin for detection of von Willebrand factor activity in animal plasmas, *J. Lab. Clin. Medicine,* **101:**74-82 (1983).
101. S. Shiau, and C. Ouyang, Isolation of coagulant and anticoagulant principles from the venom of *Trimeresurus gramineus, Toxicon,* **2:**213-220 (1965).
102. H. C. Cheng, and C. Ouyang, Isolation of coagulant and anticoagulant principles from the venom of *Agkistrodon acutus, Toxicon,* **4:**235-243 (1967).
103. C. M. Teng, Y., P. Kuo, L. G. Lee, and C. Ouyang, Effect of cobra venom phospholipase A_2 on platelet aggregation in comparison with those produced by arachidonic acid and lysophosphatidylcholine, *Thrombosis Research,* **44:**875-886 (1986).
104. C. M. Teng, Y. P. Kuo, L. G. Lee, and C. Ouyang, Characterization of the anticoagulants from Taiwan cobra (*Naja naja atra*) snake venom, *Toxicon,* **25:**201-210 (1987).
105. E. Habermann, Über das thrombinähnlich wirkende Prinzip von Jararacagift, Archiv für Experimentelle Pathologie und Pharmacologie, **234:**291 (1958).
106. A. Magalhäes, G. J. de Oliveira, and C. R. Diniz, Purification and partial characterization of a thrombin-like enzyme from the venom of the bushmaster snake, *Lachesis muta noctivaga, Toxicon,* **19:**279-294 (1981).

107. N. A. Narsh, and B. C. Whaler, Separation and partial characterization of a coagulant enzyme from *Bitis gabonica* venom, *British Journal of Haematology,* **26:**295-306 (1974).
108. W. H. Holleman, and L. J. Weiss, The thrombin-like enzyme from *Bothrops atrox* snake venom: Properties of the enzyme purified by affinity chromatography of p-aminobenzamidine substitute agarose, *Journal of Biological Chemistry,* **251:**1663-1669 (1976).
109. C. A. Bonilla, Defibrinating enzyme from Timber rattlesnake (*Crotalus h. horridus*) venom: a potential agent for therapeutic defibrination, I. Purification and properties, *Thrombosis Research,* **6:**151-169 (1975).
110. L. Andersson, Isolation of thrombin-like activity from the venom of *Trimeresurus okinavensis, Haemostasis,* **1:**31-43 (1972).
111. K. Stocker, and G. H. Barlow, The coagulant enzyme from *Bothrops atrox* venom (Batroxobin), *Methods in Enzymology,* **45:**214-223 (1976).
112. P. J. Gaffney, N. A. Marsh, and Talalak, South East Asian Journal of Tropical Medicine and Public Health, **10:**258-265 (1979).
113. D. Nolan, L. S. Hall, and G. H. Barlow, Ancrod, the coagulating enzyme from malayan pit viper (*Agkistrodon rhodostoma*) venom, *Methods in Enzymology,* **45:**205-213 (1976).
114. F. S. Markland, Crotalase, *Methods in Enzymology,* **45:**223-236 (1976).
115. L. F. Guan, X. Zhang, and C. W. Chi, Different mechanism of fibrin polymerization and fibrinopeptide release induced by human thrombin and thrombin-like enzyme (TLE) from the snake venom of *Agkistrodon halys pallas, Thrombosis and Haemostasis,* **54:**313 (1985).
116. S. S. Bajwa, F. S. Markland, and F. E. Russel, Fibrinolytic enzymes in western diamondback rattlesnake (*Crtalus atrox*) venom, *Toxicon,* **18:**285-290 (1980).
117. S. S. Bajwa, F. S. Markland, and F. E. Russell, Fibrinolytic and Fibrinogen clotting enzymes present in the venoms of western diamondback rattlesnake, *Crotalus atrox,* eastern diamondback rattlesnake, *Crotalus adamanteus,* and southern pacific rattlesnake, *Crotalus viridis helleri, Toxicon,* **19:**53-59 (1981).
118. S. S. Bajwa, H. Kirakossian, K. N. N. Reddy, and F. S. Markland, Thrombin-like and fibrinolytic enzymes in the venoms from the gaboon viper (*Bitis gabonica*), eastern cottonmouth moccasin (*Agkistrodon p. piscivorus*) and southern copperhead (*Agkistrodon c. contortrix*) snakes, *Toxicon,* **20:**427-432 (1982).
119. K. W. E. Denson, Coagulant and anticoagulant action of snake venoms, *Toxicon,* **7:**5-11 (1969).
120. K. W. E. Denson, R. Borrett, and R. Biggs, The specific assay of prothrombin using the Taipan snake venom, *British Journal of Haematology,* **21:**219-226 (1971).
121. J. Rosing, H. Speijer, J. W. P. Govers-Riemslag, G. Trans, and R. F. A. Zwaal, Purification and properties of prothrombin activators from the venom of *Notechis scutatus scutatus* and *Oxyuranus scutellatus, Thrombosis and Haemostasis,* **54:**312 (1985).
122. R. G. Macfarlane, and B. Barnett, The hemostatic possibilities of snake venom, *Lancet,* **ii:**985-987 (1934).
123. K. W. E. Denson, F. E. Russell, D. Almagro, and R. C. Bishop, Characterization of the coagulant activity of some snake venoms, *Toxicon,* **10:**557-562 (1972).
124. A. J. Quick, Thromboplastin generation: Effect of the Bell-Alton reagent and Russell's viper venom on prothrombin consumption, *American Journal of Clinical Pathology,* **55:**555-560 (1971).
125. C. Bon, and H. Hofmann, Prothrombin and Factor X activators from *Bothrops atrox* venom, *Toxicon,* **23:**553 (1985).
126.T. B. Lo, Y. H. Chen, and C. Y. Lee, Chemical studies of Formosan cobra (*Naja naja atra*) venom. Part I. Chromatographic separation of crude venom on CM-Sephadex and preliminary characterization of its components, *Journal of Chinese Chemical Society,* **13:**25 (1966).
127. C. M. Teng, Y. H. Chen, and C. Ouyang, Purification and properties of the main coagulant and anticoagulant principles of *Vipera russellii* snake venom, *Biochimica et Biophysica Acta,* **786:**204-212 (1984).
128. Z. S. Latallo, and E. Teisseyre, Evaluation of Reptilase-R and thrombin clotting time in the presence of fibrinogen degradation products and heparin, *Scandinavian Journal of Haematology (Supplement),* **13:**261-266 (1971).
129. L. L. Phillips, H. J. Weiss, and N. P. Christy, Effects of puff adder venom on the coagulation mechanism II. In Vitro, *Thrombosis et Diathesis Haemorrhagica,* **30:**499-508 (1973).

130. K. Stocker, H. Fischer, J. Meier, M. Brogli, and L. Svendsen, Characterization of the protein C activator Protac® from the venom of the southern copperhead (*Agkistrodon contortrix*) snake, *Toxicon,* **25**:239-252 (1987).
131. T. F. Huang, J. C. Holt, H. Lukasiewicz, and S. Niewiarowski, Trigramin, a low molecular weight peptide inhibiting fibrinogen interaction with platelet receptors expressed on glycoprotein IIb/IIIa complex, *The Journal of Biological Chemistry,* **262**:16157-16163 (1987).
132. T. F. Huang, C. Z. Liu and C. Ouyang, Halysin, a potent platelet aggregation inhibitor, inhibits the fibrinogen binding to the activated platelets, *Thrombosis and Haemostasis,* **62**:112 (1989).
133. T. F. Huang, C. Z. Liu, W. J. Wang and C. Ouyang, Characterization of the Trigramin-like peptides from snake venoms, the specific antagonists of the fibrinogen receptor on human platelets. Proceedings of the International Scientific Symposium on Fibrinogen, Thrombosis, Coagulation and Fibrinolysis, *Taipei,* A-054 (1989).
134. C. Ouyang, M. L. Hung and C. M. Teng, Effect of *Trimeresurus* snake venoms on platelet aggregation, Proceedings of 9th World Congress on Animal, Plant and Microbial Toxins, p. 39 (1988).
135. C. Oyuang, M. L. Hung, T. F. Huang, and C. M. Teng, Effects of Trimucytin and Triwagulerin, platelet aggregation inducers isolated from *Trimeresurus mucrosquamatus* and *Trimeresurus wagleri* snake venoms on aggregation of rabbit platelets, *Thrombosis and Haemostasis,* **62**:338 (1989).

THROMBIN-LIKE VENOM ENZYMES: STRUCTURE AND FUNCTION

Hubert Pirkle and Ida Theodor

Department of Pathology
College of Medicine
University of California,
Irvine, CA 92717, USA

INTRODUCTION

While thrombin itself has an ever-growing number of reported biochemical and biological actions, the term thrombin-like enzyme usually connotes only one of these actions — the ability to induce the clotting of fibrinogen. The phylogenetic and developmental distance between mammalian thrombin and the thrombin-like enzymes of snake venoms provides opportunities for comparative studies on the structural requirements for their highly specific coagulant actions on fibrinogen.

SOURCES

Most of the known thrombin-like enzymes come from the pit viper family, in particular the genera *Agkistrodon, Bothrops, Crotalus, Lachesis,* and *Trimeresurus* (Table 1). In additon, such enzymes have been found in the venoms of the true vipers, *Bitis gabonica* and *Cerastes vipera*, and a member of the Colubridae family, *Dispholidus typus*. There are other species, not listed here, whose whole venom seems to possess thrombin-like activity, but for which no effort toward purification has been made (Stocker, 1978).

CLOTTING OF FIBRINOGEN

In general, the thrombin-like venom enzymes display considerably less fibrinogen-clotting activity than thrombin. The most active of them is reported to have a specific activity (Table 2) a little less than half that of human thrombin (Fenton et al., 1977). The range of clotting activity among the venom enzymes appears to be quite wide, even taking into account the substantial uncertainty regarding the specific activities listed in Table 2. The sources of uncertainty are several. First, individual venom enzymes exhibit a high degree of variability in their reactivity with fibrinogens from different species, a variability that correlates poorly or not at all with the thrombins that have been used as a reference standards of NIH unitage (Csákó et al., 1975; Raw et al., 1986; Shieh et al., 1985; Wik et al., 1972). Next, the state of venom enzyme preparations with respect to active or inactive impurities and denaturation is often uncertain. Finally, assay conditions have most likely not been uniform. For these reasons the activities given in Table 2 should be taken as no more than rough approximations.

In 1958 Birger Blombäck reported the signal observation that, although the clotting of fibrinogen by thrombin entails the release of two different peptides from fibrinogen

Fibrinogen, Thrombosis, Coagulation, and Fibrinolysis, Edited by
C. Y. Liu and S. Chien, Plenum Press, New York

Table 1. Sources of Thrombin-like Venom Enzymes*

Genus	Species/Subspecies
Agkistrodon	*acutus*
	caliginosus
	contortrix contortrix
	halys blomhoffii
	halys pallas
	rhodostoma
Bitis	*gabonica*
Bothrops	*asper*
	atrox (Hoge)**
	insularis
	jararaca
	moojeni
	pictus
Cerastes	*vipera*
Crotalus	*adamanteus*
	atrox
	durissus terrificus
	horridus horridus
	viridis helleri
	viridis oreganus
Dispholidus	*typus*
Lachesis	*muta muta*
	muta noctivaga
Trimeresurus	*flavoviridis*
	gramineus
	okinavensis

*References for these venom sources can be identified from titles in the bibliography of this review.
**Termed *Bothrops atrox marajoensis* or *Bothrops marajoensis* in some publications.

(FPA and FPB), a thrombin-like venom preparation called Reptilase* induced the clotting of fibrinogen by releasing only one of these peptides, FPA. Thus, FPA rlease was evidently the trigger for fibrin polymerization. Since Reptilase fibrin seemed basically similar to thrombin fibrin, the hunt was soon under way to uncover the function of FPB release. In this effort the thrombin-like venom enzymes that relase FPA only, or strongly preferentially, became central tools in a host of studies comparing the properties of the two types of fibrin (Pirkle and Theodor, 1988). Meanwhile, Shainoff found that an enzyme from the Southern copperhead snake venom (*A. contortrix contortrix*) releases FPB preferentially (Herzig et al., 1970), almost exclusively under certain conditions (Shainoff and Welches, 1988), and that this also could be made to induce fibrin formation. This provided another probe of the fibrin polymerization process.

Table 3 summarizes the fibrinopeptide-releasing action of the various thrombin-like venom enzymes. Most of them described so far strongly preferentially release FPA. In fact, this group has generally, over the years, been thought to release no FPB. But even in his original report in 1958, Blombäck called attention to the release by Reptilase of a small amount of peptide that he considered might be FPB. And since the introduction by Kehl, Lottspeich and Henschen (1981) of a precise and sensitive HPLC method for fib-

*The source of Blombäck's Reptilase was thought to be *Bothrops jarraca* venom, but it is now known that this preparation contained also constituents of *B. atrox* (K. Stocker, personal communication).

Table 2. Fibrinogen-clotting Activity of Venom Enzymes

Venom Source	Specific Activity* ("NIH units"/mg protein)	Reference
Agkistrodon acutus	276	Ouyang et al. (1971)
A. caliginosus	I 118 II 139	Suzuki and Takahashi (1984)
A. rhodostoma	667	Nolan et al. (1976)
Bitis gabonica	45	Pirkle et al. (1986a)
Bothrops asper	350 551	Stocker and Barlow (1976) Ortiz and Gubensek (1976)
B. atrox (Hoge)	333	Stocker and Barlow (1976)
(*B. marajoensis* and *B. atrox marajoensis* in these publications)	675	Holleman and Weiss (1976)
B. insularis	656	Selistre and Giglio (1987)
B. moojeni	88 220	Stocker and Barlow (1976) Holleman and Weiss (1976)
Crotalus adamanteus	450	Bajwa and Markland (1979)
C. horridus horridus	125	Shu et al. (1983)
Lachesis muta muta	168 944	Campos et al. (1988) Silveira et al. (1989)
L. m. noctivaga	1320	Magalhães et al. (1981)
Trimeresurus gramineus	46	Ouyang and Yang (1974)
T. okinavensis	450	Andersson (1972a)

*When activity was given in other units, "NIH units" were computed according to Seegers, 1962 (1.25 Iowa units/NIH unit), according to Barlow and Devine, 1972 (3 ancrod units/NIH unit) and according to Stocker, personal communication (5.7 batroxobin units/NIH unit). The uncertain quantitative significance of these "NIH units" is explained in the text.

Table 3. Fibrinopeptide Release by Thrombin-like Venom Enzymes

I. Fibrinopeptide A
- A. Exclusive (?)
  - *Trimeresurus flavoviridis*
- B. Strongly preferential
  - *Agkistrodon rhodostoma*
  - *Bothrops atrox*
  - *Bothrops jararaca* (?)
  - *Crotalus adamenteus*
  - *Crotalus horridus horridus*
  - *Lachesis muta muta*
  - *Trimeresurus okinavensis*
- C. Moderately preferential
  - *Bitis gabonica*
  - *Agkistrodon halys blomhoffii* (?)

II. Fibrinopeptide B
- A. Strongly preferential
  - *Agkistrodon contortrix contortrix*
  - *Agkistrodon halys pallas*

Question marks are explained in the text.

rinopeptide analysis, it has become clear that most of these enzymes have at least a little FPB-releasing activity (Pirkle and Theodor, 1988). This is why a question mark appears after *T. flavoviridis*; reactions of fibrinogen with this enzyme, sufficiently long in duration to exclude FPB release, have not been reported. The question mark after *B. jararaca* refers to the mixed source used by Blombäck (see footnote). Under the heading of moderately preferential FPA release is the enzyme from *B. gabonica* which releases FPB rapidly enough to be obvious early in the reaction (Gaffney et al., 1973; Pirkle et al., 1986). Under the same heading is listed *A. halys blomhoffi* whose enzyme was mentioned briefly in a note (Suzuki, 1966) but not in sufficient detail to know where it belongs in this classification, hence another question mark. The FPB-releasing enzymes from *A. contortrix* have now been joined by a similar one from *A. halys pallas* (Guan et al., 1984). Dyr et al. (1989b) have made the fascinating observation that, in the case of one of the two *A. contortrix* thrombin-like enzymes (Dyr et al. 1989a), the preference for FPB release can be switched to FPA by calcium ion.

As a precautionary note in the use of these enzymes to probe various functions or effects of fibrinopeptide release, it is sometimes overlooked that, depending on the enzyme, other bonds in the fibrinogen molecule may be cleaved (Pirkle and Theodor, in press, b). If such secondary cleavages are not recognized, serious misinterpretations can arise.

Very low levels of fibrinogen-clotting activity have been noted for some other venom enzymes such as the kallikrein-like esterase of *C. atrox* (Bjarnason et al, 1983; Pirkle et al. 1989) and the platelet-aggregating enzyme from *B. atrox* (Niewiarowski et al., 1979).

FACTOR XIII ACTIVATION

For technical reasons, it has been fairly easy for reasearchers to note whether or not their fibrinogen-clotting preparations activate factor XIII. Many have been reported to do so, only to find that the factor XIII-activating activity can be separated from the clotting activity and, therefore, is plainly due to another venom enzyme. But there still remain a few that seem to have the intrinsic capacity to activate factor XIII. These include both of the *A. contortrix* enzymes (Dyr et al., 1989a and 1989b), the *B. gabonica* enzyme (Pirkle et al., 1986), and the platelet-aggregating enzyme from *B. atrox* (Niewiarowski et al., 1979).

OTHER BIOCHEMICAL PROPERTIES

These enzymes are evidently all serine proteinases since, in each of the many instances investigated, they were inactivated by diisopropyl phosphofluoridate or phenylmethanesulfonyl fluoride. Also like other serine proteinases, the thrombin-like venom enzymes cleave a number of synthetic esters and amides (Ascenzi et al., 1986; Exner and Koppel, 1972; Markland and Pirkle, 1977b; Simmons et al., 1985) with some preference for arginyl bonds, but otherwise little in common regarding specificity.

So far, the thrombin-like venom enzymes are single polypeptide chains with several disulfide bonds and no free sulfhydryl groups (Collins and Jones, 1974; Guan et al., 1984; Markland and Damus, 1971; Pirkle et al., 1986; Shieh et al., 1988; Shu et al., 1983; Stocker and Barlow, 1976; Suzuki and Takahashi, 1984). With the sole exception of the thrombin-like enzyme from *T. flavoviridis* (Shieh et al., 1988), they are glycoproteins whose carbohydrate content runs as high as 36% (Nolan et al., 1976). Although their molecular weights are mostly in the upper 20K to mid-30K range (Pirkle and Theodor, in press, a) a few are substantially higher, e. g. *A. contortrix*, 68K (Dyr et al., 1989a) and *D. typus*, 55-67K, depending on method (Hiestand and Hiestand, 1979). These proteins tend to be acidic, down to pI 3.2 (Ouyang et al., 1976), partly because of their sialic acid content but also due to their amino acid composition. Thus, for example, the pI of the desialylated *C. adamanteus* enzyme was 4.6 (Bajwa et al., 1979) and the carbohydrate-free enzyme from *T. flavoviridis*, 4.8 (Shieh et al., 1985 and 1988). Only the enzyme from *C. vipera* has been reported to have a pI on the alkaline side of neutrality (7.7) (Farid et al., 1989).

Inhibitors of thrombin-like enzymes have been summarized in a recent review (Pirkle and Theodor, in press, b). Of particular interest are the thrombin inhibitors, heparin and hirudin, which in most instances do not inhibit thrombin-like enzymes. Hirudin is, of course, by far the more specific, so that its capacity to inhibit the *T. flavoviridis* enzyme (Kosugi et al., 1986) is of considerable interest.

PRIMARY STRUCTURES

While some partial structures were known earlier (Pirkle et al., 1981, 1983 and 1986; Bjarnason et al., 1983), the first complete primary structure, that of the *B. moojeni* enzyme, was deduced from its cDNA sequence by Itoh, Yamashina, and coworkers (1987). In the following year the same group reported the entire gene sequence for this enzyme (Itoh et al., 1988). The gene structure of the *B. moojeni* enzyme turned out to be of considerable interest in establishing its molecular phylogenetic position among the serine proteinases. As indicated in Fig. 1, the exon/intron organization of the *B. moojeni* enzyme ("bat" for batroxobin) is identical to that of trypsin and kallikrein, in contrast to the great variation in gene organization among the serine proteinases in general. Since the thrombin-like snake venom enzymes show a high degree of homology among themselves (Fig. 2 and Table 4) it is reasonable to suppose that, like the *B. moojeni* enzyme, they all belong to the trypsin/kallikrein family, a conjecture consistent with their greater sequence similarity to trypsin and kallikrein than to thrombin (Table 4).

In an effort to find clues to the structural basis for the ability to remove FPA from fibrinogen, shared by thrombin and the venom enzymes, sequence positions were identified which are occupied by the same or similar residues in thrombin and the thrombin-like enzymes but by different residues in the non-thrombin enzymes. Such "thrombic" residues are boxed in Fig. 2. The possibility that these positions might be involved in a secondary binding site for the common fibrinogen substrate was investigated by determining their location within the as yet unpublished crystal structure of bovine thrombin (Brian Edwards, personal communication). The resulting spatial relationship of these positions to the active site of thrombin did not bear any coordinated orientation suggestive of a substrate binding function. The reason for the failure of this strategy is not immediately obvious. Perhaps these shared structural features are concerned with shared functions of a different kind which have yet to be discovered.

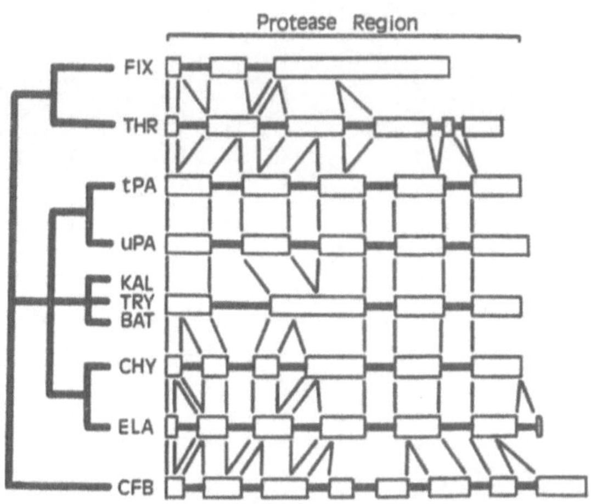

Fig. 1. Exon/intron organization of the proteinase regions of serine proteinase genes (Itoh et al., 1988). The family tree deduced from protein sequences, is given on the left. On the right, open boxes represent exons and black bars represent introns. FIX, human factor IX; THR, human thrombin; tPA, human tissue plasminogen activator; uPA, porcine urokinase; KAL, mouse kallikrein; TRY, rat trypsin; BAT, batroxobin; CHY, rat chymotrypsin; ELA rat elastase; CFB, human complement factor B. (Reproduced from Itoh and coworkers (1988) with permission of the American Society for Biochemistry and Molecular Biology.)

```
                                  10                  20                    30
Batroxobin      V I G G D E C D I N E H P F L   A F M Y Y S P R Y F   C G M T
Flavoxobin      V I G G D E C D I N E H P F L V A L Y D A W S G R F L C G G T
Crotalase       V I G G D E C N I N E H R F L V A L Y D Y W X Q X F L
Crotalus atrox  V V G G D E C N I N E H R S L V A I F V     S T E F D C G G D
Gabonase        V V G G A E C K I D G H R C L A L L Y
Thrombin        I V E G Q D A E V G L S P W Q V M L F R K S P Q E L L C G A S
Kallikrein      V V G G Y N C E M N S Q P W Q V A V Y Y F     G E Y L C G G V
Trypsin         I V G G Y T C G A N T V P Y Q V S L N       S G Y H F C G G S
                                40                  50                  60
Batroxobin      L I N Q E W V L T A A H C             N R R F             M R
Flavoxobin      L I N P E W V L T A A H C           D S K N F K           M K
Crotalus atrox  L I N V E W V L T A A H C
Thrombin        L I S D R W V L T A A H C L L Y P P W B K N F T V D D L L V R
Kallikrein      L I D P S W V I T A A H C A T D N Y Q V W L G R N N   L Y E D
Trypsin         L I N S Q W V V S A A H C         Y K S G I Q             V R
                            70                    80                  90
Batroxobin      I H L G K H A G S V A N Y D E V V R Y P K E K F I   C P N K K
Flavoxobin          L G A H S Q K V L N E D E Q I R N P K E K F I   C P N K K
Crotalase                     R S V Q F D K E Q Q R
Thrombin        I     G K H S R T R Y E R K V E K I S M L D K I Y I H P R Y N
Kallikrein      E P F A Q H R L V S Q S F P H P G F N Q D L I W N H T R Q P G
Trypsin             L G Q D N I N V V E G N Q Q F I S A S K S I V   H P S Y N
                            100                 110                 120
Batroxobin      K N V I T D K D I M L I R L D R P V K N S E H I A P L S L P S
Flavoxobin      N T E V L D K D I M L I K L D S P V S Y S E H I A P L S L P S
Crotalase                 D K D I M L I R L N K P V S Y S E H I A P L S L P S
Thrombin        W K E N L D R D I A L L K L K R P I E L S D Y I H P V C L P D
Kallikrein      D D Y S N   D L M L L H L S Q P A D I T D G V K V I D L P I
Trypsin           S N T L N N D I M L I K L K S A A S L N S R V A S I S L P T
                          130                 140                 150
Batroxobin      N P P S       V G S V C     R I M G W G A I T T     S E D T
Flavoxobin      S P P S       V G S V C     R I M G W G S I T P     V E E T
Crotalase       S P P I       V G S V C     R A M G W G Q T T S     P Q E T
Thrombin        K Q T A A K L L H A G F K G R V T G W G N R R E T W T T S V A
Kallikrein      E E P K       V G S T C     L A S G W G S I T P   D G L E L S
Trypsin         S C A S       A G T Q C     L I S G W G N T K S     S G   T S
                        160                 170                 180
Batroxobin      Y P D V P H C A N I N L     F N N T V C R E A Y N G   L P   A
Flavoxobin      F P D V P H C A N I N L     L D D V E C K P G Y P E L L P E Y
Crotalase       L P D V P H C A N I N L     L D Y E V C
Thrombin        E V Q P S V L Q V V N L P L V E R P V C K A S T R I R I T   N
Kallikrein          D D L Q C V N I D L     L S N E K C V E A H K E E V T D
Trypsin         Y P D V L K C L K A     P I L S N S S C K S A Y P G Q I T S N
                      190                 200                 210
Batroxobin      K T L C A G V L     Q G G   I D T C G G D S G G P L I C
Flavoxobin      R T L C A G V L     Q G G   I D T C G F D S G T P L I C
Crotalus atrox    T L C A G I     P E G G   L D T C G G D S G G P L I C
Thrombin        D M F C A G Y K P G E G K R G D A C E G D S G G P F V M K S P
Kallikrein      L M L C A G E M     D G G K   D T C K G D S G G P L I C
Trypsin           M F C A G Y L     E G G K   D S C Q G D S G G P V V C
                    220                 230                 240
Batroxobin          N G Q F Q   G I L S W G S D P C A E P R K P A F Y T K V F
Flavoxobin          N G Q F Q   G I V Y I G S H P C G Q S R K P G I Y T K V F
Crotalase                                         C D C K E K Y F D C W N T F
Crotalus atrox      D G K P D   G I T S
Thrombin        Y N N R W Y Q M G I V S W G   E G C D R N G K Y G F Y T H V F
Kallikrein          N G V L Q   G I T S W G F N P C G E P K K P G I Y T K L I
Trypsin             S G K L Q   G I V S W G   S G C A Q K N K P G V Y T K V C
                  250                 260
Batroxobin      D Y L P W I Q S I I A G N K T A T C   P
Flavoxobin      D Y N A W I Q S I I A G N T A A T C L P
Crotalase       K E D
Thrombin        R L K K W I Q K V I D R L G S
Kallikrein      K F T P W I K E V M K E N P
Trypsin         N Y V S W I K Q T I A S N
```

Fig. 2. Amino acid sequences of thrombin-like venom enzymes and some other serine proteinases. In order, from Itoh et al. (1987), Shieh et al. (1988), Pirkle et al. (1981, 1983), Pirkle et al. (1989), Pirkle et al. (1986), Magnusson (1975), Swift et al. (1982), and Titani et al. (1975). Batroxobin, flavoxobin, crotalase, and gabonase are trivial names for the thrombin-like venom enzymes from, respectively, *B. atrox, T. flavoviridis, C. atrox*, and *B. gabonica*. Gaps (—) have been introduced to maximize homology. Boxed residues are those that are the same or similar for thrombin-like enzymes and thrombin, but different for the other serine proteinases listed. Modified from Pirkle and Theodor (in press, a).

Table 4. Thrombin-like Enzymes and Serine Proteinases

	Sequence Identity (%)					
	Batroxobin	Flavoxobin	Crotalase	Thrombin	Trypsin	Kallikrein
Batrtoxobin	—	69	64	31	39	39
Flavoxobin	69	—	76	27	39	31
Crotalase	64	76	—	29	32	36

Batroxobin, flavoxobin, and crotalase are trivial names for the thrombin-like venom enzymes from, respectively, *B. atrox, T. flavoviridis,* and *C. adamanteus.*

ACKNOWLEDGMENTS

The work of the authors was supported in part by Grants HL-22875 and HL-31267 from the National Institute of Health.

REFERENCES

Alexander, G., Grothusen, J., Zepeda, H., and Schwartzman, R. J., 1988, Gyroxin, a toxin from the venom of *Crotalus durissus terrificus,* is a thrombin-like enzyme, *Toxicon,* **26**:953.

Andersson, L., 1972a, Isolation of thrombin-like activity from the venom of thrombin-like activity from the venom of *Trimeresurus okinavensis, Haemostasis,* **1**:31.

Andersson, L., 1972b, The action and inhibition of the fibrinogen-clotting enzyme from the venom of *Trimeresurus okinavensis, Haemostasis,* **1**:79.

Ascenzi, P., Bertollini, A., Bolognesi, M., Guarneri, M., Menegatti, E., and Amiconi, G., 1986, Primary specificity of ancrod, the coagulating serine proteinase from the Malayan pit viper (Agkistrodon rhodostoma) venom, *Biochim. Biophys. Acta,* **871**: 225.

Bajwa, S. S., and Markland, F. S., Jr., 1979, A new method for purification of the thrombin-like enzyme from the venom of the Eastern diamondback rattlesnake, *Thrombos. Res.,* **16**:11.

Bajwa, S. S., Markland, F. S., and Russell, F. E., 1981, Fibrinolytic and fibrinogen clotting enzymes present in the venoms of Western diamondback rattlesnake, *Crotalus atrox,* Eastern diamondback rattlesnake, *Crotalus adamanteus,* and Southern Pacific rattlesnake, *Crotalus viridis helleri, Toxicon,* **19**:53.

Bajwa, S. S., Kirakossian, H., Reddy, K. N. N., and Markland, F. S., 1982, Thrombin-like and fibrinolytic enzymes in the venoms from the Gaboon viper (*Bitis gabonica*), Eastern cottonmouth moccasin (*Agkistrodon p. piscivorus*) and Southern copperhead (*Agkistrodon c. contortrix*) snakes, *Toxicon,* **20**:427.

Barlow, G. H., and Devine, E. M., 1974, A study of the relationship between ancrod and thrombin clotting units, *Thrombos. Res.,* **5**:695.

Bjarnason, J. B., Barish, A., Direnzo, G. S., Campbell, R., and Fox, J. W., 1983, Kallikrein-like enzymes from *Crotalus atrox* venom, *J. Biol. Chem.,* **258**:12566.

Blombäck, B., Blombäck, M., and Nilsson, I. M., 1957, Coagulation studies on "Reptilase", an extract of the venom from *Bothrops jararaca, Thrombos. Diath. Haemorrh.,* **1**:76.

Blombäck, B., 1958, Studies on the action of thrombic enzymes on bovine fibrinogen as measured by N-terminal analysis, *Ark. Kemi,* **12**:321.

Campos, S., Escobar, E., Lazo, F., Yarleque', Marsh, N. A., Peyser, P. M., Whaler, B. C., Creighton, L. J., and Gaffney, P. J., Partial separation and characterization of thrombin-like enzyme from the venom of the Peruvian bushmaster snake, Lachesis *muta muta, in* Hemostasis and Animal Venoms, Pirkle, H., and Markland, F. S., Jr., eds., Marcel Dekker, Inc., New York, 1988, p. 107.

Collins, J. P., and Jones, J. G., 1972, Studies on the active site of IRC-50 Arvin, the purified coagulant enzyme from *Agkistrodon rhodostoma* venom, *Eur. J. Biochem.,* **26**:510.

Collins, J. P., and Jones, J. G., 1974, Identification of serine and histidine as essential amino-acid residues in the coagulation enzyme ancrod, *Eur. J. Biochem.*, **42**:81.

Csákó, G., Gazdy, E., Csernyanszky, H., and Szilagyi, T., 1975, Specificity of bovine thrombin and Reptilase for mammalian plasmas, *Blut*, **30**:283.

Damus, P. S., Markland, F. S., Jr., Davidson, T. M., and Shanley, J. D., A purified procoagulant enzyme from the venom of the Eastern diamondback rattlesnake (*Crotalus adamanteus*): In vivo and in vitro studies, *J. Lab. Clin. Med.*, **79**:906.

Dvilansky, A., Britten, A. F. H., and Leowy, A. G., 1970, Factor XIII by an isotope method. I. Factor XIII (transamidase) in plasma, serum, leukocytes, erythrocytes and platelets and evaluation of screening tests of clot solubility, *Brit. J. Haematol.*, **18**:399.

Dyr, J. E., Blombäck, B., and Kornalik, F., 1983, The fibrinogenolytic and procoagulant activity of Southern copperhead venom enzymes, *Thrombos. Res.*, **30**:185.

Dyr, J. E., Hessel, B., Suttner, J., Kornalik, F., and Blombäck, B., 1989, Fibinopeptide-releasing enzymes in the venom from the Southern copperhead snake (*Agkistrodon contortrix contortrix*), *Toxicon*, **27**:359.

Dyr, J. E., Blombäck, B., Hessel, B., and Kornalik, F., 1989, Conversion of fibrinogen to fibrin induced by preferential release of fibrinopeptide B, *Biochim. Biophys. Acta*, **990**:18.

Edgar, W., and Prentice, C. R. M., 1973, The proteolytic action of ancrod on human fibrinogen and its polypeptide chains, *Thrombos. Res.*, **2**:85.

Esnouf, M. P. and Tunnah, G. W., 1967, The isolation and properties of the thrombin-like activity from *Ancistrodon rhodostoma* venom, *Br. J. Haematol.*, **13**:581.

Ewart, M. R., Hatton, M. W. C., Basford, J. M., and Dodgson, K. S., 1970, The proteolytic action of Arvin on human fibrinogen, *Biochem. J.*, **118**:603.

Exner, T., and Koppel, J. L., 1972, Observations concerning the substrate specificity of Arvin, *Biochim. Biophys. Acta*, **258**:825.

Farid, T. M., Tu, A. T., and Farid El-Asmar, M., 1989, Characterization of cerastobin, a thrombin-like enzyme from the venom of *Cerastes vipera* (Sahara sand viper), *Biochemistry*, **28**:371.

Fenton, J. W., II, Fasco, M. J., Stackrow, A. B., Aronson, D. L., Young, A. M., and Finlayson, J. S., 1977, Human thrombins. Production, evaluation, and properties of alpha-thrombin, *J. Biol. Chem.*, **252**:3587.

Furlan, M., Seelich, T., and Beck, E. A., 1976, Clottability and cross-linking reactivity of fibrin(ogen) following differential release of fibrinopeptides A and B, *Thrombos. Haemostas.*, **36**:582.

Furukawa, Y., and Kyozo, H., 1977, Factor X converting and thrombin-like activities of *Bothrops jararaca* snake venom, *Toxicon*, **15**:107.

Gaebert, A. K., 1977, Isolation of thrombin-like factors from Malayan pit viper venom by affinity chromatography, *Toxicon*, **15**:217.

Guan, L.-F., Chi, C.-W., and Yuan, M., 1984, Study on the thrombin-like enzyme preferentially releasing fibrinopeptide B from the snake venom of *Agkistrodon halys pallas*, *Thrombos. Res.*, **35**:301.

Hatton, M. W. C., 1973, Studies on the coagulant enzyme from *Agkistrodon rhodostoma venom*. Isolation and some properties of the enzyme, *Biochem. J.*, **131**:799.

Herzig, R. H., Ratnoff, O. D., and Shainoff, J. R., 1970, Studies on a procoagulant fraction of Southern copperhead snake venom: The preferential release of fibrinopeptide B, *J. Lab. Clin. Med.*, **76**:451.

Hessel, B. and Blombäck, M., 1971, The proteolytic action of the snake venom enzymes Arvin and reptilase on N-terminal chain-fragments of human fibrinogen, *FEBS Lett.*, **18**:318.

Hiestand, P. C. and Hiestand, R. R., 1979, *Dispholidus typus* (boomslang) snake venom: Purification and properties of the coagulant principle, *Toxicon*, **17**:489.

Holleman, W. H., and Coen, L. J., 1970, Characterizatrion of peptides released from human fibrinogen by Arvin, *Biochim. Biophys. Acta*, **200**:587.

Holleman, W. H., and Weiss, L. J., 1976, The thrombin-like enzyme from *Bothrops atrox* snake venom. Properties of the enzyme purified by affinity chromatography on p-aminobenzamidine-substituted agarose, *J. Biol. Chem.* **251**:1663.

Holm, B., and Godal, H. C., 1986, Degradation of fibrin by Reptilase and thrombin, *in:* Fibrinogen and Its Derivatives, Müller-Berghaus, G., Scheefers-Borchel, U., Selmayr, E., and Henschen, A., eds., Elsevier Science Publishers B. V. (Biomedical Division), Amsterdam, pp. 45-48.

Itoh, N., Tanaka, N., Mihashi, S., and Yamashina, I., 1987, Molecular cloning and sequence analysis of cDNA for batroxobin, a thrombin-like snake venom enzyme, *J. Biol. Chem.,* **262:**3132.

Itoh, N., Tanaka, N., Funakoshi, I., Kawasaki, T., Mihashi, S., and Yamashina, I., 1988, Organization of the gene for batroxobin, a thrombin-like snake venom enzyme, *J. Biol. Chem.,* **263:**7628.

Kehl, M., Lottspeich, F., and Henschen, A., 1981, Analysis of human fibrinopeptides by high-performance liquid chromatography, *Hoppe-Seyler's Z. Physiol. Chem.,* **362:**1661.

Kirby, E. P., Niewiarowski, S., Stocker, K., Kettner, C., and Shaw, E., 1979, Thrombocytin, a serine protease from *Bothrops atrox* venom. 1. Purification and characterization of the enzyme, *Biochemistry,* **18:**3564.

Kopec, M., Latallo, Z. S., Stahl, M., and Wegrzynowicz, Z., 1969, The effect of proteolytic enzymes on fibrin stabilizing factor, *Biochim. Biophys. Acta,* **181:**437.

Kosugi, T., Ariga, Y, Nakamura, M., and Kinjo, K., 1986, Purification and some chemical properties of thrombin-like enzyme from *Trimeresurus flavoviridis* venom, *Thrombos. Haemost.,* **55:**24.

Mackessy, S. P., 1988, Isolation of a thrombin-like protease from the venom of the Northern Pacific rattlesnake, Abstracts of 9th World Congress on Animal, Plant and Microbial Toxins, Stillwater, Oklahoma, p. 37.

Maglahães, A., de Oliveira, G. J., and Diniz, C. R., 1981, Purification and partial characterization of a thrombin-like enzyme from the venom of the bushmaster snake, *Lachesis muta noctivaga, Toxicon,* **19:**279.

Magnusson, S., Peterson, T. E., Sottrup-Jensen, L., and Claeys, H., 1975, Complete primary structure of prothrombin: Isolation, structure and reactivity of ten carboxylated glutamic acid residues and regulation of prothrombin activation by thrombin, *in:* Proteases and Biological Control, Reich, E., Rifkin, D. B., and Shaw, E., eds., Cold Spring Harbor Laboratory, Cold Spring Harbor, N.Y., pp. 123-149.

Markland, F. S., and Damus, P. S., 1971, Purification and properties of a thrombin-like enzyme from the venom of *Crotalus adamanteus* (Eastern diamondback rattlesnake), *J. Biol. Chem.,* **246:**6460.

Markland, F. S., Jr., 1976, Crotalase, *Meth. Enzymol.,* **45:**223.

Markland, F. S., and Pirkle, H., 1977a, Thrombin-like enzyme from the venom of *Crotalus adamanteus* (Eastern diamondback rattlesnake), *Thrombos. Res.* **10:**487.

Markland, F. S., Jr., and Pirkle, H., 1977b, Biological activities and biochemical properties of thrombin-like enzymes from snake venoms, *in* Chemistry and Biology of Thrombin, Lundblad, R. L., Fenton, J. W., II, and Mann, K. G., eds., Ann Arbor Science Publishers, Ann Arbor, pp. 71-89.

Markland, F. S., Kettner, C., Shaw, E., and Bajwa, S. S., 1981, The inhibition of crotalase, a thrombin-like venom enzyme, by several peptide chloromethyl ketone derivatives, *Biochem. Biophys. Res. Commun.,* **102:**1302.

Markland, F. S., Kettner, C., Schiffman, S., Shaw, E., Bajwa, S. S., Reddy, K. N. N., Kirakossian, H., Patkos, G. B., Theodor, I., and Pirkle, H., 1982, Kallikrein-like activity of crotalase, a snake venom enzyme that clots fibrinogen, *Proc. Natl. Acad. Sci. USA,* **79:**1688.

Marsh, N. A. and Whaler, B. C., 1974, Separation and partial characterization of a coagulant enzyme from *Bitis gabonica* venom, *Br. J. Haematol.,* **26,** 295.

Mattock, P., and Esnouf, M. P., 1971, Differences in the subunit structure of human fibrin formed by the action of Arvin, Reptilase and thrombin, *Nature New Biol.* **223:**277.

McDonagh, J., and McDonagh, R. P., 1975, Alternative pathways for activation of factor XIII, Brit. J. Haematol., **30:**465.

Niewiarowski, S., Kirby, E. P., Brudzynski, T. M., and Stocker, K., 1979, Thrombocytin, a serine protease from *Bothrops atrox* venom. 2. Interaction with platelets and plasma-clotting factors, *Biochemistry,* **18:**3570.

Nolan, C., Hall, L. S., and Barlow, G. H., 1976, Ancrod, the coagulating enzyme from Malayan pit viper (*Agkistrodon rhodostoma*) venom, *Meth. enzymol.,* **45:**205.

Olexa, S. A., and Budzynski, A. Z., 1980, Effects of fibrinopeptide cleavage on the plasmic degradation pathways of human cross-linked fibrin, *Biochemistry,* **19:**647.

Olascoaga, M. E., Zavaleta, A., and Marsh, N. A., 1988, Preliminary studies of the effects of a Peruvian snake *Bothrops pictus* (jergon of the coast) venom upon fibrinogen, *Toxicon,* **26:**501.

Ortiz, F. A., and Gubensek, F., 1976, Isolation and some properties of blood clotting enzyme from the venom of *Bothrops asper, Bull. Inst. Pasteur,* **74:**145.

Ouyang, C., Hong, J. S., and Teng, C.-M., 1971, Purification and properties of the thrombin-like principle of *Agkistrodon acutus* venom and its comparison with bovine thrombin, *Thromb. Diath. Haemorrh.,* **26:**224.

Ouyang, C., and Hong, J.-S., 1974, Inhibition of the thrombin-like principle of *Agkistrodon acutus* venom by group-specific enzyme inhibitors, *Toxicon,* **12:**449.

Ouyang, C., and Yang, F.-Y., 1974, Purification and properties of the thrombin-like enzyme from *Trimeresurus gramineus* venom, *Biochim. Biophys. Acta,* **351:**354.

Ouyang, C., Teng, C.-M., Yang, F.-Y., and Hong, J. S., 1976, Studies of the coagulant and anticoagulant principles of Formosan crotalid venoms, *Toxicon,* **14:**415.

Ouyang, C., and Teng, C.-M., 1978, *In vivo* effects of the purified thrombin-like and anticoagulant principles of *Agkistrodon acutus* (hundred pace snake) venom, *Toxicon,* **16:**583.

Pasha, M. A. Q., Joshi, A. P., and Gangal, S. V., 1988, A study of the coagulant activity of Indian green pit viper venom, *Biochem. Int.,* **16:**219.

Pirkle, H., Markland, F. S., Theodor, I., Baumgartner, R., Bajwa, S. S., and Kirakossian, H., 1981, The primary structure of crotalase, a thrombin-like venom enzyme, exhibits closer homology to kallikrein than to other serine proteases, *Biochem. Biophys. Res. Commun.,* **99:**715.

Pirkle, H., Markland, F. S., and Theodor, I., 1983, Amino acid sequences from crotalase, a venom enzyme with thrombin-like and kallikrein-like actions, *Federation Proc.,* **42:**1993, abstr.

Pirkle, H., Theodor, I., Miyada, D., and Simmons, G., 1986, Thrombin-like enzyme from the venom of *Bitis gabonica.* Purification, properties, and coagulant actions, *J. Biol. Chem.,* **261:**8830.

Pirkle, H., and Theodor, I., 1988, Thrombin-like enzymes in the study of fibrin formation, in *Hemostasis and Animal Venoms,* Pirkle, H. and Markland, F. S., Jr., eds., Marcel Dekker, Inc., New York, p. 121.

Pirkle, H., Theodor, I., and Lopez, R., 1989, Catroxobin, a weakly thrombin-like enzyme from the venom of *Crotalus atrox.* NH_2-Terminal and active site amino acid sequences, *Thrombos. Res.,* **56:**159.

Pirkle, H., and Theodor, I., in press (a), Structure of thrombin-like snake venom proteinases, *in* Medical Use of Snake Venom Proteins, Stocker, K., ed., CRC Press, Boca Raton, FL.

Pirkle, H., and Theodor, I., in press (b), Thrombin-like enzymes from snake venoms, *in* "Handbook of Natural Toxins, Volume 5, Reptile and Amphibian Venoms", Tu, A. T., ed., Marcel Dekker, Inc., New York.

Pitney, W. R. and Regoeczi, E., 1979, Inactivation of 'Arvin' by plasma proteins, *Brit. J. Haematol.,* **19,** 67.

Pizzo, S. V., Schwartz, M. L., Hill, R. L., and McKee, p., 1972, Mechanism of ancrod anticoagulation. A direct proteolytic effect on fibrin, *J. Clin. Invest.,* **51:**2841.

Raw, I., Rocha, M. C., Esteves, M. I., and Kamiguti, A. S., 1986, Isolation and characterization of a thrombin-like enzyme from the venom of *Crotalus durissus terrificus, Brazilian J. Med. Biol. Res.,* **19:**333.

Sato, T., Iwanaga, S., Mizushima, Y., and Suzuki, T., 1965, Studies venoms. XV. Separation of arginine ester hydrolase of venom *Agkistrodon halys blomhoffii* into three enzymatic entities: "Bradykinin releasing," "clotting" and "permeability increasing," *J. Biochem.,* **57:**380.

Seegers, W. H., 1962, "Prothrombin", Harvard University Press, Cambridge.

Selistre, H. S., and Giglio, J. R., 1987, Isolation and characterization of a thrombin-like enzyme from the venom of the snake *Bothrops insularis* (jararaca ilhoa), *Toxicon,* **25:**1135.

Shainoff, J. R., and Dardik, B. N., 1979, Fibrinopeptide B and aggregation of fibrinogen, *Science,* **204:**200.

Shainoff, J., and Welches, W. R., 1988, Studies of the preferential release of fibrinopeptide B by copperhead procoagulant enzyme, *in:* Hemostasis and Animal Venoms, Pirkle, H. and markland, F. S., Jr., eds. Marcel Dekker, Inc., New York, p. 85.

Shieh, T. C., Tanaka, S., Kihara, H., Ohno, M., and Makisumi, S., 1985, Purification and characterization of a coagulant enzyme from *Trimeresurus flavoviridis* venom, *J. Biochem.,* **98:**713.

Shieh, T.-C., Kawabata, S.-I., Kihara, H., Ohno, M., and Iwanaga, S., 1988, Amino acid sequence of a coagulant enzyme, flavoxobin, from *Trimeresurus flavoviridis* venom, *J. Biochem.,* **103:**596.

Shu, Y.-Y, Moran, J. B., and Geren, C. R., 1983, A thrombin-like enzyme from timber rattlesnake venom, *Biochim. Biophys. Acta,* **748:**236.

Silveira, A. M. V., Magalhães, A., Dihiz, C. R., and de Oliveira, E. B., 1989, Purification and properties of the thrombin-like enzyme from the venom of *Lachesis muta muta, Int. J. Biochem.,* **21:**863.

Simmons, G., Bundalian, M., Theodor, I., Martinoli, J., and Pirkle, H., 1985, Action of crotalase, an enzyme with thrombin-like and kallikrein-like specificities, on tripeptide nitroanilide derivatives, *Thrombos. Res.,* **40:**555.

Stocker, K., and Barlow, G. H., 1976, The coagulant enzyme from *Bothrops atrox* venom (batroxobin), *Meth. Enzymol.,* **45:**214.

Stocker, K., 1978, Defibrinogenation with thrombin-like snake venom enzymes, *in* Fibrinolytics and Antifibrinolytics (Handbook of Experimental Pharmacology, Vol. 46), Markwardt, F., ed., Springer, Heidelberg.

Stocker, K. F., and Meier, J., 1988(a), Thrombin-like snake venom enzymes, *in* Hemostasis and Animal Venoms, Pirkle, H. and Markland, F. S., Jr., eds., Marcel Dekker, Inc., New York, p. 67.

Stürzebecher, J., Stürzebecher, U., and Markwardt, F., 1986, Inhibition of batroxobin, a serine proteinase from *Bothrops* snake venom, by derivatives of benzamidine, *Toxicon,* **24:**585.

Suzuki, T., 1966, Pharmacologically and biochemically active components of Japanese ophidian venoms, *Memorias Instituto Butantan,* **33:**519.

Suzuki, T., and Takahashi, H., 1984, Purification of two thrombin-like enzymes from the venom of *Agkistrodon caliginosus* (kankoku-mamushi), *Toxicon* **22:**29.

Swift, G. H., Dagorn, J.-C., Ashley, P. L., Cummings, S. W., and MacDonald, R. J., 1982, Rat pancreatic kallikrein mRNA: Nucleotide sequence and amino acid sequence of the encoded preproenzyme. *Proc. Natl. Acad. Sci. USA,* **79:**7263.

Tangen, O., Wik, O. K., and Berman, H. J., 1973, The effect of thrombin, Reptilase, and a fibrinopeptide B-releasing enzyme from the venom of the Southern copperhead snake on rabbit platelets, *Microvasc. Res.,* **6:**342.

Teng, C.-M., and Ko, F.-N., 1988, Comparison of the platelet aggregation induced by three thrombin-like enzymes of snake venoms and thrombin. *Thrombos. Haemostas.,* **59:**304.

Teng, C.-M., Wang, J.-P., Huang, T.-F., and Liau, M.-Y., 1989, Effects of venom proteases on peptide chromogenic substrates and bovine prothrombin, *Toxicon,* **27:**161.

Titani, K., Ericsson, L. H., Neurath, H., and Walsh, K. A., 1975, Amino acid sequence of dogfish trypsin, *Biochemistry,* **14:**1358.

Turpie, A. G. G., Prentice, C. R. M., McNicol, G. P., and Douglas, A. S., 1971, In-vitro studies with ancrod ('Arvin'), *Brit. J. Haematol.,* **20:**217.

Viljoen, C. C., Meehan, C. M., and Dawie, P. B., 1979, Separation of *Bitis gabonica* (Gaboon adder) venom arginine esterases into kinin-releasing, clotting and fibrinolytic factors, *Toxicon,* **17:**145.

Walter, M., Nyman, D., Krajnc, V., and Duckert, F., 1977, The activation of plasma factor XIII with the snake venom enzymes ancrod and batroxobin marajoensis, *Thrombos. Haemostas.,* **38:**438.

Wik, K. O., Tangen, O. T., and McKenzie F. N., 1972, Blood clotting activity of Reptilase and bovine thrombin in vitro: A comparative study on seven different species, *Brit. J. Haematol.,* **23:**37.

THROMBIN INHIBITION BY SYNTHETIC HIRUDIN PEPTIDES

John M. Maraganore* and John W. Fenton II‡

*Biogen, Inc., Cambridge, MA, and
‡Wadsworth Center for Laboratories and Research
New York State Department of Health, and
Departments of Physiology and Biochemistry
The Albany Medical College of Union University, Albany, NY, USA

INTRODUCTION

Thrombin (EC 3.4.21.5) is a central, bioregulatory enzyme in thrombosis and hemostasis (1, 2). Thrombin activities are diverse and include fibrin clot formation (3) and stabilization (4), platelet and other cellular activation (5-7), and both positive and negative control of its own zymogen activation (8, 9). When compared to other serine proteinases of the coagulation and fibrinolytic systems, thrombin is unique in lacking the majority of its propiece (prothrombin fragment F 1.2) following zymogen activation (10, 11). Accordingly, the various specificities of thrombin must be accommodated solely within the enzyme moiety consisting of A and B chains.

Examination of peptide bonds susceptible to thrombin cleavage has revealed that while exhibiting preference for arginine or lysine at the P_1 position, the enzyme exhibits no apparent pattern of specificity for amino acids neighboring the scissile bond (12). Nevertheless, thrombin action is by limited proteolysis and, thus, structures in thrombin apart from the active site pocket may serve as elements in governing specificity. Such a functional domain in thrombin is the anion-binding exosite (13) which has been implicated in α-thrombin recognition of fibrin(ogen), binding to various cells, and adsorption to negatively-charged surfaces (14-16). Thrombin's exosite may also function in interactions of enzyme with additional macromolecular substrates, cofactors, and inhibitors.

A potent and specific inhibitor for thrombin is hirudin, a 65 amino acid residue protein from the medicinal leech (17, 18). Hirudin inhibits thrombin by formation of a stoichiometric complex with K_i and K_d values reported in the pM to fM range (19, 20). Kinetic analysis of the hirudin: thrombin interaction has shown that inhibitor binds to enzyme at two separate sites: a high-affinity locus wherein binding does not alter catalytic reactivity and a locus at or near the catalytic pocket (21). The interaction of hirudin with the thrombin active site appears unique to other serine proteinase inhibitors since hirudin lacks an identifiable reactive site for interaction with the thrombin specificity pocket. Moreover, hirudin can bind to disopropylfluorophosphoryl-α-thrombin with high affinity (20), and site-directed mutagenesis of recombinant desulfatohirudin has failed to identify an amino acid residue critical for inhibitory function (20). While Lys-47 of hirudin has been implicated as an inhibitory reactive site, its absence in certain hirudin isoforms (21) and examination of site-directed mutants (20) has established that it is not essential for hirudin function. Examination of the crystallographic structure of the hirudin: thrombin complex shows further that Lys-47 plays no direct role in hirudin-thrombin interactions (A. Tulinsky, personal communication).

In addition to its interactions at the thrombin active site, hirudin contains a highly anionic C-terminal tail (Fig. 1) which appears to form a critical set of interactions with the thrombin anion-binding exosite. The functional importance of the hirudin tail was

Fibrinogen, Thrombosis, Coagulation, and Fibrinolysis, Edited by
C. Y. Liu and S. Chien, Plenum Press, New York

Fig. 1. Amino acid sequence of the hirudin C-terminal tail from residues 53-65.

identified by Chang (22) who showed that treatment of hirudin with *S. aureus* V_8 proteinase or carboxypeptidase Y, both treatments leading to removal of C-terminal structural elements, leads to significant loss of inhibitory activity. Bajusz et al. (23) had shown subsequently that short C-terminal hirudin peptides contain thrombin inhibitory activities, albeit substantially lower in potency as compared to that of hirudin. The interaction of the hirudin C-terminus with the thrombin exosite has been proposed to be predominantly electrostatic in nature (24), although certain hydrophobic residues in this segment of hirudin are known to be important for inhibitory functions (25).

Within the cluster of anionic amino acid residues at the hirudin C-terminus is Tyr-63 which is found as a sulfated derivative in isolated forms of the protein (26). An obligatory role of tyrosine sulfation for hirudin activity has been discounted since recombinant desulfato-hirudins are effective inhibitors of thrombin with specific activities comparable to those of sulfated hirudins (27-29). Also, cleavage of the sulfate ester by chemical means was found to reduce hirudin activity by only 50% (22). Nevertheless, as addressed below, Tyr-sulfation plays a critical role for antithrombin activities of synthetic C-terminal hirudin peptides.

Advances in defining structure-function relationships in the thrombin-hirudin tail interaction have derived from studies of numerous synthetic peptides. C-terminal hirudin peptides, derived from residues 45 to 64, inhibit fibrinogenolytic activities of thrombin but do not inhibit the enzyme's hydrolytic reactivity as measured in spectrophotometric assays for thrombin-catalyzed cleavage of synthetic tripeptidyl-p-nitroanilides (30, 31). As shown in Fig. 2, while hirudin can inhibit thrombin hydrolysis of tosyl-Gly-Pro-Arg-p-ntroanilide, S-Hir_{53-64}, a Tyr-sulfated dodecapeptide derived from residues 53 to 64 of hirudin, fails to inhibit the thrombin-catalyzed reaction at concentrations exceeding enzyme by as high as one million-fold. In fact, these hirudin peptides increase the rate of tripeptidyl-p-nitroanilide hydrolysis by thrombin in an uncompetitive fashion (Fig. 3). This rate enhancement, corresponding to a ~2-fold increase in k_{cat}, is found to be saturable with increasing peptide concentrations. An increase in thrombin catalytic activity toward synthetic substrates following hirudin peptide binding may derive from conformational changes at or near the thrombin active site following complex formation. Results from circular dichroism studies of the thrombin:hirudin peptide complex support this possibility (G. Villanueva, J. Fenton, and J. M. Maraganore, unpublished results).

C-terminal hirudin peptides are effective inhibitors of thrombin-catalyzed hydrolysis of fibrinogen and, thus, potent inhibitors of coagulation. In APTT assays, the minimal hirudin sequence required for maximal inhibitory activity was found to reside in peptides comprising residues 53 to 64 (31). A hirudin peptide of 20 amino acid residues in length, Hir_{45-64}, was found to have a comparable anticoagulant activity when compared to a derivative of 12 residues truncated from the N-terminus. However, subsequent removal of N-terminal amino acids yielding derivatives of 8 and 10 residues led to significant reduction in anticoagulant activities (Table 1). Likewise, treatment of a hirudin dodecapeptide, Hir_{53-64}, with carboxypeptidase A resulted of removal in C-terminal Tyr and Leu residues with concomitant loss of anticoagulant activity. When the dodecapeptide Hir_{53-64} was sulfated at the tyrosine, corresponding to Tyr_{63} in hirudin, the anticoagulant activity of the resulting analog, S-Hir_{53-64} or "hirugen", was found increased by ~10-fold (Table 1). The inhibition by hirugen of thrombin fibrinogenolytic activity is at K_i = 144 nM, while that for the corresponding unsulfated hirudin analog, Hir_{53-64}, is at K_i = 1,990 nM.

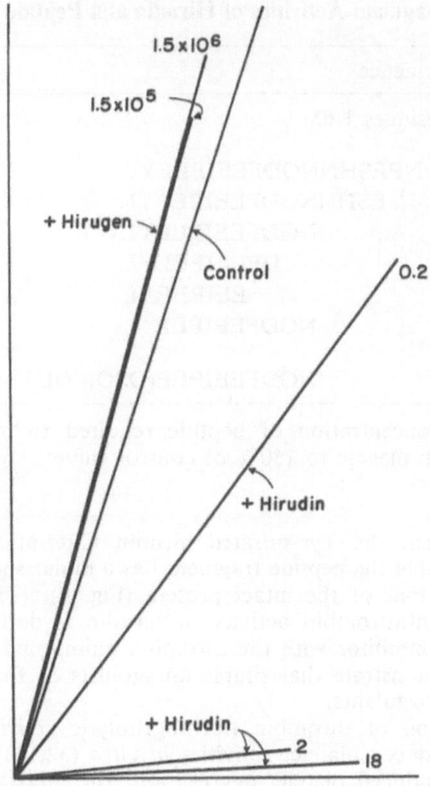

Fig. 2. Effects of hirudin and hirugen on the thrombin-catalyzed rate of Spectrozyme TH hydrolysis. The molar excess of hirudin or hirugen over thrombin is provided to the right of each curve. Assays were performed at 22-23°C using a 0.05 M Tris, pH 7.4, buffer containing 0.1 M NaCl. The rate of reaction was monitored at 405 nm.

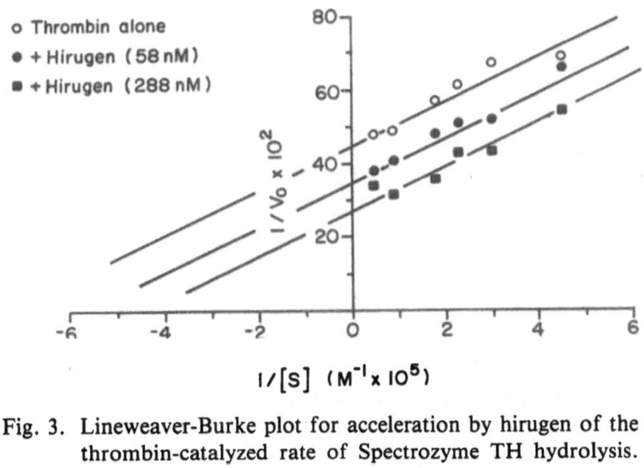

Fig. 3. Lineweaver-Burke plot for acceleration by hirugen of the thrombin-catalyzed rate of Spectrozyme TH hydrolysis.

Table 1. **Anticoagulant Activities of Hirudin and Peptide Derivatives**

Derivative	Sequence	ED_{150}[a], μM
Hirudin	Residues 1-65	0.004
Hir_{45-64}	TPNPESHNNGDFEEIPEEYL	3.4
Hir_{49-64}	ESHNNGDFEEIPEEYL	2.7
Hir_{53-64}	NGDFEEIPEEYL	2.4
Hir_{55-64}	DFEEIPEEYL	6.8
Hir_{57-64}	EEIPEEYL	>1500
Hir_{53-62}	NGDFEEIPEE	>1500
Hirugen	NGDFEEIPEE(OSO_3-Y)L	0.25

[a]ED_{150} is the concentration of peptide required to increase the APTT of human plasma to 150% of control values.

Comparisons of hirugen, the Tyr-sulfated hirudin C-terminal dodecapeptide, with natural hirudin have shown that the peptide fragment has a molar specific inhibitory activity reduced by ~50-fold from that of the intact protein (Fig. 4). This result shows that a significant component of antithrombin activity in hirudin is defined by interactions of C-terminal structure of the inhibitor with the thrombin anion-binding exosite. Moreover, the activities of hirugen demonstrate that simple antagonists of fibrinogen recognition by thrombin are effective anticoagulants.

In addition to inhibition of thrombin fibrinogenolytic activities, hirugen has been found to block thrombin-induced platelet activities *in vitro* (32, 33). As shown in Fig. 5, hirugen inhibits thrombin induced platelet aggregation, thromboxane A_2 generation, and serotonic release in a concentration-dependent manner. Since hirugen does not interfere with thrombin catalytic function, this result suggests that the anion-binding exosite participates in interactions of thrombin with platelet surface glycoprotein(s) and that antagonism of the exosite by peptides like hirugen is sufficient for inhibition of platelet activation *in vitro*. Essentially the same results have been obtained for effects of hirugen on thrombin activation of cultured human umbilical vein endothelial cells (HUVECs). Hirugen and related hirudin tail peptides are found to block thrombin-mediated HUVEC synthesis of PAF and PGI_2, and to inhibit the acquisition by endothelial cells of an adhesive surface for neutrophil binding (34). Thus, studies with hirudin C-terminal peptides show that the exosite may be a key determinant for recognition by thrombin of cellular substrate surfaces.

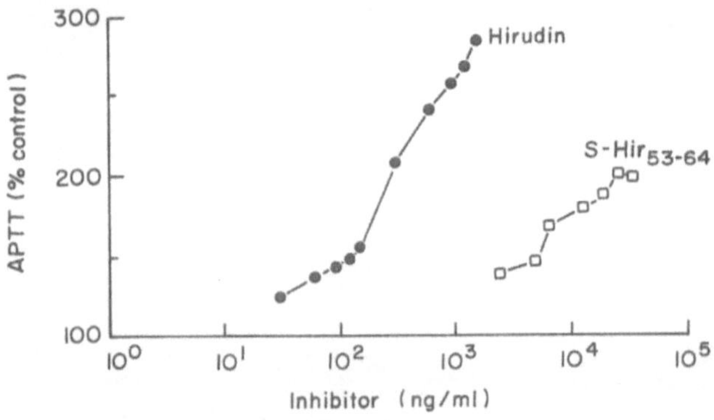

Fig. 4. Comparison of anticoagulant activities of hirudin and hirugen. Activated partial thromboplastin time assays were performed semi-automatically using pooled, normal citrated human plasma.

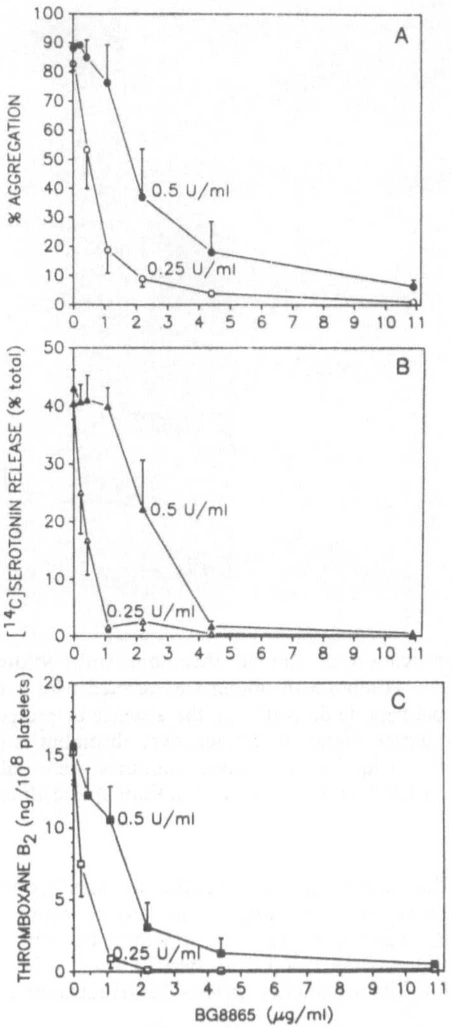

Fig. 5. Inhibition by hirugen of thrombin-induced platelet activities. Assays were performed as described previously (33).

In order to delineate the locus for hirugen binding, a N$^{\alpha}$-dinitrofluorobenzyl-peptide derivative was synthesized as affinity label for the cognate recognition site in thrombin (35, 36). The reaction of human α-thrombin with this peptide was found to yield specific covalent derivatization of enzyme (Fig. 6). The modification was determined to be stoichiometric based on SDS-PAGE analyses showing an increase in the molecular weight of the complex consistent with addition of one mole hirugen/mole thrombin. Automated Edman degradation of peptide fragments derived from the thrombin: hirugen complex allowed identification of Lys-149 as the predominant site of alkylation. Lys-149 is found in one of several loops in thrombin (in particular, loop segment 145-150) unique to this serine proteinase and, has been proposed as a locus in the exosite most proximal to the thrombin catalytic center (37). Examination of a 3-dimensional model for human α-thrombin (38) reveals that Lys-149 is separated by at least 18 Å from Ser-195 in the enzyme catalytic pocket. Thus, the site in thrombin for hirudin C-terminal peptide binding is discrete both in structure and function from the enzyme active center.

Hirugen and its derivatives comprise a unique and potentially important class of thrombin inhibitors. They have served as valuable probes for thrombin exosite function.

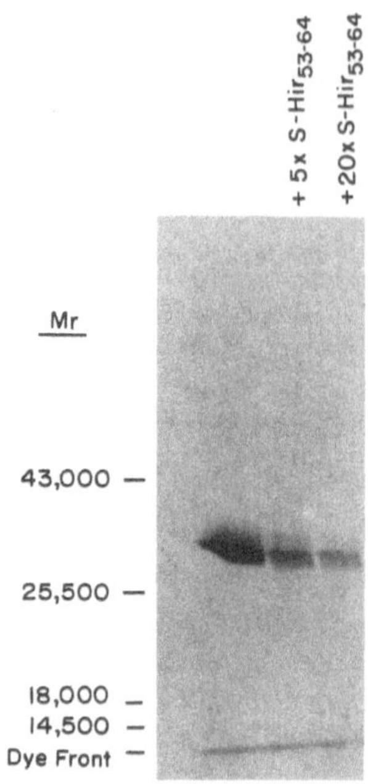

Fig. 6. Covalent modification of human thrombin with N^{α}-dinitrofluorobenzyl-[^{35}S]-hirugen peptide. Human α-thrombin was reacted with a 10-fold molar excess of the affinity-label peptide derivative in the absence or presence of non-radiolabeled hirugen. The molar excess of hirugen over thrombin is provided at the top of lanes 2 and 3. Aliquots of reaction mixtures were subjected to SDS-PAGE analysis (12.5% acrylamide, reducing conditions) and fluorographic analysis.

In addition, these peptides exhibit great potential as antithrombotic agents. Their antithrombotic efficacy has now been established in several animal models (39-41). Unlike recombinant hirudin, hirugen and its derivatives appear to exhibit a wide margin of hemostatic safety and, as short peptides, may not be limited in use by immunogenicity. Studies are warranted to explore the utility of such peptides for treatment of human thromboembolic disease.

ACKNOWLEDGMENTS

Support in part to JWF by NIH grant HL 13160-19.

REFERENCES

1. J. W. Fenton II, *Ann. NY Acad. Sci.,* **485:**5-15 (1986).
2. J. W. Fenton II, *Semin. Thrombos. Hemostas.,* **12:**200-208 (1986).
3. B. Blomback, *in:* Plasma Proteins (B. Blomback, and L. A., Hanson, eds.) pp. 223-253, Wiley, New York (1979).
4. L. Lorand, and K. Konishi, *Arch. Biochem. Biophys.,* **105:**58-67 (1964).
5. D. M. Tollefsen, J. R. Feagler, and P. W. Majerus, *J. Biol. Chem.,* **249:**2646-2651 (1974).
6. G. M. Rodgers, *FASEB J.,* **2:**116-123 (1988).
7. R. W. Colman, *Biochemistry,* **8:**1438-1445 (1969).
8. C. T. Esmon, *Science,* **235:**1348-1352 (1987).
9. C. T. Esmon, W. G. Owen, D. L. Duiguid, and C. M. Jackson, *Biochim. Biophys. Acta,* **310:**289-294 (1973).

10. S. J. F. Degen, R. T. A. MacGillivray, and E. W. Davie, *Biochemistry,* **22:**2087-2097 (1983).
11. M. J. Rabiet, A. Blashill, B. Furie, and B. C. Furie, *J. Biol. Chem.,* **261:**13210-13215 (1986).
12. J.-Y. Chang, *Eur. J. Biochem.,* **151:**217-223 (1985).
13. J. W. Fenton II, *Semin. Thrombos. Hemostas.,* **15:**265-268 (1989).
14. J. W. Fenton II, J. I. Witting, C. Pouliott, and J. Fareed, *Ann. NY Acad. Sci.,* **556:**158-165 (1989).
15. T.-L. Chang, R. D. Feinman, B. H. Landis, and J. W. Fenton II, *Biochemistry,* **18:**113-119 (1979).
16. S. D. Lewis, L. Lorand, J. W. Fenton II, and J. A. Shafer, *Biochemistry,* **26:**7597-7603 (1987).
17. J. Dodt, H. P. Muller, U. Seemuller, and J.-Y. Chang, *FEBS Lett.,* **165:**180-184 (1984).
18. F. Markwardt, *Biomed. Biochim. Acta,* **44:**1007-1013 (1985).
19. J. W. Fenton II, B. H. Landis, D. A. Walz, D. H. Bing, R. D. Feinman, M. P. Zabinski, S. A. Sonder, L. J. Berliner, and J. S. Finlayson, *in:* The Chemistry and Physiology of Human Plasma Proteins (D. H. Bing, ed.) Pergamon Press, New York, pp. 151-183.
20. S. R. Stone, and J. Hofsteenge, *Biochemistry,* **25:**4622-4628 (1986).
21. S. R. Stone, P. J. Braun, and J. Hofsteenge, *Biochemistry,* **26:**4617-4624 (1987).
22. J.-Y. Chang, *FEBS Lett.,* **164:**307-313 (1983).
23. S. Bajusz, I. Fauszt, E. Barabas, N. Diosegi, and D. Bagdy, *in:* Peptides (U. Ragnarsson, ed.) Almqvist and Wiksell Int., Stockholm, pp. 473-476 (1984).
24. G. Noe, J. Hofsteenge, G. Rovelli, and S. R. Stone, *J. Biol. Chem.,* **263:**11729-11735 (1988).
25. J. L. Krstenansky, T. J. Owen, M. T. Yates, and S. J. T. Mao, *J. Med. Chem.,* **30:**1688-1691 (1987).
26. D. Bagdy, E. Barabas, L. Graf, T. Ellebaek, and S. Magnusson, *Methods Enzymol.,* **45:**669-678 (1976).
27. J. Dodt, S. Kohler, and A. Baici, *FEBS Lett.,* **229:**87-90 (1988).
28. R. P. Harvey, E. Degryse, L. Stefani, F. Schamber, J. P. Cazenave, M. Courtney, P. Tolstoshev, and J. P. Lecocq, *Proc. Natl. Acad. Sci. USA,* **83:**1084-1088 (1986).
29. G. Loison, A. Findeli, S. Bernard, M. Nguyen-Juilleret, M. Marquet, N. Riejl-Bellon, D. Carvallo, L. Guerra-Santos, S. W. Brown, M. Courtney, C. Roitsch, and Y. Lemoine, *Biotechnology,* **6:**72-77 (1988).
30. J. L. Krstenansky, and S. J. T. Mao, *FEBS Lett.,* **211:**10-16 (1987).
31. J. M. Maraganore, B. Chao, M. L. Joseph, J. Jablonski, and K. L. Ramachandran, *J. Biol. Chem.,* **264:**8692-8698 (1989).
32. J. A. Jakubowsky, and J. M. Maraganore, *Blood,* **72:**1213 (abtract) (1988).
33. J. A. Jakubowsky, and J. M. Maraganore, *Blood,* **75:**399-406 (1990).
34. S. M. Prescott, A. R. Seeger, G. A. Zimmerman, T. M. MacIntyre, and J. M. Maraganore, *J. Biol. Chem.,* in press (1990).
35. P. Bourdon, and J. M. Maraganore, *Thromb. and Haemostas.,* **62:**533 (abstract) (1989).
36. P. Bourdon, J. W. Fenton II, and J. M. Maraganore, *Biochemistry,* in press (1990).
37. W. Bode, I. Mayr, U. Baumann, R. Huber, S. R. Stone, and J. Hofsteenge, *EMBO J.,* **8:**3467-3475 (1989).
38. B. Furie, D. H. Bing, R. J. Feldmann, D. J. Robison, J. P. Burnier, and B. C. Furie, *J. Biol. Chem.,* **257:**3875-3882 (1982).
39. T. M. Palabrica, M. J. Aronovitz, J. M. Maraganore, and M. A. Konstam, *Thrombos. and Hemostas.,* **62:**434 (abstract) (1989).
40. Y. Cadroy, J. M. Maraganore, S. R. Hanson, and L. A. Harker, *Thrombos. and Haemostas.,* **62:**188 (abstract) (1989).
41. Y. Cadroy, J. M. Maraganore, S. R. Hanson, and L. A. Harker, *Proc. Natl. Acad. Sci. USA,* submitted (1989).

EFFECTS OF STRUCTURAL MODIFICATIONS ON THE PROPERTIES OF TISSUE PLASMINOGEN ACTIVATOR (tPA)

Per Wallén, Xiang-Fei Cheng, and Per-Ingvar Ohlsson

Department of Medical Biochemistry and Biophysics
University of Umeå
90187 Umeå, Sweden

INTRODUCTION

Physiological fibrinolysis is a proteolytic degradation of polymerized fibrin. A particular fibrinolytic enzyme system is present in blood. The central reaction in this system is the activation of a proenzyme, plasminogen, to the proteolytic enzyme plasmin. The activation is triggered by highly specialized proteases, plasminogen activators. One of these, the tissue plasminogen activator (tPA), is synthesized by endothelial cells and excreted to the blood on certain stimuli. Properties, which distinguish tPA from other types of plasminogen activators are a high affinity to fibrin and the very strong stimulation exerted by fibrin on tPA induced activation (2-3 orders of magnitude). These properties have evoked the idea of an efficient ternary activation complex between tPA, plasminogen and fibrin. The work on the isolation of tPA, the determination of its primary structure and gene structure as well as elucidation of its physicochemical properties has been performed by a large number of scientists. Several reviews have been published (e.g. Collen 1980; Rånby and Wallén 1985; Danö et al. 1985; Erickson et al. 1985; Wallén 1987). Fig. 1 shows the now well established primary structure* of wild tPA.

Studies on structure-function relationships by gene technology

It has already been mentioned that tPA, which is excreted to the lumen of blood vessels from the endothelial cells, forms a specific complex with fibrin. The fact that the efficiency of the activator function of tPA furthermore is highly stimulated by fibrin, has suggested that fibrinolysis induced by tPA is a major factor in the prevention of a manifest thrombotic condition. As a consequence of these unique properties, tPA was regarded as an ideal agent for thrombolytic therapy. Early experiments to lyse thrombi performed in animals (Matsuo et al. 1981b) as well as in vitro (Mattsson et al. 1981) seemed also promising and the first attempts to use tPA in patients were made 1981 by Collen and coworkers (Weimar et al. 1981). The cloning of the tPA gene (Pennica et al. 1983) and subsequent developments made large amounts of tPA available for clinical trials. The results of these studies indicated however that tPA with the natural amino acid sequence did not come up to the expectations. The doses needed for a successful treatment were very high mainly due to the extremely short half life in the circulation (3-6 min.), largely caused by a very rapid uptake in the liver (Nilsson et al. 1984; Verstraete et al. 1985). This leads to high costs and furthermore to some undesirable side effects. During the last few years great efforts have

*The numbering of residues is based on the L-form of Jörnvall et al. (1983) and is generally used in this paper. For the sake of clarity the numbering of the shorter form of Pennica et al. (1983) is also given in some cases and then within parenthesis.

Fibrinogen, Thrombosis, Coagulation, and Fibrinolysis, Edited by
C. Y. Liu and S. Chien, Plenum Press, New York

been made to modify the tPA molecule by genetic engineering in order to increase the thrombolytic effect. These studies have also given valuable information with regard to structure-function relations.

In early publications van Zonneveld et al. (1986a, 1986b) and Verheijen et al. (1986) reported on the construction and properties of deletion mutants, which lacked one or more of the domains of the A-chain part of tPA, which counted from the N-terminus are: the Finger (F), the Epithelial Growth Factor (EGF), the Kringle 1 (K1) and the Kringle 2 (K2) domains (Fig. 1). These studies showed that two domains, F and K2, interact with fibrin. Van Zonneveld et al. demonstrated that these domaines bound differently. Variants containing the K2 domain also bound to lysine-Sepharose and could be eluted with ε-amino caproic acid (EACA). Derivatives lacking this domaine showed no affinity to lysine-Sepharose, even if they contained a F domain. The K2 dependent binding to fibrin was furthermore inhibited by EACA contrary to the F dependent binding. Thus the K2 dependent binding of tPA seems to occur by the lysine binding site whereas the F dependent binding does not. Regarding the roles of these two domains in the fibrinolytic process van Zonneveld et al. (1986b) have proposed a mechanism, according to which tPA initially binds to fibrin through the F domain forming a weak ternary complex with plasminogen and fibrin. A slight proteolysis of the fibrin network (fibrin-X monomer formation) should lead to the generation of C-terminally positioned lysine residues, to which plasminogen and tPA may bind by means of their lysine binding sites thus forming a highly efficient, second ternary activation complex. These ideas are supported by earlier findings that tPA induced fibrinolysis proceeds in two phases with distinct kinetic constants and that the switch betwen these phases occurs simultaneously with a slight degradation of fibrin (Norrman et al. 1985). They are also supported by recent findings (Larsen et al. 1988) that the clot lysis proceeds much faster for wild-type tPA than for mutants lacking the F domain, if low concentrations of the activators are used. The difference disappears however at high concentrations of the activators. A different view has been presented by Higgins and Vehar (1987). Using fibrin prepared from intact or slightly degraded fibrinogen, they found that EACA efficiently inhibits the binding of unmodified tPA to intact fibrin whereas the binding to partially degraded fibrin is largely uneffected. Their results would point to that there is two or three binding sites for fibrin in tPA. The conflicting results and different opinions regarding the mechanism for fibrin stimulated activation by tPA reflect the complexity of this reaction. Nevertheless it is now clear that at least two domains in the A chain part of tPA molecule, the Finger and the Kringle 2 domains, both are important structures interacting in the activation complexes.

A matter, which has caused interest is the question why single chain tPA in spite of formally being a proenzyme still has a very high enzymatic activity. It was first proposed that the ε-group of $lys_{280(277)}$, which is the second residue in the B chain sequence could provide the positive charge forming the salt bridge to $Asp_{480(477)}$ necessary for an active conformation of the B chain (Wallén et al. 1983). Point mutants in which Lys_{280} was replaced by a residue lacking positive charge have been produced and showed in some cases a decrease of the activity. The results have however been contradictory. A three dimensional study by means of computer-graphical methods was recently made (Heckel and Hasselbach 1988). One finding in this study was that one residue, $Lys_{419(416)}$, which so far has not been found in this position in other serine proteases, is situated near the active site and assumed to interact with Asp_{480} thus causing the conformation necessary for an active enzyme. In a recent study this residue was mutated and the product seemed to lack enzymatic activity in the single chain form but regained the activity on cleavage with plasmin (Petersen et al. 1989). It is possible that this lysine residue may be responsible for the remarkably high enzymatic activity of single chain tPA.

A main objective in the efforts to create thrombolytic agents more suitable for therapeutic use than wild tPA has been to create mutants, which as compared to native tPA have a substantially increased half life in blood by hampered cellular uptake. They should in addition have at least as high enzymatic efficiency as the wild activator. Usually three types of genetic modifications or combinations of them have been applied: 1) Deletions, in which one or more domains of the A chain are removed. 2) Point mutations, in which one or more amino acids residues are replaced. 3) Chimeras, which are constructed by exchanging domains between different proteins, e.g. connections of the A chains of plasminogen or tPA to the B chain of urokinase.

Regarding the attempts to increase the in vivo half life of tPA two types of structure have been of special interest, the E domain and the glycosylation sites. The oligosaccharides attached to tPA are formed by N-glycosylation and a number of modified forms have been

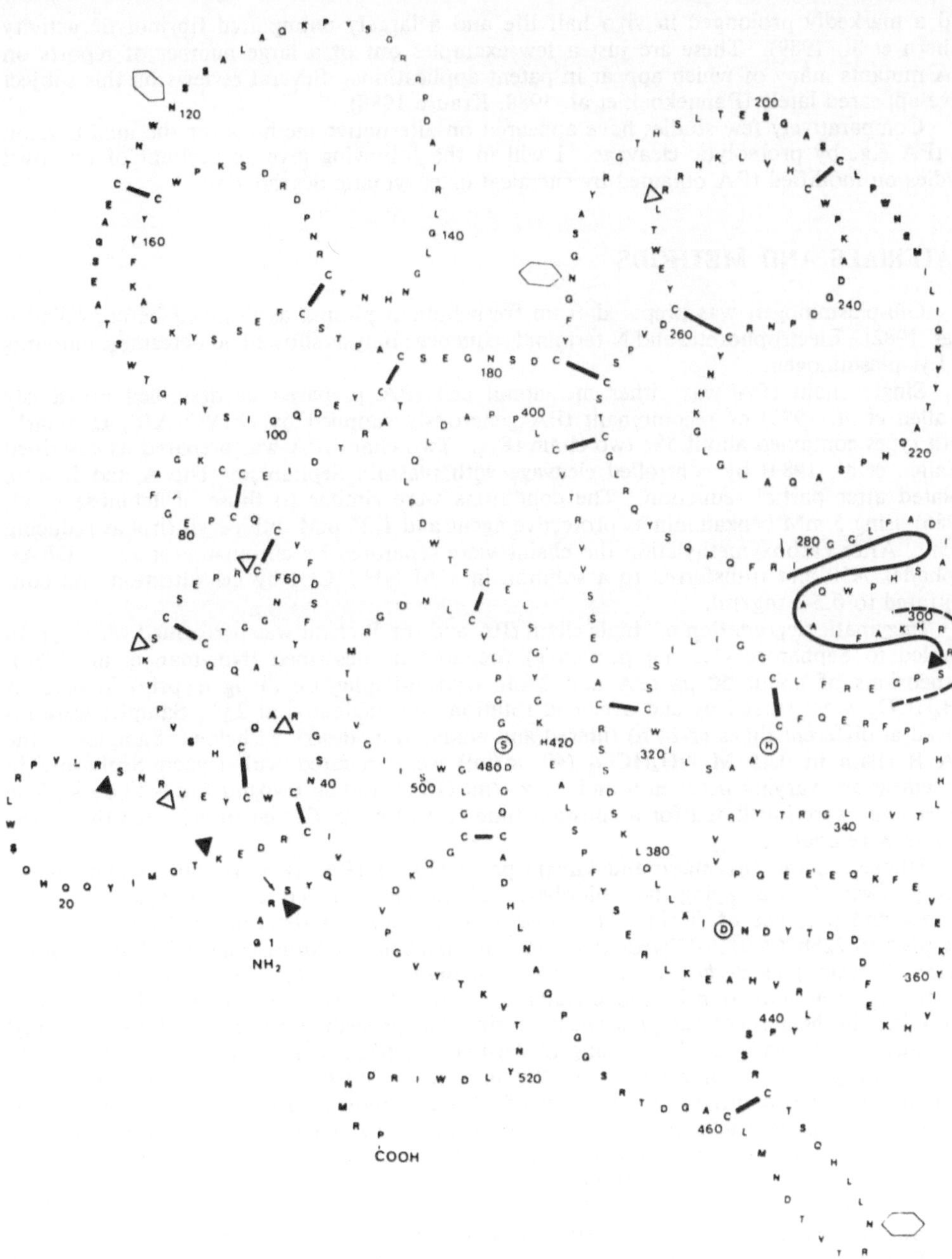

Fig. 1. The complete primary structure of single chain tissue plasminogen activator. The arrows indicate trypsin cleavages within one hour. The times refers to the experimental conditions in this study. The filled triangles indicate trypsin cleavages within four and the open triangle within 24 hours. The residue numbering is based on the L-form (Jörnvall et al. 1983).

obtained by point mutations in the glycosylation sites, Asn-X-Ser(Thr). A prolonged in vivo half life has been reported for these mutants. In a recent report the binding of tPA to hepatocytes was effectively prevented by the introduction of a new glycosylation site in the E domain by point mutation (Gething et al. 1989). In another communication point mutants with two-amino acid substitutions in the peptide link between the F and E domains had a markedly prolonged in vivo half life and a largely unimpaired fibrinolytic activity (Ahern et al. 1989). These are just a few examples out of a large number of reports on tPA mutants many of which appear in patent applications. Several reviews on this subject have appeared lately (Pannekoek et al. 1988; Krause 1988).

Comparatively few studies have appeared on alternative methods for the modification of tPA e.g. by proteolytic cleavage. I will in the following give an account of our own studies on modified tPA obtained by chemical or enzymatic degradation.

MATERIALS AND METHODS

Glu-plasminogen was prepared from fresh human plasma as desribed before (Wallén et al. 1982). Electrophoretic and N-terminal sequence analysis showed no detectable amounts of Lys-plasminogen.

Single chain tPA was either melanoma cell tPA prepared as described previously (Wallén et al. 1983) or recombinant tPA generously supplied by NOVO A/S, Denmark. Both types contained about 5% two chain tPA. Two chain tPA was prepared as described (Wallén et al. 1983) by controlled cleavage with plasmin Sepharose. The A and B were isolated after partial reduction. The conditions were similar to those of Ichinose et al. (1986) using 5 mM benzamidin as protective agent and 1.25 mM dithioerythritol as reducing agent. After carboxymethylation the chains were separated by chromatography on DEAE Sephadex A50 and transferred to a solution in 1 M NH_4HCO_3 by gel filtration and concentrated to 0.5-1 mg/ml.

Enzymatic degradation of single chain tPA and the B chain was performed with trypsin coupled to Sepharose (1.5 mg protein/g) prepared as described (Norrman et al. 1986). Suspensions of about 50 μg tPA and 2 mg trypsin-Sepharose (3 μg trypsin) in 0.15 M NH_4HCO_3, were stirred by end over end rotation and incubated at 25°. Samples were removed at different times (1-24 h) filtered and analyzed as described below. Samples of the tPA B chain in 0.15 M NH_4HCO_3 (40 μg/ml) were digested with trypsin-Sepharose in concentrations varying between 9 and 90 mg/ml (13 to 130 μg trypsin) for 1-2 hours. The suspensions were incubated for a constant time (1-2 h) at 25°C, centrifuged and the supernatants were analyzed.

Studies on the enzymatic and kinetic properties of tPA, tPA derivatives and digests were performed by assaying the amidolytic and the plasminogen activator activities in the absence and presence of fibrin. The amidolytic activity was determined using D-Ile-Pro-Arg-pNA (S-2288 KABI AB Sweden) as substrate in a Titertec Multiscan (Flow Laboratories, Irwine, Scotland) as described previously (Norrman et al.1986). The kinetic parameters for plasminogen activation by single chain tPA, two chain tPA and the tPA B chain were determined in the absence and presence of fibrin. The procedure was essentially as described previously (Norrman et al. 1985) using D-Val-Leu-Lys-pNA (S-2251, KABI AB Sweden) to monitor the generation of plasmin. In the absence of fibrin the concentration of tPA or B chain was 13 nM and the concentrations of Glu-plasminogen varied between 4 and 14 μM. When fibrin was used as stimulator the concentrations of single chain and two chain tPA were 0.02 to 0.04 nM and the range of plasminogen concentrations 0.16 to 2.5 μM. The values for the B chain determinations were 13 nM enzyme and 5 to 20 μM plasminogen. The fibrin was generated by adding a partially degraded fibrin monomer, DESAFIB-X (Biopool Sweden) to the reaction mixture to a concentration of 0.3 μM.

Trypsin cleavage sites were identified by sequence analysis (Applied Biosystem, 477A pulsed liquid phase sequencer) of peptides separated by reversed phase HPLC as described (Norrman et al. 1988) or on peptide fractions from SDS polyacrylamidgel electrophoresis collected by electroblotting onto PVDF-membranes as described (Matsudaira, 1987).

RESULTS AND DISCUSSION

Enzymatic properties of the isolated B-chain

The B-chain isolated from two chain tPA after cleavage of the interchain disulphide bridge by mild reduction had an amidolytic activity of about 1.9 mA/min/nM, which is almost as high as that of two chain tPA, about 2.3 mA/min/nM and significantly higher than the amidolytic activity of single chain tPA, about 0.4 mA/min/nM.

Table 1. Kinetic Constants for Activation of Glu-Plasminogen with Single Chain tPA, Two Chain tPA and the B-Chain from tPA. The Determinations have been Made with and Without Fibrin

	K_m (μM)	k_{cat} (s^{-1})	k_{cat}/K_m (μM^{-1}, s^{-1})
B-chain	27	0.003	0.11×10^{-3}
Tc-tPA	22	0.004	0.18×10^{-3}
Sc-tPA	60	0.003	0.05×10^{-3}
In the presence of fibrin			
B-chain	20	0.009	0.5×10^{-3}
Tc-tPA	0.14	0.06	0.43
Sc-tPA	0.1	0.08	0.80

The B-chain had a considerable plasminogen activator activity. This is indicated by determinations of kinetic constants (Table 1). In the absence of fibrin, the isolated B-chain is more effective as activator than single chain tPA but slightly less effective than two chain tPA. The variations are mainly due to differences in the K_m values, indicating a higher affinity to Glu plasminogen for the two chain tPA and the B-chain than for single chain tPA. These results agree reasonably with recently published results on isolated B-chain (Dodd et al. 1986; Rijken and Groeneveld 1986). The plasminogen activating activities of single chain tPA and two chain tPA are strongly stimulated by fibrin, mainly due to a decrease of K_m, whereas the activity of the B-chain is only slightly changed. These findings are in agreement with our present knowledge that sites in tPA interacting with fibrin are situated in the A-chain. The slightly increased activity of B-chain in the presence of fibrin may be explained by an effect on the conformation of Glu-plasminogen.

Effect of trypsin cleavage on tPA

Sites of cleavage in the single chain tPA molecule on treatment with Sepharose-trypsin were localized by sequence analysis of peptides separated by HPLC after reduction and carboxymethylation of the digests (Fig. 1). Samples were taken from the digestion mixtures at different times and the Sepharose derivative removed by centrifugation and filtration through 0.45 μm filters. The changes of enzymatic properties of tPA by this treatment are shown in Fig. 2. The rapid increase of amidolytic activity and plasminogen activator activity in the absence of fibrin observed within one hour occur according to the analysis of the structural changes simultaneously with the conversion to two chain tPA by the cleavage of a single bond, Arg_{278}-Ile_{279}. These effects are expected and well known from earlier studies on the formation of two chain tPA. (Wallén et al. 1982, Rånby, 1982). The tPA derivatives obtained on further digestion with trypsin behave differently with regard to amidolytic and-plasminogen activator activities. Whereas the amidolytic activity is largely stable during prolonged digestion the ability to activate plasminogen rapidly decreases. This indicates the presence of a plasminogen affinity site, which is destroyed by tryptic cleavage.

The rate of plasminogen activation by single chain tPA is highly increased by the addition of fibrin. As demonstrated in Fig. 2, there is no additional stimulation by the conversion to the two chain derivative. On prolonged digestion of tPA there is a gradual decrease of the fibrin stimulated plasminogen activation activity. The sequence analysis of peptides obtained on digestion revealed several bonds sensitive to trypsin in the A chain (Fig. 1). Thus four cleavages were found in the Finger domain two of which were cleaved early (Lys_{13}, Arg_{30}), cleavages were also found in the Epithelial Growth Factor (Arg_{58}) and the Kringle 2 (Arg_{252}) regions of the A chain (Fig. 1). As reviewed above studies on deletion mutants have shown that the Finger and the Kringle 2 domains of tPA are responsible for the binding to fibrin. The breaks in the A chain observed on trypsin treatment therefore explain the gradual decrease of fibrin stimulated activation of plasminogen.

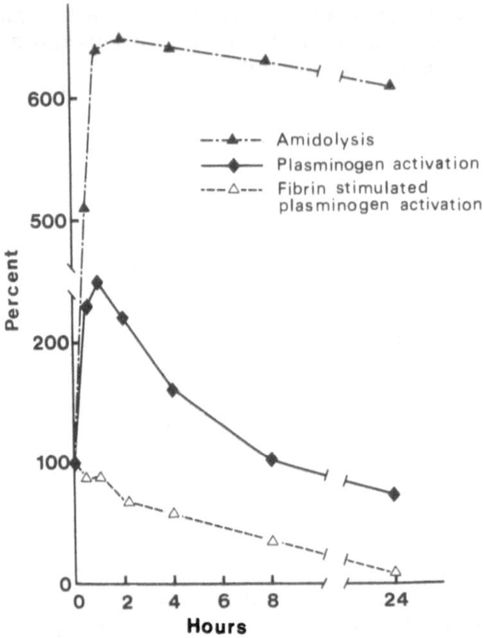

Fig. 2. Effects of tryptic digestion on enzymatic activities of the single chain tPA. Digestion of sc-tPA with trypsin-Sepharose.

In the B chain part of tPA only one bond, Arg_{302}-Ser_{303}, was cleaved at an appreciable rate. This cleavage occurred rather early, and was almost completed after digestion for 4 hours using the conditions related above. The conversion to the two chain tPA is however concluded well in advance of the appearance of the peptide starting with Ser_{383} in the digestion mixture. This indicates that the cleavage in the B chain occurs mainly in two chain tPA and implies a detachment of a peptide, $B_{1\text{-}24}$, from its covalent connection with the remainder of the tPA molecule since this peptide contains no cystein residue. It is well known from X-ray crystallographic studies that the active conformation of serin proteases such as chymotrypsin is due to an ionic bond between the N-terminus and a certain aspartic acid of the B chain sequence (Fercht, 1976). The hydrolysis of Arg_{278}-Ile_{279} is analogous to the cleavage leading to the activation of serin enzyme zymogens and the B chain of tPA highly homologous with the B chain of other mammalian serin enzymes. It is therefore likely that the same mechanism is valid for the active B chain of tPA and thus a considerable decrease of activity would be expected on removal of the $B_{1\text{-}24}$ peptide. However, even after extensive digestion (24 h) the amidolytic activity was very little affected by the cleavage at residue 24 of the B chain (Fig. 2). The main peak obtained by gel filtration of these digests on columns with Sephadex G-50 in 0.5 M ammonium hydrogen carbonate, 0.01% Tween 80, still contained most of the activity (90%) of the digest and HPLC analysis also showed it to contain the $B_{1\text{-}24}$ peptide. This showed that the peptide is bound to the remainder of the molecule by non-covalent interaction. A removal of the peptide was obtained by gel filtration in 4 M but not in 2 M guanidine hydrochloride indicating a strong binding. Peptide depleted tPA lost almost all amidolytic activity, whereas native tPA treated in the same way retained most of its activity. Small amounts of the peptide $B_{1\text{-}24}$ were isolated by HPLC and added to peptide depleted tPA. The amidolytic activity was regained in a concentration dependent manner (Norrman et al., 1988).

On the presence of a secondary plasminogen binding site in tPA

As mentioned above treatment to tPA with trypsin caused a biphasic course of non fibrin stimulated plasminogen activator activity. An early occurring stimulation due to the conversion to the two chain tPA was followed by a marked decrease. The level of amidolytic activity on the other hand showed a monophasic development with a rapid increase to a level, which was kept rather stable even after prolonged digestion, indicating that the active site of the B-chain was rather unaffected by tryptic cleavages. These findings implied that tPA may contain a secondary site for binding of plasminogen but not synthetic substrates. This should be different from the substrate binding pocket near the active site, in which Arg_{560} of plasminogen and the basic amino acid of the synthetic substrates interacts with 4 amino acid residues (Heckel and Hasselbach, 1988). Since specific properties of tPA and many other serine enzymes in general seem to be connected with structures in the A chain it was near at hand to believe that also this site should be localized to this chain. However, the fact that the activation of plasminogen by isolated B chain follows kinetics more similar to two chain than to single chain tPA suggested however that a plasminogen binding site might be localized to the B chain. This prompted us to study the effect of trypsin digestion on the amidolytic and plasminogen activator activities of tPA.

The procedure for tryptic digestion of the B chain was somewhat different from that of tPA. Instead of keeping the concentration of trypsin-Sepharose constant and use different incubation times, the time was kept constant whereas the trypsin-Sepharose concentration was varying and comparatively high, which made it possible to use a short digestion time. This procedure seemed to increase the stability of the amidolytic activity. As shown in Fig. 3 the amidolytic activity was relatively stable even at high trypsin concentration whereas the plasminogen activator activity decreased rapidly and seemed to approach a level around 10% of the original activity. The effect of fibrin on these activities (not shown) was negligible, which is expected since all structures interacting with fibrin seem to be localized in the A chain. On SDS polyacrylamide electrophoresis of the main change caused by the digestion was the generation of a fragment slightly smaller than the B chain and of about the same size as the main fragment from the B chain observed on the cleavage of tPA. The fragment from the B chain was collected from the gel by electroblotting and the amino acid sequence determined. The following sequence was obtained: Ser-Pro-Gly-Glu-Arg-Phe- . This indicates that the peptide bond at Arg_{302} is a main cleavage site also in the isolated B chain and furthermore that the petide B_{1-24} is detached but kept in position by noncovalent bonds.

These findings indicate in the first place that there is a site in tPA with affinity for plasminogen and situated in the B chain and furthermore that the binding to plasminogen is impaired by cleavage of the 24th peptide bond from the N-terminus of the B chain.

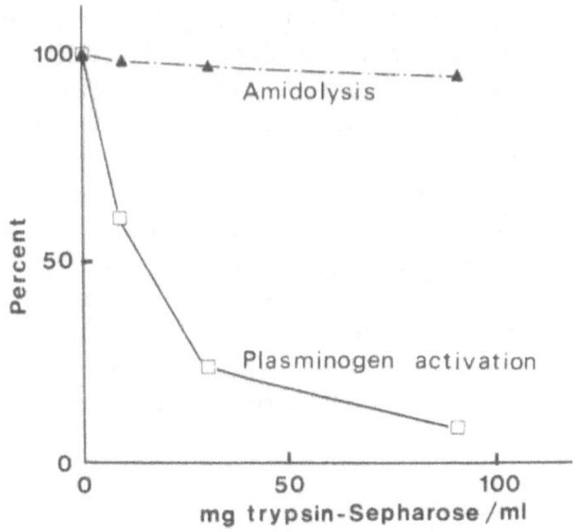

Fig. 3. Effects of tryptic digestion on the amidolytic and plasminogen activator activities of B-chain from tPA. Digestion of tPA B-chain was with trypsin-Sepharose. The determinations were performed in the absence of fibrin.

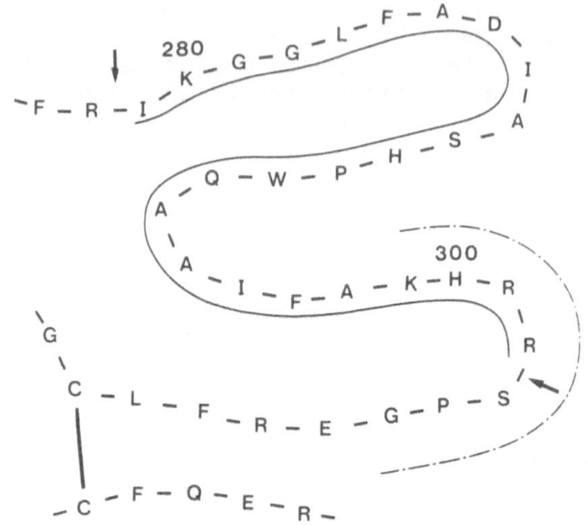

Fig. 4. Enlargement of the tPA sequence Phe_{277} to Gly_{311} of Fig. 1. This part of tPA contains the sites of cleavage at Arg_{278} and Arg_{302}. The dotted line shows the sequence in the B-chain part deleted by Madison et al. (1989).

Whether this postulated site also participates in the ternary activation complex between tPA, fibrin and plasminogen cannot be judged by the present studies. As for the localisation of a binding site it may be proposed that it is situated at the cleavage site. This part of the B chain is highly hydrophilic and probably situated at the surface of the molecule. It is, however, not possible to exclude that the cleavage of the Arg_{302} bond will cause a conformational change in the B chain, which disturbs the binding in a different part of the molecule. Recently Madison et al. (1989) reported on three variants of tPA with mutations in this region of the B chain, which all showed a varying degree of resistance against the inhibitor PAI-1. In the most resistant of these a peptide stretch of seven amino acids, $Lys_{299(296)}$-His-Arg-Arg-Ser-Pro-$Gly_{305(302)}$, was deleted. Interestingly the cleavage in the B chain reported in our present paper and proposed to cause decreased affinity to plasminogen occurs in the middle of this sequence (Fig. 4). However, using an assay, in which fibrin is used as stimulator they did not find that the plasminogen activator activity in this mutant was decreased. Although these results seem contradictory to our results, it should be stressed, that there are differences between these studies, both with regard to the types of derivatives used and to assay methods. The derivative obtained by deletion lacks all basic amino acid residues in this region except one $Arg_{307(304)}$, whereas the derivative obtained by tryptic cleavage retains all residues. With regard to assay methods Madison et al. apply a fibrin stimulated method whereas we use a method in which the plasminogen activation is unstimulated. A difference that might be even more important is that Lys-plasminogen is used as substrate in the analysis of the deletion product whereas we use the native Glu-plasminogen.

ACKNOWLEDGMENTS

Regarding the experimental part of this paper we are greatly indebted for skilful technical assistance by Mrs Kristina Jonsson and Miss Maria Brohlin. Financial support is given by Swedish Medical Research Council (Project 13X-3906), Swedish Cancer Society (Project 2515), the Medical Faculty at Umeå University and the Bergvall Foundation.

REFERENCES

Ahern, T. J., Morris, G. E., Barone, K. M., Horgan, P. G., Angus, L. B., Henson, K. S., Langer-Safir, P. R., Larsen, G. R., 1989, Distinguishing the sites in the amino-terminal region of tissue-type plasminogen activator (t-PA) required for efficient fibrinolytic activity and rapid clearance from the circulation. *Thromb. Haemostas.*, **62:**338.

Collen, D. 1980, On the regulation and control of fibrinolysis. *Thromb. Haemostas.*, **43:**77.

Danö, K., Andreasen, P. A., Gröndal-Hansen, J., Kristensen, P., Nielsen, L. S., and Skriver, L. 1985, Plasminogen activators, tissue degradation and cancer. *Advances in Cancer Research*, **44:**139.

Dodd, I., Fears, R. and Robinson, J. H. 1986, Isolation and preliminary characterization of active B-chain of recombinant tissue type plasminogen activator, *Thromb. Haemostas.*, **55:**94.

Erickson, L. A., Schleef, R. R., Ny, T. and Loskutoff, D. J., 1985, The fibrinolytic system of the vascular wall, *Clin. Haematol.*, **14:**513.

Fersht, A. R., 1972, Conformational equilibria in α- and σ-Chymotrypsin, *J. Mol. Biol.*, **64:**497.

Gething, M.-J., Sambrook, J. and McGookey, D., 1989, Addition of an oligosaccharide side-chain at an epitopic site on the EGF-like domaine of t-PA prevents binding to specific receptors on hepatic cells, *Thromb. Haemostas.*, **62:**338.

Heckel, A. and Hasselbach, K. M., 1988, Prediction of the three-dimensional structure of the enzymatic domaine of t-PA, *J. Computer-Aided Mol. Des.*, **2:**7.

Higgins, D. L. and Vehar, G. A., 1987, Interaction of one-chain and two-chain tissue plasminogen activator with intact and plasmin-degraded fibrin, *Biochemistry*, **26:**7786.

Ichinose, A., Takio, K. and Fujikawa, K., 1986, Localization of the binding site of tissue-type plasminogen activator to fibrin, *J. Clin. Invest.*, **78:**163.

Jörnvall, H., Pohl, G., Bergsdorf, N. and Wallén, P., 1983, Differential proteolysis and evidence for a residue exchange in tissue plasminogen activator suggest possible association between two types of protein microheterogeneity, *FEBS Lett.*, **156:**47.

Krause, J., 1988, Catabolism of tissue-type plasminogen activator (t-PA), its variants, mutants and hybrids, *Fibrinolysis*, **2:**133.

Larsen, G. R., Henson, K. and Blue, Y., 1988, Variants of human tissue-type plasminogen activator: Fibrin binding, fibrinolytic and fibrinogenolytic characterization of genetic variants lacking fibronectin finger-like and/or the epidermal growth factor domaines, *J. Biol. Chem.*, **263:**1023.

Madison, E. L. Goldsmith, E. J., Gerard, R. D., Gething, M.-J. and Sambrook, J. F., 1989, Serpin-resistant mutants of human tissue-type plasminogen activator, *Nature*, **339:**721.

Matsudaira, P., 1987, Sequence from picomole quantities of proteins electroblotted onto polyvinylidene difluoride membranes, *J. Biol. Chem.*, **262:**10035.

Matsuo, O., Rijken, D.C. and Collen, D., 1981a, Comparison of the relative fibrinogenolytic, fibrinolytic and thrombolytic properties of tissue plasminogen activator and urokinase in vitro, *Thromb. Haemostas.*, **45:**225.

Matsuo, O., Rijken, D. C. and Collen, D., 1981b, Thrombolysis by human tissue plasminogen activator and urokinase in rabbits with experimental pulmonary embolus, *Nature*, **291:**590.

Mattson, C., Nyberg-Arrhenius, V. and Wallén, P., 1981, Dissolution of thrombi by tissue plasminogen activator, urokinase and streptokinase in an artificial circulating system, *Thromb. Res.*, **21:**535.

Nilsson T., Wallén, P. and Mellbring, G., 1984, In vivo metabolism of human tissue-type plasminogen activator, *Scand. J. Haematol.*, **33:**49.

Norrman, B., Wallén, P. and Rånby, M., 1985, Fibrinolysis mediated by tissue plasminogen activator: Disclosure of a kinetic transition, *Eur. J. Biochem.*, **149:**193.

Norrman, B., Pohl, G., Jörnvall, H. and Wallén, P., 1986, Proteolytically induced variations in the enzymatic properties of tissue plasminogen activator, *Eur. J. Biochem.*, **159:**7.

Norrman, B., Ohlsson, P.-I. and Wallén, P., 1988, Proteolytic modification of tissue plasminogen activator: Importance of the N-terminal part of the catalytically active B-chain for enzymatic activity, *Biochemistry*, **27:**8325.

Pannekoek, H., de Vries, C. and van Zonneveld, A.-J., 1988, Mutants of human tissue-type plasminogen activator (t-PA): Structural aspects and functional properties, *Fibrinolysis*, **2:**123.

Pennica, D., Holmes, W. E., Kohr, W. J., Harkins, R. N., Vehar, G. A., Ward, C. A.,

Bennet, W. F., Yelverton, E., Seeburg, P. H., Heyneker, H. L., Goeddel, D. V. and Collen, D., 1983, Cloning and expression of tissue-type plasminogen activator cDNA in E.coli, *Nature,* **301**:214.

Petersen, L. C., Boel, E., Johannesen, M. and Foster, D., 1989, Possible involvement of a lysine residue in establishing the charge-relay system responsible for one-chain tPA activity, *Thromb. Haemostas.,* **62**:322.

Rånby, M., 1982, Studies on the kinetics of the activation of plasminogen by tissue plasminogen activator, *Biochim. Biophys. Acta,* **704**:461.

Rånby, M. and Wallén, P., 1985, Enzymatic properties of tissue-type plasminogen activator, *in*: Thrombolysis: biological and therapeutic properties of new thrombolytic agents, D. Collen, H. R. Lijnen and M. Verstraete, eds. Churchill Livingstone, Edinburgh, Vol.1:31.

Rijken, D. C. and Groeneveld, E., 1986, Isolation and functional characterization of the heavy and light chains of human tissue-type plasminogen activator, *J. Biol. Chem.,* **261**:3098.

Wallén, P., Rånby, M., Bergsdorf, N. and Kok, P., 1981, Purification and characterization of tissue plasminogen activator: on the occurrence of two different forms and their enzymatic properties, *in*: Progress in fibrinolysis, J. F. Davidson, I. M. Nilsson and B. Åstedt, eds. Churchill Livingstone, Edinburgh, Vol. 5:31.

Wallén, P., Bergsdorf, N. and Rånby, M., 1982, Purification and identification of two structural variants of porcine tissue plasminogen activators by affinity adsorbtion on fibrin, *Biochim. Biophys. Acta,* **719**:318.

Wallén, P., Pohl, G., Bergsdorf, N., Rånby, M., Ny, T. and Jörnvall, H., 1983, Purification and characterization of a melanoma cell plasminogen activator, *Eur. J. Biochem.,* **132**:681.

Wallén, P., 1987, Structure and function of tissue plasminogen activator and urokinase, *in*: Fundamental and clinical fibrinolysis, F. J. Castellino, P. J. Gaffney, M. M. Samama and A. Takada, eds. Congress Ser. 757, Elsevier Science Publisher, Amsterdam, Exerpta Medica p:1.

Weimar, W., Stibbe, J., Van Seyen, A. J., Billiau, A., De Somer, P. and Collen, D., 1981, Specific lysis of an iliofemoral thrombus by administration of extrinsic (tissue-type) plasminogen activator, *Lancet,* **2**:1018.

Van Zonneveld, A. J., Veerman, H. and Pannekoek, H., 1986a, Autonomous functions of structural domains on human tissue-type plasminogen activator, *Proc. Natl. Acad. Sci. USA,* **83**:4670.

Van Zonneveld, A. J., Veerman, H. and Pannekoek, H., 1986b, On the interaction of the finger and the kringle-2 domain of tissue-type plasminogen activator with fibrin, *J. Biol. Chem.,* **261**:14214.

Verheijen, J. H., Caspers, M. P. M., Chang, G. T. G., deMunk, G. A. W., Pouwels, P. H. and Enger-Valk, B.E., 1986, Involvement of finger domain and kringle 2 domain of tissue-type plasminogen activator in fibrin binding and stimulation of activity by fibrin, *EMBO J.,* **5**:3525.

Verstraete, M., Bounameaux, H., de Cock. F., Van de Werf, F. and Collen, D., 1985 Pharmacokinetics and systemic fibrinogenolytic effects of recombinant human tissue-type plasminogen activator in humans, *J. Pharmacol. Exp. Ther.,* **235**:506.

MOLECULAR GENETICS OF ALPHA 2 PLASMIN INHIBITOR

Nobuo Aoki

The First Department of Medicine
Tokyo Medical and Dental University
Yushima 1-5-45, Bunkyo-ku, Tokyo 113, Japan

INTRODUCTION

α_2-Plasmin inhibitor (α_2PI), also called α_2-antiplasmin, is a plasma glycoprotein which rapidly inactivates plasmin proteolytic activity (1). Its molecular weight, deduced from the cDNA sequence and the carbohydrate content (14%), is ≅58 KD (2) whereas the molecular weight estimated by SDS-gel electrophoresis was 67 KD (1). The cause of the discrepancy is not known. Its concentration in human plasma was estimated to be 6.9 ± 0.6 mg/100 ml (3), which is estimated to be ≅1.2 μM on the assumption of MW 58KD. α_2PI is a serine proteinase inhibitor (serpin) which is able to inhibit several different "serine" proteinases, but is mainly playing a role as a primary inhibitor of plasmin-mediated fibrinolysis (4). Its congenital deficiency results in a lifelong severe hemorrhagic tendency due to premature degradation of hemostatic plugs by physiologically occurring fibrinolytic process (4-7).

THREE FUNCTIONS OF α_2PI

α_2PI molecule has three functional sites: plasmin(ogen) binding site, reactive site and crosslinking site (Fig 1). α_2PI has a strong affinity for plasmin(ogen), and preferentially binds to the sites in plasmin(ogen) molecule, called lysine binding sites (LBS). LBS is the sites where fibrin is also bound. Hence, α_2PI competitively inhibits the binding of plasminogen to fibrin, thereby retarding the initiation of fibrinolytic process (8-11). The plasmin(ogen) binding site in α_2PI is located within 26 amino acid residues of the carboxyl terminal end (12, 13). Two lysine residues, Lys 436 and Lys 452 which is the carboxyl terminal end, are most likely involved in the binding (14). The plasmin(ogen) binding site also plays an important role in inhibition of plasmin. As the first step, α_2PI rapidly forms a reversible complex with plasmin through non covalent binding between LBS in plasmin and the plasmin(ogen) binding site in α_2PI (15). In the second step, a covalent bond is formed between the active site serine of plasmin and the reactive site of α_2PI, resulting in a loss of proteolytic activity of plasmin. The reactive site is located at Arg 364 (16).

Another important function of α_2PI is its crosslinking to fibrin. When blood clots, part of the α_2PI present in plasma is rapidly crosslinked to fibrin α-chain by activated coagulation factor XIII (17). While the clot retraction is progressively taking place, the fibrin-bound α_2PI becomes condensed in the clot and contributes to resistance of the clot against fibrinolysis (18). Fibrin-fibrin crosslinking is only of minor importance in endowing the clot with resistance to fibrinolysis (19). Crosslinking site in α_2PI is located at the second amino acid, glutamine, from the amino-terminus (20). The residue in fibrin(ogen) molecule where the glutamine residue of α_2PI is crosslinked was determined to be Lys 303 of Aα-chain (21). Among these three functional sites in α_2PI, the crosslinking site and plasmin(ogen) binding site are peculiar to α_2PI. No other serine protease inhibitors has a factor XIII catalyzed crosslinking activity in a physiological condition.

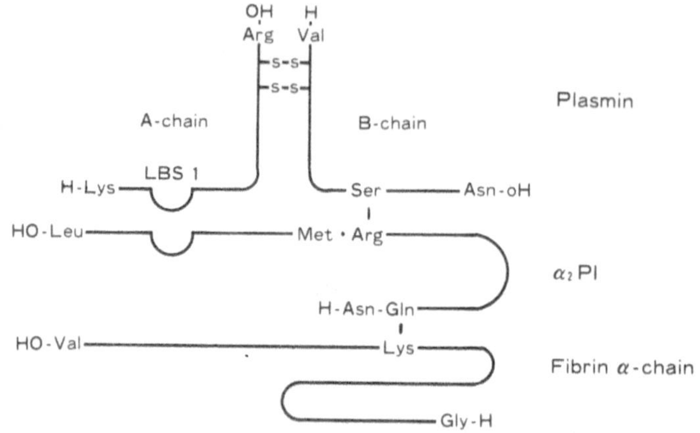

Fig. 1. A schematic model of molecular interactions in inhibition of plasmin by α_2PI cross-linked to fibrin.

When homologous amino acid sequences of α_2PI and the other members of serine protease inhibitor family (antithrombin III, α_1-antitrypsin, plasminogen activator inhibitor-1 and others) are aligned, α_2PI extends 50-52 amino acids beyond the carboxylterminal ends of the other members of the family (12). This extra 50-52 carboxyl-terminal amino acid sequence is therefore specific to α_2PI and contains the plasminogen binding site (12, 13, 14).

GENE ORGANIZATION

Studies from our laboratories (12) and those of others (2, 16) have led to the isolation of cDNA coding for human α_2PI. The cDNA was subsequently used for the isolation of overlapping genomic clones from λ phage library (22). The α_2PI gene contains 10 exons and 9 introns distributed over ∼16 kilobases of DNA (Fig 2). Number of introns is the highest among those of serine protease inhibitor gene family ever reported. All introns are located in the 5'-half of the corresponding mRNA, and the 5'-untranslated region and the leader sequence are interrupted by 3 introns totaling ∼6 kilobases. The leader sequence consisting of 39 amino acids is encoded by a part of exon II, exon III and a part of exon IV. The leader sequence consists of the prepeptide (signal peptide) of 27 amino acids and the propeptide of 12 amino acids (23). α_2PI is the only pre-pro type processing protein so

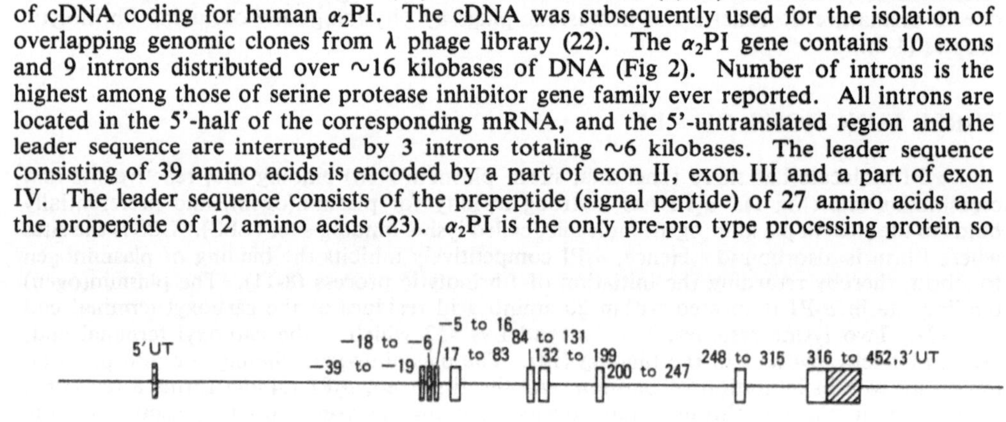

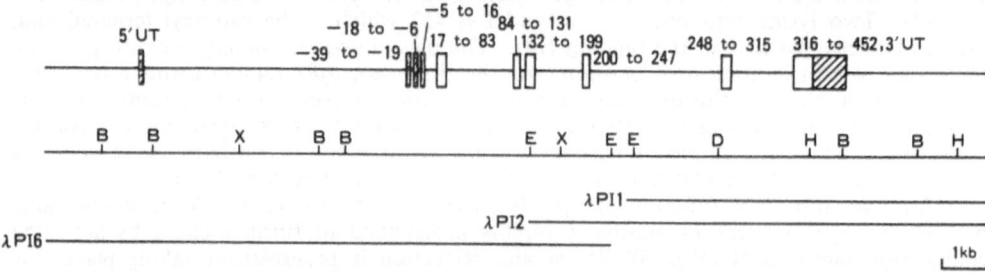

Fig. 2. Organization of the human α_2PI inhibitor gene. The first line shows the positions of exons as rectangles, and the numbers above the line indicate the amino acids at which intron-exon junctions occur. Untranslated regions (UT) are shown as hatched areas. A small 5'-untranslated region exists in the second exon. The second line indicates the positions of restriction endonuclease recognition sites. Straight lines at bottom indicate the region of the three phage clones. B, Bam HI; D, Dra I; E, Eco RI; H, Hind III; X, Xba I. (Reproduced rom Ref 22).

far reported among serine protease inhibitor family members. The NH_2-terminal region, which contains the crosslinking site, is encoded by a part of exon IV. The COOH-terminal region, which contains the reactive site and plasmin(ogen) binding site, is encoded by exon X. A TATA box sequence, multiple GC box sequences and CCAAT box-like sequence are found in the 5'-flanking region. Particularly interesting is the presence of the 16-bp sequence which is 88% similar to the 17-bp sequence in the hepatitis B virus enhancer element (24). This element displays tissue-specific activity and shows high homology with sequences in the promoter region of several liver-specific genes ; α-fetroprotein, α_1-antitrypsin and albumin.

A restriction fragment length polymorphism (RFLP) was found in α_2PI gene (25). This RFLP can be attributed to the presence of two alleles, A and B. The minor allele, B, is due to a deletion of about 720 bp in intron 8. The deletion was caused by intrastrand recombination between Alu sequences. Two alleles, A and B, are distributed with frequencies of 73.5% and 26.5%, respectively, in 66 unrelated Caucasian individuals or with frequencies of 51.0% and 49.0%, respectively, in 50 unrelated Japanese individuals.

The α_2PI gene, whose genetic symbol is PLI, was located on chromosome 18p11.1-g11.2 by in situ chromosomal hybridization (26).

MOLECULAR BASIS FOR α_2PI DEFICIENCY

The physiological importance of α_2PI has been established by discoveries of individuals with congenital deficiency of α_2PI in whom hemostatic plugs are dissolved prematurely before the restoration of injured vessels, resulting in a severe hemorrhagic tendency. So far, 10 families with congenital deficiency have been reported (27). Among these families, three were found in Japan.

Recently we analyzed the α_2PI genes from the two Japanese families with congenital α_2PI deficiency, and found that both families have a change of nucleotide sequence in an exon coding for plasma α_2PI. One family, α_2PI-Nara, has a single nucleotide insertion in exon X (28), and the other, α_2PI-Okinawa, has a trinucleotide deletion in exon VII (29).

In the case of α_2PI-Nara, a single nucleotide insertion in a region coding for the carboxyl-terminal portion caused a frameshift mutation and resulted in a production of a variant protein with alteration and elongation of the carboxyl terminal part of the molecule (28). Since the mature normal protein consists of 452 amino acids whereas the mature variant protein has extra 166 amino acids, the size of the mature variant molecule in liver cells was calculated to be approximately 37% larger than the normal molecule, disregarding the mass of carbohydrate attached to the molecules. To ensure that the frameshift mutation could cause the deficiency of α_2PI and to elucidate its mechanism, an expression vector containing the abnormal sequence or normal sequence was transfected into COS-7 cells for transient expression assay. The results indicated that ∼99% of the normal recombinant α_2PI that was synthesized in the COS-7 cells was secreted into the culture medium whereas the level of the mutant α_2PI secreted into the medium was ∼4% of the level of the normal recombinant protein secreted, although their cellular contents did not differ significantly. Probably a substantial change in its secondary and tertiary structure which was caused by the alteration and elongation of the carboxyl terminal part of α_2PI may have affected its posttranslational transport through intracellular secretory compartments. The altered conformation of the molecule may have also caused an instability of the mutant α_2PI, because no abnormal intracellular accumulation of the α_2PI was observed in spite of the apparently retarded secretion and unreduced transcript level tested with RNA blot analysis.

To confirm that the impairment in the intracellular transport of α_2PI-Nara observed in COS-7 cells is also occurring in liver cells and is the biochemical basis for the deficiency, we examined the secretory process of α_2PI-Nara in Hep G2, a hepatoma cell line synthesizing α_2PI, by taking advantage of the rare opportunity that the mutant molecule can be distinguished from the native molecule by the difference in their sizes (30). For this purpose, the expression plasmid for δ_2PI-Nara was transfected into Hep G2. A clone stably expressing α_2PI-Nara, Hep N1, was selected and the pulse-chase labeling experiments were performed. Both in Hep G2 and Hep N1, the native α_2PI was synthesized as a 63-KDa precursor form and was mostly secreted into the medium as a 72-KDa mature form 2 hours after labeling. However, a 78-KDa precursor form of α_2PI-Nara in Hep N1 remained in the cells for prolonged periods, and only a small portion of it was secreted into the medium as an 87-KDa mature form even 8 hours after labeling. The difference in size between the native α_2PI and α_2PI-Nara was in agreement with the elongation of the amino acid sequence caused by the frameshift mutation. The results obtained in this hepatoma cell line thus agree with those obtained in the COS-7 cells; thus, the impaired intracellular transport of this mutant molecule within the liver cells should be the primary cause of the α_2PI deficiency in this family.

In the case of α_2PI-Okinawa, a trinucleotide deletion in exon VII that gives rise to the deletion of Glu 137 was identified by nucleotide sequence analysis of the cloned mutant gene (29). Using the DNA samples amplified with the polymerase chain reaction, hybridization analysis by oligonucleotide probes confirmed the presence of this mutation in all the affected family members. An eukaryotic expression plasmid for α_2PI containing this mutation was constructed and transfected into COS-7 cells for transient expression analysis. The results indicated that the mutant α_2PI synthesized is mostly retained within the cells, and only a small portion of it is secreted into the medium. The region around Glu 137 is hydrophilic, and the deletion of Glu 137, a hydrophilic amino acid, transforms this region to a hydrophobic one. The significant alteration in the hydropathicity may then cause a structural change by disrupting the normal hydrogen bonding network responsible for maintaining the tertiary structure. A change of the tertiary structure or the hydropathicity of a particular region may have affected the intracellular transport and secretion of α_2PI.

To characterize α_2PIs in the cells as well as in the media in transient expression experiments, the α_2PIs were pulselabeled with ^{35}S methionine and immunoprecipitated. The immunoprecipitates were analyzed by SDS-PAGE followed by fluorography (29, 30). Part of the immunoprecipitated α_2PI was treated with endo-ß-N-acetylglucosaminidase H (endo H) or neuraminidase before SDS-PAGE to examine the state of the N-linked oligosaccharides that undergo processing along the intracellular transport pathway. The results indicated that both the normal and the mutant α_2PIs in the cells were sensitive to digestion with endo H but resistant to neuraminidase, indicating that these α_2PIs in the cells represented the precursor forms. The mutant α_2PIs which appeared late in the media in a small amount were susceptible to digestion with neuraminidase but resistant to endo H, indicating that the mutant α_2PIs in the media were mature. When chased with unlabeled methionine, the mutant α_2PIs remain within the cells as an endo H-susceptible form for a prolonged period as compared with the normal α_2PI. Only a very small portion of the mutant α_2PIs had been secreted into the media as a mature form even when intracellular precursor form became almost undetectable.

Since endo H removes N-linked oligosaccharides from polypeptides during its transit through the rough endoplasmic reticulum but has no effect on the processed oligosaccharides following transfer of the proteins to the medial stacks of the Golgi compartment (31), it is suggested that the precursor forms of the mutant α_2PIs are retained within the endoplasmic reticulum (Fig 3). And, instead of being secreted into the media, most of the mutant molecules should have undergone degradation within the cells while their transport is retarded in the intracellular transport pathway (28, 29, 30). Further studies on mechanisms leading to retention of these mutant proteins in cells are expected to help to elucidate the mechanism of the posttranslational intracellular transport and secretion pathway of secretory proteins in general.

SUMMARY

Analysis of the cDNA for α_2PI and its expression experiment indicated that the leader sequence is pre-pro peptide containing an N-terminal hydrophobic prepeptide (signal sequence) of 27 amino acids, followed by a hydrophilic propeptide of 12 amino acids. The α_2PI gene contains 10 exons and 9 introns distributed over ~16 kilobases of DNA. Two families of hereditary α_2PI deficiency were found to have changes of nucleotide sequence in an exon coding for plasma α_2PI, resulting in productions of variant proteins. These proteins are mostly retained within the cells, resulting in a deficiency of α_2PI in a circulating blood plasma.

ACKNOWLEDGMENTS

This research was supported by Grant-in-Aid for Scientific Research on Priority Areas (63637001, 63637002) and Grant-in-Aid for Scientific Research (62480260) from the Ministry of Education, Science and Culture of Japan, and a grant from Yamanouchi Foundation for Research on Metabolic Disorders.

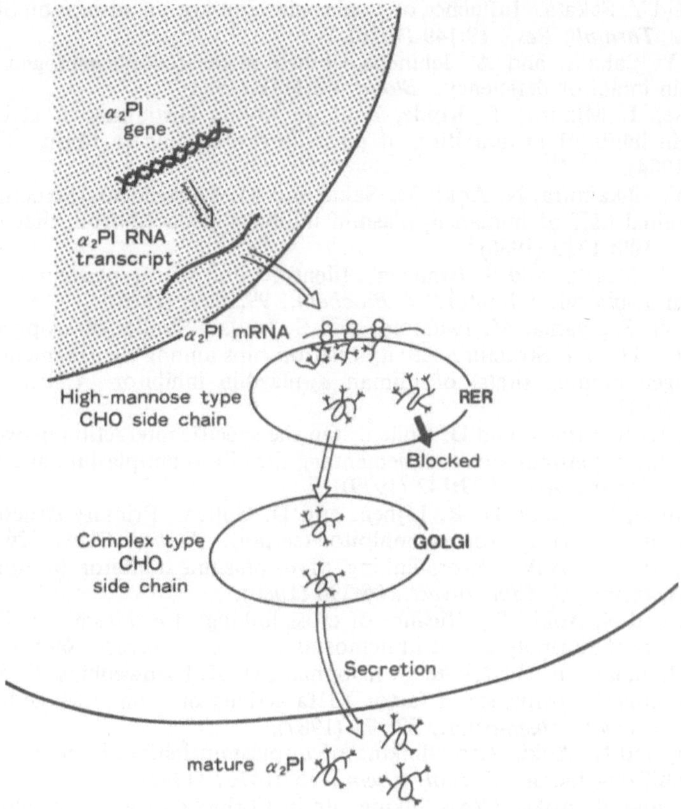

Fig. 3. Synthesis, oligosaccharide processing, and intracellular transit of α_2PI. The proteins synthesized in the rough endoplasmic reticulum (RER) are glycosylated by the addition of a high mannose type carbohydrate side chain which is Endo H-sensitive. During its transit to the Golgi apparatus, the carbohydrate side chain is converted to a complex type of oligosaccharides which is Endo H-resistant. Most of the mutant α_2PIs are retained in the RER as a Endo H-sensitive form, and their transit to the Golgi apparatus is blocked.

REFERENCES

1. M. Moroi, and N. Aoki. Isolation and characterization of α_2-plasmin inhibitor from human plasma. A novel protease inhibitor which inhibits activator-induced clot lysis. *J. Biol. Chem.*, **251:**5956 (1976).
2. M. Tone, R. Kikuno, A. Kume-Iwaki, and T. Hashimoto-Gotoh. Structure of human α_2-plasmin inhibitor deduced from the cDNA sequence. *J. Biochem.*, **102:**1033 (1987).
3. Y. Sakata, and N. Aoki. Cross-linking of α_2-plasmin inhibitor to fibrin by fibrin-stabilizing factor. *J. Clin. Invest.*, **65:**290 (1980).
4. N. Aoki, M. Moroi, M. Matsuda, and K. Tachiya. The behavior of α_2-plasmin inhibitor in fibrinolytic states. *J. Clin. Invest.*, **60:**361 (1977).
5. N. Aoki. Natural inhibitors of fibrinolysis. *Prog. Cardiovasc. Dis.*, **21:**267 (1979).
6. N. Aoki, and P. C. Harpel. Inhibitors of the fibrinolytic enzyme system. *Semin. Thromb. Hemostasis.*, **10:**24 (1984).
7. N. Aoki. Fibrinolysis: Its initiation and regulation. *J. Protein Chem.*, **5:**269 (1986).
8. N. Aoki, M. Moroi, and K. Tachiya. Effects of α_2-plasmin inhibitor on fibrin clot lysis. Its comparison with α_2-macroglobulin. *Thromb. Haemostas.*, **39:**22 (1978).

9. N. Aoki, and Y. Sakata. Influence of α_2-plasmin ininbitor on adsorption of plasminogen to fibrin. *Thromb. Res.*, **19:**149 (1980).
10. N. Aoki, Y. Sakata, and A. Ichinose. Fibrin-associated plasminogen activation in α_2-plasmin inhibitor deficiency. *Blood,* **62:**1118 (1983).
11. A. Ichinose, J. Minuro, T. Koide, and N. Aoki. Histidine-rich glycoprotein and α_2-plasmin inhibitor in inhibition of plasminogen binding to fibrin. *Thromb. Res.,* **33:**401 (1984).
12. Y. Sumi, Y. Nakamura, N. Aoki, M. Sakai and M. Muramatsu. Structure of the carboxylterminal half of human α_2-plasmin inhibitor deduced from that of cDNA. *J. Biochem.,* **100:** 1339 (1986).
13. T. Sasaki, T. Morita, and S. Iwanaga. Identification of the plasminogen-binding site of human α_2-plasmin inhibitor. *J. Biochem.,* **99:**1699, (1986).
14. T. Sasaki, N. Sugiyama, M. Iwamoto, and S. Isoda. Studies on α_2-plasmin inhibitor fragment T-11. III. Structure-activity relationships among the fragments of T-11, the plasminogen binding site(s) of human α_2-plasmin inhibitor. *Chem. Pharm. Bull,* **35:**2810, (1987).
15. B. Wiman, H. R. Lijnen, and D. Collen. On the specific interaction between the lysine-binding sites in plasmin and complementary sites in α_2-antiplasmin and in fibrinogen. *Biochim. Biophys. Acta,* **579:**142 (1979).
16. W. E. Holmes, L. Nelles, H. R. Lijnen, and D. Collen. Primary structure of human α_2-antiplasmin, a serine protease inhibitor (serpin). *J. Biol. Chem.,* **262:**1659 (1987).
17. Y. Sakata, and N. Aoki. Cross-linking of α_2-plasmin inhibitor to fibrin by fibrin-stabilizing factor. *J. Clin. Invest.,* **65:**290 (1980).
18. Y. Sakata, and N. Aoki. Significance of cross-linking of α_2-plasmin inhibitor to fibrin in inhibition of fibrinolysis and in hemostasis. *J. Clin. Invest.,* **69:**536 (1982).
19. J. W. C. M. Jansen, F. Haverkate, J. Koopman, H. K. Nieuwenhuis, C. Kluft, and Th. A. C. Boschman. Influence of factor XIIIa activity on human whole blood clot lysis *in vitro*. *Thromb. Haemostas.,* **57:**171 (1987).
20. T. Tamaki, and N. Aoki. Cross-linking of α_2-plasmin inhibitor catalyzed by activated fibrin-stabilizing factor. *J. Biol. Chem.,* **257:**14767 (1982).
21. S. Kimura, and N. Aoki. Cross-linking site in fibrinogen for α_2-plasmin inhibitor. *J. Biol. Chem.,* **261:**15591 (1986).
22. S. Hirosawa, Y. Nakamura, O. Miura, Y. Sumi, and N. Aoki. Organization of the human α_2-plasmin inhibitor gene. *Proc. Natl. Acad. Sci. USA,* **85:**6836 (1988).
23. Y. Sumi, Y. Ichikawa, Y. Nakamura, O. Miura, and N. Aoki. Expression and characterization of pro α_2-plasmin inhibitor. *J. Biochem.,* **106:**703 (1989).
24. Y. Shaul, and R. Ben-Levy. Multiple nuclear proteins in liver cells bound to hepatitis B virus enhancer element and its upstream sequences. *EMBO J.,* **6:**1913 (1987).
25. O. Miura, Y. Sugahara, Y. Nakamura, S. Hirosawa, and N. Aoki. Restriction fragment length polymorphism caused by a deletion involving Alu sequences within the human α_2-plasmin inhibitor gene. *Biochemisry,* **28:**4934 (1989).
26. A. Kato, Y. Nakamura, O. Miura, S. Hirosawa, Y. Sumi, and N. Aoki. Assignment of the human α_2-plasmin inhibitor gene (PLI) to chromosome region 18p11.1-q11.2 by in situ hibridization. *Cytogenet. Cell Genet.,* **47:**209 (1988).
27. N. Aoki. Hemostasis associated with abnormalities of fibrinolysis. *Blood Rev.,* **3:**11 (1989).
28. O. Miura, S. Hirosawa, A. Kato, and N. Aoki. Molecular basis for congenital deficiency of α_2-plasmin inhibitor. A frameshift mutation leading to elongation of the deduced amino acid sequence. *J. Clin. Invest.,* **83:**1598 (1989).
29. O. Miura, Y. Sugahara, and N. Aoki. Hereditary α_2-plasmin inhibitor deficiency caused by a transport-deficient mutation (α_2-PI-Okinawa). *J. Biol. Chem.,* **264:**18213 (1989).
30. O. Miura, and N. Aoki. Impaired secretion of mutant α_2-plasmin inhibitor (α_2PI-Nara) from COS-7 and Hep G2 cells: Molecular and cellular basis for hereditary deficiency of α_2-plasmin inhibitor. *Blood,* **75:**1092 (1990).
31. R. Kornfeld, and S. Kornfeld. Assembly of asparagine-linked oligosaccharides. *Annu. Rev. Biochem.,* **54:**631 (1985).

BIOLOGICAL PROPERTIES OF HYBRID PLASMINOGEN ACTIVATORS

P. P. Hung, J. Wilhelm, N.K. Kalyan, S.M. Cheng, H.L. James*, D. Nachowiak*, C.J. Weinheimer*, B.E. Sobel*, S.R. Bergmann*, and S.G. Lee

Wyeth-Ayerst Research, Philadelphia, PA 19101 and
*Washington University, St. Louis, MO 63110, USA

SUMMARY

A number of hybrid plasminogen activator genes were constructed from the t-PA and u-PA cDNAs and expressed using a bovine papilloma virus vector and mouse C-127 cells. Hybrid A was constructed by replacing the finger (F) and EGF domains of t-PA with the EGF and Ku domains of u-PA, while hybrids B and C had an extra Ku inserted before or after the double kringle (K1-K2) region of t-PA respectively. While all the hybrids showed comparable enzymatic activities towards a small substrate (S-2288), they had different activities in binding to fibrin clots as well in the fibrin-dependent plasminogen activation, the order of activities being: t-PA ≥ hybrid B > hybrid C > hybrid A. Carbohydrate analysis showed that while hybrid C, like rt-PA, had at least one high-mannose type sugar chain (probably at residue 117 in K1), the other hybrids had only complex-type carbohydrates suggesting that domain interaction in t-PA might influence glycan processing. Pharmacokinetic studies in dog showed that hybrid B had a significantly longer plasma half-life than rt-PA. Thrombolytic efficacies of hybrid B and rt-PA were compared in dog model using an artificially induced coronary thrombus. Complete thrombolysis was achieved with 18 mg and 50 mg dosages for hybrid B and rt-PA respectively. These data show the superior pharmacokinetic and thrombolytic properties of hybrid B compared to rt-PA.

INTRODUCTION

Blood clot disorders such as myocardial infarction, pulmonary embolism and stroke are the major causes of death and disabling diseases. The central feature of these disorders is the formation of a thrombus in a blood vessel which interrupts the flow of blood to vital organs (1, 2). Recently, tissue-type plasminogen activator (t-PA) has attracted attention for use as a thrombolytic agent for the treatment of some of these disorders (3). The most important feature of t-PA is its high binding affinity for the fibrin clot thus localizing its thrombolytic action at the clot site to avoid general bleeding complications. The enzymatic action of t-PA is to convert an inactive precursor of plasmin, called plasminogen, into active plasmin — a reaction which is greatly enhanced by the presence of fibrin (1, 4). Plasmin, in turn, proteolytically degrades the blood clot, thus achieving thrombolysis.

t-PA is a glycoprotein of Mr 65,000 and is synthesized initially as a precursor protein of 562 amino acids containing an amino terminal signal peptide which is cleaved as it is transported through the secretory pathway. The native protein is released from the endothelial cells as a single-chain form of 527 amino acids. It can, however be cleaved into two chains — heavy (H) and light (L) chains held together by a disulfide bond (5, 6). Comparison of t-PA with other blood protein sequences revealed that t-PA is organized into discrete structural domains. The carboxy-terminal L-chain displays homology with the trypsin-like serine protease family, whereas the amino-terminal H-chain contains noncatalytic domains — a finger (F), an epidermal growth factor (EGF) and two kringles (K1 and K2). Kringles, the triple disulfide-linked sequences, have been found in many proteins involved in hemodynamics — factor XII, prothrombin, plasminogen and apolipoprotein (a) (5-7). The finger and kringle domains in t-PA have been shown to be involved in binding to fibrin clot as well as fibrin dependent plasminogen activation (8).

Urokinase (u-PA), another plasminogen activator found in the urine, has a simpler structure with EGF, kringle (Ku) and protease domains. Although urokinase lacks direct binding affinity for fibrin, its single-chain form (scu-PA) shows specificity for fibrin clot in a clot lysis assay (9,10). This specificity for fibrin is thought to be due to displacement of a specific inhibitor by fibrin in a scu-PA inhibitor complex circulating in the blood (10).

In this report we present the construction and expression of chimeric PA genes made from the cDNA clones of t-PA and u-PA. Biological and biochemical properties of expressed hybrid PAs, in particular of hybrid B, will also be presented.

RESULTS

Construction and Expression of Hybrid-PA Genes

The human diploid cell line, WI-38, produces high levels of t-PA and u-PA when stimulated by endothelial cell growth factor and heparin. The cDNA library prepared from WI-38 mRNA was screened to isolate cDNA clones for t-PA and u-PA (10). From these two genes, three hybrid cDNAs were constructed by inserting urokinase Ku and/or EGF into the t-PA sequence (Fig. 1). All these manipulations were done in the NH_2-terminal non-protease part of t-PA. Hybrid A contains the urokinase EGF and Ku followed by the two kringles and the protease domain of t-PA. For hybrids B and C, the Ku was inserted either in front of (hybrid B) or behind (hybrid C) the double kringle region of t-PA.

For expression, t-PA as well as all three hybrid PA cDNAs were inserted into the BPV-vector and transfected into mouse C-127 cells. In this vector (Fig. 2), the expression of PAs is under the control of mouse metallothionein transcription promoter and the SV40 early region poly 'A' signal. An expression level of 1 to 25 μg/ml of activator proteins secreted into the media was obtained.

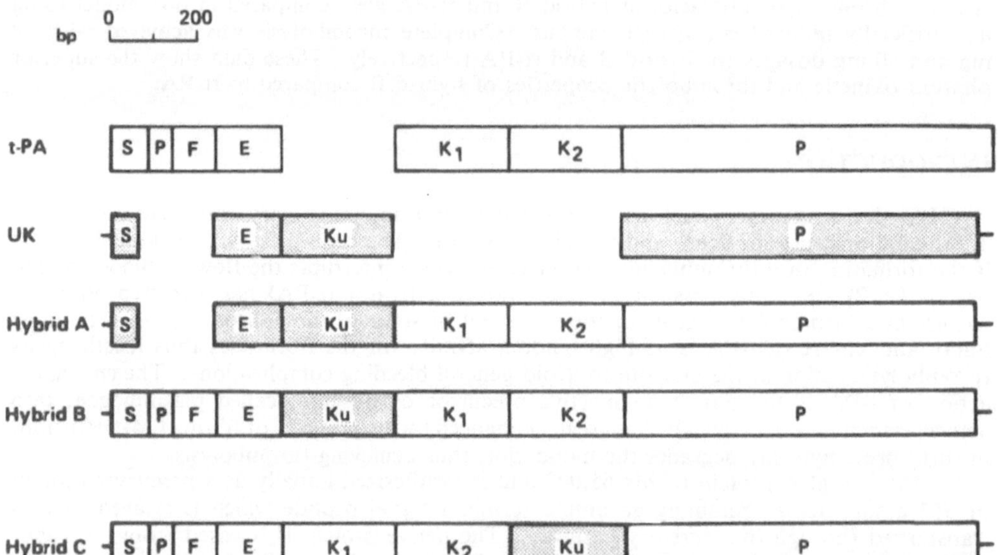

Fig. 1. Domain structure of t-PA and hybrid PAs. Sequences from u-PA are shaded. S is signal peptide.

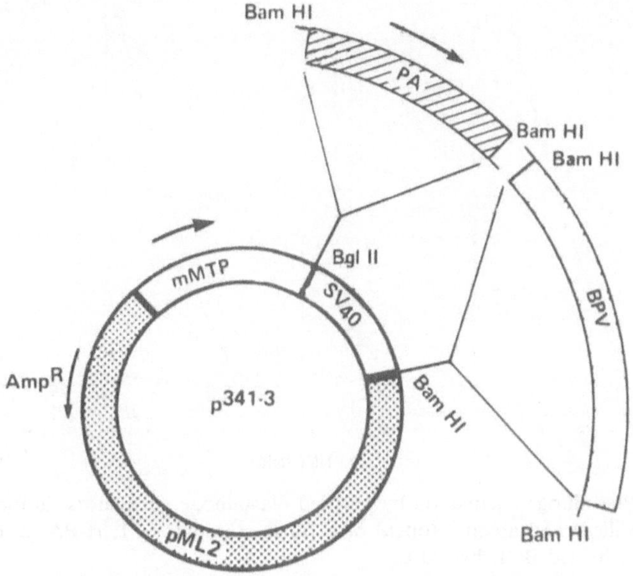

Fig. 2. Construction of expression vectors containing t-PA or hybrid PA genes.

Biochemical Characterization of Hybrid PAs

rt-PA and hybrid PAs were purified individually from serum-free conditioned media using a two-step procedure of zinc chelate- and lysine-sepharose columns. All the hybrid PAs were present mainly in single chain forms and consistent with their large sizes, showed slower mobility in SDS-polyacrylamide gel electrophoresis. As noticed by other workers (11), the PAs migrated on the gel as closely spaced doublets differing in size by approximately 5,000 Daltons. Two forms are considered to arise from partial glycosylation of one of the three glycosylated sites at residues 117, 184, and 448, most probably the one at residue 184 in K2.

Fig. 3 illustrates the ability of rt-PA and hybrid PAs to activate plasminogen in the presence or absence of fibrin and Table 1 lists the kinetic parameters of this reaction. As expected, hybrid PAs, like rt-PA, have low activity in the absence of fibrin. However, the presence of fibrin stimulates the activities of rt-PA and hybrid PAs in the following order: rt-PA > hybrid B > hybrid C > hybrid A. When binding to fibrin clots was determined, the same order of binding activity was obtained; binding of hybrid A was considerably less than of hybrids B or C. Table 1 shows that in the absence of fibrin, the Km values for glu-plasminogen of rt-PA and all hybrid PAs were more than 100 μM. Addition of fibrin lowered the Km values of hybrid PAs to 1.7 to 2.1 mM - significantly higher than for rt-PA (0.6 μM). Hybrid A showed lower catalytic efficiency (kcat/Km) than hybrid B and C.

Two interesting and related conclusions can be drawn from these data. First, the lower fibrin binding displayed by hybrid A suggests that the finger domain, which is not present in hybrid A, partially mediates fibrin binding of t-PA. Consistent with this is a recent report on deletion mutants of t-PA which shows that both F and K2 mediate fibrin binding activity (8). More recently we have shown that a t-PA mutant lacking F and EGF does not bind appreciably to fibrin (12). Second, the observation that the rt-PA and hybrid B show comparable fibrin binding activity, through hybrid B contains an extra domain between the F and K1, suggests a substantial flexibility in the interactions of these domains with fibrin. The result also shows that Ku, which does not mediate fibrin binding in its parent molecule (u-PA), clearly does not inhibit binding when inserted in a different position.

As far as catalytic properties of hybrids, are concerned, amidolytic activity towards a small substrate (S-2288, D-Ile-Pro-Arg-p-nitroanilide) was essentially unchanged suggesting the autonomous function of catalytic and noncatalytic domains of t-PA (8).

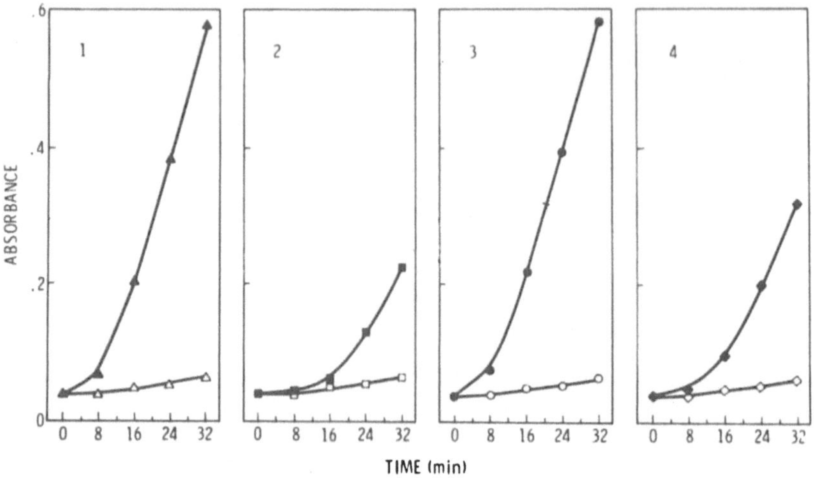

Fig. 3. Plasminogen activation by purified plasminogen activators, in the presence (filled) and absence (open) of a fibrin stimulator. 1, rt-PA; 2, hybrid A; 3, hybrid B; 4, hybrid C.

Carbohydrate Structure

Carbohydrate side-chains play an important role in the biological activity of a protein. t-PA contains three glycosylation sites present at residues 117 in K1, 184 in K2 and 448 in protease domains (13). Structural analysis of carbohydrate chains of t-PA showed that the sugar chain at residue 117 was of high-mannose type while the other two were of complex type. During our studies on structure-function analysis of t-PA, we found that one deletion mutant lacking F and EGF domain ($\triangle_{2\text{-}89}$ t-PA), had unexpectedly low activities toward fibrin binding and fibrin dependent plasminogen activation, though it had fibrin

Table 1. Kinetic Parameters for the Activation of Glu-Plasminogen by rt-PA and Hybrid PAs

System	Stimulator[b]	K_m(μM)	k_{cat}(sec^{-1})	k_{cat}/K_m (M^{-1} sec^{-1})
rt-PA	–	–	–	43[c]
Hybrid A[a]	–	–	–	42[c]
Hybrid B[a]	–	–	–	63[c]
Hybrid C[a]	–	–	–	93[c]
rt-PA	+	0.64	0.0072	11,000
Hybrid A	+	2.1	0.0041	2,000
Hybrid B	+	1.7	0.0102	6,000
Hybrid C	+	1.7	0.0052	3,100

[a]Two-chain forms

[b]The stimulator is soluble fibrin (125 μg/ml)

[c]Determined from plots of initial rates versus plasminogen concentration. Km and kcat were not individually determined because substrate saturation was not achieved at the highest concentration of glu-plasminogen used (8 μM). Km was estimated at > 100 μM as observed by other workers (11).

Table 2. Characteristics of N-linked Oligosaccharides of rt-PA and Hybrid PAs

PA	Lectin Reaction[a]		Glycosidase Susceptibility[b]	
	Man	GlcNAc/AcNeu	Neur	EndoH
rt-PA	+	+	−/+[c]	+
Hybrid A	−	+	+	−
Hybrid B	−	+	+	−
Hybrid C	+	+	−/+	+

[a]Lectins used were concanavalin A and wheat germ lectin to detect high mannose and complex-type sugar chains respectively.

[b]Neur (neuraminidase) cleaves terminal sialic acid and Endo H cleaves high mannose type sugar chains.

[c]−/+ indicate weak reaction.

binding kringle domains (12). However, these activities could be partially restored when $\triangle_{2\text{-}89}$ t-PA was treated with either neuraminidse or N-glycanase. Surprisingly, this mutant, unlike t-PA, did not have any endoglycosidase H (endo H) susceptible sugar chain (high mannose type)*.

In order to characterize carbohydrate structures of hybrid PAs, two methods were used: (i) susceptibility of PAs to specific glycosidase by observing changes in their mobility on SDS-PAGE; (ii) reaction with sugar-specific lectins after running PAs on SDS-PAGE. The results are shown in Table 2.

It is clear that hybrid C, like rt-PA, contains both high mannose type as well as complex type sugar chains, while hybrids B and C have only complex type and no high mannose type carbohydrate chains. As mentioned above, the sugar chain at residue 117 in K1 of t-PA (F-EGF-K1-K2-P) as well as of hybrid C (F-EGF-K1-K2-Ku-P) is of high mannose type. It seems likely that any changes made in the NH_2-terminal domain (as in hybrids A and B) result in altered processing of sugar chains. Similar results were obtained when deletion mutations were introduced into the EGF and/or F domains of t-PA. These results suggest that domain interactions in t-PA influence glycan processing and illustrate how protein engineering may have unexpected consequences on glycan processing.

Since hybrid B has biochemical characteristics similar to rt-PA, further comparactive studies were conducted with hybrid B and commercial rt-PA from Genentech, Inc.

Pharmacokinetics

The plasma clearance of hybrid B and rt-PA was studied in dog with a single bolus injection. Hybrid B exhibited a longer half-life than Genentech's t-PA (Fig. 4). The activity of hybrid B in blood could be detected for up to two hours after injection compared to ten minutes for rt-PA. Hybrid B and rt-PA showed biexponential disappearance from plasma. About 97% of rt-PA disappeared with t 1/2 (α) of 2 minutes and the rest with t 1/2 (β) of ~20 minutes. In contrast, 30% of hybrid B disappeared with a t 1/2 (α) of 5 minutes and 70% disappeared with t 1/2 (β) of 54 minutes. The area under the curve (AUC) for hybrid B was about 6-fold higher than for rt-PA. Thus for a given dosage, hybrid B would achieve a higher steady state level in blood than rt-PA.

Animal Studies in Dog

Thrombolytic efficacies of hybrid B and Genentech rt-PA (Activase) were compared in a dog model. Fig. 5 shows the protocol used for these studies. Coronary thrombus

*Wilhelm, J. et al., Bio/Technology 8 (1990). In press.

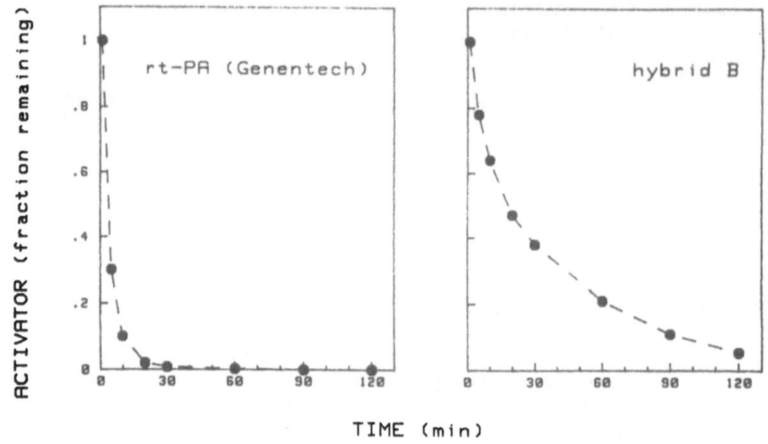

Fig. 4. Biexponential clearance of rt-PA and hybrid B in dogs.

occlusion was induced with a thrombolytic copper coil and occlusion was confirmed by arteriography. Activators were infused with an initial 10% bolus followed by 30 minute infusion of hybrid B or 60 minute infusion of rt-PA. Successful thrombolysis was defined as the complete recanalization of the artery as observed by sequential arteriography.

Initial studies with low dosages of t-PA (1 mg, 3 mg, 10 mg and 18 mg) produced no or partial lysis of the clot. However, at 50 mg dosage, t-PA showed lysis in all cases (n = 4) with a mean lysis time of 77 ± 25 (60 – 120 range) minutes. Hybid B on the other hand, lysed the clot (n = 15) at a significantly lower dosage (18 mg) with mean lysis time of 55 ± 26 (15 – 120 range) minutes. Fig. 6A shows the Mean ± S.D. peak antigen (ng/ml) and activity (IU/ml) levels in the blood when hybrid B at 18 mg (n = 12) and rt-PA at 50 mg (n = 5) were infused into the dogs. Consistent with its longer half-life, hybrid B achieved significantly higher blood levels ($p < 0.02$), although the total dosage injected was less than t-PA. The blood parameters, e.g. blood pressure, heart beat, fibrinogen and α_2-antiplasmin levels did not differ significantly in the two groups. It is interesting to note that despite the relatively high sustained levels of hybrid B, systemic activation, potentially indicative of a

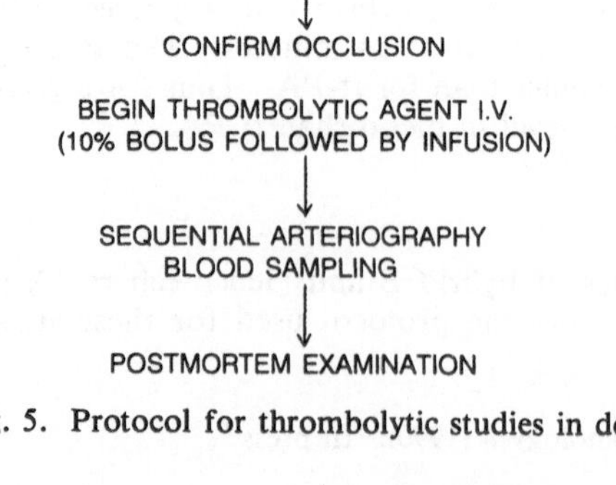

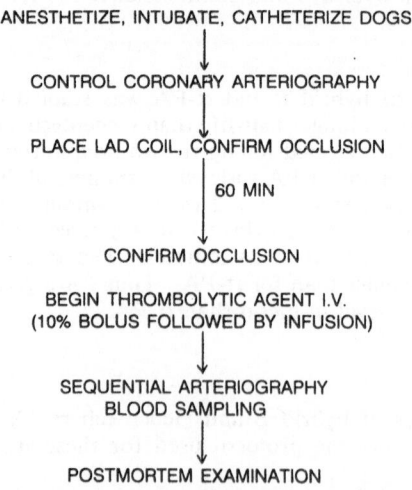

Fig. 5. Protocol for thrombolytic studies in dog.

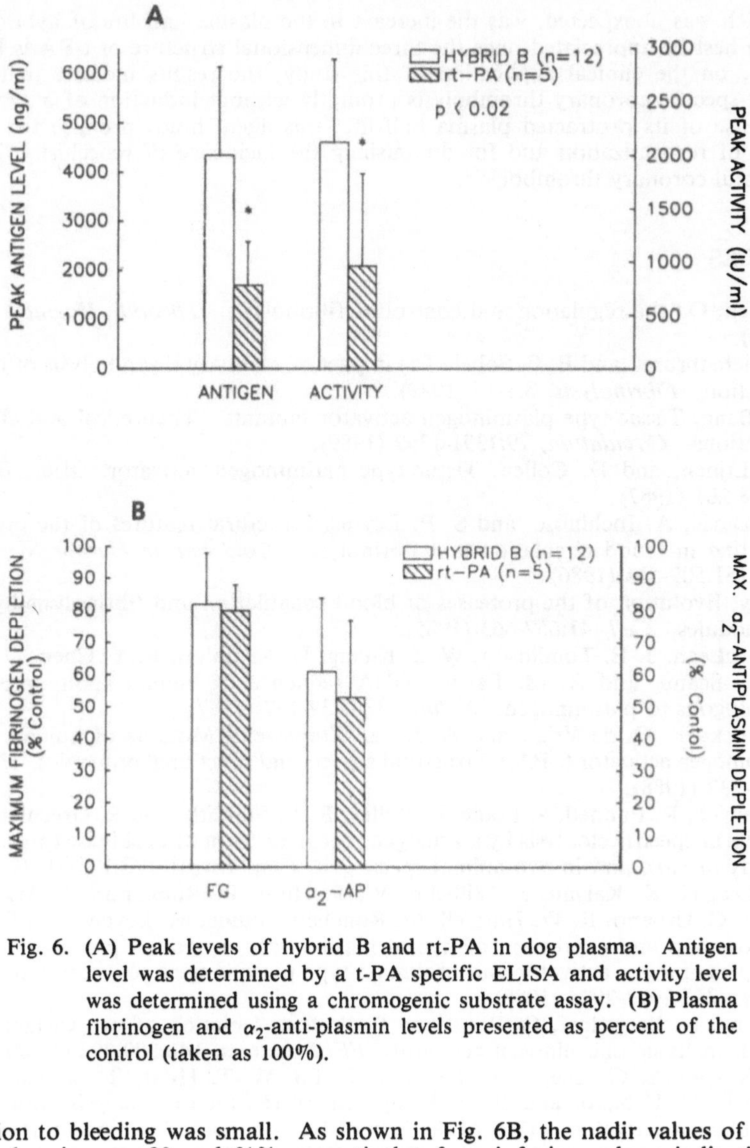

Fig. 6. (A) Peak levels of hybrid B and rt-PA in dog plasma. Antigen level was determined by a t-PA specific ELISA and activity level was determined using a chromogenic substrate assay. (B) Plasma fibrinogen and α_2-anti-plasmin levels presented as percent of the control (taken as 100%).

predisposition to bleeding was small. As shown in Fig. 6B, the nadir values of fibrinogen and α_2-antiplasmin were 82 and 61% respectively of pre-infusion values, indicative of only modest systemic effects.

CONCLUSION

Recently, a number of studies have been directed toward understanding structure-function relationship in t-PA, with a long term objective of designing a superior plasminogen activator (see ref. 2 and 8 for review). All of these studies (including the present one) show the feasibility of shuffling functional domain(s) in t-PA. Although gross functional activities are retained, some changes, probably in the structure of t-PA, do take place. For example, our study shows that the Glu-plasminogen Km values are about 3-fold higher for all three hybrids than for t-PA. Similarly, the sugar composition of hybrids A and B are quite different than for native t-PA, especially the high-mannose type sugar chain at residue 117 in K1. To our knowledge, this is the first report which shows that domain interactions in a glycoprotein may influence glycan processing and further illustrate how protein engineering may have unexpected consequences on glycan processing at distal sites. Another interesting change, which was unexpected, was the increase in the plasma half-life of hybrid B. Such changes can best be appreciated once the three-dimensional structure of t-PA is known.

Finally, on the clinical application of this study, the results indicate that hybrid B induces clot-specific coronary thrombolysis promptly without induction of a systemic lytic state. Because of its protracted plasma half-life, this agent holds promise for enhancing initial rates of recanalization and for diminishing the incidence of reocclusion in patients with successful coronary thrombolysis.

REFERENCES

1. D. Collen, On the regulation and control of fibrinolysis. *Thromb. Haemost.,* **43:**77-89 (1980).
2. A.-J. Tiefenbrunn, and B. E. Sobel, The impact of coronary thrombolysis of myocardial infarction. *Fibrinolysis,* **3:**1-15 (1989).
3. N. U. Bang, Tissue-type plasminogen activator mutants. Theoretical and clinical considerations. *Circulation,* **79:**1391-1392 (1989).
4. H. R. Lijnen, and D. Collen, Tissue-type plasminogen activator. *Ann. Biol. Clin.,* **45:**198-201 (1987).
5. E. W. Davie, A. Inchinose, and S. P. Leytus, Structural features of the proteins participating in blood circulation and fibrinolysis. *Cold Spring Harbor Symp. Quant. Biol.,* **51:**509-514 (1986).
6. L. Patty, Evolution of the proteases of blood coagulation and fibrinolysis by assembly for modules. *Cell,* **41:**657-663 (1985).
7. J. W. McLean, J. E. Tomlinson, W.-J. Kuang, D. L. Eaton, E. Y. Chen, G. M. Fless, A. M. Scanu, and R. M. Lawn, cDNA sequence of human apolipoprotein (a) is homologous to plasminogen. *Nature,* **330:**132-137 (1987).
8. H. Pannekoek, C. de Vries, and A.-J. van Zonneveld, Mutants of human tissue-type plasminogen activator (t-PA): Structural aspects and functional properties. *Fibrinolysis,* **2:**123-132 (1988).
9. V. Gurewich, R. Pannell, S. Louie, P. Kelley, R. L. Suddith, and R. Greenlee, Effective and fibrin-specific clot lysis by a zymogen precursor form of urokinase (pro-urokinase): a study *in vitro* and in two animal species. *J. Clin. Invest.,* **73:**1731-1739 (1984).
10. S. G. Lee, N. K. Kalyan, J. Wilhelm, W.-T. Hum, R. Rappaport, S.-M. Cheng, S. Dheer, C. Urbano, R. W. Hartzell, M. Ronchetti-Blume, M. Levner, and P. P. Hung, Construction and expression of hybrid plasminogen activators prepared from tissue-type plasminogen activator and urokinase-type plasminogen activator genes. *J. Biol. Chem.,* **263:**2917-2924 (1988).
11. M. Ranby, N. Bergsdorf, G. Pohl, and P. Wallen, Isolation of two variants of native one-chain tissue plasminogen activator. *FEBS Letters,* **146:**289-292 (1982).
12. N. K. Kalyan, S. G. Lee, J. Wilhelm, K. P. Fu, W.-T. Hum, R. Rappaport, R. W. Hartzell, C. Urbano, and P. P. Hung, Structure-function analysis with tissue-type plasminogen activator: effect of deletion of NH_2-terminal domains on its biochemical and biological properties. *J. Biol. Chem.,* **263:**3971-3978 (1988).
13. G. Pohl, L. Kenne, B. Nilsson, and M. Einarsson, Isolation and characterization of three different carbohydrate chains from melanoma tissue plasminogen activator. *Eur. J. Biochem.,* **170:**69-75 (1987).

THE REGULATION OF THE ACTIVATON OF THE FIBRINOLYSIS SYSTEM

A. Takada, T. Urano, and Y. Takada

Department of Physiology, Hamamatsu University
School of Medicine, Hamamatsu-shi, Shizuoka
Japan 431-31

INTRODUCTION

Fibrinolysis is a system in which fibrin clot dissolves due to the degradation of fibrin by plasmin, which is present in the blood as a precursor form, plasminogen (plg). A native form of plasminogen has glutamic acid at its N-terminus, and called as Glu-plg. Plasminogen is activated to plasmin by a group of enzymes, called plasminogen activators (PA). There are two pathways in the fibrinolysis system: the intrinsic pathway initiates upon the activation of a coagulation factor XII to XIIa in its interaction with a negatively charged foreign surface in the presence of high molecular weight kininogen and prekallikrein, and the other pathway, extrinsic pathway, initiates by PAs introduced in the blood exogenously from endothelial cells or upon the addition of bacterial activators such as streptokinase (SK).

In the present paper, we are not going to discuss PAs in the intrinsic pathway, but discuss the regulation of the activation of fibrinolysis at each step of the extrinsic pathway. Since the regulation of the activation of plasminogen by urokinase (UK) and SK has been discussed by Y. Takada in this book, I shall concentrate my discussion on the activation of plasminogen by tissue type PA (t-PA) and single chain urinary PA (scu-PA: proUK), a precursor of UK.

CONVERSION OF SINGLE CHAIN t-PA (sct-PA) AND scu-PA TO THEIR TWO CHAIN FORMS

Tissue plasminogen activator is synthesized as a single chain form in the endothelial cells (1-3), and UK is also synthesized as a scu-PA in kidney cells (4), macrophages (5) and virally transformed cells (6, 7) and tumor cells (8). Sct-PA or scu-PA was converted to their two chain form by plasmin (9-11) and kallikrein (12-14). It is well recognized that sct-PA has proteolytic activity (9, 10, 15), whereas scu-PA has little, if any, activity (6, 7, 16, 17).

Urano et al. (18) recently found that the amidolytic activity of sct-PA was enhanced in the presence of fibrinogen and values of kinetic parameters (k_{cat}/K_m) of sct-PA in the presence of fibrinogen approached to that of tct-PA, while amidolytic activity of tct-PA was not influenced in the presence of fibrinogen. We thought that increase in the amidolytic activity of sct-PA in the presence of fibrinogen may be due to the conformational changes of sct-PA upon the binding of fibrinogen. We then examined changes in kinetic parameters of sct-PA in the presence of various concentrations of fibrinogen using Lineweaver-Burk plots (Fig. 1)

Increase in the concentration of fibrinogen resulted in decrease in K_m without any change in k_{cat}. In order to get kinetic constants of sct-PA toward S-2288 in the presence

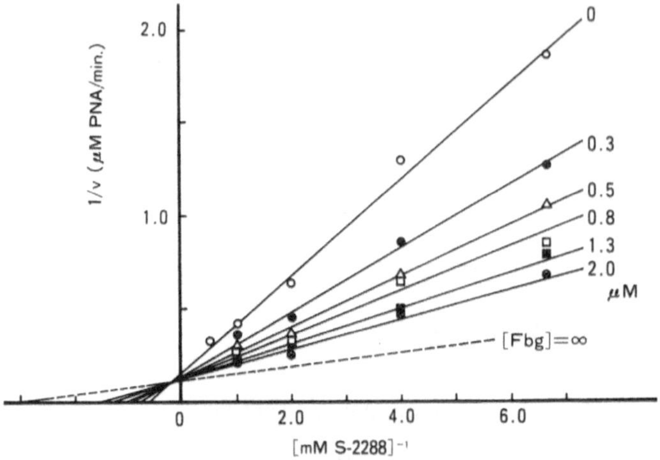

Fig. 1. Lineweaver-Burk plots of sct-PA in the presence of various concentrations of fibrinogen. Dotted line shows the velocity of the amidolytic activity of sct-PA in the presence of infinite concentrations of fibrinogen.

of the infinite concentration of fibrinogen, $[\triangle \text{ slope}]^{-1}$ value and $[\triangle \text{ intercept}]^{-1}$ value of the Lineweaver-Burk plot were calculated and plotted against the reciprocal of the concentration of fibrinogen. From the Y intercepts and X intercepts of two lines, kinetic numbers α, β and kinetically determined dissociation constant (k_a) were calculated for fibrinogen and its degradation products (Table 1).

It is shown that fibrinogen enhanced the amidolytic activity of sct-PA to the largest extent, followed by fragment X and D. Fragment E did not influence the amidolytic activity of sct-PA. It is also shown that k_a, dissociation constants between sct-PA and fibrinogen, X or D were similar.

It is well known that lysine analogues such as epsilon aminocaproic acid (EACA) or tranexamic acid enhance the rate of the activation of Glu-plg by activators due to its conformational changes induced upon the binding of lysine analogues to lysine binding sites (LBS) of plasminogen reviewed in references in 19, 20. Ichinose et al. (21) showed that the kringle 2 of t-PA has LBS to which fibrin and lysine analogues bind. We then wondered if the site of t-PA to bind with fibrinogen is identical with LBS. We tried to dissociate t-PA from fibrinogen binding by lysine analogues. Fig. 2 shows that various lysine analogues only slightly reduced a fibrinogen induced enhancement of the amidolytic activity of sct-PA.

Furthermore lysine analogues were shown not to enhance the amidolytic activity of sct-PA in contrast to the enhancement of the activation of Glu-plg by activators in their presence. Lysine analogues were also known to change the conformation of Glu-plg as measured by changes in sedimentation coefficient, the intensity of intrinsic fluorescence, and rotational relaxation time of N-terminal part. We measured changes in the intensity of tryptophan related fluorescence in the presence of various concentrations of tranexamic

Table 1. Kinetic Constants of sct-PA in the Presence of Fibrinogen and Its Degradation Products

Effectors	k_a (μM)	α	β	αK_m	βk_{cat}	β/α
Fibrinogen	2.90	0.16	1.32	0.25	6.50	8.30
Fragment X	2.30	0.36	1.14	0.56	5.60	3.20
Fragment D	2.00	0.39	1.11	0.61	5.40	2.80
Fragment E	—	1.00	1.00	1.56	4.90	1.00

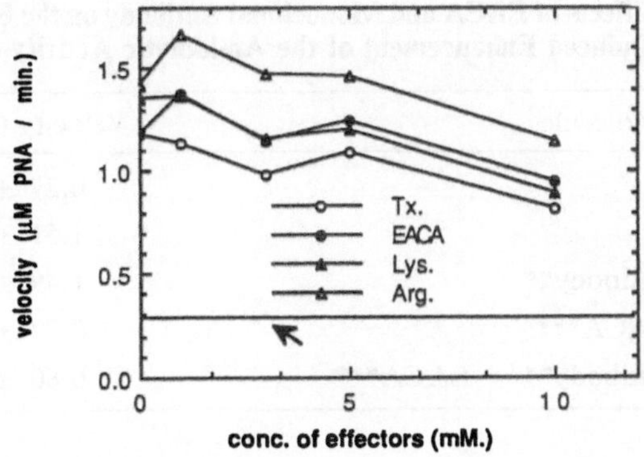

Fig. 2. Effects of lysine analogues on the fibrinogen induced enhancement of the amidolytic activity of sct-PA. 0.1 mM of S-2288, 20 nM of sct-PA and 10 μM of fibrinogen were incubated at 37°C with various concentrations of lysine analogues. Velocity in the absence of fibrinogen is shown by the arrow in the figure.

acid. Fig. 3 shows that tranexamic acid changed the conformations of sct-PA upon its binding with LBS. Since the amidolytic activity of sct-PA did not change in the presence of tranexamic acid, it may be concluded that conformational changes of sct-PA induced by tranexamic acid had nothing to do with the enhanced activity of sct-PA induced in the presence of fibrinogen.

It is also shown that not only kringle 2 but also the finger domain of sct-PA had a site to bind to fibrin (22). It is possible that fibrinogen binds to sct-PA through two sites, one in kringle 2 and the other in finger domain, and that fibrinogen still binds to sct-PA after its dissociation from kringle 2 by tranexamic acid, thereby causing both its conformational changes and enhanced amidolytic activity. We used monoclonal antibodies (Mab) specific against finger and EGF domains and EACA, and tried to dissociate fibrinogen induced enhancement of the amidolytic activity of sct-PA.

Table 2 shows that both Mab against finger and EGF domains and EACA did not reduce the enhancement, thus not dissociating fibrinogen from sct-PA. These results are in agreement with previous reports (18) in which domain-deleted mutants of sct-PA were used. The previous results also indicated that EACA did not reduce the amidolytic activity of sct-PA which lacked the finger domain.

We wanted to know if the results were true using the wild type sct-PA, since domain deleted mutants may not have the tertiary structure similar to that of the wild type t-PA.

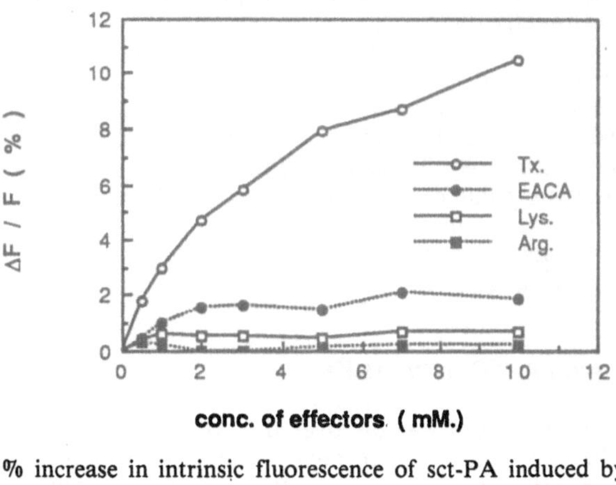

Fig. 3. % increase in intrinsic fluorescence of sct-PA induced by lysine analogues.

Table 2. **Effects of EACA and Monoclonal Antibody on the Fibrinogen Induced Enhancement of the Amidolytic Activity of sct-PA**

Effector Molecule	Velocity (μM/min)
none	0.27 ± 0.01
fbg*	1.51 ± 0.04
fbg* + antibody**	1.49 ± 0.10
fbg* + EACA***	0.87 ± 0.08
fbg* + antibody** + EACA***	0.80 ± 0.07

*10 μM
**monoclonal antibody toward EGF and finger domain, 2.5 μg/ml
***10 mM

Our results suggest that the fibrinogen binding site of sct-PA may be different from LBS or else that fibrinogen may bind to both LBS and fibrinogen binding sites, so that lysine analogues alone were not enough for dissociation of fibrinogen from sct-PA. Furthermore it is suggested that the intact conformation of fibrinogen was required for a full expression of the enhanced activity in sct-PA because slight degradation of fibrinogen by plasmin dramatically reduced the extent of the enhancement of the amidolytic activity of sct-PA. Another thing to emphasize is that fibrinogen may bind to sct-PA through its D-domain since fragment E has no enhancing effects.

ENHANCEMENT OF THE ACTIVATION OF PLASMINOGEN BY t-PA

It is well documented that the rate of the activation of plasminogen by t-PA was greatly enhanced in the presence of fibrin and its degradation products (23, 24). Hoylaerts et al. (25) proposed that t-PA forms a transient ternary complex with fibrin and plasminogen, thereby decreasing apparent Km between plasminogen and t-PA. We also wanted to measure kinetic parameters of the activation of Glu-plg by t-PA in the presence of fibrin, fibrinogen or its degradation products (26). The activation rate was most significantly enhanced in the presence of fibrin as suggested by other investigators, but it was also shown that fibrinogen or its degradation products also slightly, but significantly enhanced the activation rate, the order of enhancement being fragment D > E > fibrinogen (26). We showed that the presence of catalytic amounts of fibrin drastically decreased Km of the activation of Glu-plg by t-PA. Increase in the concentration of fibrin resulted in increase in k_{cat} with little change in K_m (Fig. 4). These results may suggest that fibrin may provide a site to which t-PA binds and plasminogen turns over rapidly there.

Fibrin, fibrinogen, or their degradation products may not be the only one group of factors to enhance t-PA induced activation of plasminogen. We got various Mabs against t-PA (27). Some of Mabs enhanced the amidolytic activity of sct-PA and tct-PA, and some of them also enhanced the rate of plasminogen activation by t-PA even in the presence of fibrin. Using immunoblotting, we found one kind of Mabs to bind to B chain of t-PA, and such Mab bound t-PA enhanced the activation of plasminogen in the presence and absence of fibrin. We have an impression that t-PA may be a molecule which changes its conformation very easily upon interaction with various molecules, one of which may be fibrin or its related substances.

CONVERSION OF scu-PA TO u-PA (UK)

When the presence of scu-PA was discovered in the urine or cell culture fluid for the first time (5, 6), scu-PA was considered to have a capacity to bind to fibrin in contrast to UK, which has little capacity to bind to fibrin (5-7, 16, 23, 24, 28). Now it seems to be established that scu-PA has as little capacity to bind to fibrin as UK itself (23, 24, 29, 30). Since then it was reported that cells synthesize and release UK as a single chain form, which

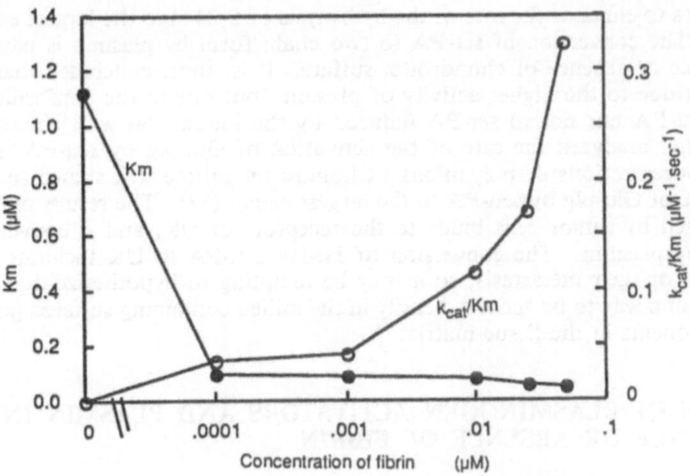

Fig. 4. Effects of fibrin concentration on catalytic efficiencies.

is converted to two chain form by plasmin and kallikrein (4-14). Although scu-PA was found in the plasma (31, 32), scu-PA is produced in immunologically activated macrophages (5), and transformed cells or tumor cells (6-8). Many cells including tumor cells have receptors for HMW (high molecular weight) UK, but not for LMW (low molecular weight) UK, which lacks EGF and kringle domains (33, 34). The hypothesis of HMW UK playing a role in tissue destruction involves the assumption that plasmin together with other proteolytic enzymes degrades the extracellular matrix. It is noteworthy that most components of the extracellular matrix can be degraded by plasmin. In addition, plasmin can activate procollagenase to collagenase (35-37), the latter facilitating the growth of tumor cells, or may digest fibrin surrounding tumor cells (33).

Since there are abundant sulfated polysaccharides in the tissue matrix such as chondroitin sulfate or heparin, we wondered if such anionic polymers may have any influence on the rate of the plasmin induced conversion of scu-PA to UK, or the scu-PA induced activation of plasminogen.

Fig. 5 shows the rate of the hydrolysis of S-2444 by UK converted from scu-PA by plasmin in the presence of chondroitin sulfate, heparin or dextran sulfate. Chondroitin

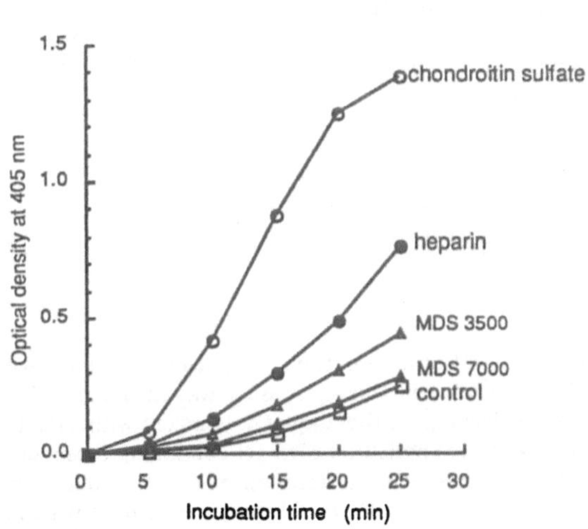

Fig. 5. Time course of the hydrolysis of S-2444 during plasmin catalyzed conversion of scu-PA to u-PA.

sulfate appears to enhance the rate of the hydrolysis of S-2444 to the largest extent, followed by heparin. The conversion of sct-PA to two chain form by plasmin is hardly influenced in the presence or absence of chondroitin sulfate. It is, thus, concluded that the enhancement was not due to the higher activity of plasmin, but due to the conformational change specific to scu-PA but not to sct-PA induced by the interaction with chondroitin sulfate.

We further analyzed the rate of the activation of Glu-plg by scu-PA in the presence of various concentrations of polyanions. Chondroitin sulfate was shown to potentiate the activation rate of Glu-plg by scu-PA to the largest extent (38). The results may suggest that scu-PA released by tumor cells binds to the receptor for UK, and is converted to its two chain form by plasmin. The conversion of HMW scu-PA to UK facilitate the spreading of tumor cells or their metastasis, so it may be tempting to hypothesize that scu-PA molecules evolved in a way to be activated easily in the milieu containing sulfated polysaccharides, normal components in the tissue matrix.

INHIBITION OF PLASMINOGEN ACTIVATORS AND PLASMIN IN THE PRESENCE OR ABSENCE OF FIBRIN

Active enzyme of the fibrinolytic system, PAs and plasmin, quickly interact with fast-acting inhibitors in the plasma. There are three PA inhibitors; PAI-1 is synthesized by endothelial cells, and released into the circulation and is considered to be responsible for the regulation of fibrinolysis in the blood (39-42). PAI-2, which was first discovered in the placenta (43, 44), is synthesized by cells of macrophage/monocyte series (5, 45). PAI-3, which was found in the urine (46), turned out to be identical with protein C inhibitor (47).

It is suggested that the inhibiton of t-PA by PAI-1 was not effective in the presence of fibrin (48), which facilitates clot lysis, but inactivates t-PA quickly once fibrin dissolves. It is also suggested that t-PA-PAI-1 complex can activate plasminogen in the presence of fibrin. In fact it is shown that fibrin lyzes at a band where t-PA-PAI-1 complex is present in fibrin autography (48). We then wanted to know if t-PA-PAI-1 complex can activate plasminogen in the absence of fibrin. We used casein zymography (49) in which plasminogen was mixed with casein and plasma added with t-PA was electrophoresed. Fig. 6 shows that

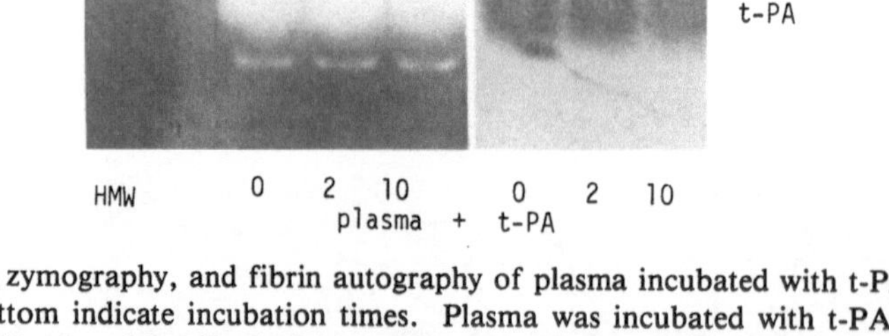

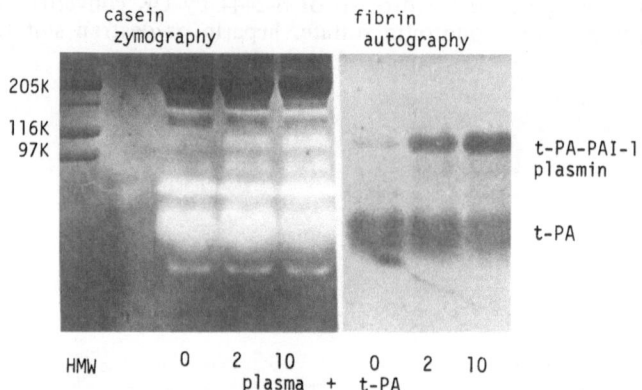

Fig. 6. Casein zymography, and fibrin autography of plasma incubated with t-PA Numbers at the bottom indicate incubation times. Plasma was incubated with t-PA at 37° for 0, 2 and 10 min. After incubation the aliquot was mixed with sample buffer. For casein zymography, samples were applied on SDS-gel containing casein and plasminogen. For fibrin autography, samples were applied on SDS-gel. Following electrophoresis gel was washed in 2.5% Triton X-100 for 90 min, then rinsed with buffer and incubated at 37°C for 3 hr. and stained with 0.25% Coomassie blue for casein zymography. For fibrin autography the gel was placed on a plasminogen-rich fibrin-agarose underlay after washing in Triton X-100.

bands of fibrinolysis and caseinolysis appeared at the site of the presence of t-PA-PAI-1 complex. These data suggest that PAI-1 is unique in the inhibition of t-PA in that the active site of t-PA did not either form a covalently bonded complex with the reactive site of PAI-1 or else that the active site of t-PA was exposed, and that t-PA and PAI-1 are complexed through a covalent bond at the site other than active and reactive sites.

ROLES OF FREE PAI-1 IN THE REGULATION OF THE FIBRINOLYTIC POTENTIAL

Now we wanted to know the roles of t-PA and PAI-1 in the regulation of the fibrinolytic potential. In order to examine these roles, we had to differentiate free PAI-1 from t-PA-PAI-1 complex. We developed a new enzyme immunoassay to measure total PAI-1, free PAI-1 and t-PA-PAI-1 complex (50).

Free PAI-1 and t-PA-PAI-1 complex were measured by enzyme immunoassay (50). Anti-PAI-1 Mab was used as first antibody, and ß-D-galactosidase labeled polyclonal anti-t-PA Fab' fragments (a-t-PA-Gal) were used as second antibody. Plasma was incubated with an anti-PAI-1 Mab immobilized silicone piece to measure amounts of t-PA-PAI-1 complex. In this case free PAI-1 and t-PA-PAI-1 complex were coupled with anti-PAI-1 Mab silicone, and only t-PA-PAI-1 complex was mesured, since the second antibody was a-t-PA Gal. Next, plasma was incubated with excess amounts of t-PA to form t-PA-PAI-1 complex from free PAI-1. This preincubated plasma was incubated with an anti-PAI-1 Mab immobilized silicone piece. Since all the PAI-1 were expected to be converted to the complex, this assay was used to measure the amounts of total PAI-1. The amounts of free PAI-1 were obtained by subtraction of t-PA-PAI-1 complex from total PAI-1.

In order to obtain a calibration line, we used a purified preparation of active PAI-1 obtained from human placenta using anhydrourokinase-Sepharose 4B (51). The amounts of PAI-1 were measured by IMULYSE (Biopool, Sweden), and diluted serially. The amounts of PAI-1 of these diluted PAI-1 preparations were measured by our method. There was a good calibration line for total PAI-1 ($y = 0.524x + 7.60$, $r = 0.993$, $p<0.001$).

In order to obtain a calibration line for t-PA-PAI-1 complex we added t-PA to PAI-1 of known concentration, and converted PAI-1 to t-PA-PAI-1 complex. The complex was also serially diluted, and measured by our method. There was a good correlation, and we could obtain a calibration line for t-PA-PAI-1 complex ($y=0.617x+7.01$, $r=0.985$, $p<0.001$).

We could also get good correlation lines for PAI-1 and t-PA-PAI-1 complex, by standardizing the concentrations of PAI-1 and t-PA-PAI-1 complex by IMULYSE. Our assay of total PAI-1 had a good correlation with data obtained using IMULYSE (52).

We compared the results obtained by our assay and using IMULYSE concerning plasma levels of total PAI-1 and free PAI-1. There was an excellent correlation between PAI-1 by IMULYSE and total PAI-1 by our assay ($y = 1.16x + 2.60$, $r = 0.946$, $p<0.001$). The y intercept of 2.60 ng/ml of total PAI-1 indicates the IMULYSE can not detect t-PA-PAI-1 complex as sensitively as our assay so that there are nearly 2 ng/ml t-PA-PAI-1 complex undetected by IMULYSE. There was also a good correlation between the concentration of PAI-1 measured by IMULYSE and those of free PAI-1 measured by our assay ($y = 0.849x - 0.122$, $r= 0.916$, $p<0.001$). This result indicates that PAI-1 measured by IMULYSE is mainly free PAI-1 and that some amounts of t-PA-PAI-1 complex could be detected by IMULYSE as suggested from x intercept (0.122/0.849). These results also suggest that there may not be much latent PAI-1, which was so abundantly shown in the supernatant of the culture of human umbilical endothelial cells (53, 54), since free PAI-1 measured by our method is an active form and the concentration of PAI-1 measured by IMULYSE is said to represent both active and latent PAI-1. The fact that the concentration of PAI-1 measured by IMULYSE is nearly the same as that measured by our method means that only small amounts of latent PAI-1 are present in the plasma.

The fibrinolytic potential of the plasma is best represented by euglobulin lysis time. We wanted to know which parameters of fibrinolytic system in the plasma are most significantly correlated to the euglobulin lysis time. It was shown that plasma free PAI-1 levels have the highest correlation coefficient to the euglobulin lysis time ($Y = 0.819x + 1.04$, $r = 0.599$, $p<0.001$), whereas the correlation to plasma levels of t-PA is low. These results suggest that free PAI-1 levels in the plasma are closely related to the fibrinolytic potential of the plasma.

ROLES OF FIBRIN IN THE PREVENTION OF THE INHIBITION OF PLASMIN BY α_2AP

Plasma is inhibited by α_2 antiplasmin (α_2AP) and $\alpha_2$macroglobulin (α_2M). α_2AP is a fast-acting inhibitor of plasmin. In order for plasmin to be inhibited quickly, LBS of plasmin must interact with plasmin binding site at the C terminal part of α_2AP (55), and then the active site of plasmin reacts with the reactive site of α_2AP. Since fibrin binds to plasmin through LBS, LBS of plasmin engaged in fibrinolysis is occupied by fibrin, so that such plasmin does not quickly interact with α_2AP. Once fibrin is degraded, LBS of plasmin is freed, then plasmin is being quickly inactivated.

Using the release of Bβ peptide from fibrin or fibrinogen by plasmin we were able to show that fibrin protects plasmin for inactivation by α_2AP. Glu-plg was incubated with UK in the presence of fibrinogen or fibrin. At various time intervals aprotinin was added to stop the reaction. After removing fibrin or fibrinogen by ethanol precipitation, the amounts of Bβ peptides (Bβ15-42 and Bβ1-42) were measured by radioimmunoassay (Fig. 7). In the absence of α_2AP, Bβ peptides were released faster from fibrinogen than from fibrin, even though Glu-plg was activated faster in the presence of fibrin than fibrinogen. The addition of α_2AP to these mixtures completely reversed the results. In the presence of α_2AP,

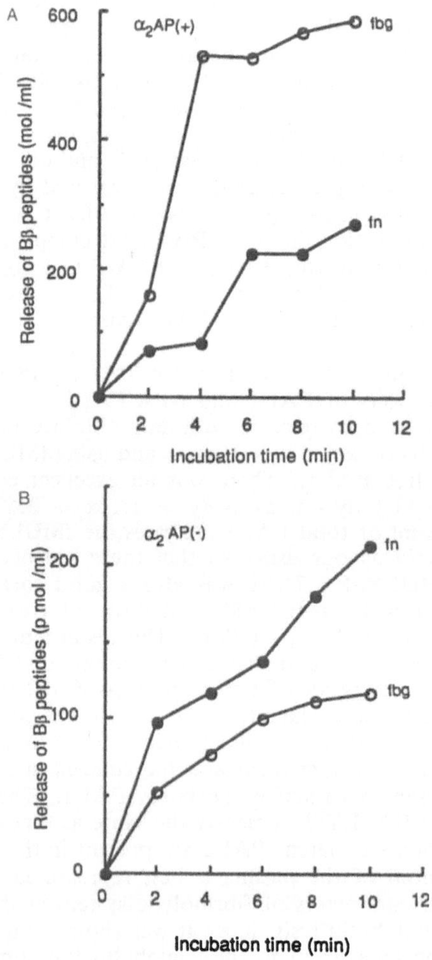

Fig. 7. Release of Bβ peptides from fibrin or fibrinogen after UK activation of Glu-plg in the absence (left) or presence (right) of α_2AP.

the release of Bβ peptides was faster from fibrin than fibrinogen, possibly due to the inactivation of α_2AP in the presence of fibrin.

We have shown that the hydrolysis of S-2251 by plasmin in the plasma clot is very fast in comparison to that in the plasma (58, 59). It can be explained that plasmin in the clot was protected from inactivation by α_2AP.

BINDING SITES ON t-PA WITH FIBRIN OR PLASMIN

It has been shown that plasminogen binds to both D and E domain of fibrin (60-62). Vali and Patthy (63) have shown that LBS of kringle 1 of plasminogen has a site to bind to E domain of fibrinogen, but D domain also binds to kringle 5 of plasminogen.

On the other hand, Nieuwenhuizen et al. (64) have shown that Aα 148-161 have site to bind to t-PA, whereas Christensen et al. (65), Norrman et al. (66), Suenson et al. (67) have indicated that lysine residues on C terminal part of Aα chain is responsible for binding with t-PA.

From these results, a scheme is presented for binding sites on t-PA and plasminogen for the formation of a trimolecular complex (Fig. 8).

We wanted to know if there is another binding site on plasminogen with t-PA. The incubation of sct-PA with plasmin in the presence of tranexamic acid resulted in the inhibition of the conversion of sct-PA to its two chain form. Fig. 9 shows the SDS-PAGE of the mixture of sct-PA and plasmin in the presence of various concentrations of tranexamic acid.

The addition of fragments of plasminogen such as K1-3 or K4 to the mixtures of sct-PA and plasmin facilitated the conversion of sct-PA to its two chain form by plasmin (Fig. 10).

These results suggest that binding of LBS (either with K1-3 or K4) facilitated the conversion of sct-PA to tct-PA by plasmin, and the binding of tranexamic acid to LBS of plasmin prevented the binding of LBS to t-PA. We, therefore, think that there is a site on t-PA molecule to interact with LBS of plasmin, which may be involved in a formation of a ternary complex between t-PA, plasminogen and fibrin.

DISCUSSION

We have worked for years on the mechanisms of the activation and regulation of fibrinolysis system. Although much attention has been paid to t-PA due to its high affinity to fibrin and its enhanced capacity to activate plasminogen, the facilitated lysis of clot is not necessarily the unique consequence caused by the activation of plasminogen by t-PA in the presence of fibrin. UK and SK also cause faster lysis of the clot by better activation of plasminogen in the presence of fibrin. These aspects are discussed in details by Y. Takada in this book. Thus clot formation itself brings about the initiation of clot lysis, which is an expression of homeostasis in vivo. Bleeding must be stopped by the platelet aggregation and fibrin clot formation. Once bleeding stops, clot must be dissolved for the recanalization of the blood vessels. Intravascular clot caused by platelet adhesion and aggregation to atherosclerotic plaques must be quickly dissolved by the activation of fibrinolysis system

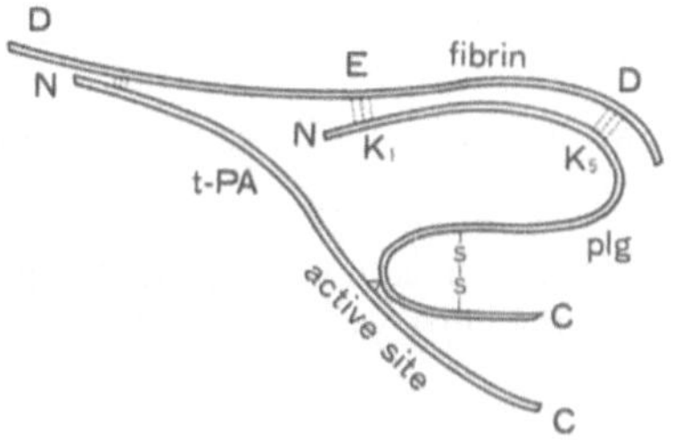

Fig. 8. A scheme of a trimolecular complex of t-PA plasminogen and fibrin.

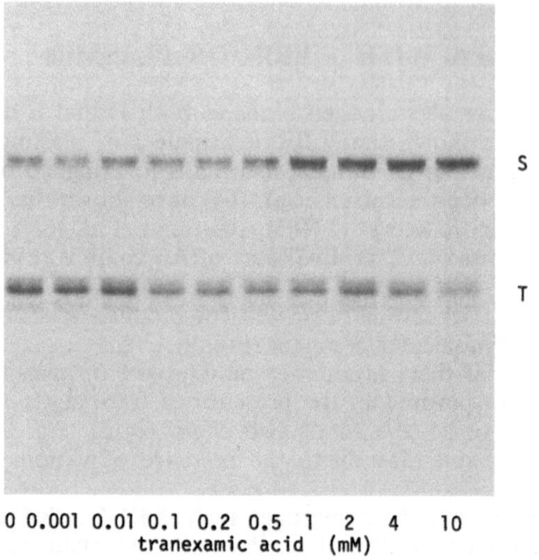

Fig. 9. Effects of tranexamic acid on the conversion of sct-PA to tct-PA by plasmin S: sct-PA, T: tct-PA.

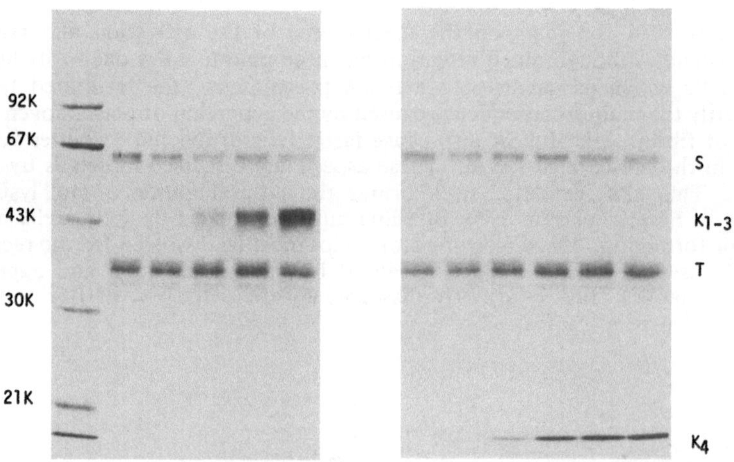

Fig. 10. Effect of K1-3 or K4 on the conversion of sct-PA by plasmin S: sct-PA, T: tct-PA, LMW: low molecular weight standards.

Otherwise blood supply to the occluded area is impaired, resulting in the necrosis of tissues, thus infarction.

We think that coagulation and fibrinolysis system have evolved to ensure the prevention of the loss of the blood by bleeding and the quick dissolution of fibrin once bleeding stops. Also intravascular clot formation is prevented as much as possible, so that many steps to safeguard against accidental coagulation have evolved such as the presence of inhibitors corresponding to active enzymes and difficulty in the initiation of coagulation in the absence of tissue injury. Characteristics of endothelial cells sometimes change so that coagulation initiates on the surface of diseased vessels. Clot must be immediately dissolved in such an occasion.

Since there are many fast acting and effective inhibitors such as α_2AP or PAI-1, active enzymes such as t-PA or plasmin are quickly inactivated. The prevention of the inhibition of active enzymes in the presence of fibrin is one of the mechanisms to facilitate the clot lysis. Secondly, sct-PA is less active in amidolytic activity than tct-PA. Glu-plg is hardly activatable by activators. The presence of fibrin changes the conformation of both sct-PA and Glu-plg, the former being easily converted to a more active two chain form, and the latter being changed to a looser and better activatable conformation. Fibrin formation, thus, signals to many steps of the fibrinolysis system so that fibrinolysis proceeds effectively.

Although we singled out each step of the regulation mechanisms of the fibrinolysis system in order to discuss effectively, each reaction in the coagulation and fibrinolysis takes place in a subtle harmony and is fine tuned to slight changes in the concentration of each factor and its conformation. We must therefore realize that the coagulation and fibrinolysis system should be analyzed as a whole in vivo. The activation of one reaction is proceeding at the time when the inhibition of the reaction has already started. Furthermore the amplification and feedback reaction are also taking place hand in hand. We should not then evaluate data using only the results of one or some reactions but always pay attention to the whole system functioning in accord.

REFERENCES

1. F. M. Booyse, J. Scheinbuks, J. Radek, G. Osikowicz, S. Fedor and A. J. Quarfoot, Immunological identification and comparison of plasminogen activator forms in cultured normal human endothelial cells and smooth muscle cells, *Thromb. Res.*, **24**:495-504 (1981).
2. G. H. Goldsmith, N. P. Ziats, and A. L. Robertson, Studies on plasminogen activator and other proteases in subcultured human vascular cells, *Exp. Mol. Pathol.*, **35**:257-264 (1981).
3. E. G. Levin, and D. J. Loskutoff, Cultured bovine endothelial cells produce both urokinase and tissue-type plasminogen activators, *J. Cell Biol.*, **94**:631-636 (1982).
4. C. Nolan, L. S. Hall, G. H. Barlow and I. I. E. Tribby, Plasminogen activator from human embryonic kidney cell cultures. Evidence of a proactivator, *Biochim. Biophys. Acta*, **496**:384-400 (1977).
5. J.-D. Vassalli, J. M. Dayer, A. Wohlwend, and D. Belin, Concomitant secretion of prourokinase and of a plasminogen activator-specific inhibitor by cultured human monocytes-macrophages, *J. Exp. Med.*, **159**:1653-1668 (1984).
6. T. C. Wun, L. Ossowski, and E. Reich, A proenzyme of human urokinase, *J. Biol. Chem.*, **257**:7262-7268 (1982).
7. L. Skriver, L. S. Nielsen, R. Stephens, and K. Danϕ, Plasminogen activator released as inactive proenzyme from murine cells transformed by sarcoma virus, *Eur. J. Biochem.*, **124**:409-414 (1982).
8. G. Markus, The role of hemostasis and fibrinolysis in the metastatic spread of cancer, *Semin. Thromb. Hemost.*, **10**:61-70 (1984).
9. M. Rånby, N. Bergsdorf, and T. Nilsson, Enzymatic properties of the one- and two-chain form of tissue plasminogen activator, *Thromb. Res.*, **27**:175-183 (1982).
10. D. C. Rijken, M. Hoylaerts, and D. Collen, Fibrinolytic properties of one-chain and two-chain human extrinsic (tissue-type) plasminogen activator, *J. Biol. Chem.*, **257**:2920-2925 (1982).
11. D. C. Stump, H. R. Lijnen, and D. Collen, Purification and characterization of single-chain urokinase-type plasminogen activator (scu-PA) from human cell cultures, *J. Biol. Chem.*, **261**:1274-1278 (1986).

12. A. Ichinose, W. Kisiel, and K. Fujikawa, Proteolytic activation of tissue plasminogen activator by plasma and tissue enzymes, *FEBS Lett.*, **175**:412-418 (1984).
13. A. Ichinose, K. Fujikawa, and T. Suyama, The activation of prourokinase by plasma kallikrein and its inactivation by thrombin, *J. Biol. Chem.*, **261**:3486-3489 (1986).
14. L. A. Miles, and J. H. Griffin, The role of molecules immunologically related to urokinase in contact system-dependent fibrinolysis, *in:* "Fundamental and Clinical Fibrinolysis," F. J. Castellino, P. J. Gaffney, M. M. Samama, and A. Takada, ed., Elsevier Science Publishers, B. V., Amsterdam, pp. 45-55 (1987).
15. F. Bachmann, and E, K. O. Kruithof, Tissue plasminogen activator: Chemical and physiological aspects, *Semin. Thromb. Haemost.*, **10**:6-17 (1984).
16. V. Gurewich, R. Pannell, S. Louie, P. Kelley, R. L. Suddith and R. Greenlee, Effective and fibrin-specific clot lysis by a zymogen precursor form of urokinase (prourokinase). A study in vitro and in two animal species, *J. Clin. Invest.*, **73**:1731-1739 (1984).
17. T. Urano, V. S. de Serrano, P. J. Gaffney, and F. J. Castellino, Activation of human [Glu^1] plasminogen by human single chain urokinase, *Arch. Biochem. Biophys.*, **264**:222-230 (1988).
18. T. Urano, V. S. de Serrano, S. Urano, and F. J. Castellino, Stimulation by fibrinogen of the amidolytic activity of single-chain tissue plasminogen activator, *Arch. Biochem. Biophys.*, **270**:356-362 (1989).
19. F. J. Castellino, B. A. K. Chibber, J. M. Beals, and V. S. de Serrano, The structure and activation of human plasminogen, *in:* "Fundamental and Clinical Fibrinolysis," F. J. Castellino, P. J. Gaffney, M. M. Samama, and A. Takada, ed., Elsevier Science Publishers, B. V., Amsterdam, pp. 19-31 (1987).
20. A. Takada, and Y. Takada, Activation mechanisms of human plasminogen by streptokinase, urokinase or tissue plasminogen activator, *in:* "Fundamental and Clinical Fibrinolysis," F. J. Castellino, P. J. Gaffney, M. M. Samama, and A. Takada, ed., Elsevier Science Publishers, B. V., Amsterdam, pp. 33-44 (1987).
21. A. Ichinose, K. Takio,and K. Fujikawa, Localization of the binding site of tissue-type plasminogen activator to fibrin, *J. Clin. Invest.*, **78**:163-169 (1986).
22. L. Banyai, A. Varadi, and L. Patthy, Common evolutionary origin of the fibrin-binding structures of fibronectin and tissue-type plasminogen activator, *FEBS Lett.*, **163**:37-41 (1983).
23. S. M. Camiolo, S. Thorsen, and T. Astrup, Fibrinogenolysis and fibrinolysis with tissue plasminogen activator, urokinase, streptokinase-activated human globulin, and plasmin, *Proc. Soc. Exp. Biol. Med.*, **138**:277-280 (1971).
24. S. Thorsen, P. Glas-Greenwalt, and T. Astrup, Differences in the binding to fibrin of urokinase and tissue plasminogen activator, *Thromb. Diath. Haemorrh.*, **28**:65-74 (1972).
25. M. Hoylaerts, D. C. Rijken, H. R. Lijnen, and D. Collen, Kinetics of the activation of plasminogen by human tissue plasminogen activator: role of fibrin, *J. Biol. Chem.*, **257**:2912-2919 (1982).
26. A. Takada, Y. Sugawara, and Y. Takada, Comparison of kinetic parameters of the activation of Glu-plasminogen by tissue plasminogen activator obtained from various sources, *Haemostasis,* **18**:117-125 (1988).
27. Y. Sugawara, Y. Takada, K. Yamamoto, and A. Takada, Kinetic analyses of the enhancement of the activities of t-PA induced by the presence of monoclonal antibody (C9-5). *Thromb. Res.*, **50**:637-646 (1988).
28. S. S. Husain, V. Gurewich, and B. Lipinski, Purification and partial characterization of a single-chain, high molecular weight form of urokinase from human urine, *Arch. Biochem. Biophys.*, **220**:31-38 (1983).
29. C. Tran-Thang, E. K. O. Kruithof, and F. Bachmann, The mechanism of in vitro clot lysis induced by vascular plasminogen activator, *Blood,* **63**:1331-1337 (1984).
30. H. R. Lijnen, C. Zamarron, M. Blaber, M. E. Winkler, and D. Collen, Activation of plasminogen by pro-urokinase. I. Mechanism, *J. Biol. Chem.*, **261**:1253-1258 (1986).
31. T. C. Wun, D. Schleuning, and E. Reich, Isolation and characterization of urokinase from human plasma, *J. Biol. Chem.*, **257**:3276-3283 (1982).
32. J. D. Tissot, P. H. Schneider, J. Hauert, M. Ruegg, E. K. O. Kruithof, and F. Bachmann, Isolation from human plasma of a plasminogen activator identical to urinary high molecular weight urokinase, *J. Clin. Invest.*, **70**:1320-1323 (1982).
33. J.-D. Vassalli, D. Baccio, and D. Belin, A cellular binding site for the Mr 55,000 form of the human plasminogen activator, urokinase, *J. Cell. Biol.*, **100**:86-92 (1985).

34. M. P. Stroppelli, A. Corti, A. Soffientini, G. Cassani, F. Blasi, and R. K. Associan, Differentiation-enhanced binding of the amino-terminal fragment of human urokinase plasminogen activator to a specific receptor on U937 monocytes, *Proc. Natl. Acad. Sci. USA,* **82:**4939-4943 (1985).
35. Y. Eeckhout, and G. Vaes, Further studies on the activation of procollagenase, the latent precursor of bone collagenase, Biochem. J., **166:**21-31 (1977).
36. M. Paranjpe, L. Engel, N. Young, and L. A. Liotta, Activation of human breast carcinoma collagenase through plasminogen activator, *Life Sci.,* **26:**1223-1231 (1980).
37. R. I. O'Grady, L. I. Upfold, and R. W. Stephens, Rat mammary carcinoma cells secrete active collagenase and activate latent enzyme in the stroma via plasminogen activator, *Int. J. Cancer,* **28:**509-515 (1981).
38. A. Rydzewski, Y. Takada, and A. Takada, Stimulation of plasmin-catalyzed conversion of single-chain to two-chain urokinase-type plasminogen activator by sulfated polysaccharides, *Thromb. Haemost.* **62:**752-755 (1989).
39. E. G. Levin, Latent tissue plasminogen activator produced by human endothelial cells in culture: Evidence for an enzyme-inhibitor complex. *Proc. Natl. Acad. Sci. USA,* **80:**6804-6808 (1983).
40. J. A. van Mourik, D. A. Lawrence, and D. J. Loskutoff, Purification of an inhibitor of plasminogen activator (antiactivator) synthesized by endothelial cells. *J. Biol. Chem.,* **259:**14914-14921 (1984).
41. M. Philips, A. -G. Juul, and S. Thorsen, Human endothelial cells produce a plasminogen activator inhibitor and a tissue-type plasminogen activator-inhibitor complex, *Biochim. Biophys. Acta,* **802:**99-110 (1984).
42. E. D. Sprengers, J. H. Verheijen, V. W. M. van Hinsbergh, and J. J. Emeis, Evidence for the presence of two different fibrinolytic inhibitors in human endothelial cell conditioned medium, *Biochim. Biophys. Acta* **801:**163-170 (1984).
43. T. Kawano, K. Morimoto, and Y. Uemura, Urokinase inhibitor in human placenta, *Nature,* **217:**175-180 (1968).
44. L. Holmberg, I. Lecander, B. Persson, and B. Åstedt, An inhibitor from placenta specifically binds urokinase and inhibits plasminogen activator released from overian carcinoma in tissue culture, *Biochim. Biophys. Acta* **544:**128-137 (1978).
45. E. K. O. Kruithof, J.-D. Vassalli, W.-D. Schleuning, R. J. Mattaliano, and F. Bachmann, Purification and characterization of a plasminogen activator inhibitor from the histiocytic lymphoma cell line U 937, *J. Biol. Chem.,* **261:**11207-11213 (1986).
46. D. C. Stump, M. Thienpont, and D. Collen, Purification and characterization of a novel inhibitor of urokinase from human urine. Quantitation and preliminary characterization in plasma, *J. Biol. Chem.,* **261:**12759-12766 (1986).
47. M. J. Heeb, F. Espana, M. Geiger, D. Collen, D. C. Stump, and J. H. Griffin, Immunological identity of heparin-dependent plasma and urinary protein C inhibitor and plasminogen activator inhibitor-3, *J. Biol. Chem.,* **262:**15813-15816 (1987).
48. E. K. O. Kruithof, C. Tran-Thang, A. Ransijn and F. Bachmann, Demonstration of a fast-acting inhibitor of plasminogen activators in human plasma, *Blood* **64:**907-913 (1984).
49. P. C. Roche, J. D. Campeau, and T. Shaw Jr., Comparative electrophoretic analysis of human and porcine plasminogen activators in SDS-polyacrylamide gels containing plasminogen and casein, *Biochim. Biophys. Acta,* **745:**82-89 (1983).
50. Y. Takada, and A. Takada, Measurements of the concentration of free plasminogen activator inhibitor (PAI-1) and its complex with tissue plasminogen activator in human plasma. *Thromb. Res.,* Suppl. VIII:15-22 (1988).
51. T.-C. Wun, M. O. Palmier, N. R. Siegel, and C. E. Smith, Affinity purification of active plasminogen activator inhibitor-1 (PAI-1) using immobilized anhydrourokinase, *J. Biol. Chem.,* **264:**7862-7868 (1989).
52. A. Rydzewski, Y. Takada, and A. Takada, Determination of plasminogen activator inhibitor-1 (PAI-1) in plasma using two different anticoagulants and methods, *Thromb. Res.* **55:**285-289 (1989).
53. C. M. Hekman, and D. J. Loskutoff, Endothelial cells produce a latent inhibitor of plasminogen activators that can be activated by denaturants, *J. Biol. Chem.,* **260:**11581-11587 (1985).
54. E. D. Sprengers, V. W. M. van Hinsbergh, and B. G. Jansen, The active and the inactive plasminogen activator inhibitor from human endothelial cell conditioned medium are immunologically and functionally related to each other, *Biochim. Biophys. Acta,* **883:**233-241 (1986).

55. T. Sasaki, T. Moita, and S. Iwanaga, Identification of the plasminogen-binding site of human α_2-plasmin inhibitor, *J. Biochem.,* **99:**1699-1705 (1986).
56. T. Urano, Y. Takada, and A. Takada, The enhanced activation of Glu-plasminogen by urokinase in the presence of fibrin or des A fibrin as measured by the release of Bβ peptide and FDP, *Thromb. Res.,* **36:**429-435 (1984).
57. A. Takada, Y. Makino, and Y. Takada, Release of N-terminal peptides from Glu-plasminogen by plasmin in the presence of fibrin, *Thromb. Res.,* **41:**819-827 (1986).
58. A. Takada, T. Urano, and Y. Takada, Influence of coagulation on the activation of plasminogen by streptokinase and urokinase, *Thromb. Haemostas.,* **42:**901-908 (1979).
59. A. Takada, T. Ito, and Y. Takada, Interaction of plasmin with tranexamic acid and $\alpha_2$plasmin inhibitor in the plasma and clot, *Thromb. Haemostas.,* **43:**20-23 (1980).
60. S. A. Cederholm-Williams, The binding of plasminogen (mol. wt. 84,000) and plasmin to fibrin, *Thromb. Res.,* **11:**421423 (1984).
61. M. A. Lucas, L. J. Fretto, and P. A. McKee, The binding of human plasminogen to fibrin and fibrinogen, *J. Biol. Chem., 258:*4249-4256 (1983).
62. A. Varadi, and L. Patthy, Location of plasminogen-binding site sin human fibrin(ogen), *Biochemistry,* **22:**2440-2446 (1983).
63. Z. Wali, and L. Patthy, The fibrin binding sites of human plasminogen: arginine 32 and 34 are essential for fibrin affinity of the kringle 1 domain, *J. Biol. Chem.,* **259:**13690-13694 (1974).
64. W. Nieuwenhuizen, M. Voskuilen, A. Vermond, G. H. Veeneman, J. H. Van Boom, E. A. Klase, and N. D. Zegers, Studies on sites in fibrin(ogen) which are involved in the acceleration of plasminogen activation, catalyzed by tissue-type plasminogen activator, *in:* "Fundamental and Clinical Fibrinolysis," F. J. Castellino, P. J. Gaffney, M. M. Samama, and A. Takada, ed., Elsevier Science publishers, B. V., Amsterdam pp. 57-65 (1987).
65. U. Christensen, The AH site of plasminogen and two C-terminal fragements. A weak lysine-binding site preferring ligands not carrying a free carboxylate function. *Biochem. J.,* **223:**413-421 (1984).
66. B, Norrman, P. Wallen, and M. Ranby, Fibrinolysis mediated by tissue plasminogen activator. Disclosure of a kinetic transition, *Eur. J. Biochem.,* **149:**193-201 (1985).
67. E. Suenson, O, Lützen, and S. Thorsen, Initial plasmin degradation of fibrin as the basis of a positive feed back mechanism in fibrinolysis, *Eur. J. Biochem.,* **140:**513-522 (1984).

THE MECHANISMS OF THE ACTIVATION OF PLASMINOGEN BY STREPTOKINASE AND UROKINASE

Yumiko Takada and Akikazu Takada

Department of Physiology, Hamamatsu University
School of Medicine, Hamamatsu-shi
Shizuoka, Japan 431-31

INTRODUCTION

Plasminogen activation to plasmin is due to enzymatic cleavage of a single peptide bond in the zymogen molecule. These enzymes are called plasminogen activators. There are three categories in plasminogen activators: streptokinase (SK), urokinase (UK) and tissue plasminogen activators (t-PA). It has been well documented that the activation rate of plasminogen by t-PA was remarkably enhanced in the presence of fibrin, and this part is reported by A. Takada in this book. On the other hand, the role of fibrin in the activation of plasminogen by UK have been controversial (1-4). Studies from several laboratories showed that fibrin and its related substances enhance the activity of SK (5-10). It was claimed that fibrinogen increases only the rate of formation of an active site in the SK-plg complex and does not have an effect on the activator activity of the SK-plg complex (8). However, other studies (5, 9, 10) found that fibrinogen also stimulates the activator activity of the complex.

In the present paper, we report on the mechanisms of the enhancement of the activation of plasminogen by SK and UK in the presence of fibrin and its related substances.

ACTIVATION OF PLASMINOGEN BY SK

Streptokinase (SK) is an extracellular protein produced by several strains of hemolytic streptococci. SK molecule is composed of 414 aminoacids and its molecular weight is 47,000-50,200 (11, 12). The amino-terminal 230 residues of SK show homology with trypsin-type of serine proteases (13). However, SK does not possess an active site serine residue and does not exhibit any of the classical properties of trypsin-like enzymes. Thus, SK can not cleave peptide bonds in proteins, has no esterase or amidase activities against synthetic substrates and it is not inhibited by diisopropyl fluorophosphate (DFP) and by pancreatic trypsin inhibitor (PTI) (14). Therefore, the mechanisms of the activation of plasminogen (plg) by SK have been controversial for years, since SK is reported to have fibrinolytic activity (15). Müllertz and Lassen (16) proposed that proactivator in the human plasma was converted by SK to activator which in turn activated plasminogen of any animal species to plasmin. Since SK was shown not to have any enzymatic activity, a following scheme has been proposed. SK reacts with plasminogen to form an equimolar complex. As a result of this complex formation a conformational change occurs in the complex and an enzyme active site is formed in the plasminogen moiety of the complex. The newly formed enzyme active site catalyzes the conversion of plasminogen to plasmin. The SK-plg complex is also transformed to the SK-plasmin complex by the newly formed active site. The conversion of the SK-plg complex to the SK-plasmin complex can occur by intramolecular cleavage of the activation peptide bond (17).

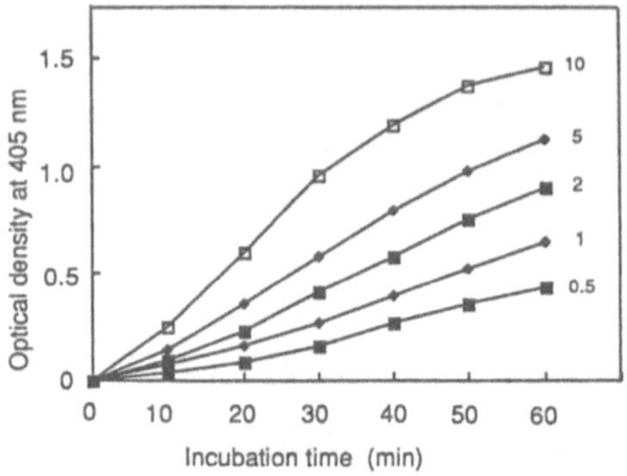

Fig. 1. Time course of the hydrolysis of S-2251 by Glu-plg and SK. Numbers in the figure indicate the concentration of SK.

Since the SK-plg complex shows amidase activity towards the plasmin-specific substrate H-D-Val-Leu-Lys pNA (S-2251) (18), we have to be careful to analyze the data of the hydrolysis of S-2251 by the mixture of SK and plasminogen. The plasminogen activator activity of the SK-plg complex is 2 to 3-fold greater than the activity of the SK-plasmin complex (19, 20).

Also, we have to be careful to use the kind of plasminogen. Native form of plasminogen has glutamic acid in its N-terminal amino residue, being called Glu-plg, whereas its proteolytically modified form has mainly lysine, partly methionine and valine, in its N-terminal, being called Lys-plg (21). Both plasminogens have two molecular forms, plg I and II, which differ from each other in their carbohydrate contents (22). Two isozymes of plasminogen are thus named Glu-plg I and II, and Lys-plg I and II.

Fig. 1 shows the time course of the hydrolysis of S-2251 (0.18 mM) by the equimolar mixture of Glu-plg (0.02 μM-0.4 μM) and SK (0.5-10 U/ml). The rate of hydrolysis of S-2251 was higher in the larger amounts of SK-plg complex, but the slope of each curve was linear for some time intervals after initial lag periods. The linearity of the curve of the hydrolysis of S-2251 means that SK-plg complex hydrolyzes S-2251.

We measured the kinetic parameters of the SK-plg complex. Fig. 2 shows the Lineweaver Burk plots of the hydrolysis of S-2251 by the mixture of Glu-plg and SK. Results indicate that Km (0.67 mM) did not change at various amounts of SK, and that Vmax increased with increases in the amounts of SK.

ENHANCEMENT OF THE ACTIVATION OF PLASMINOGEN BY SK IN THE PRESENCE OF FIBRIN OR ITS RELATED SUBSTANCES

In 1964, we have reported that there is a factor in human plasma which potentiates the rate of SK-activation of plasminogen (23). We further characterized this factor (24), which turns out to be an early degradation product of fibrinogen and named it as SK-potentiator (25, 26). SK potentiator cross-reacted with the antibody against fibrinogen, and it had a molecular weight of 240,000, consisting of ß and γ chains of fibrinogen with its α chain degraded.

When Glu-plg was mixed with SK in the presence of fibrin, SK-potentiator or other F(g)DP, and the activator activity was measured by the hydrolysis of S-2251 simultaneously added to the reaction mixture, the time course of the hydrolysis of S-2251 was shown in Fig. 3. Fibrin enhanced the activator activity to the largest extent, followed by SK-potentiator, fibrinogen and fragments D and E.

We also showed that in order to potentiate the activation of plasminogen by SK, SK-potentiator had to be present when SK and plasminogen were mixed. The addition of SK-potentiator to a preformed complex of SK and plasminogen did not result in any enhancement of the activation of plasminogen by SK (25). From these results it may be reasonable to conclude that SK forms a trimolecular complex with plasminogen and poten-

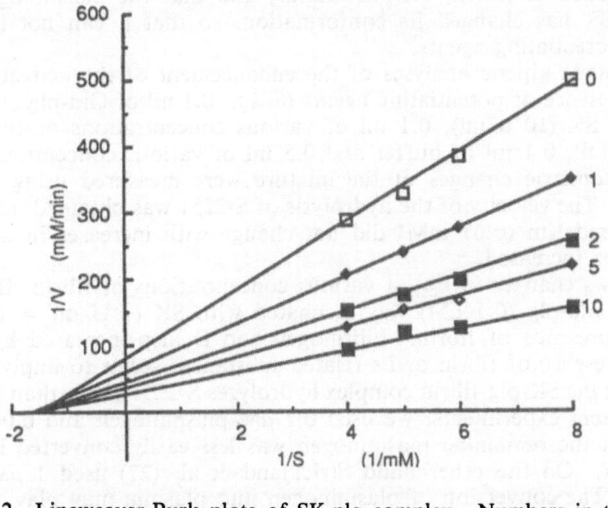

Fig. 2. Lineweaver-Burk plots of SK-plg complex. Numbers in the figure indicate the concentration of SK (u/ml). 0.1 ml of Glu-pig (0.02 μM-0.4 μM) was mixed with equimolar amounts of SK, 0.3 ml of 0.1 M Tris buffer, pH 7.4 and 0.5 ml of various concentrations of S-2251 and spectrophotometric changes of the mixture were measured using an automatic spectrophotometer.

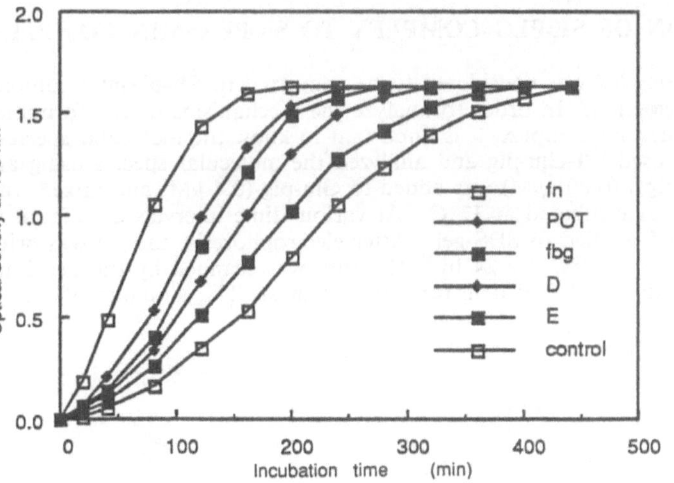

Fig. 3. Effects of fibrin or its related substances on the activation of Glu-plg by SK. "Control" indicates the time course of the hydrolysis of S-2251 (0.3 mM) by the mixture of Glu-plg (0.1 μM) and SK (0.5 u/ml: 0.02 μM). "fn", "POT", "fbg", "D" and "E" indicate the time course of the hydrolysis of S-2251 by the mixture of Glu-plg and SK in the presence of 0.1 μM of fibrin (fn), SK-potentiator (POT), fibrinogen (fbg), D and E, respectively.

tiating agents (such as fibrin, SK-potentiator) and that the plasminogen molecule in the complex with SK has changed its conformation, so that it can not form the additional complex with potentiating agents.

We then made kinetic analyses of the enhancement of the activation of plasminogen by SK in the presence of potentiating agents (6, 7). 0.1 ml of Glu-plg (1 μM) was incubated with 0.1 ml of SK (10 u/ml), 0.1 ml of various concentrations of fibrinogen, 0.1 ml of thrombin (1 u/ml), 0.1 ml of buffer and 0.5 ml of various concentrations of S-2251, and the spectrophotometric changes of the mixture were measured using an automatic spectrophotometer. The velocity of the hydrolysis of S-2251 was obtained, and Lineweaver-Burk plots were plotted Km (0.67 mM) did not change with increases in the concentration of fibrin, but Vmax increased.

Fig. 4 shows changes in k_{cat} at various concentrations of fibrin, fibrinogen, fragment D or E, when Glu-plg (0.1 μM) was incubated with SK (1 U/ml = 0.04 μM). k_{cat} was highest in the presence of fibrin. Fibrinogen and D also increased k_{cat} values. Increase in k_{cat} in the presence of fibrin or its related substances seems to imply that a trimolecular complex such as the SK-plg-fibrin complex hydrolyzes S-2251 faster than the SK-plg complex.

In the present experiments, we used 0.1 μM plasminogen and 0.04 μM SK in kinetic analyses so that the remaining plasminogen was less easily converted into plasmin by the SK-plg complex. On the other hand Strickland et al. (27) used 1 μM plasminogen and 0.004 μM SK. The conversion of plasminogen into plasmin may play a significant part in their experimental system besides the hydrolysis of S-2251. This may explain some differences in experimental results and mechanisms proposed accordingly.

Next we used Lys-plg instead of Glu-plg, and the same tendency as Glu-plg was observed. Lineweaver-Burk plots of the hydrolysis of S-2251 by the mixture of Lys-plg (0.01 μM) and SK (0.01 μM: 0.25 U/ml) and various concentrations of fibrin showed that Km (0.25 mM) did not change and Vmax increased with rise in the concentrations of fibrin.

When fibrinogen, D or E was used as a potentiating agents same tendency was observed. That means that the mechanism of the enhancement of the activation of Lys-plg by SK is similar to that of Glu-plg.

CONVERSION OF SK-PLG COMPLEX TO SK-PLASMIN COMPLEX

SK-plg complex was considered to be converted to SK-plasmin complex due to intermolecular interaction. In order to analyze the mechanisms of the formation of a dimolecular or trimolecular complex, it is important to know the molecular species in the reaction mixture. We used ^{125}I-Glu-plg and analyzed the molecular species using autoradiography.

^{125}I-Glu-plg I (0.001 μM) was added to Glu-plg (0.1 μM) and mixed with SK (0.1 μM). The mixture was incubated at 37°C. At various time intervals an aliquot was taken from the mixture and applied to SDS-gel. After electrophoresis, the gel was dried, and exposed to X-ray film at −70°C for 24 hr. The film was scanned by the densitometer. Fig. 5A and 5B show the results of densitometric tracings. The amounts of Glu-plg I started to

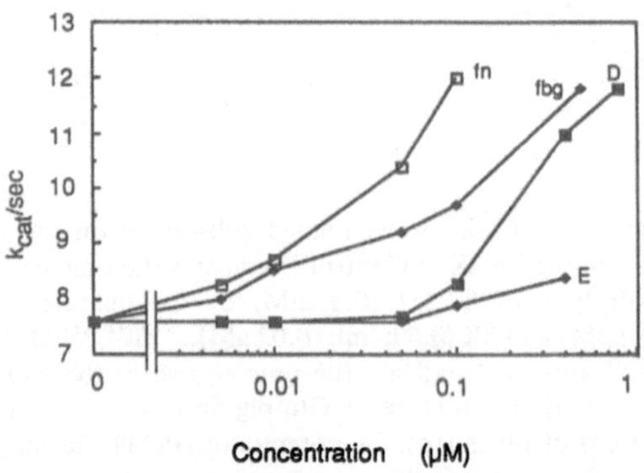

Fig. 4. k_{cat} of the mixture of Glu-plg, SK and potentiating agents.

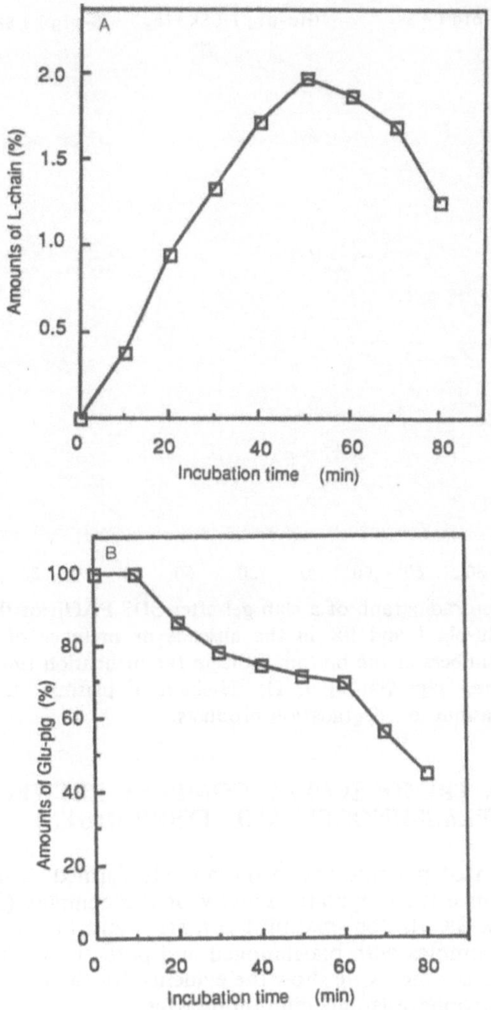

Fig. 5A. Fig. 5B. The amounts of Glu-plg I (Fig. 5A) or L-chain of plasmin (Fig. 5B) after incubation of the mixture of Glu-plg I and SK.

decline after 10 min incubation. At 20 min, the amounts of Glu-plg I was 85% of the original amounts, and nearly 70% of plasminogen was still Glu-form at 60 min (Fig. 5A).

Fig. 5B shows the changes in the amounts of L-chain after the incubation of Glu-plg with SK. The amounts of L-chain increased up to 30 min, and declined after 60 min incubation possibly due to a further degradation to peptides with smaller molecular weight. These results suggest that the main molecular species in the equimolar mixtures of SK and Glu-plg I was SK-plg complex.

Then the conversion of SK-plg complex to SK-plasmin complex in the presence of fibrin(ogen) was studied. Glu-plg I (0.1 μM) added with ^{125}I-Glu-plg I (0.001 μM) was mixed with SK (0.1 μM) in the absence or presence of fibrinogen or fibrin. Fig. 6 shows the results of autoradiography. The bands of Glu-plg disappeared faster in the presence of fibrinogen or fibrin. The bands of L-chain and smaller peptides at the bottom line appeared faster in the presence of fibrinogen or fibrin. These results show that Glu-plg was converted to plasmin faster in the presence of fibrin(ogen).

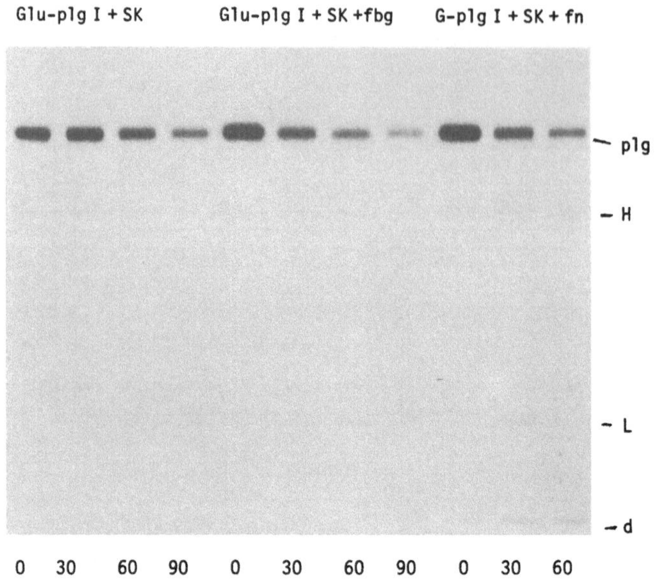

Fig. 6. Autoradiograph of a slab gel after SDS-PAGE of the mixture of Glu-plg I and SK in the absence or presence of fibrin(ogen). Numbers at the bottom indicate the incubation time of the mixture. plg: Glu-plg I, H: H-chain of plasmin, L: L-chain of plasmin, d: degradation products.

FORMATION OF A TRIMOLECULAR COMPLEX BETWEEN STREPTOKINASE, PLASMINOGEN AND FIBRIN(OGEN)

Since the addition of potentiators to the already formed complex of SK with plg (or plasmin) did not enhance the enzymatic activity of the complex (25), potentiators are not considered to enhance SK-plg (or -plasmin) complex, and kinetic studies suggest that SK forms a trimolecular complex with plasminogen and potentiators (6, 7).

In the following experiments we show the evidence for the formation of the trimolecular complex by using electrophoresis and immunoblotting.

The mixtures of Glu-plg (0.4 μM) and fibrinogen (1.6 μM), Glu-plg (0.4 μM) and SK (0.4 μM), or Glu-plg (0.4 μM), SK (0.4 μM) and fibrinogen (1.6 μM), or Glu-plg (0.4 μM), SK (0.4 μM) and fibrinogen (1.6 μM) were incubated for various time intervals at 37°C. After incubation, the mixtures were applied to polyacrylamide gel for electrophoresis. After electrophoresis, proteins were transferred to a NC membrane and reacted with anti-plg antibody or anti-fbg antibody.

The upper panel of Fig. 7 shows the results using anti-fbg antibody. The bands of fibrinogen broadened in the mixture with plasminogen. When equimolar mixtures of plasminogen and SK with fibrinogen were electrophoresed, the bands of fibrinogen migrated more in the direction of the anode. The lower panel of Fig. 7 shows the results using anti-plg antibody. The bands of plasminogen migrated nearly equal distances regardless of the simultaneous presence of fibrinogen or SK. The bands of plasminogen in the mixture of plasminogen, fibrinogen and SK after 30 min and 60 min incubation migrated more in the direction, of the anode.

These results argue against the possibility that SK activates fibrinogen bound plasminogen. As shown in Fig. 7 migration patterns of the mixture of plasminogen and fibrinogen were different from those of SK, plasminogen and fibrinogen. In the absence of SK, plasminogen and fibrinogen migrate in different sites in electrophoresis. It may, therefore, be concluded that fibrinogen complexes with plasminogen only in the simultaneous presence of SK.

We made a schematic model (Fig. 8) of a trimolecular complex considering the following points. Epsilon aminocaproic acid (EACA) did not inhibit the complex formation between SK, plasminogen and fibrinogen or related substances (28). Christensen (29) reported that

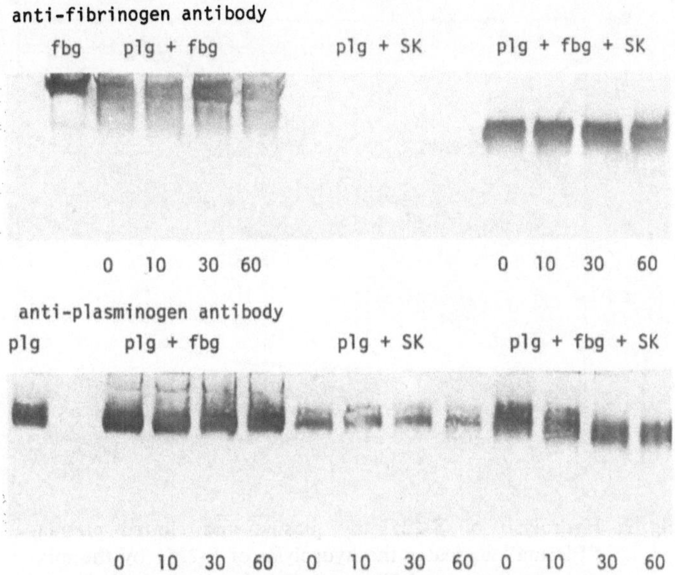

Fig. 7. Immunoblotting of fibrinogen and plasminogen after electrophoresis. Upper panel shows the bands of fibrinogen. Lower panel shows the bands of plasminogen. Numbers under the panels show the incubation times.

the aminohexyl site of plasminogen in the kringle 5 may be the site to bind to fibrin. If so, it is possible that fibrin binds not to lysine-binding sites but to the aminohexyl site of plasminogen complexed with SK. SK binds to a light chain side of plasminogen (30).

EFFECTS OF FIBRIN AND ITS RELATED SUBSTANCES ON THE ACTIVATION OF PLASMINOGEN BY UK

Roles of fibrin in the activation of plasminogen by UK had been controversial. In 1979, we indicated that the coagulation of human plasma by thrombin or calcium ions in the presence of UK resulted in the enhanced activation of native plasminogen (Glu-plg) (31). At that time, it was generally believed that the activation rate of plasminogen by t-PA

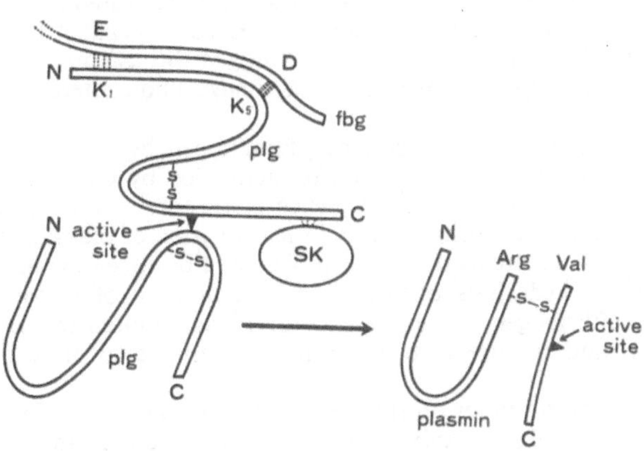

Fig. 8. Schematic model of tri-molecular complex N:N-terminus, plg:plasminogen, C:C-terminus, Fbg:fibrinogen.

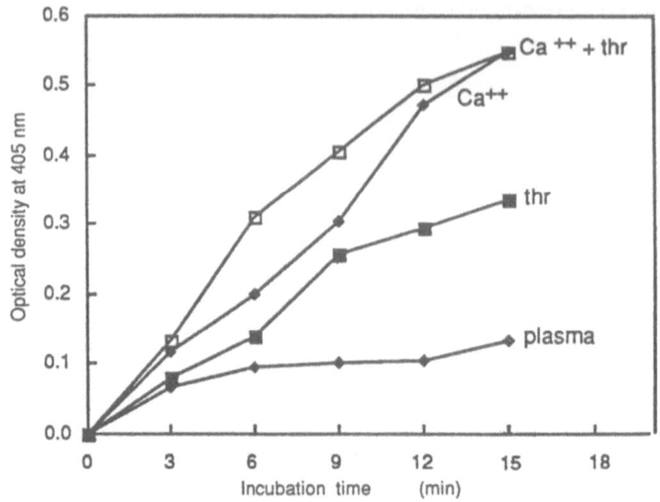

Fig. 9. Hydrolysis of S-2251 by plasma and clotted plasma. "Plasma" indicates the hydrolysis of S-2251 by the mixture of plasma and UK. "thr" indicates that of plasma, thrombin and UK. "Ca^{++}" indicates that of plasma, UK and Ca^{++}. "Ca^{++} + thr" indicates that of plasma, UK, thrombin and Ca^{++}.

was greatly enhanced in the presence of fibrin, whereas the activation of plasminogen by UK was not accelerated by the presence of fibrin. As an explanation of these findings, it was shown that UK did not bind specifically to fibrin, but t-PA was shown to be strongly adsorbed to fibrin (32). Furthermore, Glu-plg was known to bind to fibrin ineffectively, whereas Lys-plg was shown to bind well (33). From these observations, it was believed that the activation of Glu-plg by UK was not influenced in the presence of fibrin. Since our results were contradictory to those reported at that time, we first felt that the higher plasmin activity in the clotted plasma may be due to ineffective inhibition of plasmin by inhibitors in the presence of fibrin.

Also it was possible that a native form of plasminogen (Glu-plg) was activated more efficiently by UK in the presence of fibrin. It was well known that Lys-plg was activated far better than Glu-plg (34). Glu-plg was shown, however, to be activated as well as Lys-plg in the presence of appropriate concentrations of ω-amino acids (35-37). It was also well established that ω-amino acids bind to lysine binding sites (LBS) of plasminogen, and that Glu-plg, but not Lys-plg, bound with ω-amino acids, had an altered conformation (38-40), and such conformationally altered Glu-plg was shown to be activated more rapidly. Furthermore, the extent of the conformational change appeared to parallel the rate of enhanced activation of Glu-plg. Since fibrin binds to plasminogen through its LBS, it is conceivable that Glu-plg may also undergo a conformational change upon interaction with fibrin and become better activatable.

Fig. 9 shows the enhanced activation of plasminogen by UK in the clotted plasma. S-2251 was hardly hydrolyzed in the plasma after activation by UK. The significant extent of S-2251 hydrolysis was observed in the presence of thrombin, Ca^{++}, and thrombin plus Ca^{++}. These results mean that fibrin enhanced the activation of plasminogen in the plasma.

Then we used purified systems to see the effects of fibrinogen or fibrin. Fig. 10 shows the enhancement of the hydrolysis of S-2251 by the mixture of Glu-plg and UK in the presence of fibrin or fibrinogen. The presence of fibrin resulted in the largest extent of hydrolysis of S-2251, and the presence of fibrinogen resulted in more hydrolysis than in its absence.

Then, the effects of fibrin or fibrinogen on the activation of Lys-plg by UK were measured. 0.1 μM of Lys-plg was mixed with 10 U/ml of UK in the presence of 0.1 μM of fibrinogen or fibrin, and S-2251 was added. The rate of the hydrolysis was measured. In contrast to the activation of Glu-plg by UK, the presence of fibrinogen and fibrin showed little increase in the extent of the hydrolysis of S-2251.

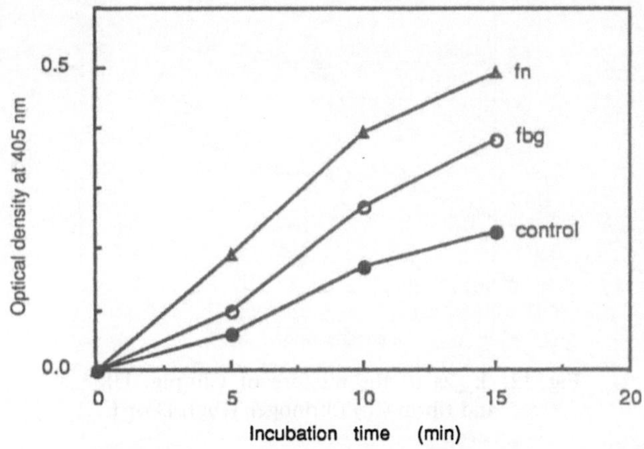

Fig. 10. Time course of the hydrolysis of S-2251 by the mixture of Glu-plg and UK in the presence of fibrinogen (fbg) or fibrin (fn). Glu-plg (1.0 μM) was mixed with UK (50 U) and 0.1 μM of fibrin or fibrinogen. The mixtures were put into curvets and spectrophotometric changes at 405 nm were measured by an automatic spectrophotometer.

Fig. 11 shows the kinetic analysis of the activation of Glu-plg by UK in the presence of fibrin. Various concentrations of Glu-plg (2 - 16 μM) were mixed with UK (10 U/ml) and S-2251 (0.6 mM) in the presence of various concentrations of fibrin. The mixture was incubated and the optical density was measured. Lineweaver-Burk plots were plotted. The addition of fibrin did not change the value of Km, but increased the value of Vmax. Addition of fibrinogen, D or E to the mixture of Glu-plg and UK showed the same tendency, that is, Km did not change and Vmax increased with increase in the concentration of fibrin or its related substances.

Fig. 12 shows changes in kcat at various concentrations of fibrin and its related substances. kcat increased most in the presence of fibrin, followed by fibrinogen, D, E.

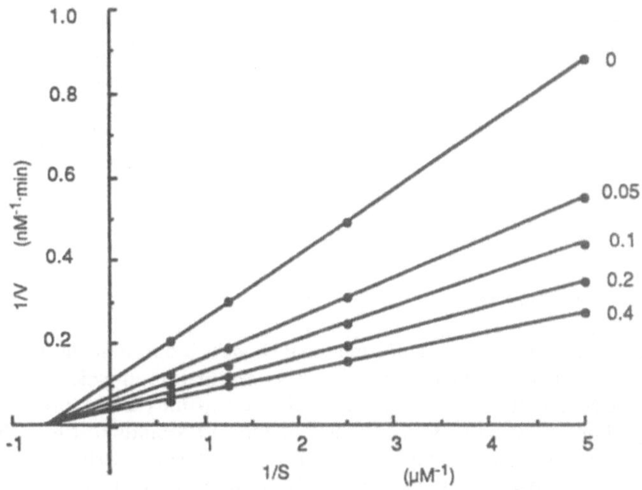

Fig. 11. Lineweaver-Burk plots of the mixture of Glu-plg, UK and fibrin.

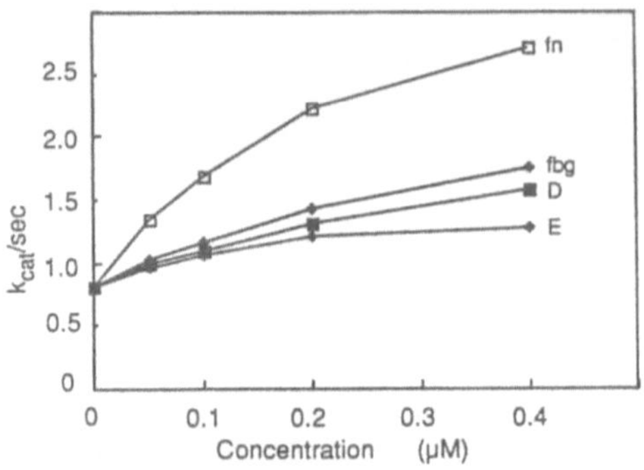

Fig. 12. k_{cat}/s of the mixture of Glu-plg, UK and fibrin (fn) fibrinogen (fbg), D or E.

From these experiments it is considered that the enhanced plasmin activity of clotted plasma is due to the enhancement of the activation of Glu-plg by UK in the presence of fibrin. In addition to this another phenomenon that helps to enhance plasmin activity is a relative inefficiency of α_2AP in inhibiting plasmin in the presence of fibrin. The LBS of plasmin must be free in order for α_2AP to interact with the active site of plasmin. Interaction of the LBS of plasmin with ω-amino acids delays the inhibition of plasmin by α_2AP. Since fibrin binds to the LBS of plasmin, it has been expected that fibrin may protect plasmin from inhibition by α_2AP. We have shown that fibrin formation really prevented plasmin from being inhibited by α_2AP (41).

It is therefore, possible that the high fibrinolytic activity (or plasmin activity) is caused by the enhanced activation of Glu-plg by UK in clotted plasma and the less effective inhibition of plasmin by α_2AP in the presence of fibrin.

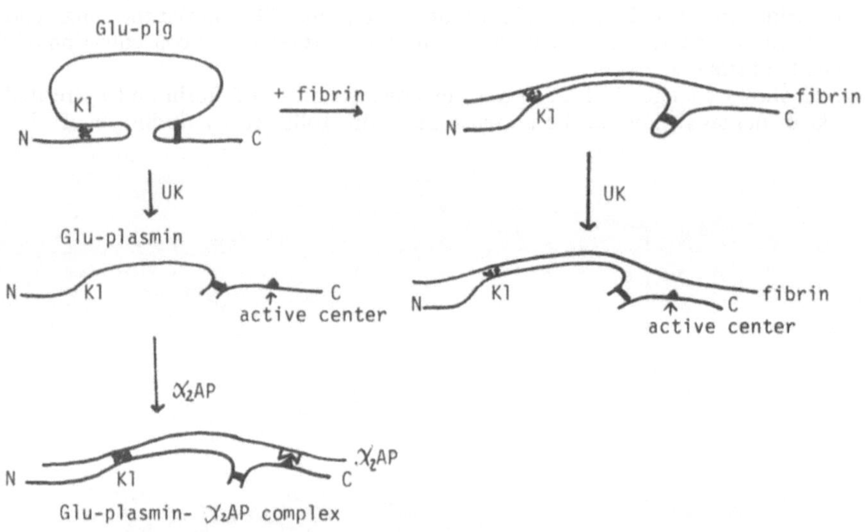

Fig. 13. Schematic model of the conformational changes induced in Glu-plg upon interaction with fibrin. Glu-plg has a tight conformation in which a lysine residue of the N-terminal portion is bound with a low affinity LBS on the first kringle. Upn interaction with fibrin, the low affinity LBS binds to fibrin, leading the loose conformation of Glu-plg. Such a loose form is easily activated by UK or other activators. When Glu-plg is activated by activators in the absence of fibrin and in the presence of α_2AP, Glu-plasmin is immediately inactivated by α_2AP, since LBS of Glu-plasmin is available for binding with α_2AP.

A model depicting the enhanced activation of Glu-plg by UK and the ineffective inhibition of plasmin by α_2AP in the presence of fibrin is shown in Fig. 13.

REFERENCES

1. S. M. Camiolo, S. Thorsen, and T. Astrup, Fibrinogenolysis and fibrinolysis with tissue plasminogen activator, urokinase, streptokinase-activated human globulin, and plasmin, *Proc. Soc. Exp. Biol. Med.,* **138:**277-280 (1971).
2. P. Wallen, Chemistry of plasminogen and plasminogen activation, *in:* "Progress in Chemical Fibrinolysis and Thrombolysis Vol. 3," J. F. Davidson, R. M. Rowan, M. M. Samama and P. C. Desnoyers, eds., Raven Press, New York, pp. 167-181 (1978).
3. J. H. Verheijen, W. Nieuwenhuizen, and G. Wijngaards, Activation of plasminogen by tissue activator is increased specifically in the presence of certain soluble fibrin(ogen) fragments, *Thromb. Res.,* **27:**377-385 (1982).
4. A. Takada and Y. Takada, Effects of fibrin or its degradation products on the activation of Glu-plasminogen by streptokinase or urokinase, *Thromb. Res.,* **30:**309-310 (1983).
5. R. Fears, M. J. Hibbs, and R. A. G. Smith, Kinetic studies in the interaction of streptokinase and other plasminogen activators with plasminogen and fibrin, *Biochem. J.,* **229:**555-558 (1985).
6. Y. Takada and A. Takada, Kinetic analyses of potentiation of plasminogen activation by streptokinase in the presence of fibrin or its degradation products, *Haemostasis,* **17:**1-7 (1987).
7. A Takada and Y. Takada, Enhanced activation activity of the mixture of streptokinase and a modified form of plasminogen (Lys-plasminogen) in the presence of fibrin: Role of conformational change of plasminogen, *Haemostasis,* **18:**106-112 (1988).
8. B. A. K. Chibber, J. P. Morris, and F. J. Castellino, Effects of human fibrinogen and its cleavage products on activation of human plasminogen by streptokinase, *Biochemistry,* **24:**3429-3434 (1985).
9. A. Takada, Y. Takada, and Y. Sugawara, The activation of Glu- and Lys-plasminogens by streptokinase: effects of fibrin, fibrinogen and their degradation products. *Thromb. Res.,* **37:**465-475 (1985).
10. S. Camiolo, G. Markus, J. L. Evers, and G. H. Hobika, Augmentation of streptokinase activator activity by fibrinogen or fibrin. *Thromb. Res.,* **17:**697-706 (1980).
11. M. Einarsson, B. Skoog, B. Forsberg, and R. Einarsson, Characterization of highly purified streptokinase and altered streptokinase after alkaline treatment, *Biochim. Biophys. Acta,* **568:**19-29 (1979).
12. T. Bilinski, T. Loch, and K. Zakrzewski, Studies on streptokinase. Purification and some molecular properties. *Acta Biochim. Pol.,* **15:**123-18 (1968).
13. K. W. Jackson and J. Tang, Complete amino acid sequence of streptokinase and its homology with serine proteases, *Biochemistry,* **21:**6620-6625 (1982).
14. F. J. Castellino, J. M. Sodetz, W. J. Brockway, and G. E. Siefring Jr., Streptokinase, *in:* "Methods in Enzymology XLV," L. Lorand, ed., Academic Press, New York, pp. 244-255 (1976).
15. W. S. Tillet and R. L. Garner, The fibrinolytic activity of hemolytic streptococci, *J. Exp. Med.,* **58:**485-502 (1933).
16. S. Müllertz and M. Lassen, An activator system in blood indispensable for formation of plasmin by streptokinase, *Proc. Soc. Exp. Biol. Med.,* **82:**264-268 (1953).
17. L. Summaria, R. C. Wohl, I. G. Boreisha, and K. C. Robbins, Specific cleavage of the arginyl-560-valyl peptide bond in the diisopropoxyphosphinyl virgin enzyme by plasminogen activators, *Biochemistry,* **21:**2056-2059 (1982).
18. G. Claeson, L. Aurell, G. Karlsson, and P. Friberger, Substrate structure and activity relationships, *in:* "Progress in Chemical Fibrinolysis and Thrombolysis Vol. 3," Raven Press, New York, pp. 299-304 (1978).
19. K. N. N. Reddy and G. Markus, Esterase activities in the zymogen moiety of the streptokinase-plasminogen complex, *J. Biol. Chem.,* **249:**4851-4857 (1974).
20. G. Markus, J. L. Evers, and J. L. Hobika, Comparison of properties of native (g/u) and modified (lys) human plasminogen, *J. Biol. Chem.,* **213:**737-739 (1978).
21. B. Wiman and P. Wallen, On the primary structure of human plasminogen and plasmin, Purification and characterization of cyanogen-bromide fragment, *Eur. J. Biochem.,* **57:**387-394 (1975).

22. M. L. Hayes, R. K. Bretthauer, and F. J. Castellino, Carbohydrate compositions of the rabbit plasminogen isozymes, *Arch. Biochem. Biophys.*, **171**:651-655 (1975).
23. A. Takada, Y. Takada, and U. Okamoto, Fractionation of plasminogen activator and proactivator in tissue and blood by gel filtration, *Keio J. Med.*, **13**:187-194 (1964).
24. A. Takada, Y. Takada, and J. L. Ambrus, Streptokinase-activatable proactivator of human and bovine plasminogen, *J. Biol. Chem.*, **245**:6389-6396 (1970).
25. A. Takada and Y. Takada, Studies on SK-potentiator of plasminogen in human plasma, *Thromb. Res.*, **13**:325-335 (1978).
26. A Takada, K. Mochizuki, and Y. Takada, Further characterization of SK-potentiators of plasminogen, *Thromb. Res.*, **19**:485-492 (1980).
27. D. K. Strickland, J. P. Morris, and F. J. Castellino, Enhancement of the streptokinase-catalyzed activation of human plasminogen by human fibrinogen and its plasminolysis products, *Biochemistry,* **21**:721-728 (1982).
28. J. H. Smith, J. P. Morris, B. A. Chibber, and F. J. Castellino, The role of the lysine binding sites of human plasminogen in the fibrinogen stimulated rate of active site formation in the streptokinase-plasminogen equimolar complex, *Thromb. Res.*, **234**:499-506 (1984).
29. U. Christensen, The AH-site of plasminogen and two C-terminal fragments, *Biochem. J.*, **223**:413-421 (1984).
30. K. C. Robbins and G. Markus, The interaction of human plasminogen with streptokinase, *in:* Fibrinolysis, P. J. Gaffney and S. Balkuv-Ulutin, ed., Academic Press, New York, pp. 61-75 (1976).
31. A. Takada, T. Urano, and Y. Takada, Influence of coagulation on the activation of plasminogen by streptokinase and urokinase, *Thromb. Haemostas.*, **42**:901-908 (1979).
32. S. Thorsen, P. Glas-Greenwalt, and T. Astrup, Differences in the binding to fibrin of urokinase and tissue plasminogen activator, *Thromb. Diath. Haemorrh.*, **28**:65-74 (1972).
33. S. Thorsen, I. Clemmensen, L. Sottrup-Jensen, and S. Magnusson, Adsorption to fibrin of native fragments of known primary structure from human fibrinogen, *Biochim. Biophys. Acta,* **668**:377-387 (1981).
34. H. Claeys and J. Vermylen, Physico-chemical and proenzyme properties of NH_2-terminal glutamic acid and NH_2-terminal lysine human plasminogen. Influence of 6-aminohexanoic acid, *Biochim. Biophys. Acta* **342**:351-359 (1974).
35. N. Alkjaersig, The purification and properties of human plasminogen, *Biochem. J.*, **93**:171-181 (1964).
36. W. J. Brockway and F. J. Castellino, Measurement of binding of anti-fibrinolytic aminoacids to various plasminogens. *Arch. Biochem. Biophys.*, **151**:194-199 (1972).
37. A. Takada and Y. Takada, Effects of ω-aminoacids and clot formation on the activation by urokinase of various plasminogen preparations, *Thromb. Res.* **18**:167-176 (1980).
38. F. J. Castellino, W. J. Brockway, J. K. Thomas, H-T. Liao, and A. B. Rawitch, Rotational diffusion analysis of the conformational alterations produced in plasminogen by certain antifibrinolytic amino acids, *Biochemistry,* **12**:2787-2791 (1973).
39. Y. Sugawara, Y. Takada, and A. Takada, Fluorescence polarization and spectropolarimetric studies on the conformational changes induced by ω-aminoacids in two isozymes of Glu-plasminogen (I and II), *Thromb. Res.*, **33**:269-275 (1984).
40. A. Takada, Y. Takada, and Y. Sugawara, Fluorescence spectrophotometric studies on the conformational changes induced by ω-aminoacids in two isozymes of Glu-plasminogen (I and II), *Thromb. Res.*, **33**:461-469 (1984).
41. A. Takada, Y. Watahiki, and Y. Takada, Effects of fibrin on the enhanced activation of plasminogen by urokinase and tissue plasminogen activator: Role of cross-link, *Thromb. Res.*, **41**:605-613 (1986).

PROTEIN C AND FIBRINOLYSIS: A LINK BETWEEN COAGULATION AND FIBRINOLYSIS

N. J. de Fouw[1,2,#], F. Haverkate[1,*], and R.M. Bertina[2]

[1]Gaubius Institute TNO, Leiden and [2]University Hospital Leiden
The Netherlands

ABSTRACT

The effect of purifed human activated protein C (APC) on fibrinolysis was studied by using *in vitro* clot lysis techniques. Clots were formed from citrated blood or plasma (supplemented with ^{125}I-labeled fibrinogen) by adding thrombin and Ca^{2+}-ions; lysis of the clots was achieved by the addition of tissue-type plasminogen activator before clot formation. The gradual release of labeled fibrin degradation products from the clot into the supernatant was taken as a measure for the lysis rate. It was demonstrated that the acceleration of clot lysis by APC added before clot formation depends on the presence of Protein S, Ca^{2+}-ions and phospholipids. These observations suggest a role of APC as anticoagulant in clot lysis, since the cofactors for the expression of its anticoagulant and profibrinolytic effect are very similar. Indeed, we could demonstrate that the profibrinolytic effect of APC *in vitro* is associated with reduction of thrombin generation through the coagulation cascade by inactivation of factor VIIIa and factor Va. For instance, APC did not accelerate the lysis of factor X deficient blood clots.

More generally, thrombin generation was associated with retarded fibrinolysis *in vitro*. Consequently anticoagulants such as APC or Heparin are profibrinolytic, whereas procoagulants such as phospholipids (in cell-free plasma) inhibit fibrinolysis through the generation of thrombin. Thrombin thus plays a crucial role as a link between coagulation and fibrinolysis. As thrombin is able to inhibit the lysis of blood and plasma clots, and not of purified fibrin clots, we hypothesize that thrombin inhibits lysis through an as yet unidentified mediator in plasma.

INTRODUCTION

Protein C is a vitamin K dependent plasma protein which can be activated by the thrombin/thrombomodulin complex to activated protein C (APC) (Esmon and Esmon, 1984). APC has been shown to exert both anticoagulant and profibrinolytic activity (Comp and Esmon, 1981; Burdick and Schaub, 1987; Colucci et al., 1984; Taylor and Lockhart, 1985a; De Fouw et al., 1986; Fulcher et al., 1984). The anticoagulant activity of APC is well documented and is a result of the inactivation of coagulation factors Va and VIIIa (Fulcher et al., 1984). For the expression of this effect Ca^{2+}-ions, phospholipid and protein S (an other vitamin K dependent plasma factor) are essential cofactors (Walker, 1981).

[#]Present address N.J. de Fouw: Unilever Research Laboratorium, Vlaardingen, The Netherlands.
*Correspondence to F. Haverkate, Gaubius Institute TNO, P.O. Box 612, 2300 AP Leiden, The Netherlands.

The mechanism by which APC enhances fibrinolysis is still unknown. It has been shown that protein S is an essential cofactor for the expression of the profibrinolytic effect *in vitro* (De Fouw et al., 1986). A plausible explanation for this profibrinolytic effect of APC could be its ability to neutralize plasminogen activator inhibitor activity present in blood by complex formation (Van Hinsbergh et al., 1985; Sakata et al., 1985; Taylor and Lockhart, 1985b; De Fouw et al., 1987; De Fouw et al., 1988a; D'Angelo et al., 1987). In a clot lysis system based on the use of purified human proteins, we and others have found that APC can indeed stimulate fibrinolysis by the inactivation of PAI-1 (De Fouw et al., 1988a; D'Angelo et al., 1987). However, relatively long incubation periods and high concentrations of APC are required to show this effect (Fay and Owen, 1989; De Fouw, 1988a). Furthermore, protein S does not influence the interaction of APC with PAI (Fay and Owen, 1989; De Fouw, 1988a). Therefore, the protein S dependent mechamism by which APC stimulates whole blood clot lysis cannot merely be explained by the protein S independent PAI-1 neutralization by APC, observed in a purified system (De Fouw, 1988a). This suggests that in whole blood (an)other mechanism(s) should exist by which APC exerts its profibrinolytic activity. In the present study we demonstrate that the profibrinolytic effect of APC in the blood and plasma clot lysis systems *in vitro* is associated with the decrease in thrombin generation, which is the result of the anticoagulant action of APC. Moreover, it is shown that thrombin plays a crucial role as a link between coagulation and fibrinolysis.

MATERIALS AND METHODS

Fibrinogen was isolated from human plasma, as described before (Van Ruijven-Vermeer and Nieuwenhuizen, 1978) and labeled with ^{125}I with the iodogen method (Fraker and Speck, 1978). The amount of radioactivity not clottable with thrombin was less than 10% of the total radioactivity. The radiolabeled fibrinogen, dissolved in 0.15 M NaCl, had a concentration of 2 mg/ml and a specific radioactivity of 75 10^6 cpm/mg.

The two chain form of t-PA was isolated from (Bowes) melanoma cells and purified as described (Rijken et al., 1979). This activator is considered to be identical with plasminogen activator after its release from the vessel wall (Rijken et al., 1980). t-PA activity was measured according to Verheijen et al. (1982). The stock solution contained 162×10^3 IU/ml and 570×10^3 IU/mg and was stored in 0.1 M Tris, 0.1% Tween-80 (pH 8.0) at $-70°C$.

Protein C was purified and activated as described previously (Bertina et al., 1982, 1984; Van Hinsbergh et al., 1985). Fibrinolytic activity could not be detected on plasminogen-rich fibrin plates (Haverkate and Brakman, 1975) nor in a t-PA assay with the chromogenic substrate S-2251 (KabiVitrum, Stockholm, Sweden).

APC was inactivated with DFP (diisopropylfluorophosphate) or with anti-protein C IgG (prepared from rabbit antiserum raised against purified human protein C as described previously (Van Hinsbergh et al., 1985). One ml of anti-protein C IgG completely neutralizes the protein C activity of 25 ml of pooled normal plasma within 15 minutes.

Factor II, VIII and X activity were measured by standard one stage coagulation methods.

α_2-Antiplasmin activity and antigen were estimated as reported by Kluft et al. (1985). The concentration of clottable fibrinogen was measured according to Clauss (1957). Inhibition of t-PA by plasma was assessed as previously described (Verheijen et al., 1982).

Pooled normal plasma. Blood was collected by venapuncture in plastic tubes containing 1/10 volume of 0.11 mol/1 sodium citrate. The blood was centrifuged at 1200 g for 10 minutes at 4°C to obtain platelet poor plasma which was subsequently centrifuged for 30 minutes at 20,000 g at 4°C to obtain platelet free plasma. The platelet free plasma of 42 healthy volunteers was pooled and stored at $-70°C$ in aliquots of 0.4 ml. This pooled normal plasma is considered to contain 1 U/ml of each of the coagulation factors.

Phospholipid preparation. Washed human brain was extracted according to Folch et al. (1957).

Factor X depleted blood. Citrated whole blood was incubated for 15 minutes at 37°C with anti-factor X IgG (2.2 mg/ml) prepared from rabit antiserum against purified human factor X and isolated by protein A-Sepharose chromatography. Under these conditions 1 ml of anti-factor X IgG (2.2 mg/ml) completely neutralizes the factor X clotting activity of 6.6 ml pooled normal plasma. Anti-factor X IgG did not influence the APC activity as was checked by an amidolytic assay using S-2366, nor did it influence the rate of the lysis of a clot prepared from factor X deficient plasma.

Factor X depleted plasma. This was prepared by passing pooled normal plasma through a column of anti-factor X IgG coupled to Sepharose 4B (8 mg antifactor X IgG/ml gel). This plasma contained less than 0.01 U/ml factor X activity and antigen.

Hereditary Factor VIII deficient plasma was isolated from the blood of patients with a severe hereditary deficiency of this factor. Factor VIII deficient plasma contained less than 0.001 U/ml activity and antigen.

Factor XIII depleted blood. Anti-factor XIII (subunit A) IgG (rabbit antiserum from Behringwerke AG, Marburg, West Germany) was partially purified and concentrated by protein A Sepharose chromotography. Blood was incubated with anti-factor XIII IgG for 15 minutes at 37°C. Under these conditions 1 ml of anti-factor XIII IgG (18 mg/ml) completely neutralizes the factor XIII activity of 25 ml plasma. This was demonstrated as described by Jansen et al. (1987).

α_2-Antiplasmin deficient blood was obtained from a patient with a severe deficiency in α_2-antiplasmin activity (< 0.03 units/ml activity; 1.0 unit/ml antigen) with a serious bleeding tendency. This patient is homozygous for an abnormal α_2-antiplasmin molecule (α_2-antiplasmin Enschede) (Kluft et al., 1987).

Whole blood clot lysis. Blood from healthy donors was collected by venapuncture in plastic tubes in 1/10 volume of 0.11 M sodium-citrate and supplemented with 7.5 μl of ^{125}I-labeled fibrinogen solution per ml, 30 or 60 international units (IU) t-PA/ml, APC and a volume of 0.15 M NaCl to obtain a final blood concentration of 80%. After rapid mixing, 250 μl aliquots were transferred to Eppendorf cups containing 3 μl of 60 μM thrombin solution (unless otherwise stated) and 2.5 μl of 1.25 M $CaCl_2$. Immediately after mixing, the cups were incubated at 37°C to allow clot formation and subsequent lysis. At different time intervals the tubes were centrifuged in a MSE Micro Centaur (Beun de Ronde, Amsterdam, Holland) centrifuge for five minutes at 11,500 g (4°C). The supernatant (100 μl) was mixed with 100 μl of a solution containing 2% sodium dodecyl sulphate (SDS), 20% glycerol, 6 M urea and 0.1 M Tris/HC1, pH 6.8, and counted for radioactivity (^{125}I FDP; fibrin degradation products) in a gamma counter (Packard Instruments, Brussels, Belgium). Clot lysis is expressed as x-b/t-b × 100%, in which x = radioactivity of the supernatant, b = radioactivity of the blank, t = total radioactivity added before clotting. The blank value (<10%) is the radioactivity of the supernatant of whole blood 30 minutes after clotting and without the addition of t-PA. One hundred percent lysis indicates complete clot lysis.

Plasma clot lysis. Citrated platelet-free plasma was supplemented with ^{125}I-fibrinogen, t-PA, phospholipid (final concentration 8 μM) and a volume of 0.15M NaCl to obtain a final plasma concentration of 80%. After rapid mixing of the suspension, 250 μl aliquots were transferred to Eppendorf cups containing 3 μl thrombin (0.8 μM, unless otherwise stated) and 2.5 μl $CaCl_2$ (2.5 M) and incubated at 37°C to allow clot formation and subsequent clot lysis. For further details see under whole blood clot lysis. Routinely, experiments were performed at least in four-fold. The standard deviation calculated from the percentage of clot lysis at different time intervals varied from 0.8 to 5%.

Plasma recalcification time. Platelet-free plasma was thawed and warmed to 37°C (about 15 minutes). Subsequently, 5 μl phospholipid (490μM) and, when indicated, 5 μl APC (435 nM) were added to 290 μl plasma and mixed. The reaction was started by adding 10 μl $CaCl_2$ (0.78 M) (no thrombin was added). The suspension was gently mixed and incubated at 37°C in a plastic tube. The recalcification time was determined with a manual tilt tube technique.

RESULTS

In previous studies of our laboratory (De Fouw et al., 1986) it has been shown that APC, added to whole blood before clot formation is induced with thrombin and Ca^{2+}-ions, greatly accelerates the lysis of the clots. The effect was shown to increase with higher APC concentrations (7.75 nM to 46.5 nM).

In the same study it was shown that APC inactivated with anti-protein C IgG or DFP no longer accelerated the lysis of blood clots, indicating that the profibrinolytic effect is APC specific and dependent on its intact active site. Further it was demonstrated that protein S is required for the expression of the profibrinolytic effect of APC.

In the present study it was investigated whether phospholipid and Ca^{2+}-ions are also cofactors for the expression of the profibrinolytic effect of APC (similar as for the expression of the anticoagulant effect of APC).

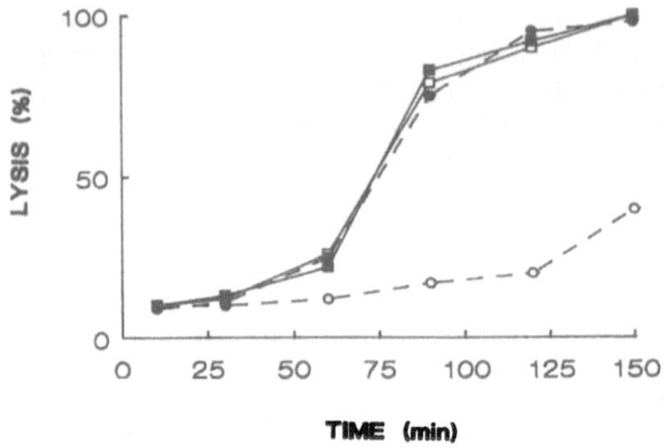

Fig. 1. Influence of APC in the absence (squares) and presence (circles) of phospholipid (8 μM) on the lysis of cell free plasma clots. To initiate clot lysis 30 IU t-PA/ml blood were added before clot formation was induced with thrombin and Ca^{2+}-ions. Curves represent lysis in the presence (closed symbols) and absence (open symbols) of APC.

The phospholipid surface which is required for the anticoagulant effect of APC can be provided by the (activated) blood cells and/or endothelial cells. To investigate the phospholipid requirement for the profibrinolytic effect of APC, cell free recalcified plasma was used instead of whole blood. When a clot lysis experiment in the absence of cells was performed, the addition of APC did not affect the lysis of the clot (Fig. 1), indicating that cells are required for the expression of the profibrinolytic effect of APC. When procoagulant phospholipid was added to this cell free-plasma, APC could again stimulate clot lysis (Fig. 1). This observation suggests that phospholipid can replace blood cells as a cofactor for the expression of the profibrinolytic activity of APC. Also Ca^{2+}-ions were required for the expression of the profibrinolytic effect of APC. This was demonstrated by measuring the lysis of clots formed from citrated blood by adding thrombin without recalcification. Addition of APC to the blood under these conditions did not affect the lysis of the clots (not shown).

To investigate the possibility that APC might enhance fibrinolysis by the inactivation of coagulation factor XIII activity, clot lysis experiments were performed in the presence and absence of APC using blood which had been preincubated with anti-factor XIIIa IgG. Fig. 2 shows that APC still accelerates whole blood clot lysis in the absence of factor XIIIa. APC was also able to accelerate whole blood clot lysis in the absence of α_2-AP using blood of a patient with a congenital deficiency in this factor (not shown). These results show that the profibrinolytic effect of APC is not associated with factor XIIIa dependent crosslinking of fibrin chains to each other or of fibrin to α_2-AP, or to an effect of APC on α_2-AP.

Since both the anticoagulant and profibrinolytic activity of APC are dependent on the presence of the cofactor protein S, calcium ions and phospholipids, we investigated whether both activities are related. As the anticoagulant activity of APC will result in an inhibition of thrombin generation, we investigated (1) whether in our lysis experiments thrombin is generated and (2) whether this generation is inhibited by APC. To that end we determined the plasma recalcification time (without adding thrombin).

Table 1 shows that under the experimental conditions used, endogenous thrombin generation resulted in a recalcification time of 9 minutes in normal plasma. Plasma deficient of F VIII or depleted of F X, used as controls in which no thrombin can be generated, did not clot within 45 minutes. APC (Table I) prolonged the recalcification time of normal plasma. These results indicate that in our lysis experiments (with exogenous thrombin) thrombin will be also generated endogenously and that APC is able to inhibit this thrombin formation.

To establish a relation between the anticoagulant and profibrinolytic activity of APC, clots of normal plasma, F VIII deficient and F X depleted plasma were lysed in the presence

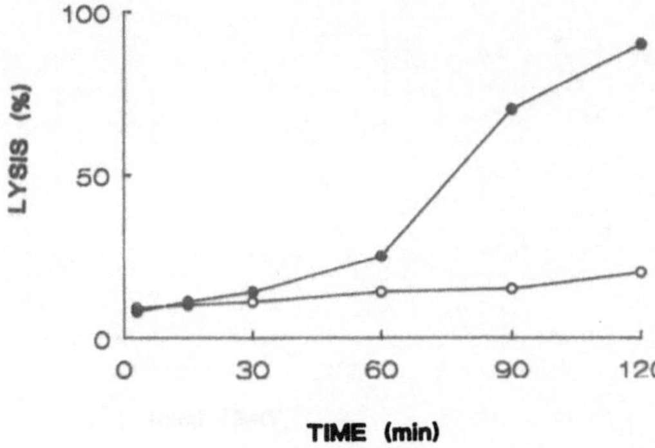

Fig. 2. Effect of the elimination of coagulation factor XIII activity on whole blood clot lysis in the presence and absence of APC. Citrated whole blood was incubated for 15 minutes at 37°C with anti-factor XIIIa IgG, recalcified, clotted and lysed as described. Closed symbols repsent lysis (initiated with 35 IU t-PA/ml blood) in the presence of APC, open symbols in the absence of 46 nM APC.

and absence of APC. Fig. 3 shows that APC accelerated lysis of normal plasma clots but not of the two other plasmas where no endogenous thrombin is generated. Moreover, it appeared that in the absence of APC lysis of normal plasma clots is highly retarded as compared with the same plasma after depletion of F X (Fig. 3). It seems, therefore, that thrombin generation is associated with retarded fibrinolysis and that APC may accelerate lysis by blocking the endogenous thrombin generation.

Also another anticoagulant, heparin, was able to accelerate lysis of normal plasma clots and not of F X depleted plasma clots (Fig. 4). This experiment shows that APC is not unique in accelerating clot lysis.

The results obtained so far indicate that clot lysis is retarded when endogenous thrombin formation occurs when compared to the lysis of clots prepared in plasmas where no endogenous thrombin generation will occur. To further investigate a possible role of thrombin in fibrinolysis, we added increasing amounts of thrombin to whole blood depleted of F X. The results (Fig. 5) show that the whole blood clot lysis is retarded when higher thrombin concentrations were added.

Table 1. Plasma recalcification times in the presence and absence of APC. Plasmas, supplemented with phospholipids (8 μM), were recalcified in the absence or presence (7 nM) of APC, and incubated at 37°C in plastic tubes. The time required for clot formation was determined using a manual tilt tube technique

	Recalcification time (min)	
Plasma	−APC	+APC
pooled normal	9	17
factor VIII def.	>45	>45
factor X def.	>45	>45

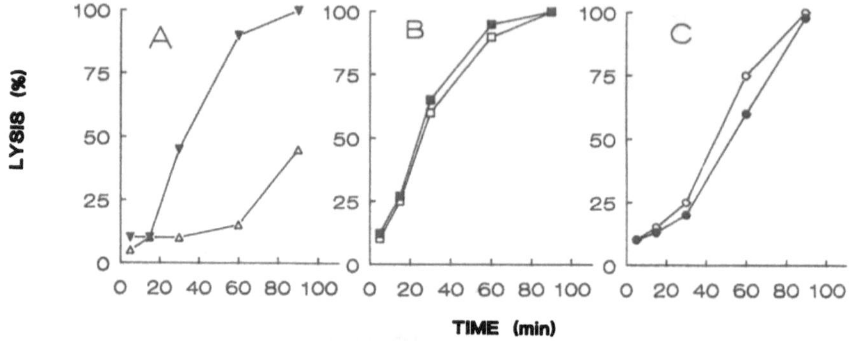

Fig. 3. Effect of APC on the lysis of clots prepared from pooled normal plasma (A), factor VIII deficient plasma (B) and factor X deficient plasma (C). Plasma was supplemented with phospholipids, recalcified, clotted and lysed as described in materials and methods. 60 IU t-PA were added prior to clot formation. Lysis in the absence (open symbols) or presence (closed symbols) of 7 nM APC.

DISCUSSION

Up till now, the mechanism by which APC stimulates fibrinolysis is unknown. Some investigators tend to believe that neutralization of plasminogen activator inhibitor activity (PAI) by APC is of crucial importance (Sakata et al., 1985; Taylor and Lockhart, 1985b; D'angelo et al., 1987), whereas others are more sceptical on a role of PAI (Van Hinsbergh et al., 1985, De Fouw et al., 1987, 1988a).

In the blood and plasma clot lysis systems a direct action of APC on PAI as a cause of the profibrinolytic effect of APC is highly unlikely for the following reasons:

1) t-PA is added in excess with respect the PAI in blood or plasma so that PAI activity is completely neutralized.
2) APC reacts relatively slowly with PAI (De Fouw et al., 1988a; Fay and Owen, 1989).
3) The reaction between APC and PAI does not depend on protein S, calcium ions and phospholipids (De Fouw et al., 1988a; Fay and Owen, 1989), which is in contrast with the cofactor requirements for the profibrinolytic effect of APC.

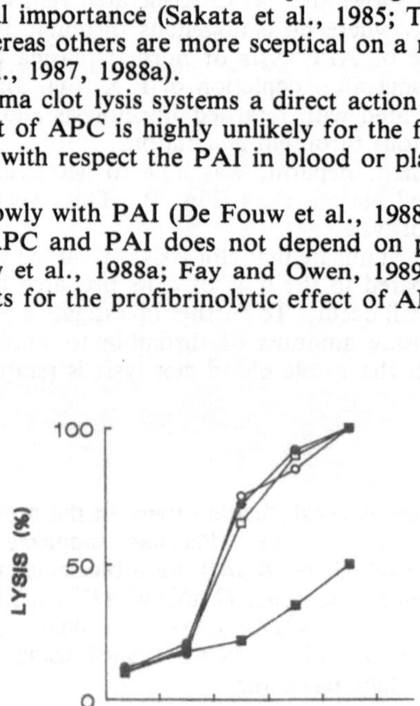

Fig. 4. Effect of heparin on plasma clot lysis. To factor X deficient plasma (circles) or pooled normal plasma (squares), both supplemented with phospholipids, heparin (0.04 U/ml, open symbols) or buffer (closed symbols) was recalcified, clotted and lysed as described under Materials and Methods. 65 IU t-PA/ml plasma were added prior to clot formation.

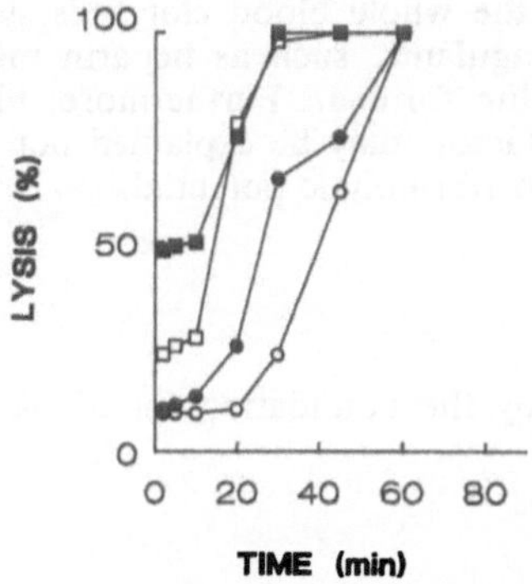

Fig. 5. Influence of thrombin concentration on whole blood clot lysis in which endogenous thrombin formation is inhibited by neutralization of factor X activity. Citrated whole blood was incubated for 15 minutes at 37°C with anti-factor X IgG (2.2 mg/ml), recalcified, clotted and lysed as described in Materials and Methods. To initiate clot lysis, 60 IU t-PA/ml blood were added before clot formation was induced. Curves represent lysis after the addition of 2.7 nM IIa (■), 5.5 nM IIa (□), 11 nM IIa (●) and 55 nM IIa (○).

Zolton and Seegers (1973) demonstrated that APC could inhibit plasmin inhibitory capacity of normal plasma. The major plasmin inhibitor was later shown to be α_2-AP. In the present study a possible involvement role of α_2-AP was excluded because APC also accelerated lysis of blood clots deficient in α_2-AP. Because APC also accelerates lysis of blood clots depleted in F XIII activity, an influence of APC on the F XIII dependent crosslinking of fibrin to fibrin or to α_2-AP, which is of crucial importance in lysis rates (Jansen et al., 1978; Sakata and Aoki, 1980), could be excluded as an explanation for the profibrinolytic effect of APC.

As the profibrinolytic and the anticoagulant activity of APC depend on the same cofactors, we concentrated our further studies on a possible relation between both types of activity. Lysis was found to be retarded if endogenous thrombin is generated and lysis is more rapid when such generation is suppressed. This points to a crucial role of thrombin in the regulation of fibrinolyis. Indeed, we confirmed that in blood and plasma clot lysis experiments thrombin is able to inhibit fibrinolysis, regardless whether human thrombin is added (Fig. 5) or generated endogenously (Fig. 3). APC accelerates fibrinolysis by virtue of its anticoagulant activity which reduces the amount of thrombin generated endogenously (Table I). In general, anticoagulants such as APC or heparin, added to the blood or plasma before clotting, act as profibrinolytic agents *in vitro*. Reversedly, procoagulants such as phospholipids in cell-free plasma (Fig. 1) inhibit fibrinolysis by accelerating endogenous thrombin formation. In line with this concept, depletion of F X which eliminates the possibility of endogenous thrombin generation is associated with an increase in lysis rates (Fig. 3).

The mechanism, by which thrombin (generation) inhibits lysis of fibrin clots, is not clear yet. As thrombin itself has no effect in a lysis system consisting of purified components (De Fouw et al., 1988b), we think that thrombin inhibits fibrinolysis via one or more, as yet, unidentified mediators in plasma and blood. It is conceivable that thrombin induces release of PAI from blood platelets (Erickson et al., 1984; Sprengers et al., 1986), thus inhibiting fibrinolysis. However, this can be only a minor part of its effect, as lysis of platelet-free plasma clots (in the presence of phospholipids, Fig. 3) is also accelerated by APC.

The amount of thrombin added to induce clot formation is approximately 5% of the amount which can be generated if all prothrombin present in plasma would be converted to thrombin. This indicates that endogenous thrombin generation is a potential source of inhibition of fibrinolysis in blood and plasma clot lysis experiments and should be considered when using these systems for lysis experiments.

Our results obtained with the whole blood clot lysis system may be extrapolated to the conditions *in vivo*. Anticoagulants, such as heparin may act also as profibrinolytic agents when blood clots are being formed. Furthermore, bleeding phenomena as caused by deficiency of a coagulation factor, may be explained not only by lack of procoagulant activity, but also by an increased fibrinolytic potential.

ACKNOWLEDGMENTS

This work was supported by the Foundation for Medical Research Fungo, grant nr. 13-30-61.

REFERENCES

Bertina, R. M., Broekmans, A. W., Van de Linden, I. K., and Mertens, K., 1982, Protein C deficiency in a Dutch family with thrombotic disease, *Thromb. Haemostas.*, **48**:1.

Bertina, R. M., Broekmans, A. W., Krommenhoek-Van Es, C., and Van Wijngaarden, A., 1984, The use of a functional and immunological assay for plasma protein C in the study of the heterogeneity of congenital protein C deficiency, *Thromb. Haemostas.*, **15**:1.

Burdick, M. D., and Schaub, R. G., 1987, Human activated protein C produces anticoagulation and increased fibrinolytic activity in the cat, *Thromb. Res.*, **45**:413.

Clauss, A., 1957, Gerinnungsphysiologische Schnellmethode zur Bestimmung des Fibrinogens, *Acta Haematol.* (Basel), **17**:237.

Colucci, M., Stassen, J. M., and Collen, D., 1984, Influence of protein C activation on blood coagulation and fibrinolysis in squirrel monkeys, *J. Clin. Invest.*, **74**:200.

Comp, P. C., and Esmon, C. T., 1981, Generation of fibrinolytic activity by infusion of activated protein C into dogs, *J. Clin. Invest.*, **68**:1221.

D'Angelo, A., Lockhart, M. S., D'Angelo, S. V., and Taylor, F. B., 1987, Protein S is a cofactor for activated protein C neutralization of an inhibitor of plasminogen activator released from platelets, *Blood*, **69**:231.

De Fouw, N. J., Haverkate, F., Bertina, R. M., Koopman, J., Van Wijngaarden, A., and Van Hinsbergh, V. W. M., 1986, The cofactor role of protein S in the acceleration of whole blood clot lysis by activated protein C *in vitro*, *Blood*, **67(4)**:1189.

De Fouw, N. J., Van Hinsbergh, V. W. M., De Jong, Y. F., Haverkate, F., and Bertina, R. M., 1987, The interaction of activated protein C and thrombin with the plasminogen activator inhibitor released from cultured endothelial cells, *Thromb. Haemostas.*, **57**:176.

De Fouw, N. J., De Jong, Y. F., Haverkate, F., and Bertina, R. M., 1988a, Activated protein C increases fibrin clot lysis by neutralization of plasminogen activator inhibitor; no evidence for a cofactor role of protein S, *Thromb. Haemostas.*, **60(2)**:328.

De Fouw, N. J., De Jong, Y. F., Haverkate, F., and Bertina, R, M., 1988b, The influence of thrombin and platelets on fibrin clot lysis rates *in vitro*. A study using a clot lysis system consisting of purified human proteins, *Fibrinolysis*, **2**:235.

Erickson, L. A., Ginsberg, M. H., and Loskutoff, D. J., 1984, Detection and partial characterization of an inhibitor of plasminogen activator in human platelets, *J. Clin. Invest.*, **74**:1465.

Esmon, C. T., and Esmon, N. L., 1984, Protein C activation, *Sem. Thromb. Haemostas.*, **10**:122.

Fay, W. P., and Owen, W. G., 1989, Platelet plasminogen activator inhibitor: purification and characterization of interaction with plasminogen activators and activated protein C, *Biochemistry*, **28**:5773.

Folch, L., Lees, M., and Sloane, S. G. H., 1957, A simple method for the isolation and purification of total lipids from animal tissues, *J. Biol. Chem.*, **226**:497.

Fraker, P. J., and Speck, J. C., 1978, Protein and cell membrane iodinations with a sparingly soluble chloroamide, 1,3,4,6-tetrachloro-3a, 6a-diphenyl-glycoluril, *Biochem. Biophys. Res. Commun.*, **80**:849.

Fulcher, C. A., Gardiner, J. E., Griffin, J. H., and Zimmerman, T. S., 1984, Proteolytic inactivation of human factor VIII procoagulant protein by activated human protein C and its analogy with factor V, *Blood*, **63**:486.

Haverkate, F., and Brakman, P., 1975, Fibrin plate assay, *in:* "Progress in Chemical

Fibrinolysis and Thrombolysis", J. F. Davidson, M. M. Samama, P. C. Desnoyers, eds., Raven Press, New York, Vol. 1, p. 151.
Jansen, J. W. C. M., Haverkate, F., Koopman, J., Nieuwenhuis, H. K. Kluft, C., and Boschman, Th. A. C., 1987, Influence of factor XIIIa activity on human whole blood clot lysis *in vitro, Thromb. Haemostas.,* **72:**171.
Kluft, C., Wijngaards, G., Jie, A. F. H., and Groeneveld, E. 1985, Appropriate milieu for the assay of α_2-antiplasmin activity with chromogenic substrates, *Haemostasis,* **15:**198.
Kluft, C., Nieuwenhuis, H. K., Rijken, D. C., Groeneveld, E., Wijngaards, G., Van Berkel, N., Dooijewaard, G., and Sixma, J. J., 1987, α_2-Anti-plasmin Enschede: dysfunctional α_2-antiplasmin molecule associated with an autosomal recessive hemorrhagic disorder, *J. Clin. Invest.,* **80:**1391.
Rijken, D. C., Wijngaards, G., Zaal-de Jong, M., and Welbergen, J., 1979, Purification and partial characterization of plasminogen activator from human uterine tissue, *Biochim. Biophys. Acta,* **580:**140.
Rijken, D. C., Wijngaards, G., and Welbergen, J., 1980, Relationship between tissue plasminogen activator and the activators in blood and vascular wall, *Thromb. Res.,* **18:**815.
Sakata, Y., and Aoki, N., 1980, Crosslinking of α_2-antiplasmin inhibitor to fibrin by fibrin-stabilizing factor, *J. Clin. Invest.,* **65:**290.
Sakata, Y., Curriden, S., Lawrence, D., Griffin, J. H., and Loskutoff, D. J., 1985, Activated protein C stimulates the fibrinolytic activity of cultured endothelial cells and decreases antiactivator activity, *Proc. Natl. Acad. Sci. USA,* **82:**1121.
Sprengers, E. D., Akkerman, J. W. N., and Jansen, B. G., 1986, Blood platelet plasminogen activator inhibitor: two different pools of endothelial cell type plasminogen activator in human blood. *Thromb. Haemostas.,* **55:**325.
Taylor Jr, F. B., and Lockhart, M. S., 1985a, Whole blood clot lysis: *in vitro* modulation by activated protein C, *Thromb. Res.,* **37:**639.
Taylor Jr, F. B., and Lockhart, M. S., 1985b, A new function for activated protein C: APC prevents inhibition of plasminogen activators by release from mononuclear leukocytes-platelet suspensions stimulated by phorboldiester, *Thromb. Res.,* **37:**155.
Van Hinsbergh, V. W. M., Bertina, R. M., Van Wijngaarden, A., Van Tilburg, N. H., Emeis, J. J., and Haverkate, F., 1985, Activated protein C decreases plasminogen activator activity in endothelial cell conditioned medium, *Blood,* **65:**444.
Van Ruijven-Vermeer, I. A. M., and Nieuwenhuizen, W., 1978, Purification of rat fibrinogen and its constituent chains, *Biochem. J.,* **169:**653.
Verheijen, J. H., Mullaart, E., Chang, G. T. G., Kluft, C., and Wijngaards, G., 1982, A simple, sensitive spectrophotometric assay for extrinsic (tissue type) plasminogen activator applicable to measurements in plasma, *Thromb. Haemost.,* **48:**266.
Walker, F. J., 1981, Regulation of activated protein C by protein S. The role of phospholipids in factor Va inactivation, *J. Biol. Chem.,* **256:**11128.
Zolton, R. P., Seegers, W. H., 1973, Autoprothrombin II-A: Thrombin removal and mechanism of induction of fibrinolysis, *Thromb. Res.,* **3:**23.

TRANSFORMATION OF PROSTACYCLIN (PGI_2) TO A BIOLOGICALLY ACTIVE METABOLITE: 5(6)-OXIDO-PGI_1 BY CYTOCHROME P450-DEPENDENT EPOXYGENASE

Patrick Y.-K. Wong

Department of Physiology and Medicine
New York Medical College
Valhalla, NY 10595, USA

SUMMARY

The renal epoxygenase has been demonstrated to be an active pathway for the conversion of PGI_2 to a new, previously unreported, metabolite. This metabolite was isolated and identified by radiogas-chromatography-mass spectrometry as 5-hydroxy-6-keto $PGF_{1\alpha}$. Its structure was further confirmed by comparison of the mass-spectra to that of the synthetic standard. The formation of 5-hydroxy-6-keto $PGF_{1\alpha}$ in the kidney suggested epoxidation of prostacyclin via the renal epoxygenase as an alternative pathway of PGI_2 metabolism.

INTRODUCTION

Metabolism of AA involves three distinct enzymatic pathways: 1: cyclooxygenase, leading to the formation of prostaglandins (PGs), thromboxane A_2 (TXA)$_2$), and prostacyclin (PGI_2); 2: lipoxygenase, leading to the formation of hydroxy and dihydroxyeicosatetraenoic acid (HETEs and diHETEs), leukotrienes (LTs); and lipoxins (LXs) and 3: cytochrome P450 dependent monooxygenase (cyto-P450), a mixed function oxidase system which is strictly dependent on molecular oxygen and NADPH, that can metabolize AA by three types of reaction: allylic oxidation leading to the formation of four different epoxyeicosatrienoic acids (EETs) which can undergo hydrolysis by epoxide hydrolase to form the corresponding diol metabolites — the dihydroxyeicosatrienoic acids, (DHET) and omega oxidation at the ω- and ω-1 position to forms the 19- and 20-HETEs, respectively (1). Since cyto-P450 exists in many forms, the predominance of one of these reactions over others may be controlled by the isoenzyme composition in each tissue or cell type. Major isoenzymes of cyto P450 that are responsible for the formation of epoxides are called EPOXYGENASE. There are two major isoenzymes of EPOXYGENASE: isoenzyme IIB_1, and isoenzyme IIB_2. The IIB_1 and IIB_2 proteins exhibit 97% amino acid similarity and have distinct chromatographic and electrophoretic properties (2). Capdevila et al. (3) had demonstrated that the liver is a rich source of cyto P450 epoxygenase, followed by the rank order of: kidney, eye and the pituitary. In rat the richest tissue content was found in the liver, followed by the kidney and lung. EETs are also found in the urine (4), platelets and endothelial cells (6). The occurrence of EETs in tissues and various cell types leads to the hypothesis that these substances may be stored in the cells by binding to protein via conjugation or by incorporation into membrane phospholipids (PL). The storage form of EETs can be released upon activation of phospholipases (phospholipase A_2, PLA_2) or phospholipase C (PLC) by neuronal, hormonal or chemical stimulation.

Fibrinogen, Thrombosis, Coagulation, and Fibrinolysis, Edited by
C. Y. Liu and S. Chien, Plenum Press, New York

Of the various systems in which epoxygenase and its epoxides (EETs) have been studied, the kidney seems to be one of the major sites of production and action of the EETs. Urine and kidney contain mainly 8, 9 and 14, 15-EET. The levels of EETs in the kidney appear to vary as a function of renal pathology (6). Sacerdoti et al. (7) had demonstrated that spontaneously hypertensive rats (SHR) developed increased renal ω- and ω-1 metabolic activities of arachidonic acid in parallel with the development of hypertension. In addition to the increased plasma PLA_2 activity found in SHR, an alteration of prostaglandin biosynthesis and catabolism had been described by Pace-Asciak et al. during the development of hypertension in SHR, which was not found in WKY rats (8). These workers describe enhanced PGI_2 production by vascular tissues of SHR as compared to their normotensive controls (9), which may have great significance in the control of systemic blood pressure. The blood pressure lowering effect of PGI_2 was found to be twice as potent as PGE_2 in the normal rats, but 3-4 fold more active in the SHR. Furthermore, Pace-Asciak et al. also reported that PGI_2 was equipotent through intravenous and intraarterial administration, indicative of the lack of pulmonary inactivation (10). In contrast, work done in our laboratory as well as that of Eling et al. (11, 12) demonstrated that PGI_2 was metabolized on transit through the lung. The increased biological action of PGI_2 infusion in hypertensive rats may be due to secondary metabolism of PGI_2 via the cyto P450 enzyme to 5(6) oxido PGI_1. Indeed, our preliminary experiment has shown that 5(6) oxido PGI_1 methylester (Me) is 2 to 3 times more potent than PGI_2 Me in lowering of blood pressure in SHR (10, 13). In this communication we summarize our recent results on the metabolism of PGI_2 by the cytochrome P450 enzyme system in the rabbit renal cortex, as well as the discovery of a new, previously unreported metabolite of PGI_2 via the cytochrome P450 enzyme system in the isolated perfused rabbit kidney.

MATERIALS AND METHODS

Radiolabeled [9-^{3}H] PGI_2 methyl ester with specific activity of 15 μCi/umole was purchased from New England Nuclear, Boston, MA. The purity of the PGI_2 methylester was established by thin-layer chromatography (TLC plate, 0.25 mm thick, 5 × 20 cm silica gel precoated plastic sheets, Brinkman) with hexane:acetone (1:1, v/v) as solvent. Radiochromatogram scans showed a single peak of PGI_2 methyl ester on TLC plates with an Rf value of 0.68. The methyl ester of [9-^{3}H] PGI_2 was converted to the PGI_2 sodium salt the day before use by mild alkaline hydrolysis and diluted with authentic PGI_2 sodium salt (Upjohn) to a specific activity of 12.76 μCi/μmole as described (9).

Kidney Perfusion

Male New Zealand rabbits were anesthetized with sodium pentobarbital (30 mg/kg). After midline laparotomy, the kidneys were exposed. The renal artery and ureter were cannulated and the kidney was flushed with Tyrode's solution, removed from the animal, and placed in a thermostatically-controlled chamber. The kidney was perfused through the renal artery with oxygenated Tyrode's solution (37°C) at a rate of 10 mg/min. 10.0 μCi of [9-^{3}H] PGI_2 Na salt (12.76 μCi/mole) was infused into the kidney through the arterial cannula over a period of 5 min. The venous and ureteral effluents were acidified with 1N HCl to pH 3.0 and extracted three times with equal volumes of ethyl acetate. The ethyl acetate extract was evaporated to dryness in vacuo.

Chromatographic Method

Thin-layer chromatography was performed with Brinkman precoated silic gel G-25 plates and solvent system A9 (organic phase of ethyl acetate: acetate acid:iso-octane:H_2O (55:10:25:50, v/v). The radioactive products were detected with a Packard model 7320 radiochromatogram scanner. High-performance liquid chromatography (HPLC) was carried out with a 6000A pump (Waters Associates, Molford, MA) system and an ultrasphere — ODS reverse-phase column maintained at a flow rate of 0.5 ml/min and monitored by a model GM770 variable wave length detector set at 192 nm. The compounds were eluted isocratically with acetonitrile: water (30:70, v/v, pH 2.95). The column effluent was collected in 0.5 ml fractions and an aliquot of 50 μl each was assessed for radioactivity.

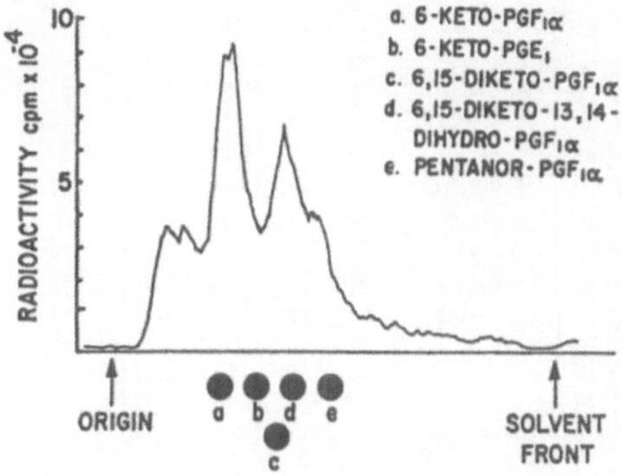

Fig. 1. Radiochromatograph scan of radioactive products extracted from the venous effluent of the rabbit kidney after infusion of [9-^{3}H]-PGI_2. Radioactive metabolites were extracted, separated and identified as described under "Methods".

Radiometric Gas Chromatography and Mass Spectrometry (GC-MS)

The radioactive products were converted to methyl ester, O-methyl-oxime trimethylsilyl ether (15) before analyses by radiometric GC coupled with MS. The final derivative was dissolved in a small aliquot of acetone before injection. GC was carried out on a Varian 2700 GC coupled with a Packard 894 radioactivity detector for simultaneous recording of mass and radioactivity. GC-MS analysis was carried out by LKB-9000 mass spectrometer. The 6-foot column [1% SE3O on Chromosorb-H (HP)] was kept at 210°C, the flash heater at 240°C, and the separator at 250°C. Electron energy was set at 22.5eV (15).

PGI_2 Metabolites in Rabbit Kidney Perfusate

Fig. 1 represents a typical radiochromatogram of the extract from rabbit kidney perfusate. Two major and one minor radioactive zones were observed. Of the total radioactive [9-^{3}H] injected, 95% was recovered in the renal effluent (average of three experiments); 34% migrated in zone 1, which corresponds to the mobility of 6-keto $PGF_{1\alpha}$, and 44% migrated to zones 2 and 3 which correspond to the mobility of pentanor $PGF_{1\alpha}$ and 6, 15-diketo-13, 14-dihydro $PGF_{1\alpha}$. However, when zones 2 and 3 were eluted from the TLC plate and subjected to radiometric GM analysis, five radioactive peaks were observed (Fig. 2). Compounds 2a and 2b were identified as 6-keto $PGF_{1\alpha}$ and 2,3-dinor $PGF_{1\alpha}$, respectively. Compounds 2c and 2d are novel metabolites and their structural characteristics were carried out by RP-HPLC analysis and GC/MS.

Gas chromatographic analysis of the trimethylsilyl derivative of compound 2c showed the radioactive peak with equivalent chain length of C-20.3. The spectrum resembled that reported for the TMS derivative of 9,11,15-trihydroxy-pentanor prosta-13-enoic acid γ-lactone or pentanor $PGF_{1\alpha}$ γ-lactone (17). Subsequently, structural comparison with the authentic standard confirmed the structural assignment as that previously reported (16). Approximately 25% of the radioactivity in both zones 2 and 3 (Fig. 2, lower tracing) was accounted for by compound 2d and was detected by radioactive GC because of incomplete separation by TLC. Compound 2d was slightly more polar than 6-keto $PGF_{1\alpha}$ as it eluted off the reverse-phase column before 6-keto $PGF_{1\alpha}$. The retention time of the methylester, O-methoxime-TMS derivative of 2d, had a C-value of 25.3. In the mass spectrum (Fig. 3), the molecular ion was found at m/e 717. Fragment ions were found at 702 [M-15], 686 [M-31], 646 [M-71], 627 [M-90], 596 [M−90+31], 556 [M−90+71], 537 [M-2×90], 516 [M-201], 514 [M−203], 466 [M−2×90+71], 203 (base peak) and 171. The molecular ion and most of the prominent fragments at the high mass end are in accord with the formation of a 6-keto-$PGF_{1\alpha}$ derivative with an additional hydroxyl group. A series of [M−71] fragments indicate that the compound is not a 19 or 20-hydroxyl compound and the spectrum differs from the corresponding spectrum from similar derivatives of 19 or 20-hydroxy-6-keto-$PGF_{1\alpha}$. The base peak at m/e 203 and the M-203 fragment at m/e 514 suggested that

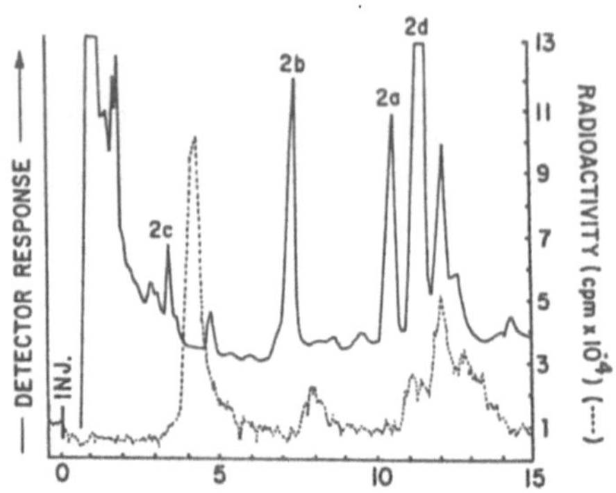

Fig. 2. Radio-gas chromatograph of [9-^{3}H]-PGI_2 metabolites recovered from zones 2 and 3; GC detector as broken line (---). Note that the radioactive detector responses were one minute later than those of the GC detector.

the most probable position of the extra hydroxyl group is at C-5 because of the favorable cleavage between the adjacent -OTMS and -NOH_3 groups. Based on this information, we concluded that compound 2d was a new metabolite of PGI_2:5-hydroxy-6-keto-$PGF_{1\alpha}$. Further, structural conformation was provided by oxidative cleavage by periodic acid. About 50 μg of compound 2d was treated with 0.5 ml of 2 mg/ml aqueous periodic acid for 16 hours. The product was converted to methyl ester TMS derivative and analyzed by GC-MS. The compound was qualitatively converted to the 15 carbon pentanor $PGF_{2\alpha}$ γ-lactone, therefore confirming the adjacent keto/alcohol functions at C-5 and C-6. 5-hydroxy-6-keto-$PGF_{1\alpha}$ was subsequently prepared by direct epoxidation of PGI_2 with m-chloroperoxybenzoic acid to form the 5,6-epoxide, followed by ring opening with diluted aqueous acid (17). The product has identical chromatographic and mass spectral characteristics as the biological material.

Similar to our previous report (15), the urinary effluent obtained from the isolated perfused kidney, during infusion of [9-^{3}H]-PGI_2 and thereafter (three experiments), did not show appreciable (less than 0.1%) radioactivity nor contain any tritiated water.

DISCUSSION

We have previously reported that renal metabolism of PGI_2 is extensive; the major metabolites in venous effluent have been isolated and identified as 7,9-dihydroxy, 4,13-diketo-dinor-$PGF_{1\alpha}$, dinor-6-keto-$PGF_{1\alpha}$, suggesting that a substantial portion of the infused [9-^{3}H]-PGI_2 was metabolized via the 15-hydroxyprostaglandin dehydrogenase followed

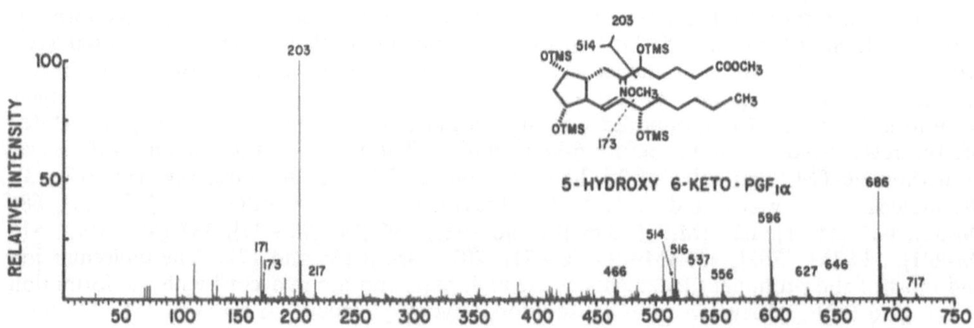

Fig. 3. Mass spectrum of methyl-o-methyl-oxamine trimethylsilyl ether derivative of 5-hydroxy-6-keto-$PGF_{1\alpha}$.

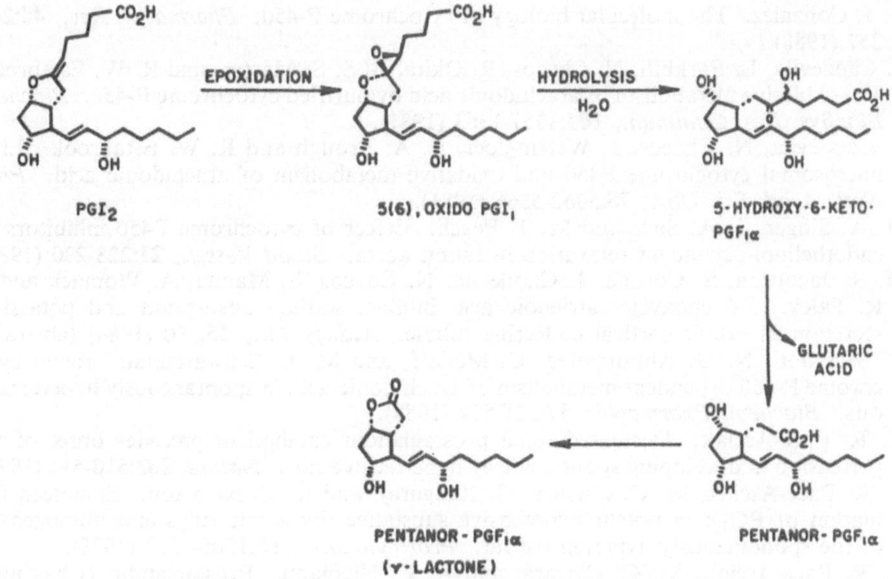

Fig. 4. Proposed metabolic pathways of epoxygenation of prostacyclin in the rabbit kidney.

by Δ 13,14 reductase, as well as by ß-oxidation (15, 17). Radioactive metabolites were not detected in the urine, suggesting that exogenous PGI_2 was not secreted by the proximal tubular secretory system. PGI_2, unlike PGE_2, is a poor substrate for the organic acid secretory system (18). In this report, we have demonstrated the identification of the new metabolite, 5-hydroxy-6-keto-$PGF_{1\alpha}$, probably generated from an unstable intermediate, "epoxide-ether": "5,(6)-oxido-PGI_1" by the renal epoxygenases. The "epoxide-ether", "5,(6)-oxido-PGI_1", is subsequently hydrolyzed to 5-hydroxy-6-keto-$PGF_{1\alpha}$ by the addition of H_2O (Fig. 4).

The occurrence of a cytochrome P450-dependent mono-oxygenase/epoxygenase system in renal microsomes has been demonstrated by Oliw et al. (16) and by Manna et al. (19). Recently, this enzyme system was demonstrated in the cells isolated from the thick ascending limb of Henle's loop (20). Drugs which caused the depletion of cytochrome P450, like $CoCl_2$, reduced the formation of the above metabolites (20). Induction of cytochrome P450 system by 3-methylcholanthrene and ß-naphthoflavone caused an increase of cytochrome P450-dependent metabolites. Since the discovery of this new epoxygenase pathway, a new group of epoxygenase products of arachidonic acid (AA) have been isolated and identified. The biological properties of these products of AA have been studied by Schwartzman et al. and Capdevila (21, 22). All of these epoxides of AA were found to be potent stimulators of polypeptide hormone release at low concentrations. Furthermore, the 5,6-epoxide of AA was found to be a potent inhibitor of ion transport as well as of Na^+/K-ATPase activity (20). In view of the potent biological effect of various unstable AA metabolites, e.g., PGI_2 and 5,6 LTA_4, the postulated intermediate, "5,(6) oxido-PGI_1", may also be biologically active (13). More recently, Oliw (23) reported that 5,(6) oxido-C-20:3 can be used as substrate for the ram seminal vesicle cyclooxygenase to generate two new products, namely 5-hydroxy-$PGI_{1\alpha}$ and 5-hydroxy-$PGI_{1\beta}$ (two isomers). In this study we demonstrated that PGI_2 can be metabolized by the renal epoxygenase enzyme system and transformed into a series of new products, including 5-hydroxy-6-keto-$PGF_{1\alpha}$. The biological significance of the products of this pathway in the kidney remains to be defined.

REFERENCES

1. A. Y. H. Lu and S. B. West. Multiplicity of mammalian microsomal cytochromes P-450. *Pharmacol. Rev.*, **31**:277-295 (1980).

2. F. J. Gonzalez. The molecular biology of cytochrome P-450. *Pharmacol. Rev.*, **40**:243-257 (1988).
3. J. Capdevila, L. Parkhill, N. Chacos, R. Okita, B. S. S. Masters and R. W. Estabrook. The oxidative metabolism of arachidonic acid by purified cytochrome P-450. *Biochem. Biophys. Res. Commun.*, **101**:1357-1363 (1981).
4. J. Capdevila, N. Chacos, J. Werringloer, R. A. Prough and R. W. Estabrook. Liver microsomal cytochrome P450 and oxidative metabolism of arachidonic acid. *Proc. Natl. Acad. Sci. USA*, **78**:5362-5366 (1981).
5. H. A. Singer, J. A. Saye and M. J. Peach. Effect of cytochrome P450 inhibitors on endothelium-dependent relaxation in rabbit aorta. *Blood Vessels*, **21**:223-230 (1984).
6. H. R. Jacobson, S. Corona, J. Capdevila, N. Chacos, S. Manna, A. Womack and J. R. Falck. 5,6 epoxyeicosatrienoic acid inhibits sodium absorption and potassium secretion in rabbit cortical collecting tubule. *Kidney Int.*, **25**:330 (1984) (abstract).
7. P. Sacerdoti, N. G. Abraham, J. C. McGiff, and M. L. Schwartzman. Renal cytochrome P-450 dependent metabolism of arachidonic acid in spontaneously hypertensive rats. *Biochem. Pharmacol.*, **37**:521-527 (1988).
8. C. R. Pace-Asciak. Decreased renal prostaglandin catabolism precedes onset of hypertension in developing spontaneously hypertensive rats. *Nature*, **263**:510-511 (1976).
9. C. R. Pace-Asciak, M. C. Carrara, G. Rangaraj, and K. C. Nicolaou. Enhanced formation of PGI_2: a potent hypotensive substance, by aortic rings and homogenates of the spontaneously hypertensive rat. *Prostaglandins*, **15**:1005-1012 (1978).
10. C. R. Pace-Asciak, M. C. Carrara, and K. C. Nicolaou. Prostaglandin I_2 has more potent hypotensive properties than prostaglandin E_2 in the spontaneously hypertensive rat. *Prostaglandins*, **15**:999-1003 (1978).
11. P. Y.-K. Wong, J. C. McGiff, F. F. Sun, and K. U. Malik. Pulmonary metabolism of prostacyclin (PGI_2) in the rabbit. *Biochem. Biophys. Res. Commun.*, **83**:731-738 (1978).
12. H. J. Hawkins, B. J. Smith, K. C. Nicolaou, and T. E. Eling. Studies of the mechanisms involved in the fate of prostacyclin (PGI_2) and 6-keto-$PGF_{1\alpha}$ in the pulmonary circulation. *Prostaglandins*, **16**:871-884 (1978).
13. X. R. He, and P. Y.-K. Wong. Transformation of prostacyclin (PGI_2) to a biologically active metabolite: 5(6)-oxido-PGI_2 by cyto-P450 dependent epoxygenase. *FASEB J.* **3**:No. 3, 2714 (1989).
14. R. A. Johnson, F. H. Lincoln, E. G. Nidy, H. P. Schneider, J. L. Thompson and U. Axen. Synthesis and characterization of prostacyclin, 6-keto-prostaglandin $F_{1\alpha}$, prostaglandin I_2, and prostaglandin I_3. *J. Amer. Chem. Soc.*, **100**:7690-7693 (1980).
15. P. Y.-K. Wong, J. C. McGiff, L. Cagen, K. U. Malik and F. F. Sun. Metabolism of prostacyclin in the rabbit kidney. *J. Biol. Chem.*, **254**:12-14 (1979).
16. E. H. Oliw, J. A. Lawson, H. R. Brash, J. A. Oates. Arachidonic acid metabolism in rabbit renal cortex: formation of two novel dihydroxyeicosatrienoic acids. *J. Biol. Chem.*, **256**:9924-9931 (1981).
17. P. Y.-K. Wong, K. U. Malik, D. M. Desiderio, J. C. McGiff and F. F. Sun. Hepatic metabolism of prostacyclin (PGI_2) in the rabbit: formation of a potent novel inhibitor of platelet aggregation. *Biochem. Biophys. Res. Commun.*, **93**:486-494 (1980).
18. M. J. S. Miller, E. G. Spokas and J. C. McGiff. Metabolism of prostaglandin E_2 in the isolated perfused kidney of the rabbit. *Biochem. Pharmacol.* **31**:2955-2960 (1982).
19. S. Manna, J. R. Falck, N. Chacos and J. Capdevile. Synthesis of arachidonic acid metabolites produced by purified kidney cortex microsomal cytochrome P450. *Tetrahedron Letters*, **24**:33-36 (1983).
20. N. R. Ferreri, M. Schwartzman, N. G. Ibraham, P. N. Chander and J. C. McGiff. Arachidonic acid metabolism in a cell suspension isolated from rabbit renal outer medulla. *J. Pharmacol. Exp. Ther.*, **231**:441-448 (1984).
21. M. Schwartzman, N. R. Ferreri, M. Carroll, N. Ibraham, R. D. Levere and J. C. McGiff. Arachidonic acid metabolism in isolated cells from the thick ascending limb of Henle's loop. *Clin. Res.*, **32**:456A (1984) (abstract).
22. J. Capdevila, N. Chacos, J. R. Falck, S. Manna, A. Negro-Vilar and S. R. Ojeda. Novel hypothalamic arachidonate products stimulate somatostatin release from the median eminence. *Endocrinology*, **113**:421-423 (1983).
23. E. H. Oliw. Metabolism of 5(6) oxidoeicosatrienoic acid by ram seminal vesicles: formation of two stereoisomers of 5-hydroxyprostaglandin I_1. *J. Biol. Chem.*, **259**:2716-2721 (1984).

STUDIES ON THE LOCALIZATION OF FIBRINOGEN BINDING SITES ON PLATELET GLYCOPROTEIN IIIa

Stefan Niewiarowski, Karin J. Norton,
Jacquelynn J. Cook and Annette Eckardt

Department of Physiology and Thrombosis Research Center
Temple University School of Medicine
Philadelphia, PA 19140, USA

INTRODUCTION

The domains of various integrins involved in the binding to adhesive proteins are poorly characterized. The approaches to localize ligand binding sites on the adhesive receptors include studies on the genetic characterization of the defective β_2 subunits in leukocyte adhesion deficiency and the defective β_1 subunit in Glanzmann thrombasthenia, studies on the chemical and photoaffinity crosslinking of the adhesive ligands to the receptors, proteolytic degradation of the integrins, attempts to recognize binding sites with monoclonal antibodies, and studies on the synthetic peptides derived from glycoprotein IIIa (GPIIIa).

OBSERVATIONS AND DISCUSSIONS

Studies performed in our laboratory indicate that fibrinogen binding ability of GPIIIa is lost during its proteolytic conversion to the "66 kDA component" in parallel with the deletion of the region of this protein spanning from residue 130 to 348. Further studies with synthetic peptides derived from this region of GPIIIa suggest that the putative fibrinogen binding site may be restricted to the residues 217-232 or 217-228 of GPIIIa.

It is well established that the binding of fibrinogen to specific receptors is essential for platelet aggregation. These receptors are associated with the glycoprotein IIb/IIIa (GPIIb/IIIa) complex that represents a major protein on platelet membranes. (1, 2). The GPIIb/IIIa complex is a member of a superfamily of receptors called integrins (3) or cytoadhesins (4) that are critical for cell-cell interaction and cell adhesion to the extracellular matrix. Each integrin forms a heterodimeric complex with a highly conserved β subunit and more variable α subunits (3, 5).

The best characterized integrins can be divided into three subgroups. First, the VLA family of heterodimers contains at least five different structures, each with a distinct α subunit, and a common β_1 subunit. (6). This subfamily of integrins, commonly known as collagen, fibronectin or laminin receptors, occur on a number of cells including fibroblasts, platelets and endothelial cells. The second subfamily (β_2) include the LFA-1, Mac-1 and p150, 95 integrins; they appear on neutrophils and macrophages (7). Finally, the vitronectin receptor on endothelial cells and the GPIIb/IIIa complex on platelets contain a common β subunit (β_3) which is identical with GPIIIa.

A number of adhesive proteins including fibrinogen, fibronectin, von Willebrand factor, vitronectin and thrombospondin bind to these integrins (2, 8). Several amino acid sequences in these adhesive proteins have been identified as cell recognition sites, the RGD sequence being most important (5). However, our knowledge of the structure/function relationship of integrins is much more limited. Kishimoto et al. (9) studied the genetic defect in leukocyte adhesion deficiency, a heritable, often fatal disease which is characterized by severe recurrent infections and deficiency of the LFA-1, Mac-1 and p150, 95 integrins. These authors demonstrated an aberrant splicing of the β_2 subunit gene with the deletion of a region spanning from amino acids 332 to 361 in this molecule. They postulate that this region is important for the formation of the heterodimer which expresses adhesive properties.

Several investigators have attempted to define the fibrinogen or RGD binding sites on the GPIIb/IIIa complex. For example, Melero and Gonzalez-Rodriguez (10) have produced a monoclonal antibody (P37) against GPIIIa which blocks platelet aggregation. This antibody recognizes a 23 kDa NH2 terminal tryptic degradation product of GPIIIa (11) which, based on molecular weight, restricts its epitope to within approximately the first 192 amino acids. In addition, on the basis of experiments with chemical crosslinking of an RGD-containing peptide to platelets, D'Souza et al. (12) hypothesized that the RGD binding site may reside in an area possibly as small as the 63 amino acid residues 109-171 of platelet GPIIIa. However, Smith and Cheresh (13), using photoaffinity labelled RGD, provided evidence that amino acid residues 61-203 are proximal to the RGD binding domain of the vitronectin receptor (placental GPIIIa). Loftus et al. (14) studied a variant of thrombasthenia (CAM family); platelets of these patients contain GPIIb/IIIa which does not bind adhesive proteins. Their experiments indicate that the single point of mutation in GPIIIa (replacement of Asp 119 with Tyr) results in the inability of this molecule to bind to RGD. Most recently, d'Souza et al. (15) provided evidence that the recognition site for the C-terminal region of fibrinogen γ chain resides within the residues 294-314 of GPIIb. This site may be in the vicinity of the fibrinogen recognition site in GPIIIa.

Along with other investigators we demonstrated that the incubation of platelet suspensions with proteolytic enzymes such as chymotrypsin, pronase, porcine pancreatic elastase and human granulocyte elastase can expose fibrinogen binding sites on the platelet surface (16-19). In 1983, we described the appearance of a component on the surface of chymotrypsin or pronase treated platelets migrating with an apparent Mr of 60 kDa in a non-reduced system and 66 kDa in a reduced system (20). Subsequently, we isolated this component from solubilized platelets pretreated with chymotrypsin (21). It migrated with an apparent Mr 62,400 in a non-reduced system with SDS polyacrylamide gel electrophoresis. In a reduced system it yielded two major subunits migrating with apparent Mr 14-17 kDa and 65 kDa. The low molecular weight component began with the NH2-terminal sequence of GPIIIa (GPNICTTR. . .) and the larger component with residue 348 of GPIIIa (GKIRSKKA. . . . .) as deduced from the cDNA clone of this glycoprotein (22, 23). The two subunits appeared to be linked by one or more S-S bridges. Fig. 1 shows the putative

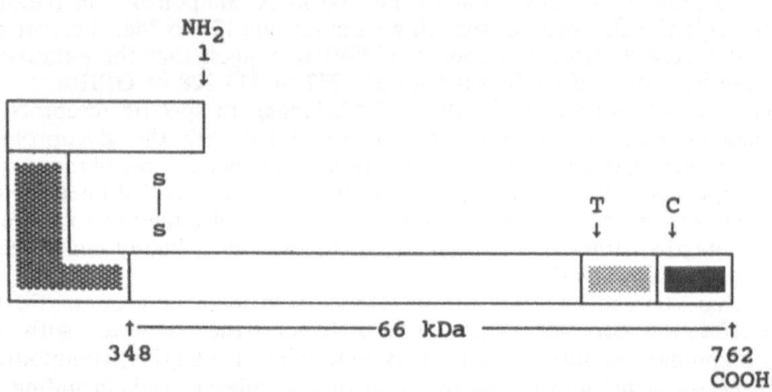

Fig. 1. Putative structure of GPIIIa and its 66 kDa domain. It is proposed that the "66 kDa" component is comprised of a light subunit near the N-terminal region (White area) and a heavy subunit near the C-terminal region (White area), the transmembrane region (light stippled area) and the cytoplasmic region (solid bar). The dark stippled area represents the region deleted from GPIIIa in the formation of 66 kDa. It is unknown how many S-S bridges are involved in maintaning the structure of the "66 kDa" component. Reprinted from S. Niewiarowski et al. Bioch. Biophys. Acta 983:91-99, 1989.

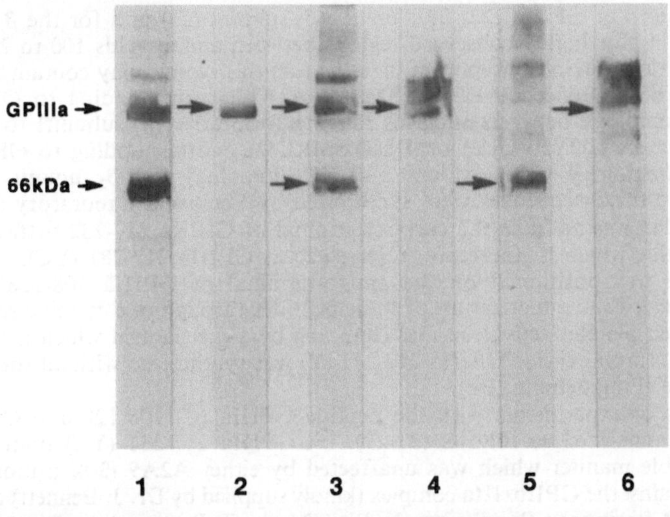

Fig. 2. Characterization of GPIIIa and the "66 kDa" component in detergent solubilized enzyme — treated platelets by means of absorption on GRGDSPK-agarose. Material obtained from 5×10^{10} platelets was applied on a 1.0 ml column of GRGDSPK-agarose. The eluates were subjected to SDS-polyacrylamide gel electrophoresis, the nitrocellulose transfers were stained with anti-GPIIIa antibody in a dilution 1:1000. Lanes 1, 3, 5 represent material which was applied to the column but did not bind. This shows the 66 kDa component identical to that seen in material prior to application to column; lanes 2, 4, 6 correspond to the material eluted from columns with 25 mM octylglucoside/0.1% RGDS buffer. Lanes 1, 2: platelets treated with chymotrypsin (ref 21); lanes 3, 4: platelets treated with porcine pancreatic elastase (ref 18); lanes 5, 6: platelets treated with human granulocyte elastase (ref 19).

structure of GPIIIa and its 66 kDa domain. The linkage of a short NH2-terminal fragment of GPIIIa to a rather remote portion of this molecule suggests that this protein is highly folded on the platelet surface. Fig. 2 shows formation of similar 66 kDa components from GPIIIa following platelet digestion with chymotrypsin, human granulocyte elastase and porcine pancreatic elastase. This figure also shows that the 66 kDa component, produced by any of the enzymes tested, is not retained on insolubilized GRGDSPK. By contrast, GPIIIa was retained and eluted with a buffer containing 25 mM octylglucoside and 1 mg/ml RGDS. In addition, the 66 kDa component did not bind to fibrinogen/agarose nor to insolubilized monoclonal antibody recognizing GPIIb/IIIa complex, while GPIIIa was retained on all columns. Furthermore, during the same course of chymotryptic digestion, the exposure of fibrinogen receptors preceded the formation of the 66 kDa component. The essential chymotryptic cleavage of GPIIIa resulting in the formation of the 66 kDa component is between Tyr-347 and Gly-348. We estimate that the proteolytic conversion of GPIIIa to the 66 kDa component results in the deletion of a domain which spans approximately amino acids 130 to 347 of the NH2-terminal of the molecule. We conclude that the fibrinogen (RGD) binding sites are deleted from this molecule (66 kDa component). Similarly, these sites are absent in the 66 kDa fragments formed from GPIIIa following digestion with human granulocyte elastase and porcine pancreatic elastase. In agreement with our study, Beer and Coller (24) presented evidence that platelet glycoprotein IIIa has a large disulfide bonded loop that is susceptible to proteolytic cleavage. Degradation products of GPIIIa formed following digestion by chymotrypsin, plasmin and trypsin gave similar patterns. From this observation they inferred that the major fragments after digestion were two chain molecules, composed of the amino terminal fragment of Mr 17 kDa to 18 kDa which is disulfide bonded to the carboxy-terminal remnant of Mr 58 to 70 kDa.

In an attempt to localize the fibrinogen binding site in the domain of GPIIIa deleted during proteolysis of this molecule we analyzed the cDNA-predicted sequences for the integrin β_3-subunit i.e. GPIIIa (22, 23), for the β_2 subunit (25) and for the β_1 subunit (26). We were searching for highly conserved regions between amino acids 100 to 250. Based on indications from all previously reported investigators this region may contain the fibrinogen binding site. The sequence DAPEGGFDAIMQATVC, corresponding to GPIIIa 217-232 displays high homology between integrins ranging from 80% (β_1 subunit) to 69% (β_2 subunit). The sequence EDYPVDIYYLMDLSYSMKDDL, corresponding to GPIIIa 108-128, displayed 90% homology with β_1 subunit and 76% homology with β_2 subunit. The peptides corresponding to these sequences were synthesized by Peninsula Laboratory (Belmont Ca). A tyrosine residue was addd to the carboxyterminal of GPIIIa 217-232 to facilitate iodination of the peptide which is, therefore, referred to as GPIIIa 217-231 (YC). In addition, a cysteine residue was positioned on the amino-terminal of GPIIIa 108-128. Preliminary experiments revealed the dimerization of these peptides through one cysteine residue in each; the dimers caused platelet activation and clumping by a mechanism which is not yet understood. For this reason peptide GPIIIa 217-231 (Y) was synthesized without the final cysteine (Wistar Institute, Philadelphia, Pa).

The results of experiments with the peptide GPIIIa (C) 108-128 and GPIIIa 217-231 (YC) can be summarized as follows (27-29). ^{125}I-GPIIIa 217-231 (YC) bound to platelets in a non-saturable manner which was unaffected by either A2A9 (30), a monoclonal antibody directed against the GPIIb/IIIa complex (kindly supplied by Dr. J. Bennett) or a polyclonal antibody against fibrinogen (kindly supplied by Dr. A. Z. Budzynski). In contrast, GPIIIa (C) 108-128 did not bind appreciably to platelets. GPIIIa 217-231 (YC) at concentrations as low as 0.01 mM significantly increased the binding of ^{125}I-fibrinogen to intact platelets. In the presence of 0.1 mM peptide, Scatchard analysis revealed a 7-8 fold increase in the number of fibrinogen binding sites over control binding sites exposed following ADP-stimulation. Furthermore, in the presence of GPIIIa 217-231 (YC), radiolabeled fibrinogen bound to thrombasthenic platelets and to normal platelets which were treated with A2A9 to block binding sites on the GPIIb/IIIa complex. In a similar manner, GPIIIa 217-231 (Y) at 0.1 mM concentration increased fibrinogen binding to platelets even in the presence of an excess of A2A9. However, the effect of this peptide was more modest with the number of binding sites of the same order of magnitude as those after ADP-induced exposure. Contrastingly, GPIIIa (C) 108-128 at 0.1 mM concentration, caused fibrinogen binding to platelets which was completely inhibited by A2A9. This suggests that the observed fibrinogen binding was secondary to platelet activation by the dimerized peptide and subsequent exposure of binding sites on GPIIb/IIIa.

In a series of preliminary experiments we tested a number of anti-GPIIb/IIIa and anti-GPIIIa antibodies for cross-reactivity with GPIIIa 108-128 and GPIIIa 217-232 using the ELISA technique (31). The results of these studies suggest that the 217-232 domain of 6PIIIa may be involved in fibrinogen binding during platelet aggregation.

Our data are consistent with the recent finding by Charo et al. (32) who reproted that GPIIIa 203-228 inhibits both fibrinogen binding to the purified receptor (GPIIb/IIIa) and platelet aggregation at low concentration of fibrinogen. Observations of these authors and our findings indicate further narrowing of the putative fibrinogen binding site to amino acid residues 217-228. Although the relationship of synthetic peptides GPIIIa 203-228 and GPIIIa 217-232 to the physiological binding sites in native GPIIIa is not established, further studies of this region of the molecule are warranted to define the amino acid residues responsible for physiological interaction of GPIIIa with fibrinogen.

ACKNOWLEDGMENT

This work was supported by NIH grants HL 15226 and HL 36579.

REFERENCES

1. E. I. B. Peerschke, The platelet fibrinogen receptor, *Seminars in Hematology,* **22**:241-259 (1985).
2. D. R. Phillips, I. F. Charo, L. V. Parise, and L. A. Fitzgerald, The platelet membrane glycoprotein IIb-IIIa complex, *Blood,* **71**:831-843 (1988).
3. R. O. Hynes, Integrins: A family of cell surface receptors, *Cell,* **48**:549-554 (1987).
4. E. F. Plow, J. C. Loftus, E. G. Levin, D. S. Fair, D. Dixon, J. Forsyth, and M. H. Ginsberg,

Immunologic relationship between platelet membrane glycoprotein GPIIb/IIIa and cell surface molecules expressed by a variety of cells, *Proc. Nat. Acad. Sci. (USA),* **83:**6002-6606 (1986).
5. E. Ruoslahti and M. D. Pierschbacher, New perspectives in cell adhesion: RGD and Integrins, *Science, (Wash., DC),* **238:**491-497 (1987).
6. M. E. Hemler, C. Thang, and L. Schwarz, The VLA protein family. Characterization of five distinct cell surface heterodimers each with a common 130,000 molecular weight subunit, *J. Biol. Chem.,* **262:**3300-3309 (1987).
7. F. Sanchez-Madrid, J. A. Nagy, E. Robbins, P. Simm, and T. A. Springer, A human leukocyte differentiation antigen family with distinct alpha subunits and a common beta subunit: the lymphocyte-function associated antigen (LFA-1), the C3bi complement receptor (OKM1/MAC-1), and the p150, 95 molecule, *J. Exp. Med.,* **158:**1785-1803 (1983).
8. J. C. Giltay, and J. A. van Mourik, Structure and function of endothelial cell integrins, *Haemostasis,* **18:**376-389 (1988).
9. T. K. Kishimoto, K. O'Connor, and T. A. Springer, Leukocyte adhesion deficiency. Aberrant splicing of a conserved integrin sequence causes a moderate deficiency phenotype, *J. Biol. Chem.,* **264:**3588-3599 (1989).
10. J. A. Melero and J. Gonzalez-Rodriguez, Preparation of monoclonal antibodies against glycoprotein IIIa of human platelets: Their effect on platelet aggregation, *Eur. J. Biochem.,* **141:**421-427 (1984).
11. J. J. Calvete, G. Rivas, M. Maurui, M. V. Alvarez, J. L. McGregor, C. L. Hew, and J. Gonzalez-Rodriguez, Tryptic digestion of human GPIIIa. Isolation and biochemical characterization of the 23 kDa N-terminal glycopeptide carrying the antigenic determinant for a monoclonal antibody (P37) which inhibits platelet aggregation, *Biochem. J.,* **250:**697-704 (1988).
12. S. E. D'Souza, M. H. Ginsberg, T. A. Burke, S. C. T. Lam, and E. F. Plow, Localization of an Arg-Gly-Asp recognition site within an integrin adhesion receptor, *Science,* **242:**91-93 (1988).
13. J. W. Smith and D. A. Cheresh, The Arg-Gly-Asp binding domain of the vitronectin receptor. Photoaffinity crosslinking implicates amino acid residues 61-203 of the beta subunit, *J. Biol. Chem.,* **263:**18726-18731 (1988).
14. J. C. Loftus, T. E. O'Toole, E. F. Plow, and M. H. Ginsberg, Identification of a GPIIa-IIIa mutation in a glanzmanns variant associated with loss of RGD binding function, *Blood,* **74 (suppl 1),** 58a (abstract) (1989).
15. S. E. D'Souza, M. H. Ginsberg, and E. F. Plow, Defining the ligand recognition site within the platelet adhesion receptor GPIIb-IIIa, *Blood,* **74 (suppl 1),** 90a (abstract) (1989).
16. J. Greenberg, J. L. Or, M. A. Packham, M. A. Guccione, E. J. Harfenist, R. L. Kinlough-Rathbone, D. W. Perry, and J. F. Mustard, The effect of pretreatment of human or rabbit platelets with chymotrypsin on their responses to human fibrinogen and aggregating agents, *Blood,* **54:**753-765 (1979).
17. S. Niewiarowski, A. Z. Budzynski, T. A. Morinelli, T. M. Brudzynski, and G. J. Steward, Exposure of fibrinogen receptor on human platelets by proteolytic enzymes, *J. Biol. Chem.,* **256:**917-925 (1981).
18. E. Kornecki, Y. H. Ehrlich, D. D. D. D. Demars, and R. H. Lenox, Exposure of fibrinogen receptors in human platelets by surface proteolysis with elastase, *J. Clin. Invest,* **77:**750-756 (1986).
19. E. Kornecki, Y. M. Ehrlich, R. Egbring, E. Gramse, R. Seitz, A. Eckardt, H. Lukasiewicz, and S. Niewiarowski, Granulocyte-platelet interactions and platelet fibrinogen receptor exposure, *Amer. J. Physiol.,* **225:**H651-H658 (1988).
20. E. G. Kornecki, G. P. Tuszynski, and S. Niewiarowski, Inhibition of fibrinogen receptor-mediated platelet aggregation by heterologous anti-human platelet membrane antibody: significance of an MR = 66,000 protein derived from glycoprotein IIIa, *J. Biol. Chem.,* **285:**9349-9356 (1983).
21. S. Niewiarowski, K. J. Norton, A. Eckardt, H. Lukasiewicz, J. C. Holt, and E. Kornecki, Structural and functional characterization of major platelet membrane components derived by limited proteolysis of glycoprotein IIIa, *Biochim. Biophys. Acta,* **983:**91-99 (1989).
22. L. A. Fitzgarld, B. Steiner, S. C. Rall, S. S. Jr., Lo, and D. R. Phillips, Protein sequence of endothelial glycoprotein IIIa derived from a cDNA clone, *J. Biol. Chem.,* **262:**3936-3939 (1987).

23. A. B. Zimrin, R. Eisman, G. F. Vilaire, E. Schwartz, J. S. Bennett, and M. Poncz, Structure of platelet glycoprotein IIIa. A common subunit for two different membrane receptors, *J. Clin. Invest.,* **81:**1470-1475 (1988).
24. J. Beer and B. S. Coller, Evidence that platelet glycoprotein IIIa has a large disulfide bonded loop that is susceptible to proteolytic cleavage, *J. Biol. Chem.,* **64:**17564-175-73 (1989).
25. S. K. A. Law, J. Gagnon, J. E. K. Hildreth, C. E. Wells, A. C. Willis, and A. J. Wong, The primary structure of the beta subunit of the cell surface adhesion glycoproteins LFA-1, CR3 and p150, 95 and its relationship to the fibronectin receptor, *EMBO Journal,* **6:**915-919 (1987).
26. J. W. Tamkun, D. W. DeSimone, D. Fonda, R. S. Patel, C. Buck, A. F. Horwitz, and R. O. Hynes, Structure of integrin, a glycoprotein involved in the transmembrane linkage between fibronectin and actin, *Cell,* **46:**271-282.
27. K. J. Norton, J. J. Cook, and S. Niewiarowski, Effect of GPIIIa 217-232 on fibrinogen binding to human platelets, *Blood,* **74:**(suppl 1), 90a (abstract).
28. J. J. Cook, K. J. Norton, and S. Niewiarowski, Immunochemical characterization of GPIIIa 108-128 and GPIIIa 217-232, *Blood,* **74:**(suppl 1), 92a (abstract).
29. K. J. Norton, "Structural and functional characterization of potential fibrinogen binding sites on glycoprotein IIIa of human platelets." Dissertation submitted to the Department of Physiology, Temple University School of Medicine in partial fulfillment of the requirements for the degree of Doctor of Philosophy, Philadelphia, PA, pp. 139 (1989).
30. J. S. Bennett, J. A. Hoxie, S. F. Leitman, G. Vilaire, and D. B. Cines, Inhibition of fibrinogen binding to stimulated human platelets by a monoclonal antibody, *Proc. Nat. Acad. Sci., (USA),* **80:**2417-2421 (1983).
31. R. L. Nachman and L. L. K. Leung, Complex formation of platelet membrane glycoproteins IIb and IIIa with fibrinogen, *J. Clin. Invest.,* **69:**263-269 (1982).
32. I. F. Charo, L. Nanninzi, D. R. Phillips, M. A. Hsu, and R. M. Scarborough, Inhibition of fibrinogen binding to GPIIb/IIIa by a GPIIIa peptide, *Blood,* **74:**(suppl 1), 135a (abstract) (1989).

AFFINITY LABELING OF NUCLEOTIDE BINDING SITES OF ENZYMES AND PLATELETS

Roberta F. Colman

Department of Chemistry and Biochemistry
University of Delaware
Newark, DE 19716, USA

INTRODUCTION

A goal of many biochemists is to identify and then to define the role of particular amino acids within the active and allosteric sites of enzymes. One approach is to chemically modify critical amino acids and to rely on the specificity of the enzyme for its substrate or normal regulatory compound to limit the extent of chemical modification to the region of the active or allosteric sites. In this approach, a reagent is designed which can react irreversibly with amino acid residues of the enzyme, but which is in addition stucturally analogous to the substrate. Such a compound may mimic the substrate in forming a reversible enzyme inhibitor complex at a particular site on the enzyme; the reactive functional group of the reagent may then form a covalent bond within the active site during the existence of this complex. This is the strategy termed affinity labeling, which can potentially lead to specific but irreversible attack within purified enzymes, or even of particular enzymes or receptors when present in a mixture of proteins.

OBSERVATIONS AND DISCUSSIONS

Examples of such compounds are the fluorosulfonylbenzoyl derivatives of nucleosides shown in Fig. 1. Compound (a) is 5'-p-fluorosulfonylbenzoyl adenosine (abbreviated 5'-FSBA), which is prepared by reaction of p-fluorosulfonylbenzoyl chloride with adenosine. This compound might reasonably be considered as an analogue of ADP, ATP or NADH. In addition to the adenine and ribose moieties, it has a carbonyl group adjacent to the 5'-position which is structurally similar to the first phosphoryl group of the naturally occuring purine nucleotides. If the molecule is arranged in an extended conformation, the sulfonyl fluoride moiety may be located in a position analogous to the terminal phosphate of ATP or to the ribose proximal to the nicotinamide ring of NADH. This sulfonyl fluoride moiety can act as an electrophilic agent in covalent reactions with several classes of amino acids, including tyrosine, lysine, histidine, serine and cysteine (1).

Structure (b) is 5'-p-fluorosulfonylbenzoyl guanosine, or 5'-FSBG, in which guanine replaces the adenine moiety in the first derivative. It might be anticipated that this purine nucleotide alkylating agent would be specifically directed toward GTP sites in proteins.

Structure (c) is the fluorescent compound 5'-p-fluorosulfonylbenzoyl-1, N^6-etheno-adenosine (or 5'-FSBεA). This nucleotide analogue, with a fluorescence emission maximum at 412 nm, may provide a means of introducing a covalently bond fluorescent probe into nucleotide sites in proteins (2).

The final compound is a new bifunctional affinity label, 5'-p-fluorosulfonylbenzoyl-8-azidoadenosine (5'-FSBAzA), containing both an electrophilic fluorosulfonyl moiety and a photoactivatable azido group (3). Following stoichiometric incorporation of reagent

Fibrinogen, Thrombosis, Coagulation, and Fibrinolysis, Edited by
C. Y. Liu and S. Chien, Plenum Press, New York

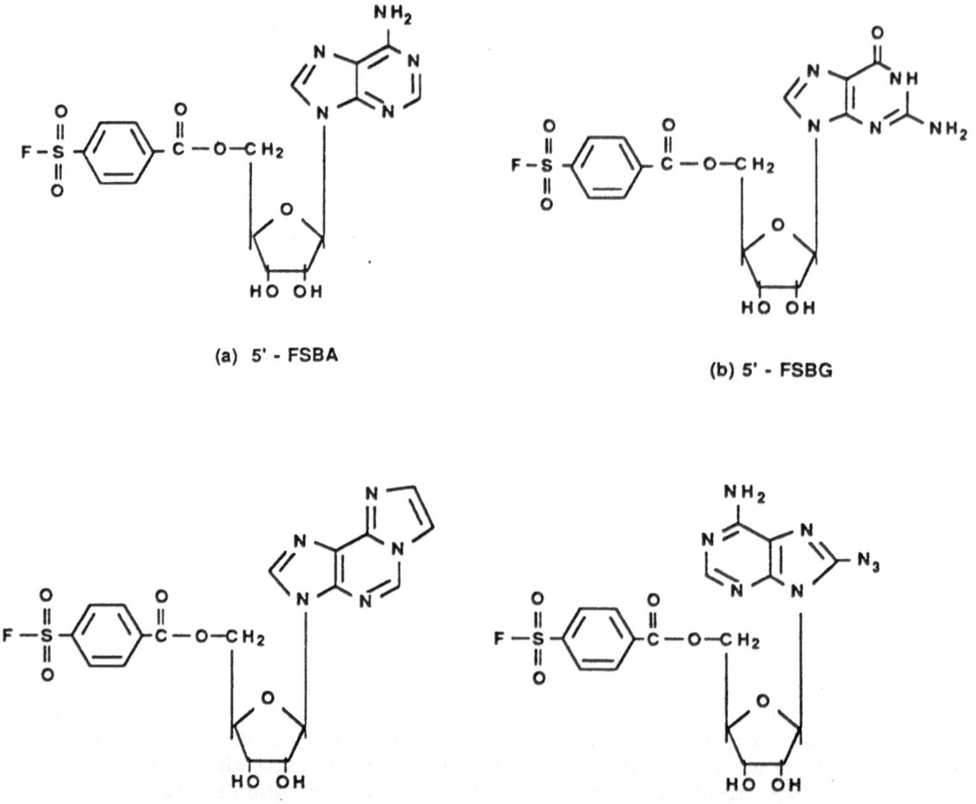

(a) 5' - FSBA

(b) 5' - FSBG

(c) 5' - FSBεA

(d) 5' - FSBAzA

Fig. 1. Fluorosulfonylbenzoyl Nucleosides.

through the fluorosulfonyl at a specific site, photolysis of the tethered molecule can lead to reaction with amino acids adjacent to the residue which is initially labeled. This reaction sequence can help to elucidate the tertiary structure of the enzyme in the region of the nucleotide site.

Two of these fluorosulfonylbenzoyl nucleosides will be used initially to illustrate the types of studies that can be conducted with purine nucleotide affinity labels. Work we have conducted on the purified enzyme bovine liver glutamate dehydrogenase will be addressed first and then our approaches to labeling aggregin, an ADP receptor protein in intact platelets, will be considered. Finally, some recent work we have carried out on human platelet cAMP phosphodiesterase using a different type of nucleotide analogue will be summarized.

Glutamate dehydrogenase catalyzes the oxidative deamination of the amino acid glutamate to form α-keto glutarate and ammonia using either NAD or NADP as its coenzyme. The activity of the bovine liver enzyme is regulated by several purine nucleotides, the most effective of which are ADP, which activates, and GTP, which inhibits. The enzyme is also inhibited by high concentrations of NADH which binds to a second non-catalytic site. Glutamate dehydrogenase is composed of six identical subunits with several nucleotide sites per subunit, including two sites for ADP, two for GTP and two for NADH (one catalytic and one regulatory). We thought that affinity labeling might provide a useful approach to distinguish among these several binding sites.

Upon incubation of glutamate dehydrogenase with 1.4 mM 5'-p-fluorosulfonylbenzoyl-1, N^6-ethenoadenosine for 200 min at pH 8 and 30°, the enzyme retains full catalytic activity as measured in the absence of any regulatory compounds; that is, the reagent does not inactivate the enzyme. It thus appears that 5'-FSBεA does not react at the active site. Native enzyme is inhibited about 90% by 1 μM GTP. A time-dependent increase is observed in the activity of the enzyme as assayed in the presence of 1 μM GTP. The rate of reacton of 5'-FSBεA with the enzyme can be monitored by this observed time-dependent desensitization to GTP inhibition. The reaction follows pseudo first order kinetics and a rate constant can be calculated from the semilogarithmic plot (4).

The effects of added substrates and regulatory compounds on the rate of reaction of glutamate dehydrogenase with 1.4 mM FSBεA have been evaluated. The rate constant is unaffected by inclusion in the reaction mixture of the substrate α-ketoglutarate or of the coenzyme NADH when these are present at concentrations high relative to their known dissociation constants. This indicates that modification by 5'-FSBεA does not occur at the active site or the NADH inhibitory site. The activator ADP also does not affect the rate constant appreciably.

The k_{obs} is decreased with increasing concentrations of GTP alone, but is most markedly altered by combinations of GTP and NADH. Indeed, complete protection is provided by saturating concentrations of these ligands, such as 25 μM GTP in the presence of 100 μM NADH. Since GTP is known to bind more tightly to glutamate dehydrogenase in the presence of reduced coenzyme, these results are consistent with reaction of 5'-FSBεA occurring at a GTP site.

The amount of reagent incorporated into the enzyme was determined by a fluorimetric technique. When excited at 310 nm, 5'-FSBεA emits light at 405 nm; however, the quantum yield is very low (about 0.01), presumably because of the stacking of the benzoyl ring on the ethenoadenosine ring system. When the ester bond between the ethenoadenosine and the fluorosulfonylbenzoyl moieties is hydrolyzed by raising the pH to 12.5, a rapid increase in fluorescence is observed because of the release of 1, N^6-ethenoadenosine, whose quantum yield is 54-fold higher than that of 5'-FSBεA. The amount of 5'-FSBεA incorporated into glutamate dehydrogenase can be determined by dissolving a known amount of the modified enzyme in 6 M urea and measuring the maximum fluorescence after raising the pH to 12.5 (5).

The extent of covalent incorporation increases with time of incubation with 5'-FSBεA and is directly proportional to the percent decrease in GTP inhibition. The data extrapolate to 1.28 moles of SBεA incorporated per subunit at 100% change in sensitivity to GTP inhibition. The catalytic and regulatory properties of modified enzyme containing about 1 mole reagent/subunit were compared with those of native enzyme to examine the effect of covalent reaction and ascertain the functional site of modification. The major difference between the two enzymes lies in their sensitivity to inhibition by GTP. Native enzyme is inhibited 96% by saturating concentrations of GTP, whereas the modified enzyme is inhibited only 70% by saturating concentrations of the inhibitory nucleotide. In addition, the affinity for GTP is about 15 times weaker in the modified enzyme. It is apparent that the sensitivity of glutamate dehydrogenase to GTP inhibition is decreased but not eliminated by reaction with 5'-FSBεA.

To determine whether FSBεA reacts at a GTP site of glutamate dehydrogenase, we measured the reversible binding of radioactive GTP to native and modified enzyme using an ultrafiltration technique. We found that native enzyme binds 2 moles of GTP in the presence of NADH, while modified enzyme binds only 1 mole of GTP. These results indicate that upon modification of glutamate dehydrogenase by 5'-FSBεA, one of the two GTP sites is eliminated. 5'-FSBεA has presumably reacted at this GTP site.

In order to identify the site on the protein which reacts with 5'-FSBεA, we prepared stable derivatives of tyrosine and lysine by reaction with the fluorosulfonylbenzoyl nucleosides. Upon acid hydrolysis, the ester linkages between the benzoyl and the nucleoside moieties are hydrolyzed and the acid stable products are carboxybenzenesulfonyl tyrosine (CBS-Tyr) and carboxybenzenesulfonyl lysine (CBS-Lys). These derivatives can be separated and quantified on an ion exchange amino acid analyzer. CBS-Lys is eluted just ahead of tyrosine, and CBS-Tyr is eluted after phenylalanine; these derivatives are thus eluted at positions distinct from each other and from standard amino acids. For glutamate dehydrogenase labeled with 5'-FSBεA, most of the incorporated reagent can be accounted for as CBS-Tyr (5).

The site of 5'-FSBεA reaction was further characterized by proteolytic digestion by trypsin and chymotrypsin of modified glutamate dehydrogenase, and fractionation of the proteolytic digest by gel filtration followed by HPLC. A single type of peptide containing modified tyrosine was isolated, with amino terminal CBS-tyrosine as determined by dansylation. This peptide has been identified as amino acids 262-265 on the basis of the known amino acid sequence of glutamate dehydrogenase, and therefore the tyrosine residue modified by 5'-FSBεA is Tyr-262 (6). This residue is thus designated as a participant in a GTP regulatory site of glutamate dehydrogenase.

Affinity labeling of purine nucleotide sites is a useful tool for tackling a very different type of problem: the probing of ADP receptor sites on the surface of an intact cell, human platelets. ADP plays two major roles in platelet activation. On one hand, ADP induces human platelets to change shape from discs to spiculated spheres, followed by platelet aggregation and storage granule secretion. On the other hand, it antagonizes the prostaglandin PGE_1 stimulation of adenylate cyclase. Action of this enzyme converts ATP to cyclic AMP. cAMP phosphodiesterase catalyzes the cleavage of cAMP to AMP. Elevation of cAMP leads to inhibition of the platelet's reaction to stimuli. Agents that increase platelet cAMP such as PGE_1, PGI_2 and adenosine, inhibit platelet shape change, adhesion, aggregation and release of granule contents. By inhibiting the stimulated adenyl cyclase, ADP reduces cyclic AMP, but this in itself cannot activate platelets. Although it is reasonable to suppose that both the effects of ADP on shape change and adenylate cyclase are mediated by the same receptor, this proves not to be the case.

Little is understood about the molecular properties of the receptors for ADP. Affinity labeling provides an approach to identifying the protein involved and to testing for the involvement of ADP in various aspects of platelet function. For this purpose, we decided to use the reactive ADP analogue, 5'-p-fluorosulfonylbenzoyl adenosine (Fig. 1a).

The utility of 5'-FSBA as a specific analogue for modification of ADP receptors on the cell surface was assessed by the ability of the compound to inhibit ADP-mediated platelet shape change from a disc to a spiny sphere. As the platelet rounds up, it absorbs more light, resulting in an increase in optical density. Washed platelets were incubated for varying time periods with 40 μM 5'-FSBA and the extent of shape change was measured on aliquots of the incubation mixture. Adenosine deaminase was included to eliminate any adenosine which might form from slow hydrolysis of 5'-FSBA. With 5'-FSBA, progressive inhibition resulted with complete inhibition at 20 min. In contrast, incubation with the solvent alone or with the guanosine derivative, 5'-FSBG, had no effect, demonstrating the specificity of 5'-FSBA.

In parallel with this experiment, we confirmed the optical density changes by direct observation using scanning electron microscopy. Platelets incubated with ADP undergo shape change as indicated by the appearance of spiny spheres with pseudopodia. Platelets incubated with the solvent alone for 40 min before being challenged with ADP still exhibit shape change. In contrast, when platelets were incubated with 5'-FSBA for 40 min and then were challenged with ADP, only 12% of the platelets show shape change as compared to 95% of those incubated with the solvent alone.

Washed intact platelets incubated with [^{3}H]-FSBA for 40 min show concentration dependent incorporation of radioactivity into platelets after extensive dialysis. The incorporation was covalent, since the platelets were solubilized in 2% sodium dodecyl sulfate, dithiothreitol and 8M urea prior to dialysis. The extent of inhibition in shape change is closely related to the covalent incorporation.

The solubilized platelets or platelet membranes prepared from the washed platelets were subjected to polyacrylamide gel electrophoresis in the presence of SDS. A single radioactive peak was found, with a molecular weight of 100,000 daltons. Labeling of this peak by radioactive 5'-FSBA was completely prevented by ADP in the reaction mixture.

Aggregation by ADP is closely related to fibrinogen binding to ADP-stimulated platelets. Almost no fibrinogen binds to intact platelets when ADP is not added. Upon exposure to ADP, [^{125}I]-labeled fibrinogen binds to platelets to an extent proportional to the concentration of fibrinogen. When platelets are preincubated with 5'-FSBA and then successively exposed to ADP and fibrinogen, no [^{125}I]-fibrinogen is bound, and shape change and aggregation are totally inhibited. Thus, the incorporation of 5'-FSBA into a single membrane protein is associated with inhibition of shape change, aggregation and fibrinogen binding.

Evidence that the action of ADP on adenylate cyclase is mediated by a different receptor that is independent of ADP-induced shape change and aggregation has been obtained using another analogue of ADP. 2-Methylthio-ADP, with a K_i of 7 nM, is 75-fold more potent in inhibiting stimulated adenylate cyclase than in causing platelet aggregation. Labeled 2-methylthio ADP binds to a single class of sites with a K_d of 10 nM. A Scatchard plot shows binding to 1500 sites per platelet when measured at concentrations of 2-methylthio-ADP below that at which platelet aggregation occurs. When platelets are incubated with 25 times the concentration of 5'-FSBA needed to block aggregation, no change in the number of binding sites for 2-methylthio-ADP is detected and there is less than a two-fold change in the dissociation constant. Thus, the affinity label 5'-FSBA provides a handle for distinguishing among several classes of ADP receptors, as well as for evaluating the participation of the 5'-FSBA-sensitive ADP receptor in various functions of intact platelets (7-9).

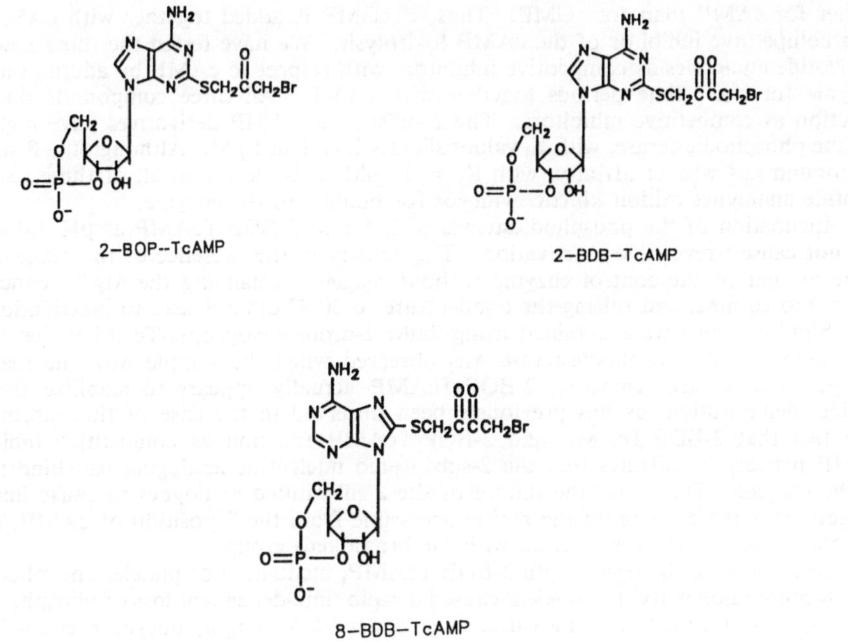

Fig. 2. Bromodioxobutyl and Bromo-oxopropyl Cyclic Nucleotides.

We have also used the approach of affinity labeling to probe the nucleotide binding site of a purified intracellular human platelet enzyme: cAMP phosphodiesterase. For this purpose, we recently synthesized three new cAMP analogues with reactive groups, shown in Fig. 2. The first of these is 2-[4-bromo-2,3-dioxobutyl]-thioadenosine 3',5'-cyclic monophosphate (2-BDB-TcAMP). This compound has the adenine, ribose and cyclic phosphate of natural cAMP. It is also water soluble and negatively charged at neutral pH. The bromoketo group is potentially reactive with several nucleophiles found in proteins, including cysteine, histidine, tyrosine, lysine, glutamic and aspartic acids. In addition, the dioxo group lends the possibility of reaction with arginine residues. Because of the location of the functional group adjacent to the 2 position, the compound might be expected to react with amino acid residues in the purine region of the cAMP binding sites of enzymes.

The second compound is 2-[3-bromo-oxopropyl]-thioadenosine 3',5'-cyclic monophosphate (2-BOP-TcAMP). This analogue is structurally identical to the first compound except that the reactive side chain is shorter by one carbonyl group. A comparison between the effectiveness of affinity labeling of an enzyme by the two top compounds in principle can yield an estimate of distance between the target amino acid and the enzyme-bound nucleotide.

The third compound is 8-[4-bromo-2,3-dioxobutyl]-thioadenosine 3',5'-cyclic monophosphate (8-BDB-TcAMP). In this analogue, the reactive BDB group is at a different position of the purine ring. The two BDB-derivatives can be used to probe distinct subregions of a cAMP binding site in an enzyme. These three compounds have now been evaluated for their potential to function as affinity labels of the active site of cAMP phosphodiesterase (10).

The enzyme that we have used is the high affinity cAMP phosphodiesterase which was purified to homogeneity from human platelets. This enzyme, when freshly prepared, exhibits a single band of 110,000 daltons upon polyacrylamide gel electrophoresis in the presence of SDS, and this is considered to be the molecular weight of the enzyme monomer. A divalent metal cation is required for activity, and this requirement can be satisfied by Mg^{++} or Mn^{++}, but not by Ca^{++}. None of the amino acid residues at the active site has yet been identifid, making this phosphodiesterase an excellent candidate for affinity labeling.

Phosphodiesterase hydrolyzes both cAMP and cGMP to their corresponding 5'-monophosphates. Although thc enzyme exhibits high affinity for both compounds (with K_m values of 0.18μM and 0.02 μM, respectively), the maximum rate of hydrolysis is 10 times higher for cAMP than for cGMP. Thus, if cGMP is added together with cAMP, it acts as a competitive inhibitor of the cAMP hydrolysis. We have tested the three new reactive nucleotide analogues as competitive inhibitors with respect to cAMP by adding them to the enzyme for short time periods together with cAMP. All three compounds do, in fact, function as competitive inhibitors. The 2-substituted cAMP derivatives have high affinity for the phosphodiesterase, with K_i values slightly less than 1 μM. Although the 8-substituted compound has weaker affinity, with K_i = 10 μM, it is clear that all of these reactive nucleotide analogues exhibit kinetic evidence for binding to the enzyme.

Incubation of the phosphodiesterase with 5 mM 2-BDB-TcAMP at pH 7.0 and 24°C did not cause irreversible inactivation. The activity in the presence of the reagent was the same as that of the control enzyme without reagent. Changing the Mg^{++} concentration from 0 to 20 mM, and raising the temperature to 30°C did not lead to inactivation.

Similar results were obtained using 5mM 2-bromo-oxopropyl-TcAMP; that is, no inactivation of the phosphodiesterase was observed when the sample with the reagent was compared to control enzyme. 2-BOP-TcAMP actually appears to stabilize the enzyme against denaturation, as has previously been observed in the case of the natural cAMP. The fact that 2-BDB-TcAMP and 2-BOP-TcAMP function as competitive inhibitors of cAMP hydrolysis indicates that the 2-substituted nucleotide analogues can bind reversibly to the enzyme. Therefore, the failure of the 2-substituted analogues to cause inactivation suggests that the enzyme, in the region accessible from the 2 position of cAMP, lacks nucleophiles appropriate for reaction with the bromoketo group.

In contrast to the results with 2-BDB-TcAMP, incubation of platelet phosphodiesterase with 8-bromodioxobutyl-thio-cAMP caused a rapid time-dependent loss of phosphodiesterase activity. The reaction rate, measured over a period of 5 min, obeyed pseudo-first order kinetics and was dependent on reagent concentration. Calculation of the second-order rate constant yielded a value of 31 min^{-1} M^{-1}. The inactivation is due to an irreversible, covalent reaction, since no recovery of activity was observed either upon dilution of the enzyme-reagent incubation mixture, or upon gel filtration to remove all the non-covalently bound reagent.

If the 8-BDB-TcAMP is inactivating the phosphodiesterase by reaction at its active site, the normal substrates and/or products should protect that enzyme against inactivation. Addition to the reaction mixture of either the substrates, cGMP or cAMP, or the product, 5'-AMP, resulted in a marked decrease in the inactivation rate (10). In contrast, NAD, which is not a normal ligand for this enzyme, did not protect against inactivation. These results suggest that the inactivation of the cAMP phosphodiesterase by 8-BDB-TcAMP is occurring by reaction at the active site of the enzyme.

The distinction between the effects of the reactive 2-substituted and 8-substituted nucleotides in causing inactivation of the phosphodiesterase may relate to the predominant conformation of the nucleotide when in solution. Nucleotides featuring substituents at the 2 position of the purine ring are known to exist predominantly in the *anti* conformtion about the purine-ribose bond; whereas nucleotides with substituents at the 8 position exist predominantly in the *syn* conformation (Fig. 2). Even if the two nucleotide analogues bind essentially to the same substrate site normally occupied by cAMP, it is likely that different amino acid side chains will be available to bromoketo groups which are tethered to either the 2 or 8 position of the purine ring. Thus, the availability of these complementary nucleotide analogues may allow systematic probing of the protein environment in the region surrounding the bound cAMP.

We have previously shown that adenosine diphosphate analogues with bromodioxobutyl groups at the 2, 6 and 8 positions of the purine ring function as affinity labels for several dehydrogenases and kinases (11, 12). In these cases, discretely modified cysteine, histidine, tyrosine and aspartate residues have been identified. It is likely that the new cAMP affinity labels will prove useful in probing the structure of the active sites of the various isozymes of cAMP phosphodiesterase, as well as of the cAMP sites of other proteins.

These examples are representative of the types of studies that we are engaged in using purine nucleotide affinity labels. These various analogues should not only be useful for our own experiments, but should also be valuable to other laboratories in exploring nucleotide sites in a variety of proteins. The 5'-p-fluorosulfonylbenzoyl adenosine which was the first of the reactive nucleotide analogues that we described, has already been found to yield specific labeling of NAD sites in several dehydrogenases and reductases. It has also labeled ATP or ADP binding sites in a large number of kinases and synthetases, in addition to providing an effective handle for examining and ADP receptor protein of platelet membranes. Furthermore, it has modified specific nucleotide sites in such diverse proteins as the ATPases, actin, myosin, luciferase and oxo-prolinase (13). We anticipate that the FSB derivatives of guanosine, ethenoadenosine and azidoadenosine, as well as the new 2- and 8-bromodioxobutyl-thio-cAMP will similarly have widespread applications in the elucidation of purine nucleotide sites in enzymes and receptors.

ACKNOWLEDGMENTS

Finally, I would like to acknowledge the major efforts of my colleagues in the work I have described here. My colleagues at the University of Delaware on the synthesis and characterization of the nucleotide analogues, as well as the work on glutamate dehydrogenase were Drs. Jerome Bailey, Dianne DeCamp and Marlene Jacobson. My collaborators at Temple University on the platelet studies were Dr. Robert W. Colman, William Figures, David Mills, Stefan Niewiarowski and Gwen Stewart.

REFERENCES

1. R. F. Colman, *Ann. Reviews of Biochem.*, **52**:67 (1983).
2. J. J. Likos, and R. F. Colman, *Biochemistry,* **20:491** (1981).
3. K. E. Dombrowski, and R. F. Colman, *Arch. Biochem. Biphys.*, **275**:302 (1989).
4. M. A. Jacobson, and R. F. Colman, *Biochemistry,* **21**:2177 (1982).
5. M. A. Jacobson, and R. F. Colman, *Biochemistry,* **22:4247** (1983).
6. M. A. Jacobson, and R. F. Colman, *Biochemistry,* **23**:6377 (1984).
7. D. C. B. Mills, W. R. Figures, L. M. Scearce, G. J. Stewart, R. F. Colman, and R. W. Colman, J. *Biol. Chem.*, **260**:8078 (1985).
8. R. W. Colman, R. N. Puri, F. Zhou, and R. F. Colman, *in:* "Platelet Membrane Receptors" ed., G. A. Jamison, Alan R. Liss, Inc., New York. 1988, p. 263.
9. R. W. Colman, W. R. Figures, Q.-X. Wu, T. A. Morinelli, G. Tuszynski, R. F. Colman, and S. Niewiarowski, *Arch. Biochem. Biophys.*, **262**:298 (1988).
10. P. Grant, D. L. DeCamp, J. M. Bailey, R. W. Colman, and R. F. Colman, *Biochemistry,* **29**:887 (1990).
11. D. L. DeCamp, and R. F. Colman, *J. Biol. Chem.*, **264**:8430 (1989).
12. Y.-C. Huang, and R. F. Colman, *J. Biol. Chem.*, **264**:12208 (1989).
13. R. F. Colman, *in:* "Proteins: Structure and Function," ed. J. L'Italien. Plenum Press, N.Y. (1987), p. 569.

THE EFFECT OF GANODERMIC ACID S ON HUMAN PLATELETS

Chuen-Neu Wang, Jia-Chyuan Chen, Ming-Shi Shiao, and Cheng-Teh Wang

Institute of Life Science, National Tsing Hua University
Hsinchu, Taiwan, ROC

SUMMARY

Incubation of gel-filtered human platelets in ganodermic acid S (lanosta-7,9(11), 24-trien-3ß, 15α-diacetoxy-26-oic acid) showed that uptake of the agent by platelets was a simple diffusion process. The agent caused platelet aggregation at concentrations above 20 μM. Above the threshold, the extent of cell aggregation was in a linear relationship to the agent concentration. Below the aggregation threshold, platelets showed neither the resynthesis of [^{32}P] phosphatidylinositol 4,5-bisphosphate ([^{32}P]PIP_2) and [^{32}P] phosphatidylinositol 4-phosphate ([^{32}P]PIP) nor the accumulation of [^{32}P] phosphatidic acid ([^{32}P]PA). The results suggested that ganodermic acid S caused the activation of PIP_2 hydrolysis. Scanning electron microscopy revealed that the morphology of platelets below the aggregation threshold appeared to be spiculate discoid shape. Above the threshold, the cells rounded up to spiculate irregular forms.

INTRODUCTION

Ganodermic acid S (lanosta-7,9(11), 24-trien-3ß-15α-diacetoxy-26-oic acid) is the major triterpenoid purified from *Ganoderma lucidum* (Fr.) Karst., which is a widely used Polyporaceae in traditional Chinese medicine (1). It has been reported that the structure analogues of ganodermic acid S can lower the cellular content of cholesterol as well as inhibit the growth of cultured hepatoma cells (2, 3). Ganodermic acid S is an amphipathic molecule with a chemical structure similar to those of detergents cholate and deoxycholate. We have shown that the incorporation of deoxycholate into platelet membrane results in an inhibition of platelet function (4). Ganodermic acid S may serve as a membrane-acting agent as well. The effects of ganodermic acid S on human platelets have been investigated in terms of 1) the uptake of ganodermic acid S by platelets; 2) the effect on biochemical events by comparison with that of thrombin stimulation; and 3) the effect on cell morphology as revealed by scanning electron microscopy (scanning EM). The results indicate that ganodermic acid S is a platelet membrane-acting agent (5).

METHODS

Gel-filtered human platelets were prepared from freshly drawn blood of healthy donors according to the method of Lages et al. (6). Platelets were incubated with various concentrations of ganodermic acid S for a certain period, and then centrifuged through a layer of silicon oil in a Laboratory centrifuge (Sigma, Model 202 CM, F.R.G.) at 13,500 x g for 1 min. The supernatant was taken to measure the remaining concentration of ganodermic

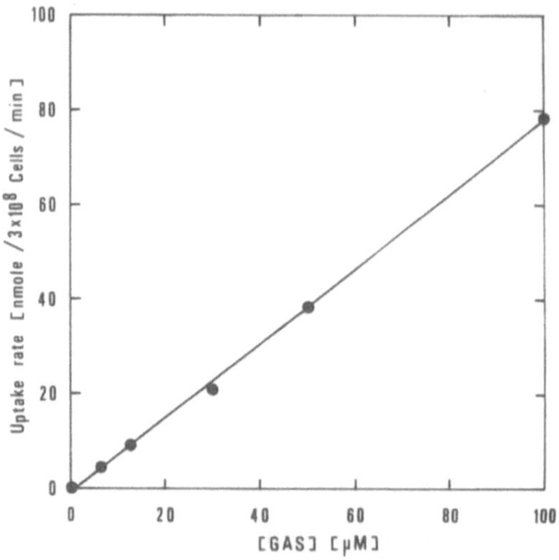

Fig. 1. A plot of the rate of uptake by platelets vs. the concentration of ganodermic acid S (GAS).

acid S in a spectrophotometer at 243 nm (Beckman, Model DU-70, U.S.A.). The sample preparation for scanning EM study was detailed in the report of Tsai et al. (7). The extent of platelet aggregation was observed in an aggrecorder (Daiichi, Model PA-3210, Japan). The study of protein phosphorylation was carried out by a modification of that described by Sano et al. (8). Fura-2/AM was employed as a probe for the estimation of the change in concentration of cytosolic free calcium ion ($[Ca^{++}]_i$). The experiment was performed according to the method of Pollock et al. with some modifications (9). Secretion of serotonin was assayed as described by Drummond and Gordon (10). Analysis of the changes in [^{32}P]-phosphoinositides and [^{32}P]PA was performed according to the method described by Holmsen et al. (11).

RESULTS

The uptake of ganodermic acid S by human gel-filtered platelets

Fig. 1 depicts a plot of the rate of uptake by platelets vs. the external concentration of ganodermic acid S. It appears to be a linear relationship. Hence, the uptake process of ganodermic acid S by platelets is a simple diffusion. If one assumes that the size of platelets in the population was homogeneous as a discoid shape with 3 μm in major diameter and 0.5 μm in thickness, then the ratio for the agent concentration was about 3×10^3 in platelets to that remaining in the aqueous phase. Therefore, the partition coefficient of ganodermic acid S in the platelet membrane should be several hundred folds of 3×10^3.

Effect on the metabolism of phosphoinositides by ganodermic acid S

Platelets appeared to aggregate in ganodermic acid S of above 20 μM. Above the threshold concentration, the extent of platelet aggregation was in a linear relationship to the concentration of ganodermic acid S. Also, platelets in various concentrations of ganodermic acid S appeared to show a concentration-dependent increase in the level of: 1) $[Ca^{++}]_i$, 2) phosphorylation of 47K and 20K proteins, and 3) serotonin release (data not shown). It implies that the incorporation of ganodermic acid S into platelet membrane might activate the turnover of phosphoinositides.

Under thrombin stimulation, the time-course profiles showed: 1) a continuous increase in [^{32}P]PA (Fig. 2A); 2) a gradual decrease in [^{32}P]phosphatidylinositol ([^{32}P]PI) (Fig. 2B); 3) a gradual increase in [^{32}P]PIP up to 1 min (Fig. 2C); and 4) an initial decrease

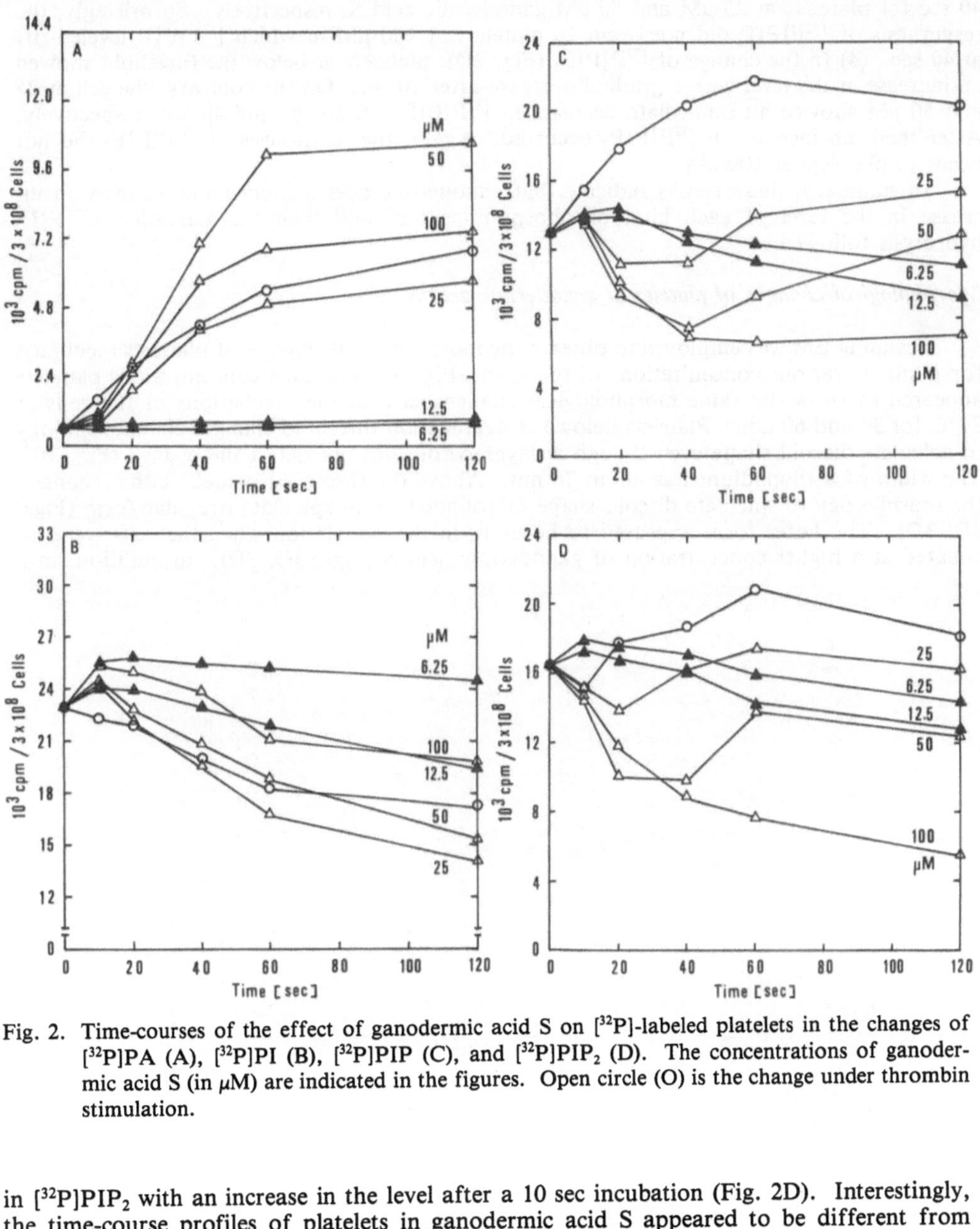

Fig. 2. Time-courses of the effect of ganodermic acid S on [^{32}P]-labeled platelets in the changes of [^{32}P]PA (A), [^{32}P]PI (B), [^{32}P]PIP (C), and [^{32}P]PIP$_2$ (D). The concentrations of ganodermic acid S (in μM) are indicated in the figures. Open circle (O) is the change under thrombin stimulation.

in [^{32}P]PIP$_2$ with an increase in the level after a 10 sec incubation (Fig. 2D). Interestingly, the time-course profiles of platelets in ganodermic acid S appeared to be different from those of control. In addition, the time-course patterns of platelets at below the aggregation threshold were different from those found at above the threshold. Specifically, (1) platelets at below the threshold showed no significant change in [^{32}P]PA (Fig. 2A). However, above the threshold, the cells showed several fold increase in [^{32}P]PA. Interestingly, at the first 10 sec, the rate of increase in [^{32}P]PA was slower than that increase observed under thrombin stimulation. After 10 sec incubation, the rate of increase in [^{32}P]PA was faster than that in thrombin. After 2 min incubation, the level of [^{32}P]PA in platelets at 25 μM was half of that in the cells at 50 μM. However, the level of [^{32}P]PA in platelets at 100 μM appeared to be 60% of that at 50 μM. (2) in the change of [^{32}P]PI (Fig. 2B), platelets at various concentrations of ganodermic acid S all showed an initial increase in the level with a decrease at a prolonged incubation. The rate of decrease was dependent on the agent concentration. Interestingly, the decreasing rate of [^{32}P]PIP appeared up to 10 sec with the appearance of a drastic decrease in [^{32}P]PIP. Then, another increase of [^{32}P]PIP showed up at 20 sec and

40 sec for platelets at 25 μM and 50 μM ganodermic acid S, respectively. Surprisingly, the resynthesis of [^{32}P]PIP did not occur to platelets at 100 μM in which [^{32}P]PIP leveled off at 40 sec. (4) In the change of [^{32}P]PIP_2 (Fig. 2D), platelets at below the threshold showed an increase in the level with a gradual decrease after 10 sec. On the contrary, the cell at 25 and 50 μM showed an immediate decrease in [^{32}P]PIP_2 till 20 sec and 40 sec, respectively. After then, an increase in [^{32}P]PIP_2 occurred. Again, the resynthesis of [^{32}P]PIP_2 did not occur to platelets at 100 μM.

In summary, these results indicate that ganodermic acid S immediately caused an increase in the level of each kind of phosphoinositide, and then the activation of PIP_2 hydrolysis followed.

Morphological changes of platelet in ganodermic acid S

Scanning EM was employed to observe the morphological changes of platelets incubated for 1 min at various concentrations of the agent (Fig. 3). For each concentration, platelets appeared to show the same morphological change between the incubations of the cells at 37°C for 30 and 60 min. Platelets below the aggregation threshold changed the morphology to spiculate discoid shape even though an aggrecorder did not detect the change (Fig. 3B). The width of a filopodium was about 70 nm. Above the threshold, platelet either changed the morphology to spiculate discoid shape or rounded up to spiculate irregular form (Figs. 3C, 3D). The latter form appeared to be more in the population when the cells were incubated at a higher concentration of ganodermic acid S (Figs. 3C, 3D). In addition, in a

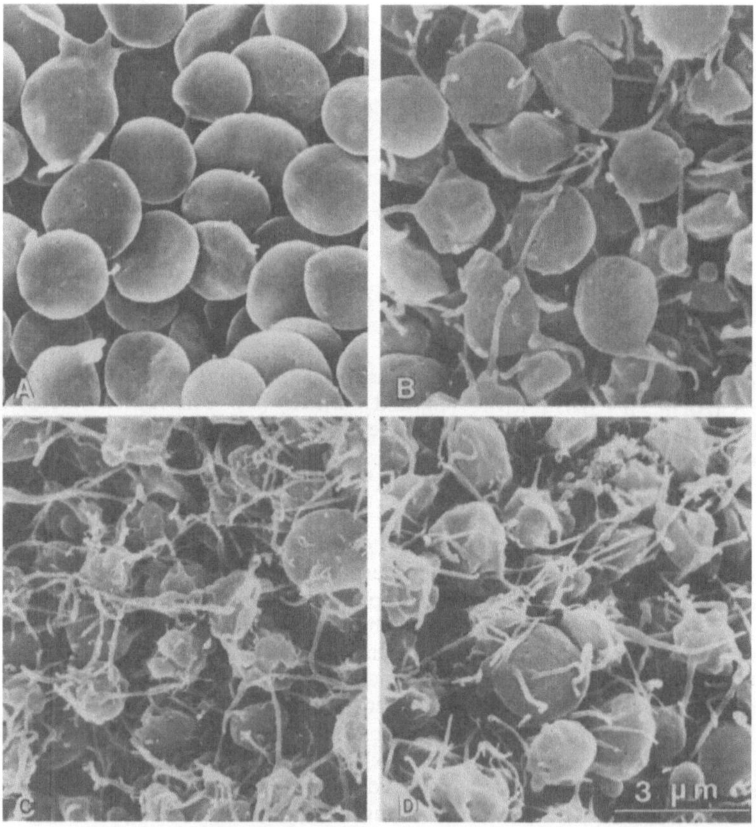

Fig. 3. Scanning electron micrographs of gel-filtered platelets in various concentrations of ganodermic acid S (X 9,000). Gel-filtered platelets were preincubated at 37°C for 60 min (A), and then incubated for 1 min with ganodermic acid S of 12.5 μM (B), 25 μM (C), and 50 μM (D).

higher concentration of ganodermic acid S, each deformed cell appeared to have a higher number of filopodia, which were longer in length (Fig. 3C, 3D).

DISCUSSION

The study indicates that ganodrmic acid S is a membrane-acting agent, since: 1) the agent is highly soluble in platelet membrane with a partition coefficient of about 10^5; and 2) it causes platelet membrane spiculation and vesiculation (Figs. 3). This kind of change in membrane morphology also occurs to platelets in anionic detergents dodecylsulfate and deoxycholate (4). According to the explanation of the bilayer couple hypothesis, the spiculation of platelet membrane may mean that the distribution of ganodermic acid S in membrane bilayer is more in the outer leaflet than in the inner one (4, 12).

The study indicates that the incorporation of ganodermic acid S into platelet membrane may exert three kinds of effect on the interconversion of phosphoinositides (Fig. 2). Firstly, the infiltration of ganodermic acid S into platelet membrane immediately causes an elevation of [^{32}P]PI from the other PI pool. Secondly, the phosphoinositide interconversion is turned on in the condition of platelets at either 25 or 50 μM ganodermic acid S. Thirdly, the interconversion of phosphoinositides may be partially blocked for platelets at a higher concentration as 100 μM ganodermic acid S. Therefore, the infiltration of various concentrations of ganodermic acid S into platelet membrane results in differential effect on the metabolism of phosphoinositides.

The study by scanning EM reveals that platelets below the aggregation threshold change the morphology to spiculate disc without change in $[Ca^{2+}]_i$ (data not shown). Hence, the formation of filopodia in ganodermic acid S may be a direct effect by the agent. Above the threshold, the change in cell volume is due to the elevation of $[Ca^{2+}]_i$, which results in the reorganization of cytoskeleton (13-15). Hence, the membrane spiculation caused by ganodermic acid S is a separate phenomenon from platelet aggregation in response to ganodermic acid S.

In summary, the study shows that the insertion of ganodermic acid S into platelets affects not only the platelet membrane morphology but also the metabolic pool of phosphoinositides in the plasma membrane.

ACKNOWLEDGMENTS

This research was supported by a grant to C.-T. Wang from the National Science Council of Republic of China (NSC79-0203-B007-03).

REFERENCES

1. M.-S. Shiao, and L.-J. Lin, *J. Nat. Prod.,* **50(5):**886-890 (1987).
2. F. R. Taylor, S. E. Saucier, E. P. Shown, E. J. Parish, and A. A. Kandutsch, *J. Biol. Chem.,* **259:**12382-12387 (1984).
3. J. O. Toth, B. Luu, and G. Ourisson, *Tetrahedron Lett.,* **24:**1081-1084 (1983).
4. Y.-J. Shiao, J.-C. Chen, and C.-T. Wang, *Biochim. Biophys. Acta,* **980:**56-68 (1989).
5. C.-N. Wang, J.-C. Chen, M.-S. Shaio, and C.-T. Wang, *Biochim. Biophys. Acta,* **986:**151-160 (1989).
6. B. Lages, M. C. Scrutton, and H. Holmsen, *J. Lab. Clin. Med.,* **85:**811-821 (1975).
7. W.-J. Tsai, J.-C. Chen, and C.-T. Wang, *Biochim. Biophys. Acta,* **940:**105-120 (1988).
8. K. Sano, Y. Takai, J. Yamanishi, and Y. Nishizuka, *J. Biol. Chem.,* **258:**2010-2013 (1983).
9. W. K. Pollock, T. J. Rink, and R. F. Irvine, *Biochem. J.,* **235:**869-877 (1986).
10. A. H. Drummond, and J. L. Gordon, *Thromb. Diath. Haemorrh.,* **31:**336 (1974).
11. H. Holmsen, C. A. Dangelmaier, and S. Rongved, *Biochem. J.,* **222:**157-167 (1984).
12. J. E. Ferrell Jr., K. T. Mitchell, and W. H. Huestis, *Biochim. Biophys. Acta,* **939:**223-237 (1988).
13. A. C. Cox, R. C. Carroll, J. G. White, and G. H. R. Rao, *J. Cell Biol.,* **98:**8-15 (1984).
14. J. E. B. Fox and D. R. Phillips, *Semin. Hematol.,* **20:**243-260 (1983).
15. L. C. Gershman, L. A. Selden, and J. E. Estes, *Biochem. Biophys. Res. Commun.,* **135:**607-614 (1986).

ROLE OF INSULIN RECEPTORS IN THE EXPRESSION OF PROSTAGLANDIN E_1 BINDING ACTIVITY IN PLATELETS

Nighat N. Kahn and A. Kumar Sinha

Division of Cardiology
Department of Medicine, Montefiore Medical Center
Albert Einstein College of Medicine
Bronx, NY 10467, USA

INTRODUCTION

Aggregation of platelets induced by agonists like thrombin, ADP, 1-epinephrine or collagen is believed to be critically important not only in the normal blood coagulation process (1), but platelet hyperactivity has been shown to be crucially important in the pathogenesis of acute ischemic heart disease (2). Studies have demonstrated increased platelet released products (3), aggregates of platelets in the circulation (4), and enhanced aggregation of platelets (2) in patients with acute myocardial infarction or unstable angina. The aggregation of platelets is counterbalanced by several humoral factors which include prostanoids like prostacyclin (PGI_2) and prostaglandin E_1 (PGE_1). These prostaglandins inhibit platelet aggregation by increasing intracellular cyclic AMP level through the activation of adenylate cyclase in these cells (5), and are believed to play a significant role in the prevention of thrombosis and atherosclerosis.

Studies of patients with diabetes mellitus have shown a close association of the disease with the incidence of premature cardiovascular complication when compared to the non-diabetic population (6). Since diabetes mellitus is a disease due to absolute or relative insulin deficiency in the system, these studies implicate the hormone as a humoral factor capable of regulating platelet function. However, no direct effect of insulin on platelet aggregation could be demonstrated.

RESULTS AND DISCUSSION

To determine the effect of insulin on platelet prostaglandin E_1 (PGE_1) receptor activity, we incubated platelet-rich plasma (PRP) with physiologic concentrations of insulin (post-prandial) for 3 hours at 23°C and the binding of [^{3}H]-PGE_1 to the washed platelets was determined (7) (Fig. 1). Treatment of PRP with increasing concentrations of insulin (0.175-2.8 nM) increased the binding of the prostanoid, and at 0.7 nM (100 μU/ml) of the hormone concentration, the specific binding of the autacoid was maximally increased by approximately 2.5 fold when compared to control. Substitution of insulin by other hormones including somatotropin, epidermal growth factor or adrenocorticotropin A showed no effect on the prostanoid binding. The time course of ^{125}I-insulin and [^{3}H]-PGE_1 binding to intact platelets demonstrated that the increased binding of the protein hormone to these cells resulted in the increased binding of the prostanoid (Fig. 2).

Scatchard plots (8) of the binding of [^{3}H]-PGE_1 to platelets in the presence and absence of insulin showed that the increased prostanoid binding by the hormone was due to the increase of prostaglandin receptor numbers without the changes of affinity (Fig. 3). In control platelet preparation, the binding of [^{3}H]-PGE_1 showed the presence of one high

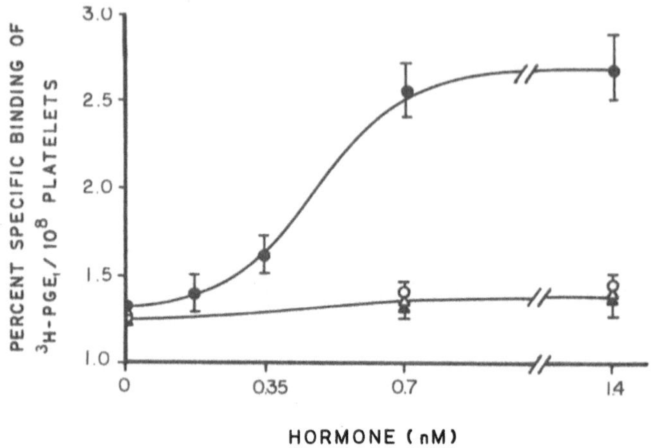

Fig. 1. Stimulation of [^{3}H]-PGE$_1$ Binding to Platelets by Insulin. PRP from normal volunteers (n = 6) prepared according to the method described before (16) were incubated with various concentrations of either insulin (●), or epidermal growth factor (○), or somatotropin (△), or adrenocorticotropin (▲) for 3 h at 23°C. After incubation platelets were separated by gel filtration and incubated with 3 nM [^{3}H]-PGE$_1$ (≈ 60,000 dpm) in total volume of 200 μl in Tyrode's buffer containing 5 mM $MgCl_2$ for 20 min to attain equilibrium (17). The platelet suspension was then filtered over a Whatman glass fiber filter (GF/C), presoaked in the same buffer under mild vacuum and washed twice with 5.0 ml of the buffer. The platelets were adsorbed on the filters which were than dried and the radioactivity was determined in a liquid scintillation counter (7). The nonspecific binding was measured by adding 15 μM unlabelled PGE$_1$. The specific binding was calculated as described before (7). Results shown are mean ± S. E. of six experiments in triplicate (Ref. 7, with permission of the publisher).

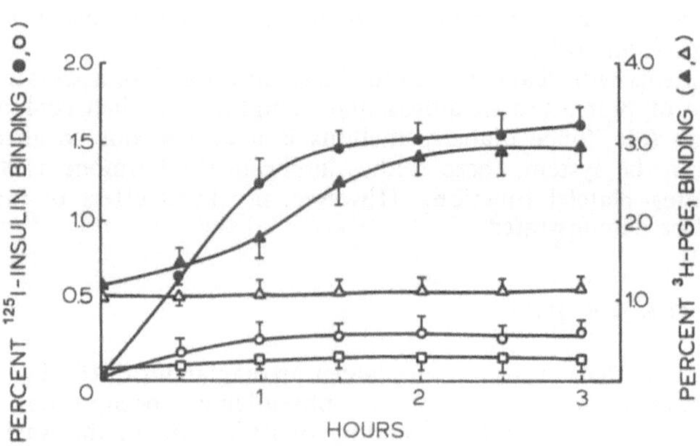

Fig. 2. Time Course of Binding of ^{125}I-Insulin and [^{3}H]-PGE$_1$ to Platelets. PRP was incubated with 0.7 nM ^{125}I-insulin for 2.5 h at 23°c. After incubation for various times platelet suspension was filtered over a Whatman glass fiber filter (GF/C) to separate the free ligand from the mixture as described in Fig. 1. The radioactivity of the filters was then determined in a gamma counter. Nonspecific binding was determined by adding 0.7 μM unlabelled insulin to the platelet suspension. This value was substracted from the total ^{125}I-insulin binding to obtain the specific binding. The solid circle (●) indicate the total ^{125}I-insulin binding and the hollow circles (○) indicate the nonspecific binding. The total [^{3}H]-PGE$_1$ binding is shown by solid triangle (▲) and the nonspecific prostaglandin binding is represented by hollow squares (□). The hollow triangles (△) show the binding of [^{3}H]-PGE$_1$ to platelets. Results shown are the mean ± S. E. of three experiments each in triplicate (Ref. 7, with permission of the publisher).

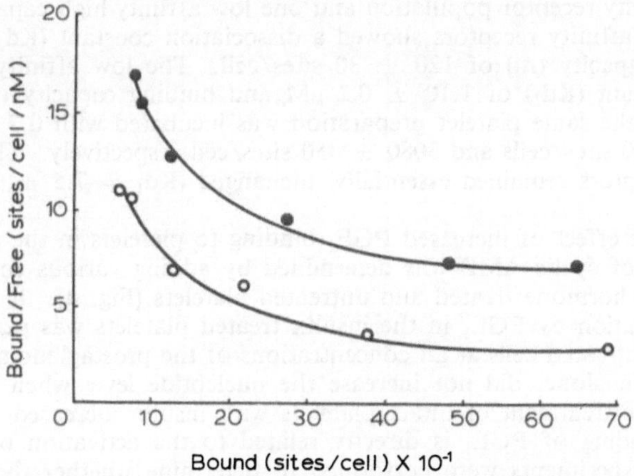

Fig. 3. Scatchard Plots of the Binding of [^{3}H]-PGE_1 to Platelets Incubated in the Presence and Absence of Insulin. PRP was incubated in the presence or absence of insulin (0.7 nM) as described above. The platelets were then isolated and the binding of [^{3}H]-PGE_1 was determined by adding 0-15 μM PGE_1 containing 3 nM [^{3}H]-PGE_1 in Tyrode's buffer as described in Fig. 1. Scatchard plots of the [^{3}H]-PGE_1 binding to platelets were analyzed by computer as described before (17). (●), represents the binding of [^{3}H]-PGE_1 in the presence of insulin and (○), indicates control PRP.

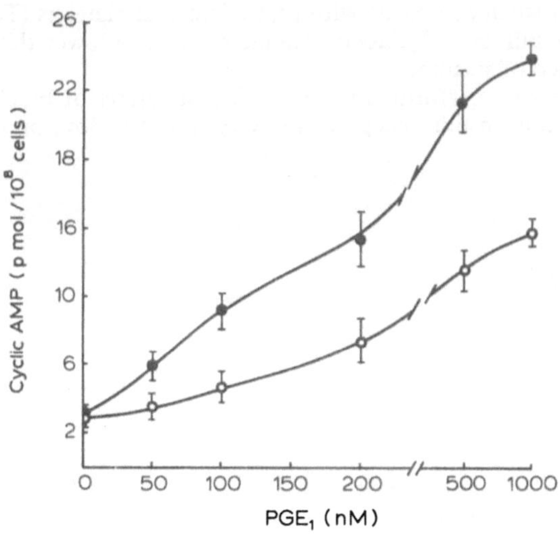

Fig. 4. Stimulation of Cyclic AMP Formation in Platelets by PGE_1 in the Presence of Insulin. PRP was incubated with 0.7 nM insulin for 2.5 h at 23°C. After incubation 10 mM theophylline was added and incubated for 2 min at 23°C, followed by the addition of various amounts of PGE_1 as indicated. The reaction was terminated by adding ice-cold trichloroacetic acid (5%) and the concentration of cyclic AMP in the trichloroacetic acid extract was determined (18). Insulin + PGE_1 (●); PGE_1 alone (○). Results are the mean ± S.E. of three experiments (Ref. 7, with permission of the publisher).

affinity-low capacity receptor population and one low affinity-high capacity receptor population. The high affinity receptors showed a dissociation constant (Kd_1) of 9.0 ± 1.2 nM with a binding capacity (n_1) of 120 ± 30 sites/cell. The low affinity receptors had the dissociation constant (Kd_2) of 1.10 ± 0.2 μM and binding capacity (n_2) of 1400 ± 250 sites/cell. When the same platelet preparation was incubated with 0.7 nM insulin, n_1 and n_2 increased to 260 sites/cells and 3080 ± 450 sites/cell respectively. The dissociation constants of the receptors remained essentially unchanged (Kd_1 = 7.5 ± 1.0 nM and Kd_2 = 0.8 ± 0.20 μM).

To assess the effect of increased PGE_1 binding to platelets in the presence of insulin, the cellular level of cyclic AMP was determined by adding various concentrations of the prostanoid to the hormone treated and untreated platelets (Fig. 4). It was found that the cyclic AMP formation by PGE_1 in the insulin treated platelets was persistently 2-2.5 fold higher than the untreated cells at all concentrations of the prostaglandin tested. Treatment of PRP by insulin alone, did not increase the nucleotide level when compared with the control. Since the treatment of intact platelets with insulin increased [^{3}H]-PGE_1 binding, and since the binding of PGE_1 is directly related to the activation of adenylate cyclase in platelets (9), experiments were performed to determine whether the insulin stimulated prostanoid binding would also stimulate the activation of adenylate cyclase. Treatment of PRP with increased amounts of insulin (0.175-14 nM) was found to result in the increased activation of adenylate cyclase by the prostaglandin when compared with the control (Fig. 5). At 0.7 nM insulin concentration, the platelet membrane adenylate cyclase was maximally stimulated by more than 2-fold by 0.1 μM PGE_1 compared to the untreated membranes.

Since the biological effects of insulin is mediated through the insulin receptor interaction in target tissues (10), the relation between the binding of insulin and prostaglandin E_1 to their respective receptors on the cell surface was next investigated using platelets from the patients with noninsulin-dependent diabetes mellitus (NIDDM). Scatchard analysis of the binding of [^{125}I]-insulin to normal platelets showed that these cells contain 96 ± 10 fmol/10^8 cells with dissociation constant of 1.45 ± 0.30 nM (Fig. 6). In contrast, the same analysis showed that platelets from the patients with NIDDM contained significantly ($P < 0.001$) lower number of insulin receptors (56 ± 12 fmol/10^8 cells) with similar dissociation constant (1.58 ± 0.41 nM) (Fig. 6). Platelets from the diabetic patients showed not only lower number of insulin binding sites/platelets, these cells also showed lower number of PGE_1 receptors/cell when compared with that of normal platelets (Table 1). The stimulation of PGE_1 receptor number of diabetic platelets was also lower than that of the normal platelets in the presence of insulin.

Experiments were next performed to determine the effect of blocking insulin receptors in normal platelets by anti-insulin receptor antibody (IgG fraction, prepared from the serum

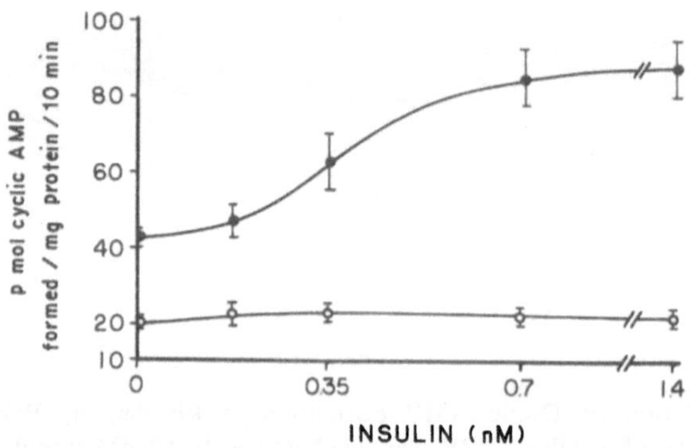

Fig. 5. Activation of Adenylate Cyclase by PGE_1 in Membranes from Platelets Incubated in the Presence or Absence of Insulin. PRP was incubated in the presence of insulin (0.7 nM) as described above. After incubation platelet membrane were prepared and the activation of adenylate cyclase by PGE_1 (100 nM) was determined (9). ●, PRP + insulin; ○, PRP alone. Results are the mean ± S.E. of six experiments (Ref. 7, with permission of the publisher).

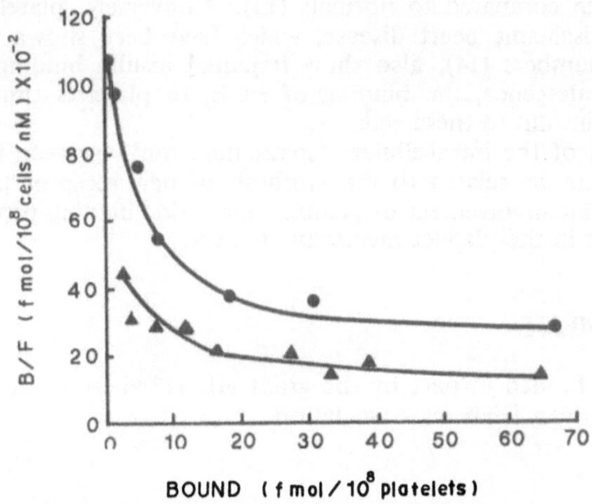

Fig. 6. Scatchard Plots of the Binding of ^{125}I-Insulin to Platelets from Normal Volunteers and NIDDM Patients. The binding of ^{125}I-insulin to platelets from normal volunteers and the diabetic patients was determined by the method described before (7). Each point represents the mean of 10 experiments each in triplicate in normal platelets (●), and 7 experiments, also in triplicate, in diabetic patients (▲).

of a diabetic patient) on the subsequent [^{3}H]-PGE_1 binding to these cells. It was found that incubation of normal PRP with the antibody (80 μg/ml) for 3 hours at 23°C not only blocked the binding of insulin to these cells (46 ± 5 fmol/10^8 cells), but also resulted in the inhibition of PGE_1 binding (160 ± 10 fmol/10^8 cells). In contrast, normal human IgG up to 100 μg/ml under similar conditions, did not inhibit the binding of either ^{125}I-insulin (110 ± 20 fmol/108 cells) or [^{3}H]-PGE_1 (265 ± 26 fmol/10^8 cells) when added to PRP.

These studies demonstrate that insulin at physiologic concentration increases the binding of PGE_1 to platelets and consequently increases the sensitivity of these cells to the prostanoid through increased formation of cyclic AMP. However, it is not the hormone alone that increases the binding of PGE_1 to platelets, the prostanoid also increases the binding of insulin to platelets (11) and to erythrocyte membranes (12). These results, taken together, indicate the possible existence of intracellular "communication" between the receptors of insulin and PGE_1 in that the interaction of one of the agonists with its own receptors, evokes the stimulation of cellular response of the second agonist through upregulation of the corresponding receptors. The interaction of insulin with its own receptors not only increased the prostanoid binding, but blockade of the hormone receptors by antireceptor antibody decreased even the basal PGE_1 binding. That the decrease of insulin receptor numbers would also decrease the prostaglandin binding, had been corroborated by the diminished PGE_1 binding in NIDDM platelets where the insulin receptor numbers are known

Table 1. Binding of [^{3}H]-PGE_1 to Normal and NIDDM Platelets in the Presence or Absence of Insulin

Platelets	Addition	PGE_1 Binding (fmol/10^8 cells) Mean ± S.D.
Normal	None	275 ± 41
Normal	Insulin (0.7 nM)	605 ± 65
NIDDM	None	160 ± 30
NIDDM	Insulin (0.7 nM)	238 ± 65

to be decreased when compared to normals (13). Conversely, platelets from nondiabetic patients with acute ischemic heart disease, which have been shown to contain decreased PGE_1/I_2 receptors numbers (14), also show impaired insulin binding (15). Furthermore, during cardiac convalescence, the binding of PGE_1 to platelets simultaneously improves with the binding of insulin to these cells.

The mechanism of the intracellular "communication" between these receptors is not known, and could not be related to the synthesis of new receptor proteins. Preliminary results indicate possible involvement of guanine nucleotide binding protein in the expression of "spare" receptors in the platelet membrane bilayer.

ACKNOWLEDGEMENT

This work was funded in part by the grant HL-41386 from the National Institute of Health and the American Diabetes Association.

REFERENCES

1. G. V. R. Born and R. M. Hardisty, *in:* "Human Blood Coagulation, Haemostasis and Thrombosis." R. Biggs, ed. Blackwell Scientific Publication, London. pp. 159-187 (1972).
2. M. A. DeWood, J. Spores, R. Notske, L. T. Mouser, R. Burroughs, M. S. Golden, and H. T. Lang, *New Eng. J. Med.,* **303:**897-902 (1980).
3. A. C. DeBoer, A. G. C. Turpie, R. W. Butt, R. V. Johnston, and E. Genton, *Circulation,* **66:**327-33 (1982).
4. M. Sobel, E. W. Salzman, G. C. Davies, R. I. Handin, J. Sweeney, J. Ploetz, and G. Kurland, *Thromb. Haemost.,* **40:**66-72 (1978).
5. G. A. Robison, A. Arnold, and R. C. Hartman, *Pharmacol. Res. Commun.,* **1:**325-32 (1969).
6. P. J. Palumbo, L. J. Melton, and R. G. Tancredi, *in:* "Clinical Diabetes Mellitus," J. K. Davidson, ed. Thieme Inc. New York pp. 349-60 (1986).
7. N. N. Kahn, and A. K. Sinha, *J. Biol. Chem.,* **265:**4976-81 (1990).
8. G. Scatchard, *Ann. N. Y. Acad. Sci.,* **51:**660-72 (1949).
9. N. N. Kahn and A. K. Sinha, *Biochim. Biophys. Acta.,* **972:**45-53 (1988).
10. M. P. Czeck, *Annu. Rev. Biochem.,* **46:**359-84 (1977).
11. N. N. Kahn and A. K. Sinha, (unpublished).
12. T. K. Ray, A. K. Dutta-Roy, and A. K. Sinha, *Biochim. Biophys. Acta,* **856:**421-27 (1986).
13. M. Udvardy, G. Pfliegler, and K. Rak, *Experientia,* **41:**422-33 (1985).
14. N. N. Kahn, H. S. Mueller, and A. K. Sinha, *Circulation Res.* (in press) (1990).
15. N. N. Kahn, W. A. Bauman, H. S. Mueller, and A. K. Sinha, *Clinical Res.,* **37:**269A (1989).
16. N. N. Kahn, and A. K. Sinha, *Biochim. Biophys. Acta,* **984:**113-118 (1989).
17. A. K. Dutta-Roy and A. K. Sinha, *J. Biol. Chem.,* **262:**12685-91 (1987).
18. A. G. Gilman, *Proc. Natl. Acad. Sci. USA,* **67:**305-12 (1970).

PLATELET-FIBRIN INTERACTION IN THE SUSPENSION AND UNDER FLOW CONDITIONS

C. J. Jen, S. J. Hu, H. J. Wu, T. S. Lin*, and C. W. Mao*

Department of Physiology and Department of Electrical Engineering*,
National Cheng-Kung University,
Tainan, Taiwan, ROC

ABSTRACT

Interactions between platelets and fibrin are important in hemostasis but often confused with platelet-fibrinogen interactions. A stirred mixture of solubilized fibrin and washed platelets at neutral pH range showed drastic reduction in turbidity and concomitant platelet adhesion onto newly formed fibrin strands. This platelet-fibrin interaction did not require platelet activation nor did it cause platelet aggregation. A device consisting of a parallel-plate flow chamber mounted on a fluorescence microscope has been constructed to allow direct visualization and recording of platelet-fibrin interaction under flow conditions. Platelets in whole blood adhered to the fibrin-coated portion but not to the uncoated portion of the flow chamber. Slow motion playback of video tapes indicated that the adhesion phenomenon was a dynamic process that involved attaching, detaching, relocation and transient contact. The fibrin coating influenced platelet adhesion both by increasing the number of cells making short-term attachments to the surface and by increasing the duration of cells attached to the surface. These observations provided basic characteristics of platelet-fibrin interaction.

INTRODUCTION

Platelets and fibrin are the two major components in a clot. The final product of tne coagulation system is a crosslinked fibrin network, while the final stage of platelet activation is in the fom of platelet aggregates. That is to say, a clot may be considered as a platelet-incorporated fibrin network. Two concepts concerning platelet-fibrin(ogen) interaction in the broad sense have been well documented. First, the plasma membrane of platelets can serve as the catalytic site of the coagulation system (1). Second, fibrinogen is an essential component for platelet aggregate formation (2). Both phenomena require platelet activation and neither deals with direct interaction between platelets and fibrin. Recently several studies on direct interaction between platelets and fibrin have been carried out (3-7). However, most of these studies deal with only activated platelets. Our studies in the past have shown that the interaction between resting platelets and fibrin can occur under certain experimental conditions (8,9). This report provides some characteristics of such interaction both in the suspension and under flow conditions.

MATERIALS

Thrombin, apyrase, hirudin, bovine serum albumin, creatine phosphokinase, 5'-p-fluorosulfonylbenzoyladenosine, acetylsalicylic acid (aspirin), and dipyridamole were pur-

Fibrinogen, Thrombosis, Coagulation, and Fibrinolysis, Edited by
C. Y. Liu and S. Chien, Plenum Press, New York

chased from Sigma Chemical (St. Louis, MO, USA). Prostaglandin E_1 (PGE_1) was a gift from Dr. W.C. Chang, National Cheng-Kung Univeristy. Acridine red was obtained from Tokyo Kasei Kogyo (Tokyo, Japan). All other chemicals were reagent graded.

Human fibrinogen was purchased from Kabi Vitrum (Stockholm, Sweden) and subsequently purified by DEAE-cellulose chromatography to remove plasminogen, fibronectin and factor XIII (10). The method for fibrin preparation has been described previously (8). Briefly, fibrinogen was converted into fibrin by incubating with thrombin overnight. The resulting fibrin clot was blotted and redissolved in a NaBr/acetic acid buffer (pH 5.3). Hirudin was included to neutralize any remaining thrombin.

Blood samples were drawn into sodium citrate from healthy donors who had not taken any medication for at least two weeks. Washed platelets were prepared by repeated centrifugation and suspended in a Tyrode/HEPES buffer (0.128M NaCl, 2.7×10^{-3}M KCl, 5×10^{-4}M $MgCl_2$, 3.6×10^{-4}M NaH_2PO_4, 1.2×10^{-3}M $NaHCO_3$, 0.01M HEPES, 5.5×10^{-3}M glucose, and 3.5 mg/ml albumin, pH 7.4). Washed red cells were also prepared by centrifugation and suspended in the same buffer. Blood cell counts were perfomed using electronic cell counters (Cell Dyn 100 and 400, Metertech, Taipei, Taiwan). For fluorescence microscopy studies, platelets in whole blood were labeled with 2 μM acridine red and monitored using the light filter set for rhodamine observations.

RESULTS

Platelet-Fibrin Interaction in the Suspension

Fibrin was polymerized when the acidity of a fibrin monomer solution was neutralized. Without stirring and in the absence of cells, this polymerization phenomenon could be traced by monitoring the light-scattering intensity change or by assaying the amount of fibrin trapped on a filter (8). With stirring, the polymerization process followed a sequence of visible events. First, the solution became turbid and gradually formed small particles. Then these particles grew in size and became filamentous via collision with one another. At the end, almost all filamentous fibrin entangled together to form a gel-like material surrounding the stirring bar.

When platelets were present, the mixture was turbid at the beginning. As the fibrin polymerization process went on, platelets became associated with fibrin. Eventually $<1\%$ of platelets were left in the suspension, and this process was accompanied with a large increase of the sample transmittance. An aggregometer could be conveniently used to trace the turbidity/light transmittance change associated with platelet-fibrin interaction in the suspension (Fig. 1). Under these conditions, however, little platelet aggregation was present when the sample was observed by use of a light microscope. Therefore, the drop in sample turbidity was due to platelets adhering to fibrin, not due to platelet aggregation. The addition of 1 mM EGTA to the reaction mixture reduced the turbidity change to the minimum, indicating at least submicromolar levels of Ca^{2+} was necessary for such platelet-fibrin interaction. If platelets in the mixture were substituted by red cells, little transmittance change could be detected (Fig. 1). The gel-like end product of fibrin was slightly pinkish and $>95\%$ of red cells could be recovered from the suspension at the end of experiment.

To further elucidate the specificity, we also prepared reaction mixtures containing both platelets and red cells. The turbidity of this type of samples did not change appreciably during fibrin polymerization. However, scanning electron microscopic observations of the gel-like material showed numerous platelets with few red cells attached to the fibrin mesh (Fig. 2). Moreover, these platelets individually adhered to fibrin rather than formed aggregates.

Various platelet activation inhibitors were used to test whether this platelet-fibrin interaction required preactivation of platelets or not. The inhibitors used in our study included those blocking the effect of ADP from outside (creatine phosphate-creatine phosphokinase, and 5'-p-fluorosulfonyl-benzoyladenosine) and those blocking arachidonic acid metabolism from inside (PGE_1 and aspirin). None of these inhibitors was effective in blocking this platelet-fibrin interaction (Table 1). As a matter of fact, formalin-fixed platelets behaved like control platelets in this system. We also prepared ADP-activated platelets that were subsequently fixed and washed. When these platelets were mixed with polymerizing fibrin, the sample transmittance increased slightly faster than that in control samples. Little difference in respective plateau values was observed (curves not shown here).

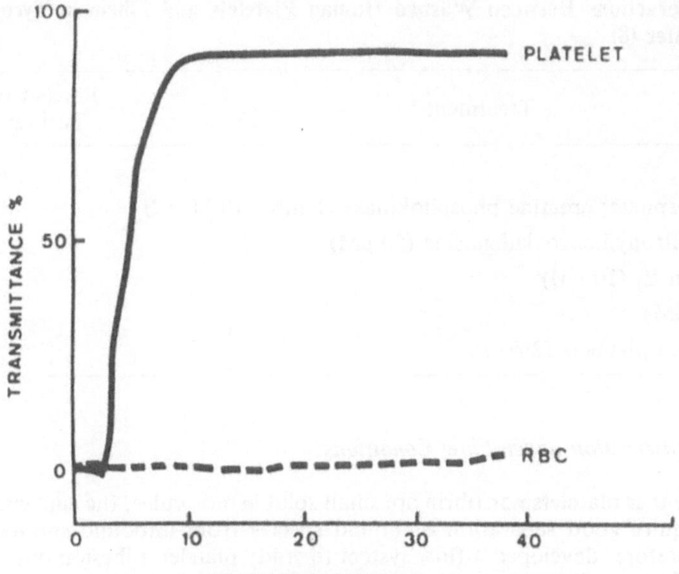

Fig. 1. Cell-fibrin interaction as indicated by light transmittance change (8). Fibrin (12 mg/ml) in NaBr/acetic acid (pH 5.3) was added to 19 volumes of Tyrode/HEPES buffer (pH 7.35) containing either washed platelets or washed red cells. Sample turbidity was continuously monitored by an aggregometer (NKK Hematracter 2) under stirred conditions. To make the initial turbidities almost the same, the sample either contained 3×10^8 platelets/ml or 5×10^7 red cells/ml. At 40 minute there were 2×10^6 platelets/ml or 4.85×10^7 red cells/ml left in the respective suspension.

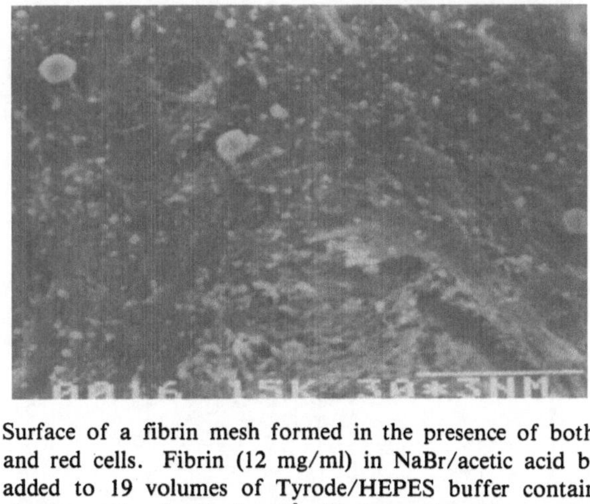

Fig. 2. Surface of a fibrin mesh formed in the presence of both platelets and red cells. Fibrin (12 mg/ml) in NaBr/acetic acid buffer was added to 19 volumes of Tyrode/HEPES buffer containing both platelets and red cells (3×10^8 cells /ml each). The reaction mixture was stirred for 5 minutes and the resulting materials were washed, fixed, critical-point dried, and metalshadowed. This sample was viewed under a Hitachi S405 scanning electron microscope. Numerous platelets and a few red cells were associated with the fibrin mesh (background).

Table 1. Interactions Between Washed Human Platelets and Fibrin in Tyrode/HEPES Buffer (8)

Treatment	Percent of Maximal Turbidity Change
Control	100
Creatine phosphate; creatine phosphokinase (1 mM; 10 U/ml)	98
5'-p-Fluorosulfonylbenzoyladenosine (50 μM)	96
Prostaglandin E_1 (10 μM)	96
Aspirin (10 μM)	94
Formalin-fixed platelets (2%)	97

Platelet-Fibrin Interaction under Flow Conditions

Because neither platelets nor fibrin are small soluble molecules, the conventional binding studies that require good separation of bound species from unbound species are difficult to do. We, therefore, developed a flow system to study platelet adhesion onto fibrin-coated surface under flow conditions.

A device consisting of a parallel-plate flow chamber mounted on an epifluorescence microscope, similar to the one developed by Hubble and McIntire (11), has been assembled to allow direct visualization of the process of platelet adhesion (Fig. 3). The flow chamber consisted of four parts: a steel cover plate, a cover glass, a silicone rubber gasket, and a plastic distributor. The dimension of the chamber cavity was $32.5 \times 12.5 \times 0.13$ mm^3. Blood was drawn through the flow chamber by a syringe pump. Since our observation point was close to the center of cover glass, the flow there was laminar, fully-developed with trivial edge effect. Under these circumstances we selected experimental flow rates of 0.15 ml/min and 0.94 ml/min, corresponding to surface shear rates of 70 s^{-1} and 445 s^{-1} respectively. The fluorescence dye, acridine red, could label both platelets and leukocytes but not red

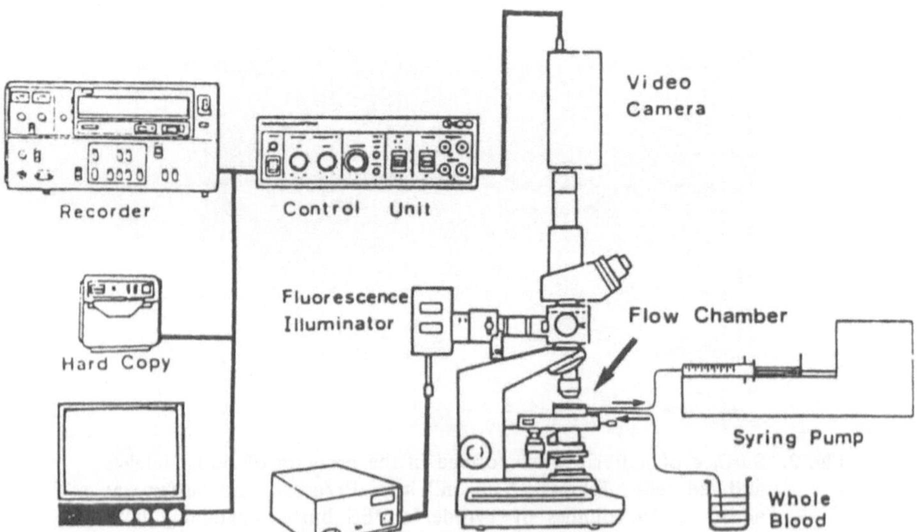

Fig. 3. A diagram of fluorescence video microscopy and flow system setup. Details of the flow chamber are described in the text.

cells. Therefore, platelets in whole blood became distinguishable under a fluorescence microscope.

At the beginning of an experiment, a cover glass coated with a small droplet of fibrin (1 mg/ml) in the middle was mounted on the flow chamber. The upstream edge of this droplet was located under the microscope. The flow system was first flushed with Tyrode/HEPES buffer without glucose and albumin. Then the fluorescence-labeled blood was drawn through the flow chamber at a predetermined flow rate and the whole process of platelet adhesion was recorded on a video tape.

When fluorescence-labeled blood went through the flow chamber, many bright particles were observed to move along the flow direction (Fig. 4). The numerous small particles represented platelets, while a few large bright spheres represented white cells. At 70 s^{-1} surface shear rate, these particles showed visible tails since they moved rather slowly in the periphery of the surface. In contrast, moving cells were too faint to be individually recognized at 445 s^{-1} (Fig. 5). Cells farther away from the surface were not visible since they moved very rapidly and were off-focused. Cells adhered on the surface showed clear bright images. As blood flowing through the chamber, more and more small bright dots (platelets) accumulated at the portion of surface that had been previously coated with fibrin. Few platelets adhered to the uncoated portion of glass surface. In the presence of 5 mM EGTA, only occasional transient contacts (see definition below) of platelets with the surface could be obvserved with no net accumulation throughout the experiment.

When the adhesion process was examined in detail by slow motion playback of the video tape, it consisted of at least four different types of events: attaching, detaching, relocation, and transient contact. The definitions of these terms are as following:

Attaching: a cell from the bulk flow making a surface contact and remaining at the contact location for at least 1/5 second (6 frames).

Detaching: a previously attached cell leaving the surface to the bulk flow or leaving the observational area.

Relocation: a previously attached cell moving to a new location at least 3 μm apart without joining the bulk flow.

Transient Contact: a cell from the bulk flow making a surface contact and remaining at the contact location for less than 1/5 second but longer than 1/30 second (1 frame).

When three adjacent blocks (1000 μm^2 each) along the flow direction were assigned, such as in Fig. 5, the events of platelet adhesion could be quantified. Table 2 represents

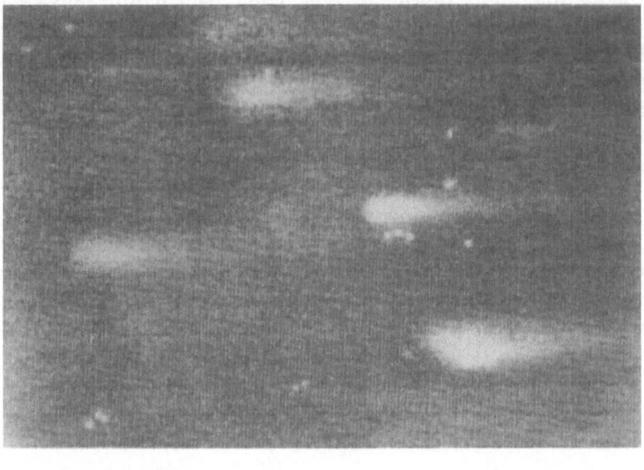

Fig. 4. Platelet deposition onto fibrin-coated surface under 70 s^{-1} surface shear rate and at 10 second flow time. The flow direction was from right to left. Scattered small bright dots represented adhered platelets. The big bright objects were white cells. Their faint, horizontal shadows indicated that they were not adhered. Platelets flowing by could be observed on the tape playback as faint white dots with horizontal tails as well. However, they were too faint to leave a distinguishable image on a single frame as shown here.

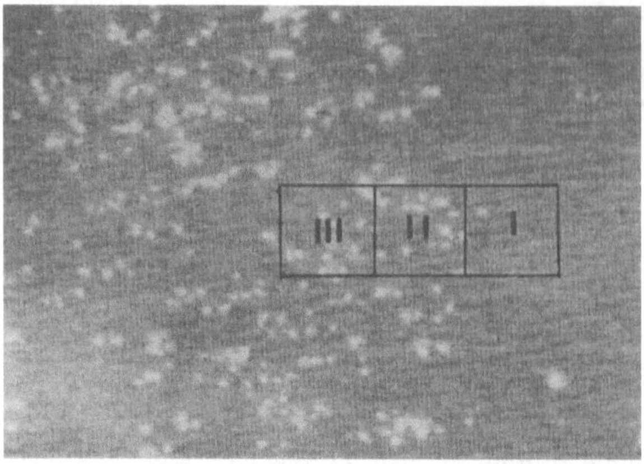

Fig. 5. Platelet deposition onto fibrin-coated surface under 445 s^{-1} surface shear rate and at 60 second flow time. The flow direction was from right to left. Under this flow condition, most white cells only left diffused bright image at the background unless they bounced off the surface. Moving platelets were not visible anymore. Only adhered platelets produced clear image. As flow time increased, more and more platelets adhered to the coated region to make the coating boundary distinguishable. Three adjacent squares, 1000 μm^2 each, were marked on a transparent sheet that was taped on the monitor screen. Individual events happened in each square could be registered during slow-motion playback of the video tape. Note: Square I was in the uncoated region while the other two squres were in the fibrin-coated region.

the results from the first 2 minutes of a 445 s^{-1} experiment. Most of the cells attached to uncoated area (block I) departed eventually while cells attached to fibrin-coated area (block II & III) tended to stay longer. Consequently, a net accumulation of adhered platelets in fibrin-coated area happened. Nevertheless, most of the attached cells did not remain at their initial attachment sites for more than 1 minute. This was reflected by high frequencies of relocation and detaching. As a matter of fact, some cells would make several relocations before they finally settled or detached. Although cells making transient contacts did not actually stay at the surface longer than 1/5 second, they could still be regarded as making short-term attachments. Cells merely bounced off the surface could not be registered by our current setup. Our results indicated that fibrin coating influenced platelet adhesion both by increasing the number of cells making short-term attachments to the surface and by increasing the duration of cells attached to the surface.

Recently we have developed a computer program that allows us to identify those adhered platelets. The accuracy of identification was better than 90%. By using this program we have been able to make large area (20,000μm^2) quantification. Fig. 6 shows a frame with platelets identified and marked by small dots at their centers. Since the location of each identified platelet was registered in computer, we could compare platelet location files from frames grabbed every 5 seconds. In addition to the accumulative growth curve, the curves for staying, newly attached and detached cells could be also obtained (Fig. 7) However, this computation did not detect transient contacts. Moreover, it classified relocation as a pair of attaching and detaching events, if the evacuated site of a relocation process was not reoccupied within the time interval (5 seconds).

DISCUSSION

This report provides the following characteristics of platelet-fibrin interaction: i) platelets in the form of washed platelets, platelet rich plasma or whole blood can be associated with fibrin; ii) this interaction does not induce platelet aggregation either in the suspension or

Table 2. Classification of Platelet-Surface Interaction During the First 2 Minutes of Blood Flowing Through the Flow Chamber at a Surface Shear Rate of 445 s^{-1}

Time Course (sec)		0–10	10–20	20–30	30–40	40–50	50–60	60–70	70–80	80–90	90–100	100–110	110–120	Total
	I	–	2	1	2	1	1	1	–	–	–	–	–	8
Attaching	II	1	1	2	5	9	8	6	1	9	4	2	2	50
	III	1	1	3	3	3	8	4	3	2	2	3	3	36
	I	–	2	–	1	1	1	1	–	–	–	–	–	6
Detaching	II	–	–	1	–	2	2	1	1	1	1	–	2	11
	III	–	–	1	–	1	–	–	1	1	1	2	–	7
	I	–	–	1	–	–	–	–	–	–	–	–	–	1
Relocation	II	–	1	–	1	4	5	3	2	2	2	1	1	22
	III	–	–	2	2	2	1	3	2	1	2	–	2	17
Transient	I	–	2	–	–	–	–	1	–	–	–	–	1	4
Contact	II	–	2	3	1	3	1	1	–	1	2	1	2	17
	III	1	–	3	3	–	3	1	–	1	2	1	2	17
Net	I	–	–	1	1	–	–	–	–	–	–	–	–	2
Change*	II	1	1	1	5	7	6	5	–	8	3	2	–	39
	III	1	1	2	3	2	8	4	2	1	1	1	3	29

*Net Change = Attaching — Detaching

under flow conditions; iii) this interaction requires calcium; iv) this interaction cannot be blocked by platelet activation inhibitors; v) although platelets gradually accumulate on fibrin-coated surface, the adhesion phenomenon is a dynamic process that involves attaching, detaching, relocation and transient contact.

Recent studies, notably those from Hantgan and coworkers, have clearly demonstrated that fibrin and fibrin oligomers prepared in the presence of excess amounts of fragment D or GPRP can bind to ADP activated platelets (5-7). However, whether this binding is similar to the fibrinogen binding to platelets that involves platelet surface glycoprotein II_b-III_a complex is still controversial. On one hand, some monoclonal antibodies against the glycoprotein II_b-III_a complex inhibit the binding of fibrin oligomers to ADP stimulated platelets (6). On the other hand, some other antibodies against different epitopes of the same glycoprotein complex and RGD-containing peptides as well as fibrinogen $\gamma^{400\text{-}411}$ peptide increase clot tension, indicating augmented platelet-polymerizing fibrin interaction (12). It is possible that platelet-fibrin interaction involves at least two components, one is related to the platelet glycoprotein II_b-III_a complex while the other is not. According to Harfenist et al. (4) high levels of nonspecific binding of GPRP solubilized fibrin (i.e., those not displaced by fibrinogen) have been observed with stimulated as well as non-stimulated platelets. This two-component-interaction hypothesis can explain that under certain experimental conditions, such as those reported in our previous study (8) and by Brown et al. (13), platelets treated with various inhibitors or formalin showed similar degree of association with fibrin. Moreover, the nonspecific binding can also explain that platelets in whole blood or platelet rich plasma, both containing large amounts of fibrinogen, can be associated with fibrin in the suspension and under flow conditions. It would be very interesting to compare platelet-fibrin interaction with platelet-fibrinogen interaction under the same experimental conditions.

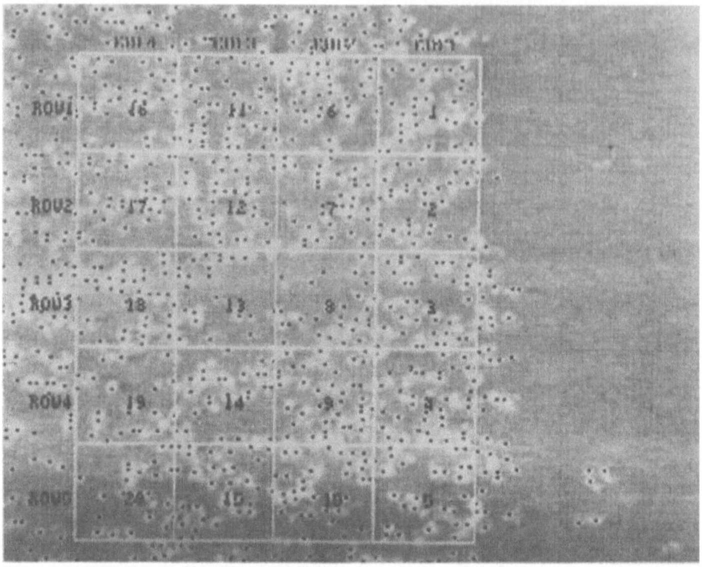

Fig. 6. Platelet deposition onto fibrin-coated surface under 445 s^{-1} surface shear rate and at 180 second flow time. The flow direction was from right to left. Adhered platelets have been marked with small black dots. A total counting area of 20,000 μm^2 have been marked in the fibrin-coated region with one boundary arranged close to the leading edge of coated region.

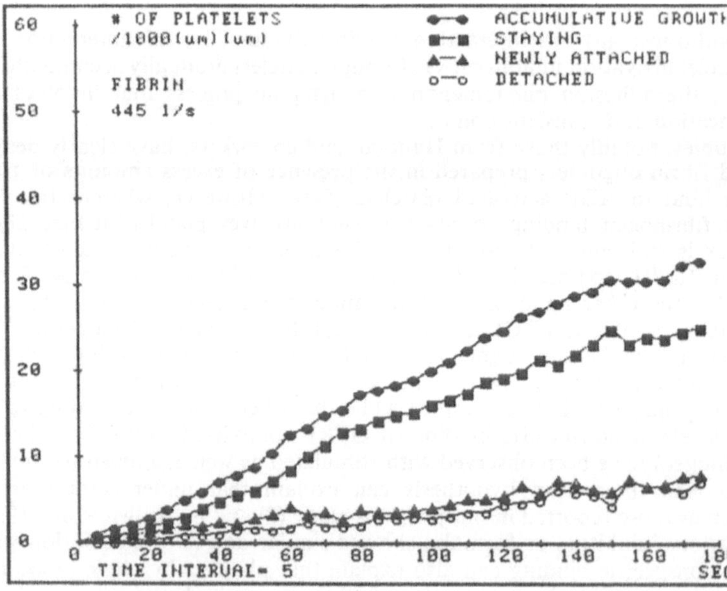

Fig. 7. Data analysis of the first three minutes in a 445 s^{-1} surface shear rate experiment.

We are currently using our flow system to investigate the platelet adhesion process to fibrinogen-coated and fibrin-coated surfaces (14).

ACKNOWLEDGMENTS

The authors are grateful to Dr. J.A. Hubble for helpful suggstions in constructing the flow system and to Mr. H.M. Wang and Mr. C.C. Tai for performing some pilot studies. This study was supported by National Science Council Grants NSC77-0412-B006-08 and NSC77-0404-E006-12.

REFERENCES

1. P. N. Walsh, Platelet-mediated coagulation protein interactions in hemostasis, *Semin. Hematol.,* **22:**178-186 (1985).
2. E. I. B. Peerschke, The platelet fibrinogen receptor, *Semin. Hematol.,* **22:**241-259 (1985).
3. G. P. Tuszynski, E. Kornecki, C. Cierniewski, L. C. Knight, A. Koshy, S. Srivastava, S. Niewiarowski, and P. N. Walsh, Association of fibrin with platelet cytoskeleton, *J. Biol. Chem.,* **259:**5247-5254 (1984).
4. E. J. Harfenist, M. A. Packham, and J. F. Mustard, Comparison of the interactions of fibrinogen and soluble fibrin with washed rabbit platelets stimulated with ADP, *Thromb. Haemostasis,* **53:**183-187 (1985).
5. R. R. Hantgan, R. G. Taylor, and J. C. Lewis, Platelets interact with fibrin only after activation, *Blood,* **65:**1299-1311 (1985).
6. R. R. Hantgan, Fibrin protofibril and fibrinogen binding to ADP-stimulated platelets: evidence for a common mechanism, *Biochim. Biophys. Acta,* **968:**24-35 (1988).
7. J. C. Lewis, C. Johnson, P. Ramsamooji, and R. R. Hantgan, Orientation and specificity of fibrin protofibril binding to ADP-stimulated platelets, *Blood,* **72:**1992-2000 (1988).
8. C. J. Jen and S. J. Hu, Direct platelet-fibrin interaction that does not require platelet activation, *Am. J. Physiol.,* **253:**H745-H750 (1987).
9. Y. L. Chiu, Y. L. Chou, and C. J. Jen, Platelet deposition onto fibrin-coated surfaces under flow conditions, *Blood Cells,* **13:**437-447 (1988).
10. E. J. Harfenist, G. Raychaudhuri, M. A. Packham, and J. F. Mustard, An investigation into the role of coagulation factor XIII in ADP-induced aggregation and fibrinogen binding with rabbit platelets, *Blood,* **60:**905-911 (1982).
11. J. A. Hubble and L. V. McIntire, Technique for visualization and analysis of mural thrombogenesis, *Rev. Sci. Instrum.,* **57:**892-897 (1986).
12. I. Cohen, D. L. Burk, and J. G. White, The effect of peptides and monoclonal antibodies that bind to platelet glycoprotein II_b-III_a complex on the development of clot tension, *Blood,* **73:**1880-1887 (1989).
13. R. S. Brown, S. Niewiarowski, G. J. Stewart, and M. Millman, A double-isotope study on incorporation of platelets and red cells into fibrin, *J. Lab. Clin. Med.,* **90:**130-140 (1977).
14. C. J. Jen, H. J. Wu, and T. S. Lin, Direct observation of platelet adhesion onto fibrinogen-coated and fibrin-coated surfaces, *Thromb. Haemostasis,* **62:**418 (1989).

AN INDEPENDENT HAEMOSTATIC MECHANISM: SHEAR INDUCED PLATELET AGGREGATION

J. R. O'Brien and G. P. Salmon

Central Laboratory
St. Mary's Hospital
Portsmouth, Hants., UK

ABSTRACT

We have published (1) evidence indicating that high shearing forces alone applied to platelets expose and activate a unique domain on glycoprotein IIb/IIIa (GPIIb/IIIa) at the platelet surface. In the presence of von Willebrand's factor (vWf) and divalent cations the platelets will aggregate. This paper reviews the extensive literature on high shear effects. It describes a device in which high shear produced by forcing heparinised whole blood through a complex filter normally results in platelet activation; the platelets aggregate and then block the filter. This system is inhibited by antibodies to GPIIb/IIIa and to vWf: fibrinogen is apparently not involved. The same antibodies to GPIIb/IIIa and vWf prevent high shear induced thrombosis occurring in vivo in animal models. The filter blockage is not influenced by aspirin, heparin and ticlopidine and so involves a different mechanism from the aspirin sensitive mechanisms involved in clinical thrombosis prevention in vivo in man. While there are a number of unexplained phenomena in this global test nevertheless this filter model is a simple way of studying a recently recognised pathway which is almost certainly involved in thrombogenesis in man.

INTRODUCTION

Sections of a thrombus will usually show extraordinary plemorphism. In any one area there may be a deposit predominantly of platelets or fibrin, or red cells, or polymorphs. Often there is evidence of streaming in the fibrin deposition, indicating that shearing forces influence the mechanisms leading to thrombus formation at a particular site at a particular time. Evidently many mechanisms are involved. However this article will focus only on the activation of platelets by shearing forces.

Early Evidence of Shear

The first evidence obtained by the author (2) was when blood was forced through a column of glass beads and the platelets were counted before and after passage through the column. At low pressure — low shear — the platelet retention in this column was normal in von Willebrand's (vW) patients relative to the controls. At high pressure, however, the retention in vW patients was relatively grossly decreased. It is of interest that very similar results (Fig. 1) were obtained 20 years later by Badimon et al. (3) who measured platelet deposition on de-endothelialized porcine aorta using vW and normal blood in a perfusion chamber at high and low pressures.

The next model chosen for studying the effect of high shear involved forcing blood through a nuclepore filter with 5 μ holes (4). Native blood was forced through the filter

Fibrinogen, Thrombosis, Coagulation, and Fibrinolysis, Edited by
C. Y. Liu and S. Chien, Plenum Press, New York

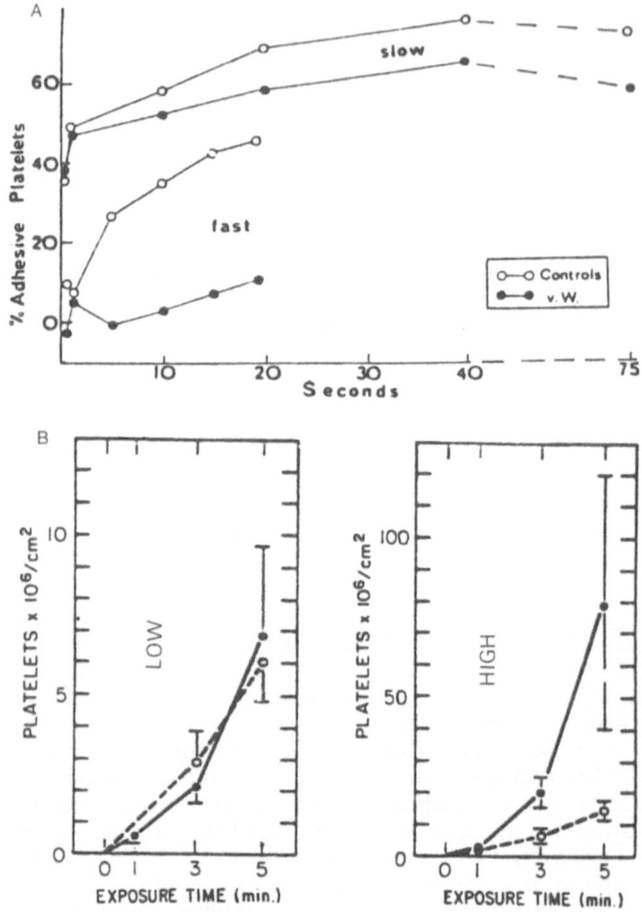

Fig. 1.

which was subsequently washed and stained. It could be seen that very few platelets were retained on the upstream surface of the filter, indicating that the filter material was relatively inert. Downstream, below the filter, great masses of platelets were deposited indicating that the shearing forces alone had activated the platelets.

The Literature

The extensive literature referring to the effect of high shear will not be reviewed in any detail. Shear can be produced very precisely in a *viscometer* (5, 6, 7, 8) but the blood is usually retained in the viscometer for many seconds or minutes and any observed changes may have been modified by metabolic changes subsequent to an initial event.

Another method of studying the effects of shear employs the *Baumgartner technique* (9) in which rings of de-endothelialized blood vessels are everted over a glass rod. This rod is then inserted into a narrow tube through which blood is forced (10, 11, 12). The shearing forces and the deposition of platelets on the sub-intima can then be monitored. The third method involving high shear induced by stenosis in vivo will be reviewed at the end of this article.

A very approximate summary of the conclusions of all these works suggest that high shearing forces acting on platelets expose and "activate" the membrane glycoprotein IIb/IIIa (GPIIa/IIIb). This GPIIb/IIIa in the presence of von Willebrand factor (vWf) acting as the ligand will cause platelet aggregation. There are some reports suggesting that glycoprotein Ib (5, 6) and fibrinogen and ADP may also be involved. However differences in

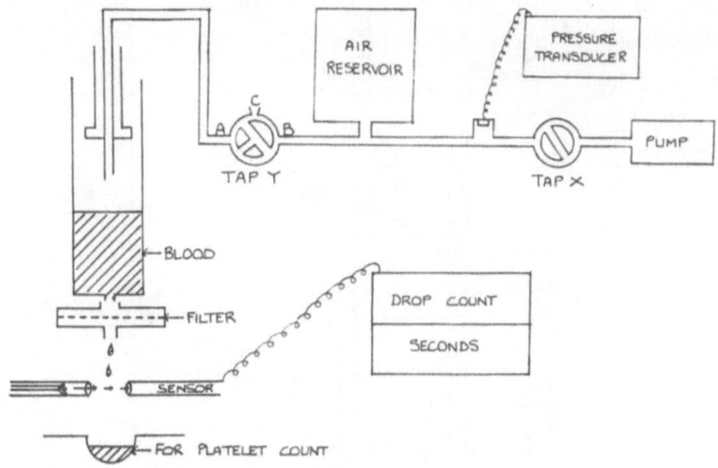

Fig. 2.

techniques may be responsible for the differing results. There are also several reports that shear induced phenomena are unaffected by aspirin.

The in Vitro Filter Test

Basically the author's device (1, 13) (Fig. 2) which has similarities to that of Uchiyama (14) consists of a reservoir of air pumped up to a predetermined pressure (usually 40 mmHg) and measured by a pressure transducer. A routine 5 ml syringe is filled with native or anticoagulated blood and fixed on top of a nuclepore filter holder, in which the chosen filter is placed; usually this is a PALL U100 which retains particles larger than 10 μ in diameter. On turning the tap Y, the high pressure air is delivered on top of the blood which is forced through the filter. The first drop falling through a light path triggers a drop counter and a timer. Arbitrarily we record the cumulative drop count every 5 seconds. In addition the difference between the initial platelet count and the count after passage through the filter must indicate the number of platelets retained in the filter. This, expressed as a percentage of the initial count is referred to as the "percentage platelet retention". Under most circumstances platelet retention and filter blockage are closely correlated.

The nature of the filter surface does not seem to have an important influence on the result. The positively charged PALL U100 filter, made of fine glass fibres which is hydrophilic, and a negatively charged glass filter give similar results; as does a filter made of lipophilic polypropylene fibres.

The speed of transit through the filter is remarkably fast. Heparinised platelet rich plasma pours in a stream through the filter initially, but blockage usually occurs in about 40 seconds; thus platelets are activated, aggregate and are retained. The initial transit time has been calculated to be 8 ms. This must be one of the shortest exposure times of any platelet studies to date (8), and this agrees with the in vivo situation when blood is pouring out of cut capillaries. Thus this device permits the study of native or anticoagulated blood pouring through capillary sized holes. Damaged endothelium, "tissue juice", collagen and vessel contraction seem to be the only factors missing.

Fig. 3 shows the effect of a pressure of 40 mmHg on normal heparinised blood. Initially the cumulative drop count increases rapidly but by 40 seconds virtually no more drops come through the filter which EM shows to be blocked by platelet aggregates. It will be seen that initially (0 - 3 secs) platelet retention is some 30% and, as the filter progressively blocks, up to 80% of the platelets are retained. A completely different situation exists if the pressure is reduced to 5 mmHg. The tracing of the drops is a straight line indicating that there is no progressive blocking. The percentage of platelets retained initially is relatively high (30%) but subsequently few platelets are retained (see below). With EDTA blood at 40 mmHg there is very little slowing in the drop count and after an initial retention of some 30% of platelets, few additional platelets are retained. Heparinised normal platelet

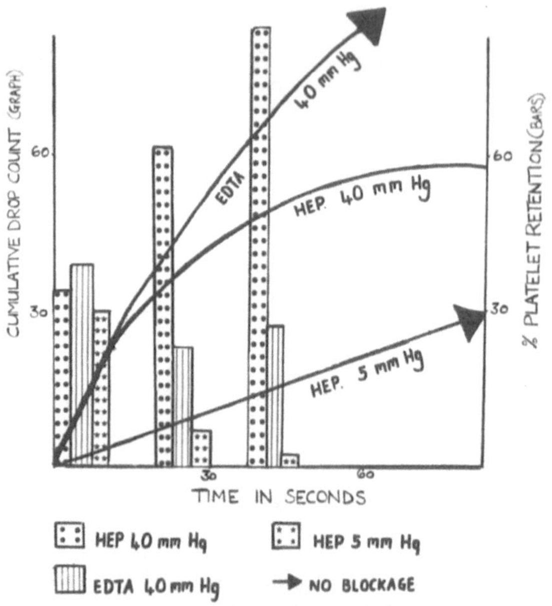

Fig. 3.

rich plasma, initially pours through the filter, but blockage with platelet aggregates is complete in about 40 secs. These results emphasise that high pressure is necessary. Divalent cations are necessary. Red cells affect the rate of flow but in their absence filter blockage can cocur. This, anticipating subsequent results, shows that red cell haemolysis is not a vital part of this process.

In studies with antibodies a concentration of antibody known to cover or block a particular domain on, for example the GP IIb/IIIa complex on the platelet membrane is added one minute before testing the blood in the filter. The percentage of platelets retained in the presence of the antibody is compared with retention occurring in its absence. Thus a ratio of 100 indicates that there is no change in platelet retention; the addition has had no effect. A ratio of 25 indicates that only 25% of the platelets that would normally be retained were retained in the presence of the antibody, i.e. gross inhibition of retention. Table 1 gives a typical result with one of 6 antibodies tested known to block a domain on GPIIb/IIIa. It will be seen that platelet retention was reduced to 34% relative to the controls; thus marked platelet inhibition occured. Two antibodies to GPIIb/IIIa (15, 16) are of particular interest because they have been shown to cover only the domain which apparently is unique for the adhesion of vWfactor and do not prevent fibrinogen binding. One of the antibodies also inhibits platelet retention. Patients with Glanzmann's thrombasthaenia, with a deficiency or abnormality of GPIIb/IIIa do not block the filter and retention is abnormally low. Thus GPIIb/IIIa is required.

Synthetic peptides containing the amino acid sequence arg-gly-asp, namely GRGDS (17) and RGDT (18) will cover the site on the IIb/IIIa which is required for von Willebrand's factor (and fibrinogen) to bind platelets to platelet or platelet to collagen. They also prevent platelet aggregation and filter blocking.

Table 1 also gives typical results when one of 6 antibodies to von Willebrand's factor was tested. Again the ratio is low indicating marked inhibition of platelet retention. The blood from all 19 vW patients when tested poured through the filters; there was low retention and no blackage. Thus vWf is required for filter blockage. It should be noted that both in Glanzmann's thrombasthaenia and in WV disease the plasma fibrinogen is normal, yet for different reasons both these clinical syndromes are completely abnormal in this in vitro filter test. Evidently fibrinogen is not essential, and vWf is. The one patient with Bernard Soulier Syndrome (abnormal glycoprotein Ib) and one antibody to glycoprotein Ib behaved almost normally in the filter suggesting that glycoprotein Ib is not of major importance.

Table 1. Comparison of Platelet Retention at 40 mmHg by Different Bloods with and without Additions

Blood Tested	n	% Platelet Retention 10 – 20 seconds	% of normal
Normal	80	67 ± 17	100
von Willebrand's Disease	19	24 ± 11**	36
Normal + Anti-vWF (88H8)	5	19 ± 6** (71 ± 8)	27
Glanzmann's Thrombasthaenia	2	12 ± 2	18
Normal + Anti-GPIIb/IIIa (LJP5)	4	36 ± 6** (64 ± 12)	56
Normal + 300 μM RGDT	4	27 ± 6* (60 ± 22)	45
Bernard Soulier Syndrome	1	67	100
Normal + Anti-GPIb (6DI)	4	55 ± 12 (69 ± 12)	80

n = Number of blood studied. The figures in brackets are the appropriate controls. % of normal: platelet retention of the blood tested is expressed as a percentage of the appropriate control. *$p < 0.05$, **$p < 0.01$ treated blood/appropriate control.

Table 2 demonstrates that the same critical concentration of an antibody to GPIIb/IIIa is less inhibitory at 100 mm/Hg than at 40 mm/Hg. A concentration of anti-vWf is equally inhibitory at the two pressures. This suggests that high shear stress exposes more GPIIb/IIIa sites, while vWf is not influenced by shear (5).

The Hypothesis

Using native or heparinized blood, and even in this global test, these observations are fully compatible with the following hypothesis: the shearing forces generated within the filter activate and expose a unique (15, 16) domain of the GPIIb/IIIa complex at the platelet surface. This domain in the presence of divalent cations binds with sites on vWf containing the Arg-Gly-Asp sequence. This process results in platelet aggregation and filter blockage. Fibrinogen (19), fibrinectin and vitronectin are apparently not essential or perhaps not even involved. This is all the more remarkable since there are 300 molecules of fibrinogen to one of vWf (20).

Complications

While these conclusions seem to be precise and gain support from many different studies in the literature, there may be other factors involved in this global situation. One

Table 2. Comparison of Platelet Rentention at 40 with 100 mmHg with Monoclonal Antibodies to vWf and GPIIb/IIIa

Blood Tested	n	% Platelet Retention 10 – 20 second: 40 mmHg	100 mmHg
Normal	5	71 ± 8	77 ± 5
Normal + Anti-vWf (88H8)	5	19 ±6	15 ± 6
Normal	4	64 ± 12	78 ± 3
Normal + Anti-GPIIb/IIIa (LJP5)	4	36 ± 6	68 ± 3**

n = number of blood studied.
**$p < 0.01$ compared to the same sample at 40 mmHg.

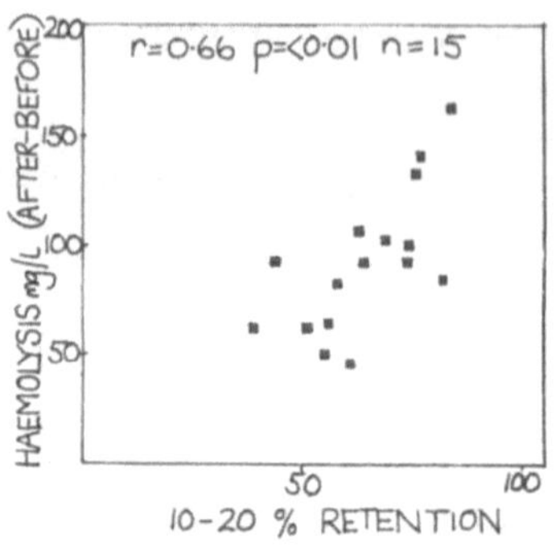

Fig. 4.

of these concerns the unexplained difference between the first and the second stage. The first stage is defined as events occurring within the first 3 seconds of blood passing through the filter. From Fig. 3 it will be seen that at low pressure (5 mmHg) and in the presence of EDTA; normal numbers — some 30% — of platelets are retained in the first stage. Thereafter in both these situations few platelets are retained. Nevertheless vWf and GPIIb/IIIa appear necessary in the first stage. Further studies are required to elucidate this apparent discrepancy between the factors necessary for the first and second stage.

While it is clear that vWf is necessary for filter blockage, if this molecule is assayed using the antibody employed in the standard VIII R:Ag Elisa assay, the correlation between vWf and platelet retention is curious. Within the 18 patients with vW disease (low R:Ag and low retention) there is no correlation between R:Ag and platelet retention. Similarly within 80 normal controls, again there is no correlation; similarly in bloods from 10 pregnant women, in which there is a high von Willebrand's factor and high retention, again there is no correlation. Yet if all these 3 separate groups are correlated together there is a significant correlation overall with an r value of 0.6. These findings are compatible with a further, as yet unidentified, factor(s) confusing these measurements. Perhaps the R:Ag assay is measuring an inappropriate epitope on von Willebrand's factor. Indeed Nichols et al (21) found that an antibody to vWf given to pigs prolonged the bleeding time and prevented a from of thrombosis but bad little effect on levels of VIIIC or vWf measured as R:Ag.

Other events may be also be confusing the correlation between vWf and retention. Perhaps ADP liberated from haemolysed red cells may contribute. Fig. 4 shows that the mean haemolysis caused by the shearing forces in the filter is correlated with platelet retention suggesting ADP liberated from haemolysed red cells plays a part. Conversely when retention is high perhaps lysis occurs secondarily. It is relevant that normal platelet rich plasma blocks the filter rapidly, and vW disease platelet rich plasma does not. Thus vWf is essential for blocking and ADP from lysed red cells can be irrelevant. But some contribution from shear induced platelet lysis is not excluded.

A Test for Red Cell Fragility

During a study of pregnancy (22) we were able to show that at all pressures the amount of haemolysis occurring in the pregnant blood was greater than that occurring in normal blood (Table 4) (and pregnant blood with high vWf and fibrinogen blocks faster than normal) (8). This is in line with increased fragility reported by Majid (23). Thus it appears that this method may be appropriate for investigating even minor degrees of red cell fragility.

Clinical Relevance

The clinical uses of this test have not been extensively explored but some uses are listed in Table 5 (and see Uchiyama (14)). The injection of cryoprecipitate or of DDAVP which mobilizes large vWf multimers (24) produces an immediate and marked increase in platelet retention (25). A failure of response to DDAVP in this test is reported (26); there was a failure of normal haemostasis and a transfusion was required.

Table 3. The First and Second "Stage" Retention Ratio

	n	First 0 – 3 secs	Second 10 – 20 secs
Normal Heparin Blood 40 mmHg	80	100	100
Normal Heparin Blood 10 mmHg	4	114	52
Normal EDTA Blood 40 mmHg	4	100	21
Normal Blood + RGDT 300 μM	4	105	45
von Willebrand's Disease	19	53	36
Normal Blood + Anti vWf (88 Hg)	5	45	27
Glanzmann's Thrombasthenia 1		61	21
Glanzmann's Thrombasthenia 2		104	15
Normal Blood + Anti-GPIIb/IIIa HP1/ID	4	61	20

n = number studied.
Retention ratio: the mean value of the calculated retention divided by the retention of the normal bloods with no addition × 100.

Table 4. Comparison of Haemolysis in Normal Controls and Pregnancy by Passage through the Filter

Pressure	Haemoglobin Released mg/L			
	n	Normal Controls	n	Pregnancy
Pre-filter	40	70 ± 55	8	72 ± 21
10 mmHg	2	42	4	1379 ± 580
20 mmHg	2	186	4	3614 ± 833
40 mmHg	2	186	4	3614 ± 1912
100 mmHg	2	1937	–	–

Table 5. A Clinically Useful Test

Platelet retention is *decreased* and the filter does not block in:

1. von Willebrand's disease.
2. Glanzmann's disease.
3. Severe thrombocytopenia ($<50{,}000/10^9/1$).
4. Myeloproliferative thrombocythaemia.
5. Early in cardio-pulmonary bypass surgery.

It is *increased* (abnormally rapid blocking) in:

1. Reactive thrombocythaemia.
2. In pregnancy.
3. At the end of cardio-pulmonary bypass surgery.
4. A dramatic improvement in retention and the rate of filter blocking occurs immediately after an infusion of DDAVP or cryoprecipitate.

It may be asked how relevant are these in vitro findings to the in vivo situation and to clinical haemostasis and thrombosis. It has been shown that there are 2 stages in filter blockage (Fig. 3), the first occuring even in the presence of low shear or EDTA and is then non-progressive. It is just possible that this non-progressive deposition is related to the observation that in vivo removal of the endothelium by balloon catheter but without stenosis, results in the deposition of only a monolayer of platelets.

It is noticeable that the four treatments generally accepted to contribute to the inhibition of thrombosis, namely oral anticoagulants, heparin, aspirin and ticlopidine all carry the risk of bleeding. Thus it may be asked whether vW patients with a long bleeding time are protected from thrombosis. Information on this point is not complete but this argument raises the suggestion that artificially induced von Willebrand's factor deficiency might prevent thrombosis.

Shear and Thrombosis in Vivo

Only when there is intimal damage and 70 — 90% stenotic obstruction causing high shear does platelet deposition progress to thrombosis (27, 28, 29). Nicholls et al (21) have produced a thrombosis model in the pig coronary artery and have shown that this type of thrombosis does not occur in pigs with vW disease. Bellinger (29) reports that thrombosis in normal pigs is inhibited by an antibody to vWf which also causes prolongation of the bleeding time.

Coller et al (28) has used a similar model involving endothelial damage and stenosis to produce high shear in the carotid artery in the monkey. Normally this produces recurrent episodes of platelet thrombi forming distal to the stenosis. In this model thrombosis is inhibited by an antibody to GP IIb/IIIa. The deposition of platelets and thrombus formation in a dacron arterio-venous shunts in baboons was also inhibited by antibodies to GPIIb/IIIa. Significantly aspirin and heparin had no effect on this model (30).

The experimental prevention of thrombosis in vivo by antibodies to vWf and to GP IIb/IIIa and by RGD containing peptides in situations where there is endothelial damage and high shear very strongly suggests that this mechanism is one, presumably amongest many, which can predispose to thrombosis. The importance of this shear induced mechanism in clinical practice is not known but it is universally accepted that, with or without plaque rupture (37), coronary artery stenosis is almost always present in patients with myocardial infarction. Clearly the filter device is one way of studying this process.

Aspirin, heparin and ticlopidine are clinically beneficial and so must affect mechanisms involved in thrombosis. In the filter model none of these drugs influence filter blocking. This contrast indicates that high shear induced thrombosis involves different pathways.

These observations suggest that in any future studies of the prothrombotic state vWf appropriately measured and the anisotropy (32) and rigidity of platelets and their response to shearing forces with the exposure of GPIIb/IIIa may be rewarding. The use of antibodies may not be the only way of inhibiting shear induced platelet aggregation and its contribution to thrombus formation.

ACKNOWLEDGEMENTS

We thank Drs. B. Coller and Z Ruggeri and W. Bowie for antibodies and Pall Europe Ltd., for the filters. Mrs. Bastable for secretarial assistance and Librarian Mrs. B. Pitman. We thank the publishers of Journal of Laboratory and Clinical Medicine and Journal of Clinical Pathology for permission to publish the graphs in Fig. 1 and the publisher of Blood for Fig. 2.

REFERENCES

1. J. R. O'Brien, and G. P. Salmon, Shear stress activation of platelet glycoprotein IIb/IIIa plus von Willebrand factor causes aggregation: filter blockage and the long bleeding time in von Willebrand's disease, *Blood,* **70(No5)**:1354-61 (1987).
2. J. R. O'Brien, and J. B. Heywood, Some interactions between human platelets and glass: von Willebrand's disease compared with normal, *J. Clin. Path.,* **20**:56-64 (1967).
3. L. Badimon, J. J. Badimon, J. Rand, V. T. Turitto, and V Fuster, Platelet deposition on von Willebrand factor-deficient vessels. Extracorporeal perfusion studies in swine

with von Willebrand's disease using native and heparinized blood, *J. Lab. Clin. Med.*, **110(No5):**634-47 (1987).
4. J. R. O'Brien, M. D. Etherington, and P. Weir, Platelet aggregation inhibitors: a 5μ nuclepore filter "bleeding time", *in:* Sixth International Congress on Thrombosis of the Mediterranean League against Thromboembolic Diseases, Monte Carlo, 307 T (1980).
5. J. L. Moake, N. A. Turner, N. A. Stathopoulos, L. Nolasco, and J. D. Hellums, Shear-induced platelet aggregation can be mediated by vWf released from platelets, as well as by exogenous large or unusually large vWF multimers, requires adenosine diphosphate, and is resistant to aspirin, *Blood,* **71(No 5):**1366-74 (1988).
6. D. M. Peterson, N. A. Stathopoulos, T. D. Giorgio, J. D. Hellums and J. L. Moake, Shear-induced platelet aggregation requires von Willebrand factor, and platelet membrane glycoproteins Ib and IIb-IIIa, *Blood,* **69(No 2):**625-628 (1987).
7. Y. Ikeda, M. Murata, and Y Araki, et al, Importance of fibrinogen and platelet membrane glycoprotein IIb/IIIa in shear-induced platelet aggregation, *Thromb. Res.*, **51:**157-163 (1988).
8. L. J. Wurzinger, R. Opitz, P. Blasberg, and H. Schmid-Schonbein, Platelet and coagulaon parameters following millisecond exposure to laminar shear stress, *Thromb. Haemost.*, **54(2):**381-386 (1985).
9. H. R. Baumgartner, The role of blood flow in platelet adhesion, fibrin deposition, and formation of mural thrombi, *Microvasc. Res.*, **5:**167-79 (1973).
10. H. R. Baumgartner, T. B. Tschopp, and D. Meyer, Shear rate dependent inhibition of platelet adhesion and aggregation on collagenous surfaces by antibodies to human factor VIII/von Willebrand factor, *Br. J. Haematol.*, **44:**127-39 (1980).
11. K. S. Sakariassen, P. F. E. M. Nievelstein, B. S. Coller, and J. J. Sixma, The role of platelet membrane glycoproteins Ib and IIb-IIIa in platelet adherence to human artery subendothelium, *Br. J. Haematol.*, **63:**681-91 (1986).
12. L. Badimon, J. J. Badimon, V. T. Turitto, and V. Fuster, Role of von Willebrand factor in mediating platelet-vessel wall interaction at low shear rate; the importance of perfusion conditions, *Blood,* **73(No 4):**961-67 (1989).
13. J. R. O'Brien, and G. P. Salmon, Unpublished results (1989).
14. S. Uchiyama, M. L. Bach, P. Didisheim and E. J. W. Bowie, Clinical evaluation of a new test of haemostasis: the filter bleeding time. *Thromb. Res.*, **34:**397-405 (1984).
15. S. Berliner, K. Niinga, J. R. Roberts, B. A. Houghton, and Z. M. Ruggeri, Generation and characterization of peptide-specific antibodies that inhibit vW factor binding to glycoprotein IIb/IIIa without interacting with other adhesive molecules, *J. Biol. Chem.*, **263:**7500-05 (1988).
16. T. V. Lombardo, E. Hodson, J. R. Roberts, T. J. Kunicki, T. S. Zimmerman, and Z. M. Ruggeri, Independent modulation of von Willebrand factor and fibrinogen binding to the platelet membrane glycoprotein IIb/IIIa complex as demonstrated by monoclonal antibody, *J. Clin. Invest.*, **76:**1950-58 (1985).
17. E. F. Plow, M. D. Pierschbacher, E. Ruoslahti, G. A. Marguerie, and M. H. Ginsberg, The effect of arg-gly-asp-containing peptides on fibrinogen and von Willebrand factor binding to platelets, *Proc. Natl. Acad. Sci. USA,* **82:**8057-61 (1985).
18. Y. Cadroy, R. A. Houghten, and S. R. Hanson, RGDT peptide selectively inhibits platelet-dependent thrombus formation in vivo. Studies using a baboon model, *J. Clin. Invest.*, **84:**939-44 (1989).
19. H. J. Weiss, J. Hawiger, Z. M. Ruggeri, V. T. Turitto, and P. Thiagarajan, T. Hoffmann. Fibrinogen-independent platelet adhesion and thrombus formation on subendothelium mediated by glycoprotein IIb-IIIa complex at high shear rate, *J. Clin. Invest.*, **83:**288-97 (1989).
20. D. F. Mosher, Influence of proteins on platelet-surface interactions, *in:* Interactions of the blood with natural and artificial surfaces, E. Salzman, ed., Marcel Dekker, N.Y. 55-101 (1972).
21. T. C. Nichols, D. A. Bellinger, T. A. Johnson, M. A. Lamb, T. R. Griggs, von Willebrand's disease prevents occlusive thrombosis in stenosed and injured porcine coronary arteries, *Circulation Res.*, **59(No 1):**15-26 (1986).
22. J. R. O'Brien, G. P. Salmon, and R. V. Majer, Abnormalities in the in vitro "filter bleeding time" in pregnancy and in pre-eclamptic toxaemia (PET). *Thromb. and Haemostas.*, **62:**424 (1989).
23. S. M. Magid, M. Perlin, and E. L. Gottfried. Increased erythrocyte osmotic fragility in pregnancy, *Am. J. Obstet. Gynecol.*, **144:**910-914 (1982).

24. P. M. Mannucci, Desmopressin: a nontransfusional form of treatment for congenital and acquired bleeding disorders, *Blood,* **72(no. 5):**1449-55 (1988).
25. J. R. O'Brien, and G. P. Salmon, Heat treatment in von Willebrand's disease, *Br. Med. J.,* **291:**409 (1985).
26. J. R. O'Brien, P. J. Green, G. P. Salmon, Desmopressin and sheared platelets: a test, *Lancet,* **i:**655 (1988).
27. S. R. Hanson, F. I. Pareti, and Z. M. Ruggeri, et al. Effects of monoclonal antibodies against the platelet glycoprotein IIb/IIIa complex on thrombosis and hemostasis in the baboon, *J. Clin. Invest.,* **81:**149-58 (1988).
28. B. S. Coller, J. D. Folts, L. E. Scudder, and S. R. Smith, Antithrombotic effect of a monoclonal antibody to the platelet glycoprotein IIb/IIIa receptor in an experimental animal model, *Blood,* **68(No 3):**783-86 (1986).
29. D. A. Bellinger, T. C. Nichols, and M. S. Read, et al, Prevention of occlusive coronary artery thrombus by a mural monochromal antibody to porcine von Willebrand factor, *Proc. Natl. Acad. Sci. USA,* **84:**8100-04 (1987).
30. V. Fuster, B. Stein, L. Badimon, and J. H. Chesebro, Antithrombotic therapy after myocardial reperfusion in acute myocardial infarction, *J. Am. Coll. Cardiol.,* **12 (No 6):**78A-84A (1988).
31. M. J. Davies, A. C. Thomas, P. A. Knapman, and J. R. Hangartner, Intramyocardial platelet aggregation in patients with unstable angina suffering sudden ischemic cardiac death, *Circulation,* **73(No 3):**418-27 (1986).
32. S. Raha, C. Opper, and W. Wesemann, Correlation of membrane anisotropy with function in subpopulations of human blood platelets, *Br. J. Haemat.,* **72:**397-401 (1989).

REGULATION OF EICOSANOID BIOSYNTHESIS IN ENDOTHELIAL CELLS: CRITICAL ROLE OF DE NOVO SYNTHESIS OF PROSTAGLANDIN ENDOPEROXIDE SYNTHASE

Kenneth Kun-Yu Wu

Department of Internal Medicine and Vascular
Disease Research Center, University of Texas
Medical School at Houston, Houston, TX 77030, USA

INTRODUCTION

Prostaglandins are important autacoids participating in a myriad of physiologic and pathophysiologic processes. Biosynthesis of prostaglandins and eicosanoids is regulated at several enzymatic steps (1). As shown in Fig. 1, liberation of arachidonic acid from membrane phospholipids upon cell activation is catalyzed by phospholipases. This is a rate limiting step. Once arachidonic acid is liberated, it is metabolized via the cyclooxygenase and lipoxygenase pathways. The enzyme that is responsible for catalysis of arachidonic acid into endoperoxides is prostaglandin endoperoxide synthase (prostaglandin G/H synthase, prostaglandin H synthase or cyclooxygenase). This molecule possesses two enzymic activities: cyclooxygenase which catalyzes the oxygenation of arachidonic acid into prostaglandin G_2 (PGG_2) and peroxidase converting PGG_2 into PGH_2 (2). Prostaglandin H synthase is upregulated by peroxides including PGG_2 but once PGG_2 is converted to PGH_2, oxygen radicals that are generated appear to cause irreversible inactivation of the cyclooxygenase activity (3-5). This leads to limited synthesis of PGH_2. PGH_2 is the common precursor for prostacyclin (PGI_2), PGE_2, $PGF_{2\alpha}$, PGD_2 and thromboxane A_2 (TXA_2). With compromised PGH_2 synthesis, production of these biologically active metabolites is consequently self-limited. This enzymatic step, hence, plays a central role in controlling eicosanoid biosynthesis.

Several studies on endothelial cells have shown that agonists such as thrombin, histamine and ionophore stimulate prostacyclin production rapidly and PGI_2 synthesis reaches plateau at 10-30 min and then declines abruptly (6, 7). The rapid rise and fall of PGI_2 formation in endothelial cells in response to these agonists are consistent with the anticipated autoinactivation of cyclooxygenase. It is, hence, intriguing to note that eicosanoid biosynthesis induced by cytokines and growth factors follows a time course quite different from that of PGI_2 stimulation by thrombin, histamine or ionophore A23187. The onset of eicosanoid synthesis is delayed but once eicosanoid synthesis is started, it is sustained for a prolonged period of time lasting for several hours (8-11). The eicosanoid synthesis is blocked by pretreating the cells with cycloheximide or actinomycin D indicating that the sustained eicosanoid formation depends on protein synthesis. Recent studies reveal that the mechanism by which these factors induce eicosanoid biosynthesis is mediated through the stimulation of de novo synthesis of prostaglandin H synthase. The purpose of this paper is to provide an overview of this important mechanism of eicosanoid biosynthesis.

Prostaglandin H (PGH) synthase

PGH synthase is a membrane-associated enzyme. It was purified to homogeneity from ram seminal vesicles in 1970s (12-14). The purified enzyme appeared to be a homodimer

Fibrinogen, Thrombosis, Coagulation, and Fibrinolysis, Edited by
C. Y. Liu and S. Chien, Plenum Press, New York

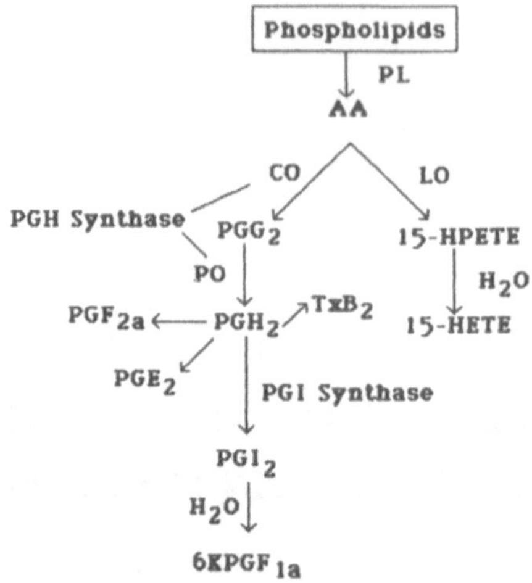

Fig. 1. Arachidonic acid metabolism in endothelial cells. The metabolic cascade serves to illustrate the enzymic regulation of prostacyclin biosynthesis. Abbreviations used are as follows: PL: phospholipase A_2; CO: cyclooxygenase activity; PO: peroxidase activity; LO: lipoxygenase; AA arachidonic acid; TXB_2 thromboxane B_2, 15-HPETE, 15-hydroperoxyeicosatetraenoic acid, 15-HETE, 15-hydroxyeicosatetraenoic acid, PGG_2, PGH_2, $PGF_{2\alpha}$, PGE_2 are prostaglandin G_2, H_2, $F_{2\alpha}$ and E_2 respectively. $6KPGF_{1\alpha}$ is 6-keto-$PGF_{1\alpha}$.

and the apparent molecular mass of each subunit was estimated to be 70 kDa. This has been corroborated by the amino acid sequence derived from the full-length cDNA recently cloned from ram seminal vesicle (15, 16). The full-length cDNA consists of 1,800 nucleotides encoding a protein of 600 amino acids including a signal sequence of 24 amino acids. There are 3 potential sites for N-glycosylation. A small hydrophobic region is present which is considered to be the membrane-associated site. The serine reported to be acetylated by aspirin (17) is at the serine 506 residue. As aspirin causes irreversible inactivation of the cyclooxygenase activity by acetylating this residue, it has been suggested that the active site of the enzyme is at or near the serine 506. Work is now in progress to test the validity of this assumption by using molecular biology techniques.

PGH synthase is widely distributed in human cells. It remains unclear, however, whether the PGH synthase in each cell type is identical nor is it clear whether there is specific difference. Earlier experiments showed that the enzyme in various cell types from a wide variety of mammalian species exhibited differential sensitivities to inhibition by nonsteroidal anti-inflammatory agents (18). It was further shown that the human platelet PGH synthase is 2 orders of magnitude more sensitive to aspirin than the vascular endothelial cell enzyme (19). It is possible that this enzyme may be present in isozyme forms.

Cyclooxygenase is inactivated rapidly after arachidonic acid is bound to the enzyme and converted to PGH_2 (3-5). Its half-life at the "activated" state is estimated to be around 5 min (20). Inactivation appears to be related to the generation of free radicals when PGG_2 is converted to PGH_2 (3-5). The exact molecular mechanism by which irreversible inactivation of cyclooxygenase occurs remains to be investigated. Recent work from the in vitro experiments on the purified enzyme has provided clues as to the potential site of enzyme degradation. Limited trypsin digestion of the purified enzyme results in the cleavage of the enzyme at arginine 253 yielding 2 major fragments with an apparent molecular mass of 38 kDa and 33 kDa respectively (21, 22). It seems likely that this may also be the site of intracellular proteolytic degradation.

Endothelial cell PGH synthase turnover

Two distinct types of PGI_2 stimulation have been observed in cultured endothelial cells. One type is the rapid but limited stimulation by thrombin, histamine, ionophore A23187 and arachidonic acid and the other is the delayed but sustained stimulation by cytokines. We postulated that these 2 types of stimulation are critically related to the regulation of PGH synthase turnover by these agonists. To test this hypothesis, we first determined the steady state level of PGH synthase in resting and stimulated cultured human umbilical endothelial cells (HUVEC) by Western blot analysis using an affinity-purified polyclonal antibodies raised in rabbits agonist PGH synthase purified from ram seminal vesicle (23). PHG synthase is present as 70 kDa band on Western blot in resting HUVEC. The 70 kDa band was significantly enhanced by cytokines (IL-1 and IL-2). In contrast, histamine, arachidonic acid and ionophore A23187 did not increase the 70 kDa band but instead they caused degradation fragments visible on the Western blot. Enhancement of the 70 kDa band by cytokines was time and concentration dependent. The enhancement was apparent after 1 hr of incubation, reached plateau approximately 6 hr. Increment in the enzyme mass shown by Western blot is accompanied by a corresponding rise of the enzyme activity. These data indicate that cytokines increase the steady state level of PGH synthase whereby they induce sustained eicosanoid biosynthesis.

Increased steady state level of PGH synthase may be due to enhanced enzyme synthesis, retarded enzyme degradation or both. To test these possibilities, we determined the methionine incorporation by a reverse immunoblot technique recently described (24). The methionine incorporation in IL-1 treated cells was increased over that in resting cells by approximately 2-fold. The degradation rate was evaluated by pulse and chase experiments. Cells were prelabeled with L-[^{35}S]methionine for 40 min and then chased with cold methionine. The decay of methionine was determined over 60 min periods by reverse immunoblot. The decay curves of resting and IL-1 treated cells were similar. They showed a single exponential decay. The half-life of the enzyme is 5.3 min and 6.3 min in IL-1 treated and resting cells respectively. It is interesting to note that a constant level of enzyme appears to decay very slowly regardless whether the cells are at the quiescent or stimulated state. It is speculated from these data that there may be 2 pools of PGH synthase.

Biologically active phorbol esters such as phorbol 12-myristate 13-acetate (PMA) are known to stimulate eicosanoid synthesis (25). PMA exerted a dose- and time-dependent stimulation of endothelial cell eicosanoid biosynthesis by a process requiring protein synthesis (24). We have shown that PMA enhances the 70 kDa PGH synthase band on the Western blot and the time course of stimulation is similar to that of IL-1 stimulation (24). Like IL-1, PMA boosts the methionine incorporation without altering the degradation rate (24). These results indicate that phorbol esters stimulate endothelial cell prostacyclin synthesis by a mechanism similar to IL-1. In fact, it has been suggested that the action of IL-1 on stimulating PGH synthase production is signaled by activation of protein kinase C (26).

These results indicate that stimulation of de novo synthesis of PGH synthase is an important mechanism for prostacyclin and other eicosanoid biosynthesis. This may be a common mechanism by which cytokines, growth factors and phorbol esters induce sustained formation of PGI_2, PGE_2 and other biologically active eicosanoids. The schematic view of this mechanism is shown in Fig. 2. These stimulating factors bind to their respective receptors, trigger signal transduction whereby 2 concurrent events occur. On one hand, the phospholipase A_2 is activated, arachidonic acid liberated and metabolized into PGI_2, PGE_2 and other prostaglandins. This process is short-lived because of autoinactivation of cyclooxygenase. On the other hand, the cyclooxygenase pool is constantly renewed because of stimulation of de novo synthesis of PGH synthase. This permits the sustained eicosanoid biosynthesis. The mechanism by which de novo synthesis of PGH synthase occurs is not entirely clear. Recent studies suggest that it may be due to enhanced message expression of PGH synthase gene.

Pathophysiological implications

In summary, PGI_2 biosynthesis may be regulated by 3 mechanisms: (1) short-lived, limited production stimulated by physiological agonists such as thrombin, histamine, bradykinin, etc.; (2) sustained formation elicited by cytokines and growth factors and (3) cell-cell cooperation in the formation of PGI_2. This is exemplified by shuntinig of PGH_2 generated by stimulated platelets into endothelial cells, vascular smooth muscle cells or lymphocytes which is converted into PGI_2 (27-29). These mechanisms work in concert to provide an adequate quantity of PGI_2 for defending agonist platelet thrombi and vasospasm.

Sustained stimulation of PGE_2 and PGI_2 formation by cytokines has important implications in inflammation and tissue injury. It has recently been shown that in joints afflicted

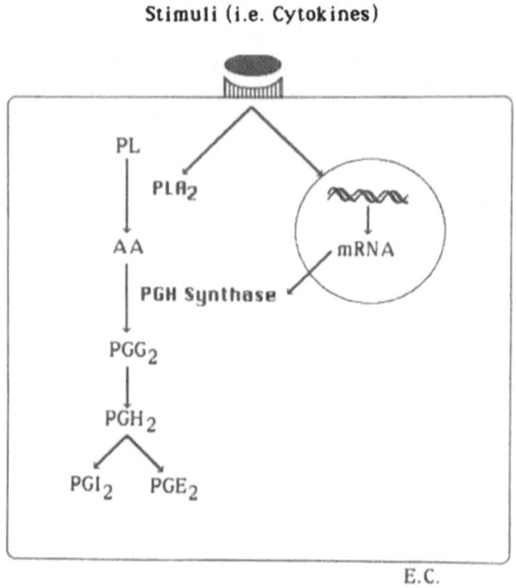

Fig. 2. Schematic view of cytokine effect on arachidonic acid metabolism in endothelial cells.

with rheumatoid arthritis there is a large quantity of IL-1 generated accompanied by a high level of PGE_2 in the synovial fluid. PEG_2 is thought to be produced by microvascular endothelial cells, synovial cells and fibroblasts. PGE_2 is considered to be a major inflammatory mediator. Similar processes may take place in ischemic or injured tissues where IL-1, growth factors and eicosanoids may interplay to propel a perpetual cycle of tissue damage and infarction. Further investigations into this area of research will provide a better understanding of diagnosis and treatment of these important disorders.

The author wishes to thank Nancy Fernandez for preparing this manuscript. The work is supported in part by a program project grant from the U.S. National Institutes of Health (NS-23327).

REFERENCES

1. W. E. M. Lands, The biosynthesis and metabolism of prostaglandins, *Ann. Rev. Physiol.*, **41**:633-652 (1979).
2. S. Ohki, N. Ogino, S. Yamamoto, and O. Hayaishi, Prostaglandin hydroperoxidase an integral part of prostaglandin endoperoxide synthetase from bovine vascular gland microsomes, *J. Biol. Chem.*, **254**:829-836 (1979).
3. W. L. Smith and W. E. M. Lands, Oxygenation of polyunsaturated fatty acids during prostaglandin biosynthesis by sheep vesicular gland, *Biochemistry*, **11**:3276-3285 (1972).
4. R. W. Egan, J. Paxton, and F. A. Kuehl, Mecahnism for irreversible self-deactivation of prostaglandin synthetase, *J. Biol. Chem.*, **251**:7325-7335 (1976).
5. M. E. Hemler and W. E. M. Lands, Evidence for a peroxide-initiated free radical mechanism of prostaglandin biosynthesis, *J. Biol. Chem.*, **255**:6253-6261 (1980).
6. R. S. Kent, S. L. Diedrich, and J. Whorton, Regulation of vascular prostaglandin synthesis by metabolites of arachidonic acid in perfused rabbit aorta, *J. Clin. Invest.*, **72**:455-465 (1983).
7. T. M. McIntire, K. S. Zimmerman, and S. M. Prescott, Cultured endothelial cells synthesize both platelet-activating factor and prostacyclin response to histamine, bradykinin, and adenosine triphosphate, *J. Clin. Invest.*, **76**:271-280 (1985).
8. V. Rossi, G. Breviario, P. Ghezzi, E. Dejana, and A. Mantovani, Prostacyclin synthesis induced in vascular cells by interleukin-1, *Science*, **229**:174-176 (1985).
9. R. R. Schleef, M. P. Bevilacqua, M. Sawdey, M. A. Gimbrone, and D. J. Loskutoff, *J. Biol. Chem.*, **263**:5797-5803 (1988).

10. P. J. Whiteley and P. Needleman, Mecahnism of enhanced fibroblast arachidonic acid metabolism by mononuclear cells, *J. Clin. Invest.*, **74**:2249-2253 (1984).
11. C. R. Albrightson, N. L. Baenziger, and P. Needleman, Exaggerated human vascular cell prostaglandin biosynthesis mediated by monocytes: Role of monokines and interleukin 1, *J. Immunol.*, **135**:1872-1877 (1985).
12. M. Hemler and W. E. M. Lands, Purification of the cyclooxygenase that forms prostaglandins, *J. Biol. Chem.*, **251**:5575-5579 (1976).
13. T. Miyamoto, N. Ogino. S. Yamamoto, and O. Hayaishi, Purification of prostaglandin endoperoxide synthetase from bovine vesicular gland microsomes, *J. Biol. Chem.*, **251**:2629-2636 (1976).
14. F. J. Van der Ouderaa, M. Buytenhek, D. H. Nugteven, and D. A. Van Dorp, Purification and characterization of prostaglandin endoperoxide synthetase from sheep vesicular glands, *Biochim. Biophys. Acta,* **487**:315-331 (1977).
15. D. L. DeWitt and W. L. Smith, Primary structure of prostaglandin G/H synthase from sheep vesicular gland determined from complementary DNA sequence, *Proc. Natl. Acad. Sci. (USA),* **85**:1412-1416 (1988).
16. J. P. Merlie, D. Fagan, J. Mudd, and P. Needleman, Isolation and characterization of the complementary DNA for sheep seminal vesicle prostaglandin endoperoxide synthase, *J. Biol. Chem.*, **263**:3550-3553 (1988).
17. G. J. Roth, N. Stanford, and P. W. Majerus, Acetylation of prostaglandin synthase by aspirin, *Proc. Natl. Acad. Sci. (USA),* **72**:3073-3076 (1975).
18. R. J. Flower, Drugs which inhibit prostaglandin biosynthesis, *Pharmacol. Reviews,* **26**:33-67 (1974).
19. J. W. Burch, N. L. Baenziger, N. Stanford, and P. W. Majerus, Sensitivity of fatty acid cyclooxygenase from human aorta to acetylation by aspirin, *Proc. Natl. Acad. Sci. (USA),* **75**:5181-5184 (1978).
20. J. M. Fagan and A. L. Goldberg, Inhibitors of protein and RNA synthesis cause a rapid block in prostaglandin production at the prostaglandin synthase step, *Proc. Natl. Acad. Sci. (USA),* **83**:2771-2775 (1986).
21. Y.-N. Chen, M. J. Bienkowki, and L. J. Marnett, Controlled tryptic digestion of prostaglandin H synthase, *J. Biol. Chem.*, **262**:16892-16899 (1987).
22. R. J. Kulmacz and K. K. Wu, Topographic studies of microsomal and pure prostaglandin H synthase, *Arch. Biochem. Biophys.*, **268**:502-515 (1989).
23. K. Frasier-Scott, H. Hatzakis, D. Seong, C. M. Jones, and K. K. Wu, Influence of natural and recombinant IL-2 on endothelial cell arachidonate metabolism, *J. Clin. Invest.*, **82**:1877-1883 (1988).
24. K. K. Wu, H. Hatzakis, S. S. Lo, D. C. Seong, and S. K. Sanduja, Stimulation of de novo synthesis of prostaglandin G/H synthase in human endothelial cells by phorbol ester, *J. Biol. Chem.*, **263**:19043-19047 (1988).
25. G. A. Beaudry, M. Waite, and L. W. Daniel, Regulation of arachidonic acid metabolism in Madin-Darby canine kidney cells: Stimulation of synthesis of the cyclooxygenase system by 12-0-tetradecanoly-phrobol-13-acetate, *Arch. Biochem. Biophys.*, **239**:242-247 (1985).
26. A. Raz, A. Wyche, and P. Needleman, Temporal and pharmacological division of fibroblast cyclooxygenase expression into transcriptional and translational phases, *Proc. Natl. Acad. Sci. (USA),* **86**:1657-1661 (1989).
27. P. A. Needleman, A. Wyche, and A. Raz, Platelet and blood vessel arachidonate metabolism and interactions, *J. Clin. Invest.*, **63**:345-349 (1979).
28. A. J. Marcus, B. B. Weksler, E. A. Jaffe, and M. J. Broekman, Synthesis of prostacyclin from platelet derived endoperoxides by cultured human endothelial cells, *J. Clin. Invest.*, **66**:979-986 (1980).
29. K. K. Wu, A. C. Papp, C. E. Manner, and E. R. Hall, Interaction between lymphocytes and platelets in the synthesis of prostacyclin, *J. Clin. Invest.*, **79**:1601-1606 (1987).

MODULATION OF ENDOTHELIAL FUNCTION BY HYPOXIA: PERTURBATION OF BARRIER AND ANTICOAGULANT FUNCTION, AND INDUCTION OF A NOVEL FACTOR X ACTIVATOR

Satoshi Ogawa, Revati Shreeniwas, Caesar Butura,
Jerold Brett, and David M. Stern

Department of Physiology and Cellular Biophysics
College of Physicians and Surgeons of Columbia University
New York, NY 10032, USA

ABSTRACT

Exposure of the vessel wall to hypoxemia is a central feature of ischemic cardiovascular disease. This led us to examine the perturbation of endothelial cell properties under hypoxia. An atmosphere of pO_2 of 12 mmHg is not lethal to the endothelial cells for up to five days, but barrier function was impaired. Increased passage of macromolecule tracers were observed in time- and dose-dependent manner and electron microscopy demonstrated small gaps (0.5 — 1.0 μm) between cells. Expression of the anticoagulant cofactor thrombomodulin was also perturbed: thrombomodulin activity and antigen decreased in parallel. Northern blots showed almost complete suppression of thrombomodulin in hypoxic culture. Furthermore, synthesis of other proteins, such as fibronectin, was slightly enhanced under hypoxia. In addition to the suppression of these anticoagulant cofactor, hypoxic endothelial cell displayed a noval procoagulant activity distinct from tissue factor. Further study revealed that hypoxic endothelial cultures directly activated Factor X, as assessed by functional assays and SDS-PAGE. In addition to this no activation of Factor IX or prothrombin was observed. The hypoxia-induced Factor X activator was membrane-associated, required calcium to form Factor Xa, was inhibited by $HgCl_2$ but not by PMSF, and had Km $\approx$ 25 μg/ml. Co-incubation of hypoxic cultures with cycloheximide prevented the expression of this activity, suggesting that protein synthesis is required for its expression. These functional perturbations of endothelial cells were reversible following reoxygenation. These data indicate that hypoxia imposes a selective perturbation on endothelial cell function, suggesting the possible contribution of hypoxemia to vascular dysfunction in ischemia.

INTRODUCTION

Hypoxemia is a frequent accompaniment of disorders of the circulatory system, especially those associated with ischemic vascular disease. The effect of hypoxia on cellular functions is quite complex, including changes in cellular functions, and induction of certain proteins (oxygen regulated proteins) along with suppression of others (1, 2). Endothelial cells, which form the luminal vascular surface, are directly exposed to changes in environmental conditions in the intravascular space. Of the many changes in vascular function caused by hypoxemia, two are of special clinical importance; increased vascular permeability (3, 4) and a prothrombic tendency (5). Twenty years ago, Sevitt (6) indicated that thrombi formed in venous valve pockets, an area subject to stasis and hypoxia (5). Morever, the observation that the clot was attached to vessel surface of the valve cusp raised the speculation that hypoxia-induced perturbation of endothelial function might play an important role in the pathogenesis of thrombosis (7). Based upon theses considerations, we have exposed cultured bovine endothelial cells to low oxygen concentrations, and assessed perturbation of its two central functions, barrier function and regulation of intravascular coagulation.

MATERIAL AND METHOD

1) Cultured endothelial cells and conditions to achieve hypoxia.

Bovine aortic endothelial cells were grown from newborn calf aortas in minimal essential medium containing penicillin-streptomycin (50 U/ml, 5 μg/ml), glutamine and 10% fetal calf serum (Hyclone, Logan, Utah) as described (8). Cultures were characterized by morphologic criteria, and immunofluorescence for von Willebrand factor and thrombomodulin (9). Cells were separated for subculture with trypsin/EDTA, and cells from passages 3-12 were grown to confluence in different size culture dishes (Becton Dickinson Labware, Lincoln Park, NJ). Permeability studies employed cultures grown, as described previously (10), on 6.5 mm diameter polycarbonate membranes, mounted on polystyrene inserts (Transwell, Coaster, Cambridge, MA). Cultures were characterized based on cell density, determined by Coulter Counter (Model ZM; Coulter Electronics, Luton, England), time in culture, and labelling index using Amersham kit (Arlington Hts., Ill). After achieving confluence (7-10 days after planting, labelling index < 1%, 2×10^5 cells/cm^2) in an ambient atmosphere, endothelial cells were transferred to the hypoxia chamber for further study.

Confluent endothelial cultures were made hypoxic by transferring them to an incubator attached to the hypoxia chamber, which could maintain a humidified atmosphere with the desired low oxygen tensions (Coy Laboratory Products, Ann Arbor, MI). At intervals throughout these experiments, the oxygen tension of culture medium bathing the cells was repeatedly analyzed by gas analyzer ABL-2 (Radiometer, Sweden). Values shown in the figures are the partial pressure of oxygen in the medium. During the course of these experiments, pH of the medium was constant. In some experiments, either cycloheximide (0.1 μl/ml for 48 hr; Sigma) or Warfarin (1 μg/ml for 48 hr; Sigma) was added to the growth medium under hypoxic conditions. Endothelial monolayers were incubated with recombinant human tumor necrosis factor and procoagulant activity was studied as described below.

2) Determination of endothelial monolayer permeability.

Confluent endothelial monolayers grown to confluence on Transwell inserts were transferred to hypoxia chamber, and after the indicated time of hypoxa, a radiolabelled marked was then added, either ^{3}H-inulin (3 μg/ml; 24 Ci/g New England Nuclear), ^{3}H-sorbitol (38 ng/ml; 24 Ci/mmole, New England Nuclear) or ^{125}I-albumin (150 ng/ml; 5000 cpm/ng), to upper chamber. Radiolabelled albumin was prepared by the lactoperoxidase method (11) using Enzymobeads (Biorad, Sacremento, CA), desalted with a Sephadex G25 column (Pharmacia, Piscataway, NJ), and dialyzed exhaustively versus Hanke's balanced salt solution. Transport of tracers from the inner to outer chamber, i.e. across the endothelial monolayer, was determined by dividing radioactivity emerging in the outer well by radioactivity remaining in the inner well.

3) Assessment of endothelial coagulant properties.

Endothelial thrombomodulin activity and antigen were assessed after exposing cultures to hypoxia, as follows: cultures exposed to different oxygen tensions, washed three times with Hanke's balanced salt solution, and then incubated for 60 min at 37°C in HEPES (10 mM, pH = 7.45), NaCl (137 mM), glucose (11 mM), KCl (4 mM), $CaCl_2$ (2 mM) and bovine serum albumin (1 mg/ml) containing protein C (100 μg/ml) and thrombin (0.1 U/ml). Formation of activated protein C was terminated by adding antithrombin III (100 μg/ml), and the amount of the enzyme formed was determined using a chromogenic assay, hydrolysis of the substrate (Spectrozyme, American Diagnostica, NY). Enzyme concentration was determined by comparison with a standard curve made in the presence of known amounts of activated protein C (12).

Total thrombomodulin antigen was assessed by radioimmunoassay of endothelial cell detergent lysates (1% NP-40) prepared in the presence of protease inhibitors, PMSF (2 mM; Sigma) and leupeptin (0.3 mM; Boehringer Mannheim, Houston, TX). The radioimmunoassay was carried out according to the method described previously (13). The sensitivity of this radioimmunoassay was 10 ng/ml, which corresponded to 80% binding on the standard curve.

Levels of thrombomodulin were compared with total protein synthesis and fibronectin production. The effect of hypoxia on overall protein biosynthesis was assessed by evaluating incorporation of ^{3}H-leucine (60 Ci/mmole; New Englane Nuclear) into trichloroacetic acid precipitable material, as described by Madri et al. (14). Endothelial elaboration of fibronectin was assessed by obtaining aliquot of culture supernatant, and determining their fibronectin content by radioimmunoassay. The radioimmunoassay for fibronectin was carried out by the same procedure described for thrombomodulin except that tracer was ^{125}I-bovine fibronectin, the antibody was polyclonal anti-bovine fibronectin (Calbiochem San Diego, CA), and incubation of first antibody was done for 8 hrs at 25°C.

Northern blots to assess levels of thrombomodulin and fibronectin mRNA were carried out by the procedure reported previously (15, 16). cDNA probes for thrombomodulin (generously provided by Dr. E. Sadler, Washington Univ., St. Louis, Mo) (17) and fibronectin (generously provided by Dr. R. Hynes, M.I.T., Cambridge, Mass) (18) were labelled using random labelling (Boehringer Mannheim random primed DNA labelling kit, IN) and hybridization was performed at 42°C as described previously (19).

Study of endothelial cell procoagulant properties was carried out using coagulant assays. The coagulant assay was performed on cells scraped with a rubber policeman to obtain a suspension ($\approx 10^6$ cells/ml). The cells were washed three times with veronal buffer and then 0.2 ml of the same buffer was added along with the citrated bovine plasma (0.2 ml) and $CaCl_2$ (20 mM; 0.2 ml), and the mixture was incubated at 37°C for the first visual evidence of fibrin clot. The Fctor Xa clotting assay was performed by incubating purified Factor X (50 μg/ml) with hypoxic or normoxic endothelial cell suspensions or monolayers for indicated time, removing aliquot (60 μl) and adding them to factor VII/X deficient plasma (60 μl; Sigma) along with cephalin (60 μl) and $CaCl_2$ (20 mM; 60 μl). Enzyme concentration was determined by comparison with a standard curve made with known amounts of Factor Xa. In some experiments, after the initial exposure to hypoxia, endothelial cells were further incubated with either antibody to tissue factor (10 μg/ml; 1 hr at 37°C) (blocking anti-bovine tissue factor monoclonal antibody was generously provided by Dr. R. Bach, Mt. Sinai School of Medicine, NY) (20), mercury chloride (0.1 mM), or PMSF (1 mM, for 30 min at room temp.) prior to carry out the assay. Activation of Factor X by hypoxic endothelium was also studied by examining cleavage of ^{125}I-Factor X on reduced SDS-PAGE. Factor X was radiolabelled by the lactoperoxidase method using Enzymobeads as described previously for albumin. ^{125}I-Factor X was activated by Russell's viper venom (generously supplied by Dr. Richard Hart, American Diagnostica, NY).

To examine modulation of vessel wall coagulant properties using fresh native endothelium, an experiment was carried out with calf aortic segments obtained immediately after sacrifice (Max Cohen, Inc., NY). The vessel segments were incubated for 24 hr in complete endothelial growth medium and exposed to either normoxic or hypoxic ($pO_2 \approx 14$ mmHg) conditions for 24 hr, then washed with Hanke's balanced sale solution, placed in a template device, and functional assays for thrombomodulin and the Factor X activator were carried out as described above. As described previously (21), vessel segments were placed between the two sheets of the template such that water tight wells with exposed native endothelium were formed.

4) Morphologic studies.

For immunocytologic studies, cell monolayers grown on coverslips were exposed to normoxia or hypoxia, and fixed in that atmosphere in phosphate-buffered saline, pH 7.2, containing 3.5% formalin and 0.1% NP-40 for 5 min and washed in phosphate-buffered saline. For visualization of F-actin, coverslips were incubated with rhodamine conjugated phalloidin (Molecular Probes, Junction City, OR) for 30-45 min, washed in phosphate-buffered saline and mounted in Gelvatol containing 1 mg/ml p-phenylenediamine. Mounted coverslips were examined in a Leitz Dialux 20 microscope with a 2.4 Ploempak filter block and water immersion fluorite objectives, and recorded on Kodak Tri-X film. The protocol for visualization of protein S has been described previously (22) and was identical to that used for thrombomodulin and von Willebrand factor. In each case, primary antibody was visualized with fluorescein-conjugated goat anti-rabbit IgG antibody (Sigma).

For scanning electron microscopy, monolayers grown on glass coverslips were exposed to hypoxia or normoxia and fixed in that atmosphere in cacodylate buffer (0.1 M, pH 7.2) containing glutaraldehyde (2.5%) and sucrose (2%) for 45 min, and then washed three times in buffer followed by post-fixation in buffer containing osmium tetroxide (2%). Fixed monolayers were dehydrated in ethanol, transferred to Pel-Dry (Ted Pella Inc., Redding CA) and dried by sublimation in a fume hood. Dried monolayers were coated with gold palladium and viewed in a Joel T-300 scanning electron microscope.

RESULTS

1) General properties and barrier function of hypoxic endothelial cultures.

Consistent with a previous report (23), endothelial monolayers tolerated the hypoxic environment (as low as $pO_2 \approx 14$ mmHg, up to 5 days) well with no evidence of cell death, based on trypan blue exclusion and LDH release. However, cells exposed to hypoxia for 24 hrs appeared flatter, their peripheral bands of circumferential stress fibers appeared diffuse, and in the central cytoplasm reticular networks of thin actin filaments developed, compared with normoxic controls. These changes became more apparent by 48 hr, and were reversible following return to the ambient atmosphere (Fig. 1A-C). Consistent with this, scanning electron microscopy revealed the presence of small gaps between cells (Fig. 1D), suggesting that the barrier function would be perturbed. Using the Transwell membrane system, cultures shifted to hypoxia ($pO_2 \approx 14$ mmHg) demonstrated an increase in the passage of macromolecular tracers across the monolayers (Fig. 2A-C). The hypoxia-induced diffusional defect was observed in time- and dose-dependent manner, and restitution to an ambient air atmosphere led to a reversal of these alterations in monolayer permeability within 48 hr (data not shown), consistent with restoration of normal morphology of the monolayer.

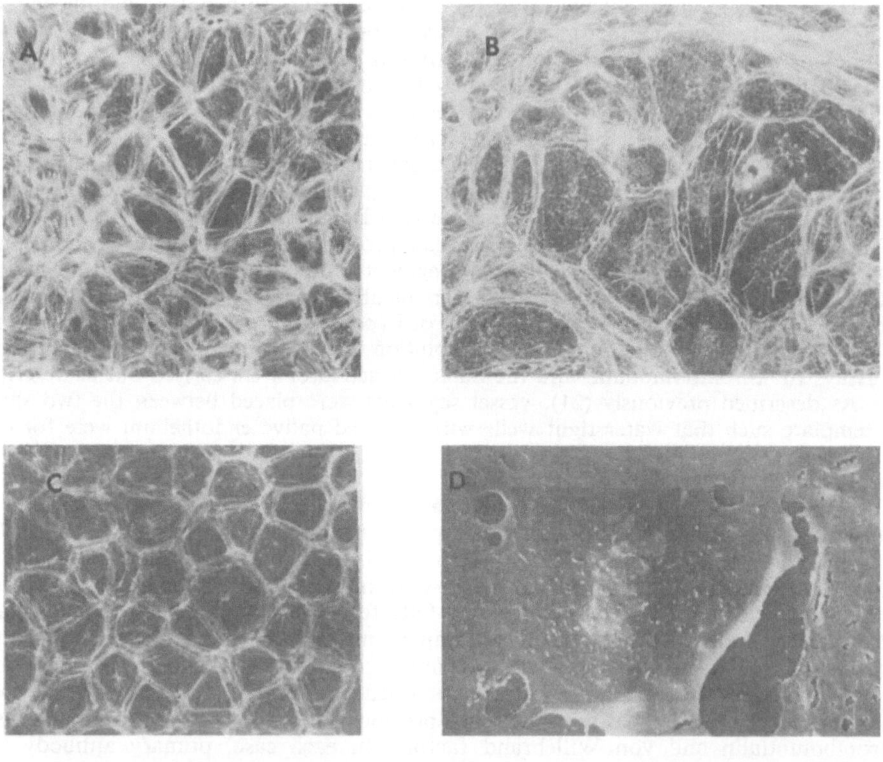

Fig. 1. Morphologic features of hypoxic endothelial cells. A-C: Changes in the actin based cytoskeleton visualized by rhodamine phalloidin staining. (A) Control, (B) 48 hr in hypoxia, (C) 48 hr recovery after hypoxia. (D) Scanning electron micrograph showing the intercellular gaps (X6500).

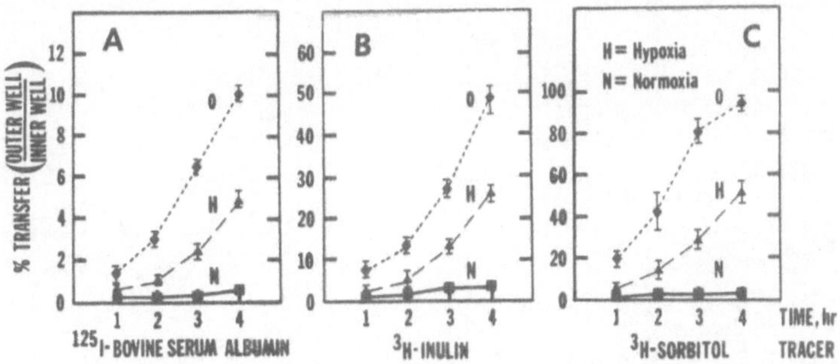

Fig. 2. Permeability of endothelial cells (72 hr in hypoxia) assessed by transfer of three tracers, (A) ^{125}I-BSA, (B) ^{3}H-inulin, (C) ^{3}H-sorbitol. In each case, H, N, and C, represents the passage of tracers across hypoxic monolayer, normoxic monolayer, and membranes without cells.

2) Hypoxia-mediated suppression of endothelial thrombomodulin.

Exposure of endothelium to an atmosphere with $pO_2 \approx 14$ mmHg led to a fall in cell surface thrombomodulin activity and antigen (Fig. 3A), reaching a maximum by 72 hr, at which time cell surface thrombomodulin activity and antigen had declined by about 80%. In addition to this, immunohistochemical studies revealed an apparent decrease in intensity of thrombomodulin antigen on hypoxic cell surface (Fig. 3B). Hypoxia-mediated suppression of thrombomodulin was also dependent on the oxygen concentration (data not shown) and reversible: levels of thrombomodulin activity/antigen recovers within 48 hr of reoxygenation. Based on the ^{3}H-leucine incorporation study, endothelial protein synthesis fell with a different time course and showed a decline of only about 35% compared with normoxic controls, suggesting that there might be a selective suppression of thrombomodulin under hypoxia. Consistent with this, another macromolecular product of endothelium, fibronectin, showed a slight increase in hypoxia (data not shown). Finally, Northern blots of endothelial mRNA hybridized with cDNA probes for thrombomodulin and fibronectin showed a marked decrease in thrombomodulin, but not in fibronectin (Fig. 3C).

3) Hypoxia-mediated induction of endothelial procoagulant activity.

Exposure of endothelial cultures to hypoxia induced clot-promoting activity, assessed by a shortening of the recalcification time of plasma. Expression of cell surface procoagulant activity was dependent on the time cultures were maintained in hypoxia (Fig. 4), and also depended on the oxygen concentration, especially at lower oxygen tensions (data not shown). Previous observations, indicating that endothelial cells incubated with certain cytokines express tissue factor (12, 24), led us to examine the possible induction of tissue factor procoagulant activity by hypoxia. A blocking monoclonal antibody to bovine tissue factor, which could reverse the procoagulant activity of tumor necrosis factor treated cells, failed to affect the clotting time of hypoxic endothelium (Table 1), suggesting that procoagulant activity other than tissue factor was responsible for the shortening of clotting time of hypoxic endothelium. The possibility that hypoxic endothelium could directly activate Factor IX or prothrombin was not supported by the results of coagulant and functional assays, and cleavage studoes of radiolabelled tracers on SDS-PAGE. However, studies with ^{125}I-Factor X revealed direct cleavage of the coagulation factor with formation of the heavy chains of Factor Xaα and Xaβ (Fig. 5, lane A) (25) on reduced SDS-PAGE. In contrast, neither buffer controls (Fig. 5, lane A)nor normoxic cultures (Fig. 5, lane B) cleaved Factor X. Control activation studies of the same iodinated Factor X by Russell's viper venom demonstrated cleavage of Factor X, mainly with Factor Xaα heavy chain (Fig. 5 lane D), as previously reported (26).

Further studies using Factor Xa coagulant assays showed that the induction of this procoagulant activity on hypoxic endothelial cells was dependent on the incubation time in hypoxia and the oxygen concentration (Fig. 6A-B). Neither antibody to tissue factor, factor VII, Factor IX, nor Factor VIII had any effect on Factor X activation (data not shown), indicating that the hypoxia-induced Factor X activator was distinct from classical intrinsic and extrinsic components.

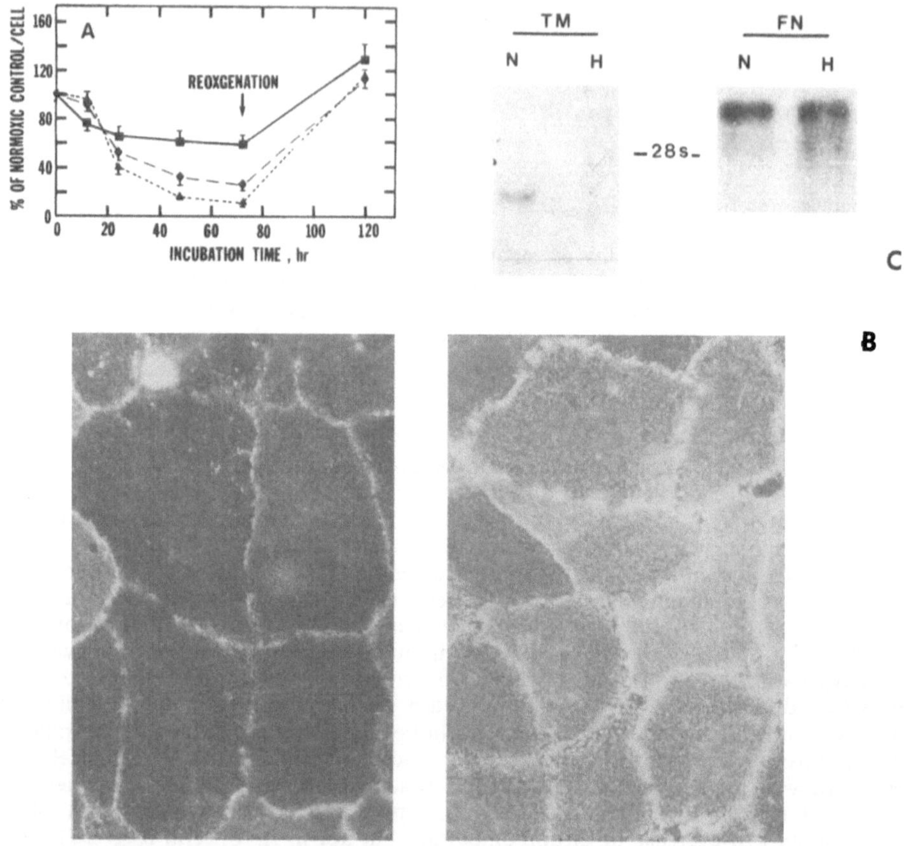

Fig. 3. (A) Hypoxia induced suppression of endothelial protein synthesis assessed by ^{14}C-leucine incorporation (I), thrombomodulin antigen (II), surface thrombomodulin activity (III). (B) Immunofluorescence for cell surface thrombomodulin in hypoxic (I) and normoxic (II) cultures. (C) Northern blots of 24 hr hypoxic (H) and normoxic (N) endothelial cell mRNA hybridized with cDNA probes with thrombomodulin (TM) and fibronectin (FN).

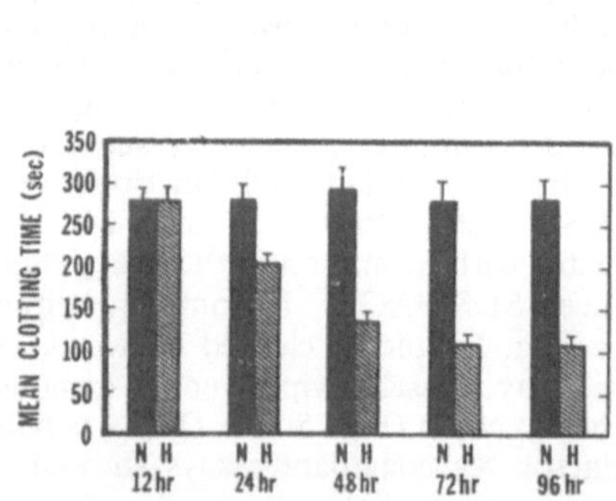

Fig. 4. Time course of hypoxia-induced expression of procoagulant activity assessed by the clotting time of recalcified plasma (H). (N) denotes normoxic controls.

Table 1. Effect of Blocking Tissue Factor Monoclonal Antibody on the Procoagulant Activity of Hypoxic and Tumor Necrosis Factor-Treated Endothelial Cultures

Cell Treatment	Clotting Time (Sec.)
Normoxia	255 ± 25
Hypoxia	130 ± 15
Hypoxia + Antibody	128 ± 17
TNF (1 μM)	125 ± 16
TNF + Antibody	278 ± 24

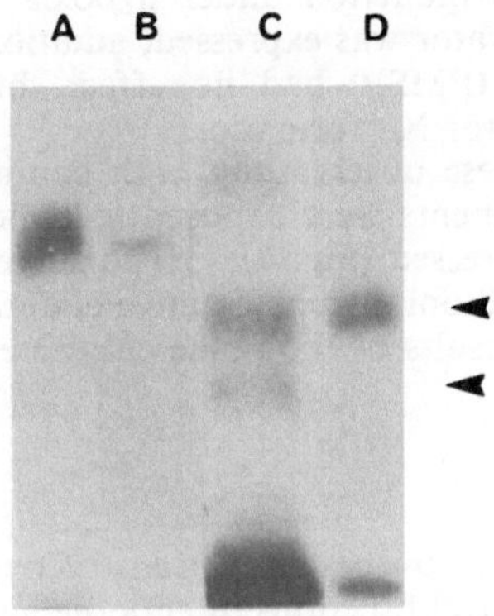

Fig. 5. Activation of ^{125}I-Factor X by hypoxic endothelial cells (72 hr in hypoxia) seen on reduced SDS-PAGE. Each lane represents ^{125}I-Factor X incubated with buffer only (A), normoxic endothelial cells (B), hypoxic endothelial cells (C), or Russell's Viper venom. Arrows correspond to the heavy chains of Factor Xaα and Xaβ.

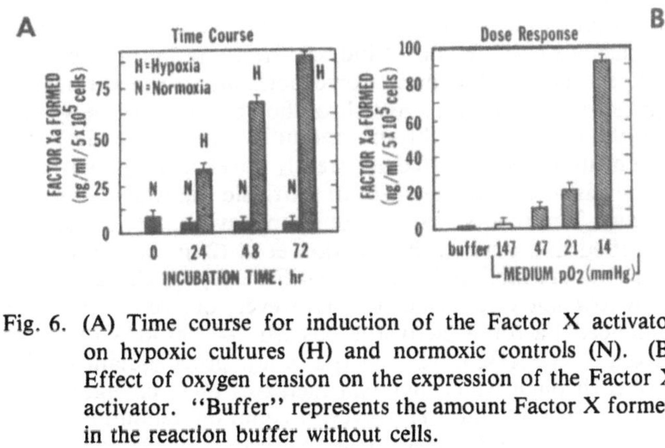

Fig. 6. (A) Time course for induction of the Factor X activator on hypoxic cultures (H) and normoxic controls (N). (B) Effect of oxygen tension on the expression of the Factor X activator. "Buffer" represents the amount Factor X formed in the reaction buffer without cells.

Table 2. Characterization of the Hypoxia-Induced Endothelial Activator of Factor X

1.	Localized on surface of hypoxic endothelial cells.
2.	Requires Ca^{++} to form Factor Xa (Maximal effect at 5 mM).
3.	Km ≈ 25 μg/ml and Vmax ≈ 1.1 ng/ml/min at pH = 7.4.
4.	Expression blocked by cycloheximide but not by warfarin.
5.	Activity inhibited by $HgCl_2$ but not by PMSF.

To characterize better the nature of this hypoxia-induced Factor X activator, several further studies were done and the results were summarized in Table 2. In brief, this activator was localized on the cell surface of hypoxic endothelium and required calcium to form Factor Xa (maximal effect at ≈ 5 mM). Activation of Factor X was saturable and appeared to fit Michael-Menten kinetics, with Km ≈ 25 μg/ml and Vmax 1.1 ng/ml/min at pH = 7.4. Expression of the Factor X activator was cycloheximide sensitive, suggesting that protein synthesis was necessary for its induction under hypoxia, although warfarin was without effect. Once the Factor X activator was expressed, addition of the serine protease inhibitor phenylmethylsulfonyl fluoride (PMSF) had no effect, but mercury chloride, a cysteine protease inhibitor, blocked Factor Xa formation.

In order to extrapolate these observations with cultured endothelium to native endothelium in situ, calf aortic segments were exposed to hypoxia and their ability to activate protein C and Factor X was assessed (Fig. 7). Hypoxic vessel segments showed a decrease in thrombomodulin activity, and only hypoxic native endothelium had the ability to activate Factor X, consistent with the results of the tissue culture experiments.

DISCUSSION

This study clearly show that low concentrations of oxygen, though not lethal to endothelial cells, can perturb two central functions of endothelium. In our experiments, increased permeability of endothelial monolayers, which was evident after 48 hr in hypoxia, was related to the formation of small intercellular gaps, which could form a paracellular pathway for fluid to escape. Although cultured endothelial monolayer used in this and previous studies (10, 27) are more permeable than native endothelium, this system provides a simple and useful model to examine the pathophysiology of increased vessel permeability, which can be seen in high altitude pulmonary edema or cerebral edema which accompanies ischemia (3, 4). Consistent with our observations, Stelzner et al. (3) demonstrated increased fluid in the lungs within 24-48 hr after exposing rats to hypoxia.

Another striking effect of hypoxia on the endothelial cell function was the perturbation of endothelial procoagulant properties. Thrombomodulin, a central anticoagulant cofactor expressed on the endothelial cell surface, is also reported to decrease following incubation of endothelial cells with tumor necrosis factor. In the latter case, increased degradation and/or decreased synthesis plays the main role in the decrease of thrombomodulin activity on endothelial cell surface (28). Hypoxia-induced decrease in thrombomodulin antigen and activity, and disappearance of the band corresponding to the thrombomodulin message, suggests that at lower oxygen tensions less thrombomodulin may be synthesized.

In addition to the reduction of thrombomodulin synthesis, endothelial cells exposed to hypoxia expressed a membrane-associated apparently novel procoagulant, a Factor X activator, which appears to be distinct from intrinsic or extrinsic pathways. From the properties of this Factor X activator identified so far, it would appear to most closely resemble the tumor procoagulant, identified and purified by Gordon et al (29): both are expressed on the cell surface membrane, both are inhibited not by PMSF but by $HgCl_2$, and both can be extracted from cell surface by low salt veronal buffer. However, our recent pilot studies have shown that the hypoxiainduced Factor X activator has Mr. ≈ 100 kDa (based on elution of the purified protein from SDS-PAGE), whereas the tumor procoagulant has Mr ≈ 68 kDa (29).

The mechanisms through which hypoxia mediates these changes in cell surface coagulant activities is unclear at this time. However, a common denominator of hypoxia-induced changes in cellular function could relate to a redirection of protein biosynthesis with sup-

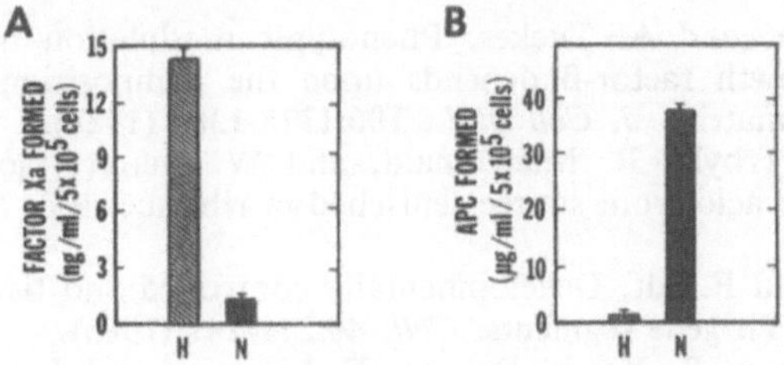

Fig. 7. (A) Activation of Factor X on native endothelium of aortic segments incubated for 24 hr in hypoxia (H) or normoxia (N). (B) Thrombomodulin activity on native endothelium of aortic segments incubated for 24 hr in hypoxia (H) or normoxia (N). Thrombomodulin activity was assessed by the activation of protein C.

pression of certain proteins and induction of others (1, 2) i.e. suppression of thrombomodulin synthesis and the induction of the Factor X activator. Although the physiologic significance of these findings in tissue culture must be extended to the in vivo situation, they provide new insights into possible mechanisms underlying the pathogenesis of vascular dysfunction in hypoxemia.

REFERENCES

1. J. Sciandra, J. Subject, and C. Hughes, Induction of glucoseregulated proteins during anaerobic exposure and of heat-shock proteins after reoxygenation, *PNAS (USA)*, **81**:4842-4847 (1984)
2. C. Heacock and R. Suhterland. Induction of characteristics of oxygen regulated proteins, *J. Radiat. Oncol. Biophys.*, **12**:1287-1290 (1986).
3. T. Stelzner, R. O'Brien, K. Sato, and J. Weil, Hypoxia-induced increase in pulmonary transvascular protein escape in rats, *J. Clin. Invest.*, **82**:1840-1847 (1988).
4. G. Kinasewitz, L. Groome, W. Marshall, W. Leslie, and H. Diana, Effect of hypoxia on permeability of pulmonary endothelium of canine visceral pleura, *J. Appl. Physiol*, **61**;554-560 (1986).
5. J. Hamer, P. Malone, and I. Silver, The pO_2 in venous pockets: its possible bearing on thrombogenesis, *Brit. J. Surg.*, **68**:166-170 (1981).
6. S. Sevitt, The acutely swollen leg and deep vein thrombosis, *Brit. J. Surg.*, **54**:886-890 (1967).
7. P. Malone, A hypothesis concerning the aetiology of venous thrombosis, *Med. Hypothesis*, **5**:189-201 (1977).
8. S. Schwartz, Selection and characterization of bovine aortic endothelial cells, *In Vitro*, **14**:966-984 (1978).
9. D. Stern, J. Brett, K. Harris, and P. Nawroth, Participation of endothelial cells in the protein C-protein S anticoagulant pathway: the synthesis and release of protein S. *J. Cell. Biol.*, **102**:1971-1978 (1986).
10. J. Brett, H. Gerlach, P. Nawroth, S. Steinberg, G. Godman, and D. Stern, Tumor necrosis factor/cachectin increases permeability of endothelial cell monolayers by a mechanism involving regulatory G proteins. *J. Exp. Med.*, **169**:1977-1991 (1989).
11. G. David and R. Reisfeld, Protein iodination with solid state lactoperoxidase, *Biochemistry*, **13**:1014-1020 (1974).
12. P. Nawroth and D. Stern, Modulation of endothelial cell hemostatic properties by tumor necrosis factor, *J. Exp. Med.*, **163**:740-745 (1986).
13. C. Esposito, H. Gerlach, J. Brett, D. Stern, and H. Vlassara, Endothelial receptor-mediated binding of glucose-modified albumin is associated with increased monolayer permeability and modulation of cell surface coagulant properties. *J. Exp. Med.*, **170**:1387-1408 (1989).

14. J. Madri, B. Pratt, and A. Tucker, Phenotypic modulation of endothelial cells by transforming growth factor-ß depends upon the composition and organization of the extracellular matrix. *J. Cell Biol.,* **106:**1375-1384 (1988).
15. J. Chirgwin, R. Przbyla, R. MacDonald, and W. Rutter, Isolation of biologically active ribonucleic acid from sources enriched in ribonuclease, *Biochemistry,* **18:**5294-5299 (1979).
16. G. Yancopoulos and F. Alt, Developmentally controlled and tissue-specific expression of unrearranged Vh gene segments, *Cell,* **40:**271-281 (1985).
17. D. Wen, W. Dittman, R. Ye, L. Deaven, P. Majerus, and J. Sadler, Human thrombomodulin: Complete cDNA sequence and chromosome localization of the gene, *Biochem.,* **26:**4350-4357 (1987).
18. J. Schwarzbauer, J. Tamkin, I. Lemischka, and R. Hynes, Three different fibronectin mRNA's arise by alternative splicing within the coding region, *Cell,* **35:**421-431 (1983).
19. M. Reth and F. Alt, Novel immunoglobulin heavy chains are produced from DJh gene segment rearrangements in lymphoid cells, *Nature,* **312:**418-423 (1984).
20. R. Bach, Initiation of coagulation by tissue factor, *CRC Critical Review in Biochem.,* **23:**339-368 (1988).
21. D. Stern, P. Nawroth, W. Kisiel, M. Handley, M. Drillings, and J. Bartos, A coagulant pathway on bovine aortic segments leading to generation of Factor Xa and thrombin, *J. Clin. Invest.,* **74:**1910-1921 (1984).
22. J. Brett, P. Steinberg, P. deGroot, P. Nawroth, and D. Stern, Norepinephrine down-regulated the activity of protein S on endothelial cells, *J. Cell Biol.,* **106:**2109-2118 (1988).
23. S. L. Lee and B. Fanburg, Glycolytic activity and enhancement of serotonin uptake by endothelial cells exposed to hypoxia/anoxia, *Circ. Res.,* **60:**653-688 (1987).
24. M. Bevilacqua J. Pober, G. Majeau, R. Cotran, and M. Gimbrone, Recombinant TNF induces procoagulant activity in endothelium, *PNAS (USA),* **83:**4533-4537 (1984).
25. K. Fujikawa, M. Coan, M. Legaz, and E. Davie, The mechanism of activation of bovine Factor X by intrinsic and extrinsic pathways, *Biochemistry,* **13:**5290-5299 (1974).
26. B. Furie, A. Gottlieb, and W. Williams, Activation of bovine Factor X by the venom coagulant protein of Vipera Russeli: Complex formation of the activation fragments, *Biochem. Biophys. Acta,* **365:**121-132 (1974).
27. M. Clark, M. J. Chem, S. Crooke, and J. Bomalaski, Tumor necrosis factor induces phospholipase A_2-activating protein in endhothelial cells, *Biochem. J.,* **250:**125-130 (1988).
28. K. Moore, C. Esmon, and N. Esmon, Tumor necrosis factor leads to the internalization and degradation of thrombomodulin from the surface of bovine aortic endothelial cell in culture, *Blood,* **73:**159-165 (1989).
29. A. Falanga and S. Gordon, Isolation and characterization of cancer procoagulant: a cystine proteinase from malignant tissue, *Biochemistry,* **24:**5558-5567 (1985).

FIBRIN AND THE VESSEL WALL

K. L. Kaplan[1,*], A. Bini[1,#], J. Fenoglio[1], Jr., and B. Kudryk[2]

Columbia University College of Physicians and Surgeons[1] and
The New York Blood Center[2]
New York, NY, USA

ABSTRACT

Fibrin is a major component of atherosclerotic plaques, and there may also be situations in which intravascular fibrin is formed in contact with the endothelium. The studies to be presented describe the distribution of fibrinogen/fibrin I, fibrin II, and fragments D and D-dimer in normal vessels and atherosclerotic plaques of increasing severity and also describe some functional effects of fibrin on normal endothelium. Immunohistochemical studies using three specific monoclonal antibodies with the avidin-biotin complex immunoperoxidase technique demonstrated that little fibrinogen/fibrin I or fibrin II and no D/D-dimer were detected in normal aortas. In early lesions and in fibrous plaques, fibrinogen/fibrin I and fibrin II were distributed in long threads and around vessel wall cells. D/D-dimer was not seen in early lesions. In advanced plaques all three molecular forms were detected in areas of loose connective tissue, in thrombi, and around cholesterol crystals. Thus increased fibrin formation and degradation may be associated with progression of atherosclerotic disease. Additionally, the presence of fibrin II around vessel wall cells suggests that these cells may be involved in the fbgn to fibrin transition within the vessel wall.

The second aspect of the work to be presented concerns effects of fibrin on vascular endothelium. Fibrin formed on the surface of cultured human umbilical vein endothelial cells stimulated production of prostacyclin and tissue plasminogen activator by the cells in a time- and dose-dependent manner. Stimulation of prostacyclin was completely inhibited by indomethacin and partially inhibited by actinomycin D, cycloheximide, and trifluoperazine, while stimulation of t-PA synthesis was completely inhibited by actinomycin D and cycloheximide and partially inhibited by cytochalasin D, vinblastine, and trifluoperazine. Neither collagen polymerization on the cells nor thrombin added in subclotting concentrations was stimulatory, and Reptilase fibrin caused only a slight increase in t-PA. Soluble fibrin II generated in the presence of gly-pro-arg-pro failed to stimulate production of prostacyclin or t-PA. Thus these studies demonstrate that polymerized fibrin II is necessary for stimulation. Such stimulation could limit vascular occlusion by inhibiting platelet activation and stimulating fibrinolysis via tissue plasminogen activator.

INTRODUCTION

Thrombosis is a major clinical complicaton of atherosclerosis, and thrombosis on the surface of a plaque is responsible for acute events in patients with atherosclerosis, i.e. myocardial infarctions and many strokes. However, in addition to precipitating acute ischemic events, thrombosis may be involved in atherosclerotic disease in another way, i.e. as a factor in the

*Present address: Mt. Sinai Services, City Hospital at Elmhurst, 79-01 Broadway, Elmhurst, NY 11373.
#Present address: Mario Negri Sud Institute, Via Nazional 66030 S. Maria Imbaro, Chieti, Italy

Fibrinogen, Thrombosis, Coagulation, and Fibrinolysis, Edited by
C. Y. Liu and S. Chien, Plenum Press, New York

development of atherosclerotic plaques. There is a great deal of evidence that platelet activation during thrombus formaton contributes to plaque development by secretion of the platelet-derived growth factor (1), but that topic will not be considered further in this paper.

The other major component of the thrombus is fibrin. There has been considerable interest in fibrin as a component of atherosclerotic plaques for more than 100 years. As long ago as 1852 Rokitansky (2) hypothesized that atherosclerotic lesions develop by organization of intravascular fibrin deposits. Beneke (3), in 1874, stated that organization of thrombi led to connective tissue formation that thickened lesions. Mallory in 1912 (4) and Clark et al. (5) in 1936 noted that plaque morphology suggested a thrombotic component in their development. In 1948, Duguid (6) repeated Rokitansky's original assertion that plaques develop through organization of thrombi. Many other investigators in the intervening years have examined the question of fibrin in atherosclerotic plaques, using biochemical, histochemicl, immunohistochemical, and electron microscopic techniques to identify fibrinogen and/or fibrin. However, these techniques have left unanswered questions regarding fibrinogen-related protein in the vessel wall:

Is this protein fibrinogen or fibrin?
Is it present incidentally?
Does fibrin form within the vessel wall?
Does it enter by endothelialization of a mural thrombus?
Does it contribute to the development of a plaque?
Is the fibrin covalently linked to other proteins within the plaque?

This paper will present recent results from our laboratory that provide answers to some of these questions.

Immunochemical studies of fibrinogen-related protein in atherosclerotic plaques and thrombi (7)

Some definitions are required before the results of these studies can be presented. First, the different types of atherosclerotic plaques will be described. Fatty plaques consist of focal accumulations of smooth muscle cells and macrophages that contain and are surrounded by lipid. Fibrous plaques consist of accumulations of lipid-laden smooth muscle cells and macrophages, extracellular lipid, collagen, elastic fibers, and cell debris, covered by a fibrous cap. Finally, complicated plaques develop from fibrous plaques that have undergone hemorrhage, calcification, cell necrosis, and/or mural thrombosis.

Secondly, several fibrinogen/fibrin-related proteins and peptides must be defined. Fibrinogen itself is a dimer consisting of two Aα chains, two Bβ chains, and two γ chains. Fibrin I is formed from fibrinogen by thrombin cleavage of the Aα 16-17 bonds to remove FPA. Fibrin II is formed from fibrinogen by cleavage of the Aα 16-17 bonds to remove FPA and the Bβ 14-15 bonds to remove FPB. It can also be formed from fibrin I by cleavage of the Bβ 14-15 bonds. Fragment X is formed from fibrinogen or fibrin by plasmin cleavage at β 42-43 and at multiple α chain sites. Fragment D is a late plasmin cleavage product of fibrinogen or fibrin arising from the C-terminal portion of the molecule. When it arises from plasmin cleavage of crosslinked fibrin, it occurs as a dimer (D-dimer). The N-terminal disulfide knot (NDSK) and fragment E are N-terminal digestion products resulting from CNBr or plasmin digestion respectively.

Tissue samples from surgery or post-mortem were obtained and digested with CNBr. Several radioimmunoassays were then used to quantitate the various fibrinogen-related proteins in the CNBr digests. These are shown in Table 1.

Table 1

Antigen	Assay
Total fibrinogen-related protein	NDSK
Fibrinogen	FPA (thrombin-releasable)
Fibrin I	FPB (thrombin-releasable) Bβ 1-42
Fibrin II	β 15-42

Thrombi were examined first to confirm the methods. Both acute and organized thrombi were studied, with similar results. Only trace amounts of fibrinogen and fibrin I were detected in these specimens. 75-80% of the fibrinogen-related protein (NDSK) was identified as Fibrin II. The fragment X content was calculated as the difference between the NDSK and the total of fibrinogen, fibrin I, and fibrin II. In the thrombi fragment X comprised 20-25% of the NDSK.

Aortic samples were then examned. Normal aortas contained predominantly fibrinogen, with about 15% fibrin I and 5% frgment X. Fatty plaques contained about 50% fibrinogen, 25% fibrin I, and small amounts of fibrin II and fragment X. Fibrous plaques were similar to fatty plaques except that they contained somewhat less fibrinogen and fibrin I but more fibrin II and fragment X. When complicated plaques were examined, they were found to have a distribution of fibrinogen-related proteins very similar to that of thrombi, although the fibrinogen-related protein in the complicated plaques was only 6% of the total protein as compared with 50-70% in the thrombi. Thus these studies provided evidence that fibrinogen was present in normal aortas and in early plaques in the absence of fibrin. As plaques became more severe, the proportions of fibrinogen, fibrin I, fibrin II, and fragment X changed, suggesting that transformation might be occurring within the vessel wall.

Immunohistochemical studies of thrombi and plaques (8, 9)

Subsequent studies were then carried out to identify the different forms of fibrinogen-related antigen within the vessel wall. To accomplish this, the avidin biotin immunoperoxidase technique was employed, using three monoclonal antibodies that specifically identified fibrinogen/fibrin I (antibody 18C6, reactive with Bβ 1-42 but not with FPB or β 15-42), fibrin II (antibody T2G1, reactive with the N-terminal β chain of fibrin), and fragments D and D-dimer (antibody GC4, reactive with fragments D and D-dimer but not with intact fibrinogen or fibrin). Polyclonal antibodies to fibrinogen and albumin were also used.

Examination of an organizing thrombus from a femoral plaque showed intense staining of the histologically more recent area of the thrombus with the antibody that recognized fibrinogen/fibrin I, but the older areas of the thrombus stained weakly. Staining with this and the other fibrinogen-related antibodies occurred either in parallel, wave-like layers or as a fibrin mesh. Staining for fibrin II was diffuse throughout the thrombus. Fragments D and D-dimer were stained intensely in the histologically older, more central area of the thrombus. The polyclonal antibody to fibrinogen recognized most of the areas identified by the three monoclonals, and albumin was present diffusely throughout the thrombus.

Most histologically normal sections of aortas contained no detectable fibrinogen-related antigens. However, focal staining of fibrinogen/fibrin I was detected in the intimal and subintimal layers of one specimen with diffuse staining in the adventitia, and diffuse staining of all three layers was found in a second specimen, with traces of fibrin II in the media and adventitia.

An early coronary plaque revealed fibrinogen/fibrin I present on the luminal surface and deeper into the media. Some fibrin II was present on the luminal surface. No fragment D was detected in this lesion. The polyclonal antibody to fibrinogen stained the areas stained by the monoclonals. The distribution of albumin staining was similar to that for fibrinogen, while IgG staining was distinct, appearing in association with connective tissue in the intima and in the adventitia.

Another early coronary plaque, with a normal-appearing intimal surface, contained a cluster of foam cells surrounded by fibrinogen/fibrin I. Staining was also seen as short threads around cells that appeared to be smooth muscle cells or macrophages. Only traces of fibrin II were found in this lesion, as small flecks around cells. A small area of staining for fragments D and D-dimer was visible deep within the intima. Albumin staining was intense deep within the intima, in a distribution similar to that seen with the polyclonal antifibrinogen antibody.

Examination of a fibrous plaque from a popliteal artery revealed fibrinogen/fibrin I distributed in small flecks associated with smooth muscle cells and macrophages. In another area of the same lesion, fibrinogen/fibrin I was found as longer threads and as diffuse staining in an area of loose connective tissue. In this latter area, fibrin II staining was more prominent and was also seen around foam cells. A large focus of fragments D and D-dimer was also noted in this area. The albumin antibody stained similarly to the fibrin II and fragment D monoclonals, and albumin staining was widely distributed.

An advanced lesion with cholesterol crystals showed fibrinogen/fibrin I in the central

Table 2

Lesions	Fibrinogen/Fibrin I	Fibrin II	Fragments D & D-dimer
Normal aortas	2/12	1/12	—
Early plaques	6/7	3/7	1/7
Fibrous plaques	5/6	4/6	4/6
Advanced plaques	7/8	7/8	6/8
Thrombi	16/18	18/18	17/17

area of the lesion. Fibrin II was more widespread throughout the lesion, and fragments D and D-dimer were distributed similarly to fibrin II. Albumin staining was intense in the areas where fibrinogen-related antigen staining was seen.

The antibody reactivity of the lesions examined is summarized in Table 2.

Overall, the patterns of fibrin staining of atherosclerotic plaques can be considered as luminal staining or as staining within the vessel wall. Luminal staining was either associated with surface thrombus or occurred as a thin layer on the luminal surface. Staining within the vessel wall occurred as bundles of short threads, as a net-like pattern, or as small flecks surrounding vessel wall cells presumed to be smooth muscle cells or macrophages or surrounding cholesterol crystals or Ca^{++}. The bundles of threads and the net-like pattern are likely to represent incorporated thrombus, but the flecks of fibrin surrounding vessel wall cells may represent fibrin that is formed within the atherosclerotic lesion.

Effects of fibrin on the endothelium (10)

In occasional lesions examined immunohistochemically, fibrin was noted to be present on the endothelium. Previous studies by others have demonstrated deleterious effects of fibrin on endothelial morphoiogy and function, i.e. rapid disorganization of the monolayer (11, 12), altered migration (11, 13), increased pinocytosis (14), and increased rate of DNA synthesis (14). Fibrin has also been reported to stimulate rapid release of von Willebrand factor from cultured endothelial cells (15). Because the studies described earlier had suggested the possibility that fibrin formation might occur within the vessel wall, it was of interest to determine whether fibrin formed on the endothelium might alter endothelial cell hemostatic function.

To carry out these experiments, normal citrated plasma or plasma diluted with serum or with tissue culture medium was added to dishes of cultured cells. Fibrin was then formed by the addition of calcium to generate thrombin. In the initial experiment, it was found that marked stimulation of prostacyclin production occurred over a 22 hour incubation when cells were incubated with recalcified plasma, but that there was no increase when cells were incubated with serum that had been prepared in a test tube and then added to the cells or with plasma in the absence of recalcification. Additional experiments demonstrated that tissue plasminogen activator production was similarly stimulated in a time-dependent manner. The shape of the curve varied in different experiments, with significant stimulation by 6 hours in some experiments but not until 20 hours in others. Stimulation was also found to be fibrin-dose dependent. Prostacyclin synthesis was maximal at approximately 1 mg fibrin/35 mm dish, whereas tissue plasminogen activator synthesis increased at least up to 5 mg fibrin/dish.

To define the mechanism by which stimulation occurred, a series of inhibitors were tested. Stimulation of prostacyclin production was inhibited about 50% by actinomycin D and cycloheximide, suggesting that protein synthesis was required. Presumably this reflected a requirement for synthesis of one or more enzymes in the prostacyclin synthetic pathway. Indomethacin completely inhibited stimulation of prostacyclin synthesis, as expected. Neither cytochalasin D nor vinblastine significantly affected stimulation of prostacyclin synthesis, indicating that cytoskeletal function was not required. Trifluoperazine, an inhibitor of calcium-dependent reactions involving calmodulin and protein kinase C and of phospholipase A_2, caused slight inhibition of stimulation of prostacyclin synthesis. Which of the potential actions of trifluoperazine was actually responsible for the inhibition is not known.

When these same inhibitors were tested for their ability to prevent fibrin stimulation of tissue plasminogen activator synthesis, different results were obtained. Both actinomycin D and cycloheximide completely inhibited stimulation and decreased background synthesis, indicating that protein and RNA synthesis were absolutely required. Indomethacin had no effect. Cytochalasin D caused 60% inhibition of stimulation and vinblastine caused 40% inhibition, demonstrating that in this case normal cytoskeletal function was required for stimulaton. As with prostacyclin, there was slight inhibition of stimulation of tissue plasminogen activator synthesis by trifluoperazine.

The possibility that stimulation of the cells by fibrin was a nonspecific effect due to polymerization of a large molecule on the surface of the cells was then considered. However, when soluble collagen was added to the cultures and allowed to polymerize, there was no stimulation of either prostacyclin or tissue plasminogen activator synthesis.

Because thrombin was generated in the culture dishes in order to form fibrin, it was possible that the stimulation seen was due to thrombin and not to fibrin, since others had previously reported stimulation of tissue plasminogen activator (16, 17) and prostacyclin (18, 19) production by thrombin. When thrombin was added to cells at a concentration of 1 U/ml, which in this system did not induce clot formation, no stimulation of prostacyclin or tissue plasminogen activator production occurred. When Ca^{++} was added alone or together with thrombin, stimulation was seen, indicating that sub-clotting levels of thrombin were not stimulatory.

Additional experiments were carried out to determine the molecular requirements of fibrin for stimulation. Reptilase was added to generate fibrin I (desFPA fibrin), and this failed to stimulate prostacyclin production and caused much less stimulation of tissue plasminogen activator synthesis than did fibrin II (desFPAdesFPB fibrin).

The requirement for polymerization of fibrin in order to effect stimulation of prostacyclin and tissue plasminogen activator production was examined by carrying out experiments in the presence and absence of gly-pro-arg-pro, a peptide that inhibits fibrin polymerization by binding to the site complementary to the Aα chain polymerization site (20). Typical stimulation was seen in the absence of gly-pro-arg-pro, but stimulation was completely abolished in the presence of this peptide.

The role of secreted platelet products in the stimulation of prostacyclin and tissue plasminogen activator production was examined by comparing the stimulatory activity of fibrin formed from platelet rich-plasma with that formed from platelet-poor plasma (Table 3). It was evident that the presence of platelets or secreted platelet products did not affect the endothelial cells under these conditions.

These studies have demonstrated that fibrin formed on the surface of cultured endothelial cells stimulates the cells to produce increased amounts of prostacyclin and tissue plasminogen activator in a time and dose-dependent manner. Although thrombin was generated in these experiments, it did not appear to be responsible for the stimulation since when added directly to cells it had no effect unless it was added in sufficient concentration to generate fibrin. Additionally, thrombin formed in the presence of the fibrin polymerization inhibitor gly-pro-arg-pro failed to stimulate the cells, whereas the same amount of thrombin did stimulate when fibrin was allowed to polymerize. It also appeared that fibrin II (desFPAdesFPB fibrin) was necessary for stimulation since little or no stimulation occurred when Reptilase was used to form fibrin I (desFPA fibrin). The requirement for fibrin II may be related to the fact that fibrin II is a better substrate for factor XIII than is fibrin I and raises the possibility that fibrin may be crosslinked to the cells. Finally, different cell mechanisms appeared to be involved in stimulation of prostacyclin and tissue plasminogen activator synthesis since different effects of inhibitors were noted.

Table 3

Condition	6-keto-$PGF_{1\alpha}$	t-PA
Platelet-poor plasma	1.49	0.57
Platelet-rich plasma	1.55	0.54
Platelet-poor plasma + Ca^{++}	7.56	0.95
Platelet-rich plasma + Ca^{++}	7.88	0.94

REFERENCES

1. R. Ross, J. Glomset, and L. Harker. A platelet-dependent serum factor that stimulates the proliferation of arterial smooth muscle cells in vitro. *Proc. Natl. Acad. Sci. USA,* **71**:1207-1210, (1974).
2. C. V. Rokitansky. A manual of pathologic anatomy. The Syndenham Society, *London,* **4**:265-275, (1852).
3. F. W. Beneke. Grundlinien der Pathologie des Stoffwechsels. *Berlin.,* (1874).
4. F. B. Mallory. The Infectious Lesions of Blood Vessels, *in*: The Harvey Lectures, New York (1912), pp. 150-166.
5. E. Clark, I. Graef, and H. Chasis. Thrombosis of the aorta and coronary arteries. *Arch. Pathol.,* **22**:183-212, (1936).
6. J. B. Duguid. Thrombosis as a factor in the pathogenesis of aortic atherosclerosis. *J. Pathol. Bacteriol.,* **60**:57-61, (1948).
7. A. Bini, J. J. Fenoglio, Jr., J. Sobel, J. Owen, M. Fejgl, and K. L. Kaplan. Immunochemical characterization of fibrinogen, fibrin I and fibrin II in human thrombi and atherosclerotic lesions. *Blood,* **69**:1038-1045, (1987).
8. A. Bini, J. J. Fenoglio, Jr., R. Mesa-Tejada, B. Kudryk, and K. L. Kaplan. Identification and distribution of fibrinogen, fibrin, and fibrin(ogen) degradation products in atherosclerosis. Use of monoclonal antibodies. *Arteriosclerosis,* **9**:109-121, (1989).
9. A. Bini, R. Mesa-Tejada, J. J. Fenoglio, Jr., B. Kudryk, and K. L. Kaplan. Immunohistochemical characterization of fibrin(ogen)-related antigens in human tissues using monoclonal antibodies. *Lab. Invest.,* **60**:814-821, (1989).
10. K. L. Kaplan, T, Mather, L. DeMarco, and S. Solomon. Effect of fibrin on endothelial cell production of prostacyclin and tissue plasminogen activator. *Arteriosclerosis,* **9**:43-49, (1989).
11. J. L. Kadish, C. E. Butterfield, and J. Folkman. The effect of fibrin on cultured vascular endothelial cells. *Tissue Cell,* **11**:99-108, (1979).
12. K. Watanabe and K. Tanaka. Influence of fibrin, fibrinogen and fibrinogen degradation products on cultured endothelial cells. *Atherosclerosis,* **48**:57-70, (1983).
13. R. R. Schleef and C. R. Birdwell. The effect of fibrin on endothelial cell migration *in vitro. Tissue Cell,* **14**:629-636, (1982).
14. R. R. Schleef and C. R. Birdwell. Biochemical changes in endothelial cell monolayers induced by fibrin deposition *in vitro. Atherosclerosis,* **4**:13-20, 1984.
15. J. A. Ribes, C. W. Francis, and D. D. Wanger. Fibrin induces release of von Willebrand factor from endothelial cells. *J. Clin. Invest.,* **79**:117-123, (1987).
16. E. G. Levin, U. Marzec, J. Anderson, and L. A. Harker. Thrombin stimulates tissue plasmnogen activator release from cultured human endothelial cells. *J. Clin. Invest.,* **74**:1988-1995, (1984).
17. E. G. Levin, D. M. Stern, P. P. Nawroth, R. A. Marlar, D. S. Fair, J. W. Fenton 2nd, and L. A. Harker. Specificity of the thrombin-induced release of tissue plasminogen activator from cultured human endothelial cells. *Thromb. Haemost.,* **52**:115-119, (1986).
18. B. B. Weksler, C. W. Ley, and E. A. Jaffe. Stimulation of endothelial cell prostacyclin production by thrombin, trypsin, and the ionophore A23187. *J. Clin. Invest.,* **62**:923-930, (1978).
19. S. L. Hong. Effect of bradykinin and thrombin on prostacyclin synthesis in endothelial cells from calf and pig aorta and human umbilical vein. *Thrombosis Res.,* **18**:787-795, (1980).
20. A. P. Laudano and R. F. Doolittle. Studies on synthetic peptides that bind to fibrinogen and prevent fibrin polymerization. Structural requirements, number of binding sites, and species differences. *Biochemistry,* **19**:1013-1019, (1980).

INTERACTIONS BETWEEN FIBRIN, COLLAGEN AND ENDOTHELIAL CELLS IN ANGIOGENSIS

H. Mei Liu, Danny Ling Wang, and Chung Yuan Liu

Institute of Biomedical Sciences
Academia Sinica
Nankang, Taipei, Taiwan, ROC

ABSTRACT

The role of fibrin in the generation of new blood vessels was examined in this study. Using a wound chamber model, we investigated the sequential interactions between endothelial cells and the extracellular matrix during angiogenesis. Silicone tubes 5 mm long and 1.4 mm in internal diameter were sutured to the cut ends of thigh muscles in the rats. The contents of the chamber were removed at intervals for histological, immunohistochemical and electron microscopic studies. We observed an initial phase of fluid accumulation in the wound chamber followed by formation of a fibrin/fibronectin clot. Migration of endothelial cells, macrophages and fibroblasts into the clot occurred after the 1st week. The subsequent phase of fibrinolysis was accompanied by deposition of collagen and organization of endothelial cells into capillary tubes. These findings support the view that angiogenesis is the product of interactions between endothelial cells and a changing extracellular matrix (ECM) and requires the participation of soluble and immobilized plasma proteins and local ECM factors. Our findings indicate that fibrin is intimately involved in both hemostasis and angiogenesis; these are sequential steps in the initial phase of wound healing. Thus, fibrin/fibrinogen occupies a central position and provides a vital link in the initiation of the cascade event of wound healing.

INTRODUCTION

The formation of a fibrin clot results from the polymerization of soluble fibrionogen mediated by thrombin. Subsequent to injury and hemorrhage, the fibrin/fibrinogen clot promotes platelet aggregation and plays a key role in hemostasis and wound healing. Following intravascular thrombosis, the fibrin clot undergoes spontaneous lysis and is replaced by newly-formed capillaries. This leads to reestablishment of the circulation, a process known as "recanalization" (Kumar 1984, Taussig 1984). While a great deal is known about the cellular and molecular mechanisms of thrombosis and fibrinolysis, their physiological roles in angiogenesis and in the overall process of wound healing has not been clearly defined.

Angiogenesis plays important role in many physiological and pathological conditions such as embryonic development (Patten, 1971), wound healing (Gimbrone et al., 1976, Ross et al., 1977, Liu, 1988) and tumor growth (Ausprunk and Folkman, 1977, Folkman et al., 1971, Zetter, 1980). Endothelial cells in adult organisms are normally quiescent but are activated by pathological stimuli. These cells undergo proliferation, migration and formation of new blood vessels. The highly vascular granulation tissue paves the way for subsequent steps of cell growth and tissue regeneration in a wide variety of organs (Kurkinen et al., 1980, Hollund et al., 1982).

Fibrinogen, Thrombosis, Coagulation, and Fibrinolysis, Edited by
C. Y. Liu and S. Chien, Plenum Press, New York

Information on the morphological and biochemical events of angiogenesis is largely derived from studies of cultured endothelial cells (Folkman et al., 1979, Maciag, 1984, Montesano et al., 1985). The relationship of the endothelial cells with the extracellular matrix (ECM) under the in vivo wound situation has not been studied in great detail. In this communication, we report our experience in the use of a wound chamber model to study the sequential morphological steps of angiogenesis (Liu et al., 1989). The interactions between endothelial cells with plasma-derived components such as fibrin/fibronectin and with ECM factors such as collagen were the foci of this study.

METHODS

Adult rats weighing 250-300 gm were used. The rats were anesthetized by an intraperitoneal injection of pentobarbital (40 mg/kg). Skin incisions were made on the lateral aspect of the thigh and the muscles were split to allow the implantation of impermeable transluscent silicone tubes, 5 mm long and 1.4 mm in internal diameter (Mentor Co, Goleta,

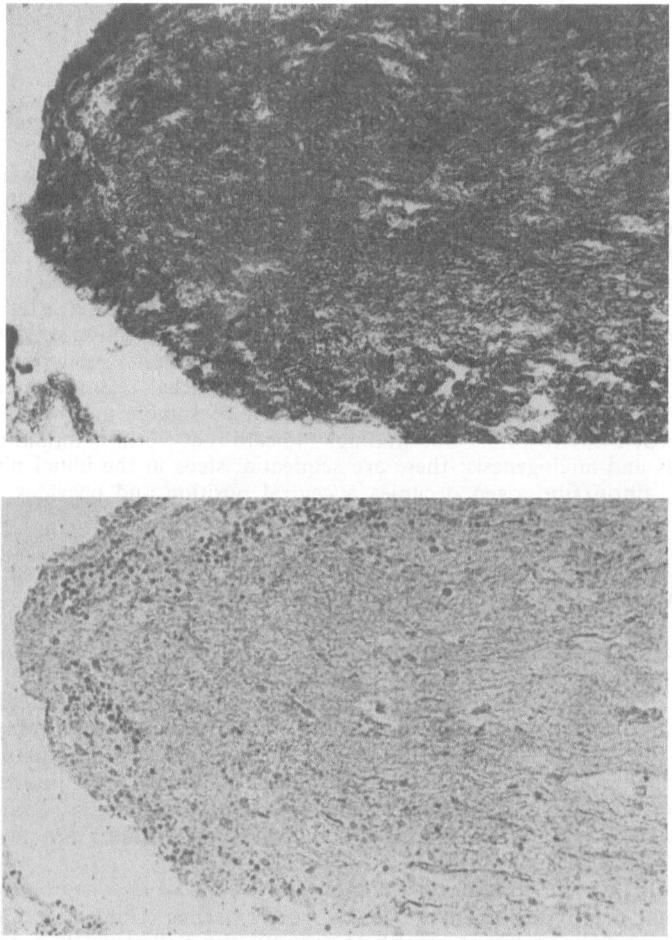

Fig. 1. Immunocytochemical (PAP) staining of wound chamber clot on the 7th day showing positive and diffuse staining for fibrinogen (top). The bottom side was a negative control in which the primary antibody was omitted (bottom) X 80. The clot showed a similar reactivity to fibronectin (not shown).

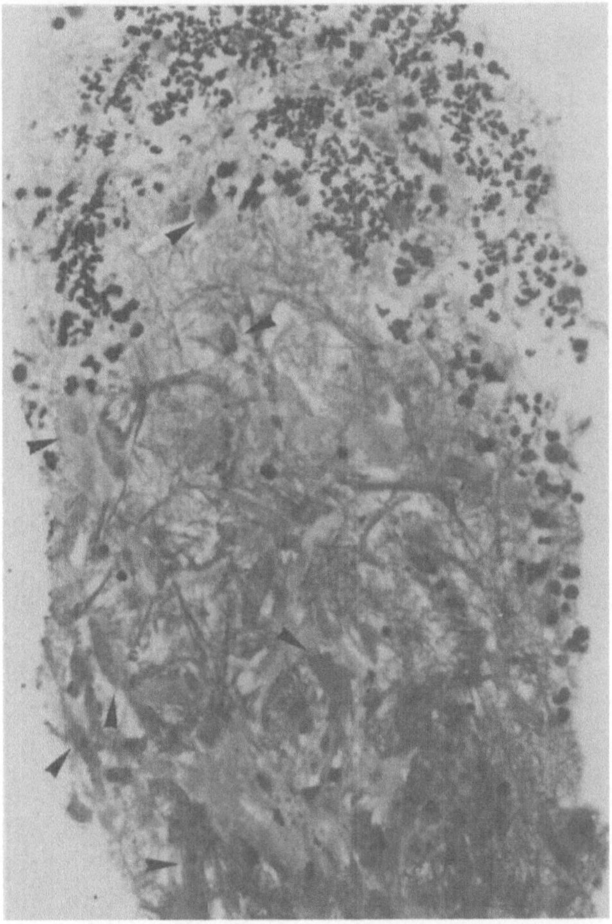

Fig. 2. Light microscopic view of wound chamber clot on the 7th day showing a network of fibrin containing scattered cells (arrows) and collections of red blood cells at one end. Toluidine blue stain. X 80.

CA). The ends of the tubes were sutured to the cut ends of the leg muscles and the skin incisions were closed with metal clips. The rats were sacrifised on 3, 5, 7, 10, 14, 18 and 21 post-operative days. Under deep pentobarbital anesthesia, the rats were perfused intracardiacally with 200 ml of phosphate buffered saline (PBS) followed by 400 ml of 3% paraformaldehyde in PBS. The material present in the tubes were removed and processed for histological, electron microscopic and immunohistochemical studies.

For light microscopy, the tissues were fixed for an additional 12 hours in 5% buffered formalin, serially dehydrated and embedded in paraffin. Sections cut at 4 micron were stained with H&E stain for orientation. For immunohistochemical study, deparaffinized sections were first incubated in 3% H_2O_2 for 10 minutes to eliminate the endogenous peroxidase, then in 1% albumin solution for 20 minutes to block the non-specific binding. Sections were incubated successively in the following reagents with 15 minutes of washing in PBS between incubations: polyclonal rabbit antibodies against fibrinogen and fibronectin 1:100 dilution for 60 minutes; swine anti-rabbit IgG, 1:50 dilution for 30 minutes and rabbit peroxidase anti-peroxidase (PAP) 1:50 dilution for 30 minutes. The slides were finally reacted with chromogen in 0.03% H_2O_2, mounted in glycerin and examined under a light microscope.

For electron microscopic study, the tissues were fixed for 2 hours in 3% glutaraldehyde and cut into 1-2 mm fragments. After rinsing in cocodylate buffer, the samples were post-fixed in osmium tetraoxide, serially dehydrated and embedded in epon. Care was taken

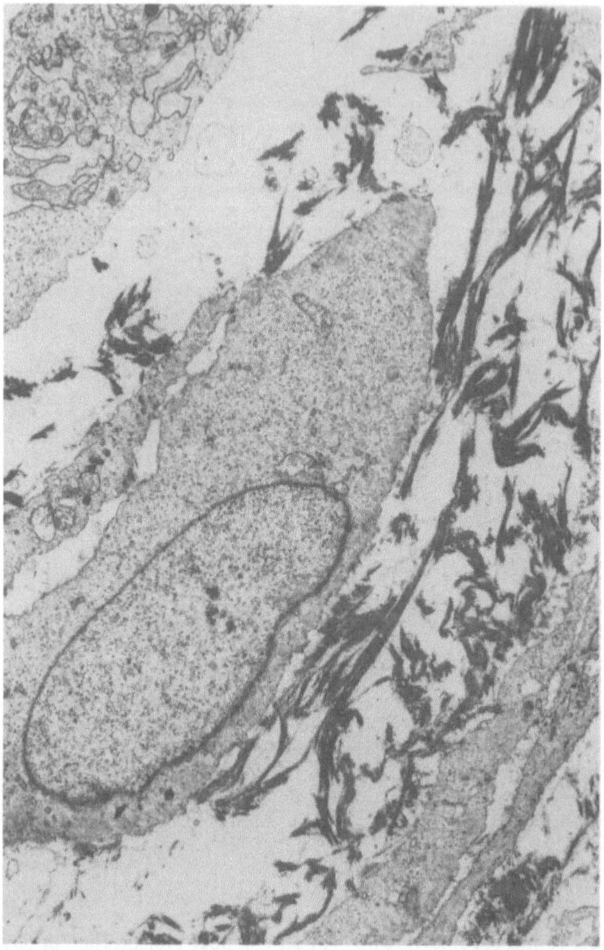

Fig. 3. Electron microscopy of wound chamber clot removed on the 7th day showing undifferentiated cells whose surfaces were covered by fibrin threads. The cells had smooth outline and the cytoplasm contained scanty organelles. X 24,000.

to orient the tissues so that they were cut either crossly or parallel with the axis of the wound chamber. Thick section of 1 micron were stained with 1% toluidine blue. Ultrathin sections were stained with uranyl acetate and lead citrate and examined under a Jeol 1200 XT electron microscope.

FINDINGS

The morphological events occurring in the wound chamber can be divided into the following stages.

(1) Initial stage (1st week): fluid accumulation and clot formation — during the first few days, there was accumulation of bloody fluid in the wound chambers as the result of extravasation of formed and soluble blood components. By the 5th day, a soft clot composed of interlacing fibrin strands enclosing groups of red blood cells, platelets and leukocytes was formed within the wound chamber. Immunohistochemical study showed that the clots reacted positively and diffusely to antisera against fibrinogen and fibronectin (Fig. 1).

(2) Early stage (2nd week): Migration and proliferation of cells. By 7th day, many undifferentiated spindle-shaped cells and macrophages were found within the fibrin clot

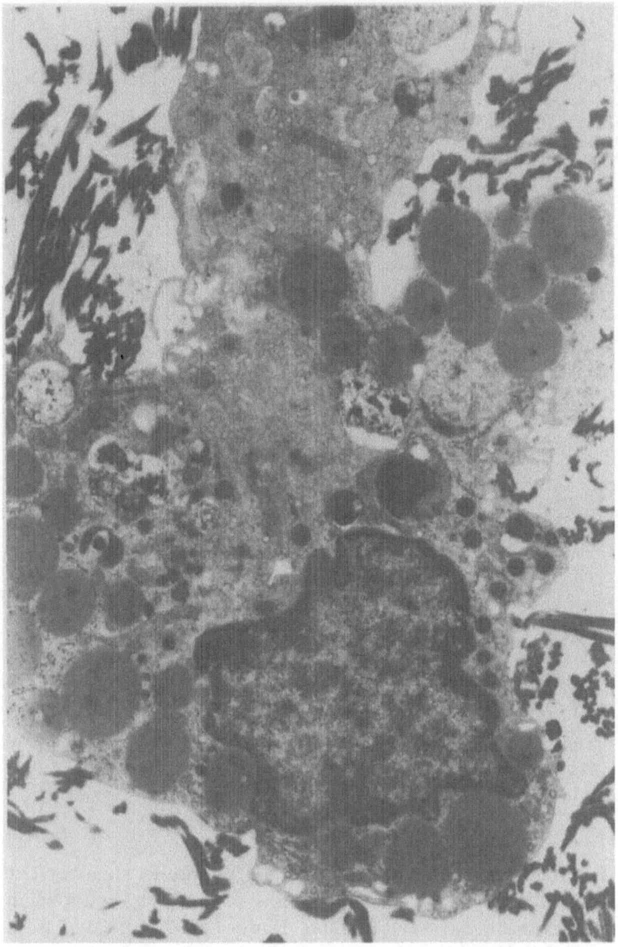

Fig. 4. A macrophage surrounded by fibrin was present in the wound chamber on 10th day. The cytoplasm was filled with lipid granules and lysosomes containing phagocytosed debris. X 24,000.

(Fig. 2). Under electron microscopy, individual cells were enclosed within the fibrin network and the cell surfaces were studded with the fibrin strands. The spindle cells had smooth external surface and their elongated cytoplasms were oriented along the axis of the wound chamber (Fig. 3). The cytoplasm contained scanty organelles such as rough endoplasmic reticulum (RER) and mitochondria. The macrophages contained abundant lipid droplets and phagosomes (Fig. 4).

(3) Intermediate stage (2-3rd week): Fibrinolysis, collagen deposition and endothelial cell organization. During the 2-3 weeks, the endothelial cells showed signs of differentiation and organization. The cytoplasm was enlarged and contained an increased amount of RER. The cells elaborated numerous cytoplasmic processes which had converged toward one another (Fig. 5). By this time, the fibrin in the intercellular space had virtually disappeared and thin collagen fibrils appeared next to the endothelial cells.

(4) Late stage (3rd week): Capillary tube formation. During the 3rd week, there was increasing deposition and realignment of collagen. The wound chamber was made up of a collagen matrix containing islands of new blood vessels in various stages of development (Fig. 7). Initially, open capillaries were relatively few and were surrounded by solid cell clusters referred to as "endothelial buds." These were highly organized structures composed of tightly packed overlapping and interdigitating endothelial cell processes separated

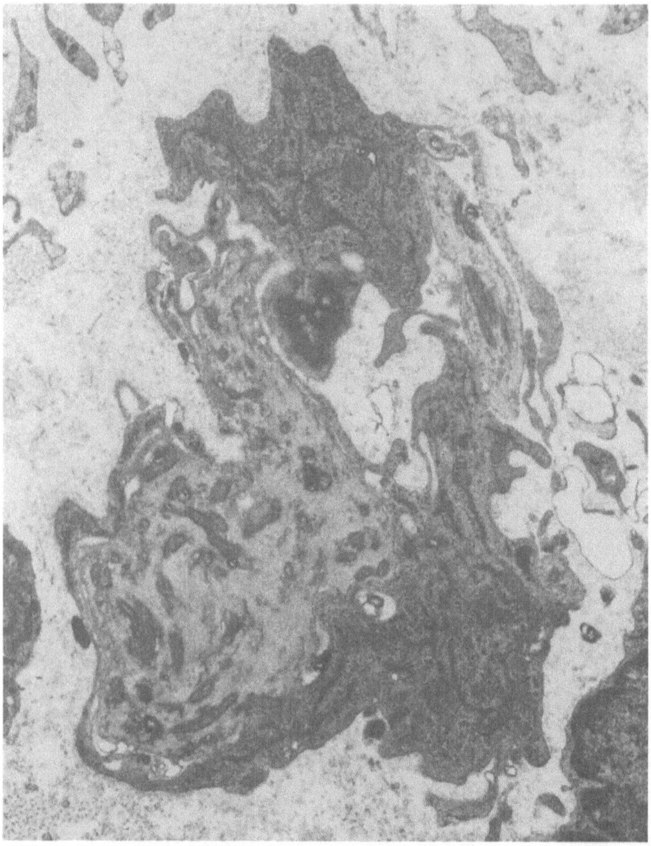

Fig. 5. Electron microscopy of the material in wound chamber clot on the 10th day showing aggregation of multiple endothelial cell processes but no basal lamina. Fibrin had disappeared from the extracellular space where scattered thin collagen fibrils were seen. X 16,000.

from the collagen matrix by basement membrane (Fig. 6a). Slit-like lumens can be distinguished in the center of these structures. The endothelial cells were characterized by the presence of cytoplasmic inclusions (Weibel-Palade bodies), large lipid droplets and abundant cytoplasmic filaments and microtubules (Fig. 6b).

By 21 days, the wound chamber was filled with a network of interconnecting open capillaries embedded on a collagenous matrix (Fig. 8). Figure 9 is a schematic drawing to demonstrate the sequence of events occurring in the wound chamber during the 3 week period of observation.

DISCUSSION

The formation of a network of new capillaries in the wound chamber is the product of a series of interactions between endothelial cells and various soluble and immobilized extracellular factors. During the initial phase, there is exudation of plasma components, red and white blood cells and platelets. This is followed by the formation of a clot in which fibrin and fibronectin co-distributed. The hematogenous origin of the wound chamber fluid has been verified by the fact that it had a similar protein profile as the rat serum on one- and two-dimensional electrophoresis (Liu et al., to be published).

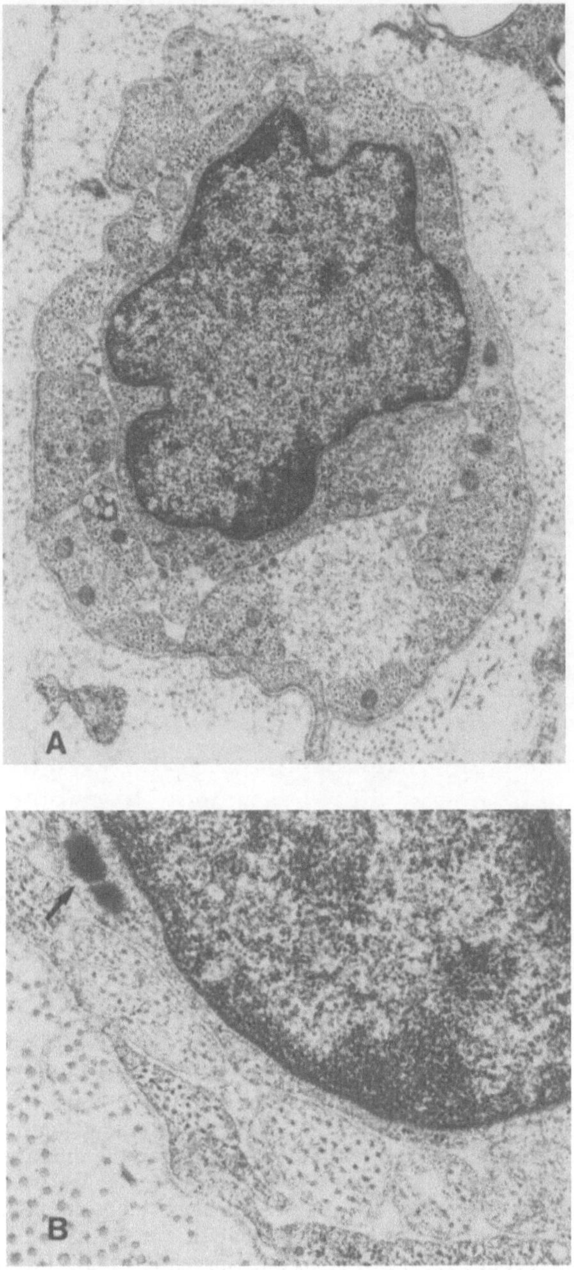

Fig. 6A. An "endothelial bud" composed of multiple closely packed endothelial cell processes enclosed within a common basal lamina. A small lumen containing flocculent material was present. X 18,000.

Fig. 6B: Close view of an endothelial cell comprising the capillary bud. The cytoplasm contained dense bodies (Weibel-Palade bodies) and abundant microtubules. X 24,000.

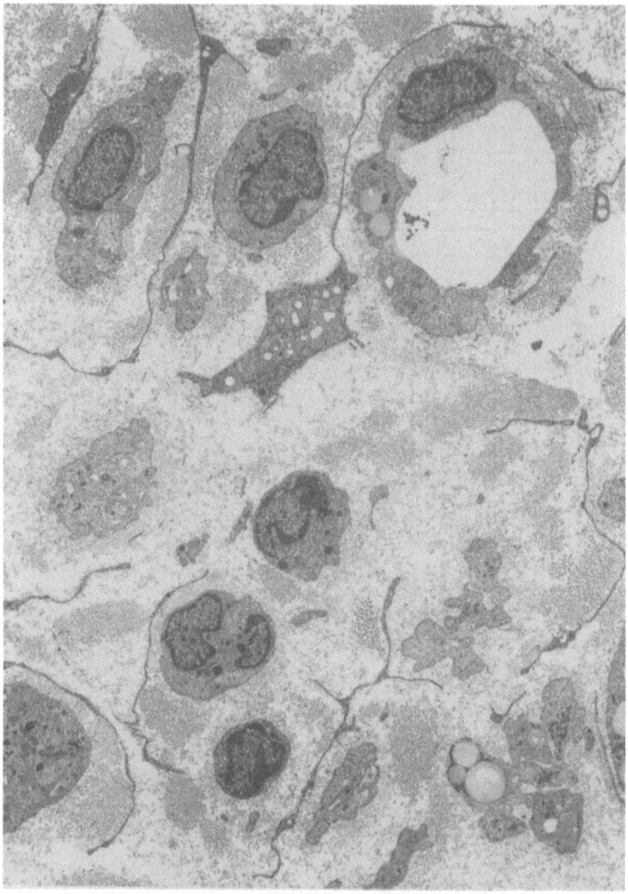

Fig. 7. Wound chamber clot removed on the 14th day showing several endothelial aggregates (capillary buds) and one open capillary (top right). The capillary buds were embedded on a well organized collagen matrix and each capillary bud was encircled by thin, elongated fibroblast processes. X 12,000.

The migration of endothelial cells, fibroblasts and macrophages into the fibrin/fibronectin clot is the result of a chemical interaction and adhesion between cell surface and certain adhesive regions on the fibrinogen molecule (Cheresh, 1987). Endothelial cells exclusively recognize an Arg-Gly-Asp (RGD) tripeptide sequence at the C-terminal on the α-chain of fibrinogen through a RDG-directed receptor, integrin, and the interaction can be inhibited by synthetic peptide containing this sequence. Fibronectin, von Willebrand factor, collagen and laminin are among the group of adhesive proteins containing the RGD tripeptide (Ruoslahti et al., 1989). Fibronectin and its proteolytic fragments are also chemotactic for endothelial cells (Vaheri et al., 1983, Bowersox and Sorgente, 1982). Fibronectin is covalently linked to fibrin by factor XIIIa during normal clotting process (Mosher, 1970). The fibrin/fibronectin complex in concert with other soluble growth and trophic factors may be instrumental in bringing about the endothelial cell proliferation and migration into the wound chamber.

Endothelial cells in tissue culture respond to different kinds of growth factors either alone or in combination (Auerbach, 1981, Guilino, 1981). The well known examples are endothelial cell growth factor or ECGF (Maciag, 1984, Burgess et al., 1985) and fibroblast growth factor or FGF (Gospadarowicz and Lui, 1981, Thomas et al., 1984) both of which were isolated from brain and pituitary. Tumor tissues elaborate endothelial cell growth

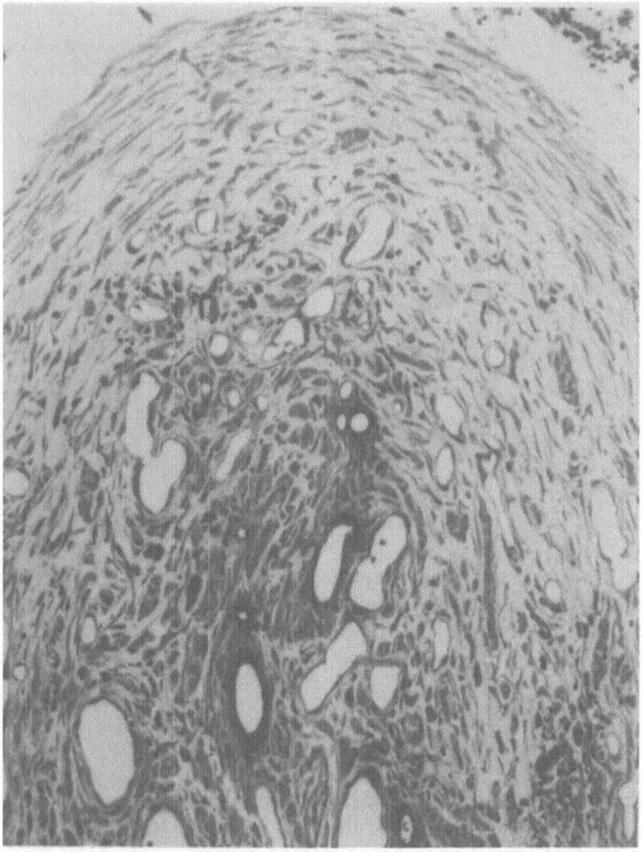

Fig. 8. Light microscopy of a wound chamber clot removed on the 21th day. The clot was filled with a collagen matrix containing many open capillaries. X 80.

factors (Gimbrone et al., 1973) one of which is transforming growth factor or TGF (Roberts et al., 1986). Within the present wound chamber, soluble growth factors such as the PDGF produced by the platelets (Johnsson et al., 1982) and by macrophages (Thakral et al., 1979, Martinet et al., 1986) may contribute to the proliferation of endothelial cells in the clot. It is also possible that other factors including plasma components, fibrin degradation products (Thompson et al., 1985) and hydrolytic enzymes such as proteases and plasminogen activator may act synergistically to stimulate the proliferation of the endothelial cells.

At a time when the fibrin clots in the wound chamber were occupied by proliferated endothelial cells and fibroblasts, mechanisms were triggered which led to fibrinolysis and collagen deposition. The appearance of collagen fibrils in the extracellular matrix coincided with the beginning of organization of endothelial cells into endothelial buds. The observed fibrinolysis is probably mediated by plasmin generated from plasminogen through the action of plasminogen activator produced by the endothelial cells and fibroblasts in the wound chamber. Our findings suggest that one of the physiological roles of fibrinolysis is to promote the deposition of collagen which in turn promotes capillary organization.

Clonal endothelial cells spontaneously form tubes after one month in tissue culture and the process can be accelerated from one month to two weeks by the addition of collagen gel or fibrin and collagen gel or laminin gel over the monolayers (Madri et al., 1983, Montesano et al., 1985, Delvos et al., 1982, Grant et al., 1989). It is well known that the differentiation and organization of epithelial cells and endothelial cells depend on the presence of extracellular matrix (ECM) molecules such as collagen, proteoglycans, laminin and fibronectin (Bernfield et al., 1972, Hall et al., 1982). The nature of the signaling agent and the transducing

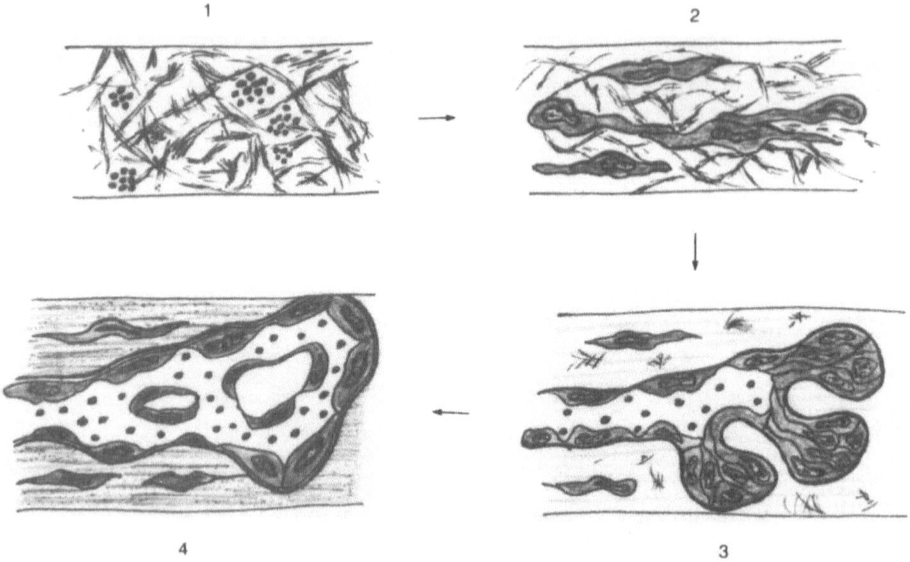

Fig. 9. Schematic drawing showing the sequence of events in the genesis of new capillaries. (1) Initial stage of fibrin clot formation. (2) Migration of endothelial cells and fibroblasts into the chamber. (3) Fibrinolysis, deposition of collagen and organization of endothelial cells into "capillary bunds." (4) Formation of a network of open capillary tubes.

system which switch the cells from a phase of proliferation and migration to that of differentiation is not clear. It has been postulated, and there is evidence to support it, that variation in the balance of tensile strength generated by the cytoskeletal contractile elements in their resistance to the extracellular attachment points may play a regulatory role (Ingber et al., 1989). High ECM densities resist cell-generated tensile strength and promote cell spreading and growth while low densities permit cell shortening and aggregation.

Another factor involved in tissue organization is the basal lamina which appeared during the time of formation of endothelial buds. The endothelial cells are anchored to the basal lamina which in turn anchores to the collagen matrix. This serves to establish the inside-outside position and cell polarity which initiates normal function.

The present study has demonstrated that angiogenesis is the result of a series of interaction between endothelial cells with a changing ECM factors. The fibrin/fibronectin clot together with soluble growth factors provides the substratum for the migration and proliferation of the endothelial cells. The collagen and basal lamina which are deposited after fibrinolysis may provide the signals necessary for the endothelial cells to change from a spreading and proliferative form to an aggregating form which eventually results in the formation of capillary tubes.

Findings from the present study suggest that fibrin, in addition to being the key figure in thromobosis and hemostasis, has an important role in angiogenesis as well. Angiogenesis is an important initial event in wound healing (Shoshan, 1981, Clark et al., 1982). The formation of vascular granulation tissue provides the substratum for subsequent steps of cell growth and tissue regeneration in all types of organs regardless of the nature of the injury. We have postulated that the newly formed capillaries are abnormally permeable thus allowing the entrance of growth-related factors from blood into the wound tissues (Liu 1988a,b,c).

From a pathophysiological point of view, the survival of the organisms depends on their ability to cope with injuries, a process known as "reaction to injury" (Forbus 1943). The grave consequences of injuries are hemorrhage, infection and cell loss. In order to counteract these adversities, various defensive measures have evolved and these include hemostasis, defensive and immune mechanisms, and wound healing (Fig. 10). Fibrin, by

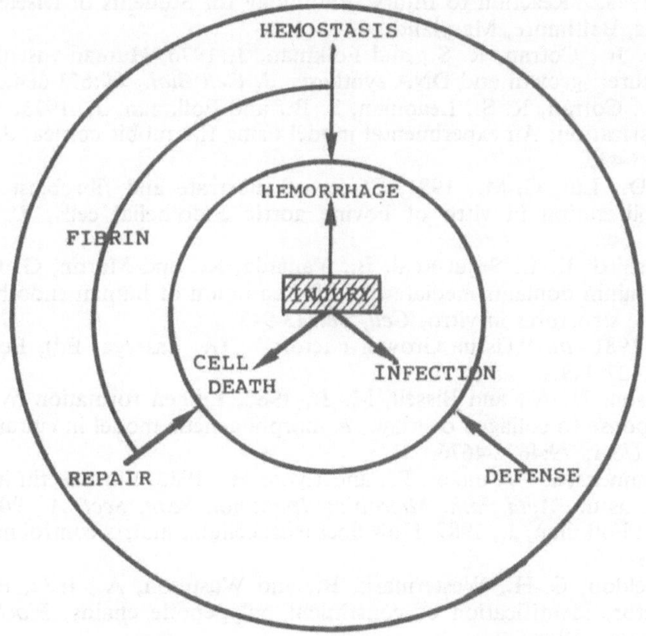

Fig. 10. A diagram showing the dual roles of fibrin in the overall picture of the organism's reaction to injury.

virtue of its involvement in both hemostasis and angiogenesis, occupies a central position in wound healing and in the overall picture of reaction to injury.

REFERENCES

Auerbach, R., 1981, *in:* "Lymphokines". (E. Pick, Ed), *Academic Press,* **pp:**69-79.

Ausprunk, D. H., and Folkman, J., 1977, Migration and proliferation of endothelial cells in preformed and newly formed blood vessels during tumor angiogenesis, *Microvas Res.,* **14:**53-65.

Bernfiled, M. R., Banerjee, S. D., and Cohn, R. H., 1972, Dependence of salivary gland morphology and branching morphogenesis upon acid mucopolysaccharide-protein (proteoglycan) at the epithelial surface, *J. Cell. Biol.,* **52:**674-689.

Bowersox, J. C., and Sorgente, N., 1982, Chemotaxis of aortic endothelial cells in response to fibronectin, *Cancer Res.,* **42:**2547-2551.

Burgess, W. H., Mehlman, T., Friesel, R., Johnson W. V., and Maciag, T., 1985, Multiple forms of endothelial cell growth factor. Rapid isolation and biological and chemical characterization, *J. Biol. Chem.,* **260:**11389-11392.

Cheresh, D. A., Berliner, S. A., Vicente, V., and Ruggeri, Z. M., 1989, Recognition of distinctive adhesive sites on fibrinogen by related integrins on platelets and endothelial cells, *Cell,* **58:**945-953.

Clark, R. A. F., DellaPelle, P., Manseau, E., Lanigan, J. M., Dvorak, H. F., and Colvin, R. B., 1982, Blood vessel fibronectin increases in conjunction with endothelial cell proliferation and capillary ingrowth during wound healing, *J. Invest. Dermat.,* **79:**269-276.

Delvos, U., Gajdusek, H., Sage, L. A., Harker, A., and Schwartz, M., 1982, Interactions of vascular wall cells with collagen gels, *Lab. Invest.,* **46:**61-72.

Folkman, J., Merter, E., Abernathy, C., and Williams, G., 1971, Isolation of a tumor factor responsible for angiogenesis, *J. Exp. Med.,* **133:**275-288.

Folkman, J., Haudneschild, C., and Zetter B. R., 1979, Longterm culture of capillary endothelial cells, *Proc. Natl. Acad. Sci. USA,* **76:**5217-5221.

Forbus, W. D., 1943, "Reaction to Injury, Pathology for Students of Disease." Williams and Wilkins, Baltimore, Maryland.
Gimbrone, M. A. Jr., Cotran, R. S., and Folkman, J, 1976, Human vascular endothelial cells in culture: growth and DNA synthesis, *J. Cell Biol.,* **60:**673-684.
Gimbrone, M. A., Cotran, R. S., Leapman, S. B., and Folkman, J., 1973, Tumor growth and vascularization; An experimental model using the rabbit cornea, *J. Natl. Cancer Inst.,* **52:**413-427.
Gospadarowicz, D., Lui, G.-M., 1981, Effect of substrate and fibroblast growth factor on the proliferation in vitro of bovine aortic endothelial cells, *J. Cell Physiol.,* **109:**69-81.
Grant, D. S., Tashiro, K. I., Segui-Real, B., Yamada, K., and Martin, G. R., 1989, Two different laminin domains mediate the differentiation of human endothelial cells into capillary-like structures in vitro, *Cell,* **58:**933-943.
Guillino, P. M., 1981, *in:* "Tissue Growth Factors". (R. Baserga, Ed), Berlin, Springer-Verlag. pp:427-449.
Hall, H. G., Farson, D. A., and Bissell, M. J., 1982, Lumen formation by epithelial cell lines in response to collagen overlay: A morphogenetic model in culture, *Proc. Natl. Acad. Sci, USA,* **79:**4672-4676.
Hollund, B., Clemmensen, I., Junker, P., and Lyon, H., 1982, Fibronectin in experimental granulation tissue, *Acta Path. Microbiol. Immunol. Scan. Sect. A.,* **90:**159-165.
Ingber, D. E., and Folkman, J., 1989, How does extracellular matrix control morphogenesis? *Cell,* **58:**803-805.
Johnsson, A., Heldon, C.-H., Westermark, B., and Wasteson, A., 1982, Platelet-derived growth factor; identification of constituent polypeptide chains, *Biochem. Biophys. Res. Commun.,* **104:**66-74.
Kumar, V., 1984, *in:* "Pathological Basis of Disease." (S. L. Robbins and R. S. Cotran, Eds), 2nd ed., pp:91-102.
Kurkinen, M., Vaheri, A., Roberts, P. J., and Stenman, S, 1980, Sequential appearance of fibronectin and collagen in experimental granulation tissue, *Lab. Invest.,* **43:**47-51.
Liu, H. M., Schmid, K., 1988a, The nerve growth promoting activity of human plasma α_1-acid glycoprotein, *J. Neurosci. Res.,* **20:**60-72.
Liu, H. M., 1988b, Neovasculature and blood-brain barrier in ischemic brain infarct, *Acta Neuropathol.,* **75:**422-426.
Liu, H. M., and Sturner, W. Q., 1988c, The extravasation of plasma proteins in traumatic brain lesions, *Forensic Sci. International,* **38:**285-295.
Liu, H. M., Wang, D. L., and Liu, C. Y., 1989, Angiogenesis in a wound chamber model, *Submitted for publication.*
Maciag, T., 1984, Angiogenesis, in: "Progress in Hemostasis and Thrombosis." Vol 7, Grune and Stratton, pp:167-182.
Madri, J. A., Williams, S. K., Wyatt, T., and Mezzio, C., 1983, Capillary endothelial cell cultures: phenotypic modulation by matrix components, *J. Cell Biol.,* **97:**153-165.
Martinet, Y., Bitterman, P. B., Mornex, J., Grotendorst, G. R., Martin, G. R., and Crystal, R. G., 1986, Activated human monocyte express the c-sis proto-oncogene and release a mediator showing PDGF-like activity, *Nature,* **319:**158-160.
Montesano, R., Mouron, P., and Orci, L., 1985, Vascular outgrowth from tissue explants embedded in fibrin or collagen gels: A simple in vitro model of angiogenesis, *Cell Biology International Reports,* **9:**869-875.
Mosher, D. F., 1971, Cross-linking of cold-insoluble globulin by fibrin-stabilizing factor, *J. Biol. Chem.,* **250:**6614-6616.
Patten, B. M., 1971, The early embryology of the chick. New York, McGraw-Hill.
Roberts, A. B., Sporn, M. B., Assoian, R. K., Smith, J. M., Roche, N. S., Wakefield, L. M., Heine, U. I., Liotta, L. A., Falanga, V., Kehrl J. H., and Fauci, A. S., 1986, Transforming growing factor type B:Rapid induction of fibrosis and angiogenesis in vivo and stimulation of collagen formation in vitro, *Proc. Natl. Acad. Sci. USA,* **83:**4167-4171.
Ross, R., Glomset, J., and Harker, L., 1977, Response to injury and atherogenesis, *Am. J. Path,* **86:**665-674.
Ruoslahti, E., and Pierschbacher, M. D., 1987, New perspective in cell adhesion: RGD and integgrins, *Science,* **238:**491-497.
Shoshan, S., 1981, Wound Healing, *in:* "International Review of Connective Tissue Research, vol. 9., D. A. Hall and D. S. Jackson (eds)., Academic Press, pp:1-26.

Taussig, M. J., 1984, Processes in Pathology and Microbiology, Blackwell-Scientific Publishers, Oxford, London. pp:640-646.
Thakral, K. K., Goodson, W. H., and Hunt, T. K., 1979, Stimulation of wound blood vessel growth by wound macrophages, *J. Cell Physiol,* **96:**203-213.
Thompson, W. D., Campbell, R., and Evans, A. T., 1985, Fibrin degradation and angiogenesis: quantitative analysis of the angiogenic response in the chick chorioallantoic membrane, *J. Pathol.,* **145:**27-37.
Thomas, K. A., Candelore, M. R., Rios-Candelore, M., and Fitzpatrick, S., 1984, Purification and characterization of acidic fibroblast growth factor from bovine brain, *Proc. Natl. Acad. Sci.,* **81:**357-365.
Vaheri, A., Salonen, E.-M., Varito, T., Hedman, K., and Stenman, S., 1983: Fibronectin and tissue injury. *In:* "Biology and Pathology of the Vessel Wall." Neville Woolf, Praeger, East Sussex, U.K. pp:161-171.
Zetter, B. R., 1980, Migration of capillary endothelial cells is stimulated by tumor-derived factors, *Nature,* **285:**41-44.

NEW DEVELOPMENTS IN THROMBOLYTIC THERAPY

D. Collen and H. K. Gold

Center for Thrombosis and Vascular Research
University of Leuven, Belgium and
Massachusetts General Hospital
Harvard Medical School, Boston, MA, USA

SUMMARY

Thrombotic complications of cardiovascular disease are a main cause of death and disability and, consequently, thrombolysis could favorably influence the outcome of such life-threatening diseases as myocardial infarction, cerebrovascular thrombosis and venous thromboembolism.

Thrombolytic agents are plasminogen activators that convert plasminogen, the inactive proenzyme of the fibrinolytic system in blood, to the proteolytic enzyme plasmin. Plasmin dissolves the fibrin of a blood clot, but may also degrade normal components of the hemostatic system and predispose to bleeding. Currently, five thrombolytic agents are either approved for clinical use or under clinical investigation in patients with acute myocardial infarction. These include streptokinase, urokinase, recombinant tissue-type plasminogen activator (rt-PA), anisoylated plasminogen streptokinase activator complex (APSAC) and single chain urokinase-type plasminogen activator (scu-PA, prourokinase). The first generation thrombolytic agents, streptokinase (and probably also urokinase), are only moderately efficacious and their administration is associated with extensive systemic fibrinogen breakdown. In comparative studies performed in patients with acute myocardial infarction, recombinant tissue-type plasminogen activator (rt-PA) is a more effective and fibrin-specific thrombolytic agent than streptokinase. The acylated plasminogen streptokinase activator complex (APSAC) has a profile of thrombolytic efficacy and fibrin-specificity that is similar or somewhat better than that of streptokinase, but has the advantage that it can be administered by bolus injection. Single chain urokinase-type plasminogen activator is more fibrin-specific than urokinase. Comparative data on the efficacy and safety of this agent are limited as it is in the early stage of clinical investigation.

Reduction of infarct size, preservation of ventricular function and/or reduction in mortality has been observed with streptokinase, rt-PA and APSAC. Therefore, thrombolytic therapy will probably become routine therapy for early acute myocardial infarction.

In patients with acute myocardial infarction, intravenous streptokinase recanalizes 40-45 percent of occluded coronary arteries and reduces mortality by 25 percent; it costs approximately $200 for a therapeutic dose of 1,500,000 units. Recombinant tissue-type plasminogen activator (rt-PA) is more potent for coronary arterial thrombolysis, producing both more rapid and more frequent (65-70 percent) reperfusion, but it costs over $1,000 for a therapeutic dose of 100 mg. Side effects (mainly bleeding) and the incidence of reocclusion associated with the use of streptokinase and rt-PA are not markedly different. Whether the higher efficacy of rt-PA will translate into a comparably larger reduction of mortality remains to be determined in large comparative clinical trials. Both agents are available for clinical use. The choice of agent for the treatment of acute myocardial infarction at present must be based on considerations of lower cost of streptokinase versus the higher efficacy for coronary recanalization of rt-PA.

Recent reviews of thrombolytic agents have reached apparently contradictory conclusions with respect to the comparative properties of thrombolytic agents (1-3). In particular, the data on the relative efficacy for coronary thrombolysis, the speed of reperfusion, the frequency of reocclusion, the occurrence of bleeding complications, and the impact on mortality have been presented and interpreted differently.

All available thrombolytic agents still suffer significant shortcomings, including the need for large doses to be therapeutically efficient, a limited fibrin-specificity and residual toxicity in terms of bleeding complications. New developments towards further improved efficacy and fibrin-specificity of thrombolytic therapy include the use of combinations of synergistic thrombolytic agents, mutants of t-PA and scu-PA, chimeric t-PA/scu-PA molecules, antibody-targeted thrombolytic agents, and/or combinations of fibrin-dissolving agents with anti-platelet strategies.

In this communication, we will briefly review the components of the fibrinolytic system, the mechanism of fibrin-specific thrombolysis, the present state of clinical trials with thrombolytic agents in acute myocardial infarction, and finally, new trends in thrombolytic therapy.

THE FIBRINOLYTIC SYSTEM

A. Components of the fibrinolytic system

The fibrinolytic system, schematically represented in Fig. 1, contains a proenzyme, plasminogen, which can be converted to the active enzyme plasmin by the action of several different types of plasminogen activators. Plasmin is a serine protease that digests fibrin to soluble degradation products. Natural inhibition of the fibrinolytic system occurs both at the level of the plasminogen activator and at the level of plasmin (4-6).

Plasminogen is a single chain glycoprotein consisting of 790 amino acids that is converted to plasmin by cleavage of the Arg560-Val561 peptide bond. The plasminogen molecule contains structures, called lysine binding sites, which mediate its binding to fibrin and accelerate the interaction between plasmin and its physiological inhibitor, α_2-antiplasmin. The lysine binding sites play a crucial role in the regulation of fibrinolysis. Plasminogen binds to platelets and endothelial cells, and this binding enhances the conversion to plasmin.

Plasminogen activators are serine proteases with a high specificity for plasminogen. The hydrolysis of the Arg560-Val561 peptide bond of plasminogen yields the active enzyme, plasmin.

Streptokinase is a nonenzyme protein with molecular weight of 47,000 that is produced by beta-hemolytic streptococci. Streptokinase activates the fibrinolytic system indirectly by forming a 1:1 stoichiometric complex with plasminogen. The formation of this complex exposes an active site in the modified plasminogen moiety, whereby the complex becomes a potent plasminogen activator.

APSAC (anisoylated plasminogen streptokinase activator complex) is an inactive derivative of the plasminogen-streptokinase activator complex obtained by acylation of its active site serine. Spontaneous deacylation at physiological pH with a t½ of approximately 35 min promotes its reactivation.

Urokinase is a trypsin-like serine protease composed of two polypeptide chains, connected by a disulfide bridge, which activates plasminogen directly to plasmin. Single chain

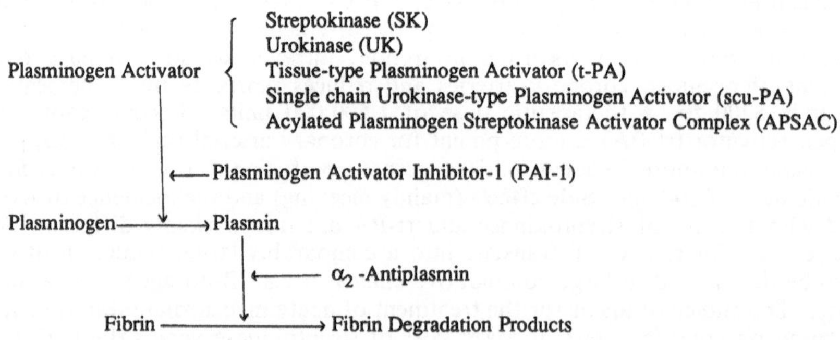

Fig. 1. Schematic representation of the fibrinolytic system.

urokinase-type plasminogen activator (scu-PA) or pro-urokinase is a single chain glycoprotein containing 411 amino acids, which is converted to urokinase by hydrolysis of the Lys158-Ile159 peptide bond. scu-PA has very little activity towards low molecular weight substrates in vitro but does posses intrinsic plasminogen activating potential. The catalytic efficiency of scu-PA is approximately two orders of magnitude lower than that of urokinase.

Tissue-type plasminogen activator (t-PA) is a trypsin-like serine protease composed of 527 amino acids. It occurs either as a single chain glycoprotein or as a two chain proteolytic derivative, but both forms have comparable enzymatic properties. t-PA is a poor plasminogen activator in the absence of fibrin, but it binds specifically to fibrin and activates plasminogen at the fibrin surface several hundred-fold more efficiently than in the circulation. t-PA binds to endothelial cells and is cleared through a saturable, probably hepatic mechanism.

α_2-antiplasmin is a glycoprotein of the serine protease inhibitor (Serpin) superfamily, composed of 452 amino acids with Arg364-Met365 as the reactive site. α_2-Antiplasmin reacts very rapidly with plasmin, first to form a reversible but inactive complex, which is then slowly converted into an irreversible complex. The rapidity of the first step of the reaction is dependent on the presence of free lysine binding sites and a free active center in the plasmin molecule.

Plasminogen activator inhibitor-1 (PAI-1) is a fast acting inhibitor of t-PA and urokinase occurring at very low concentration in the blood, but which may be significantly increased in several disease states including venous thromboembolism and ischemic heart disease. It is a Serpin, composed of 379 amino acids with Arg346-Met347 as the reactive site.

B. Mechanism of fibrin-specific thrombolysis

Plasmin, the proteolytic enzyme of the fibrinolytic system has a low substrate specificity. In purified systems it will degrade fibrinogen almost as well as fibrin and, when circulating freely in the blood, it will degrade a number of plasma proteins including fibrinogen and the blood coagulation factors V and VIII. Plasmin generated in blood will, however, rapidly be neutralized by α_2-antiplasmin via the mechanism detailed above. Extensive systemic activation of the fibrinolytic sysem will result in α_2-antiplasmin depletion, excess free plasmin generation, and secondary homostatic breakdown. In contrast, plasmin that is generated at the fibrin surface has both its lysine binding sites and active center occupied and thus is only very slowly inactivated by α_2-antiplasmin (4).

Clot-specific thrombolysis will, in view of these interactions, require plasminogen activation at, or in the vicinity of, the fibrin clot. The two physiological plasminogen activators, t-PA and scu-PA, exert clot-specificity in a plasma environment via entirely different molecular mechanisms (5). t-PA is relatively, but not totally, inactive in the absence of fibrin. It binds specifically to fibrin, thereafter acquiring a high affinity for plasminogen, resulting in preferential plasminogen activation on the fibrin clot. In a plasma milieu, relatively fibrin-specific clot lysis can be obtained within a rather narrow concentration range with scu-PA. In the absence of fibrin, no significant plasminogen activation occurs. Apparently, plasma exerts a competitive inhibitory effect that is reversed by fibrin (5). In addition, fibrin-bound plasminogen may be more sensitive to activation by scu-PA and scu-PA may be converted to the more active urokinase at the clot surface (6).

THROMBOLYTIC THERAPY IN ACUTE MYOCARDIAL INFARCTION

In the late 1970's, it was recognized that thrombosis within the infarct related coronary artery played a major role in the pathogenesis of acute myocardial infarction (7). It was further recognized that early administration of either intracoronary or intravenous streptokinase resulted in reperfusion of 75 and 45 percent of patients, respectively (8). This has formed the basis for several large scale studies employing intravenous thrombolytics with the intention to 1) define the real impact of early coronary artery reperfusion on patient survival and 2) establish patterns of efficacy and safety for new and potentially improved thrombolytic agents. The agents used for thrombolytic therapy in humans can be classified in two main categories: (a) the "classical" thrombolytic agents, streptokinase and urokinase, and (b) the "second-generation" thrombolytic agents, t-PA, scu-PA, and acylated plasminogen streptokinase activator complex (APSAC).

A. Placebo-controlled trials with mortality end-points

The results of the major placebo controlled clinical studies with intravenous thrombolytic therapy in patients with acute myocardial infarction are shown in Table 1.

Streptokinase

The first major study employing intravenous streptokinase in early acute myocardial infarction and demonstrating relative improvement in survival was carried out in 11,806 patients and reported as the GISSI Study (9, 10). Life-threatening bleeding was only observed in 0.3 percent of patients receiving streptokinase and judged not to be a limitation for therapy. This study confirmed and extended the observations with intracoronary streptokinase (11). The positive results of the GISSI trial were not confirmed in a very similar but somewhat smaller study of the ISAM study group (12). Furthermore, in the ISAM-study, life-threatening bleeding was more common, occurring in 0.8 percent of patients. However, the ISIS-2 study has confirmed and extended the GISSI findings in 17,189 patients with an inhospital mortality reduction from 11.7 to 8.9 percent in patients receiving streptokinase (13). This study also showed a significant reduction of mortality with aspirin and that the effects of streptokinase and aspirin are additive. Two additional studies (14, 15) have provided further evidence of the effect of streptokinase therapy on mortality reduction and on its relative safety.

In aggregate, these studies are suggestive of two conclusions. First, the intravenous administration of streptokinase early (preferably within 4 hours) in acute myocardial infarction is of benefit to both early and late patient survival. Second, the incidence of life-threatening side effects, predominantly major bleeding and intracranial hemorrhage, is not negligible but occurs consistently in less than I percent of the patients treated with streptokinase.

Tissue-type plasminogen activator

Following the demonstration of the potential of natural t-PA as a thrombolytic agent, an intensive effort was launched to enhance its production by recombinant DNA technology (16). Subsequently, improved left ventricular function (17-19) and reduced mortality (19, 20) were demonstrated with recombinant t-PA (rt-PA) in placebo controlled trials (Table 1).

A major recent concern with the use of rt-PA has been the occurrence of intracerebral bleeding. This complication occurred in 1.6 percent of 1,014 patients that received 150 mg of rt-PA combined with heparin and aspirin (21). This frequency has been reduced to 0.6 percent in 1,452 patients of the TIMI trial with the use of 100 mg of rt-PA (21). In the large scale placebo-controlled ASSET trial (20), the stroke rates in the rt-PA and placebo groups were, however, similar.

Thus, initial evaluation of rt-PA in myocardial infarction has been promising, suggesting its potential role as an improved thrombolytic agent. However, its use is not risk-free despite its relatively higher clotselectivity than streptokinase.

Acylated plasminogen streptokinase activator complex

In controlled studies comparing intravenous APSAC with intravenous or intracoronary streptokinase, APSAC was found to cause reperfusion at a frequency intermediate between that of intravenous and intracoronary streptokinase (22, 23). A marked reduction in mortality was obtained with APSAC in the AIMS Study (24).

B. Comparative biological properties of thrombolytic agents

The relevant biological properties of thrombolytic agents in terms of their benefit/risk ratio in acute myocardial infarction comprise their efficacy for coronary artery recanalization, the speed of reperfusion, the incidence or reocclusion, the frequency and severity of bleeding complications and their effect on mortality.

At present, insufficient results of comparative clinical trials are available to allow definitive quantitative comparison of the relative efficacy and safety of the various thrombolytic agents, although such attempts have recently been made (3, 25). At the present time, sufficient information is available from randomized controlled trials or from uncontrolled

Table 1. Major Placebo Controlled Clinical Trials with Intravenous Thrombolytic Therapy in Acute Myocardial Infarction

Study	Patient number	Adjunctive Therapy		Major Bleeding (%)		Neurological Event Rate (%)			Early Mortality Rate (%)			
		Aspirin	Heparin	Placebo	Active	Definition	Placebo	Active	Endpoint	Placebo	Active	P
A. Streptokinase												
GISSI (9, 10)	11,806	14%	21%	?	0.3	Stroke	0.9	1.1	21d	13.0	10.7	0.002
						SK-CVA		0.2				
ISAM (12)	1,741	+	+	?	0.8	ICB	0	0.5	21d	7.1	6.3	NS
W. Wash (14)	368	?	+	(0.7)	(13)	ICB	0	0.5	14d	9.7	6.3	NS
N. Zealand (15)	219	+	+	0	1	ICB	0	0	30d	12.9	2.5	0.012
ISIS-2 (13)	17,189	+/–	–	0.2	0.6	Stroke	0.9	0.8	35d	11.7	8.9	0.001
						SK-ICB		0.1-0.2				
B. rt-PA												
ECSG (19)	721	+	+	1.9	3.7	Stroke	0.5	2.0	14d	5.7	2.8	0.053
						rt-PA-ICB	?	1.4				
ASSET (20)	5,111	–	+(24 hr)	0.4	1.4	Stroke	1.0	1.1	30d	9.8	7.2	0.002
C. APSAC												
AIMS (24)	1,004	?	+	?	?	Stroke	1.0	0.4	30d	12.2	6.4	0.0016

Abbreviations: *ICB: intracerebral bleeding; CVA: cerebrovascular accident; SK-CVA: CVA associated with streptokinase therapy; SK-ICB: ICB associated with SK therapy; rt-PA ICB: ICB associated with rt-PA therapy.

trials with comparable endpoints, to allow a semi-quantitative comparison of recombinant t-PA with streptokinase and APSAC with streptokinase. For urokinase and scu-PA, such comparisons can only be made in a preliminary and tentative manner. The comparative properties of thrombolytic agents, to the extent that they can at present be evaluated, are summarized in Table 2, and will be further detailed below.

Efficacy for coronary thrombolysis

The efficacy of plasminogen activators for coronary thrombolysis in patients with acute myocardial infarction has been established by coronary angiography in "reperfusion" trials, in which patients with occluded coronary arteries, selected by angiography *before* treatment, were subjected to angiography at different time intervals after the start of thrombolytic therapy. Alternatively, "patency" trials have only studied patients by angiography at different time intervals *after* the start of thrombolytic therapy. Furthermore, coronary thrombolysis, either spontaneous or in association with thrombolytic therapy, is a time dependent phenomenon. Even without thrombolytic therapy, more than half of the patients with acute myocardial infarction will eventually (i.e. within several days) have a patent infarct-related vessel (7, 19, 26). The time window after the onset of symptoms within which reperfusion is useful for myocardial salvage is limited to several hours, which reduces the time for relevant comparisons of the efficacy of reperfusion to a couple of hours after the start of the therapy.

Tables 3 and 4 summarize the results of available efficacy trials with intravenous streptokinase, rt-PA, APSAC and urokinase strictly grouped as "reperfusion" and "patency" trials. Only trials with reasonably comparable endpoints, i.e., angiography around 90 minutes (60 to 180 minutes) after the onset of the infusion are included in the analysis. Results of reperfusion trials, with a mean time from onset of symptoms to treatment of more than 4 hours, show a 60 percent higher efficacy of rt-PA as compared to streptokinase (69 vs. 43 percent), and an intermediate value for APSAC (56 percent reperfusion). Patency trials with a mean time from onset of symptoms to treatment of usually less than 3 hours show a 50 percent higher efficacy for rt-PA (75 percent). The lower frequencies of open arteries in the reperfusion trials as compared to the patency trials (43 vs. 52 percent for streptokinase, 69 vs. 75 percent for rt-PA and 56 vs. 77 for APSAC) may be due either to inclusion of some patients with initially open arteries in the patency trials or to a somewhat higher efficacy of these agents towards fresher coronary clots. In the study, patency rates of 75 percent at 90 minutes obtained with 1 mg/kg of rt-PA could not be further increased with the combined use of high doses of rt-PA (1 mg/kg over 1 hr) and of urokinase (2,000,000 units) (70) suggesting that a significant proportion of occluded coronary arteries may be refractory to reperfusion with thrombolytic agents and that the maximally obtainable reperfusion frequencies have been approached with rt-PA at the currently recommended dosage.

Two randomized trials have been reported directly comparing the efficacy of streptokinase and rt-PA by angiography 90 minutes after the start of the infusion, namely the TIMI-I trial (33, 71) and the first trial of the European Cooperative Study Group (ECSG) (45). The TIMI-I trial was a reperfusion trial, whereas the ECGS trial was a patency trial. However, patency data from the TIMI-I trial have also recently been reported, allowing pooling of the patency results of these two comparative trials (48). The coronary patency at 90 minutes after the start of the infusion, substratified for patients treated within 3 hours of the onset of symptoms, and for patients treated between 3 and 6 hours, are summarized in Table 5. The patency at 90 minutes was significantly higher in both subgroups of patients treated with rt-PA as compared to patients treated with streptokinase. The overall frequency of patent coronary arteries with rt-PA (143 of 204 patients, 70 percent) relative to streptokinase (95 of 207 patients, 46 percent) is 52 percent higher.

The results summarized in Tables 3 to 5 thus show that, when measured within a time frame useful for salvaging myocardial tissue, streptokinase is a less efficient agent for coronary thrombolysis, reperfusing at best 1 out of 2 occluded arteries, whereas rt-PA is significantly more efficient, reperfusing at least 2 out of 3 occluded arteries. Thus, the predictions based on the biochemical properties of rt-PA have indeed translated into a higher efficacy for coronary thrombolysis.

The relative efficacy for coronary thrombolysis of APSAC has been determined in controlled comparative trials with both intravenous streptokinase (22) and with intracoronary streptokinase (23). Brochier et al. (22) determined angiographic patency at a mean of 105

Table 2. Comparative Properties of Thrombolytic Agents in Acute Myocardial Infarction

	Streptokinase	APSAC	Urokinase	scu-PA	t-PA
Coronary thrombolysis					
Reperfusion	+ (43%)	+ or ++	+ or ++	?	+++ (68%)
Patency	+ (52%)	++	+ or ++	++ or +++	+++ (75%)
Speed of reperfusion	+	+ or ++	?	+ or ++	++
Reocclusion	++	+ or ++	+ or ++	+ or ++	++
Fibrin-specificity	–	–	– or +	+	++
Bleeding complications	++	++	?	+ or ++	+
Strokes	+ (1%)*	+	?	–	+ or ++ (?)
Allergic side effects	+	+	–	–	–
Duration of infusion (min)	60	5	30–60	60	180
Dose (currently recommended or most frequently used)	1,500,000 units (~ 15 mg)	30 mg	3,000,000 units (~ 30 mg)	70–80 mg	100 mg
Early mortality reduction	+ (25%)	+ or ++	?	?	+ or ++

At present there is insufficient data from comparative controlled trials with thrombolytic agents to allow definitive quantitative conclusions on their relative properties. In this table an attempt is made to grade relative properties semiquantitatively on the basis of results of controlled trials or of uncontrolled trials with sufficiently comparable design and endpoints. When sufficient data are available, quantitative estimates are included within brackets.

*The stroke rate in the absence of thrombolytic therapy is approximately 1 percent.

Table 3. Reperfusion Demonstrated by Coronary Angiography at 60-180 Minutes after the Start of Therapy in Patients with Acute Myocardial Infarction

Study	Dose*	Mean time to treatment (hrs)	Reperfusion (percent)	Angiographic endpoint (**)
A. Streptokinase				
Neuhaus et al. 1983 (27)	1.7 M, 60 min	3.7	60 (24/40)	60 min
Schröder et al. 1983 (28)	0.5 M, 30 min	3.8	52 (11/21)	60 min
Rogers et al. 1983 (29)	1 M, 45 min	6.5	44 (7/16)	75 min
Alderman et al. 1984 (30)	0.725 M, 85 min	3.4	62 (8/13)	90 min
Spann et al. 1984 (31)	0.85-1.5 M, 60 min	3.5	49 (21/43)	60 min
Hillis et al. 1985 (32)	1.5 M, 60 min	4.5	32 (11/34)	90 min
TIMI-I Phase I 1985 (33)	1.5 M, 60 min	4.8	31 (37/119)	90 min
de Marneffe et al. 1985 (34)	1.5 M, 30 min	3.6	80 (8/10)	60 min
Total		4.3	43 (127/296)	
B. rt-PA				
Collen et al. 1984 (35)	TC, 0.5-0.75 mg/kg 30-180 min	4.7	75 (25/33)	90 min
TIMI-I 1985 (33)	TC, 50 mg/90 min	4.8	62 (70/113)	90 min
Williams et al. 1986 (36)	TC, 50 mg/90 min	4.8	68 (25/37)	90 min
Gold et al. 1986 (37)	TC, 0.4-0.75 mg/kg 60-120 min	3.0	83 (24/29)	60 min
TIMI-B 1987 (38)	SC, 70 mg/90 min	4.6	71 (59-83)	90 min
TIMI-C 1987 (38)	SC, 100 mg/90 min	4.6	68 (42/62)	90 min
Total		4.4	69 (245/357)	
C. APSAC				
Bonnier et al. 1987 (39)	30 U	2.5	64 (23/36)	90 min
De Wilde et al. 1985 (40)	30 U	(<3)	67 (10/15)	90 min
Marder et al. 1986 (41)	30 U	4.1	60 (9/15)	90 min
Timmis et al. 1987 (42)	30 U	3.3	53 (8/15)	90 min
Anderson 1988 (23)	30 U	(<6)	51 (59/115)	90 min

*M: mega units; TC: two-chain; SC: single-chain.

**Studies with angiography performed many hours or days after therapy have not been included in this analysis.

min after administration of APSAC or intravenous streptokinase and found patency frequencies of 72 percent in 42 patients treated with APSAC and of 56 percent in 44 patients given intravenous streptokinase, a difference that was not yet statistically significant. Anderson et al. (23) measured reperfusion rates in patients with proven coronary artery occlusion treated with APSAC or intracoronary streptokinase. When the authors compared reperfusion with APSAC at 90 minues and with intracoronary streptokinase at 60 minutes, the reperfusion rates were 51 percent and 60 percent respectively, a nonsignificant difference. The cumulative rate of reperfusion with intracoronary streptokinase was however markedly higher than that with APSAC at any given time point after the onset of therapy. In aggregate these comparative studies suggest that the efficacy for coronary thrombolysis of APSAC is comparable or somewhat higher than that of intravenous streptokinase, but lower than that of intracoronary streptokinase.

In the recently reported PRIMI trial (72) in which intravenous rscu-PA (80 mg over 1 hour) was compared with intravenous streptokinase in a double blind protocol, a trend ($p = 0.15$) towards higher patency rates at 90 minutes was observed for rscu-PA (72 percent) as compared to streptokinase (64 percent).

Table 4. Patency Demonstrated by Coronary Angiography at 60-180 Minutes after the Start of Therapy in patients with Acute Myocardial Infarction

Study	Dose*	Mean time to treatment (hrs)	Reperfusion (percent)	Angiographic endpoint (**)
A. Streptokinase				
Taylor et al. 1984 (43)	0.85 M, 60 min	3.2	77 (12/22)	60 min
Schwartz et al. 1984 (44)	1.5 M, 90 min	2.5	45 (25/55)	90 min
Verstraete et al. 1985 (45)	1.5 M, 60 min	2.6	55 (34/62)	90 min
Cribier et al. 1986 (46)	1.5 M, 60 min	3.1	52 (11/21)	60 min
Brochier et al. 1987 (22)	1.5 M, 60 min	2.8	56 (24/43)	105 min
Monnier et al. 1987 (47)	1.5 M, 60 min	2.5	64 (7/11)	150 min
Chesebro et al. 1988 (48)	1.5 M, 60 min	4.7	42 (61/146)	90 min
Stack et al. 1988 (49)	1.5 M, 60 min	3.0	44 (95/216)	90 min
Lopez-Sendón et al. 1988 (50)	1.5 M, 60 min	2.9	58 (14/24)	90 min
Vogt et al. 1988 (67)			72 (21/31)	
PRIMI 1989 (72)	1.5 M, 60 min	<4	64 (124/194)	90 min
Total		3.0	52 (433/825)	
B. rt-PA				
Verstraete et al. 1985 (52)	TC, 0.75 mg/kg, 90 min	3.4	61 (38/60)	90 min
Verstraete et al. 1985 (45)	TC, 0.75 mg/kg, 90 min	3.0	70 (43/61)	90 min
Verstraete et al. 1987 (53)	TC, 40 mg/90 min	2.5	66 (78/119)	90 min
Topol et al. 1987 (54)	SC, 1.25 mg/kg, 3 hr	4.0	71 (27/38)	90 min
Topol et al. 1987 (55)	SC, 150 mg, 6-8 hr	3.9	75 (288/386)	90 min
Topol et al. 1987 (56)	SC, 150 mg, 6-8 hr	3.8	67 (60/89)	90 min
		2.1	81 (43/53)	90 min
Topol et al. 1988 (57)	SC, 1.5 mg/kg, 3 hr	2.8	79 (104/132)	90 min
Simoons et al. 1988 (58)	SC, 100 mg, 3 hr	2.6	89 (160/180)	90 min
Johns et al. 1988 (59)	SC, 1 mg/kg, 90 min	(<6)	76 (52/68)	90 min
TIMI-IIA 1988 (60)	SC, 100 mg, 90 min	2.8	76 (47/62)	120 min
	SC, 70 mg, 90 min	3.2	75 (97/130)	120 min
McNeill et al. 1988 (61)	SC, 100 mg, 90 min	2.3	82 (14/17)	90 min
Neuhaus et al. 1988 (62)	SC, 70 mg, 90 min	(<6)	69 (43/62)	
Neuhaus et al. 1988 (63)	SC, 100 mg, 90 min	2.8	86 (30/35)	90 min
Total		3.0	75 (1,124/1,492)	
C. APSAC				
Kasper et al. 1986 (64)	30 U	2.5	84 (42/50)	90 min
Brochier et al. 1987 (22)	30 U	(<6)	72 (28/39)	90 min
Monassier et al. 1987 (65)	30 U	(<6)	74 (37/50)	90 min
Relik-Van Wely et al. 1988 (66)	30 U	(<4)	76 (75/99)	90 min
Lopez-Sendon et al. 1988 (50)	30 U	2.9	90 (18/20)	90 min
Vogt et al. 1988 (67)	30 U	2.6	76 (23/30)	90 min
Total		2.7	77 (223/288)	
D. Urokinase				
Mathey et al. 1988 (68)	2 M bolus		60 (30/50)	90 min
Neuhaus et al. 1988 (69)	1.5 M bolus plus 1.5 M over 90 min		63 (40/63)	90 min
Total			62 (70/113)	

*M: mega units; TC: two-chain; SC: single-chain.

**Studies with angiography performed many hours or days after therapy have not been included in this analysis.

Table 5. Relative Efficacy of rt-PA and Streptokinase for Coronary Thrombolysis in Patients with Acute Myocardial Infarction: Randomized Trials

Time from onset of symptoms to therapy	rt-PA	SK	P
Less than 3 h			
ECSG	23/29 (79%)	20/35 (57%)	0.06
TIMI	11/13 (85%)	11/21 (52%)	0.06
Combined	34/42 (81%)	31/55 (55%)	<0.01
3 to 6 h			
ECSG	20/32 (62%)	14/26 (54%)	0.51
TIMI	89/130 (69%)	50/125 (40%)	<0.001
Combined	109/162 (67%)	64/151 (42%)	<0.001
Overall combined	143/204 (70%)	95/207 (46%)	

Data from (48).

Relative speed of reperfusion

The average time to reperfusion in patients with successful thrombolysis with intravenous streptokinase, rt-PA or APSAC is approximately 45 minutes (23, 29, 58, 71). However, available data suggest that there is a linear relationship between the cumulative percent of patients reperfused at any given time point and the time elapsed since the start of the infusion.

Consequently, the speed of reperfusion is proportional to the relative efficacy of the thrombolytic agents. In as much as rt-PA has a 50 percent higher efficacy for coronary thrombolysis than streptokinase, it will have a comparably higher speed of reperfusion. The relative speed of reperfusion with APSAC is less clear to the extent that its efficacy relative to intravenous streptokinase is not yet adequately established. In the PRIMI trial (72), patency rates at 60 minutes were significantly higher ($p < 0.001$) with rscu-PA (71 percent) as compared to streptokinase (48 percent).

Reocclusion

Reocclusion following successful reperfusion has variably been estimated on the basis of recurrent ischemic pain, ECG changes, release of cardiac enzymes and coronary angiography. Furthermore, reocclusion has been determined at different times after thrombolytic therapy, either during the early hours, the first days or at the end of the hospital stay.

Comparable studies with rigorously defined endpoints to assess reocclusion are scarce. Some authors have reported incidences of reocclusion as high as 30 percent or more both for streptokinase and for rt-PA, whereas most frequently reocclusion rates of around 15 percent have been reported for both agents (73).

Two prospective studies with rt-PA have been caried out that specifically address the issue of reocclusion, using angiographic endpoints at 6-24 hours (53) or at 24-48 hours (74). The frequency of reocclusion was 7 percent of 73 patients in the first study, although 3 additional patients out of 60 reoccluded subsequently during the hospital course (53). In the second study (74), the angiographically documented reocclusion rate was only 5 percent in 78 patients. In the largest available study with APSAC, the angiographically documented reocclusion frequency, between 90 min and 24 hours after treatment, was 8 percent in a subgroup of 1/3 of patients treated, which is however an underestimate due to the fact that about 1/3 of the patients had undergone additional interventional procedures (23). In the PRIMI trial (72), reocclusion rates at 24-36 hours with both rscu-PA and streptokinase were low, but some selection may have occurred due to frequent coronary interventions. Thus, presently available data suggest that reocclusion occurs to a comparable extent with all thrombolytic agents.

Bleeding complications

Bleeding complications in association with thrombolytic therapy are most likely due to the combined actions of the thrombolytic agents on blood coagulation components, on the vessel wall and on the hemostatic plug. In addition, demographic characteristics of the patient and adjunctive anticoagulant or antiplatelet therapy may contribute to a bleeding tendency. Quantitative and qualitative evaluation of a bleeding tendency and of spontaneous bleeding from results of non-comparative studies is very difficult, especially in association with highly variable frequencies of invasive cardiovascular procedures. Consequently, bleeding complications can only validly be compared in randomized controlled trials. In the four comparative trials with streptokinase and rt-PA, the frequency of bleeding complications was comparable in two trials (33, 75), and less frequent in the rt-PA group than in the comparative streptokinase group in the other two trials (45, 76). Although available data from small randomized trials suggest a somewhat lower bleeding incidence with rt-PA as compared to streptokinase, the markedly higher fibrin-selectivity of rt-PA has not resulted in a proportional reduction of hemorrhagic complications.

The contribution to bleeding of adjunctive anticoagulant therapy with heparin, which has been demonstrated for thrombolytic therapy with streptokinase (77), will need to be further explored in the combination with rt-PA. In this context, the finding that bleeding times do not consistently prolong in patients during rt-PA infusion (no significant prolongation in 60 percent of the patients) and that there is no significant correlation between bleeding times and reperfusion (78) might be relevant and indicative of some degree of selectivity of rt-PA for a thrombus relative to a hemostatic plug. No data are available on serial bleeding times following streptokinase therapy.

A major recent concern with the use of rt-PA has been the occurrence of intracerebral bleeding in association with the use of 150 mg of rt-PA combined with heparin, in 1.6% of 1,014 patients (21). This frequency was found to be lower (0.6 percent) in 1,452 patients of the TIMI trial with the use of 100 mg of rt-PA (21), and to 0.4% in 3,768 patients treated with 80-120 mg up to the summer of 1987 (79). Nonetheless, the possibility of intracerebral bleeding due to thrombolytic therapy remains a concern.

In the comparative trials with APSAC and streptokinase bleeding complication rates were found to vary widely, depending on the nature of the study (invasive or non-invasive). In the largest comparative study of APSAC with intravenous streptokinase (22) the bleeding complication rate was similar, whereas in the invasive study of Anderson et al. (23) bleeding and blood loss was more pronounced in the APSAC group than in the intracoronary streptokinase group. In the PRIMI trial (72), bleeding complications were less frequent with rscu-PA than with streptokinase.

Evaluation of the relative risk of cerebral bleeding in association with thrombolytic therapy is most important, but it is unfortunately confounded by several intervening factors. Firstly, acute myocardial infarction, even in the absence of thrombolytic therapy, is associated with an incidence of cerebrovascular accidents that has been estimated to be approximately 1 percent in the recent large clinical trials with streptokinase and rt-PA (10, 13, 20). Secondly, it is difficult to discriminate between hemorrhagic and thrombotic strokes without the use of CT scans. Thirdly, thrombolytic therapy may convert a thrombotic stroke into a cerebral hemorrhage, without necessarily deteriorating the clinical outcome. Finally, adjunctive therapy may cause an increase of intracranial bleeding above the intrinsic and unavoidable frequency associated with thrombolytic therapy or specific thrombolytic agents. Therefore, the definitive answer to the question of the relative frequency of intracerebral bleeding associated with the various thrombolytic agents will require careful comparative studies with very large numbers of patients. Table 6 summarizes the results of neurological events during hospital follow-up in the large placebo controlled trials with streptokinase or rt-PA. Although some evidence for treatmentinduced cerebral bleeding has been observed, the frequency of neurological events is not significantly higher than in the control groups. Consequently, preliminary results of placebo controlled trials do not suggest that, at the recommended doses, streptokinase (1.5 megaunits) or rt-PA (100 mg) will significantly increase the combined frequency of thrombotic and hemorrhagic cerebrovascular accidents in patients with acute myocardial infarction.

Reduction of mortality

Mortality reduction in placebo-controlled trials has now been demonstrated for strep-

Table 6. Neurological Event Rate in Hospital Following Thrombolytic Therapy with Streptokinase or rt-PA

Study	Number of patients	Adjunctive therapy Aspirin	Heparin	Neurological Event Rate Definition	Group I	Group II
A. Streptokinase vs. placebo					Placebo	Streptokinase
ISAM (12)	1,741	+	+	ICB	0/882	4/859
GISSI (10)	11,806	?	?	Stroke	54/5,852	63/5,860
				SK-CVA	–	(0.2%)
ISIS-2 (13)	17,189	+/–	?	Stroke	75/8,381	66/8,377
				SK ICB	—	(0.1 – 0.2%)
TOTAL					129/15,115	133/15,096
B. rt-PA vs. placebo					Placebo	rt-PA
Van de Werf (19)	721	+	+	CVA	2/366	7/355
				rt-PA	ICB	(1.4%)
ASSET (20)	5,111	–	+ (24hr)	CVA	25/2,493	28/2,512
TOTAL					27/2,859	35/2,867
C. Streptokinase vs. rt-PA					SK	rt-PA
ECSG (45)	129	+	+	CVA	1/64	0/65
TIMI-I (71)	316	+	+	ICB	0/159	0/157
White (75)	270	?	?	ICB	1**/135	1/135
PAIMS (76)	171	–	+	ICB	1/85	0/86
TOTAL					2/443	1/442

**During retreatment with rt-PA after reinfarction, not included in total.

tokinase, rt-PA and APSAC (Table 1). Streptokinase reduces overall mortality at 14-30 days by an estimated 25 percent. However, in the individual trials, mortality rates in the control groups vary between 6.5 and 13 percent and reductions in early (2-4 weeks) mortality with streptokinase vary between 18 and 81 percent. Clearly, the large variability in mortality in the control groups and the impact of streptokinase thereon are influenced by patient selection, by adjunctive therapy including anticoagulant and antiplatelet drugs, and by mechanical coronary interventions. In this context, it is virtually impossible to compare the in-hospital mortalities of uncontrolled trials after rt-PA treatment in patients with acute myocardial infarction, which average around 6 percent in the more than 5,000 patients included in published studies, with those of streptokinase which average about 10 percent in the several tens of thousands of patients, treated with streptokinase (3). The in-hospital mortality rates of the two placebo-controlled rt-PA trials (19, 20), also do not provide definitive answers concerning the relative impact of streptokinase and rt-PA on mortality. Indeed, in the ASSET trial, in the absence of aspirin, patients treated with rt-PA within 3 hours from the onset of symptoms had a reduction in mortality of 25.7 percent, whereas in the corresponding subgroup of the ISIS-II trial, the reduction in mortality with streptokinase was 27.6 percent. However, in the trial of the European Cooperative Study Group (19), in the presence of adjunctive therapy with aspirin (and heparin), the reduction in mortality in patients treated within 3 hours from the onset of symptoms was 82 percent, whereas in the corresponding subgroup of ISIS-II it was 40.3 percent. Furthermore, in the ISAM-study, with a control group mortality and adjunctive therapy similar to that of the European Cooperative Study Group trial, streptokinase had a markedly lower impact on mortality. In aggregate, these studies show that valid conclusions on the impact of thrombolytic agents will only be obtained from results of careful prospective, and large comparative trials, such as the ongoing G1SS1-2 and the planned ISIS-3 trial. Similar restrictions hold for the interpretation of results obtained with APSAC.

If the clinical benefit of thrombolytic therapy in patients with acute myocardial infarction is proportional to the efficacy for coronary thrombolysis, the size of randomized clinical trials required to established differences between thrombolytic agents can be cal-

Table 7. Comparative Studies of Streptokinase and rt-PA in Patients with Acute Myocardial Infarction

Study	Patient number	Early Mortality Rate SK	rt-PA
ECSG (45)	129	3/64	3/64
TIMI-I (71)	316	14/159	12/157
White (75)	270	10/135	5/135
PAIMS (76)	171	7/85	4/86
		34/443	24/442

culated. On the basis of controlled clinical trials of streptokinase versus placebo, in over 35,000 patients with acute myocardial infarction, the reduction in hospital mortality in the treatment group was found to be approximately 25 percent. A hypothetical thrombolytic agent with a 50 percent higher efficacy of coronary thrombolysis than streptokinase would thus be anticipated to reduce early mortality by 37.5 percent. Assuming a control mortality of 9 percent in the absence of thrombolytic therapy, the mortality with streptokinase treatment would be reduced to 6.75 percent and that with the hypothetical more potent agent to 5.6 percent. In order to establish such a difference with a statistical power of 0.8 and a significance level of 0.05, more than 10,000 patients would have to be entered into a randomized trial. These numbers illustrate the tremendous task to be undertaken in order to translate efficacy for coronary thrombolysis into the most relevant clinical endpoint, mortality.

In the meantime, it might well be of interest to review the available data from the four small comparative trials with streptokinase and rt-PA, carried out to date (45, 71, 75, 76), which are summarized in Table 7. Cumulative in-hospital mortalities were 36/443 (8.1 percent) in patients randomized to streptokinase and 24/442 (5.4 percent) in patients allocated to rt-PA. It should however be stressed that these results are derived from small studies, albeit randomized, which were not prospectively designed for a mortality endpoint. However, they agree remarkably well with the values estimated on the basis of the hypothesis that the clinical outcome is primarily determined by the efficacy of the thrombolytic agent.

NEW TRENDS IN THROMBOLYTIC THERAPY

Thrombolytic therapy for acute myocardial infarction, in its present form, is based on the premise that dissolution of the fibrin component of a coronary thrombus is necessary and sufficient for recanalization. However, this approach, even with the most potent thrombolytic agents presently available, fails in approximately 25 percent of the patients. In addition, large doses are required and the treatement is associated with unpredictable bleeding and with recurrent post-treatment ischemic events. Therefore the quest for further improved thrombolytic agents or alternative therapeutic regimens continues.

Acute ischemic coronary syndromes occur nearly always in patients with atherosclerotic coronary artery disease and in association with plaque rupture (80,81) which triggers intraluminal thrombosis (7). The composition of the intraluminal thrombus formed in association with the rupture of an atherosclerotic plaque is heterogeneous, consisting of platelet-rich material contiguous to the area of plaque rupture, and erythrocyte-rich material extending both proximally and distally. This suggests two alternative and potentially complementary targets for coronary thrombolysis: the erythrocyte-rich whole blood clot and the platelet-rich thrombus. While the potential and limitations of fibrinolytic agents for the dissolution of whole blood clots is well known, the potential of pharmacological dispersion of platelet clumps and platelet-rich thrombus has not been fully explored. Furthermore, the mechanism of bleeding and strategies to reduce it need to be elucidated.

A. Attempts to improve the fibrin-dissolving potency of fibrinolytic agents

Present research in this area is carried out along several lines. In addition to the studies described below, several groups have reported in vitro data on synergism of thrombolytic agents, mutants of t-PA or scu-PA, and hybrids of t-PA and scu-PA at the recent Congress of the International Society on Thrombosis and Haemostasis, Brussels, Belgium, July 9-10, 1987, with abstracts published in Thrombosis and Haemostasis, Volume 58, #1, 1987 and at the Internatioal Congress on Fibrinolysis, Amsterdam, the Netherlands, June 28-July 1, 1988, with abstracts published in Fibrinolysis, Volume 2, Suppl. 1, 1988.

Synergism

The intrinsic fibrin selectivity of t-PA and scu-PA is mediated by entirely different molecular mechanisms. It is therefore not unreasonable to expect that, if administered in combination, the effect on clot dissolution would be more than additive. Synergism, if significant, would also allow a reduction of the total dose while obtaining a therapeutic effect that is equal to that achieved with higher doses of either agent alone. As a result, the potential for hemostatic side effects may be significantly reduced. In an in vitro system, composed of a ^{125}I-fibrin-labeled plasma clot immersed in citrated plasma, t-PA, scu-PA, and urokinase in molar ratios between 4:1 and 1:4 did not display significant synergism (82-84). In vivo, however, in a jugular vein thrombosis model in the rabbit, significant synergism between t-PA and scu-PA and between t-PA and urokinase for thrombolysis was observed (85). When t-PA and scu-PA were infused in a molar ratio of approximately 1:3, the specific thrombolytic activity of the mixture was approximately threefold higher than was anticipated on the basis of additive effects of both agents. The synergistic effect of t-PA and urokinase was only of borderline significance. A synergistic effect of at least a factor of two between t-PA and scu-PA was also found on coronary arterial thrombolysis in dogs (86).

Preliminary results in patients with acute myocardial infarction (87, 88) suggest that t-PA acts synergistically with scu-PA and urokinase in humans as well. Indeed, combining t-PA and scu-PA at approximately one-fifth of their individual thrombolytic doses produced coronary artery reperfusion in patients with acute myocardial infarction without associated systemic fibrinogen breakdown. Although these results are still preliminary and need to be confirmed in larger studies, they are potentially of significant clinical importance. The synergistic effect of t-PA and urokinase on coronary reperfusion has, however, not been confirmed in a large-scale study, although the use of the combination was associated with a significant reduction of the frequency of reocclusion (70).

Mutants of t-PA and scu-PA

We have attempted to produce mutants of t-PA and scu-PA with improved fibrin-selectivity with the use of recombinant DNA techniques.

Deletion of specific domains of the t-PA molecule has been achieved. One of these t-PA mutants, called t-PA-△FE3X, a truncated t-PA molecule without the finger-like and growth factor domains and without carbohydrate side chains, appears to have an extended half-life and a slower plasma clearance rate *in vivo* and has retained most of its thrombolytic potency (89).

Since scu-PA has intrinsic albeit limited plasminogen activating potential, attempts were made using recombinant DNA technology to prevent its conversion to two-chain urokinase. Two new constructs were produced by replacement of the lysine at position 158 with either glycine or glutamic acid. The kinetics of plasminogen activation with these mutants have been reported (90).

Although these mutants do activate plasminogen to plasmin, their catalytic efficiency is only about 10 percent of that of the natural form of scu-PA. A clear dose-response relationship was seen when such mutants were injected in rabbits with an experimental jugular vein thrombus (91). However, the specific thrombolytic activity of these mutants in vivo is also reduced by several fold. Thus, engineering of the scu-PA molecule, in order to prevent its conversion from a one-chain to a two-chain molecule, has not yielded improved molecules for thrombolysis.

Chimers of t-PA and scu-PA

Another approach has been to construct chimeric (hybrid) molecules that contain domains of both the t-PA and scu-PA molecules. The resultant molecule might combine the fibrin-affinity of t-PA (which is responsible for its concentration at the clot surface) with both the enzymatic properties of scu-PA (which is responsible for its stability in plasma) and the fibrin selectivity of scu-PA. One such chimer consisting of amino acids Ser1 through Thr263 of t-PA fused to amino acids Leu144 to Leu411 of scu-PA, has been studied in detail (92). Although this chimer has a higher fibrin affinity than scu-PA, its affinity is not as high as that of intact t-PA. Studies of the chimer *in vivo* indicate that it has maintained most of the thrombolytic potential of scu-PA, but does not appear to be a superior agent for thrombolysis (93). The chimeric approach remains however promising, because it might enable the development of an agent that has increased fibrin selectivity at a reduced dose.

Antibody-targeted thrombolytic therapy

Several alternatives to target the action of thrombolytic agents towards the thrombus with the use of fibrin-specific antibodies are presently being investigated (94, 95). These include chemical conjugates of fibrin-specific antibodies with thrombolytic agents or recombinant fusion proteins comprising a fibrin-specific antibody site and the B-chain of t-PA. Alternatively, chemical conjugates between a fibrin-specific and a t-PA-specific antibody, and biosynthetically produced heteroduplex antibodies that are both fibrin and t-PA-specific, could bind to fibrin and localize endogenous or exogenous t-PA. These conjugates display significantly enhanced clot-specific lysis in vitro and, in an animal model, in vivo.

We have recently constructed a conjugate of rscu-PA with a murine monoclonal antibody directed against human fibrin fragment D-dimer that had a 8-fold increased specific thrombolytic activity in vitro (96) and a 8-fold higher thrombolytic potency in a rabbit jugular vein thrombosis model (97).

B. Attempts to interfere with platelet-rich thrombus

The role of platelet aggregation in the pathogenesis of intraluminal coronary arterial thrombosis in patients with acute cardiac ischemic syndromes appears firmly established as reviewed above. Indeed, pathological examination has consistently revealed platelet-rich zones contiguous with plaque rupture in patients with acute myocardial infarction, unstable angina and sudden ischemic death. In addition aspirin appears to be efficient in the primary and secondary prevention of ischemic coronary events (98).

Platelet disaggregation

Although the mechanisms of platelet aggregation and pharmacological approaches to interfere with the process are well known (99), no efficient means are available to disperse preformed platelet-rich thrombus. In a plasma milieu in vitro, concentrations of thrombolytic agents that will rapidly dissolve whole blood clots, are virtually inactive towards preformed platelet aggregates (100) unless the platelets are pretreated with aspirin. rt-PA disaggregates platelet clumps but only at very high concentrations (100), an order of magnitude higher than those currently achievable by intravenous infusion. This suggested to us that platelet-rich regions of coronary clot might be more resistant to lysis than erythrocyterich zones and that predominance of platelet-rich zones would limit the efficacy of intravenous thrombolytic therapy. Consequently, one approach to increased efficacy and speed of coronary artery reperfusion might consist of the pharmacological dispersion of platelet-rich thrombus, in combination with fibrin-dissolving therapy. Alternatively, specific interference with platelet deposition during fibrinolysis might accelerate thrombolysis and prevent reocclusion.

At present three approaches towards platelet disaggregation appear feasible. The first consists of the use of high dose intracoronary administration of plasminogen activators capable of disaggregating platelets. However, the logistic restrictions to intracoronary administration precludes its widespread use. The second approach involves the use of combinations of plasminogen activators with potent antiplatelet agents (101). Thus, the combination of plasminogen activators with the anti-platelet GPIIB/IIIa antibody in animal models has resulted in accelerated lysis and elimination of reocclusion at markedly reduced doses of the fibrinolytic agent. In a third approach the thrombolytic agent could be targeted to platelet aggregates by means of specific monoclonal antibodies.

Prevention of platelet aggregation

Platelet aggregation clearly contributes to and may be the primary mechanism underlying reocclusion. This is supported by our findings that monoclonal antibodies against platelet GPIIb/IIIa efficiently abolish reocclusion in animal models with intensive thrombogenic stimulus (102). Finally, preliminary studies have indicated that arterial occlusion by platelet-rich thrombus cannot be prevented by heparin, but that it is efficiently prevented by local infusion of a synthetic thrombin inhibitor (103). These results confirm and extend recent observations with a synthetic chloromethyl ketone (104).

C. Attempts to reduce the bleeding tendency

The observation that bleeding also occurs in association with fibrinspecific thrombolytic agents has led to the hypothesis that these agents cannot distinguish between the fibrin of a pathologic thrombus and that of a hemostatic plug (2,3). However, in a recent study with rt-PA in 50 patients with acute myocardial infarction, resulting in coronary artery patency in 83 percent, serial template bleeding times remained normal throughout the infusion period in 60 percent of the patients, despite full heparin anticoagulation. No correlation was observed between the template bleeding time and success or failure of reperfusion of the occluded coronary artery (78). These observations are indicative of significant discrimination between hemostatic plugs and occlusive thrombi. The residual hemorrhagic tendency associated with the use of rt-PA despite its fibrinspecificity might relate to its interference with platelets (100) or with the endothelial cells (105) which may mediate enhanced local activation of plasminogen. Consequently, mutants of rt-PA with reduced affinity for endothelial cells might be associated with a reduced bleeding tendency. Elimination of the structures in rt-PA responsible for the interaction with endothelial cells, while maintaining those mediating the fibrin-specific enzymatic properties, might be possible with the use of recombinant DNA technology.

CONCLUSIONS

The beneficial effect of thrombolytic therapy in acute ischemic coronary syndromes, and particularly in acute myocardial infarction is now well established. The limited efficacy and potentially life-threatening side effects of thrombolytic agents remain a major problem. The present limitations of thrombolytic therapy can be explained on the basis of the heterogeneity of coronary arterial thrombus, consisting of both erythrocyte-rich and platelet-rich zones, and knowledge of the mechanism of fibrindissolution and platelet disaggregation. This unified concept suggests alternative and complementary pharmacological approaches to coronary artery recanalization. Available evidence suggests that the efficacy of coronary thrombolysis may be augmented either by improvement of the potency and specificity of fibrin-dissolving agents, by dispersion of aggregated platelets or by a combination of both. An at least partial discrimination between a pathological occlusive thrombus and a physiological hemostatic plug has been achieved with rt-PA and may be accessible to further modification. Continued investigations along several new research lines will provide new insights and promote progress towards the development of the ideal thrombolytic therapy, characterized by maximized stable coronary arterial thrombolysis with minimized bleeding.

REFERENCES

1. D. Collen, D. C. Stump, H. K. Gold, Thrombolytic therapy. *Ann. Rev. Med.,* **39:**405-423 (1988).
2. V. J. Marder, S. Sherry, Thrombolytic therapy: current status (first of two parts). *N. Engl. J. Med.,* **318:**1512-1520 (1988).
3. V. J. Marder, S. Sherry, Thrombolytic therapy: current status (second of two parts). *N. J. Engl. Med.,* **318:**1585-1595 (1988).

4. D. Collen, On the regulation and control of fibrinolysis. *Thromb. Haemost.,* **43**:77-89 (1980).
5. D. Collen, Molecular mechanisms and clinical applications of thrombolysis. Les Cahiers de la Fondation Louis Jeantet de Médicine, **1**:41-54 (1986).
6. F. Bachmann. Fibrinolysis, *in*: Thrombosis and Haemostasis. Eds: M Verstraete, J. Vermylen, R. Lijnen, J. Arnout, Leuven University Press, Leuven, Belgium, p. 227-265 (1987).
7. M. A. De Wood, J. Spores , R. Notske, et al., Prevalence of total coronary occlusion during the early hours of transmural myocardial infarction. *N. Engl. J. Med.,* **303**:897-902 (1980).
8. K. P. Rentrop, Thrombolytic therapy in patients with acute myocardial infarction. *Circulation,* **71**:627-631 (1985).
9 Gruppo Italiano per lo studio della streptochinasi nell'infarto, miocardico (GISSI). Effectiveness of intravenous thrombolytic treatment in acute myocardial infarction. *Lancet,* **1**:397-402 (1986).
10. F. Rovelli, C. De Vita, G. A. Feruglio, et al., GISSI trial: early results and late follow up. *J. Am. Coll. Cardiol.,* **10**:33B-39B (1987).
11. M. L. Simoons, M. van de Brand, C. de Zwaan, et al., Improved survival after early thrombolysis in acute myocardial infarction. *Lancet,* **2**:578-581 (1985).
12. The ISAM Study Group, A prospective trial of intravenous streptokinase in acute myocardial infarction (I.S.A.M.). Mortality, morbidity, and infarct size at 21 days. *N. Engl. J. Med.,* **314**:1465-1471 (1986).
13. ISIS-2 (Second International Study of Infarct Survival) Collaborative Group, Randomized trial of intravenous streptokinase, oral aspirin, both or neither among 17,187 cases of suspected acute myocardial infarction: ISIS-2, *Lancet,* **2**:349-360 (1988).
14. J. W. Kennedy, G. V. Martin, and K. B. Davis, et al., The Western Washington intravenous streptokinase in acute myocardial infarction randomized trial. *Circulation,* **77**:345-352 (1988).
15. H. D. White, R. M. Norris, and M. A. Brown, et al., Effect of intravenous streptokinase on left ventricular function and early survival after acute myocardial infarction. *N. Engl. J. Med.,* **317**:850-855 (1987).
16. D. Collen, Human tissue-type plasminogen activator: from the laboratory to the bedside. *Circulation,* **72**:18-20 (1985).
17. A. D. Guerci, G. Gerstenblith, and J. A. Brinker, et al., A randomized trial of intravenous tissue plasminogen activator for acute myocardial infarction with subsequent randomization to elective coronary angioplasty. *N. Engl. J. Med.,* **317**:1613-1618 (1987).
18. M. O'Rourke, D. Baron, and A. Keogh, et al., Limitation of myocardial infarction by early infusion of recombinant tissue-type plasminogen activator. *Circulation,* **77**:1311-1315 (1988).
19. F. van de Werf, A. E. R. Arnold, Intravenous tissue plasminogen activator and size of infarct, left ventricular function, and survival in acute myocardial infarction. *Br. Med. J.,* **297**:1374-1379 (1988).
20. R. G. Wilcox, G. von der Lippe, C. G. Olsson, G. Jenssen, A. M. Skene, and J. R. Hampton, Trial of tissue plasminogen activator (rt-PA) for mortality reduction in acute myocardial infarction: The Anglo-Scandinavian study of early thrombolysis (ASSET). *Lancet,* **2**:525-530 (1988).
21. E. Braunwald, G. L. Knatterud, E. Passamani, T. L. Robertson, and R. Solomon, Update from the thrombolysis in myocardial infarction trial. *J. Am. Coll. Cardiol.,* **10**:970 (1987).
22. M. L. Brochier, L. Quillet, and H. Kulbertus, et al., Intravenous anisoylated plasminogen streptokinase activator complex versus intravenous streptokinase in evolving myocardial infarction. *Drugs 3,* **(Suppl 3)**:140-145 (1987).
23. J. L. Anderson, R. L. Rothbard, and R. A. Hackworthy, et al., Multicenter reperfusion trial of intravenous anisoylated plasminogen streptokinase activator complex (APSAC) in acute myocardial infarction: Controlled comparison with intracoronary streptokinase. *J. Am. Coll. Cardiol.,* **11**:1153-1163 (1988).
24. AIMS Trial Study Group. Effect of intravenous APSAC on mortality after acute myocardial infarction: preliminary report of a placebo-controlled clinical trial. *Lancet,* **1**:545-549 (1988).
25. S. Sherry, Appraisal of various thrombolytic agents in the treatment of acute myocardial infarction. *Am. J. Med. 83,* **Suppl. 2A**:31-46 (1987).

26. K. P. Rentrop, F. Feit, and H. Blanke, et al., Effects of intracoronary streptokinase and intracoronary nitroglycerin infusion on coronary angiographic patterns and mortality in patients with acute myocardial infarction. *N. Engl. J. Med.*, **311**:1457-1463 (1984).
27. K. L. Neuhaus, U. Tebbe, G. Sauer, G. Rahlf, H. Kreuzer, and H. Kostering. Hochdosierte intravenose Kurzinfusion von Streptokinase beim acuten Myocardinfarkt, *in*: G. Trübestein, F. Etzel, eds. Fibrinolytische Therapie. FK Schattauer Verlag Stuttgart/New York, 475-480 (1983).
28. R. Schröder, G. Biamino, and E. R. Leitner, et al., Intravenous short-term infusion of streptokinase in acute myocardial infarction. *Circulation,* **67**:536-548 (1983).
29. W. J. Rogers, J. A. Mantle, and W. P. Hood, et al., Prospective randomized trial of intravenous and intracoronary streptokinase in acute myocardial infarction. *Circulation,* **68**:1051-1061 (1983).
30. E. L. Alderman K. R. Jutzy, and L. E. Berte, et al., Randomized comparison of intravenous versus intracoronary streptokinase for myocardial infarction. *Am. J. Cardiol.*, **54**:14-19 (1984).
31. J. R. Spann, S. Sherry, and B. A. Carabello, et al., Coronary thrombolysis by intravenous streptokinase in acute myocardial infarction: acute and follow-up studies. *Am. J. Cardiol.*, **53**:655-661 (1984).
32. L. D. Hillis, J. Borer, and E. Braunwald, et al., High dose intravenous streptokinase for acute myocardial infarction: preliminary results of a multicenter trial. *J. Am. Coll. Cardiol.*, **6**:957-962 (1985).
33. The TIMI Study Group, Special Report. The thrombolysis in myocardial infarction (TIMI) trial. *N. Engl. J. Med.*, **312**:932-936 (1985).
34. M. de Marneffe, E. Van Thiel, and M. Ewalenko, et al., High-dose intravenous thrombolytic therapy in acute myocardial infarction: efficiency, tolerance, complications and influence on left ventricular performance. *Acta. Cardiol.*, **40**:183-198 (1985).
35. D. Collen E. J. Topol, and A. J. Tiefenbrunn, et al., Coronary thrombolysis with recombinant human tissue-type plasminogen activator: a prospective, randomized, placebo-controlled trial. *Circulation,* **70**:1012-1017 (1984).
36. D. O. Williams, J. Borer, and E. Braunwald, et al., Intravenous recombinant tissue-type plasminogen activator in patients with acute myocardial infarction: a report from the NHLBI thrombolysis in myocardial infarction trial. *Circulation,* **73**:338-346 (1986).
37. H. K. Gold, R. C. Leinbach, and H. D. Garabedian, et al., Acute coronary reocclusion after thrombolysis with recombinant human tissue-type plasminogen activator: prevention by a maintenance infusion. *Circulation,* **73**:347-352 (1986).
38. H. S. Mueller, A. K. Rao, and S. A. Forman., Thrombolysis in myocardial infarction (TIMI): comparative studies of coronary reperfusion and systemic fibrinogenolysis with two forms of recombinant tissue-type plasminogen activator. *J. Am. Coll. Cardiol.*, **10**:479-490 (1987).
39. H. J. R. M. Bonnier, R. F. Visser, H. C. Klomps, and H. J. M. L. Hoffmann, Comparison of intravenous anisoylated plasminogen streptokinase activator complex and intracoronary streptokinase in acute myocardial infarction. *Am. J. Cardiol.*, **62**:25-30 (1988).
40. P. Dewilde, Y. Taeymans, D. Demoor, L. Huygehens, and P. Block, Intravenous thrombolysis with BRL 26921 in acute myocardial infarction. Presented at the International Symposium on Cardiovascular Pharmacotherapy, Geneva, (1985) Abstract 96.
41. V. J. Marder, R. L. Rothbard, P. G. Fitzpatrick, and C. W. Francis, Rapid lysis of coronary artery thrombi with anisoylated plasminogen: streptokinase activator complex. Treatment by bolus intravenous injection. *Ann. Intern. Med.*, **104**:304-310 (1986).
42. A. D. Timmis, B. Griffin, J. C. P. Crick, J. S. Flax, and E. Sowton, An interim report of a double-blind placebo controlled recanalisation study of anisoylated plasminogen streptokinase activator complex in acute myocardial infarction. *Drugs,* **33 (Suppl 3)**:146-50 (1987).
43. G. J. Taylor, F. L. Mikell, and H. W. Moses, et al., Intravenous versus intracoronary streptokinase therapy for acute myocardial infarction in community hospitals. *Am. J. Cardiol.*, **54**:256-260 (1984).
44. F. Schwartz, M. Hofmann, G. Schuler, K. von Olshausen, R. Zimmermann, and W. Kübler, Thrombolysis in acute myocardial infarction: effect of intravenous followed by intracoronary streptokinase application on estimates of infarct size. *Am. J. Cardiol.*, **53**:1505-1510 (1984).

45. M. Verstraete, R. Bernard, and M. Bory, et al., Randomised trial of intravenous recombinant tissue-type plasminogen activator versus intravenous streptokinase in acute myocardial infarction. *Lancet,* **1:**842-847 (1985).
46. A. Cribier, J. Berland, N. Saoudi, M. Redonnet, N. Moore, and B. Letac, Intracoronary streptokinase, OK! ..., Intravenous streptokinase, first? Heparin or intravenous streptokinase in acute infarction: preliminary results of a prospective randomized trial with angiographic evaluation in 44 patients. *Haemostasis,* **16 (Suppl 3):**122-129 (1986).
47. P. Monnier, U. Sigwart, and A. Vincent, et al., Anisoylated plasminogen streptokinase activator complex versus streptokinase in acute myocardial infarction. Preliminary results of a randomised study. *Drugs,* **33 (Suppl 3):**175-178 (1987).
48. J. Chesebro, G. Knatterud, and E. Braunwald, Correspondence section. *N. Engl. J. Med.,* **319:**1544 (1988).
49. R. S. Stack, C. M. O'Connor, and D. B. Mark, et al., Coronary perfusion during acute myocardial infarction with a combined therapy of coronary angioplasty and high-dose intravenous streptokinase. *Circulation,* **77:**151-161 (1988).
50. J. Lopez-Sendón, R. SeabraıGomes, and F. Martin Santos, et al., Intravenous anisoylated plasminogen streptokinase activator complex (APSAC) versus intravenous streptokinase (SK) in myocardial infarction (AMI). A randomized multicenter trial. *Eur. Heart J.,* **9 (Suppl A):**10 (1988).
51. H. I. Miller, A. Roth, A. Parades, B. Shagarodsky, G. Barabash, and S. Laniado, A comparison of early thrombolytic therapy with streptokinase and tissue plasminogen activator in acute myocardial infarction. *Eur. Heart J.,* **9 (Suppl A):**215 (1988).
52. M. Verstraete, W. Bleifeld, and R. W. Brower, et al., Double-blind randomised trial of intravenous tissue-type plasminogen activator versus placebo in acute myocardial infarction. *Lancet,* **2:**965-969 (1985).
53. M. Verstraete, A. E. R. Arnold, and R. W. Brower, et al., Acute coronary thrombolysis with recombinant human tissue-type plasminogen activator: initial patency and influence of maintained infusion on reocclusion rate. *Am. J. Cardiol.,* **60:**231-237 (1987).
54. E. J. Topol, D. C. Morris, and R. W. Smalling, et al., A multicenter, randomized, placebo-controlled trial of a new form of intravenous recombinant tissue-type plasminogen activator (Activase) in acute myocardial infarction. *J. Am. Coll. Cardiol.,* **9:**1205-1213 (1987).
55. E. J. Topol, R. M. Califf, and B. S. George, et al., A randomized trial of immediate versus delayed elective angioplasty after intravenous tissue plasminogen activator in acute myocardial infarction. *N. Engl. J. Med.,* **317:**581-588 (1987).
56. E. J. Topol, E. R. Bates, and J. A. Jr. Walton, et al., Community hospital administration of intravenous tissue plasminogen activator in acute myocardial infarction: improved timing. Thrombolytic efficacy and ventricular function. *J. Am. Coll. Cardiol.,* **10:**1173-1177 (1987).
57. E. J. Topol, B. S. George, and D. J. Kereiakes, et al., A multicenter, randomized, controlled trial of intravenous tissue plasminogen activator and early intravenous heparin in acute myocardial infarction. *J. Am. Coll. Cardiol.,* **11:**232A (1988).
58. M. L. Simoons, A. E. R. Arnold, and A. Betriu, et al., Thrombolysis with tissue plasminogen activator in acute myocardial infarction: no additional benefit from immediate percutaneous coronary angioplasty. *Lancet,* **1:**197-203 (1988).
59. J. A. Johns, H. K. Gold, and R. C. Leinbach, et al., Prevention of coronary artery reocclusion and reduction in late coronary artery stenosis after thrombolytic therapy in patients with acute myocardial infarction. A randomized study of maintenance infusion of recombinant human tissuetype plasminogen activator. *Circulation,* **78:**546-556 (1988).
60. The TIMI Research Group, Immediate vs delayed catheterization and angioplasty following thrombolytic therapy for acute myocardial infarction. TIMI IIA results. *J. Am. Med. Assoc.,* **260:**2849-2858 (1988).
61. A. J. McNeill, J. S. Shannon, and S. R. Cunningham, et al., A randomised dose ranging study of recombinant tissue plasminogen activator in acute myocardial infarction. *Br. Med. J.,* **296:**1768-1771 (1988).
62. K. L. Neuhaus, U. Tebbe, and M. Gottwik, et al., Intravenöse Infusion von recombinant tissue plasminogen activator (rt-PA) and Urokinase beim akuten Myokardinfarkt: Zwischenergebnisse der G.A.U.S.-Studie (German Activator Urokinase Study). *Klin. Wschr.,* **66:**102-108 (1988).

63. K. L. Neuhaus, W. Fuerer, U. Tebbe, S. Jeep-Tebbe, and A. Vogt, Efficacy of a 90-minute infusion of 100 mg tissue plasminogen activator (rt-PA) in acute myocardial infarction. *Eur. Heart J.,* **9 (Suppl A):**9 (1988).
64. W. Kasper, T. Meinertz, and H. Wollschläger, et al., Coronary thrombolysis during acute myocardial infarction by intravenous BRL 26921, a new anisoylated plasminogen-streptokinase activator complex. *Am. J. Cardiol.,* **58:**418-421 (1986).
65. J. P. Monassier, M. Brochier, and B. Charbonnier, et al., EMINASE versus streptokinase à la phase aiguë de l'infarctus du myocarde: étude randomisée (étude I.R.S. II). 14e Congrès de Cardiologie de langue française; 25-30 (1987).
66. L. Relik-Van Wely, J. M. J. van der Pol, and R. F. Visser, et al., A preliminary report, on the angiographic assessed patency and reocclusion in patients treated with APSAC for acute myocardial infarction (AMI). A Dutch Multicentre Study. *Eur. Heart J.,* **9 (Suppl A):**8 (1988).
67. P. Vogt, M. D. Schaller, P. Monnier, and U. Kaufamann, et al., Systemic thrombolysis in acute myocardial infarction: Bolus injection of APSAC versus infusion of streptokinase. *Eur. Heart J.,* **9 (Supppl A):**213 (1988).
68. D. G. Mathey, J. Schofer, F. H. Sheehan, H. Becher, V. Tilsner, and H. T. Dodge, Intravenous urokinase in acute myocardial infarction. *Am. J. Cardiol.,* **55:**878-882 (1985).
69. K. L. Neuhaus, U. Tebbe, and M. Gottwik, et al., Intravenöse Infusion von recombinant tissue plasminogen activator (rt-PA) and Urokinase beim akuten Myokardinfarkt: Zwischenergebnisse der G.A.U.S.-Studie (German Activator Urokinase Study). *Klin. Wschr.,* **66:**102-108 (1988).
70. E. J. Topol, R. M. Califf, and B. S. George, et al., Coronary arterial thrombolysis with combined infusion of recombinant tissue-type plasminogen activation and urokinase in acute myocardial infarction. *Circulation,* **77:**1100-1107 (1988).
71. J. H. Chesebro, G. Knatterud, and R. Robert, et al., Thrombolysis in myocardial infarction (TIMI) trial. Phase I: a comparison between intravenous tissue plasminogen activator and intravenous streptokinase. *Circulation,* **76:**142-154 (1987).
72. Primi trial study group, Randomised double-blind trial of recombinant pro-urokinase against streptokinase in acute myocardial infarction. *Lancent,* **1:**863-867 (1989).
73. P. L. Raynaud, and B. Desveaux, Réocclusion après traitement par l'Actilyse. *Arch. Mal. Coeur.,* **81 (Suppl 1):**25-32 (1988).
74. R. S. Kent, A. G. Batson, and J. K. LittleJohn, Thrombolytic effect of tissue-type plasminogen activator, *in:* Controversies in Thrombolysis. Eds: R. Schroder, S. Sherry, K. L. Mettinger. Current Medical Literature Ltd, London (in press).
75. H. D. White, J. T. Rivers, and A. H. Maslowski, et al., Effect of intravenous streptokinase as compared with that of tissue plasminogen activator on left ventricular function after first myocardial infarction. *N. Engl. J. Med.,* **320:**817-821 (1989).
76. B. Magnani, for the PAIMS Investigators, Plasminogen activator Italian multicenter study (PAIMS). Comparison of intravenous recombinant single-chain human tissue-type plasminogen activator (rt-PA) with intravenous streptokinase in acute myocardial infarction. *J. Am. Coll. Cardiol.,* **13:**19-26 (1989).
77. S. Schulman, D. Lockner, S. Granqvist, G. Bratt, C. Paul, and D. Nyman, A comparative randomized trial of low dose versus high dose streptokinase in deep vein thrombosis of the thigh. *Thromb. Haemost.,* **51:**261-265 (1984).
78. L. W. Gimple, H. K. Gold, and R. C. Leinbach, et al., Bleeding time measurement predicts spontaneous bleeding during thrombolysis with recombinant tissue-type plasminogen activator (rt-PA). *J. Am. Coll. Cardiol.,* **11:**231A (1988).
79. F. D. A. Summary Basis for Approval of Alteplase (86-0236). (1987) p. 15.
80. M. Friedman, The coronary thrombus: its origin and fate. *Human Pathol.,* **2:**81-128 (1971).
81. M. J. Davies, and A. C. Thomas, Plaque fissuring — the cause of acute myocardial infarction, sudden ischaemic death, and crescendo angina. *Br. Heart J.,* **53:**363-373 (1985).
82. H. R. Lijnen, C. Zamarron, M. Blaber, M. E. Winkler, and D. Collen, Activation of plasminogen by pro-urokinase. I. Mechanism. *J. Biol. Chem.,* **261:**1253-1258 (1986).
83. D. Collen, F. De Cock, E. Demarsin, H. R. Lijnen, and D. C. Stump, Absence of synergism between tissue-type plasminogen activator (t-PA), single chain urokinase-type plasminogen activator (scu-PA) and urokinase on clot lysis in a plasma milieu in vitro. *Thromb. Haemost.,* **56:**35-39 (1986).

84. V. Gurewich, and R. Pannell, Synergism of tissue-type plasminogen activator (t-PA) and single-chain urokinase-type plasminogen activator (scu-PA) on clot lysis in vitro and a mechanism for this effect. *Thromb. Haemost.*, **57**:372 (1987).
85. D. Collen, J. M. Stassen, D. C. Stump, M. Verstraete, Synergism of thrombolytic agents in vivo. *Circulation*, **74**:838-842 (1986).
86. A. A. Ziskind, H. K. Gold, T. Yasuda, M. Kanke, and J. L. Guererro, et al., Coronary thrombolysis in dogs with synergistic combinations of human tissue-type plasminogen activator (r-PA) and single chain urokinase-type plasminogen activator (scu-PA). *Clin. Res.*, **35**:337A (1987).
87. D. Collen, D. C. Stump, and F. Van de Werf, Coronary thrombolysis in patients with acute myocardial infarction by intravenous infusion of synergic thrombolytic agents. *Am. Heart J.*, **112**:1083-1084 (1986).
88. D. Collen, and F. van de Werf, Coronary arterial thrombolysis with low-dose synergistic combinations of recombinant tissue-type plasminogen activator (rt-PA) and recombinant single-chain urokinase-type plasminogen activator (rscu-PA) for acute myocardial infarction. *Am. J. Cardiol.*, **60**:431-444 (1987).
89. P. Cambier, F. Van de Werf, G. R. Larsen, and D. Collen, Pharmacokinetics and thrombolytic properties of a nonglycosylated mutant of human tissuetype plasminogen activator, lacking the finger and growth domains, in dogs with copper coil-induced coronary artery thrombosis. *J. Cardiovasc. Pharmacol.*, **11**:468-472 (1988).
90. L. Nelles, H. R. Lijnen, D. Collen, and W. E. Homes, Characterization of recombinant human single chain urokinase-type plasminogen activator mutants produced by site-specific mutagenesis of lysine 158. *J. Biol. Chem.*, **262**:5682-5689 (1987).
91. D. Collen, J. Mao, and J. M. Stassen, et al., Thrombolytic properties of Lys-158 mutants of recombinant single chain urokinase-type plasminogen activator in rabbits with jugular vein thrombosis. *J. Vasc. Med. Biol.*, **1**:46-49 (1989).
92. L. Nelles, H. R. Lijnen, D. Collen, and W. E. Holmes, Characterization of a fusion protein consisting of amino acids 1 to 263 of tissue-type plasminogen activator and amino acids 144 to 411 of urokinase-type plasminogen activator. *J. Biol. Chem.*, **262**:10855-10862 (1987).
93. D. Collen, J. M. Stassen, E. Demarsin, L. Kieckens, H. R. Lijnen, and L. Nelles, Pharmacokinetics and thrombolytic properties of chimaeric plasminogen activators consisting of the NH_2-terminal region of human tissue-type plasminogen activator and the COOH-terminal region of human single chain urokinase-type plasminogen activator. *J. Vasc. Med. Biol.*, **1**:234-240 (1989).
94. E. Haber, and M. Runge, et al., Antibody targeted fibrinolysis. *Thromb. Haemost.* **57**:253 (1987).
95. E. Haber, T. Quertermous, G. R. Matsueda, and M. S. Runge, Innovative approaches to plasminogen activator therapy. *Science*, **243**:51-56 (1989).
96. M. Dewerchin, H. R. Lijnen, B. Van Hoef, F. De Cock, and D. Collen, Biochemical properties of conjugates of urokinase-type plasminogen activator with a monoclonal antibody specific for crosslinked fibrin. *Eur. J. Biochem.*, **185**:141-149 (1989).
97. D. Collen, M. Dewerchin, J. M. Stassen, L. Kieckens, and H. R. Lijnen, Thrombolytic and pharmacokinetic properties of conjugates of urokinase-type plasminogen activator with a monoclonal antibody specific for cross-linked fibrin. *Fibrinolysis*, **3**:197-202 (1989).
98. L. A. Harker, Clinical trials evaluating platelet-modifying drugs in patients with atherosclerotic cardiovascular disease and thrombosis. *Circulation*, **73**:206-223 (1986).
99. V. Fuster, L. Badimon, J. Badimon, P. C. Adams, V. Turitto, and J. H. Chesebro, Drugs interfering with platelet functions: mechanisms and clinical relevance, *in*: Thrombosis and Haemostasis. Eds: M. Verstraete, J. Vermylen, R. Lijnen, J. Arnout, Leuven University Press, p. 349-418 (1987).
100. J. Loscalzo, and D. E. Vaughan, Tissue plasminogen activator promotes platelet disaggregation in plasma. *J. Clin. Invest.*, **79**:1749-1755 (1987).
101. T. Yasuda, H. K. Gold, and R. C. Leinbach, et al., Tissue plasminogen activator (t-PA) resistant platelet rich white thrombus (WT) and combination treatment of t-PA and antiplatelet antibody to GPIIb/IIIa receptor (7E3). *Circulation*, **78**:II-15 (1988).
102. H. K. Gold, B. S. Coller, and T. Yasuda, et al., Rapid and sustained coronary artery recanalization with combined bolus injection of recombinant tissue-type plasminogen activator and monoclonal antiplatelet GPIIb/IIIa antibody in a canine preparation. *Circulation*, **77**:670-677 (1988).

103. I. K. Jang, A. A. Ziskind, H. K. Gold, R. C. Leinbach, J. T. Fallon, and D. Collen, Prevention of arterial platelet occlusion by selective thrombin inhibition. *Circulation,* **78:**II-311 (1988).
104. S. R. Hanson, and L. A. Harker, Interruption of acute platelet-dependent thrombosis by the synthetic antithrombin D-phenylalanyl-L-prolyl-L-arginyl chloromethyl ketone. *Proc. Natl. Acad. Sci. USA,* **85:**3184-3188 (1988).
105. K. A. Hajjar, N. M. Hamel, P. C. Harpel, and R. L. Nachman, Binding of tissue plasminogen activator to cultured human endothelial cells. *J. Clin. Invest.,* **80:**1712-1719 (1987).

CORONARY THROMBOSIS: PATHOGENESIS AND PREVENTION

G. V. R. Born

Pathopharmacology Unit
The William Harvey Research Institute
St. Bartholomew's Hospital Medical College
Charterhouse Square, London EC1M 6BQ, UK

SUMMARY

Acute myocardial infarction is most commonly initiated by fissuring of an atheromatous plaque. Through such fissures the blood is exposed to thrombogenic constituents of the intima, causing thrombotic obstruction of the coronary artery. Why plaque fissuring occurs is not known. Our investigation is to establish which types of plaque undergo fissuring by relating their mechanical with their cellular and biochemical properties; and to quantify the distribution of fissures. Results so far indicate that fissures occur predominantly in plaques with lipid pools in one segment of intima, and that the commonest single site of fissuring is that of maximal stress concentration as predicted by computer modelling. The results also suggest that arterial spasm at the immediate site of fissuring is not involved, as more than half the fissures occur at sites where there is no residual medial smooth muscle. Obstructive coronary thrombosis is initiated in most cases by plaque fissure with local haemorrhage which induces intravascular platelet aggregation. Recent observations with novel techniques have provided evidence that platelet aggregation *in vivo* is initiated by ADP and potentiated by thromboxane A_2 and thrombin, with actual contribution of exposed collagen still undetermined. These observations provide an explanation for the limited effectiveness of any simple platelet-inhibiting drug, including Aspirin, by itself whenever arterial, eg. coronary or cerebral thrombosis is initiated by haemorrhages into atheromatous plaques. On the other hand, Aspirin *is* significantly effective when myocardial infarction follows unstable angina and when strokes follow transient episodes of cerebral ischaemia. This partial effectiveness can be explained through an action of Aspirin on platelets by assuming that, in such cases, their thrombo-embolic aggregation is initiated by haemodynamic effects of atheromatous lesions. Recently we discovered that the uptake of atherogenic low-density lipoprotein is accelerated by noradrenaline at its physiological blood concentrations (Shafi, Cusack & Born, 1989: *Proc. Roy. Soc. B.*, 235, 289). This may help to explain the increased incidence of coronary thrombosis in conditions associated with elevated blood noradrenaline, including cigarette smoking and stress; and it may open new therapeutic approaches.

INTRODUCTION

The principal pathological facts of obstructive coronary thrombosis are as follows (Born, 1979): (1) Thrombi do not form in normal arteries, but in atherosclerotic arteries. (2) Atherosclerosis increases slowly, whereas thrombosis occurs rapidly and is individually unpredictable; therefore, atherosclerotic arteries must be subject to sudden, unpredictable events. (3) Most occlusive thrombi are associated with fissures in underlying atheromatous plaques. (4) The central portion of occlusive thrombi consists mainly of aggregrated platelets.

OBSERVATIONS AND DIŞCUSSIONS

What is the mechanism responsible for initiating platelet aggregation in an atherosclerotic artery, as an apparently random event in time? Close serial sectioning (Friedman & Byers, 1965; Constantinides, 1966) and reconstruction of occluded segments of coronary arteries (Davies & Thomas, 1981, 1984) established that the central platelet-rich segment of an obstructive thrombus is usually, if not invariably, associated with recent haemorrhage into an underlying atherosclerotic plaque. Such haemorrhages occur through fissures or fractures in the plaque, and it is reasonable to assume that the sudden appearance of such a fissure or fracture is the random, individually unpredictable event affecting coronary arteries that has to be assumed to account for the clinical onset of acute myocardial infarction (Born, 1979). Why such a defect should develop at a particular moment is uncertain. Perhaps it is analogous to the sudden appearance of fine cracks in the wings of jet air craft that is ascribed to the cumulative effects of variable stresses on metal known as fatigue failure (Born, 1979).

How does haemorrhage into a ruptured plaque trigger platelet thrombogenesis? This can be regarded as part of the general question of how platelets are caused to aggregate by haemorrhage. An explanation commonly put forward is that the process is initiated by platelets adhering to collagen that is exposed where damaged vessels walls are denuded of endothelium (Mustard et al., 1977). Adhering platelets then release other agents, including thromboxane A_2 and adenosine diphosphate, that in turn are responsible for the adhesion of more platelets as growing aggregates. This is unlikely, however, to be the complete explanation, for the following reasons. First, haemostatic and thrombotic aggregates of platelets grow very rapidly and without delay (Hugues, 1959; Born and Richardson, 1980). In contrast, platelet aggregation by collagen begins, ever under optimal conditions for rapid reactivity, only after a delay or lag period of several seconds (Silner, Nossell & LeRoy, 1968). Secondly, platelets tend to aggregate as mural thrombi when anticoagulated blood flows through plastic vessels (Didisheim, Pavlovsky & Kobayashi, 1972) for example in artificial organs such as oxygenators or dialysers (Richardson, Galletti & Born, 1976) that contain neither collagen nor anything else capable of activating platelets similarly. This implies that there are conditions under which platelets are activated in the blood by something other than, or in addition to, the collagen in the walls of living vessels.

Recent *in vivo* experiments on three mammalian species, one of them man, indicate that the haemostatic aggregation of platelets is initiated by adenosine diphosphate (Zawilska, Born & Begent, 1982) which is released from injured cells in the blood vessels (Born, Görög & Kratzer, 1981; Born & Kratzer, 1984). It is reasonable to assume that cellular injury associated with the cracking of atheromatous plaques releases enough adenosine diphosphate locally to initiate thrombotic platelet aggregation in coronary arteries. The effect of this adenosine diphosphate, which is very rapid, is augmented first by thromboxane A_2 and later by much more adenosine diphosphate released from the platelets themselves. When a haemorrhage occurs through an atheromatous fissure into the arterial walls, the extravasated blood remains comparatively static; this condition can be presumed to favour the appearance of thrombin which initiates fibrin formation and contributes to platelet aggregation.

In this situation, therefore, platelets are apparently exposed simultaneously to several potent aggregating agents, only some of which are produced by the platelets themselves through their release reaction, which is inhibited by Aspirin. These considerations can therefore in principle account for the comparative ineffectiveness of Aspirin in clinical trials of the secondary prevention of myocardial infarction; but they leave open the question why the drug is apparently effective when myocardial infarction is associated with unstable angina. Could it be that this type of angina points to a pathogenetic mechanism which differs from other antecedents of myocardial infarction and is more similar to the mechanism underlying cerebrovascular disturbances?

Through an unusal and interesting development it has recently become possible to propose a pathogenetic mechanism for unstable angina as a result of a therapeutic success. Over several years there have been extensive and expensive clinical trials of drugs potentially effective against the most serious complications of atherosclerotic cardiovascular disease, namely cerebral thrombosis which causes stroke and coronary thrombosis which causes heart attacks. In several large controlled trials of Aspirin for the secondary prevention of myocardial infarction involving a total of over 13,000 patients, the drug produced no significant benefit, although some of the trials showed a trend in that direction (May et al., 1982). However, in two recently reported trials, Aspirin was very significantly effective in preventing myocardial infarction and death when the selection of patients was limited to those with unstable angina (Lewis et al., 1983; Cairns et al., 1984). Controlled trials of Aspirin for the prevention of stroke and of two clinical disorders which commonly precede stroke, namely transient ischaemic attacks and visual disturbances, have also demonstrated significant benefit (Fields, Lemak & Frankowski, 1977). This divergence suggests differences in the pathogenesis of these diseases. These differences may become understandable after considering how thrombogenesis may differ in atherosclerotic carotid as against atherosclerotic coronary arteries.

There is increasing evidence that in carotid as opposed to coronary arteries, haemodynamic disturbances alone can initiate the formation of embolising platelet thrombi. This conclusion is based mainly on non-invasive, ultrasound techniques that can be applied to carotid arteries but not to coronary arteries (Lusby et al., 1981, 1982; see also Born, 1985). In over 90% of patients affected by prestroke syndromes (characteristically transient ischaemic attacks and visual disturbances) two complementary imaging techniques demonstrated atherosclerotic lesions usually at the carotid bifurcation, that is, extracranially. In most of these cases the lesions constrict the arterial lumen severely, so that continuous vortices are established in the blood flow. At constant blood pressure the flow of blood is faster through the constriction than elsewhere in the artery. Therefore, high flow and wall shear rates are no hindrance to the aggregation of platelets as thrombi (Born, 1977). Indeed, the question arises of whether thrombogenic platelet aggregation can be brought about by abnormal haemodynamic conditions alone.

Evidence of increased platelet aggregation brought about by the operation of haemodynamic factors was provided by experiments in which blood was made to flow through branching channels in extra-corporeal shunts (Rowntree & Shionoya, 1927; Mustard et al., 1962). Deposits of platelets formed consistently on the shoulders of a bifurcation in the flow chamber, but nowhere else in the channels. When the chambers were perfused, not with blood, but with platelet-rich plasma, no deposit was formed, showing that red cells were also essential if deposition were to take place. The dependence that the deposition of platelets from flowing blood has on the red cells that surround and outnumber them could be caused by physical or chemical mechanisms; or by both acting together. A physical mechanism is contributed by the flow behaviour of the erythrocyte, which increases the diffusion of platelets in whole blood over that in plasma by up to two orders of magnitude (Turitto & Baumgartner, 1975). Thus, regions of flow separation and delays are evidently capable, as seen in similar flow in artificial vessels (Mustard et al., 1962), of causing platelet aggregates to form in the blood stream that are then carried as emboli into the cerebral circulation.

The exact mechanism that induces platelets to aggregate under these conditions is still uncertain. The established therapeutic effectiveness of Aspirin in a high proportion of these cases would suggest that the platelets' release reaction is essential. Release of aggregating agents from platelets has long been assumed to subserve a "chain reaction" or positive feedback mechanism (Born, 1965) that could, in principle, account for platelet aggregation in haemostasis and thrombosis. This assumption was based mainly on *in vitro* experiments that left considerable uncertainty about the contribution of the release reaction to the initiation of aggregation *in vivo*. The rapidity of the process, and the presence of other tissues, make it impossible to follow the release reaction quantitatively *in vivo* by the methods that permit its observation *in vitro*. Because it is the platelet reaction that is inhibited by Aspirin, we adopted a novel *in vivo* approach (Born, Gorog & Kratzer, 1981). With quantitative electron microscopy we showed that haemostatic aggregation of platelets can get well under way without participation of the release reaction, that is, the reaction which is inhibited by Aspirin in controlled clinical trials for the secondary prevention of myocardial infarction. The question remains how the release reaction may be triggered haemodynamically, eg, through collisions with red cells or through their reversible distortion with the release of ADP (Born, 1977 and 1979).

Thus, in spite of suggestive experimental evidence, the question remains why Aspirin is to some extent clinically effective against primary myocardial infarction in unstable angina patients. It may be, as has also been pointed out elsewhere, (Lewis et al., 1983) that the pathological conditions causing these cerebral and cardiac manifestations produce similar haemodynamic effects. For the present it must be assumed further than haemodynamic disturbances suffice to induce platelet aggregation in a way which is uninhibitable by Aspirin. Clearly, much work remains to be done to find out whether or not this is so. Thus, a puzzling but important question is why in both of the recent trials Aspirin was effective in almost exactly half of the cases (Lewis et al., 1983; Cairns et al., 1984). Does this indicate again two different pathogenetic mechanisms, both manifesting themselves as a consequence of unstable angina, but only one of them involving the postulated haemodynamic effects? Whatever the answer to this, recent evidence for the clinical effectiveness of at least one drug against the commonest and most dangerous consequence of unstable angina is very encouraging.

REFERENCES

Born, G. V. R., and Richardson, P. D., Activation time of blood platelets, *J. Membr. Biol.,* **57**:87-90 (1980)

Born, G. V. R., Görög, P., and Kratzer, M. A. A., Aggregation of platelets in damaged vessels, *Phil. Trans. R. Soc. Lond. B.,* **294**:241-250 (1981).

Born, G. V. R., Arterial thrombosis and its prevention, *in:* Proc. VIII World Congress, Cardiology. S. Hayase, S. Murao (eds). Amsterdam: Excerpta Medica 81-91 (1979).

Born, G. V. R., Fluid-mechanical and biochemical interactions in haemostasis, *Br. Med. Bull.,* **33**:193-197 (1977).

Born, G. V. R., Platelets in thrombogenesis: mechanism and inhibition of platelet aggregation, *Ann. R. Coll. Surg. Engl.,* **36**:200-206 (1965).

Born, G. V. R., and Kratzer, M. A. A., Source and concentration of extracellular adenosine triphosphate during haemostasis in rats, rabbits and man., *J. Physiol Lond,* **354**:419-429 (1984).

Born, G. V. R., The carotid plaque, *in:* Extracranial Cerebrovascular Disease: Diagnosis and Management, F. Robicsek (ed). New York: Macmillan (in press) (1985).

Cairns, J., Gent, M., Singer, J., Finnie, K., Froggatt, G., Holder, D., Jablonsky, G., Kostuk, W., Melendez, L., Myers, M., Sackett, D., Sealey, B. Q., and Tanser, P., A study of Aspirin (ASA) and sulfinpyrazone (S) in unstable angina, *Circulation,* **70**:suppl 2, Abstract 1659 (1984).

Constantinides, P., Plaque fissures in human coronary thrombosis, *J. Atheroscler. Res.,* **6**:1-17 (1966).

Davies, M. J., and Thomas, T., The pathological basis and microanatomy of occlusive thrombus formation in human coronary arteries, *Phil. Trans. R. Soc. Lond. B.,* **294**:225-229 (1981).

Davies, M. J., and Thomas, A., Thrombosis and acute coronary-artery lesions in sudden cardiac ischaemic death, *New Engl. J. Med.,* **310**:1137-1140 (1984).

Didisheim, P., Pavlovsky, M., and Kobayashi, I., Factors that influence or modify platelet function, *Ann. N. Y. Acad. Sci.,* **201**:307-315 (1972).

Fields, W. S., Lemak, R. F., and Frankowski, R. F., Controlled trials of Aspirin in cerebral ischaemia, *Stroke,* **8**:301-328 (1977).

Friedman, M., and Byers, S. O., Induction of thrombi upon preexisting arterial plaques, *Amer. J. Path.,* **46**:567-75, (1965).

Hugues, J. C., Agglutination précoce des plaquettes au cours de la formation du clou hémostatique, *Thromb. Diathès. haemorrh.,* **3**:177-186 (1959).

Lewis, Jr., H. D., Davies, J. W., Archibald, D. G., Steinke, W. E., Smitherman, T. C., III Doherty, J. E., Schnaper, H. W., LeWinter, M. M., Linares, E., Maurice Pouget, J., Sabharwal, S. C., Chesler, E., and DeMots, H., Protective effects of Aspirin against acute myocardial infarction and death in men with unstable angina, *New Engl. J. Med.,* **309**:396-403 (1983).

Lusby, R. J., Ferrell, L. D., Ehrenfeld, W. K., Stoney, R. J., and Wylie, E. J., Carotid plaque haemorrhage: its role in production of cerebral ischaemia, *Arch. Surg.,* **117**:1479-1488 (1982).

Lusby, R. J., Machleder, H. I., Jeans, W., Skidmore, R., Woodcock, J. P., Clifford, P. C., and Baird, R. N., Vessel wall and blood flow dynamics in arterial diseases, *Phil. Trans. R. Soc. Lond. B.,* **294**:231-239 (1981).

May, G. S., Eberlein, K. A., Furberg, C. D., Passamain, E. R., and DeMets, D. S., Secondary prevention after myocardial infarction: a review of long-term trials, *Prog. Cardiovas. Dis.,* **24**:331-352 (1982).

Mustard, J. F., Moore, S., Packham, M. A., Kinlough Rathbone, R. L., Platelets, thrombosis and atherosclerosis, *Proc. Biochem. Pharmacol.,* **13**:312-325 (1977).

Mustard, J. F., Murphy E. A., Rowsell, H. C., and Downie, H. G., Factors influencing thrombus formation *in vivo, Amer. J. Med.,* **33**:621-647 (1962).

Richardson, P. D., Galletti, P. M., and Born, G. V. R., Regional administration of drugs to control thrombosis in artificial organs, *Trans. Am. Soc. Artif. Intern. Organs.*, **22**:22-29 (1976).

Rowntree, L. G., and Shionya, T., Studies in experimental extracorporeal thrombosis. Part I: Methods for the direct observation of extra-corporeal thrombus formation, *J. Exp. Med.*, **46**:7-12 (1927).

Turitto, V. T., and Baumgartner, H. R., Platelet interaction with subendothelium in a perfusion system: physical role of red blood cells, *Microvasc. Res.*, **5**:167-79 (1975).

Wilner, G. D., Nossell, H. L., and LeRoy, E. C., Aggregation of platelets by collagen, *J. Clin. Invest.*, **47**:2616-2621 (1968).

Zawilska, K. M., Born, G. V. R., and Begent, N. A., Effect of ADP-utilising enzymes on the arterial bleeding time in rats and rabbits, *Br. J. Haematol.*, **50**:317-325 (1982).

IMPLICATIONS OF THE TIMI TRIALS

Allan M. Ross

George Washington University
Washington, D. C., USA

ABSTRACT

The TIMI Trials consist of three primary studies plus numerous ancillary observations. TIMI I was the direct reperfusion comparison between streptokinase and TPA. The central observations were: Higher reperfusion rates with TPA; Equivalent frequencies of bleeding complications; No differences in clinical outcome parameters; Only early reperfusion produced LV function improvement. TIMI IIA compared routine post-IV lytic therapy (TPA) immediate PTCA with a delayed interventional approach at 18-48 hours post lysis. The study demonstrated: Equivalent outcome LV function; Equivalent mortality rates; More emergency CABG in the "immediate" group; More hemorrhagic complications in the immediate group. TIMI IIB compared delayed but routine interventional post-lytic care (PTCA) with a more conservative strategy. Additionally patients were secondarily randomized to acute administration of IV beta blocker (metoprolol) or delayed administration of beta blocker, starting on Day 6. The primay results of this investigation include: A very low early (2 week) mortality rate with either strategy (about 5%); preserved low mortality at the one year follow up for both groups; no difference (routine intervention versus conservative strategy) in outcome LV function; no difference in comparative mortality rates; less recurrent ischemia in the routine PTCA group; 13% need for PTCA in the conservative group; no difference in LV function based upon beta blockade assignment but some decrease in mortality in acute beta blockade subgroups.

INTRODUCTION

In 1983 the United States' National Institutes of Health established a multicenter study group to investigate the value of thrombolysis in myocardial infarction. This collaborative effort has come to be known by the acronym "T.I.M.I.".

The first proposed trial for the group was to have been a randomized, placebo controlled trial of IV lytic therapy versus placebo with endpoints of outcome left ventricular function and mortality (since at that time no such study of this important question had been undertaken or at least completed). Only the plasminogen activators streptokinase (SK) and urokinase were generally available and the study group initially planned to select one of these agents for the trial. At about the same time however, the second generation and relatively fibrin specific recombinant form of tissue plasminogen activator (rt-PA) was entering initial clinical trials and showed promise of being a more efficient material for effecting coronary artery clot lysis. Were that expectation to prove fact, there was concern that a study using a less effective drug might turn out to be of limited relevance by the time a large and several years' duration trial had been completed and analyzed. Consequently the group decided to perform a direct comparison of the lytic efficacy of SK versus rt-PA.

Fibrinogen, Thrombosis, Coagulation, and Fibrinolysis, Edited by
C. Y. Liu and S. Chien, Plenum Press, New York

OBSERVATIONS AND DISCUSSIONS

TIMI — I

The resultant protocol, now commonly referred to as TIMI phase one (1) was undertaken as a double blind controlled comparison of the two agents, in angiographically defined infarct arteries, with the primary endpoint of successful recanalization or reperfusion at 90 minutes after the start of drug infusion.

The trial design was to take acute myocardial infarction patients within seven hours of the onset of their ischemic-type pain (of 30 minutes or more duration) and with ST segment elevation $\geq$ 1 mm in at least two contiguous ECG leads directly to coronary angiography. After definition of the perfusion status of the infarct related artery two intravenous infusions were begun. Patients were randomized to receive either 1.5 million units of SK over one hour or double chain rt-PA given as 40 mg in the first hour then 20 mg per hour for the next two hours. Patients received one of the active medications and simultaneously a placebo of the other. Clinical site investigators and angiographers were blinded as to which active agent was infused, and were required to perform repeat coronary angiograms of the infarct artery every 15 minutes until 90 minutes after drug initiation. Films were sent to a remote central reading laboratory for independent blinded analysis of successful reperfusion. A specific grading system for coronary perfusion was created in which grade 3 meant normal antigrade filling and emptying of the affected artery, grade 2, was complete filling of the vessel but with somewhat delayed flow. Grades 0 and 1 were functionally reperfusion failure.

The results of this trial in essence were that rt-PA was successful in fostering reperfusion (grades 2 or 3) by 90 minutes after drug infusion onset twice as often (62%) than was SK (31%) if the artery had been initially closed (Table 1). It was further observed that a relatively high rate of reperfusion was achieved using rt-PA independent of the duration of the infarct symptoms whereas the success rate with SK, modest up to 4 hours, fell off to quite low rates beyond four hours.

No major difference in any other measured clinical parameters (bleeding, reocclusion, global ventricle function outcome or death) was seen comparing the two agents.

The population studied in TIMI — 1 has been followed out to a year past infarction and several interesting additional observations have been made chief amongst these is the demonstration that early artery patency determines a very favorable outcome. Ejection fraction (by contrast ventriculography) increases from the day of infarcton to a 7-10 day repeat study were seen only in patients with grade 2 or 3 perfusion by 90 minutes after the start of treatment and was greatest in those treated earliest after the start of pain (2). Even

Table 1. Reperfusion, rt-PA Versus SK

	Successful Reperfusion (grades 0,1 becoming grades 2,3) rt-PA	SK
All Patients	70/113 (62%)	37/119 (31%)
Pts by hours from pain onset to treatment		
1–2	2/3 (67%)	0/1 (0%)
2–3	6/7 (86%)	8/17 (47%)
3–4	17/31 (55%)	10/23 (43%)
4–5	19/29 (66%)	8/31 (26%)
5–6	7/13 (54%)	8/29 (28%)
6–7	11/18 (61%)	1/8 (13%)
>7	8/12 (67%)	2/10 (20%)

Table 2. Mortality Rate As a Function of Infarct Artery Patency at 90 Minutes

Mortality at:	Patency	Occluded
21 days	3.1%	7.0%
6 months	5.6%	12.5%
12 months	8.1%	14.8%

using regional wall motion analysis methodolgy, no consistent improvement was observed in patients whose infarct artery was not patent by four hours after pain onset.

A further analysis of ventricular function has been undertaken utilizing pre-discharge radionuclide ventriculograms (3). These studies confirmed regional wall motion benefits related to successful recanalized and therefore more evident in rt-PA treated patients than in those who received SK.

Finally, when clinical follow up was performed up to a year after entry into TIMI-I. A mortality rate benefit for those with open arteries was seen versus those in whom thrombolysis had not been successful (Table 2).

TIMI STUDIES A — C

The next undertaking of the TIMI investigator group was necessitated by changes in the manufacturing processes of rt-PA. In the TIMI I studies the dominant form of the molecule was "double chain". When the production method for large quantity output was changed to the suspension cell method, the rt-PA produced became predominantly a single chain (SC) molecule with different dose-response characteristics. This led directly to open label angiographically controlled dose finding investigations now known collectively as TIMI A thru C (5). Single chain rt-PA proved less potent (per mg) than the original material. Doses of 80 mg of SC rt-PA given in a similar fashion to the protocol followed in TIMI-I produced 90 minutes reperfusion in only 45% of patients. Subsequent increases in SC doses produced 71% reperfusion with a total dose of 100 mg.

Further increasing the dose to 150 mg produced earlier recanalization but not a higher final rate, and more fibrinolysis but not more overall clinical bleeding. Thus for a brief period, 150 mg SC rt-PA was selected for subsequent clinical efficacy trials.

rt-PA and Intracerebral Bleeding

The TIMI-II studies (to be discussed subsequently) began with the higher (150 mg) dose of SC rt-PA. After an initial experience with this regimen an escalating frequency of intra cerebral hemorrhage was noted (1.6%). The dose was therefore decreased to a total of 100 mg in subsequent patients and the overall experience in these trials with 100 mg has been associated with a hemorrhagic CVA rate of 0.5% in a large cohort of patients (Table 3).

By the completion of the activities reviewed above, the original intention of the TIMI group to perform a large placebo controlled trial required review. By then, several such investigations had been completed by other groups or were at least well along in their recruitment phases. It was becoming clear that early intravenous thrombolytic therapy in acute myocardial infarction was associated with definite benefits in the major endpoints of

Table 3. rt-PA Dose and Intracerebral Bleeding (ICB)

Total Dose	Number of Patients Treated – Frequency	Percent with ICB
150 mg	1014	1.6%
100 mg	1172	0.5%

outcome left ventricular function and early mortality rates. Consensus of the group at that time was that an additional placebo controlled trial was not warranted. Instead TIMI turned toward the study of the emerging major clinical uncertainty i.e. defining the most appropriate post thrombolytic management strategy.

TIMI IIA

Clinical practice, at least in the United States and much of Western Europe increasingly followed a strategy of earliest possible administration of a lytic drug intravenously followed by immediate coronary angiography and usually percutaneous transluminal coronary angioplasty (PTCA). This approach was designed firstly to reap the benefits of early lysis but then to augment that benefit by mechanical reperfusion (angioplasty) of lytic drug failures, known to be at least 20% of the treated population. Further goals of immediate angiography and PTCA was the reduction in severity of the underlying atherosclerotic plaque in the infarct attery which was thought to offer reduced rates of subsequent reocclusion and greater myocardial salvage together producing better final outcome left ventricular function. Additionally it was expected that stenosis reduction by early PTCA would lower the substantial frequency of post infarction ischemia seen in patients given intravenous lytic therapy.

TIMI II consisted of two investigative arms quantitating outcome differences for patients who received immediate rt-PA (plus aspirin and heparin) followed by immediate angioplasty, delayed angioplasty, or a conservative, noninterventional post thrombolytic course.

In TIMI IIA the primary comparison was of angiography and PTCA immediately after IV rt-PA versus PTCA delayed for 18 to 48 hours. The issue was of considerable importance considering the major logistical differences between an around the clock emergency capability versus a treatment approach that allowed for more scheduled, elective procedures. This question, it must be stressed, was asked at a time when it had been widely assumed (but unproven) that angiography and PTCA were probably advisable for most or all post thrombolytic patients to maximize beneficial outcome. To be interrogated in TIMI IIA was primarily the timing of PTCA. (Actually there was also a third, conservative treatment arm TIMI IIA but those patients were analyzed subsequently within the TIMI IIB trials).

In TIMI IIA (6) patients received rt-PA with aspirin and heparin then were randomized to go to angiography and angioplasty (PTCA) immediately or within an 18 to 48 hour post lytic time window. Those who had procedures in the "immediate" group had PTCA attempts for both reperfused arteries with residual stenoses in excess of 60% in diameter, and persistently obstructed arteries (so-called "salvage PTCA"). Patients in the 18 to 48 hour delayed group had PTCA only of stenoses, not persistent occlusion as it was believed that late reperfusion was not of clinical benefit.

The trial results included a high post rt-PA patency rate (75% in the immediate group, 83% if delayed) and a high PTCA success rate (84 and 93% respectively).

Primary endpoints were group mean ejection fraction in the two groups as well as clinical parameters including complication rates. No differences were seen comparing the two strategies in terms of outcome ejection fraction but the immediate PTCA group had more bleeding complications and a more frequent need for emergency bypass operation (Table 4).

Table 4. TIMI IIA: Comparing Immediate Angiography/PTCA to the Same Procedures Deferred to 18-48 Hours After Lytic

	Immediate	Delayed
Abrupt (procedure related) reocclusion	7.8%	2.8%
Urgent CABG	4.3%	1.9%
21 day mortality	7.2%	5.7%
Cath complications	5.2%	1.1%
Pre-discharge left ventricular ejection fraction	50%	49%

Table 5. TIMI IIB Results

	Invasive Strategy	Conservative Strategy
Pts who had angioplasty	60%	13%
Group mean 6 weeks resting ejection fraction	50%	50%
6 weeks mortality rate	3.7%	3.6%
Reinfarction	4.3%	6.3%
Ischemic exercise test	12.8%	17.7%

The trialists concluded that immediate post lytic angiography and PTCA offered no major group advantage and could be deferred in favor of the less resource consuming delayed approach. Although not a predefined primary endpoint, even mortality rate favored (was lower with) the delayed approach, 5.7% vs 7.2%, (p = NS).

TIMI IIB

The most recently completed TIMI study was a large and comprehensive comparison of routine post lytic angiography and PTCA in the delayed time frame, i.e., 18 to 48 hours after rt-PA versus a conservative approach after thrombolysis. The "invasive" approach was identical to that described for the 18 to 48 hour delayed PTCA treatment arm of TIMI IIA. The "conservative" approach consisted of rt-PA, aspirin and heparin followed by close clinical observation for a week, then exercise testing. If and when patients in the "conservative" arm developed spontaneous or exercise induced post infarction ischemia, then they had coronary angiography and if anatomically appropriate, subsequent PTCA. The sample size was large, 3262 patients; treatment with rt-PA was early, 2.6 hours from symptom onset; and PTCA success rate was high when technically appropriate, 93%. The major endpoint of the study were death, recurrent non-fatal infarction and outcome left ventricular function at hospital discharge and at six weeks follow-up (7).

The results of this comparison are shown in Table 5. Comparing the invasive and the conservative strategy there were no significant differences in mortality rates at six weeks (5.2 vs 4.7% respectively), non-fatal reinfarction (5.9% vs 5.4%) or outcome left ventricular ejection fraction (50.5% vs 49.9%) were seen. The invasively treated group was disadvantages in terms of any adverse outcome, 13% vs 10.6%, $p < 0.04$ but had fewer predischarge ischemic exercise tests, 12.8% vs 17.7%, $p < 0.001$.

The interpretation of these results by the investigators was that routine catheterization and usually PTCA was unnecessary for most MI patients who receive early effective lytic therapy, provided that the procedures could be rapidly performed if and when recurrent ischemia became manifest.

A second smaller randomization was also done in eligible patients (with no contraindications) between acute beta blockade (Metoprotol 5 mg intravenously every 2 minutes X3) and delayed beta blockade (orally, day 6). There 1390 patients in this portion of the study with endpoints of left ventricular function, mortality, reinfarction and recurrent ischemic events.

Of the above, outcome left ventricular ejection fraction (EF) was prestated as the primary endpoint: in this parameter there was no difference (50 percent hospital discharge EF, both acute and deferred beta blocker groups, both at hospital discharge and at 6 weeks follow-up). Nonfatal reinfarction and early recurrence of ischemia however did appear to be reduced by early intravenous beta blockade as did total mortality in a surprising subgroup of low risk patients. These observations, while perhaps not conclusive, are of sufficient interest to warrant further study and hopefully confirmation.

TIMI III

The most recent of the TIMI trials, just entering active patient recruitment, is designed to evaluate the role of both thrombolysis and an aggressive interventional strategy in unstable angina and non ST segment elevation infarct patients. Those two conditions with often indistinguishable initial presentation are thought to have similar pathophysiology, i.e., unstable atherosclerotic plaque plus (usually) nonocclusive superimposed thrombus. Previous studies of thrombolysis have either systematically excluded such patients from evaluation, included them but within the overall group of infarcts rather than evaluating them separately, or been pilot-type trials only, with small sample size and inconclusive results.

In TIMI III a large cohort will be randomized to aggressive medical therapy and rt-PA or its placebo. One arm of the investigation will also randomize to early PTCA or a multiple drug medical approach. Of the many endpoints in TIMI III, angiographic improvement, death, uncontrolled ischemic pain and enzymatically documented myocardial necrosis will be of primary interest.

REFERENCES

1. J. H. Chesebro, G. Knatterud, R. Roberts, J. Borer, L. S. Cohen, J. Dalen, H. T. Dodge, C. K. Francis, D. Hillis, P. Ludbrook, J. E. Markis, H. Mueller, E. R. Passamani, E. R. Powers, A. K. Rao, T. Robertson, A. Ross, T. J. Ryan, B. E. Sobel, J. Willerson, D. O. Williams, B. L. Zaret, and E. Braunwald: Thrombolysis in myocardial infarction (TIMI) trial, phase I: a comparison between intravenous tissue plasminogen activator and intravenous streptoknase. *Circulation,* **76:**142-154 (1987).
2. F. H. Sheehan, E. Braunwald, P. Canner, H. T. Dodge, J. Gore, P. Van Natta, E. R. Passamani, D. O. Willams, B. Zaret, and Co-Investigators: The effect of intravenous thrombolytic therapy on left ventricular function: a report on tissuetype plasminogen activator and streptokinase from the thrombolysis in myocardial infarction (TIMI Phase I) trial. *Circulaton,* **75:**817-829 (1987).
3. F. J. Wackers, M. L. Terrin, D. S. Kayden, G. Knatterud, S. Forman, E. Braunwald, B. L. Zaret, and the TIMI Investigators: Quantitative radionuclide assessment of regional ventricular function after thrombolysis therapy for acute myocardial infarction: results of phase I thrombolysis in myocardial infarction (TIMI) trial. *JACC,* **13:**998-1005 (1989).
4. J. E. Dalen, J. M. Gore, E. Braunwald, J. Borer, R. J. Goldberg, E. R. Passamani, S. Forman, G. Knatterud, and the TIMI Investigators: Six- and twelve-month follow-up of the phase I thrombolysis in myocardial infarction (TIMI) trial. *AJC,* **62:**179-185 (1988).
5. H. S. Mueller, A. K. Rao, S. A. Forman, and the TIMI Investigators: Thrombolysis in myocardial infarction (TIMI): comparative studies of coronary reperfusion and systemic fibrinogenolysis with two forms of recombinant tissue-type plasminogen activator. *JACC,* **10:**479-490 (1987).
6. The TIMI Research Group: Immediate vs delayed catheterization and angioplasty following thrombolytic therapy for acute myocardial infarction. *JAMA,* **260:**2849-2858 (1988).
7. The TIMI Study Group: Comparison of invasive and conservative strategies after treatment with intravenous tissue plasminogen activator in acute myocardial infarction. *N. Eng. J. Med.,* **320:**618-627 (1989).

THROMBOTIC MICROANGIOPATHY

Hau C. Kwaan

Department of Medicine, Northwestern University Medical School and
Section of Hematology/Oncology, VA Lakeside Medical Center,
Chicago, IL 60611-4494, USA

THROMBOTIC MICROANGIOPATHY

In 1924, Moschcowitz (1) reported the occurrence of an "acute febrile pleiochromic anemia" in a 16 year old girl who died 13 days after presenting with fever, malaise and upper extremity weakness. On autopsy, widespread hyaline microthrombi were found in the terminal arterioles and capillaries (1, 2). Since then, this particular syndrome has been better characterized and is now known as thrombotic thrombocytopenic purpura (TTP) (3) or Moschcowitz disease (4). This syndrome is seen in a wide variety of conditions and is referred to as "Thrombotic Microangiopathy" (TM) (5-7). They all have in common a microangiopathic hemolytic anemia, thrombocytopenia, and the presence of a microvascular thrombotic lesion. The microvascular lesion is usually generalized. Its heavy involvement in certain locations results in the expression of dysfunction of specific organs, commonly the kidney, central nervous system, heart and lung. The various conditions associated with TM are listed in Table 1 and include hemolytic uremic syndrome, postpartum renal failure and other microangiopathic hemolytic anemias as seen in connective tissue disease, cancer, infection, and as a result of drug toxicity. A complete review on this topic was recently published (8).

THROMBOTIC THROMBOCYTOPENIC PURPURA (Moschocwitz Disease) (TTP)

Though the classic form of this disorder is most common with an estimated annual incidence rate of 1-20 per million population (9, 10), it merits attention because of its high mortality (40% or higher) if unrecognized (11), but significantly reduced if aggressively treated (12).

The disease is characterized by the widespread occurrence of a vascular lesion in arterioles and capillaries, sparing the venules. A microthrombus is present in the lumen while the vessel wall shows no cellular infiltration or other signs of inflammatory changes (Fig. 1). Microaneurysmal changes in some vessels have been described (13) but are generally considered nonspecific. Within the lumen, a hyaline thrombus is present, often partially, rather than completely occluding the blood flow. Light microscopy shows that the thrombus is PAS and Giemsa stain-positive with a finely granular appearance. It does not stain for iron or hemoglobin, thereby disproving Moschcowitz's original belief (1, 2) that it might have been composed of agglutinated red blood cells. Immunofluorescent studies reveal that it is composed of fibrin (14, 15) and platelets (16, 17), but complement and immunoglobulins may occasionally be found (17-20). The structure of the hyaline thrombi is confirmed by ultrastructural observations of intraluminal aggregates of platelets in various stages of degranulation (21). Proliferating endothelial cells are seen covering the thrombus. Depending on the age of the lesion, the endothelial hyperplasia may be so profuse that the thrombus appears to be subendothelial.

Postmortum examination reveals widespread involvement with adrenal glands most severely affected, with varying degrees in the lung, gastrointestinal tract, gall bladder,

Table 1. Classification of Thrombotic Microangiopathy

Primary	
	Classic thrombotic thrombocytopenic purpura (Moschcowitz's Disease)
	Hemolytic uremic syndrome
Secondary	
	Thrombotic microangiopathy associated with:
	Pregnancy and puerperium
	Cancer
	Tissue transplantation
	Renal allograft
	Bone marrow allograft
	Infection, including HIV
	Connective tissue disorder
	Immune complex disease
	Drugs and toxins

skeletal muscles, retina, pituitary gland, ovaries, uterus, and testes. Antemortem biopsies of the gingiva, skin and bone marrow may yield the diagnostic lesion in 30-50% of cases (22-25). Focal areas of hemorrhage are frequently present adjacent to an involved vessel. Consequently, in choosing the site of a skin biopsy, a petechial spot is preferred (25). More extensive hemorrhage is seen in the brain, often the cause of a fatal outcome. Ischemic changes and infarction of the organ are not common except in the pancreas, kidney and brain. The lack of severe ischemic changes may be due to sparing of venous channels, thus allowing some collateral circulation (26).

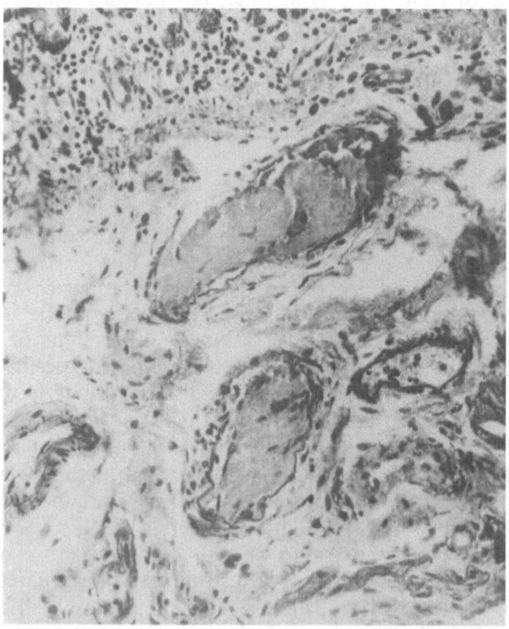

Fig. 1. Microthrombi seen in the submucosal vessels in the small intestines in TTP. Hematoxylin and eosin stain. (original magnification × 400), (with permission, Kwaan, H. C., *Sem. Hemat.* **24:**71-81, 1987).

PATHOGENESIS

Although many hypotheses have been put forth, the etiology of this disease is still uncertain. Much is known, however, of the pathogenesis of the characteristic microvascular thrombi which are rich in platelets, and believed to be the result of platelet aggregation. This can be brought about by either vascular endothelial changes or platelet aggregation in the circulation, but both conditions may indeed be present. Various investigators have observed evidence of vascular and endothelial damage, including impaired fibrinolytic activity of involved vessels (18, 25), impaired prostacyclin production (27, 28), and the presence of immune complexes in vessel wall (29). On the other hand, other etiologic factors may be derived from circulating blood including a plasma platelet aggregating protein (30), the lack of an inhibitor of platelet aggregating protein (31), the presene of large von Willebrand factor multimers (32, 33), antiendothelial cell antibody (34), or platelet-associated IgG (35). This platelet aggregating factor (PAF) is not always present. One investigator reported that it was found only in 4 out of 26 samples (33). Thus, other etiologic factors must be sought. Recently, abnormalities of von Willebrand factor (vWF) in the form of a large multimer was reported in the plasma of TTP patients (32). Using a sensitive labelled platelet serotonin release method, Kelton et al. (33) confirmed that an abnormal platelet aggregating (and release) factor was present in 41 out of 48 patients. The positive test was also found to be correlated with the activity of the disease. They further found that the presence of vWF increases the likelihood of positivity of the test so that the large vWF multimer may be a cofactor for the aggregating action of PAF. Subsequently, they reported that a calcium-dependent cysteine protease was present in the sera of TTP patients and that it fulfills all the characteristics previously described for a PAF (36). This exciting development is shedding new light on the etiology of TTP. What prompted this abnormality of the homeostasis of the calcium-dependent cysteine protease remains to be shown.

As a result of the microthrombi, three major events occur in this condition: hemolysis, thrombocytopenia, and altered functions of various visceral organs. The partial occlusion of the arterioles and capillaries allows the passage of blood through altered hemodynamic conditions. Red blood cells flowing through the partially occluded vessels are traumatized and fragmented, forming schistocytes resulting in hemolysis (37). The thrombocytopenia is the result of excessive platelet consumption due to aggregation and platelet thrombi formation.

CLINICAL FEATURES

Vascular occlusive lesions are dispersed throughout the body sparing practically no organs. Thus, a myriad of manifestations can occur depending on the severity of the interruption of blood flow to a particular area (Table 2). The most common presenting symptoms are those of neurologic disorders, hemorrhage of various types, and fever. The triad of hemolytic anemia, thrombocytopenia and neurologic manifestations was present in 74% and the additional features of fever and renal disorders, forming a pentad, was found in 40% of 258 patients reviewed by Ridolfi and Bell (10). Since the pathognomonic microthrombotic lesion is widespread in the body, signs and symptoms referable to any viscera may be found, including cardiac arrhythmias (38, 39), pulmonary failure (40), acute pancreatitis (41), ischemic bowel disease (42), and ocular changes (43, 44).

Laboratory findings show a picture of microangiopathic hemolytic anemia with the invariable presence of the characteristic schistocytes or fragmented red blood cells in the peripheral blood, and the associated stigmata of hemolysis such as increased unconjugated bilirubin and lactic dehydrogenase levels in blood. The direct antiglobulin test is negative. The thrombocytopenia is moderately severe with platelet counts averaging around 20,000/uL with a wide day-to-day fluctuation. The bone marrow picture is one of compensatory hyperplastic erythropoiesis and increased megakaryocytosis. In contrast to disseminated intravascular coagulation (DIC), rarely are there any coagulation abnormalities. The normal values of fibrinogen, prothrombin time, partial thromboplastin time, and low titers of fibrin degradation products are particularly helpful in differentiating TTP from DIC.

If untreated, a grave outcome is encountered with variously reported mortality rates of 60-90% and a mean survival of 8.5 days (8, 45). In recent years, since the advent of plasmapheresis and plasma exchange, the survival rate has significantly improved to around 82% (12). Those with progressive renal failure in spite of exchange transfusion, a falling

Table 2. Presenting Manifestations in TTP in 557 Patients Reported by Four Different groups of Investigators

Type	Amorosi, Ultman (35) (246 Cases)	Ridolfi, Bell (10) (225 Cases)	Petitt (9) (38 Cases)	Kennedy et al. (36) (48 Cases)
Neurologic	60%	52%	92%	71%
Hemorrhagic	44%	38%	NS	74%
Malaise, Weakness Fatigue	25%	29%	NS	27%
Nausea, Vomiting	24%	24%	NS	14%
Fever	20%	NS	87%	14%
Pallor	17%	NS	NS	6%
Abdominal Pain	11%	14%	NS	20%
Jaundice	9%	2.5%	NS	10%
Arthralgia, Myalgia	7%	6%	NS	NS

hematocrit, and failure of correction of thrombocytopenia following exchange transfusion, have a poor chance of survival (46). Other poor prognostic factors include being a member of the black race, having an initial high BUN and creatinine level (45), and a failure to improve within 72 hours after initiation of treatment (9). Relapses are seen in approximately 7% of the patients with complete clinical and hematologic remission between episodes. These patients have an unfavorable prognosis (9, 10).

TREATMENT

Once the diagnosis is established, the management is immediately directed to attempting to remove the putative platelet agglutinating factor by plasma exchange through plasmapheresis. The additional findings of the lack of an inhibitor of the platelet aggregating proteins in TTP plasma further support the rationale for plasma exchange and plasma infusion. Dramatic improvement in the severity of hemolysis and thrombocytopenia can often be seen within 24-48 hours of aggressive pheresis with exchange of 2-4 L/day. Recent histologic observations of tissue samples of patients who had received therapeutic plasmapheresis gave the impression of improvement in their microvascular lesions with smaller microthrombi and less endothelial proliferation (47, 48). Other means of inhibiting platelet aggregation, such as the use of aspirin and dipyridamole, may also be helpful, though the results are less consistent. It is clear from many reports that platelet transfusion is often followed by exacerbation and death and is therefore absolutely contraindicated (49). Since there is ample evidence of endothelial damage in TTP with a lack of fibrinolytic activity in the microthrombotic lesions, unlike those seen in DIC, anticoagulation would not be of benefit in the resolution of these thrombi and the use of heparin or warfarin is not recommended. Though there are occasional reports of immunologic abnormalities, the corticosteroids have also not been proven to be useful and are not routinely recommended except in acute fulminating cases.

HEMOLYTIC UREMIC SYNDROME (HUS)

This syndrome is referred to as Gasser's Syndrome (50) and shares with TTP the features of acquired microangiopathic hemolytic anemia, thrombocytopenia, acute renal failure, and cerebral symptoms. The clinical picture in TTP and in HUS frequently overlaps with the two conditions often being indistinguishable. There are, however, many features that may point to their differences. HUS more commonly affects infants and children rather than adults. The microvascular lesion is not as widespread as in TTP, and is largely confined to the kidneys. Although HUS has been found worldwide, it is of interest that more cases are reported in Argentina than elsewhere (51).

Acute hemolysis associated with renal failure as well as constitutional symptoms of malaise, weakness and fever, occur after a brief prodromal illness of gastroenteritis or upper respiratory tract infection of 10 days to 3 weeks. In contrast to TTP, the fever is mild and often transient and some mild hemorrhagic manifestations are usually present as petechiae and mucosal bleeding. Hematuria, oliguria or anuria may be present associated with hemolytic signs of encephalopathy, seizure and alterations of consciousness. Concomitant hypertension will add to the encephalopathy and may also lead to cardiac complications. Abdominal pain is present. The pathologic change is one of a microthrombotic lesion most commonly seen in the glomerular capillaries leading to varying renal dysfunction.

As in TTP, the etiology of the vascular lesion is unknown. Endothelium as well as the cell membrane of red blood cells and platelets may be damaged (52). The protective effect of vascular prostacyclin may be lacking in HUS (27).

MANAGEMENT

If prolonged anuria, seizures and hypertension are present, close monitoring of the circulating status is necessary to avoid fluid overload and dialysis. An ongoing prospective clinical trial is currently evaluating the benefits of early peritoneal dialysis.

The use of heparin is controversial (53, 54). While there is no advantage to thrombolytic therapy (55), the evaluation of a regimen of heparin plus dipyridamole yielded inconclusive results (54). On the other hand, plasmapheresis may be of benefit and should be used in patients with a relentless progression of their disease (56). With adequate supportive measures, the overall progress is believed to show a mortality of 6% with complete recovery of the disease in 64% of the cases (56, 58-60).

SECONDARY THROMBOTIC MICROANGIOPATHY

As mentioned in the introduction, the syndrome of TM is observed in many disorders other than TTP and HUS. An underlying condition can usually be identified without much difficulty. This group of TM syndromes is best viewed as a secondary TM complicating the primary disorders. The most commonly recognized ones are listed in Table 1. Each has characteristics of its own though all have the presence of microangiopathic hemolytic anemia, thrombocytopenia and renal failure in common.

Pregnancy and Puerperium (61, 62)

The syndrome of TM occurring during pregnancy and puerperium has been described variously as TTP or HUS. In a review of 65 cases of TTP during pregnancy (62), it was found that 89% occurred antepartum with 58% at or before 24 weeks of gestation. The pregnancy was clearly jeopardized by this complication with a high fetal death rate of 80% and only 25% of the mother-infant pair surviving. Aggressive plasma therapy with or without exchange by pheresis determines the maternal outcome with 68% mortality without, and a dramatic reduction to 0% with this treatment modality ($P>0.05$). Successful treatment allows the pregnancy to reach full term with the newborn baby having normal platelet counts, suggesting that the TTP platelet aggregating proteins do not cross the placental barrier.

The TM syndrome may occur postpartum and in women taking oral contraceptives. In these settings, various terms have been used in the past to describe this syndrome including: "irreversible postpartum renal failure" (63), "malignant nephrosclerosis" (64), "postpartum intravascular coagulation" (65), and "postpartum hemolytic uremic syndrome" (66). The most common pathologic feature of these cases is the subendothelial deposition of "mucinous" material in the renal glomerular vessels and arterioles. Electron-microscopic and immunohistochemical studies of biopsy material suggest that the material consists of fibrin in an atypical form or as incompletely polymerized fibrin, and of platelets. The glomeruli show varying degrees of damage, ranging from focal necrosis to total infarction. The resulting glomerulosclerosis with associated tubular atrophy may be present in those patients with irreversible renal impairment. Additional changes, present in the afferent arterioles and intralobular arteries of the kidney, may be seen with similar histologic appearance within the glomerular capillary loops. The differentiation from the pre-eclampsia/eclampsia syndrome should be relatively easy in mild cases since thrombocytopenia is uncommon with only 7% pre-eclamptics having platelet counts less than 150,000 ul (66). However, in severe cases, distinction may be impossible. The low plasma antithrombin III level in pre-eclampsia (67, 68) along with a picture of DIC may be helpful to separate it from TTP/HUS.

Thrombotic Microangiopathy in Cancer Patients (69)

The association between cancer and a "hypercoagulable" state predisposes a cancer patient to thromboembolic complications as well as DIC. These complications should not be confused with a distinct clinical entity of TM in cancer patients. This occurs in patients with widely metastatic carcinoma mostly derived from primary lesions in the stomach, lung and breast, presenting with more severe anemia and more frequent bleeding manifestations due to their underlying cancer and myelosuppressed state. The outcome depends on the responsiveness of the cancer to either hormonal or chemotherapeutic antitumor agents.

Another group of cancer patients with TM are those associated with chemotherapy. The most common agent implicated is mitomycin, though cases of TM have also been reported following the administration of cisplatinum, bleomycin, daunorubicin, cytosine arabinoside, chlorozotocin, neocarcinostatin and nitrosourea. Endothelial damage from the antitumor drug is believed to play an important role in the pathogenesis.

Infection, Especially that by HIV

Until the last few years, TM occurred mostly in bacterial infection, especially by enteric organisms. In the case of E. coli sepsis, the presence of a verotoxin has been found to induce platelet aggregation. Since the first report in 1984 (70), many more cases of TTP in HIV infected patients have been reported (71, 72). In addition, it has also been reported in HTLV-1 infection (73). Endothelial cell damage by HIV may be important in the etiology of the TM. Since immune complexes have been found in HIV infections, these complexes are also believed to play a role in the pathogenesis of TM.

MISCELLANEOUS

The other causes of secondary TM, shown in Table 1, include connective tissue disease, especially systemic lupus erythematosus, immune complex disease, a miscellaneous group of drugs (additional to the chemotherapeutic agents discussed above), and allograft rejection (74). These various types of secondary TM are distinct from the classical TTP/HUS in many ways. For one, TM is a rare complication of these disorders and, if it occurs, is frequently not clinically recognized until histologic studies are made. The severity of the TM varies from acute fulminating TTP with fatal outcome, to limited renal involvement with spontaneous recovery. The pathogenesis of TM may be different from that of TTP in that a likely causative agent of endothelial damage can be identified in most instances, such as a bacterial toxin in the case of TM associated with infection or chemotherapeutic drug toxicity as in that of mitomycin-associated TM. The prompt removal of the offending agent may be sufficient for the reversal of the abnormal vascular lesion.

In light of our knowledge of the pathogenesis of TTP indicating the importance of platelet aggregation, platelet transfusion is not advisable (47). Whether plasma exchange by plasmapheresis will offer additional therapeutic benefit is uncertain due to our limited experience. Likewise, it is not known whether or not antiplatelet aggregating agents are useful.

REFERENCES

1. E. Moschcowitz, Hyaline thrombosis of the terminal arterioles and capillaries: A hitherto undescribed disease. *Proc. N.Y. Pathol. Soc.,* **24**:21-224, (1924).
2. E. Moschcowitz, An acute febrile pleiochromic anemia with hyaline thrombosis of the terminal arterioles and capillaries. An undescribed diseae. *Arch. Int. Med.,* **36**:89-93, (1925).

3. K. Singer, F. P. Bornstein, and A. Wiles, Thrombotic thrombocytopenic purpura. *Blood,* **2**:542-554, (1947).
4. M. Bernheim, J. Roget, and F. Larbre, et al., Purpura thrombocytopenique thrombotique aigu avec anemie hemolytique. Microangiopathic (Maladie de Moschcowitz). *Ann. Pediatr. (Paris),* **33**:359-366, (1957).
5. W. C. Symmers, Thrombotic microangiopathic haemolytic anemia. *Br. Med. J.,* **2**:897-903, (1952).
6. E. C. Rossi, F. A. Carone, and F. del Greco, Hemolytic uremic syndrome and platelet-endothelial interactions, *in:* "Hemostasis, Prostaglandins and Renal Disease. (eds) G. Remuzzi, G. Mecca, G. deGaetano. Raven Press, NY, 1980, pp 321-329, 370-371.
7. H. C. Kwaan, Thrombotic Microangiopathy. *Sem. Hematol.,* **24**:69-70, (1987).
8. Thrombotic Microangiopathy I, II, *in:* "Seminars in Hematol." (Ed.) H. C. Kwaan, **24**:69-201, (1987).
9. R. M. Petitt, Thrombotic thrombocytopenic purpura. A thirty-year review. *Sem. Thromb. Hemost.,* **6**:350-355, 1980.
10. R. L. Ridolfi, and W. R. Bell, Thrombotic thrombocytopenic purpura: Report of 25 cases and a review of the literature. *Medicine,* **60**:413-428, (1981).
11. H. C. Kwaan, Thrombotic thrombocytopenic purpura. *JAMA,* **247**:3119-3120, (1982).
12. R. M. Bukowski, J. S. Hewlett, and J. W. Harris, et al., Exchange transfusion in the treatment of thrombotic thrombocytopenic purpura. *Sem. Hematol.,* **13**:219-232, (1976).
13. J. L. Orbison, Morphology of thrombotic thrombocytopenic purpura with demonstration of aneurysms. *Am. J. Pathol.,* **28**:129-135, (1952).
14. J. B. Craig, and D. Gitlin, The nature of the hyaline thrombi in thrombotic thrombocytopenic purpura. *Am. J. Pathol.,* **33**:251-258, (1957).
15. R. Komori, Thrombotic microangiopathy in patients with malignant tumors. *Acta. Pathol. JPN,* **12**:379-405, (1962).
16. J. R. Carter, Generalized capillary and arteriolar platelet thrombosis. *Am. J. Med. Sci.,* **213**:585-592, (1947).
17. H. C. Kwaan, G. Gallo, and Potter, et al., The nature of the vascular lesion in thrombotic thrombocytopenic purpura. *Ann. Int. Med.,* **68**:1169-1170, (1968).
18. H. C. Kwaan, The pathogenesis of thrombotic thrombocytopenic purpura. *Sem. Thromb. Hemost.,* **5**:184-198, (1979).
19. M. J. Mant, M. N. Couchi, and G. Medley, Thrombotic thrombocytopenic purpura; report of a case with possible immune etiology. *Blood,* **40**:416-421, (1972).
20. D. D. Weisenberger, M. L. O'Conner, and M. H. Hart, Thrombotic thrombocytopenic purpura with C3 vascular deposits: Report of an unusual case. *Am. J. Clin. Pathol.,* **67**:61-63, (1977).
21. J. D. Feldman, M. R. Mardiney, and E. R. Unanue, et al., The vascular pathology of thrombotic thrombocytopenic purpura: An immunohistochemical and ultrastructural study. *Lab. Invest.,* **15**:927-946, (1966).
22. T. E. Blecher, and A. B. Roper, Early diagnosis of thrombotic microangiopathy by paraffin sections of aspirated bone marrow. *Arch. Dis. Child.,* **42**:158-162, (1967).
23. R. Bukowski, Thrombotic thrombocytopenic purpura: A review. *Prog. Hemost. Thromb.,* **6**:287-337, (1982).
24. A. Goodman, R. Ramos, and M. Petrelli, et al., Gingival biopsy in thrombotic thrombocytopenic purpura. *Ann. Int. Med.,* **89**:501-504, (1978).
25. H. C. Kwaan, Role of fibrinolysis in thrombotic thrombocytopenic purpura. *Sem. Thromb. Hemost.,* **6**:395-400, (1980).
26. A. I. Bernheim, Widespread capillary and arteriolar platelet thrombi. *J. Mt. Sinai. Hosp.,* **10**:287-291, (1943).
27. G. Remuzzi, R. Misiani, and G. Mecca, et al., Thrombotic thrombocytopenic purpura: A deficiency of plasma factors regulating platelet-vessel wall interactions? *N. Eng. J. Med.,* **299**:311 (1978) (letter).
28. G. Remuzzi, R. Misiani, and D. Marchesi, et al., Treatment of the hemolytic uremic syndrome with plasma. *Clin. Nephrol.,* **12**:279-284, (1979).
29. P. D. Neame, Immunologic and other factors in thrombotic thrombocytopenic purpura (TTP). *Sem. Thromb. Hemost.,* **6**:416-429, (1980).
30. F. A. Siddigui, and E. C. Y. Lian, Novel platelet-agglutinating protein from a thrombotic thrombocytopenic purpura plasma. *J. Clin. Invest.,* **76**:1330-1337, (1985).

31. E. C. Y. Lian, P. T. K. Mui, and F. A. Siddique, et al., Inhibition of platelet aggregating activity in thrombotic thrombocytopenic purpura plasma by normal adult immunoglobulin G. *J. Clin. Invest.*, **73**:548-555, (1984).
32. J. J. Moake, C. K. Rudy, and J. H. Troll, et al., Unusually large plasma factor VIII: von Willebrand factor multimers in chronic relapsing thrombotic thrombocytopenic purpura. *N. Eng. J. Med.*, **307**:1432-1435, (1982).
33. J. G. Kelton, J. Moore, and A. Santos, et al., Detection of a platelet-aggregating factor in thrombotic thrombocytopenic purpura. *Ann. Intern. Med.*, **101**:589-593, (1984).
34. R. T. Wall, and L. A. Harker, The endothelium and thrombosis. *Ann. Rev. Med.*, **31**:361-371, (1980).
35. J. Morrison, and R. McMillan, Elevated platelet-associated IgG in thrombotic thrombocytopenic purpura. *JAMA*, **235**:1944, (1977).
36. W. G. Murphy, J. C. Moore, and J. G. Kelton, Calcium-dependent cysteine protease activity in the sera of patients with thrombotic thrombocytopenic purpura. *Blood*, **70**:1678-1683, (1978).
37. M. C. Brain, J. V. Dacie, and D. Hourihane, et al., Microangiopathic hemolytic anemia: The possible role of vascular lesions in pathogenesis. *Br. J. Haematol.*, **8**:358-374, (1962).
38. T. N. James, and R. W. Monto, Pathology of the cardiac conduction system in thrombotic thrombocytopenic purpura. *Ann. Intern. Med.*, **65**:37-43, (1966).
39. R. L. Ridolfi, G. M. Hutchins, and W. R. Bell, The heart and conduction system in thrombotic thrombocytopenic purpura. *Ann. Int. Med.*, **91**:357-363, (1979).
40. R. C. Bone, J. E. Henry, and J. Petterson, et al., Respiratory dysfunction in thrombotic thrombocytopenic purpura. *Am. J. Med.*, **65**:262-270, (1978).
41. H. N. Harrison, Thrombotic thrombocytopenic purpura associated with pancreatic islet cell necrosis. *Arch. Int. Med.*, **102**:124-130, (1958).
42. H. R. Hellstrom, E. C. Nash, and E. R. Fischer, Thrombotic thrombocytopenic purpura as a cause of massive gastrointestinal hemorrhage, report of a case. *Gastroenterol*, **36**:132-136, (1959).
43. D. R. Lewellen, and L. J. Singerman, Thrombotic thrombocytopenic purpura with optic disk neovascularizatin vitreous hemorrhage, retinal detachment and optic atrophy. *Am. J. Ophthalmol.*, **89**:840-844, (1980).
44. S. P. B. Percival, Ocular findings in thrombotic thrombocytopenic purpura (Moschcowitz's disease). *Br. J. Ophthalmol.*, **54**:73-78, (1970).
45. S. S. Kennedy, L. R. Zacharski, and J. R. Beck, Thrombotic thrombocytopenic purpura: Analysis of 48 unselected cases. *Sem. Thromb. Hemost.*, **6**:341-349, (1980).
46. A. V. Pisciotta, and J. L. Gottschall, Clinical features of thrombotic thrombocytopenic purpura. *Sem. Thromb. Hemost.*, **6**:330-340, (1980).
47. L. Gordon, H. C. Kwaan, and E. C. Rossi, The deleterious effect of platelet transfusion and recovery thrombocytosis in thrombotic microangiopathy. *Sem. Hematol.*, **24**:194-201, 1987.
48. H. C. Kwaan, and P. H. Domer, Fatal thrombotic thrombocytopenic purpura despite plasmapheresis. A case report of autopsy findings. *Haemostasis*, **19**:344-347, (1989).
49. C. Gasser, C. Gautier, and A. Steck, et al., Hamolytisch-uramische syndrome. Bilaterale nierenindennekrosen bei akuten erworbenen hamolytischen anamien. *Schweiz. Med. Woschenschr.*, **85**:905-909, (1955).
50. C. A. Gianantoino, M. Vitacco, and F. Mendilaharzu, et al., The hemolytic-uremic syndrome. *Nephron*, **11**:174-192, (1973).
51. S. O'Regan, R. W. Chesney, and B. S. Kaplan, et al., Red cell membrane phospholipid abnormalities in the hemolytic uremic syndrome. *Clin. Nephrol.*, **15**:14-17, (1980).
52. W. Proesmans, and R. Eeckels, Has heparin changed the prognosis of the hemolytic-uremic syndrome. *Clin. Nephrol.*, **2**:169-173, (1974).
53. B. S. Kaplan, P. D. Thomson, and J. P. deChadarevian, The hemolytic-uremic syndrome. *Pediatr. Clin. N. Am.*, **23**:761-777, (1976).
54. C. Loirat, F. Beafils, and E. Sonsino, et al., Traitement du syndrome hemolytique et uremique de l'enfant par l'urokinase. Essai controle cooperatif. *Arch. Fr. Pediatr.*, **41**:15-19, (1984).
55. W. Proesmans, R. Eckels, and B. Van Damme, et al., Antithrombotic therapy in childhood haemolytic uraemic syndrome. A randomized prospective study. *in:* "Paediatric Nephrology." (eds) J. Brodehl, J. H. H. Ehrich. Springer Verlag, Berlin, pp. 285-288 (1984).

56. B. S. Kaplan, and P. D. Thomson, Hyperuricemia in the hemolytic uremic syndrome. *Am. J. Dis. Child.*, **130**:854-856, (1976).
57. B. S. Kaplan, and W. Proesmans, The hemolytic uremic syndrome of childhood and its variants. *Sem. Hematol.*, **24**:148-160, (1978).
58. R. S. Trompeter, R. Schwartz, and C. Chantler, et al., Haemolytic uraemic syndrome: analysis of prognostic features. *Arch. Dis. Child.*, **58**:101-105, (1983).
59. C. Loirat, E. Sonsino, and A. V. Moreno, et al., Hemolytic-uremic syndrome: An analysis of the natural history and prognostic features. *Acta. Paediatr. Scand.*, **73**:505-514, (1984).
60. H. C. Kwaan, and L. V. Gratkins, Thromboembolism in obstetrical patients. *in*: "Thrombosis. (Eds) H. C. Kwaan, E. J. W. Bowie. W. B. Saunders, Philadelphia, (1982), p. 168-174.
61. C. P. Weiner, Thrombotic microangiopathy in pregnancy and the postpartum period. *Sem. Hematol.*, **24**:119-129, (1987).
62. A. C. Schoolwerth, R. S. Sandler, and S. Klahr, Nephrosclerosis postpartum and in women taking oral contraceptives. *Arch. Int. Med.*, **136**:178-185, (1976).
63. A. Segonds, N. Louradour, and J. M. Suc, et al., Postpartum hemolytic uremic syndrome: a study of three cases with a review of the literature. *Clin. Nephrol.*, **12**:229, (1979).
64. M. L. Schwartz, and W. E. Brenner, The obfuscation of eclampsia by thrombotic thrombocytopenic purpura. *Ann. J. Obstet. Gynecol.*, **131**:18, (1978).
65. J. A. Pritchard, P. C. Cuuningham, and R. A. Mason, Coagulation changes in eclampsia: Their frequency and pathogenesis. *Am. J. Obstet. Gynecol.*, **124**:855, (1975).
66. C. P. Weiner, and J. K. BRandt, Plasma antithrombin III activity: An aid in the diagnosis of preeclampsia-eclampsia. *Am. J. Obstet. Gynecol.*, **64**:46-49, (1984).
67. C. P. Weiner, H. C. Kwaan, and C. Xu, et al., Antithrombin III activity in women with hypertension during pregnancy. *Obstet. Gynecol.*, **65**:301-309, (1985).
68. A. J. Murgo, Thrombotic microangiopathy in the cancer patient including those induced by chemotherapeutic agents. *Sem. Hematol.*, **24**:161-177, (1987).
69. R. V. Boccia, E. P. Gelmann, and C. C. Baker, et al., A hemolytic uremic syndrome with the acquired immunodefinicency syndrome. *Ann. Int. Med.*, **101**:716-717, (1984).
70. J. Jokela, T. Flynn, and K. Henry, Thrombotic thrombocytopenic purpura in a human immunodeficiency virus (HIV)-seropositive homosexual man. *Am. J. Hematol.*, **25**:341-343, (1987).
71. A. N. Leaf, L. J. Laubenstein, and B. Rephael, et al., Thrombotic thrombocytopenic purpura associated with human immunodeficiency virus Type 1 (HIV-1) infection. *Ann. Int. Med.*, **109**:194-197, (1988).
72. A. C. Dixon, D. W. Kwock, and J. M. Makamura et al., Thrombotic thrombocytopenic purpura and human T-lymphotrophic virus, Type 1 (HTLV-1). *Ann. Int. Med.*, **110**:93-94, (1989).
73. H. C. Kwaan, Miscellaneous Secondary Thrombotic Microangiopathy. *Sem. Hematol.*, **24**:141-147, (1987).

CLINICAL TRIALS WITH ALTEPLASE (RT-pa) IN ACUTE MYOCARDIAL INFARCTION

David P. de Bono

Department of Cardiology
University of Leicester, Glenfield General Hospital
Leicester LE3 9QP, UK

INTRODUCTION

Alteplase (recombinant human tissue plasminogen activator, rt-PA) first became available for clinical trials in acute myocardial infarction in 1984 (1). At this time there was already considerable experience with thrombolysis using intracoronary streptokinase (2), and a consensus that the benefits of coronary thrombolysis operated through the early restoration of coronary patency. The potential benefits of rt-PA compared with streptokinase as seen in 1984 are listed in table 1 — the principal hope being that a "thrombus specific" agent such as rt-PA would provide more effective and rapid thrombolysis with a minimum of bleeding complications. The endpoint in the earliest trials (TIMI 1 and ECSG 1) was angiographic coronary patency — the "hardest" endpoint realistically attainable with the limited amounts of rt-PA then available. The results of these studies were favourable to rt-PA in comparison both with placebo (ECSG-1) and streptokinase (ECSG-2 and TIMI-1) but it was clear that larger studies would be needed both to provide adequate safety data for registration, and to provide more clearcut evidence concerning "outcome" measurements such as mortality and left ventricular function. Moreover the TIMI-1 study had raised a worry concerning the apparent high rate of reocclusion following rt-PA thrombolysis (3,4). The question of reocclusion was specifically addressed in ECSG-3, with reassuring results. Nevertheless, concern about reocclusion undoubtedly influenced subsequent trials both in terms of the use of a prolonged rt-PA infusion and a feeling that results might be further improved by combining thrombolysis with early angioplasty.

After the publication of the GISSI-1 trial in 1986 (5) with its convincing demonstration of mortality reduction by early streptokinase thrombolysis the TIMI investigators felt it was no longer ethically viable to conduct placebo-controlled studies of thrombolysis in the United States of America, and subsequent TIMI studies concentrated on dose-ranging and on the evaluation of angioplasty and of adjunctive therapy such as beta-blockade. A similar line was independently pursued by the TAMI investigators.

From an European viewpoint, placebo controlled studies were still regarded as desirable, and the ECSG-5, and ASSET studies were set up. These have now clearly shown that coronary thrombolysis with intravenous rt-PA is effective both in preserving left ventricular function and in reducing mortality compared with "conventional" therapy not involving thrombolysis.

Meanwhile however both the AIMS study with anistreplase and the ISIS-2 study with streptokinase and aspirin have shown substantial mortality reductions with these agents, at the cost of only a modest rate of complications. Inspection of the entry criteria, baseline characteristics and placebo-group mortality rates of different trials shows such variability that direct comparison on the basis of percentage mortality reduction compared to placebo is a very unsatisfactory way of comparing different agents. In view of the considerable price differential between rt-PA and streptokinase a direct comparison between these agents

Fibrinogen, Thrombosis, Coagulation, and Fibrinolysis, Edited by
C. Y. Liu and S. Chien, Plenum Press, New York

Table 1. Advantages of rt-PA Compared with Streptokinase

Fibrin specific:	So less systemic fibrinolysis: So less bleeding?
	Potentially wider therapeutic margins?
Non antigenic:	So can be used repeatedly
	Dose need not be adjusted because of previous exposure
Short half life:	So thrombolysis can be readily stopped if problems arise

in a large trial with "outcome" endpoints is clearly needed, and such trials are already in progress (GISSI-2) or being planned (ISIS-3).

The European Cooperative Study Group "Family" of Trials

The Europan cooperative study group for rt-PA in myocardial infarction (ECSG) was established in 1984 to conduct multicentre studies with rt-PA in myocardial infarction in Europe. The "family" of trials is summarised in table 2.

ECSG 1 This (actually the second trial to be published) was a comparison of Genentech-Boehringer Ingelheim rt-PA (0.75mg per Kg infused intravenously over 90 minutes) compared with placebo (6). All patients also received heparin. The endpoint was coronary patency at 90 minutes after the start of infusion. There was no pre-treatment angiogram. Centrally-assessed coronary patency was 61% for the rt-PA group ($n=63$) and 18% for the placebo group ($n=65$.) This difference was highly significant ($p<0.01$). Bleeding complications, mostly a consequence of the early angiography, were more common in the rt-PA group.

ECSG 2 compared the same dose of rt-PA with streptokinase 1,500,000 units given over 60 minutes (7). The trial had identical entry criteria to ECSG 1 but was conducted in different centres. The endpoint was again angiographic patency at 90 minutes. This was 70% in the rt-PA group ($n=62$) and 55% in the streptokinase group ($n=62$). The 95% confidence intervals for the difference were +32% to −2%, $p = 0.054$. Major bleeding complications were rare in both groups. Minor bleeding complications mainly related to the angiography were relatively common, with a trend towards increased frequency in the streptokinase group (haematoma 19 vs 12, prolonged bleeding at puncture site 10 vs 3, blood transfusion 4 vs 2).

Table 2. The European Cooperative Study Group "Family" of rt-PA Trials

Trial	Year	Comparison	Endpoint	Reference
ECSG 1	1985	rt-PA vs Placebo	Patency	6
ECSG 2	1985	rt-PA vs SK	Patency	7
ECSG 3	1986	short vs long rt-PA infusion	Reocclusion	8
ECSG 4	1988	rt-PA vs rt-PA + PTCA	LV Function Infarct Size	9
ECSG 5	1988	rt-PA vs Placebo	LV Function Infarct Size Mortality	10
ECSG 6	1989	rt-PA vs rt-PA + heparin	Patency	

ECSG 3 This trial was designed to look at the rate of reocclusion, as assessed by serial angiography, in patients shown to have a patent infarct related coronary artery after infusion of 40 mg of rt-PA who were then randomised to receive either further rt-PA (40 mg over 6 hours) or placebo, both groups receiving aspirin and heparin (8). 123 patients entered the study, and 86 had a patent vessel on angiography after 90 minutes. Subsequent reocclusion rates were low in both the further rt-PA and placebo groups (3 vs 2) and the pooled reocclusion rate was 7% (95% confidence interval 2-15%).

ECSG 4 In this study (9), patients with clinical and ECG evidence of myocardial infarction and symptoms of less than 5 hours duration (n = 367) all received heparin, aspirin and an infusion of 100 mg rt-PA (10 mg bolus, then 50,20,20 mg per hour). Patients were randomised by telephone to either a policy of "non-invasive" management (angiography and intervention only if clinically indicated) or to immediate angiography with percutaneous angioplasty of the infarct related vessel if there was more than a 60% residual stenosis. 180 actually underwent early angiography: because this was done as soon as possible after entry to the trial it was possible to plot a time course of vessels becoming patent by plotting patency rate against time to angiography. The estimated patency rate at 90 minutes was 89%. 72 patients had a completely occluded vessel at the time of angiography, and in 57 of these PTCA secured patency. Of 96 patients with a patent but stenosed vessel, post PTCA reocclusion occurred in three. Despite the success in securing earlier and more complete patency in the PTCA group, there was no benefit in terms of either enzymatic infarct size or left ventricular function. Although there was a higher proportion of patients in the rt-PA group who had an occluded vessel at the time of hospital discharge (20 vs 12) this alone probably does not wholly account for the lack of benefit of early PTCA: however other explanations such as enhancement of reperfusion injury are speculative.

ECSG 5 This was a placebo controlled trial with the same entry criteria as ECSG 4 in which patients were randomized to receive either 100 mg rt-PA (same dose schedule as ECSG 4) or placebo (10). All patients received aspirin and heparin. There was a small but significant benefit to the treated patients in terms of a smaller enzymatic infarct size and better left ventricular function. The overall mortality rate was very low, but the 3 month mortality rate was lower in the rt-PA than in the placebo group (4.8% vs 7.9%, 95% confidence interval for risk ratio 0.34 to 1.07).

TIMI Studies The TIMI (Thrombolysis In Myocardial Infarction) study group was established in 1983. The results of the first phase TIMI trial (TIMI-1) were published in preliminary form in 1985 (3) and definitively in 1987 (11). 290 patients with clinical and ECG features of myocardial infarction of less than 7 hours duration underwent pretreatment angiography, and 232 with occluded infarct related vessels received thrombolytic treatment. 113 patients received rt-PA (80 mg as 40, 20, 20 mg in successive hours) and the 90 minute reperfusion rate was 62%. 119 patients received streptokinase (1,500,000 units) and the 90 minute reperfusion rate was 31% ($p<0.001$). In a subset of patients who had repeat angiography before discharge, reocclusion occurred in 24% of 62 patients treated with rt-PA, and in 14% of 29 patients given streptokinase. If deaths or clinical reinfarction are also attributed to reocclusion, then reocclusion rates become similar at 29% in the rt-PA and 30% in the streptokinase group.

Open-label TIMI studies have been conducted to evaluate the most effective dose regimes of rt-PA. The early TIMI (TIMI-1) and ECSG (ECSG 1,2,3) studies were conducted with predominantly double-chain rt-PA (Genentech batch no. G 11021). A subsequent change in manufacturing process produced a predominantly single chain product (G 11034) which is slightly less effective on a weight for weight basis, but significantly more fibrin selective. This agent was used in subsequent TIMI and ECSG trials. The open lable TIMI studies suggested that a dose of 150 mg of rt-PA would give the most rapid coronary thrombolysis, and this dose was chosen for the phase 2 TIMI studies. However a higher than expected incidence of cerebral bleeding in the first 290 patients caused a subsequent decision to reduce the dose to 100 mg (12).

The TIMI II trial evaluated the role of elective angiography with a view either to angioplasty or coronary artery bypass grafting as compared to a "conservative" policy after rt-PA thrombolysis. In contrast to the ECSG 4 study, angiography was not performed immediately, but a mean of 32 hours after presentation. There was no difference in mortality, left ventricular function or reinfarction rates in the "invasive" as compared with the "conservative" randomisation group.

TAMI Trials

The TAMI (Thrombolysis and Angioplasty in Myocardial Infarction) group have conducted a number of studies with rt-PA in myocardial infarction. TAMI-1 (14) randomised patients to immediate angiography and subsequent angioplasty if a suitable lesion were present, or to non-invasive therapy. There was no significant difference in outcome, but more patients in the PTCA group required emergency coronary bypass surgery, and more in the "conservative" group needed urgent or emergency PTCA. TAMI-3 is one of the few studies to have compared coronary patency after rt-PA with heparin to patency after rt-PA without concomitant heparin for the four hours of rt-PA infusion. There was no difference in patency rate.

ASSET The Anglo-Scandinavian study of Early Thrombolysis is a comparison of rt-PA with placebo designed to have mortality as the sole endpoint (15). The principal entry criterion was a clinical diagnosis of myocardial infarction with onset within the preceding 5 hours, and ECG criteria were not specified. The rt-PA dose was 100 mg (10 mg bolus, 50 mg in the first hour, then 20 mg for each of the next two hours). All patients were given 5000 units of heparin, and an infusion of 1000 u heparin per hour for 21 hours. No subsequent anticoagulation or aspirin therapy was permitted. 2516 patients were allocated to rt-PA and 2495 to placebo. The one month fatality rates were 7.2 and 9.8% respectively (relative reduction 26%, 95% confidence interval 11-39%). Bleeding complications were seen in 6.3% of patients given rt-PA (classified as major in 1.4%) and in 0.8% of those given placebo (major in 0.4%).

Other Trials Placebo controlled trials in Australia by O'Rourke and Colleagues (16) and the National Heart Foundation of Australia study group (17) have shown a beneficial effect on Left ventricular function after rt-PA therapy. Guerci and colleagues (18) randomised patients to rt-PA or placebo, performed angiography 1-6 hours after the end of the 3 hour rt-PA or placebo infusion, and further randomised patients with patent vessels to angioplasty on day 3 or to conservative therapy. Patency was higher (64%) in the rt-PA group than after placebo (24%). The rt-PA group also had a higher left ventricular ejection fraction. Angioplasty had no effect on resting ejection fraction, but did improve ejection fraction after exercise, and diminished the incidence of post infarct angina.

Overview of Results with Respect to Efficacy

There is no agreed absolute index of efficacy for coronary thrombolytic agents. Measurements based on the proportion of patent coronary vessels will be affected by the time at which angiography is undertaken, and the proportion of occluded vessels in the original sample. Attempts to measure the true rate of reperfusion by performing pretreatment angiography mean that the thrombolytic drugs are being tested against "old" thrombus which may be more resistant to lysis. Efficacy also has to take account of total dose, dose schedule, and the incidence of unacceptable side effects. Probably the most realistic measure of efficacy is the patency rate at 90 minutes using the maximum dose which will have an acceptable level of side effects. For rt-PA, the dose is approximately 100 mg and the patency rate 75 + 10%. On the evidence currently available, this is rather better than the efficacy of intravenous streptokinase 1,500,000 units and similar to the claimed efficacy of APSAC 30 mg. Clinical efficacy also has to take account of the risks of reocclusion, reperfusion damage, immunogenicity and long term side effects. Many of these will depend as much on the follow-up or ancillary therapy regime as on the thrombolytic agent — elegantly demonstrated, in the case of aspirin, by the ISIS-2 study (19). The ASSET and ECSG V studies have demonstrated that rt-PA is effective in reducing post-MI mortality compared with placebo. However the placebo group mortality in ECSG V was also very low (perhaps because all patients received heparin and aspirin) and the lack of follow-up therapy in ASSET may have prevented patients deriving the maximum benefit. Better data on comparative efficacy of different agents will come from the GISSI-2 and ISIS-3 studies, but these will still be limited to a comparison of a limited number of dose schedules and follow-up regimes. A summary of presently-available data on the efficacy of rt-PA is given in table 3.

Overview of Results with Respect to Safety

rt-PA has been remarkably free of adverse haemodynamic or allergic side effects.

Table 3. Efficacy of rt-PA

1) Coronary Patency	vs. placebo	ECSG 1
		Nat. Heart Found.Aust
		Guerci et al.
	vs. SK	TIMI 1
		ECSG 2
2) LV Function	vs. placebo	ECSG 5
		Nat.Heart Found.Aust
		O'Rourke et al.
		Guerci et al.
3) Mortality	vs. placebo	ECSG 5
		ASSET

Virtually the only deleterious consequence of rt-PA thrombolysis is haemorrhage. The hope that rt-PA therapy could be administered without a fall in plasma fibrinogen has been partially realised: present dosage schedules cause an average maximum fall in plasma fibrinogen of only 33% as compared with over 90% for streptokinase, and this may account for the lower incidence of bleeding complications relative to streptokinase in the ECSG 2 study. However overall bleeding complications with rt-PA are significantly higher than with placebo. The incidence of cerebrovascular complications in the ASSET study was similar in the rt-PA and placebo groups (1.1 vs 1%) but the distribution was different, with more haemorrhage in the rt-PA group and more cerebral thrombosis and embolism in the placebo group. The risk of cerebral haemorrhage with rt-PA seems to be related both to rt-PA dosage and to patient weight with a fixed rt-PA dosage — the haemorrhage rate increases with a dose greater than 1.5 mg rt-PA per Kg.

False Paths and Red Herrings

After four years clinical experience with rt-PA, the clinical efficacy and safety profile of the drug are becoming better defined. If therapeutic efficacy is a function of the rate of restoration of coronary patency (and there is no convincing evidence to the contrary) rt-PA is likely to be the most effective of the current generation of thrombolytic agents. Its clinical superiority to, say, streptokinase in terms of survival and ventricular preservation remains however to be demonstrated formally. In retrospect, perhaps this point was treated too lightly in the early trials of rt-PA: however at the time there were insufficient supplies to mount large scale trials against streptokinase, whose true efficacy has in any event only been confirmed relatively recently by the GISSI and ISIS-2 studies. The early TIMI studies raised a worry concerning reocclusion which perhaps diverted attention towards both prolonged (and therefore complex) dose schedules and a preoccupation with early angioplasty. In retrospect pehaps it would have been fruitful to pay more attention to alternative dose schedules and to the best form of follow up therapy.

The Future: GISSI-2 and ISIS-3

The lack of comparative large scale trials of rt-PA against other effective thrombolytic agents is likely to be remedied in the near future by two trials. GISSI-2, organised from Milan, which has been recruiting since early 1988, is comparing rt-PA with streptokinase. ISIS-3, organised from Oxford, is planned to compare streptokinase with (Wellcome) rt-PA and probably also with APSAC. Both studies plan to recruit 20-30,000 patients, but even with this number a pure "mortality" endpoint may not be reached. Convincing and accurate information should however be available concerning long and short term side effects, and it may be possible to reach a definitive verdict concerning the "best" of the current generation of agents.

REFERENCES

1. F. van de Werf, P. A. Ludbrook, S. R. Bergmann et al. Coronary thrombolysis with tissue type plasminogen activator in patients with evolving myocardial infarction. *N. Engl. J. Med.,* **310:**609-13 (1984).
2. R. Schroeder, G. Biamino, E. R. Leitner et al. Intravenous short term infusion of streptokinase in myocardial infarction. *Circulation,* **67:**536-48 (1983).
3. The TIMI study group. The thrombolysis in myocardial infarction (TIMI) trial. Phase 1 findings. *N. Engl. J. Med.,* **312:**932-36 (1985).
4. D. O. Wiiliams, J. Borer, E. Braunwald et al. Intravenous recombinant tissue plasminogen activator in patients with myocardial infarction: a report from the NHLBI thrombolysis in myocardial infarction trial. *Circulation,* **73(2):**338-46 (1986).
5. Gruppo Italiano per lo Studi deela Streptochinasi nell' Infarto miocardico (GISSI). Effectiveness of intravenous thromboytic therapy in acute myocardial infarction. *Lancet,* **i:**397-401 (1986).
6. M. Verstraete, W. Bleifeld, R. W. Brower et al. Double blind randomized trial of intravenous recombinant tissue type plasminogen activator versus placebo in acute myocardial infarction. *Lancet,* **ii:**965-969 (1985).
7. M. Verstraete, R. Bernard, M. Bory et al. Randomized trial of intravenous recombinant tissue type plasminogen activator versus intravenous streptokinase in acute myocardial infarction. *Lancet,* **i:**842-847 (1985).
8. M. Verstraete, A. E. R. Arnold, R. W. Brower et al. Acute coronary thrombolysis with recombinant human tissue type plasminogen activator — initial patency and influence of a maintained infusion on reocclusion rate. *Am. J. Cardiol.,* **60:**231-7 (1987).
9. M. L. Simoons, A. E. R. Arnold, A. Betriu et al. Thrombolysis with rt-PA in acute myocardial infarction: no additional benefit of immediate PTCA. *Lancet,* **i;**(1988).
10. F. van de Werf, A. E. R. Arnold, and the European Cooperative Study Group for Recombinant Tissue Plasminogen Activator. Effect of intravenous tissue plasminogen activator on infarct size, left ventricular function and survival in patients with acute myocardial infarction. *Br. Med. J.,* (1988).
11. J.H. Chesebro, G. Knatterud, R. Roberts et al. Thrombolysis in myocardial infarction (TIMI) trial, Phase I: a comparison between intravenous tissue plasminogen activator and intravenous streptokinase. *Circulation,* **76:**142-54 (1987).
12. E. Passamani. Thrombolysis in myocardial infarction: the NHLBI experience, *in*: Sobel B.E., Collen D., Grossbard EB (eds). Tissue Plasminogen Activator in Thrombolytic Therapy. M. Dekker, New York/Basel, pp75-86 (1986).
13. TIMI Study Group. Comparison of invasive and conservative strategies after treatment with intravenous tissue plasminogen activator in acute myocardial infarction: results of the thrombolysis in myocardial infarction (TIMI) phase II trial. *New Engl. J. Med.,* **320:**618-26 (1989).
14. E. J. Topol, R. M. Califf, B. S. George et al. A randomized trial of immediate versus delayed elective angioplasty after intravenous tissue plasminogen activator in acute myocardial infarction. *N. Engl. J. Med.,* **317:**581-88 (1987).
15. R. G. Wilcox, G. von der Lippe, C. G. Olsson, G. Jensen, A. M. Skene, J. R. Hampton. Trial of tissue plasminogen activator for mortality reduction in acute myocardial infarction: Anglo Scandinavian Study of early thrombolysis (ASSET). *Lancet,* **ii:** 525-30 (1988).
16. M. O'Rourke, D. Baron, A. Keogh et al. Limitation of myocardial infarction by early infusion of recombinant tissue type plasminogen activator. *Circulation,* **77(6):**13311-5 (1988).
17. National Foundation of Australia Coronary Thrombolysis Group. Coronary thrombolysis and myocardial salvage by tissue plasminogen activator given up to 4 hours after onset of myocardial infarction. *Lancet,* **i:**203-8 (1988).
18. A. D. Guerci, G. Gerstenblith, J. A. Brinker, et al. A randomized trial of intravenous tissue plasminogen activator for acute myocardial infarction with subsequent randomization to elective coronary angioplasty. *N. Engl. J. Med.,* **317:**1613-18 (1987).
19. ISIS-2 (Second International Study of Infarct Survival) Collaborative Group. Randomised trial of intravenous streptokinase, oral aspirin, both or neither among 17,187 cases of suspected myocardial infarction: ISIS-2. *Lancet,* **ii:**349-60 (1988).

THE EFFECTS OF STREPTOKINASE AND TISSUE PLASMINOGEN ACTIVATOR ON LEFT VENTRICULAR FUNCTION

Harvey D. White

Cardiovascular Research
Green Lane Hospital
Auckland, New Zealand

INTRODUCTION

One of the major questions remaining in thrombolysis is which drug we should be using. In terms of "head to head" comparisons there are few data. Streptokinase and tissue plasminogen activator have been compared in terms of patency and tissue plasminogen activator has been shown to be superior (1, 2). However, patency cannot be directly extrapolated to patient benefit and direct comparisons of the effects of these drugs on left ventricular function and mortality are required. Left ventricular function can be considered as a surrogate for mortality but in addition the assessment of left ventricular function also evaluates the capacity of a patient to exercise, and relates to quality of life. In this study the effects of streptokinase and tissue plasminogen activator on left ventricular function were compared.

PATIENT POPULATION

Only patients with first myocardial infarctions were randomised in this trial. This was in order to attempt to ensure that baseline left ventricular function was similar. Consecutive patients were randomised aged < 70 years who presented within three hours after the onset of 30 minutes of typical ischaemic chest pain. ST-segment elevation of 1 mm or more was required in limb leads and V_4 to V_6, and ST-elevation of 2 mm or more in leads V_1 to V_3. Other exclusions were any history of a previous stroke, a confirmed peptic ulcer during the previous six months, history of a bleeding disorder, history of genitourinary bleeding, current administration of anticoagulants, continuing hypertension (systolic pressure above 200 mm unable to be readily lowered with medical therapy), recent surgery or trauma within two weeks, left bundle branch block or other associated severe non-cardiac disease that might limit survival. Patients with cardiogenic shock were included and the study was approved by the Ethics Committees in Auckland.

TREATMENT

Medications were given in a double-blind manner using a double-dummy technique. Patients were randomly given intravenous streptokinase (1.5×10^6 units over 30 minutes) plus rt-PA placebo over three hours, or streptokinase placebo infused over 30 minutes plus rt-PA infused over three hours. The rt-PA was given as a 10 mg bolus followed by 50 mg over the first hour and 20 mg over each of the subsequent two hours. A combination of 25 mg of aspirin and 200 mg of dipyridamole was given orally on admission to the coronary care unit and continued twice daily. Heparin was given 30 minutes after beginning the blind infusion and continued for 48 hours. The initial dose was 1000 iu per hour and this was subsequently adjusted to keep the activated partial thromboplastin time within the range of 90-100 seconds. Intravenous beta-blockers were not given but oral bata-blockers were given on the third day to patients without contraindicatons. Calcium-channel-blockers and nitrates were prescribed only when indicated for post-infarction angina. Cardiac catheterisation was not performed before the planned three-week angiogram unless post-infarction angina occurred which was either associated with marked ST-segment changes or did not respond readily to medical therapy. This was in order to minimise the confounding effects of myocardial stunning. Angioplasty and coronary artery surgery were only performed for severe medical symptoms and not just for coronary anatomic appearances with the exception of a left main coronary artery stenosis of > 50% diameter. Reinfarction was defined as chest pain associated with a rise in creatine kinase to twice the upper limit of normal.

DATA ANALYSIS

Left ventriculography was performed by simultaneous biplane cineventriculography. Ventricular volumes and ejection fractions were calculated from the right anterior oblique ventriculogram by means of an integration method with correction factors based on a comparison between true and calculated volumes of radiopaque casts of the left ventricle that had been made of hearts at autopsy. The infarct-related artery was identified by localisation of Q-waves and ST-segment changes on the electrocardiogram. The patency of the infarct-related artery was classified according to the criteria of the Thrombolysis in Myocardial Infarction trial (TIMI).[1] Analysis was performed at a central laboratory by experienced cardiac radiologists blinded to the treatment assignments.

STATISTICAL ANALYSIS

Sample size calculations were based on our previous trials and the sample size required 216 ventriculograms to show a 4% difference in ejection fraction with a power of 80% and a level of significance (two-tailed) of 0.05. All analyses were made on an intention-to-treat basis.

RESULTS

Two hundred and seventy patients were randomised and ventriculograms were obtained in 89% of all patients. Mean time from the onset of chest pain to the beginning of the infusion was 2.5 ± 0.6 hours in both groups. Mean ages were 56 ± 9 years for the streptokinase group and 55 ± 9 years for the rt-PA group. The baseline characteristics were similar for both randomised groups.

Table 1. Left Ventricular Function at Three Weeks After Infarction

	Streptokinase (n = 116)	rt-PA (n = 124)
Ejection fraction %	58 ± 12	58 ± 12
Anterior	55 ± 15	54± 14
Inferior	59 ± 11	60 ± 10
End-systolic volume (ml)	61 ± 29	66 ± 31
Anterior	67 ± 33	75 ± 32
Inferior	59 ± 26	60 ± 29

Left Ventricular Function

The primary end-point of this trial was left ventricular function. The ejection fraction was exactly the same (58%) for both groups. There was also no difference for patients with anterior or inferior infarctions. With respect to end-systolic volumes, the volumes were slightly larger in the patients treated with rt-PA but this was not significantly different (Table 1). Similar numbers of patients required diuretic therapy for heart failure, 7% of patients treated with streptoknase and 9% of patients treated with rt-PA.

Patency of the Infarct-Related Artery

In this study patency was not assessed acutely, but at three weeks. Patency for the streptokinase-treated patients was 75% (TIMI Grade II or higher) and 76% for the rt-PA patients.

Cardiac Enzymes

There was no difference in the peak CK, 2635 ± 1996 streptokinase vs 2565 ± 1911 rt-PA iu/l. The time to peak CK was also similar, 14 ± 7 hours for the streptokinase group and 14 ± 8 hours in the rt-PA-treated patients.

Mortality

Survival measured by actuarial methods at a mean of 9 ± 6 months showed no significant difference between the two groups. There were 12 deaths in the streptokinase-treated patients and eight deaths in the rt-PA-treated patients, $p = 0.34$. For 30-day survival there was also no significant difference with ten deaths in the streptokinase group (one patient died from a cerebral haemorrhage after receiving rt-PA for threatened reinfarction) and five deaths in the rt-PA group.

Requirement for Angioplasty or Surgical Intervention

A conservative strategy was used in this study. In the first 30 days only four patients (one with left main disease) had interventions in the streptokinase group and three patients (two with left main disease) in the rt-PA group.

Reinfarction

Despite the conservative indications for angioplasty or surgery, reinfarction was infrequent, with the incidence in both groups at 30 days being 5%.

Adverse Effects

One patient had minor melena following streptokinase treatment. Three patients had major bleeding following rt-PA treatment. Two of these died from intracranial haemorrhage and a third patient required splenectomy after the development of a subcapsular haematoma. Minor bleeding occurred in 11% of each group and minor allergy in 2% of the streptokinase-treated patients. Hypotension (a fall in the blood pressure to < 80 mmHg in the first three hours after administration of blinded therapy) occurred in 34 streptokinase patients and 24 rt-PA-treated patients, p = ns.

DISCUSSION

Clinical trials in acute coronary thrombosis can measure a number of end-points. The efficacy of an agent in opening occluded infarct-related arteries is clearly an important primary evaluation. From the patient's point of view, however, the more important evaluations are whether survival and quality of life are improved. In this study we compared two thrombolytic agents using left ventricular function as the end-point measured in patients with first infarctions. Left ventricular function relates to both survival and to quality of life (3). The findings of this study are that rt-PA and streptokinase have similar effects on the preservation of left ventricular function.

The sizing calculations for this trial showed that 216 patients would be required to be randomised. The trial at the 95% confidence limits excludes a difference between the groups of greater than a 3% benefit for each agent. A difference of less than this is unlikely to be of clinical importance. The numbers of patients required to assess differences in ejection fraction are much smaller than for mortality trials. This is because ejection fraction can be measured as a continuous variable in most patients whereas mortality may be measured in fewer than 5% of patients after thrombolytic therapy. Therefore to show a 20% reduction in mortality to 3% requires large numbers of patients; of the order of 15,000 patients (4). Clearly conclusions cannot be drawn about mortality from small left ventricular function trials. However, this trial can answer the question for which it was designed.

From the only previous comparative study, the TIMI trial (1), improvement of ejection fraction from baseline at ten days was not different; an increase in ejection fraction of 0.8% with tissue plasminogen activator vs 1.4% with streptokinase. In this trial only 50% of patients had ventriculograms and in addition intervention was delayed for up to seven hours. Few patients were randomised and had angiography within three hours; 28 patients. This study, along with our current study (5), shows that the agents are very similar in terms of preserving myocardial function. If rt-PA achieves higher patency rates than streptokinase, then the question is raised as to why this is not translating into left venticular function benefit. It is of interest that there was a nonsignificant trend for end-systolic volumes to be lower in the streptokinase-treated patients as we have shown that end-systolic volume is the most important long-term prognostic factor after recovery from myocardial infarction (3).

Despite a conservative strategy for angioplasty and surgery (3% intervention rate at 30 days) the reinfarction rate was only 5%. These rates may relate to the use of low-dose aspirin and they confirm that the "watchful waiting" strategy is appropriate after thrombolytic therapy.

CONCLUSION

Several trials have shown that both survival and left ventricular function are improved with thrombolytic therapy (5-7) and left ventricular function remains the most important prognostic factor after myocardial infarction. In three thrombolytic trials ejection fraction has been shown to be a more important prognostic factor than patency of the infarct-related artery (8-10). This study shows that streptokinase and rt-PA, when given wthin three hours of the onset of first myocardial infarction, have similar effects on preservation of left ventricular function. It remains to be seen whether there are substantial differences between these agents in terms of mortality.

REFERENCES

1. J. H. Chesebro, G. Knatterud, R. Roberts et al, Thrombolysis in myocardial infarction (TIMI) trial, phase I: a comparison between intravenous plasminogen activator and intravenous streptokinase, *Circulation,* **76**:142-54 (1987).
2. M. Verstraete, R. Bernard, M. Bory et al, Randomised trial of intravenous recombinant tissue-type plasminogen activator versus intravenous streptokinase in acute myocardial infarction, *Lancet,* **I**:842-7 (1985).
3. H. D. White, R. M. Norris, M. A. Brown, P. W. T. Brandt, R. M. L. Whitlock, and C. J. Wild, Left ventricular end-systolic volume is the major determinant of survival after recovery from myocardial infarction, *Circulation,* **76**:44-51 (1987).
4. R. M. Norris and H. D. White, Therapeutic trials in coronary thrombolysis should measure left ventricular function as the primary end-point of treatment, *Lancet,* **I**:104-6 (1988).
5. H. D. White, R. M. Norris, M. A. Brown et al, Effect of intravenous streptokinase on left ventricular function and early survival after acute myocardial infarction, *N. Engl. J. Med.,* **317**:850-5 (1987).
6. P. W. Serruys, M. L. Simoons, H. Suryapranata et al, Preservation of global and regional left ventricular function after early thrombolysis in acute myocardial infarction, *J. Am. Coll. Cardiol.,* **7**:729-42 (1986).

7. F. Van der Werf, and A. E. R. Arnold for the European cooperative study group for recombinant tissue type plasminogen activator (rt-PA), Intravenous tissue plasminogen activator and size of infarct, left ventricular function, and survival in acute myocardial infarction, *Br. Med. J.*, **297**:1374-9 (1988).
8. M. L. Simoons, P. W. Serruys, M. van den Brand et al., Improved survival after early thrombolysis in acute myocardial infarction, *Lancet,* **II**:578-81 (1985).
9. M. L. Stadius, C. Maynard, J. K. Fritz et al. Coronary anatomy and left ventricular function in the first 12 hours of acute myocardial infarction: the Western Washington randomised intracoronary streptokinase trial, *Circulation,* **72**:292-301 (1985).
10. E. J. Topol, R. M. Califf, B. S. George et al, Insights derived from the thrombolysis and angioplasty in myocardial infarction (TAMI) trials, *J. Am. Coll. Cardiol.*, **12**:24A-31A (1988).

MODERN STRATEGIES FOR TREATMENT OF ACUTE MYOCARDIAL INFARCTION: SIGNIFICANCE OF HAEMOSTASEOLOGICAL AND RHEOLOGICAL FINDINGS

G. Pindur, C. Sen*, E. Wenzel, F. Jung, C. Özbek*, H. Schwerdt*, H. Schieffer*, L. Bette*, and C. Miyashita

Abteilung für Klinische Haemostaseologie und Transfusionsmedizin, und Medizinische Universitätsklinik und Poliklinik, Innere Medizin III*, Universitätskliniken Homburg/Saar, FRG

ABSTRACT

In spite of equivalent clinical efficacy of various thrombolytic agents for the treatment of acute myocardial infarction there is evidence of different drug-depending influences on the haemostatic and fibrinolytic system. In the present study 40 patients with acute myocardial infarction have been investigated. 20 patients received 750,000 and 1.5 mio U streptokinase (SK), respectively, 10 patients 30 mg anisoylated plasminogen streptokinase activator complex (BRL 26921) and 10 patients a combination of 200,000 U urokinase (UK) and 4.5 mio U pro-urokinase (PUK) as a short-term intravenous treatment. Reperfusion of coronary arteries has been achieved in 70 to 100 percent. Major not fatal bleedings occurred in 2 patients. One patient died within 72 hours after beginning of the myocardial infarction. The longest duration of fibrinolytic activity was observed in the BRL 26921 group (half-disappearance time close to 2 h). It was significantly shorter in the SK groups showing a dose-dependency. Plasma concentration of fibrinogen dropped beyond normal levels following SK and BRL 26921, but not under UK/PUK. Plasma viscosity correlated with fibrinogen decrease and displayed a dose-depending relationship with the presence of SK. Haemorheological effects are suggested to be important for the clinical efficacy of thrombolytic therapy in myocardial infarction.

INTRODUCTION

Fibrinolytic therapy is generally accepted as a strategy for the treatment of acute myocardial infarction (AMI), since limitation of the infarct size and improvement in ventricular function is achieved by recanalization of occluded coronary arteries, thus reducing immediate mortality and improving long-term survival.

Meanwhile, there is general evidence that intracoronary fibrinolytic treatment is not feasible for widespread use because of the limited capacities of coronary laboratories (Rentrop, 1985). Moreover, the time-delay of several hours before institution of intracoronary therapy is of disadvantage for the clinical outcome in patients with evolving myocardial infarction.

Recently, it has been shown in a randomized trial, that the intravenous application of streptokinase in a dosage of 1,500,000 units during 1 hour allows to reduce acute and late mortality (GISSI, 1986, 1987), and is feasible to be carried out in about 50% of the patients suffering from AMI.

It is assumed that the injection of high doses of fibrinolytic agents leads to a rapidly established and increased fibrinolytic activity. Thus, studies were conducted where the

Fibrinogen, Thrombosis, Coagulation, and Fibrinolysis, Edited by
C. Y. Liu and S. Chien, Plenum Press, New York

effects of such regimens on the coronary reperfusion and clinical outcome in patients with AMI are compared in order to find out the disappearance rate of the fibrinolytic activities (Doenecke, 1986, Koehler, 1987; Wenzel 1987). To gain experience on the relevance of haemorheological changes during fibrinolytic therapy in some cohorts rheological as well as immunhaemostaseological methods were used. In this paper we will focus our interest especially on the validity of these findings for improving fibrinolytic therapy in AMI.

PATIENTS, MATERIAL AND METHODS

Laboratory tests and statistical analysis

Activated partial thromboplastin time (APTT), thromboplastin time (PT), thrombin time (TT), and clottable fibrinogen (Clauss method), and alpha$_2$-antiplasmin were measured using test kits from Boehringer Mannheim. Plasminogen activity was determined using a chromogenic substrate from Behringwerke Marburg. The fibrinolytic activity was measured in euglobulin precipitates using calibrated fibrin plates as described by Doenecke (1986) and Köhler (1987). Concentrations of plasminogen and of alpha$_2$-antiplasmin were measured by immunochemical methods (for details see Köhler, 1987). Thrombin-antithrombin-III-complex (TAT) was measured with a commercially available test kit by ELISA (Behringwerke Marburg) and fibrin split products (D-Dimers) by ELISA using reagents from Boehringer Mannheim.

The Mann-Whitney test was used to compare the laboratory data on the level of $p < 0.05$. The disappearance rate of fibrinolytic activities was evaluated by least square regression analysis (see Doenecke 1986 and Köhler 1987).

Haemorheological parameters such as erythrocyte aggregation (EA), and plasma viscosity (PV) were measured according to Jung (1986).

Patients and protocols (see Table 1)

Thirty patients with clinical evidence and electrocardiographic signs of AMI were included in the study. Patients with a history of bleeding disorders and with angina pectoris symptoms lasting more than 6 hrs. or with a first creatine kinase (CK)-concentration exceeding 140 units per ml were excluded.

If the diagnosis of AMI was confirmed by the clinical course, ECG-changes and CK-course, the patients were treated and monitored in the coronary care unit. Coronary angiography (CA) was not performed prior to treatment.

The recognition of reperfusion (and reocclusion) was based upon clinical, electrocardiographic and enzymatic criteria. However, verification by CA followed 1 day to 4 weeks later.

In the first group, 20 consecutive patients alternatively received either 750,000 units or 1,500,000 units streptokinase (SK) within 5 — 10 minutes (Table 1). In the second group 10 consecutive patients received 30 mg acylated SK-plasminogen complex (BRL 26921, equivalent to 1.1×10^6 U SK) intravenously within 2 minutes (Table 1).

Table 1. Overview on Clinical Data in Patients with Myocardial Infarction Treated with Various Dose Regimens of Fibrinolytic Agents: Bolus Injections of SK, BRL 26921, Bolus Injection of UK Followed by Continuous Infusion of PUK Over 40 min. Intravenously, N = 10 per Group

Group	IA	IB	II	III
Agent	SK	SK	BRL 26921	UK/PUK
Dosage (U)	750,000	1.5 mio.	30 mg	UK: 200,000 PUK: 4.5 mio.
Mean age (yrs.)	61+/−9	57+/−10	54+/−14	56+/−12
Reperfusions	8/10	8/10	10/10	7/10
Reocclusions	2/10	2/10	0/10	1/10
Major bleeding	0/10	0/10	1/10	1/10
Mortality (<72 h)	0/10	1/10	0/10	0/10

Additionally, 10 patients with acute myocardial infarction were treated with 200,000 U urokinase (UK) given as a bolus intravenously followed by the infusion of 4.5 mio U pro-urokinase (PUK) over a period of 45 min.

In the first studies (IA, IB), cortisone and heparin treatment were not standardized. Heparin therapy was started 0 to 18 hrs. after SK injection when a shortening of the APTT was observed (up to 2,500 IU per hr.). In the second study the patients received 250 mg Prednisolon and heparin 2 hrs. after bolus injection of BRL 26921 (1,000 IU per hr.). In both studies the subsequent heparin therapy was adjusted in relation to an increased APTT of 60 to 80 sec. (normal values 30 — 40 sec.) after 6 to 12 hrs.

RESULTS

Clinical outcome and side effects of thrombolytic therapy (see Table 1)

The efficacy of fibrinolytic therapy for recanalization of infarct related arteries was verified by coronary angiography, additionally laboratory markers, such as CK were studied. A higher mortality rate (2 out of 10) was observed in group IB, receiving 1,500,000 U SK.

A higher frequency of reocclusion was clinically and angiographically observed in the SK groups IA and IB and missing recanalization in 3 out of 10 patients in the urokinase and pro-urokinase group contrary to the absence of reocclusion in patients of the BRL 26921 group. The highest rate of reperfusion, as estimated from indirect or clinical parameters, was also achieved in the group treated with BRL 26921. The cohorts of both studies (IA, IB, II) did not exhibit further differences with regard to the non-invasive markers, such as frequency and duration of "reperfusion arrhythmias," rapid resolution of ST-segment elevations, and sudden relief of chest pain (Doenecke, 1986).

Disappearance rate of fibrinolytic activity and haemorheological and coagulation studies

An increase in fibrinolytic activity in the PUK group was recorded as early as 5 min after the beginning of the therapy, maximum levels were measured 1 hr. later. It was followed by a rapid decrease in activity over the next one hr. The decrease in fibrinogen concentrations was small and showed a minimum of 81 +/− 18% (mean +/− SD) compared to the preinfusion values just about 2 hrs. after beginning of therapy. Values below normal range were not observed (Fig. 1). Plasminogen activity decreased during the treatment and yielded the lowest value (73 +/− 17% of baseline) just 1 hr. after onset of therapy and returned to the initial level over the following 24 hrs. The most pronounced changes were observed in the activity of alpha-2-antiplasmin which decreased to a minum of

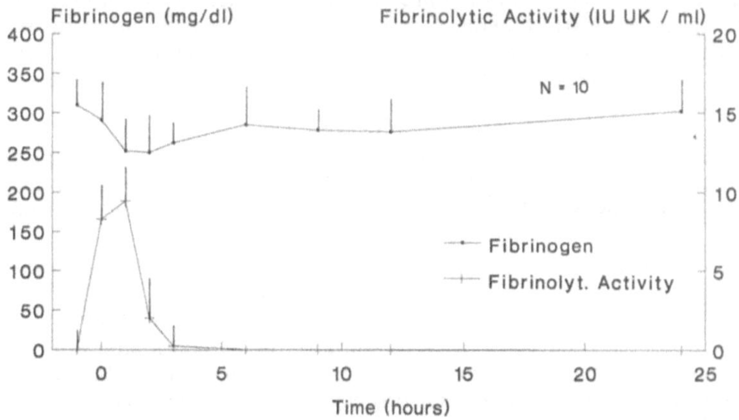

Fig. 1. PUK/UK in patients with AMI fibrinogen and fibrinolytic activity. Time course of fibrinogen and fibrinolytic activity (mean +/− SD) in the plasma of patients with acute myocardial infarction treated with 200,000 U UK and 4.5 mio U PUK (N = 10).

Table 2. Haemorheological Changes in Patients with AMI (N = 10 per Group) Undergoing Fibrinolytic Therapy with SK and UK/PUK

	erythrocyte aggregation	plasma viscosity	fibrinogen	group
Decrease* (%)	18.4	5.3	59	IA (SK)
	+/−27.2	+/−2.6	+/−40.4	
Decrease* (%)	8.0	3.9	37	III (UK/PUK)
	+/− 7.7	1.3	+/−12.8	

*comparing the values before and 1 hour after beginning of fibrinolytic therapy (mean +/− SD).

20 +/− 22% of the initial value 1 hr. after the beginning of therapy. This was followed by a slow increase, yet, the initial value failed to be reached when measured 24 hrs. later.

The kinetic of fibrinolytic activity after SK bolus injection was clearly biphasic. The half-disappearance time of the fibrinolytic activity was 31 and 15 minutes after 750,000 U SK and 1,500,000 U SK, respectively. After ultra-short infusion of BRL 26921 the fibrinolytic activity disappeared more slowly, and the biphasic decrease was less pronounced. The half-disappearance time for BRL 26921 was 112.5 min as evaluated by a double exponential mathematical model.

TAT and fibrin split products (D-Dimers) increase significantly after onset of fibrinolytic activity as studied in patients receiving UK and PUK. In the patient cohort with early increase in fibrinolytic activity and early drop in fibrinogen (SK, BRL 26921, UK/PUK) we could observe a relation between plasma viscosity and behaviour of fibrinogen (Table 2). In the UK/PUK group we observed a correlation between decreased erythrocyte aggregation and D-Dimers. These findings could be predictive for the risk of reocclusion despite the fact of high interindividual variability of the haemorheological findings.

DISCUSSION

Clinical outcome, validity of haemostaseological and haemorheological findings

Coronary artery reperfusion was achieved in 70 — 100 percent of the patients with AMI studied in the present trial using SK, acylated plasminogen-SK complex (BRL 26921) or UK/PUK respectively. These findings are in accordance with the great majority of studies of fibrinolytic agents in AMI showing successful vascular recanalization in more than 50 percent (Marder, 1988). Although highest efficacy was recognized in the patients who were given BRL 26921, the size of the groups studied is too small for clear evaluation of the different thrombolytic agents of this trial.

In this study haemorrhagic complications were generally confined to haematoma at the site of the vessel puncture. Local measures were sufficient to control these problems. Only 1 out of 30 patients receiving SK or BRL 26921 suffered from major bleeding complications and needed blood replacement therapy. This patient had taken aspirin prior to admission.

Treatment by intravenous administration of fibrinolytic agents has been studied in many randomized prospective trials in the past. In these studies, various dose regimens were used. However, the recent trend in systemic thrombolytic therapy has been to use very high doses (500,000 — 1,500,000 U) of intravenous streptokinase administered over 30 to 60 minutes. This causes a significant decrease in clottable fibrinogen as well as a marked increase in fibrinogen degradation products. It should be taken into consideration that systemic fibrinogenolysis has also a favourable influence on microcirculation and on the reduction of myocardial necrosis (Jan. 1975; Wenzel, 1983).

There exist a number of experimental studies in animals indicating that fibrinogenolysis as well as haemodilution prevent damage and necrosis of tissue in case of experimentally induced hypoxemia.

On the other hand, haemorheological parameters are proven to change in relation to the increasing hypoxemia in patients suffering from coronary heart disease and to predict cerebral vascular complications as well as thromboembolic complications in patients suffering from arteriosclerosis and AMI (Schmidt-Schönbein, 1977; Theiss, 1980, Dodds, 1980; Jan. K.-M., 1975). The increase in plasma vascosity seems to be related to the increase in high molecular plasma proteins, e.g. fibrinogen and acute phase proteins. The decreased flow properties of blood seem to be mainly connected to the increase in haematocrit, especially in patients suffering from hyperlipidemia and hypertension.

In 1975 Wende elaborated an animal model (canines) to study the fibrinogenolytic effect of high dose streptokinase infusion. In this study damage to the myocardial tissue induced by experimentally coronary artery ligature and monitored by epicardial ECG could be diminished significantly by administration SK prior to vessel occlusion.

This finding suggest that fibrinogenolysis actually prevents damage to tissue following arterial occlusion even if it does not significantly influence the macrocirculation, and even if thrombolysis does not occur.

Disappearance rate of fibrinolytic activities

Systemic fibrinogenolysis was observed both after SK and BRL 26921. The differences between these subgroups were small and not significant. A comparatively higher drop of fibrinogen was observed in the BRL 26921 group. The depletion of alpha$_2$-antiplasmin activity lasted longer after BRL 26921 than after 1,500,000 U SK.

The initial rapid decrease of the fibrinolytic activity with half-disappearance times of 15 and 31 minutes is in agreement with the findings of Auel (1975) and Martin (1983). These authors reported a half-disappearance time of 18 and 30 minutes after SK infusion of 100,000 and 1.5 million U/per hr. respectively. In contrast, the half-life of fibrinolytic activity after BRL 26921 was found to be 70 minutes, when the first 165 minutes after administration were investigated (Staniforth, 1983).

No relevant differences in the decrease in plasminogen were found in the cohorts of our studies when streptokinase was used. Marked changes observed after 30 mg BRL 26921 are in agreement with the findings of Hoffmann et al. (1985) and Matsuo et al. (1981).

In the UK/PUK group fibrinolytic activity increased as early as 5 min. after bolus injection of urokinase, however, declined rapidly over the next one hr. The decrease in fibrinogen and plasminogen activity was small in comparison to the other agents. A slight but significant fibrinogenolysis could be observed following therapy. The clinical outcome and reperfusion rate in the patient cohort treated with PUK and UK is comparable with SK and BRL 26921 which exhibit high fibrinolytic activities and significant decrease of fibrinogen. This seems to be in accordance with the hypothesis on the protective effect of the favorable influence of systemic fibrinogenolysis on microcirculation and on the reduction of myocardial infarction size.

In conclusion the efficacy of high dose infusion (bolus injection) of fibrinolytic agents is satisfying with regard to the high coronary reopening rate and minor adverse effects, when administered during an early stage of myocardial infarction.

It must be pointed out that the fibrin selective thrombolytic agents seem not to be superior to high-dose SK, BRL 26921 or PUK combined with UK. It should be taken into consideration that a systemic fibrinogenolysis appears to have a favorable influence on microcirculation and on the reduction of myocardial necrosis. Additionally, it may help to prevent reocclusion of the coronary arteries.

REFERENCES

Dodds, A. J., Boyd, M. J., Allen, J., Bennet, E. D., Flute, P. T., Dormandy, J. A., 1980, Changes in red cell deformability and other haemorheological variables after myocardial infarction. *Br. Heart. J.,* **44:**508-11.

Doenecke, P., Schwerdt, H., Hellstern, P., Wenzel, E., Bette L., 1986, Die Bolusinjektion von anisoyliertem Plasminogen-Streptokinase-Aktivator-Komplex (BRL 26921) als alternatives Konzept der systemischen Lyse bei akutem Myokardinafarkt. *Klin. Wschr.,* **64:**682-687.

Fletcher, A. P., Alkjaersig, N., Sherry S., The clearance of heterologous protein from the circulation of normal and immunized man. *J. Clin. Invest.,* **37:**1306-1315.

GISSI Study group, 1986. Effectiveness of intravenous thrombolytic treatment in acute myocardial infarction. *Lancet i,* 397-401.

Hoffmann, J. J. M. L., van Rey, F. J. W., Bonnier, J. J. R. M., 1985, Systematic effects of BRL 26921 during thrombolytic treatment of acute myocardial infarction. *Thromb. Res.,* **37**:567-572.

Jan, K.-M., Chien, S., Bigger, J. T., 1975, Observations on blood viscosity changes after acute myocardial infarction. *Circulation,* **51**:1079-1084.

Jung, F., Kiesewetter, H., Roggenkamp, H. G., Ringelstein, E. B., Gerhards, M., Kotitschke, G., Wenzel, E., Zeller H., 1986, Bestimmung der Referenzbereiche rheologischer Parameter: Studie an 653 zufällig ausgewählten Probanden im Kreis Aachen. *Klin. Wochenschr.,* **64**:375-381.

Köhler, M., Hellstern, P., Doenecke, P., Schwerdt, H., Özbek, C., Miyashita, C., Winter, R., von Blohn, G., Bette, L., Wenzel E., 1987, High-dose systemic streptokinase and acylated streptokinase-plasminogen complex (BRL 26921) in acute myocardial infarction: Alterations of the fibrinolytic system and clearance of fibrinolytic activity. In: *Haemostasis,* **17**:32-39.

Marder, V. J., and Sherry, S, 1988, Thrombolytic therapy: current status. *New Engl. J. Med.,* **318**:1512-1520.

Matsuo, O., Collen, D., Verstraete, M., 1983, On the fibrinolytic and thrombolytic properties of active-site p-anisoylated streptokinase-plasminogen complex (BRL 26921). *Thromb. Res.,* **24**:347-358.

Pindur, G., Sen, S., Köhler, M., Miyashita, C., Heiden, M., Dyckmans, J., Schieffer, H., Wenzel. E., 1988, Parameters of fibrinolysis during combination therapy of acute myocardial infarction with pro-urokinase/urokinase. *Haemostasis,* **18**:191-192.

Rentrop, K. P., Blanke, H., Karsch, K. R., Wiegend, V., Köstering, H., Rahlf, G., Oster, H., Leitz, K., 1979, Wiedereröffnung des Infarktgefäßes durch transluminale Rekanalisation und intrakoronare Streptokinase-Applikation. *Dtsch. Med. Wschr.,* **104**:1438-1440.

Rentrop, K. P., 1985, Thrombolytic therapy in patients with acute myocardial infarction. *Circulation,* **71**:627-631.

Schmid-Schönbein, H., Rieger, H., Hess H., 1977, Quantitative Erfassung der Effekte fibrinolytischer Therapie auf das Fließverhalten des Blutes. *Klin. Wschr.,* **55**:111-119.

Staniforth, D. H., Smith, R. A. G., Hibbs, M., 1983, Streptokinase and anisoylated streptokinase plasminogen complex. Their action on haemostasis in human volunteers. *Eur. J. Clin. Pharmacol.,* **24**:751-756.

Theiss, W., Volger, E., Wirtzfeld, A., Keisel, I., Blömer H., 1980, Gerinnungsbefunde und rheologische Messungen bei Streptokinasebehandlung des akuten Myokardinfarktes. *Klin. Wochenschr.,* **58**:607-615.

Wende, W., Stühlen, H. W., Meyer, J., Bleifeld, W., Holzhüter, H., Wenzel E., 1975, Größe des akuten tierexperimentellen Herzinfarktes unter Streptokinase-indizierter Fibrinolyse. *Klin. Wschr.,* **53**:755-760.

Wenzel, E., Kiesewetter, H., Hellstern, P., Radtke, H., Jung, F., Köhler, M., von Blohn G., 1983, Hämorheologische Wirkungen der Fibrinolysetherapie. *In:* Ehrly, A. M., (Ed.), Therapie mit hämorheologisch wirksamen Substanzen, Verhandlungsbericht 2, Kongreß der DGKH, W. Zuckschwerdt Verlag, München, Bern, Wien, pp. 58-68.

Wenzel, E., Özbek, C., Schwerdt, H., Dyckmans, J., Blohn, G. V., Miyashita, C., Doenecke, P., Bette L., 1987, Thrombolytic therapy of acute myocardial infarction by intravenous bolus injection of fibrinolytic agents: a critical review of published findings based on new observations, *in:* Castellion et al. (eds.), Fundamental and clinical fibrinolysis, Elsevier Science Publishers B. V., New York, Amsterdam, London, pp. 205-214.

CLINICAL APPLICATION OF TISSUE PLASMINOGEN ACTIVATOR THERAPY IN ACUTE MYOCARDIAL INFARCTION

Shih-Pu Wang, Mau-Song Chang and Benjamin N. Chiang

Division of Cardiology
Veterans General Hospital and Yang-Ming Medical College
Taipei, Taiwan, ROC

INTRODUCTION

Acute myocardial infarction is related to plaque rupture and thrombosis formation in the infarct-related coronary artery. Salvage of ischemic area with thrombolysis in the early stage of infarction has proved to be the best way to limit infarct size, preserve left ventricular function and reduce mortality in the majority of victims. Tissue plasminogen activator (tPA) is the most promising thrombolytic agent used to salvage ischemic myocardium; the efficacy has been systematically demonstrated in previous studies (1-4).

The purpose of this open label trial is to examine the different dosage of tPA in the treatment of acute myocardial infarction. The efficacy of treatment was monitored by serial serum CK-MB changes in the first 24 hours, by electrocardiographic changes, and by cardiac catheterization three hours after tPA infusion. Safety was monitored by serial examination of coagulation factors and hematologic changes.

PATIENTS AND METHODS

This prospective study was conducted from March 1st 1987 to August 30th 1988. Patients of acute myocardial infarction, age between 21 and 70 years old, who came to the emergency room were included in the study. Further inclusion criteria were severe chest pain lasting more than 15 minutes but less than 5 hours, elevation of the ST segment on ECG by more than 0.2 mV in at least two leads, and unresponsive to nitroglycerin.

The exclusion criteria were cardiogenic shock, previous myocardial infarction or coronary bypass graft surgery, bleeding tendency, valvular disease or malignancy. Patients with any one of the above mentioned conditions were excluded from the study.

Once the diagnosis of acute myocardial infarction was confirmed, patients who gave informed consent to join the study and had subsequent catheterization, received intravenous injection of heparin 5,000 IU followed by tPA (Actilyse). Two different regimes of dosage were used: Group A received a total tPA dose of 100 mg in three hours; Group B 70 mg in three hours. Actilyse is a product of Genetech Inc, supplied by Boehringer Ingelheim International GmbH.

In group A patients, the first 10 mg was given by a bolus injection in the emergency room, followed with 50 mg infusion in one hour, then 20 mg per hour for two hours. In Group B patients, the first 10 mg bolus was the same, but the first-hour dose was reduced to 40 mg, and then 10 mg per hour for two hours. All patients were then observed in the coronary care unit. Coronary arteriography was performed, whether day or night, at the third hour following tPA infusion.

Hematological and Coagulational Assessments: On admission, the following parameters were assessed for each individual patient: Blood group and cross-matching; full blood count including platelets; fibrinogen and fibrinogen degradation products (FDP); FDP D-dimer which is specific for fibrin degradation products (FbDP); prothrombin time (PT); activated partial thromboplastin time (APTT); reptilase time (RT); α_2-antiplasmin (AAP); euglobulin lysis time (ELT); and antithrombin III (AT III). These examinations were performed at 1, 2, 4, 6, 12, 24, 48 hours after tPA infusion in order to evaluate the safety of thrombolytic therapy. The difference between groups A and B were compared for possible significance.

Cardiac Enzymes: The serum creatine phosphokinase (CK) and iso-enzyme MB were also examined before admission and after tPA infusion at 1, 2, 3, 4, 8, 12, 16, 24, 36, 48, 72 hours to assess the maximal value and peaking time after treatment.

Clinical Evaluation: Plain chest X-Ray film was taken before admission to the coronary care unit, and thereafter daily three times. 12 leads electrocardiogram was done on admission and every four hours for the first day; then every day for one week.

Blood pressure, pulse rate, respiration rate and chest pain were recorded hourly for the first 24-36 hours and thereafter as needed according to the patient's condition.

Killip classification, clinical complications or any untoward side effects observed during and after infusion were recorded in the coronary care units daily and during the hospitalization.

Coronary Arteriography and Hemodynamics: After the patients had been started on tPA therapy, the catheterization was prepared for three hours later, approached by femoral puncture with Judkin's method; various parameters were recorded, including arterial pressure, left ventricular end-diastolic pressure (LVEDP), and left ventriculogram. Coronary arteriography was done with multiple projections, including right anterior oblique (RAO) 30 degrees and left anterior oblique view (LAO) 60 degrees, to demonstrate the stenosis of coronary tree.

RESULTS

For the 18-month period from March 1987 to August 1988, 51 consecutive patients with acute myocardial infarction were studied. There were 29 cases with anteroseptal myocardial infarction and 22 cases with inferior myocardial infarction. There were 50 males and one female, with an average age of 60.4 ± 7.8 years old. 31 cases were in group A, and 20 cases in group B. The body weights were 66.53 ± 9.36 Kg in group A, 67.83 ± 9.42 Kg in group B; there was no significant statistical difference. Patients in group A received 100 mg tPA, with an average dosage of 1.54 ± 0.24 mg/Kg; group B patients received 70 mg, with an average dosage of 1.05 ± 0.14 mg/Kg; the dosages in two groups were significantly different ($p<0.05$).

TPA was administrated for an average of 3.55 ± 1.0 hours, one case in group A expired from cardiogenic shock within two hours during the therapy; no catheterization was performed, his necropsy revealed three heavily atherosclerotic but patent coronary arteries associated with apical hemorrhagic infarction without rupture. Two cases in group B expired before catheterization and had no autopsy. Inferior myocardial infarction was found with significantly lower systolic and diastolic blood pressures and heart rate than anteroseptal myocardial infarction.

Chest pain was divided into four grades according to the severity as in the Canadian Heart Association classification (5). The patients arrived at the emergency room, early or late, were associated with severe nausea and cold sweating in the majority. The Killip's classification of pulmonary basal rales and hypotension were recorded as: 78% in class 1, 20% in class 2, 2% in class 3, no patients in class 4. The was no significant difference between the two groups.

Coagulation parameters revealed decreases of plasminogen, fibrinogen, AAP and Euglobulin lysis time from one hour after therapy, with the maximal decrease found between 1-4 hours. FDP, FDP Dimer, RT, PT, APTT were increased. The decreases of plasminogen, fibrinogen and AAP in the first six hours were significantly greater in group A patients; but not Euglobulin lysis time. The increases of FDP and RT were significantly higher in group A patients between 4-12 hours; but not FDP D-dimer and APTT. All the changes related to thrombolysis recovered in 48 hours after tPA infusion. The increases of both FDP D-dimer and FDP after infusion indicate that tPA can induce not only local clot thrombolysis, but also has a systemic fibrinolytic action.

The hemorrhagic complications from the thrombolytic therapy were relatively few in the 51 patients, most of whom had had catheterization and the average blood samples drawn were more than 200 ml each during the therapeutic course. Two cases had local hematoma at the artery puncture site for catheterization. One case, with gastrointestinal bleeding, required transfusion of two units of packed red cells. Another case was found to have anemia after thrombolytic therapy, and had one unit of packed cells without any apparent bleeding site. The average red blood cells, hemoglobin and platelets count were significantly decreased, but not hematocrit. Some degree of hemoconcentration might be present during first 24 hours.

There were 48 cases who had catheterization, 30 in group A and 18 in group B. All the patients received catheterization about three hours after starting tPA therapy. 19 cases had single vessel disease, 17 had double vessel disease, 12 had triple vessel disease, and two cases had left main disease. The infarct-related vessels involved left anterior descending artery in 26 cases, right coronary artery in 18 cases, and left circumflex artery in 4 cases only. The patency rate of infarct-related vessels was 90% in group A and 72% in group B; there was no significant statistical difference. Good antegrade flow of grade 2 or 3 was gained in 26/30 or 86.7% in Group A and 13/18 or 72% in group B; there was no significant difference between two groups because of the rather small number of cases. Patients in group B, however, had significantly less grade 3 antegrade flow and a higher percentage of severe stenosis, which might be an indirect implication of inadequate thrombolysis.

Left ventricular ejection fraction was well preserved in a majority of the patients; only 19/48 cases or 39.6% fell below 50%. The average ejection fraction was as high as 51.4 ± 13.1% on the left ventriculogram, which was slightly higher in group B patients; and there was no significant difference between inferior or anteroseptal myocardial infarction. The degree of antegrade flow showed no significant influence on immediate left ventriculogram. Regarding the segmental motion of left ventricle, 3 cases were normal, 21 cases had hypokinesia, 26 cases had akinesia, including two cases who had both hypokinesia and akinesia.

The serum creatine kinase (CK) and its iso-enzyme MB showed rapid elevation after admission. The time of peaking was at 6 hours in almost all cases. Cases with antegrade grade 3 flow had significantly less CK and MB release than the other three groups. The CK and MB increased more in the anteroseptal than inferior infarction. The peaking time was slightly earlier in the anteroseptal infarction.

Arrhythmias, including short runs of ventricular tachycardia, accelerated idioventricular tachycardia, and ventricular premature beats, were common manifestations in both inferior and anteroseptal infarction. These arrhythmias were usually self-limiting, requiring no more than xylocaine therapy.

DISCUSSION

Intravenous thrombolytic therapy is now a well accepted therapeutic procedure for acute myocardial infarction. The majority of the infarct-related vessels in the early hours of infarction are totally occluded by newly developed clots superimposed on the stenotic segment. Ischemic injury and necrosis rapidly develop in the non-perfused myocardium beyond the occluded coronary artery. Shock, acute heart failure, recurrent ventricular arrhythmia and even sudden death could precipitate unless the occluded artery was opened. Intravenous administration of thrombolytic agent in the early stage of infarction provides a hope to open the occluded artery and interrupt the subsequently adverse effect of occlusion, thus giving a chance of better recovery.

In comparison to the effectiveness of thrombolysis from streptokinase and tPA, it has become well known that the infarct-related coronary artery, studied within the early hours of acute myocardial infarction, is totally occluded in 87% of the cases (6). Intravenous thrombolytic therapy has given a favorable response in many studies the world over, using streptokinase which remains a major thrombolytic agent. Streptokinase, in a comparative study with tPA, was effective for opening infarct-related vessels in 55-57% of acutely infarcted patients, but the performance of tPA was even better. This has been proved in vast majority of patients described in the literature (7-8). In the present study, tPA opened 72-90% of the infarct-related vessels when examined 180 minutes after infusion; good antegrade flow could be obtained in the majority of these patients. Previous coronary arteriograms have usually been examined at 90 minutes after tPA infusion. Delaying the coronary arteriogram to 180 minutes after infusion should improve the patency rate of infarct-related vessels, as shown in a recent report (10). The good patency rate reported here is compatible with the findings of previous comparative studies. Early coronary arteriography at 90 minutes is not necessary. Delay of the procedure to 180 minutes is not only safe, but also generates better results.

The coronary patency rate obtained this way represents the combination of spontaneous reperfusion and thrombolytic effect of tPA. It may not completely represent the therapeutic efficacy of tPA. However, its effect to the patient is the same. Coronary arteriography is essential in assessing patency rate in the study, but it may be considered not so necessary in clinical practice. When the procedure is done about 180 minutes after therapy, it may not reflect the patency of the result at 90 minutes, or the perfusional status of early reperfusion. However, our study shows those cases with TIMI grade 3 antegrade flow at 180 minutes had significant less CK and MB release, as compared with those with poor antegrade flow. The meaning is clear that those with better antegrade coronary flow had less myocardial damage, and the patency of coronary artery at 180 minutes does reflect that the myocardium has been salvaged. The therapeutic efficacy of tPA in acute myocardial infarction and the result of late coronary arteriography is still valid.

The effect of intravenous tPA on thrombolysis has a linear correlation with the infusion rate and dosage. The patency of infarct-related artery was is 70-75% with tPA given at 0.75 mg/Kg over 90 minutes. In our study, tPA given at the dosage 1.0 mg/Kg (total 70 mg) the patency rate is 72% in three hours; which is similar with previous reports (11). When the dosage is increased to 1.54 mg/Kg, the patency rate is up to 90%, the antegrade flow is better and stenosis is less. This evidence supports a higher adequate dosage for patients in need.

This series of study included more anteroseptal than inferior myocardial infarction. The infarct-related vessels after thrombolytic therapy all showed significant residual stenotic lesions. The distribution of vessels involved in myocardial infarction showed that the most commonly involved artery was the left anterior descending artery which is 54%; followed by the right coronary artery in the second place, 37.5%; left circumflex artery was the least involved coronary artery of the three, only 8%. The frequencies among single-vessel, double-vessel and triple-vessel diseases were almost the same, which is compatible with previous reports. The thrombolytic therapy in this study saved left ventricular function in more than half of the infarction patients, the ejection fraction was more than 50% in 60% of the cases. However, the left ventriculogram did not show any significant difference of ejection fraction in those with better antegrade flow. Some other studies showed local wall motion abnormalities did get better after a successful thrombolytic therapy (12).

This clinical trial has proved that tissue plasminogen activator is therapeutically effective in opening infarct-related vessels, with relatively few bleeding complication. The patency rate showed no significant difference with dosage range from 70 mg to 100 mg. However, the higher dose of tPA (100 mg) is better than the lower dosage regime (70 mg). The thrombolytic therapy could provide emergency management in the early course of acute myocardial infarction. It is recommended to do coronary arteriography at 180 minutes rather than 90 minutes after tPA infusion; the patency rate at 180 minutes is better than previously reported.

REFERENCES

1. S. R. Bergmann, K. A. A. Fox, M. M. Ter-Pogossian, and D. Sobel. Clot-selective coronary thrombolysis with tissue-type plasminogen activator. *Science* **220**:1181-3 (1983).
2. B. Sobel, Early intervention in acute myocardial infarction: One center's perspective, *Am. J. Cardiol.,* **69**:983-990 (1984).
3. B. E. Sobel, R. W. Gross, and A. K. Robinson, Thrombolysis, clot selectivity and kinetics, *Circulation,* **70**:160-164 (1984).
4. A. J. Tiefenbrunn, A. K. Robinson, P. B. Kurnik, P. A. Ludbrook and B. E. Sobel, Clinical pharmacology in patients with evolving myocardial infarction of tissue-type plasminogen activator produced by recombinant DNA technology, *Circulation,* **71**:110-116 (1985).
5. L. Campeau, Grading of angina pectoris, *Circulation,* **54**:522 (1976).

6. M. A. DeWood, J. Spores, R. Notske, L. T. Mouser, R. Burroughs, M. S. Golden, and H. T. Lang, Prevalence of total coronary occlusion during the early hours of transluminal myocardial infarction, *N. Engl. J. Med.,* **303**:897 (1980).
7. The TIMI study group, The thrombolysis in myocardial infarction (TIMI) trial, *N. Engl. J. Med.,* **312**:923-6 (1985).
8. M. Verstraete, M. Bory, and D. Collen, et al., Randomized trial of intravenous recombinant tissue-typed plasminogen activator versus intravenous streptokinase in acute myocardial infarction, *Lancet,* **2**:965-969 (1985).
9. J. H. Chesebro, G. Knatterud, and R. Roberts, et al., Thrombolysis in myocardial infarction (TIMI) trial, phase I: a comparison between intravenous tissue plasminogen activator and intravenous streptokinase, clinical findings through hospital discharge, *Circulation,* **76**:142-154 (1987).
10. A. K. Rao, C. Pratt, and A. Berke, et al., Thrombolysis in myocardial infarction (TIMI) trial- Phase I: Hemorrhagic manifestation and changes in plasma fibrinogen and the fibrinolytic system in patients treated with recombinant tissue plasminogen activator and streptokinase, *J. Am. Coll. Cardiol.,* **11**:1-11 (1988).
11. D. O. Williams, J. Borer, E. Braunwald, et al. Intravenous recombinant tissue-typed plasminogen activator in patients with acute myocardial infarction: A report from the NHLBI thrombolysis in myocardial infarction trial. *Circulation* **73**:338-346 (1986).
12. F. H. Sheehan, D. F. Mathey, J. Schofer, et al., Limitations in the interpretation of rest-exercise ejection fraction changes after early thrombolytic therapy during acute myocardial infarction. *Am. J. Cardiol.,* **61**:743-748 (1988).

FIBRINOGEN PROTEOLYSIS AND COAGULATION SYSTEM ACTIVATION DURING THROMBOLYTIC THERAPY

John Owen, Betty Grossman, Joan Sobel, and Bohdan Kudryk

The Bowman Gray School of Medicine, Wake Forest University
Winston-Salem, NC, The College of Physicians and Surgeons
Columbia University and the Lindsley E. Kimball
Research Institute of the New York Blood Center
New York, NY, USA

INTRODUCTION

Thrombolytic treatment of patients with acute myocardial infarction is now well accepted as the treatment of choice in most settings (24). At this time most data has been accumulated with the use of streptokinase, and this drug has been convincingly demonstrated to reduce mortality (5). Reperfusion is achieved in 45- 75% of patients, with direct intracoronary administration being slightly more effective than administration by the intravenous route (25). Some problems remain, bleeding is significant and rethrombosis continues to be an obstacle (7).

The problem with bleeding was seen to be dose related, with most bleeding being associated with evidence of systemic fibrinogenolysis (10, 20). Tissue plasminogen activator held the promise of active thrombolysis with little or no fibrinogen proteolysis (1). This selectivity, together with the promise of unlimited supply thanks to biotechnology (15), naturally lead to a clinical trials (2, 4, 20-23). In the United States, one study was the TIMI trial (Thrombolysis In Myocardial Infarction) (19). This study was a double blind prospective test of streptokinase versus tissue plasminogen activator. The data reported here were obtained from an ancillary study carried out on patients at Columbia Presbyterian Hospital who were enrolled in the TIMI trial (13, 14).

Streptokinase was to be administered as a continuous intravenous infusion at a dose of 1,500,000 units over one hour. The dose of t-PA used was 80 mg; 40 mg in the first hour, followed by 20 mg/hr for two hours by continuous intravenous infusion. At this dose of t-PA, we calculated that there was a risk of significant fibrinogen proteolysis, and hence of bleeding complications. Prediction of risk is difficult, neither fibrinogen depletion nor total degradation product concentration are good indices (10). Following the lead of Marder (10), we postulated that the circulating concentration of fragments X could have a direct effect on the tendency to bleed. Early fragment X is clottable, and can incorporate into fibrin clots (18). These fragment X containing clots have reduced tensile strength and increased susceptibility to further plasmic degradation (11, 16). In order to further examine this postulate we measured the plasma levels of fibrinopeptide Bβ1-42 (17). This peptide is cleaved from the amino terminus of the Bβ chain of fibrinogen or B fibrin (Fibrin I) by plasmin. The concentration of fibrinopeptide Bβ1-42 thus serves as an index of in vivo fragment X formation (12).

Fig. 1 shows the mean plasma concentrations of fibrinopeptide Bβ1-42 as a function of time for the two groups of patients. The patterns are markedly different, streptokinase induces an immediate, massive release of fibrinopeptide Bβ1-42. In contrast, t-PA shows a distinct lag phase before significant effect. In order to translate these fibrinopeptide Bβ1-42 concentration data into amount of fragment X formed we used standard pharmacokinetics.

Fibrinogen, Thrombosis, Coagulation, and Fibrinolysis, Edited by
C. Y. Liu and S. Chien, Plenum Press, New York

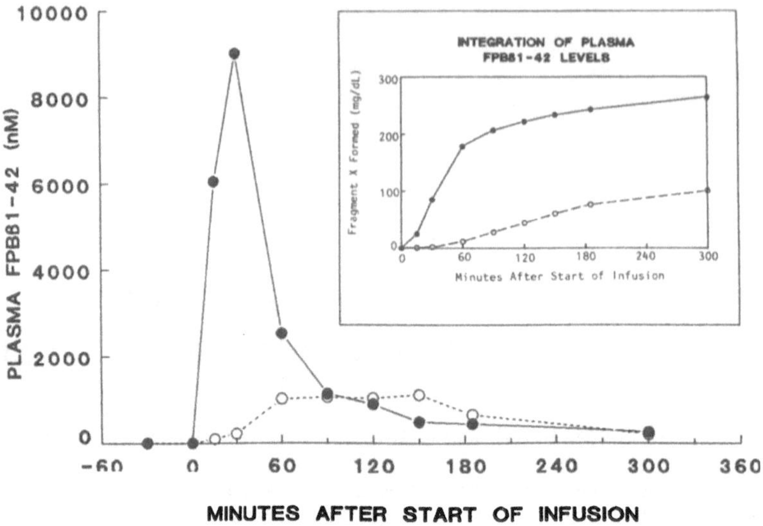

Fig. 1. Plasma levels of fibrinopeptide Bβ1-42 before, and for 5 hours after the start of thrombolytic therapy. The insert shows the cumulated amount of fibrinopeptide Bβ1-42 released, expressed as the amount of fibrinogen needed to release this amount of peptide. Data have been grouped at each time point for the two treatments. Solid symbols and continuous line refer to streptokinase treated patients, open symbols and broken line refer to rt-PA treated patients.

The relationship is given by:

$$\text{Cumulative proteolysis} = FCR_{fp} \int_{t=0}^{t=t_{max}} [fp]t\, dt$$

FCR = Fractional Catabolic Rate = $\ln(2)/t_{1/2}$
[fp] = fibrinopeptide concentration

The results obtained are shown graphically in the insert in Fig. 1, the units have been converted to mg/dL to facilitate comparison with fibrinogen concentrations. The infusion of streptokinase induces immediate formation of fragment X which continues to be rapid during the period of drug infusion. At the end of the infusion plasmin action continues at an accelerated rate for the duration of the study. The data for patients treated with t-PA show a distinct lag, then steady fibrinogen proteolysis for the duration of drug infusion. Similar to streptokinase, t-PA induces a prolonged phase of plasmin action following the cessation of drug infusion. At the end of the 5 hour study period, the cumulated fibrinogen proteolysis was some 250 mg/dL for streptokinase and 100 mg/dL for t-PA. These values are very similar to the maximum measured values for serum FDP, 230 mg/dL for streptokinase and 120 mg/dL for t-PA.

The circulating concentration of fragment X reflects the balance between formation and removal. In this case removal would likely be due to further proteolysis by plasmin with formation of fragments Y, D and E. In order to assess this we carried out immunoblotting studies. The samples were subjected to PAGE in 7.5% gels, electrophoretically transferred to nitrocellulose and probed with rabbit anti-human fibrinogen. This antibody reacts strongly with the D domain of fibrinogen and its degradation products, and contains antibody against the carboxy terminal regions of the Aα chain. Representative immunoblots are shown in Fig. 2 for streptokinase treatment, and in Fig. 3 for t-PA treatment.

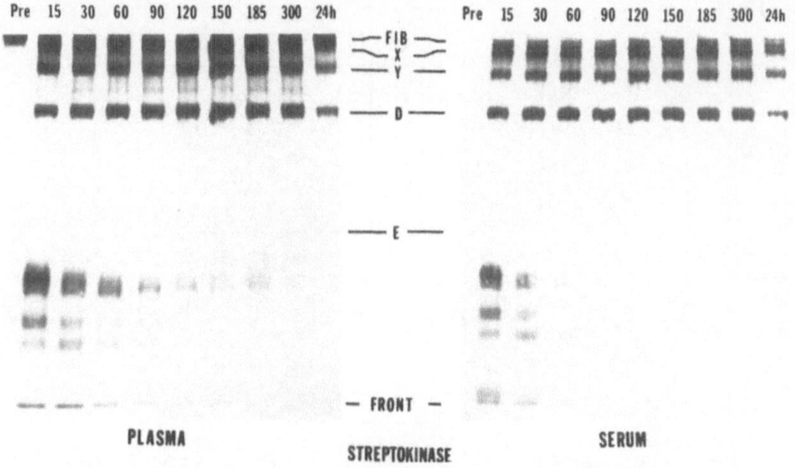

Fig. 2. Immunoblots after 7% PAGE of fibrinogen antigen in plasma and serum of a patient receiving streptokinase. The time in minutes after starting treatment is shown at the top of each lane. This figure has been reproduced from the Journal of Clinical Investigation 79:1642-1649, 1987 by copyright permission of the American Society for Clinical Investigation.

It is easily seen that there is considerable fibrinogen proteolysis with either drug, though streptokinase is faster and leads to more extensive degradation. Fragment E is not visualized using this antibody, its expected location is shown for orientation. The low molecular size bands seen below the position of fragment E are derived from the Aα chain appendage, and are particularly prominent in the t-PA treated patient. Note also that this patient shows a distinct D-dimer band. With regard to the concentration of fragment X it is impossible to give firm figures. Streptokinase induces more fragment X formation, but progressive proteolysis reduces the concentration. On the other hand t-PA induces less fragment X formation, but is less efficient and causing further degradation. These factors combine to

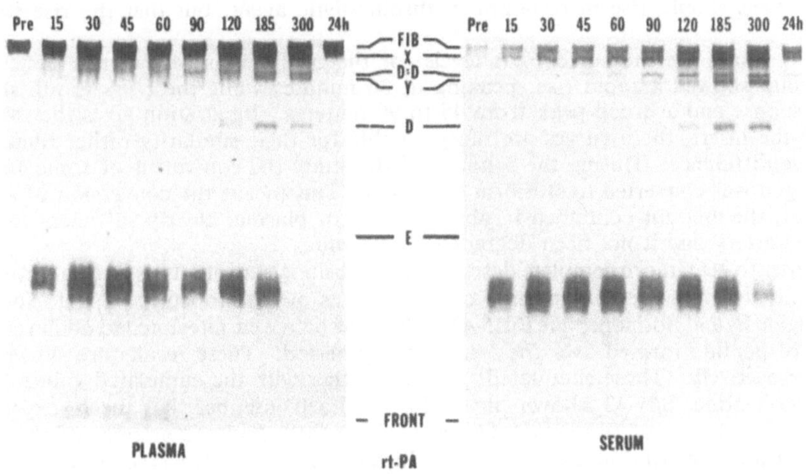

Fig. 3. Immunoblots after 7% PAGE of fibrinogen antigen in plasma and serum of a patient receiving rt-PA. The time in minutes after starting treatment is shown at the top of each lane. This figure has been reproduced from the Journal of Clinical Investigation 79:1642-1649, 1987 by copyright permission of the American Society for Clinical Investigation.

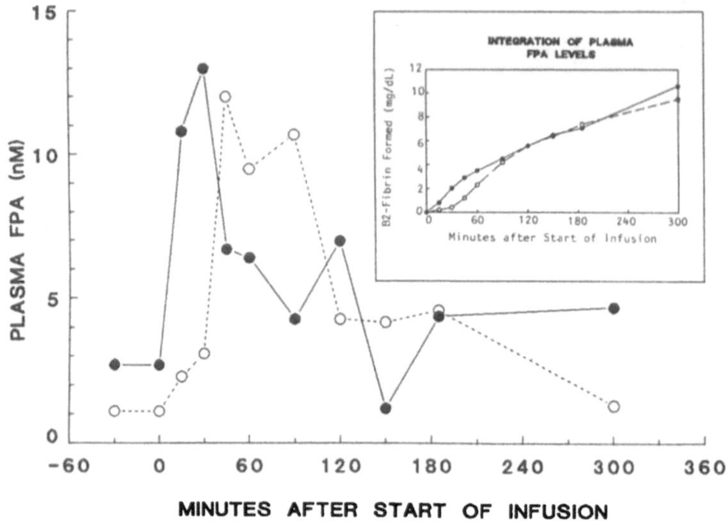

Fig. 4. Plasma levels of fibrinopeptide A. Other details are as in Fig. 1.

make the plasma fragment X concentrations more alike for the two treatments. This similarity is in keeping with the observed similarity in the frequency of bleeding complications (19), and lends support to the postulate that fragment X is a significant factor in the bleeding diathesis.

Turning to the problem of rethrombosis, we can distinguish three time periods of concern, immediate, short term and long term. It is the first of these that is of concern here. Immediate rethrombosis is a significant problem in thrombolysis, the thrombus frequently reappearing immediately after the cessation of drug infusion (6). We interpret this as evidence for procoagulant activity balanced only by pharmacologic plasmin activity. This view is supported by the observation that, for t- PA, the problem is minimized by prolonged infusion of drug (6). Further support for excess thrombin activity can be drawn from the partial protection offered by heparin administration. Thrombus formation requires fibrin formation, and fibrin formation occurs with the release of fpA by thrombin. We postulated that fpA levels should rise in response to thrombolytic agent, but that the rise should be transient.

Fig. 4 shows the measured FPA levels for the two groups of patients. The streptokinase group showed a rapid rise, peaking at 30 minutes while the. t-PA group showed a slower response and a broad peak from 45 to 90 minutes. Integration gives the two curves shown in the insert, these curves are most notable for their similarity rather than for any substantive difference. During the 5 hours of the study the equivalent of some 10 mg/dL of fibrinogen was converted to B Fibrin (Fibrin I). This means the conversion of some 250 mg/patient, the amount contained in about 70 mL of plasma; clearly sufficient to occlude a coronary artery had it not been degraded by plasmin.

In order to get a more complete description of fibrinogen proteolysis during thrombolytic therapy, we also measured the plasma concentrations of fibrinopeptide Bβ1-13 (desarginyl fibrinopeptide B) and fibrinopeptide Bβ15-42 (3, 8). The data were integrated to obtain cumulated amounts of peptide formed over the 5 hour study period. These results are shown in Fig. 5 and 6 respectively. These cumulated values, together with the cumulated values for FPA and fibrinopeptide Bβ1-42 shown in table 1. Each number has been expressed as the equivalent amount of fibrinogen needed to form the calculated amount of peptide.

The pathways of fibrinogen proteolysis which could contribute to the plasma fibrinopeptide concentrations are shown in Fig. 7. From this figure it is clear that the fibrinopeptide Bβ15-42 could well have been derived from fibrinopeptide Bβ1-42 by thrombin action rather than from Fibrin (Fibrin II) by plasmin action. This means that, during thrombolytic therapy, an elevated level of fibrinopeptide Bβ15-42 does not necessarily mean thrombolysis. At the bottom of each diagram is shown the release of D-dimer from the cross linked thrombus. Conceptually, this moiety should reflect the thrombolysis, we have not been impressed by

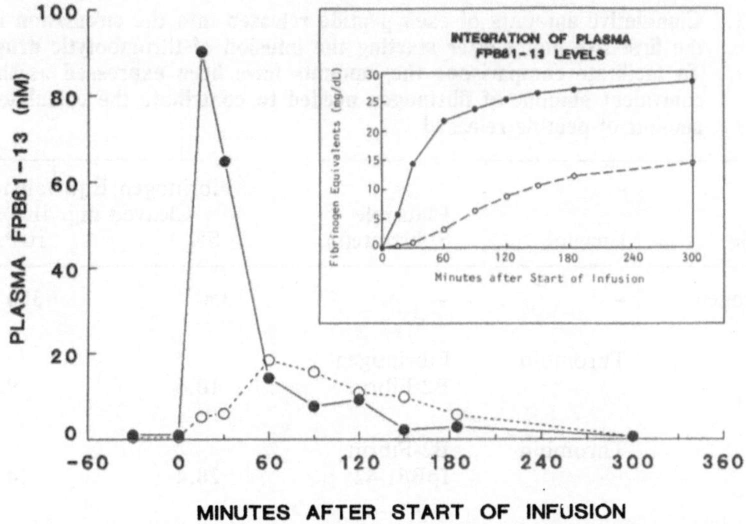

Fig. 5. Plasma levels of fibrinopeptide Bβ1-13. Other details are as in Fig. 1.

any obvious relationship between the appearance of D-dimer on the immunoblot and successful thrombolysis. At this stage we believe the search to be still underway for a suitable blood test for efficacy.

In summary, we have used serial measurements of the plasma concentration of a range of fibrinopeptides to quantitate fibrinogen proteolysis during thrombolytic therapy. A consistent picture has emerged which points the way to future developments. There clearly is a need for a facile and direct assay for fragment X, we are working on this and also on assays for the carboxy Aα chain peptides (9). The induction of thrombin activity which is heparin resistant suggests the need for alternative and perhaps novel inhibitors. Lastly, we believe that patients undergoing thrombolytic therapy are in essence undergoing a replay of the events leading to thrombus formation. As such, these patients provide a unique

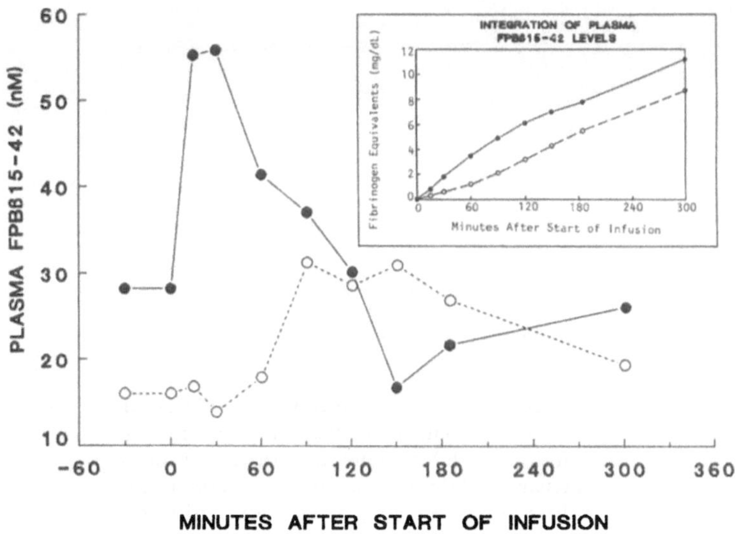

Fig. 6. Plasma levels of fibrinopeptide B—15-42. Other details are as in Fig. 1.

Table 1. Cumulative amounts of each peptide released into the circulation in the first five hours after starting the infusion of thrombolytic drug. To facilitate comparisons the amounts have been expressed as the equivalent amount of fibrinogen needed to contribute the calculated amount of peptide released

Peptide	Enzyme	Plausible Substrate(s)	Fibrinogen Equivalents Cleaved mg/dL SK	rt-PA
Fibrinogen	-	-	350	350
FPA	Thrombin	Fibrinogen B2-Fibrin	10.6	9.5
FPB	Thrombin	B2-Fibrin fpBβ1-42	28.4	14.5
Bβ1-42	Plasmin	Fibrinogen B2-Fibrin	264	101
Bβ15-42	Plasmin Thrombin	Fibrin fpBβ1-42	11.2	8.7
FDP	Plasmin	All fibrinogen derivatives	225	120

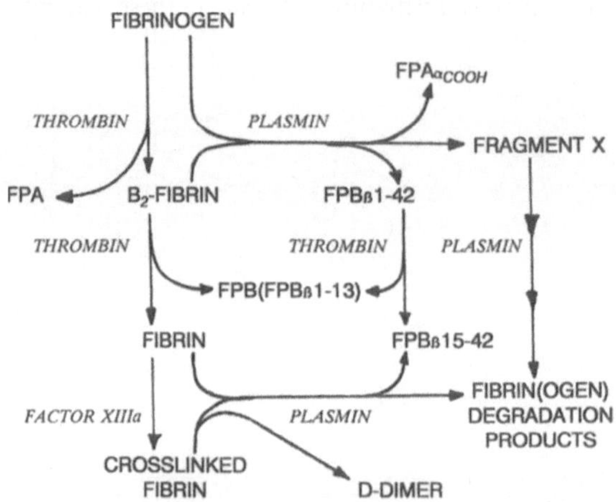

Fig. 7. Major pathways of proteolysis of fibrinogen involving thrombin and plasmin. The terms B_2-Fibrin and Fibrin refer to species previously designated Fibrin-I and Fibrin-II respectively. These terms were adopted by the Scientific and Standardization Committee of the International Society on Thrombosis and Hemostasis.

opportunity to study thrombogenesis, an opportunity not available in the typical patient with acute myocardial infarction.

REFERENCES

1. S. R. Bergmann, K. A. Fox, M. M. Ter Pogossian, B. E. Sobel, and D. Collen, Clot-selective coronary thrombolysis with tissue-type plasminogen activator, *Science,* **220:**1181-1183 (1983).
2. D. Collen, E. J. Topol, A. J. Tiefenbrunn, H. K. Gold, M. L. Weisfeldt, B. E. Sobel, R. C. Leinbach, J. A. Brinker, P. A. Ludbrook, I. Yasuda, N. H. Bulkley, A. K. Robison, A. M. Jr. Hutter, W. R. Bell, J. J. Jr. Spadaro, B. A. Khaw, and E. B. Grossbard, Coronary thrombolysis with recombinant human tissue-type plasminogen activator: a prospective, randomized, placebo- controlled trial, *Circulation,* **70:**1012-1017 (1984).
3. T. Eckhardt, H. L. Nossel, A. Hurlet Jensen, K. S. LaGamma, J. Owen, and M. Auerbach, Measurement of desarginine fibrinopeptide B in human blood, *J. Clin. Invest.,* **67:**809-816 (1981).
4. W. Flameng, F. Van de Werf, J. Vanhaecke, M. Verstraete, and D. Collen, Coronary thrombolysis and infarct size reduction after intravenous infusion of recombinant tissue-type plasminogen activator in nonhuman primates, *J. Clin. Invest.,* **75:**84-90 (1985).
5. GISSI. Effectiveness of intravenous thrombolytic treatment in acute myocardial infarction, *Lancet,* 397-401 (1986).
6. H. K. Gold, R. C. Leinbach, H. D. Garabedian, T. Yasuda, J. A. Johns, E. B. Grossbard, I. Palacios, and D. Collen, Acute coronary reocclusion after thrombolysis with recombinant human tissue-type plasminogen activator: prevention by a maintenance infusion, *Circulation,* **73:**347-352 (1986).
7. H. K. Gold, R. C. Leinbach, I. F. Palacios, T. Yasuda, P. C. Block, M. J. Buckley, C. W. Akins, and W. M. Daggett, Coronary reocclusion after selective administration of streptokinase, *Circulation,* **68:**150-154 (1983).
8. B. Kudryk, A. Rohoza, M. Ahadi, J. Chin, and M. E. Wiebe, Specificity of a monoclonal antibody for the NH2-terminal region of fibrin, *Mol. Immunol.,* **21:**89-94 (1984).
9. C. Y. Liu, J. H. Sobel, J. I. Weitz, K. L. Kaplan, and H. L. Nossel, Immunologic identification of the cleavage products from the A-alpha and B-beta chains in the early stages of plasmin digestion of fibrinogen. *Thrombos Haemostas,* **56:**100-106 (1986).
10. V. J. Marder, N. R. Shulman, and W. R. Carroll, The importance of intermediate degradation products of fibrinogen in fibrinolytic hemorrhage, *Trans. Assoc. Am. Physicians,* **80:**156-167 (1967).
11. R. P. McDonagh, J. McDonagh, and F. Duckert, The influence of fibrin crosslinking on the kinetics of urokinase-induced clot lysis. *Brit. J. Haematol.,* **21:**232-332 (1971).
12. H. L. Nossel, J. Wasser, K. L. Kaplan, K. S. LaGamma, I. Yudelman, and R. E. Canfield, Sequence of fibrinogen proteolysis and platelet release after intrauterine infusion of hypertonic saline, *J. Clin. Invest.,* **64:**1371-1378 (1979).
13. J. Owen, K. D. Friedman, B. Grossman, C. Wilkins, A. D. Berke, and E. R. Powers, Quantitation of fragment X formation during thrombolytic therapy with streptokinase and with tissue plasminogen activator, *J. Clin. Invest.,* **79:**1642-1649 (1987).
14. J. Owen, K. D. Friedman, B. A. Grossman, C. Wilkins, A. D. Berke, and E. R. Powers, Thrombolytic therapy with tissue plasminogen activator or streptokinase induces transient thrombin action. *Blood,* **72:**616-620 (1988).
15. D. Pennica, W. E. Holmes, W. J. Kohr, R. N. Harkins, G. A. Vehar, C. A. Ward, W. F. Bennett, E. Yelverton, P. H. Seeburg, H. L. Heyneker, D. V. Goeddel, and D. Collen, Cloning and expression of human tissue-type plasminogen activator cDNA in E. coli, *Nature,* **301:**214-221 (1983).
16. S. V. Pizzo, M. L. Schwartz, R. L. Hill, and P. A. McKee, The effect of plasmin on the subunit structure of fibrinogen, *J. Biol. Chem.,* **247:**636-645 (1972).
17. P. Rentrop, H. Blanke, K. R. Karsch, W. Rutsch, M. Schartl, W. Merx, R. Dorr, D. Mathey, and K. Kuch, Changes in left ventricular function after intracoronary streptokinase infusion in clinically evolving myocardial infarction, *Am. Heart J.,* **102:**1188-1193 (1981).
18. L. L. Shen, R. P. McDonagh, J. McDonagh, and J. Hermans, Early events in the plasmin digestion of fibrinogen and fibrin: effects of plasmin on fibrin polymerization, *J. Biol. Chem.,* **252:**6184-6189 (1977).

19. The TIMI Study Group, The thrombolysis in myocardial infarction (TIMI) trial: phase 1 findings, *New. Eng. J. Med.,* **312**:932-936 (1985).
20. F. Van de Werf, S. R. Bergmann, K. A. Fox, H. de Geest, C. F. Hoyng, B. E. Sobel, D. Collen, and B. E. Sobel, Coronary thrombolysis with intravenously administered human tissue-type plasminogen activator produced by recombinant DNA technology, *Circulation,* **69**:605-610 (1984).
21. F. Van de Werf, P. A. Ludbrook, S. R. Bergmann, A. J. Tiefenbrunn, K. A. Fox, H. de Geest, M. Verstraete, D. Collen, and B. E. Sobel, Coronary thrombolysis with tissue-type plasminogen activator in patients with evolving myocardial infarction, *N. Engl. J. Med.,* **310**:609-613 (1984).
22. M. Verstraete, R. Bernard, M. Bory, R. W. Brower, D. Collen, D. P. de Bono, R. Erbel, W. Huhmann, R. J. Lennane, J. Lubsen, D. Mathey, J. Meyer, H. R. Michels, W. Rutsch, M. Schartl, W. Schmidt, R. Uebis, and R. von Essen, Randomized trial of intravenous recombinant tissue-type plasminogen activator versus intravenous streptokinase in acute myocardial infarction, *Lancet,* **1**:842 (1985).
23. M. Verstraete, W. Bleifeld, R. W. Brower, B. Charbonnier, D. Collen, D. P. de Bono, A. J. Dunning, R. J. Lennane, J. Lubsen, D. G. Mathey, P. L. Michel, Ph. Raynaud, J. Schofer, A. Vahanian, J. Vanhaeke, G. A. van de Kley, F. Van de Werf, and R. von Essen, Double-blind randomized trial of intravenous tissue-type plasminogen activator versus placebo in acute myocardial infarction, *Lancet,* **2**:965 (1985).
24. M. Verstrate, Intravenous administration of thrombolytic agent is the only realistic approach in evolving myocardial infarction, *Eur. Heart J.,* **6**:586 (1985).
25. S. Yusef, R. Collins, R. Peto, C. Furberg, M. J. Stampfer, S. Z. Goldhaber, and C. H. Hennekens, Intravenous and intracoronary fibrinolytic therapy in acute myocardial infarction: overview of results on mortality, reinfarction and side effects from 33 randomized controlled trials, *Eur. Heart J.,* **6**:556 (1985).

ALTERED RHEOLOGICAL PROPERTIES OF BLOOD FOLLOWING ADMINISTRATIONS OF TISSUE PLASMINOGEN ACTIVATOR AND STREPTOKINASE IN PATIENTS WITH ACUTE MYOCARDIAL INFARCTION

Kung-ming Jan, Eric Powers*, Walter Reinhart, Andrew Berke*, Allen Nichols*, Rita Watson*, Dennis Reison*, Allan Schwartz*, and Shu Chien

Department of Physiology and Cellular Biophysics and
*Department of Medicine, College of Physicians and Surgeons
Columbia University, New York, NY 10032, USA

ABSTRACT

Tissue blood flow is determined by rheological properties of blood as well as by vascular resistance. In acute myocardial infarction patients who participated in the TIMI I trial, we compared the effects of recombinant tissue plasminogen activator (rt-PA) and streptokinase (SK) on blood rheological properties and plasma fibrinogen concentration. Blood viscosity was determined by using a coaxial cylinder viscometer at shear rates, γ, of 0.01-200 sec^{-1}. Red blood cell (RBC) deformability was studied by filtration through polycarbonate microsieves with pore size of 3 and 5 μm. Therapy with rt-PA resulted in slight decreases but statistically significant in blood viscosity from 5.2 ± 0.5 to 4.9 ± 0.4 cP (γ = 52 sec^{-1}), plasma viscosity from 1.36 ± 0.09 to 1.32 ± 0.06 cP, and plasma fibrinogen from 0.26 ± 0.04 to 0.21 ± 0.03 g/dl. SK therapy resulted in reductions in blood viscosity from 5.1 ± 0.5 to 4.6 ± 0.3 cP, plasma viscosity from 1.26 ± 0.10 to 1.16 ± 0.03 cP, and fibrinogen from 0.26 ± 0.06 to 0.10 ± 0.05 g/dl. Changes observed with SK were significantly greater than those observed with rt-PA (all $p<0.05$), and the differences persisted at 10 days after thrombolytic therapy. RBC deformability was similar in the two groups. The greater reduction of blood viscosity after SK than rt-PA suggests that, for a given degree of arterial patency, myocardial blood flow may be better maintained with SK than rt-PA in patients with acute myocardial infarction.

INTRODUCTION

It has been well established that acute myocardial infarction (AMI) is usually associated with an acute thrombotic occlusion of an atherosclerotic coronary artery (1, 2). Recent randomized clinical trials of intravenous thrombolytic therapy in patients early in the course of AMI have demonstrated clear benefits in survival, improvement of left ventricular function and limitation of infarct size in treated patients (3-6). As the results of clinical trials continue to accumulate, early intravenous thrombolytic therapy has become the standard treatment for most patients with AMI. Recent studies have focused on the search for thrombolytic agents with high fibrin specificity (7) because of the expectation that agents which are relatively fibrin-specific will cause less depletion of circulating fibrinogen and other coagulation factors and will, therefore, result in less bleeding complications than non-fibrin-specific agents.

Tissue blood flow is determined by the arteriovenous pressure drop and flow resistance. The flow resistance, in turn, is the product of vascular hindrance and blood viscosity. The major determinants of blood viscosity are plasma viscosity, hematocrit, cell aggregation, cell deformation, and flow rate (8). In patients with AMI, the blood viscosity is significantly increased (9); there is an initial rise in hematocrit followed by an increased plasma viscosity and enhanced red cell aggregation, both of which result from elevations of fibrinogen and α-2 globulin in the blood plasma. In these studies, red cell deformability was found to be normal. Dormandy *et al.* (10), however, reported that a reduction of red cell deformability occurred immediately following AMI and lasted 24 hours, attributable to an unknown washable plasma factor. At low flow rates, a primary factor for elevation of blood viscosity is red cell aggregation (8). Thus, in coronary arteries with significant stenosis in which flow rates are low, enhanced red cell aggregation and elevated blood viscosity may be significant contributors to coronary flow resistance. As a corollary, factors which can reduce blood viscosity may have beneficial effects on myocardial blood flow (11).

Thrombolytic agents cause restoration of tissue blood flow in patients with AMI and may cause reductions in hematocrit and plasma fibrinogen concentration, both of which are important determinants of blood viscosity. Therefore, it is likely that changes in blood rheology can influence tissue blood flow following successful thrombolytic therapy. Furthermore, since the non-fibrin-specific agent SK would be expected to deplete fibrinogen to a greater extent than the fibrin-specific agent rt-PA, SK might be expected to have greater effects on blood viscosity than rt-PA. The objective of the present investigation is to compare the changes in the rheological properties of blood in patients with AMI following administrations of tissue plasminogen activator (rt-PA) or streptokinase (SK). Blood rheological measurements were performed in the patients enrolled in Phase I of the Thrombolysis in Myocardial Infarction (TIMI) trial at the Presbyterian Hospital in the City of New York.

MATERIALS AND METHODS

The patient selection and the investigational protocols of TIMI have been described elsewhere (3, 12-14). In brief, the studies were carried out on patients who were identified as having AMI with chest pain of at least 30 minutes but less than 7 hours in duration, diagnostic electrocardiographic (ECG) ST elevation, and no contraindication to thrombolytic therapy. Cardiac catheterization was performed immediately after enrollment and before treatment. Patients who demonstrated significant obstruction of the infarct-related artery received an intravenous infusion of either rt-PA or SK (50 mg of a total of 80 mg given as 40 mg the first hour then 20 mg per hour for two hours of rt-PA or 1.5×10^6 units of SK given over one hour). In those patients who achieved reperfusion, intravenous heparin was also given (a bolus of 5,000 U followed by 1,000 U/h). In each patient, 20 ml of heparinized blood was obtained before and 90 minutes after initiation of thrombolytic therapy. Follow-up cardiac catheterization was performed at 7 to 10 days, and another heparinized blood sample was taken.

1. Patients

This study was performed on 19 patients (15 males and 4 females) with AMI. The mean age was 57.8 years (range 40 to 70). All patients fulfilled the entry criteria of the TIMI I trial. The duration of chest pain averaged 4.8 ± 1.8 hours. The ECG patterns revealed 11 patients with anterior wall infarction and 8 with inferior wall infarction. When the study was performed, patients received either rt-PA or SK under double blind assignment. When the study was finished, opening of the codes revealed that 10 patients had received rt-PA and 9 had received SK. During the period of study, mild bleeding complications were noted in 7 of the rt-PA and 5 of the SK treated patients; transfusion was given to 2 patients in each of the two groups. Measurements on each blood sample included the following:

2. Hematological Measurement

The hematocrit (Hct) of each sample was determined by centrifugation at 15,000 g for 5 min and corrected for plasma trapping (15). Red blood cell (RBC) and white blood cell counts were performed in a Coulter counter and the hemoglobin (Hb) concentration was determined using the cyanmethemoglobin method. Mean corpuscular volume (MCV) and mean corpuscular hemoglobin concentration (MCHC) were calculated from the measured values of Hct, RBC count and Hb.

3. Plasma Protein Analyses

The total plasma protein concentration of each sample was determined by using a refractometer (Carl Zeiss, Inc.) (16). Plasma fibrinogen concentration was determined by the method of Ratnoff and Menzie (17), and the plasma protein fractions were analyzed by means of microzone electrophoresis on cellulose-acetate membrane.

4. Measurements of Blood Viscosity

The viscometer used in this study was an air-bearing coaxial cylinder viscometer with the sensitivity and precision needed for the measurement of blood viscosity over a wide range of shear rates. The components and the principle of operations in this viscometer have been described elsewhere (9, 18). In this study, the viscosity measurements were performed at a temperature of 37°C and over a shear rate range of 0.01 to 208 sec^{-1}. The present rheological studies were designed to analyze the relative role of each of the four determinants of blood viscosity at a given temperature, i.e., RBC concentration, plasma viscosity, RBC aggregation and RBC deformability (8). RBC concentration and plasma viscosity were obtained from direct measurements. Because RBC aggregation primarily occurs at low shear rates, e.g., 0.5 sec^{-1}, low-shear blood viscosity corrected for RBC concentration and plasma viscosity has been used to indicate the degree of RBC aggregation. Accordingly, measurements of blood viscosity were also made on blood with Hct adjusted to $45.0 \pm 0.1\%$ by using autologous packed red cells or plasma, and the blood viscosity value was then divided by plasma viscosity to obtain the relative viscosity, or the viscometric aggregation index (VAI), which is the relative viscosity of blood at Hct = 45% at shear rate of 0.52 sec^{-1} (19):

$$\mathrm{VAI} = \eta_B'/\eta_p. \quad (1)$$

To assess RBC deformability from viscosity measurement, it is necessary to eliminate the effects of Hct, plasma viscosity, and RBC aggregation. Therefore, an aliquot of RBCs was washed and resuspended in Ringer's solution (pH = 7.4, containing 12 mM tris buffer and 0.25 g/dl human serum albumin) at Hct adjusted to 45%, and the viscosity of such an RBC suspension in Ringer's solution reflects RBC deformability. In summary, the measurements in the present studies included: 1) the viscosity of the original blood at shear rates ranging from 0.01 to 208 sec^{-1} (η_B); 2) the viscosity of blood at a low shear rate of 0.52 sec^{-1} with Hct adjusted to $45.0 \pm 0.1\%$ by adding either autologous packed cells or plasma (η_B'); 3) the viscosity of RBC suspensions in Ringer's solution at a Hct of 45% (η_R); and 4) the plasma viscosity (η_p). Because the viscosity of plasma is independent of shear rate, the values of plasma viscosity obtained at shear rates of 5.2 and 0.52 sec^{-1} were averaged.

5. Filtration Through Polycarbonate Sieves for Determination of RBC Deformability

In order to assess RBC deformability in passage through narrow pores, RBCs in autologous plasma at a hematocrit of 10% were subjected to filtration through polycarbonate sieves. The filtration apparatus used in this study has been described elsewhere in detail (20). The RBC suspensions were placed in a glass syringe and pumped at a constant flow rate of 0.82 ml/min and a temperature of 37°C through polycarbonate filters (Nucleopore Corp., Pleasanton, CA) with pore diameters of 3 and 5 μm. The filtration pressure during constant flow was measured on the upstream side close to the filter with a pressure transducer connected to an amplifier and a recorder. Prior to the filtration of RBC suspensions, plasma was filtered to obtain the filtration pressure for the suspending medium (Po). The initial pressure rise (Pi), which reflects the deformability of RBCs in suspension, was determined. The RBC filtration resistance index ß value which is defined as the resistance for a single RBC passing through a pore as compared to that of cell-free suspending medium, was calculated as:

$$\beta = 1 + [Pi/Po) - 1]V/h \quad (2)$$

where V is the fraction of the pore volume occupied by the RBC, i.e., the ratio of MCV to the filter pore volume, and h is the fractional volume of RBC (= Hct/100) in suspension (21).

6. Statistical Analysis

Non-paired Student t testing was used to evaluate the significance of differences between the measurements in rt-PA and SK treated patients. ANOVA was used to evaluate the significance of differences between the initial measurement and the subsequent measurements. All values are expressed as mean ± 1 SD. A p value <0.05 was considered to be statistically significant.

RESULTS

Hematological Data

The initial Hct values prior to thrombolytic therapy were 43.2±4.5% and 42.4±2.0% for rt-PA and SK patients, respectively. Immediately following thrombolytic therapy, these values decreased slightly to 41.8±4.2% and 41.1±1.0%, respectively ($p>0.05$). After 10 days of hospitalization, the Hct values fell progressively to 38.9±4.8% and 31.7± %, respectively, (Fig. 1). The initial values of MCV were 92.0±7.0 and 91.5±8.0 μm^3 for rt-PA and SK treated patients, respectively, while the values of MCHC were 33.2±1.5 and 33.5±1.9g%, respectively. These values did not change significantly in subsequent measurements.

Plasma Protein Concentrations

Significant changes in the concentrations of plasma protein fractions were seen only in fibrinogen and α_2-globulin (Fig. 1). The initial concentration of fibrinogen was not different between rt-PA and SK groups (0.26±0.04 vs. 0.26±0.06 g/dl). Ninety minutes following the beginning of thrombolytic therapy, the plasma fibrinogen concentrations decreased to 0.21±0.03 g/dl (80% of pretreatment value) in the rt-PA group and to 0.10±0.05 g/dl (less than 40%) in the SK group; the decrease with SK was significantly greater than the decrease with rt-PA ($p<0.05$). Ten days after therapy, compared to the initial values, the rt-PA group showed a significant increase in fibrinogen concentration to above the pretreatment level (0.40±0.1 g/dl), while the SK group showed-only a slight rise in the fibrinogen concentration to reach to pretreatment level (0.28±0.06 g/dl). For α_2-globulin, there were no changes in concentrations after thrombolytic therapy (rt-PA: 0.67±0.07 and SK: 0.68±0.16 g/dl prior to therapy; rt-PA: 0.69±0.08 and SK: 0.62±0.09 g/dl at 90 minutes), but at 10 days the concentration of α_2-globulin was higher ($p<0.05$) in the rt-PA group (1.0±0.17 g/dl) than in the SK group (0.55±0.09 g/dl) (Fig. 1). Other plasma protein fractions, i.e., albumin and α_1, β_1, β_2, and γ-globulins, did not show significant differences or changes between the two groups during the course of the study. The total protein concentrations were not significantly different between the two groups before and immediately after therapy. At 10 days, however, the rt-PA patients showed a slightly higher total protein concentration than the SK patients (Fig. 1).

Viscosity Studies

The results of measurements on blood viscosity (η_B) are summarized in Fig. 2. The viscosity of blood was determined over a wide range of shear rate. Values at two shear rates, i.e., 52 and 0.52 sec^{-1}, are given to represent blood viscosities at high and low shear rates, respectively. Immediately following thrombolytic therapy, both groups showed significant decreases in η_B at both high and low shear rates. At 10 day, η_B in the rt-PA group returned to the initial values, while η_B in the SK group remained decreased. Plasma viscosity (η_B) decreased immediately following thrombolytic therapy, and a greater reduction in η_p was seen in the SK group than in the rt-PA group. The viscometric aggregation index (VAI) of blood showed a significant reduction 90 minutes following the onset of thrombolytic therapy; the decrease was greater in the SK than in the rt-PA group ($p<0.05$). At 10 days these values were significantly higher in the rt-PA group than in the SK group

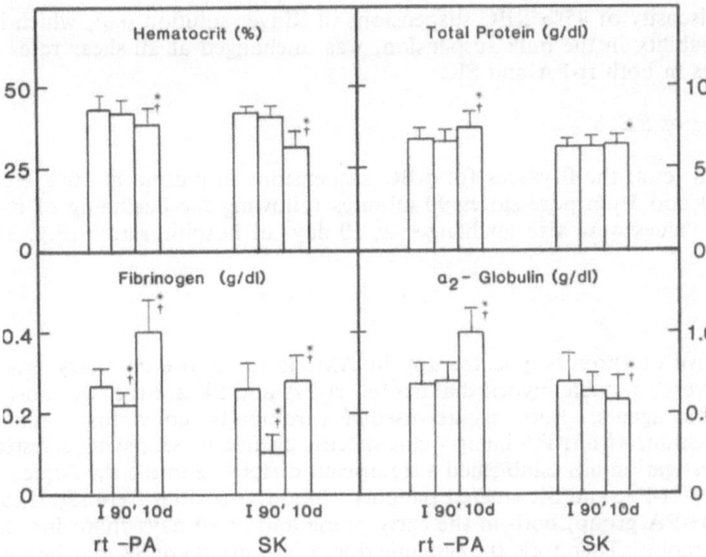

Fig. 1. Effects of thrombolytic therapy with tissue plasminogen activator (rt-PA) and streptokinase (SK) on hematocrit and plasma concentrations of total protein, fibrinogen and α_2-globulin. I, 90' and 10d represent before (initial measurements), 90 min after, and 10 days after thrombolytic therapy, respectively. Asterisks denote significant differences between rt-PA and SK, and crosses denote significant differences from the initial measurements. The vertical lines are standard deviations.

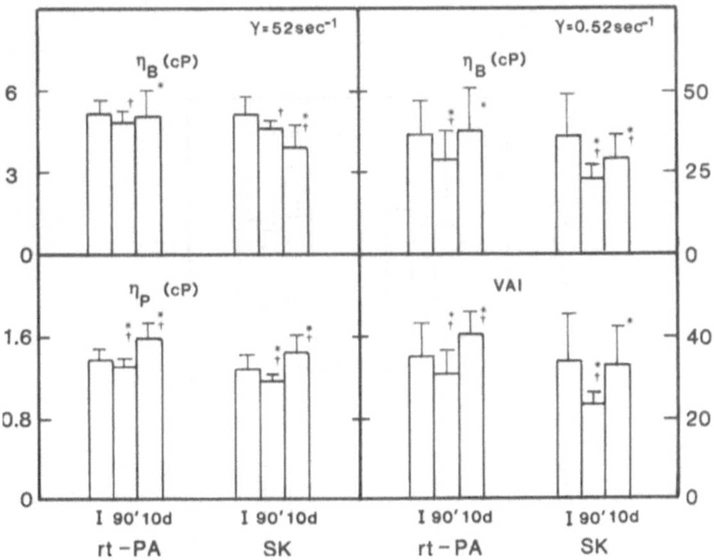

Fig. 2. Effects of thrombolytic therapy with tissue plasminogen activator (rt-PA) and streptokinase (SK) on whole blood viscosity (η_B) at a high shear rate (γ = 52 sec^{-1}) and a low shear rate (γ = 0.52 sec^{-1}), plasma viscosity (η_p), and the viscometric aggregation index (VAI). I, 90' and 10d represent before (initial measurements), 90 min after, and 10 days after thrombolytic therapy, respectively. Asterisks denote significant differences between rt-PA and SK, and crosses denote significant differences from the initial measurement. The vertical lines are standard deviations.

(Fig. 2). The viscosity of 45% RBC suspensions in Ringer solution (η_R), which is an index of RBC deformability in the bulk suspension, was unchanged at all shear rates during the course of studies in both rt-PA and SK.

Filtration Studies of RBCs

In filtration tests, the ß values for RBC suspensions in plasma at 10% Hct were unchanged using 3 and 5 μm pore sieves 90 minutes following the beginning of rt-PA or SK therapy. These values were also unchanged at 10 days of hospitalization (Fig. 3).

DISCUSSION

The objective of thrombolytic therapy in AMI is to restore coronary artery patency and oxygen delivery to viable myocardial tissue. rt-PA and SK are the two most frequently used thrombolytic agents. Both agents dissolve thrombi by converting circulating plasminogen into plasmin, with rt-PA being a clot-specific agent and SK having a systemic effect. The present investigation has established a significant difference in the rheological properties of blood between rt-PA and SK treated patients: blood viscosities were significantly lower in the SK than rt-PA group, both in the early phase and at 10 days following the onset of thrombolytic theraphy. Therefore, the systemic fibrinolytic effects of SK may be advantageous because of the consequent reduction of blood viscosity and prevention of further thrombus formation.

Oxygen delivery to tissue is the product of oxygen content in blood and the blood flow; the latter is determined by perfusion pressure and flow resistance. For a given vascular geometry, the flow resistance is primarily determined by blood viscosity. Perfusion pressure to the myocardium can be increased by increasing aortic pressure but this occurs at the cost of increasing cardiac work and cardiac oxygen consumption. Only by reducing vascular resistance or blood viscosity is it possible to increase coronary blood flow without increasing myocardial oxygen consumption. Thus, thrombolytic agents may increase tissue perfusion

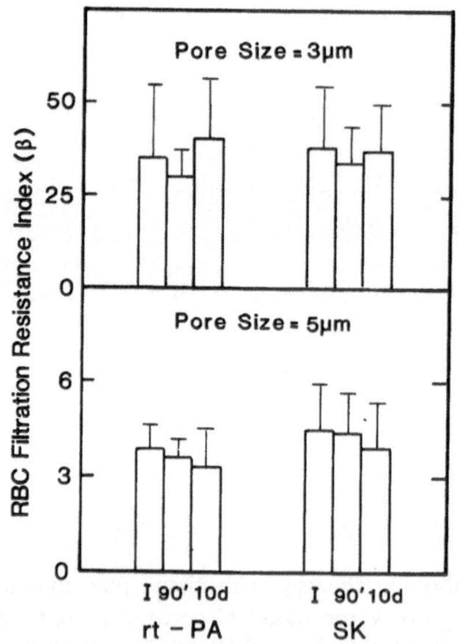

Fig. 3. Effects of thrombolytic therapy with tissue plasminogen activator (rt-PA) and streptokinase (SK) on RBC deformability indicated by the RBC filtration resistance index (ß) determined by using polycarbonate sieves with pore sizes of 3 μm (top) and 5 μm (bottom), respectively. I, 90' and 10d represent before (initial measurements), 90 min after, and 10 days after thrombolytic therapy, respectively. The vertical lines are standard deviations.

and oxygen delivery not only by decreasing the severity of coronary stenosis but also by reducing the blood viscosity.

Blood is a suspension of cells in plasma, and blood viscosity depends on the cell concentration and plasma viscosity. At given levels of cell concentration and plasma viscosity, blood viscosity varies with cell deformation and cell aggregation, both of which are processes depending on the shearing condition during flow (8). At high shear rates, e.g., above 50 sec^{-1}, red cells are subjected to shear deformation and shear dispersion, resulting in a relatively low blood viscosity. At low shear rates, e.g., below 10 sec^{-1}, however, red cells are less deformed and tend to form rouleaux, leading to an elevation of blood viscosity. In the normal circulation, the shear rates in arterie and capillaries are probably above 200 sec^{-1} and it is of the order of 20 sec^{-1} in most venules and small veins (22). The shear rate can approach zero in stagnant parts of the circulation such as in some veins of the lower extremities (23). In pathological conditions, such as shock or the partial occlusion of vessels, the conditions of low flow result in a reduction of shear rates in the vascular bed. Therefore, in the presence of a coronary stenosis, local shear rate is reduced to raise the blood viscosity, and this may become a significant factor in elevating the flow resistance.

Following an AMI, there are elevations of whole blood and plasma viscosities and RBC aggregation mainly due to changes in plasma fibrinogen concentration (9). The increase in plasma fibrinogen concentration in patients with AMI, has been correlated with an increased risk of death and reinfarction (24). Mathematical modeling of the effect of changes in blood viscosity on myocardial blood flow has shown that the increases in plasma fibrinogen observed following a myocardial infarction are sufficient to cause a marked reduction in oxygen delivery to the subendocardial myocardium of the left ventricle (25). The present investigation has shown that intravenous administration of rt-PA or SK reduces plasma fibrinogen concentration, leading to decreases in plasma and blood viscosities and in RBC aggregation. The decrease in plasma viscosity reduces blood viscosity at all shear rates, while the inhibition of RBC aggregation causes a disproportionate reduction in blood viscosity at low shear rates. This reduction in viscosity would be expected to substantially increase myocardial blood flow and oxygen delivery, particularly in the presence of stenosed coronary arteries and at subendocardium where the shear rate is low.

Neuhof *et al.* (26) measured blood pressure, cardiac output, and cardiac oxygen consumption in AMI patients treated with SK. The total peripheral resistance was found to be significantly lower in the SK treated than in control patients. In addition, cardiac output in the SK treated patients increased by 20% above baseline level without a corresponding rise in myocardial oxygen consumption. A decrease in peripheral resistance was also seen in a large randomized controlled study performed by the European Cooperative Study Group (27). Systolic blood pressure was reduced by 13-18 mmHg, and diastolic pressure by 7-10 mmHg for 48 hours after the start of SK administration. In a subset of these patients, cardiac output was measured and found to increase, showing that the decrease in blood pressure due to a reduction in peripheral resistance rather than a pump failure. This decrease in resistance which improves cardiac performance may in part be the result of a decrease in blood viscosity.

If, because of a coronary thrombus, blood flow to the infarct zone ceases completely, removal of the obstruction offers the only hope of salvaging myocardial tissue. However, in the majority of patients with an AMI and total coronary occlusion, blood flow does not cease completely but rather residual blood flow to the infarct zone probably occurs. Davies *et al.* (28) angiographically demonstrated a cyclical pattern of occlusion and patency in AMI patients. This observation supports the theory of Gasser (29) that the pulsatile change in myoglobin release, S-T segment elevation, and CK-MB release seen in AMI are due to intermittent occlusion of the coronary artery. Blanke *et al.* (30) showed that in the large majority of patients suffering from an AMI, there was residual blood flow to the ischemic area via either a partially patent vessel or via collaterals, and that this residual blood flow could be correlated with limitation of the infarct area and prevention of deterioration in myocardial function. A decrease in blood viscosity induced by a thrombolytic agent would enhance blood flow and oxygen delivery to the infarct zone and help to salvage the endangered myocardium even if vascular patency was not affected by treatement. Thus, thrombolytic agents and SK in particular may be of benefit in the absence of reperfusion due to decreases in peripheral resistance and improved myocardial oxygen delivery.

At a constant aortic pressure, coronary artery perfusion is dependent on blood viscosity, and this dependence is increased as the lumen of the coronary arteries decreases and the shear rate is lowered, as in severe coronary artery disease. While high shear rate viscosity is probably an important determinant of normal coronary blood flow, blood flow through partially occluded vessels or small diameter collaterals is likely to be more dependent on low-shear viscosity (25). Depletion of fibrinogen leads to disproportionately greater decreases in low-shear viscosity, because it reduces the tendency of RBCs to aggregate. Thus, the viscosity reduction may enhance coronary blood flow in the partially occluded vessels or small diameter collaterals. Blood flow through the latter has been shown to be important in limiting the extent of myocardial infarction (30).

Direct evidence of a beneficial effect of decreased blood viscosity on myocardial function is available from an experimental study in dogs. S-T segment elevation induced by ligation of branches of the left descending coronary artery could be reduced by SK administration. Since the ligatures completely occluded flow through the coronary arterial branches, enhancement of collateral flow was the only mechanism by which oxygen delivery could be increased (31). Clinical observations on the effect of SK in deep vein thrombosis and arterial occlusion also support the idea that an improvement in blood viscosity leads to an increased collateral circulation. When SK is given in deep vein thrombosis, improvement in pain occurs before there is thrombophlebographic evidence of lysis of the clot (32). Similarly, in patients with chronic obstruction of the pelvic artery or arteries of the lower leg, SK improved walking distance even though there was almost no evidence of an improvement in the obstruction, and these improvements were attributed to a decrease in blood viscosity (33).

In summary, at the doses studied, SK therapy caused a significantly greater decrease in blood viscosity than rt-PA, primarily due to greater reductions in fibrinogen concentration and plasma viscosity. These differences persisted up to 10 days after thrombolytic therapy. The greater decrease of blood viscosity after SK than rt-PA suggests that, for a given degree of arterial patency, myocardial blood flow in AMI patients may be better maintained with SK than rt-PA.

REFERENCES

1. M. A. DeWood, J. Spores, R. Notske, L. T. Mouser, R. Burroughs, M. S. Golden, and H. T. Lang, Prevalence of total coronary occlusion during the early hours of transmural myocardial infarction. *N. Engl. J. Med.,* **303:**897 (1980).
2. E. Braunwald, The path to myocardial salvage by thrombolytic therapy. *Circulation,* **76 (Suppl II):**II2 (1987).
3. The TIMI Study Group. The thrombolysis in myocardial infarction (TIMI) trial: Phase I findings. *N. Engl. J. Med.,* **312:**932 (1985).
4. Gruppo Italiano per lo Studio della Streptochinasi nell'Infarto Miocardico (GISSI): Effectiveness of intravenous thrombolytic treatment in acute myocardial infarction. *Lancet,* **1:**397 (1986).
5. The ISAM Study Group. A prospective trial of intravenous streptokinase in acute myocardial infarction (ISAM). *N. Engl. J. Med.,* **314:**1465 (1986).
6. G. V. Martin, M. L. Stadius, K. B. Davis, J. L. Ritchie, F. K. Sheehan, C. Maynard, and J. W. Kennedy. The Western Washington intravenous streptokinase trial; effects of intravenous streptokinase on vessel patency and left ventricular function. *Circulation,* **74 (Suppl II):**II-367 (1986).
7. M. Verstraete. The search for the ideal thrombolytic agent. *J. Am. Coll. Cardiol.,* **10:**4B (1987).
8. S. Chien. Biophysical behavior of red cells in suspension. *In*: D. M. Surgenor, editor: The Red Blood Cells, 2nd edition, New York, 1975, Academic Press, vol 2, p. 1031.
9. K. M. Jan, S. Chien, and J. T. Bigger Jr. Observation on blood viscosity changes after acute myocardial infarction. *Circulation,* **51:**1079 (1975).
10. J. Dormandy, M. Boyd, and E. Ernst. Red cell filterability after myocardial infarction. *Scand. J. Clin. Lab. Invest.,* **156:**195 (1981).
11. W. Bell. Objectives of thrombolytic therapy in acute myocardial infarction. *Am. J. Med.,* **83(Suppl 2A):**11 (1987).
12. F. H. Sheehan, E. Braunwald, P. Canner, H. T. Dodge, J. Gore, P. Van Natta, E. R. Passamani, D. O. William, and B. Zaret. The effect of intravenous thrombolytic therapy on left ventricular function: a report on tissue plasminogen activator and streptokinase from the Thrombolysis in Myocardial Infarction (TIMI Phase I) trial. *Circulation,* **75:**817 (1987).

13. J. H. Chesebro, G. Knatterud, R. Roberts, J. Borer, L. S. Cohen, J. Dalen, H. T. Dodge, C. K. Francis, D. Hillis, P. Ludbrook J. E. Markis, H. Hueller, E. R. Passamani, E. R. Powers, A. K. Rao, T. Robertson, A. Ross, T. J. Ryan, B. E. Sobel, J. Willerson, D. O. Williams, B. L. Zaret, and E. Braunwald. Thrombolysis in Myocardial Infarction (TIMI) trial, Phase I: a comparison between intravenous plasminogen activator and intravenous streptokinase. Clinical findings through hospital discharge. *Circulation,* **76:**142 (1987).
14. A. K. Rao, C. Pratt, A. Berke, A. Jaffe, I. Ockene, T. L. Schreiber, W. R. Bell, G. Knatterud, T. L. Robertson, and M. L. Terrin. Thrombolysis in Myocardial Infarction (TIMI) trial — Phase I: hemorrhagic manifestations and changes in plasma fibrinogen and the fibrinolytic system in patients treated with recombinant tissue plasminogen activator and streptokinase. *J. Am. Coll. Cardiol.,* **11:**1 (1988).
15. S. Chien, R. J. Dellenback, S. Usami, and M. I. Gregersen. Plasma trapping in hematocrit determination: Differences among animal species. *Proc. Soc. Exp. Biol. Med.,* **119:**1155 (1965).
16. B. S. Neuhausen, and D. M. Rioch. The refractometric determination of serum protein. *J. Biol. Chem.,* **55:**353 (1923).
17. O. D. Ratnoff, and C. Menzie. A new method for the determination of fibrinogen in small sample of plasma. *J. Lab. Clin. Med.,* **37:**316 (1951).
18. S. Chien, S. Usami, R. J. Dellenback, and C. A. Bryant. Comparative hemorheology-hematological implications of species differences in blood viscosity. *Biorheology,* **8:**35 (1971).
19. S. Chien, and K. M. Jan. Ultrastructural basis of the mechanism of rouleaux formation. *Microvasc. Res.,* **5:**155 (1973).
20. W. H. Reinhart, S. Usami, E. A. Schmalzer, M. M. L. Lee, and S. Chien. Evaluation of red blood cell filterability test: Influence of pore size, hematocrit level, and flow rate. *J. Lab. Clin. Med.,* **104:**501 (1984).
21. E. A. Schmalzer, R. Skalak, S. Usami, M. Vayo, and S. Chien. Influence of red cell concentration on filtration of blood cell suspensions. *Biorheology,* **20;**29 (1983).
22. S. Chien. Blood rheology and its relation to flow resistance and transcapillary exchange, with special reference to shock. *Advances in Microcir.,* **2:**89 (1969).
23. A. D. McLachlin, J. A. McLachlin, T. A. Jory, and E. C. Rawling. Venous stasis in the lower extremities. *Ann. Surg.,* **152:**678 (1960).
24. I. Weinberger, J. Fuchs, A. Tobul, Z. Rotenberg, and J. Agmon. Plasma viscosity changes during acute myocardial infarction. *Circulation,* **72(Suppl III):**III-417 (1985).
25. R. J. Gordon, G. K. Snyder, H. Tritel, and W. J. Taylor. Potential significance of plasma viscosity and hematocrit in myocardial ischemia. *Am. Heart J.,* **87:**175 (1974).
26. H. Neuhof, D. Hey, E. Glaser, H. Wolf, and H. G. Lasch. Hemodynamic reactions induced by streptokinase therapy in patients with acute myocardial infarction. *Eur. J. Intensive Care Med.,* **1:**27 (1975).
27. The European Cooperative Study Group. Streptokinase in acute myocardial infarction extended report of the European Cooperative Trial. *Acta. Med. Scand, Supple,* **648:**7 (1981).
28. G. J. Davies, S. Chierchia, and A. Maseri. Prevention of myocardial infarction by very early treatment with intracoronary strepkinase. Some clinical observation. *N. Engl. J. Med.,* **311:**1488 (1984).
29. R. N. A. Gasser, F. Dienstl, B. Puschendorf, S. Hauptlorenz, M. Moll, and E. Dworzak. New perspectives on the function of coronary artery spasm in acute myocardial infarction: the thromboischemic reentry mechanism. *Angiology* **37:**880 (1986).
30. H. Blanke, M. Cohen, K. R. Karsch, R. Fagerstrom, and K. P. Rentrop. Prevalence and significance of residual flow to the infarct zone during the acute phase of myocardial infarction. *J. Am. Coll. Cardiol.,* **5:**827 (1985).
31. W. Wende, H. W. Stühlen, J. Meyer, W. Bleifeld, H. Holzhüter, and E. Wenzel. Influence of a streptokinase-induced fibrinolysis on the extent of the acute experimental myocardial infarction. *Klin. Wochenschr.,* **53:**755 (1975).
32. A. M. Ehrly. Rheological changes due to fibrinolytic therapy. *In:* K. Messmer, H. Schmid-Schönbein, editor: Hemodilution. Theoretical Basis and Clinical Application. Basel, 1972, S. Karger, p. 289.
33. H. D. Bruhn, P. Jipp, and I. Hagenah. Thrombolytic treatment of chronic obturations of the pelvic arteries and the arteries of the lower extremities. *Med. Klin.,* **68:**1430 (1973).

MONOCLONAL ANTIBODIES FOR THE DETECTION OF THROMBOSIS

P. J. Gaffney, P. S. Gascoine, L. J. Creighton, and P. M. Tymkewycz

National Institute for Biological Standards and Control
Blanche Lane, Potters Bar
Hertfordshire EN6 3QG, UK

INTRODUCTION

The early detection of vascular occlusions or the propensity to such occlusions is of primary importance in human medicine. Since fibrin is an important structural component in both venous and arterial thrombi (1), the detection of fibrin and its fragments in vivo or in the circulating blood has underpinned a number of strategies for the early diagnosis of thrombosis.

Fibrin becomes incorporated into thrombi following the thrombin-mediated conversion of fibrinogen to fibrin by hydrolytic cleavage of fibrinopeptides A and B (FPA and FPB). During the subsequent polymerisation of fibrin, plasmin is generated on the surface of the forming fibrin polymer (2), leading to the formation of fibrin degradation products (FDP) following the degradation of fibrin-containing thrombi in the vasculature (3). During these competing interactions of plasmin and thrombin for the fibrinogen-fibrin system of molecules, many new distinct epitopes are revealed on fibrin and its fragments. While excellent reviews of the immunobiology of fibrinogen/fibrin are available (4, 5), it is proposed here to comment only on immunological probes, mostly expressed as monoclonal antibodies (mabs), which have proved clinically valuable in following the behaviour of the fibrinogen-fibrin system as it relates to the detection and monitoring of thrombosis in man.

SUBUNIT COMPOSITION OF FIBRINOGEN/FIBRIN FRAGMENTS

The structure and shape of fibrinogen dictates the sequence of degradative reactions by which plasmin attacks the molecule. Despite the fact that plasmin has a general hydrolytic affinity for most arginines and lysines, only a limited number are available in fibrinogen because it contains three major disulphide-rich inaccessible core regions. The carboxy terminal (40,000 MW) of the two A-alpha chains of fibrinogen are first digested by plasmin followed by segregation of one disulphide-rich region of the molecule known as fragment D. As shown in Fig. 1, there is a consensus that the digestion of fibrinogen to its plasmin-resistant core fragments D and E takes place through an asymmetric sequence which involves the generation of an intermediate fragment called Y, which subsequently degrades to its two core fragments, D and E. The detail of this sequence of degradation reactions has been reviewed elsewhere (7) and Fig. 1 is adequate here to demonstrate that quite a number of new epitopes are generated during the degradation sequence. Monoclonal antibodies to some of these new structures have been developed (4, 5). The interaction of thrombin with fibrinogen generates a half-stagger polymeric fibrin (Fig. 2) which becomes crosslinked rather rapidly between the gamma chains of adjacent fibrin subunits and later between the alpha chains of these subunits (for review see Ref 8). When crosslinked fibrin is digested by plasmin the same domainal fragmentation takes place as shown in Fig. 1

Fibrinogen, Thrombosis, Coagulation, and Fibrinolysis, Edited by
C. Y. Liu and S. Chien, Plenum Press, New York

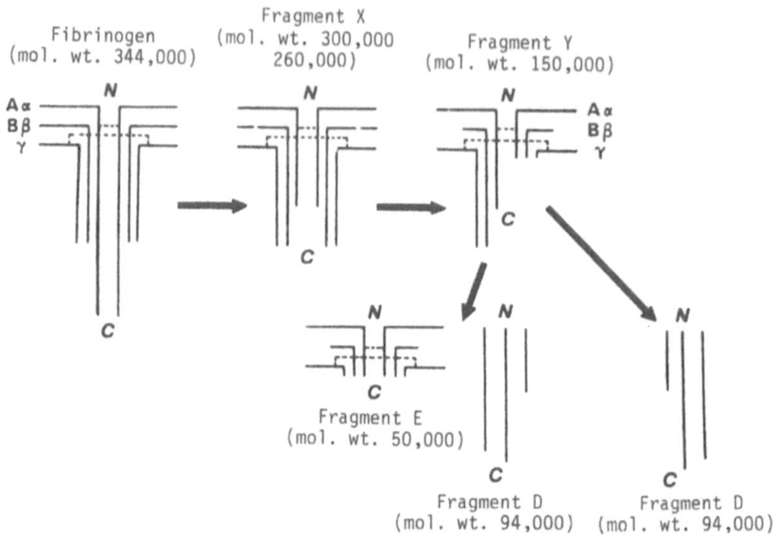

Fig. 1. Schematic diagram of the plasmin-mediated conversion of fibrinogen to its core fragments D and E showing the intermediate fragments X and Y. The polypeptide chain origin of the constituent subunits of each fragment can be related to those of the preceding fragment and to the polypeptide chains of the originating fibrinogen molecule — A-alpha, B-beta and gamma. The amino (N)- and carboxy (C)-terminal ends of the fragments are shown, to aid in locating subsequent sites of structural breakdown. The range of molecular weights of fragment X depends on the amount of peptide material removed from the carboxyl ends of the A-alpha chains and whether or not peptides have been hydrolysed from the N-terminal ends of the B-beta chains. (Reproduced with permission from ref 6).

for fibrinogen. Thus it is not surprising that in crosslinked fibrin digests there is a large variety of X, Y, D and E structural combinations held together by Factor XIII mediated gamma chain crosslinks from the originating fibrin. Fig. 2 is an attempt to domainally characterise some of this heterogeneous group. A general formula for all these crosslinked fragments is given as:-

$$(Y \text{ or } D - X_n - Y \text{ or } D)_2$$

ASSAY OF FIBRINOGEN AND FIBRIN DEGRADATION PRODUCTS (FgDP and FnDP)

Ultimately antibodies may be developed which have specificity for each component within the heterogeneous group of crosslinked fibrin degradation products (XL-FnDP) shown in Fig. 2. Currently efforts to generate monoclonal antibodies have been guided by a negative reaction with fibrinogen and a positive reaction with one or other of a rather mixed group of FDP found in plasma. The exception has been in the case of D dimer, which has been for many years (9) regarded as a potentially valuable marker of in vivo fibrin formation and degradation. The first such assay was developed by Rylatt et al. (10) and involved in the use of a D dimer-specific mab (coded DD3B6) as a catcher antibody on a polyvinyl 96-well plate and the tagging of the immunologically-immobilised D dimer fraction with a pan-specific mab. Using a two-site ELISA system the assay can detect various crosslinked FDP fractions which include D dimer structures, crosslinked Y-D and the heterogeneous X-oligomer fraction. This D dimer assay has demonstrated a mean concentration of D dimer-like fragments in normal plasma of 75 ng/ml, with an upper limit

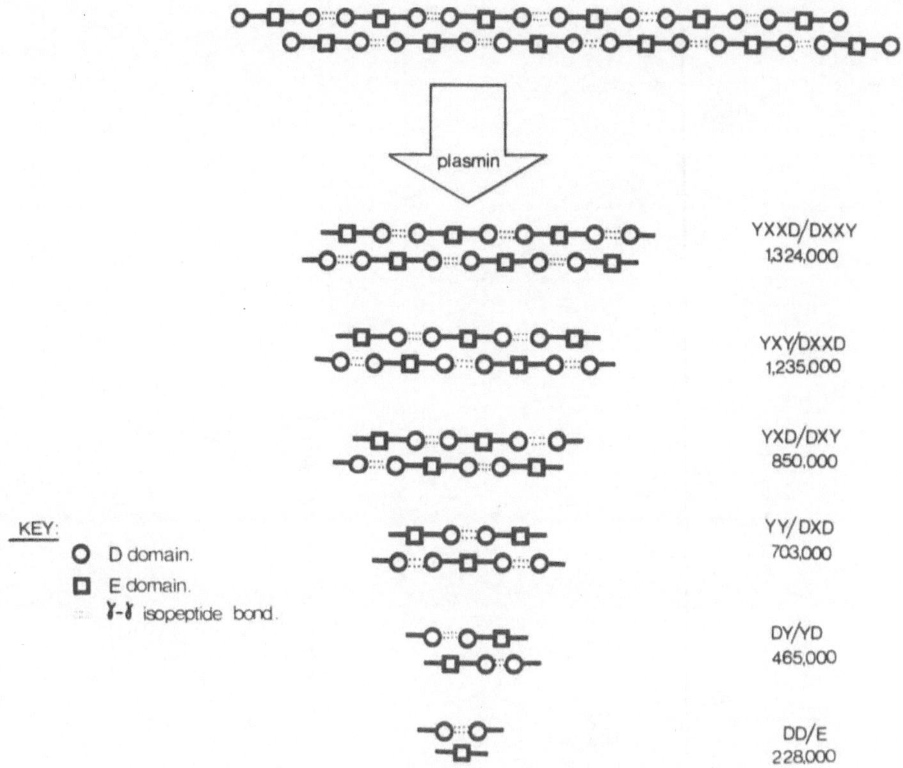

Fig. 2. Plasmin mediated digestion of XL-FN. Top of figure depicts the half-stagger two chain structure of fibrin, showing crosslinks between the D domains of adjacent fibrin sub-units. The action of plasmin generates various molecular weight structures, all having various X, Y, D, E domains associated by the gamma chain crosslinks in the originating fibrin. Lysis by plasmin of only COOH-A-alpha regions and the coiled segments between D and E domains dominates the sequence of lytic events.

of about 140 ng/ml. Concentrations in the plasma of patients with pulmonary embolism (PE), deep vein thrombosis (DVT), aterial thrombo-embolism and DIC were raised in all cases (11, 12). Another similar assay involves a two-site ELISA system using IgM mabs (NIBn 52) coated on PVC 96-well plates as a catcher and horseradish peroxidase (HRP)-labelled NIBn 178 as the tag antibody (13). This assay system has a high specificity for the X-oligomer fraction, while no reaction was observed with fibrinogen, D dimer or NXL-FDP. Using this assay in conjunction with a lyophilised X-oligomer standard, Gaffney et al. (14) have shown that patients with PE, DVT, myocardial infarction (MI) and peripheral vascular disease (PVD) had elevated levels of X-oligomer (Fig. 3), while the same assay has indicated that individuals following MI have consistently elevated levels (15). Normal plasma contained levels of 10- 200 ng/ml and the assay was reproducible in samples which had been snapfrozen and thawed many times. Two mabs with specific affinity for X-oligomer have been developed and each is directed towards a different conformational epitope on the complex X-oligomer structure (13). These mabs are all IgM class because of the unique intrasplenic method of immunizing the BALB/c mice prior to splenectomy and hybridisation procedures (16). However, mabs of the IgG class have also been developed and these react in a similar manner to the above described NIBn 52 in a two-site ELISA assay (17).

While crosslinked fibrin degradation products (XL-FDP) are detected by mabs to both D dimer and X-oligomer, it is well to point out that NIBn 52 reacts only with X-oligomer while D dimer mabs react with all the major crosslinked FDP's. This allows X-oligomer and D dimer mabs to be used in tandem not only to monitor thrombolysis in vivo but also to assess the extent of systemic fibrinolysis generated during the thrombolytic regime (18).

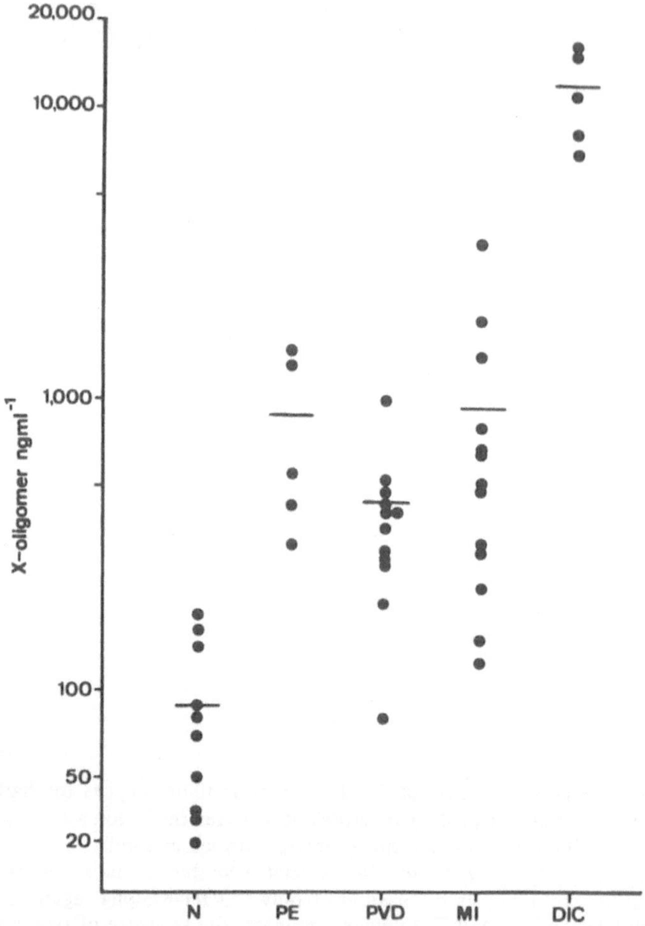

Fig. 3. Levels of X-oligomer in the plasmas of normal subjects (N) and patients with diagnosed pulmonary embolism (PE), peripheral vascular disease (PVD), myocardial infarction (MI) and disseminated intravascular coagulation (DIC). (Taken from ref 13).

Since X-oligomers are the initial fragments released from thrombi during lysis, any subsequent degradation to the smaller sized Y-D and D-D fragments may be regarded as a marker of systemic fibrinolysis.

It can be argued that the production of fibrin in the circulation is a prerequisite for the activation of plasminogen and thus the formation of NXL- and XL-FDP; however, the third assay system to be described here has been developed to allow the assessment of the total FDP fraction and the subfraction derived from both fibrinogen (FgDP) and fibrin (FnDP) (19, 20). Mabs specific for the E domain of fibrinogen (FDP 14) have been used to immobilise the total FDP (TDP) in plasma, and the fibrinogen and fibrin-derived FDP fractions can be individually identified using, respectively, a HRP-labelled mab for fibrinopeptide A (Y-18) and a HRP-labelled mab for crosslinked FDP (DD-13). Should only the TDP level in plasma be required then the second antibody used would be a polyclonal to FDP. Whether primary fibrinogen digestion in the circulation is a frequent event is questionable. However using the specific assay for FgDP it has been demonstrated that high levels exist in plasma during arvin induced defibrination (Hampton et al., unpublished data). Whether this plasma FgDP fraction originates from circulating fibrinogen or from fibrinogen bound to fibrin in the circulation is unknown; however the latter seems most probable. It remains to be seen whether this procedure has any clinical value in diagnosis. However, there is little doubt that the assay of XL-FDP in plasma is a valuable marker of an advanced state of hypercoagulability, whether transient, transient/local, disseminated, or otherwise. All the assays described above have the capacity to measure the XL-FDP fraction in plasma. Interlaboratory studies have been conducted under the auspices of the European Concerted Action on Thrombosis (ECAT) to compare the data obtained by the above three assays in plasma from patients with a variety of cardiovascular related diseases. Studies such as these and the extensive use of these mab-specific assays in many laboratories will be the final arbiter of the value of these assays in evaluating the importance of activated coagulation/fibrinolysis in health and disease.

IMAGING THROMBI: VALUE OF FIBRIN SPECIFIC ANTIBODIES

It has been evident for some time that there is the need for a relatively non-invasive procedure to detect vascular occlusions and to assess the extent of such occlusions. Bosnjakoovic et al. (21) were the first to report a useful thrombus-imaging approach using a polyclonal antibody to fibrin: however, no further development of this work has been reported. With the advent of monoclonal antibodies a number of promising approaches have appeared in the literature and most can trace their conceptual origins to the pioneering work of Plow and Edgington (22, 23) from which emerged the promise of quite unique and interesting neoantigens associated with the physiologically based cleavage of both fibrinogen and fibrin. Since fibrin in both arterial and venous thrombi is inevitably crosslinked (1) early efforts to establish unique antibodies for clot imaging were directed towards regions of conformational and primary structure associated with the crosslinked sites on the alpha and gamma chains of fibrin (24, 25). This approach has not been successful to date in producing antibodies with a specific affinity for fibrin, since it seems that these strategic crosslinked sections of gamma chains of fibrin are buried in the fibrin structure as was initially suggested from the internalization of the gamma chain in whole fibrinogen (26). Two new strategies have been adopted with a view to developing fibrin-specific antibodies the first utilizing clearly defined amino acid sequences of fibrin and the second approach depending mainly on the neoantigenic concepts associated with conformation as outlined by Plow and Edgington (22).

To date several fibrin-specific mabs have been developed, some of which have been used successfully for imaging in animals and patients. Rosebrough et al. (27) have used monoclonal antibodies to the thrombin-treated NH_2-terminal disulphide knot (T-NDSK) of human fibrinogen to image canine thrombi. They found that these mabs (denoted T2G1) crossreacted quite well with dog and human fibrin, while showing no affinity for either dog or human fibrinogen. This allowed successful imaging experiments to be conducted with T2G1 and its $F(ab')_2$ fragment in a dog model for venous thrombosis. The seeming interspecies compatability of the epitope(s) to which T2G1 is directed suggests conservation of sequential/conformational regions of fibrin which are related to clot formation and are hidden in the native fibrinogen molecule. This needs consideration when proposing animal models for imaging both autologus and heterologous thrombi. Another series of imaging experiments using T2G1 in the carine model has been carried out by Knight et al. (28). This group have used $^{99}Tc^{m}$ labelled antibody and its fab' fragment. Another group have used $^{99}Tc^{m}$ labelled fab' fragment of T2G1 to image deep vein thrombi in patients (29). Further successful fibrin probing has been performed using a mab to the approximate hexapeptide region from the NH_2 terminus of the ß chain of fibrin (30), while Feitsma et al. (31) have used a mab to fibrinogen fragment Y (Y22) in successful human clot imaging experiments. Fig. 4 shows imaging data in a human thrombus/rabbit jugular vein model using a mab to high molecular weight crosslinked fibrin fragments (X-oligomer), NIBn 123 and a mab to D dimer, DD-3B6. These imaged the human thrombus optimally after about 18 hours using an ^{131}-I label (32).

In addition our laboratory has developed a number of fibrin specific mabs using a variety of immunogens. One of these (denoted NIBn 12B3) has been described elsewhere (33) and this antibody has been tested in preliminary imaging studies using the rabbit human thrombus model. Uptake of this antibody onto the human clot was demonstrated. This antibody was raised against the denatured chains of human fibrinogen. Using solid-phase enzyme immuno-assay and immunoblot procedures this antibody was found to recognise an epitope on the denatured A-alpha chain of fibrinogen which had a high affinity for fibrin (10^9 M) and did not react with fibrinogen in solution. NIBn 12B3 reacts with X-

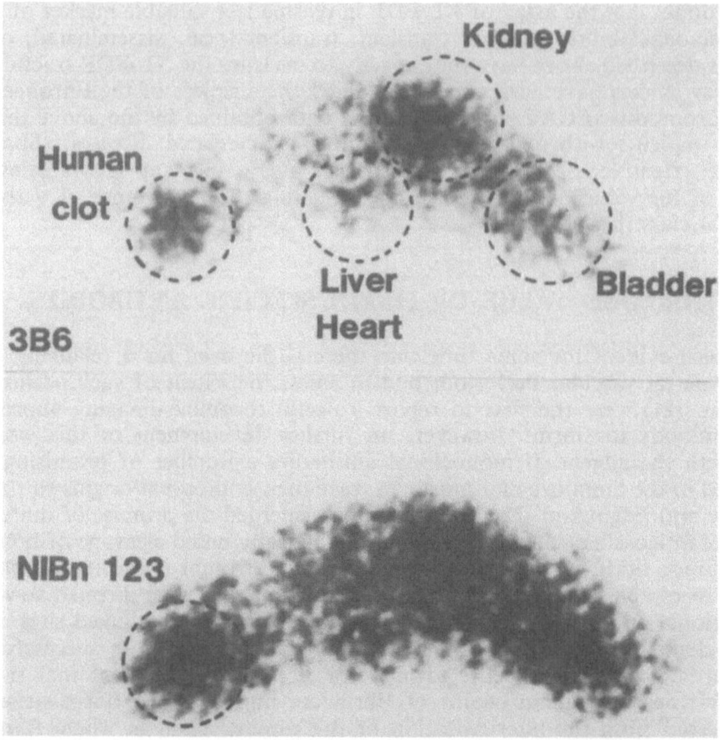

Fig. 4. Nuclear images 24 hours after injection of ^{131}-iodine labelled DD3B6 (upper image) and NIBn 123 (lower image) into the ear vein of rabbit with human jugular vein thrombus. (Taken in part from ref 31.)

oligomers but not with D dimer, D, E and T-NDSK. This is probably due to the fact that large portions of the domainal structures in fibrin are present in the X-oligomer structure. During the transition of crosslinked fibrin to the X-oligomers a 40 kd peptide is cleaved from the carboxy-terminal end of the A-alpha chain (see Fig. 1) suggesting that the epitope lies in the 27 kd peptide remaining. The inability of this antibody to bind to T-NDSK suggests that the antibody recognises a site between amino acid 51 and 214 on this A-alpha chain. The lack of cross reactivity of NIBn 12B3 with fragments D and E suggests that the epitope lies between residues 78 to 104 on the A-alpha chain (see Fig. 5). Further delineation of the epitope was carried out whereby A-alpha chains were treated with trypsin, plasmin and cyanogen bromide. Both plasmin and trypsin digests of the A-alpha chains lost their ability to bind to the antibody in solution while binding was retained in the CNBR digest, suggesting that the epitope lies in one of the following A-alpha sequences, 78-91, or 92-104.

CONCLUDING REMARKS AND FUTURE PERSPECTIVES

There is little doubt that the direct measurement in plasma of fibrin degradation products (FnDP) are a reliable marker of prior fibrin deposition in the vasculature. It remains to be seen what clinical significance can be derived from these assays.

The generation of fibrin specific monoclonal antibodies must have a broad application in vivo and in in vitro pathology. The immediate possibility of imaging fibrin rich thrombi and malignant tumours comes to mind. However it must be stressed that animal models used in some experiments are less than satisfactory because of the presumed lack of immunocross-reactivity of the various mab probes in most cases with circulating animal fibrinogen. The

1	Ala	Asp	Ser	Gly	Glu	Gly	Asp	Phe	Leu	Ala	Glu	Gly	Gly	Gly	Val	Arg	Gly	Pro	Arg	Val	Val	Glu	Arg	His	Gln
26	Ser	Ala	Cys	Lys	Asp	Ser	Asp	Trp	Pro	Phe	Cys	Ser	Asp	Glu	Asp	Trp	Asn	Tyr	Lys	Cys	Pro	Ser	Gly	Cys	Arg
51	Met ↓	Lys	Gly	Leu	Ile	Asp	Glu	Val	Asn	Gln	Asp	Phe	Thr	Asn	Arg	Ile	Asn	Lys	Leu	Lys	Asn	Ser	Leu	Phe	Glu
76	Tyr	Gln	Lys ↓↓	***ASN***	***ASN***	***LYS***	***ASP***	***SER***	***HIS***	***SER***	***LEU***	***THR***	***THR***	***ASN***	***ILE***	***MET***	***GLU***	***ILE***	***LEU***	***ARG***	***GLY***	***ASP***	***PHE***	***SER***	***SER***
101	***ALA***	***ASN***	***ASN***	***ARG*** ↓↓↓	Asp	Asn	Thr	Tyr	Asn	Arg	Val	Ser	Glu	Asp	Leu	Arg	Ser	Arg	Ile	Glu	Val	Leu	Lys	Arg	Lys
126	Val	Ile	Gln	Lys	Val	Gln	His	Ile	Gln	Leu	Leu	Gln	Lys	Asn	Val	Arg	Ala	Gln	Leu	Val	Asp	Met	Lys	Arg	Leu
151	Glu	Val	Asp	Ile	Asp	Ile	Lys	Ile	Arg	Ser	Cys	Arg	Gly	Ser	Cys	Ser	Arg	Ala	Leu	Ala	Arg	Glu	Val	Asp	Leu
176	Lys	Asn	Tyr	Glu	Asp	Gln	Gln	Lys	Gln	Leu	Glu	Gln	Val	Ile	Ala	Lys	Asp	Leu	Leu	Pro	Ser	Arg	Asp	Arg	Gln
201	His	Leu	Pro	Leu	Ile	Lys	Met	Lys	Pro	Val	Pro	Asn	Leu	Val	Pro	Gly	Asn	Phe	Lys	Ser	Gln	Leu	Gln	Lys	Val

Fig. 5. Part of the A-alpha chain indicating amino acid sequence location (large type) of the epitope to which NIBn 12B3 is directed. Single arrow indicates carboxy terminal cleavage point (Met 51) of amino terminal disulphide knot (N-DSK) while double and triple arrows show the plasmin-mediated amino terminal cleavage (lys 78) of fragment E and the carboxy terminal cleavage of fragment D (asp 105) respectively.

ideal imaging model is the human model and is as yet unclear which radiolabel is best, ^{131}Iodine, ^{111}Indium, ^{99}Tc^{m} or other. Considerable work is still required before the ideal thrombus imaging agent will be available.

Apart from thrombus imaging, fibrin specific mabs may be useful in targeting lytic therapy to occlusive thrombi. Ito et al. (34) have used fibrin specific mabs coupled to streptokinase as a potent thrombolytic complex while Bode et al. (35) have used a fibrin specific IgG and/or its Fab fragment coupled to urokinase as an efficient thrombus-specific lytic agent. More recently the group of Collen (36-38) have coupled a fibrin specific mab (15C5) with urokinase and various recombinant single chain urokinases and shown that these coupled antibody-enzyme constructs evidenced enhanced clot lysis in the rabbit thrombus model. Experiments such as these are of limited value in lytic therapy because of the supplies available of the various mablytic agent complexes. The concept of using recombinant DNA procedures to express gene constructs of an antibody (59D8) and the active site chain of tissue plasminogen activator lytic agent has been attempted (39). While the fibrin affinity and the plasminogen activation properties of the expressed fusion protein were realised, the poor expression demands further work in order to supply adequate quantities for human and animal experimentation. This approach of cloning gene constructs which express a hybrid protein possessing both lytic effect and fibrin or thrombus affinity seems to indicate the way forward.

REFERENCES

1. P. J. Gaffney, M. Brasher, K. Lord, C. J. L. Strachan, A. R. Wilkinson, V. V. Kakkar, and M. F. Scully, Fibrin subunits in venous and arterial thromboembolism. *Cardiovasc. Res.,* **10**:421 (1976).
2. D. Collen, On the regulation and control of fibrinolysis. *Thromb. Haemost.,* **43**:77 (1980).
3. P. J. Gaffney, Fibrinolysis, *in:* "Thrombosis and Haemostasis." A. L. Bloom, D. P. Thomas (eds) Churchill Livingstone, Edinburgh p. 223 (1987).
4. E. F. Plow and T. S. Edgington, Surface markers of fibrinogen and its physiological derivatives revealed by antibody probes, *Sem. Thromb. Haemost.,* **8**:36 (1982).
5. B. Kudryk, Z. D Grossman, J. G. McAfee and S. F. Rosebrough, Monoclonal antibodies as probes for fibrin(ogen) proteolysis, *in:* "Monoclonal Antibodies in Immunoscintigraphy." J.-F. Chatal (Ed) CRC Press, Boca Raton, FL., Ch. 19, 365 (1989).
6. P. J. Gaffney, The biochemistry of fibrinogen and fibrin degradation products, *in:* "The Biochemistry, Physiology and Pathology of Haemostasis. D. Ogston and B. Bennet (Eds) John Wiley, New York P. 105 (1977).
7. C. S. Francis and V. J. Marder, A molecular model of plasmic degradation of cross-linked fibrin. *Sem. Thromb. Haemost.,* **8**:25 (1981).

8. C. G. Curtis, Plasma factor XIII, *in:* "Thrombosis and Haemostasis." A. L. Bloom, D. P. Thomas (Eds) Churchill Livingstone, Edinburgh pp. 216 (1987).
9. P. J. Gaffney and F. D. P., *Lancet* **2:**1422 (1972).
10. D. B. Rylatt, A. S. Blake, L. E. Cottis, D. A. Massingham, W. A. Fletcher, P. P. Masci, A. N. Whitaker, M. Elms, I. Bunce, A. J. Webber, D. Wyatt and P. G. Bundesen, An immunoassay for human D-dimer using monoclonal antibodies, *Thromb. Res.,* **31:**767 (1983).
11. A. N. Whitaker, M. Elms, P. P. Masci, P. G. Bundesen, D. B. Rylatt, A. J. Webber and I. Bunce, Measurement of crosslinked fibrin derivatives in plasma: an immunoassay using monoclonal antibodies, *J. Clin. Path.,* **37:**882 (1984).
12. M. Elms. I. H. Bunce, P. G. Bundesen, D. B. Rylatt, A. J. Webber, P. P. Masci and A. N. Whitaker, Measurement of crosslinked fibrin degradation products — an immunoassay using monoclonal antibodies, *Thromb. Haemost.,* **50:**591 (1983).
13. P. J. Gaffney, L. J. Creighton, M. J. Perry, M. Callus, R. Thorpe and M. Spitz, Monoclonal antibodies to crosslinked fibrin degradation products (XL-FDP) I: Characterisation and preliminary evaluation in plasma, *Br. J. Haematol.,* **68:**83 (1988).
14. P. J. Gaffney, L. J. Creighton, M. Callus and R. Thorpe, Monoclonal antibodies to crosslinked fibrin degradation products (XL-FDP) II: Evaluation in a variety of clinical conditions, *Br. J. Haematol.,* **68:**91 (1988).
15. S. Rogers, P. M. Sweetman, M. J. Perry and P. J. Gaffney, Plasma levels of fibrin fragments in men with myocardial infarction, *Thromb. Res.,* **43:**389 (1985).
16. R. Thorpe, M. J. Perry, M. Callus, P. J. Gaffney and M. Spitz, Single shot intrasplenic immunisation: an advantageous procedure for production of monoclonal antibodies to human fibrin fragments, "Hybridoma" **3:**381 (1984).
17. P. J. Gaffney, L. J. Creighton, A. Curry, B. Mahon and R. Thorpe, Monoclonal antibodies to crosslinked fibrin degradation products: comparison of specificities of IgM and IgG subclasses, *in:* "Fibrinogen: Biochemistry, Physiology and Clinical Relevance." C. D. O. Lowe, J. T. Douglas, C. D. Forbes, A. Henschen (Eds) Elsevier, Amsterdam pp. 193 (1987).
18. P. J. Gaffney, L. J. Creighton, R. Harris and M. J. Perry, *in:* "Monoclonal antibodies (mabs) to crosslinked fibrin fragments; their characterisation and potential clinical use, "Fibrinogen and its Derivatives." G. Muller-Berghaus, U. Scheefers-Borchel, E. Selmayr, A. Henschen (Eds) Elsevier, Amsterdam pp. 273 (1986).
19. J. Koopman, F. Haverkate, P. W. Koppert, W. Nieuwenhuizen, E. J. P. Brommer and W. G. L. van der Werf, New immunoassay of fibrin(-ogen) degradation products in plasma using a monoclonal antibody, *J. Lab. Clin. Med.,* **109:**75 (1987).
20. P. W. Koppert, W. Kuipers, B. Hoegee-de Nobel, E. J. P. Brommer, J. Koopman and W. Nieuwenhuizen, A quantitative enzyme immunoassay (EIA) for primary fibrinogenolysis products in plasma, *Thromb. Haemost.,* **57:**25 (1987).
21. V. B. Bosnjakovic, B. D. Jankovic, J. Horvat and J. Cvoric, Radiolabelled anti-human fibrin antibody: a new thrombus-detecting agent, *Lancet,* **1:**452 (1977).
22. E. Plow and T. S. Edgington, Discriminatory neoantigenic differences between fibrinogen and fibrin derivatives, *Proc. Natl. Acad. Sci. USA,* **70:**1169 (1973).
23. E. Plow and T. S. Edgington, Surface markers of fibrinogen and its physiological derivatives revealed by antibody probes, *Sem. Thromb. Haemost.,* **8:**35 (1982).
24. J. H. Sobel, S. Birken, P. Ehrlich, R. Friedman, Z. Moustafa and R. E. Canfield, Characterisation of a crosslink-containing fragment derived from the x-polymer of human fibrin and its application in immunologic studies using monoclonal antibodies, *Thromb. Haemost.,* **46:**240 (abstract) (1981).
25. B. Lahiri, J. A. Koehn, R. E. Canfield, S. Birken and J. Lewis, Development of an immunoassay for the COOH-terminal region of the gamma chains of human fibrin, *Thromb. Res.,* **23:**103 (1981).
26. P. J. Gaffney and P. Dobos, A structural aspect of human fibrinogen suggested by its plasmin degradation, *FEBS Lett,* **15:**13 (1971).
27. S. F. Rosebrough, B. Kudryk, Z. D. Grossman, J. G. McAfee, G. Subramanian, C. A. Ritter-Hrncirik, L. S. Witanowski and G. Tillapaugh-Fay, Radioimmunoimaging of venous thrombi using iodine-131 monoclonal antibody, *Radiology,* **156:**515 (1985).
28. L. C. Knight, Imaging thrombi with radiolabelled anti-fibrin monoclonal antibodies, *Nucl. Med. Comm.,* **9:**823 (1989).
29. P. Broadhurst, R. Wilkins and A. Lahiri, Fibrin monoclonal antibodies in diagnosis of deep venous thrombosis, *Lancet,* **3:**272 (1989).

30. K. Y. Hui, E. Haber and G. R. Matsuda, Monoclonal antibodies to a synthetic fibrin-like peptide bind to human fibrin but not fibrinogen, *Science,* **222:**1129 (1983).
31. R. I. J. Feitsma, D. Blok, M. N. J. M. Wasser, W. Nieuwenhuizen and E. K. J. Pauwels, A new method for $^{99}Tc^{m}$ — labelling of proteins with an application to clot detection with an antifibrin monoclonal antibody, *Nucl. Med. Comm.,* **8:**771 (1987).
32. P. M. Tymkewycz, L. J. Creighton, P. S. Gascoine, G. D. Zanelli, P. M. Webbon and P. J. Gaffney, Imaging of human thrombi in the rabbit jugular vein: comparison of two fibrin specific monoclonal antibodies, *Thromb. Res.,* **54:**411-421 (1989).
33. P. M. Tymkewycz, P. S. Gascoine and P. J. Gaffney, A Monoclonal antibody (mab) to an A-alpha chain epitope which is exposed only on the surface of fibrin, *Thromb. Haemost.,* **62:**478 (1989).
34. R. K. Ito, G. L. Davis, E. J. Yunis, A Houranieh and B. E. Statland, Immunofibrinolysis: human fibrin-specific monoclonal antibodies as carriers for fibrinolytic agents, *Thromb. Haemost.,* **54:**276 (1985).
35. C. Bode, G. R. Matsueda, K. Y. Hui and E. Haber, Antibody-directed urokinase: a specific fibrinolytic agent, *Science* **229:**765 (1985).
36. D. Collen, M. Dewerchin, J. M. Stassen, L. Kieckens and H. R. Lijnen, Thrombolytic and pharmacokinetic properties of conjugates of urokinase-type plasminogen activated with a monoclonal antibody specific for crosslinked fibrin, *Fibrinolysis,* **3:**197 (1989).
37. M. Dewerchin, H. R. Lijnen and D. Collen. Characterisation of a chemical conjugate between a low molecular weight form of recombinant single chain urokinase-type plasminogen activator (comprising Leu^{144} through Leu^{411}) and $F(ab')_2$ — fragments of a fibrin D dimer-specific monoclonal antibody. *Fibrinolysis,* **4:**11 (1990).
38. M. Dewerchin, H. R. Lijnen, B. van Hoef, F. De Cock and D. Collen, Characterisation of conjugates of thrombin-treated single chain urokinase type plasminogen activator with a monoclonal antibody specific for crosslinked fibrin, *Fibrinolysis,* **4:**19 (1990).
39. J. M. Schnee, M. S. Runge, G. R. Matsueda, N. W. Hudson, J. G. Seidman, E. Haber and T. Quertermous, Construction and expression of a recombinant antibody-targeted plasminogen activator. *Proc. Natl. Acad. Sci. USA,* **84:**6904 (1987).

THE PHARMACOLOGY AND CLINICAL PHARMACOLOGY OF DEFIBROTIDE: A NEW PROFIBRINOLYTIC, ANTITHROMBOTIC AND ANTI-PLATELET SUBSTANCE

O.N. Ulutin, S. Balkuv-Ulutin, M.S. Ugur, T. Ulutin,
Y. Özsoy, and G. Çizmeci

Haemostasis and Thrombosis Research Center*
Cerrahpasa Medical School of Istanbul University
Istanbul, Turkey

ABSTRACT

Defibrotide, a deoxypolyribonuclide, has been found to modulate endothelial cell function causing increase in t-PA and decrease in PAI levels and also increase in PGI_2 production. In addition, it increases platelet c-AMP levels and decreases MDA and TXB_2 formation in human. Defibrotide inhibits platelet aggregate formation in vitro experiments as well as end-to-end anostomosis in rats. So, defibrotide inhibits the activation of platelets. Besides an increase of protein C and S levels a synergic action of heparin was observed in animal experiments. A strong antithrombotic effect has been observed in animal models.

The drug has a beneficial effect in the cases of DVT, POVD, stroke and thromboembolism. Through its action we may say that the drug acts in a novel fashion in contrast to the other drugs used in this area.

Defibrotide is a single-stranded polydeoxyribonucleotide obtained from deoxyribonucleic acid of mammalian lungs by controlled depolimerization. Since 1981 in our laboratory and in the clinical department we have been investigating a newly developed agent defibrotide *in vitro* experiments, animal experiments, and also its clinical pharmacology and clinical application.

Some of our findings are already published and compared with literature (40, 43, 46). Because of the limited space we are not going to review the literature in detail but we are going to summarize our observations on this compound in the following order. I — *in vitro* experiments, II — Animal experiments, III — clinical pharmacology in human.

In vitro studies

In a study we performed in our laboratory (10, 11) we demonstrated that after collagen and AA induction a significant increase in the PGI_2 and TXA_2 production in isolated leucocyte suspensions occurred (Fig. 1). On the other hand defibrotide inhibited these significantly (Table 1).

This inhibition was more significant in $PGF_{1\alpha}$, after AA and collagen inductions in isolated leucocyte suspensions.

Di Perri et al. (16, 17) demonstrated that in activated polymorphonuclear leucocyte suspension defibrotide inhibited lysozyme release and superoxide anion generation. In this study it has been shown that defibrotide has a modulating effect on the entrance and avail-

*Haemostasis and Thrombosis Research Center is affiliated with the Technical and Scientific Research Council of Turkey.

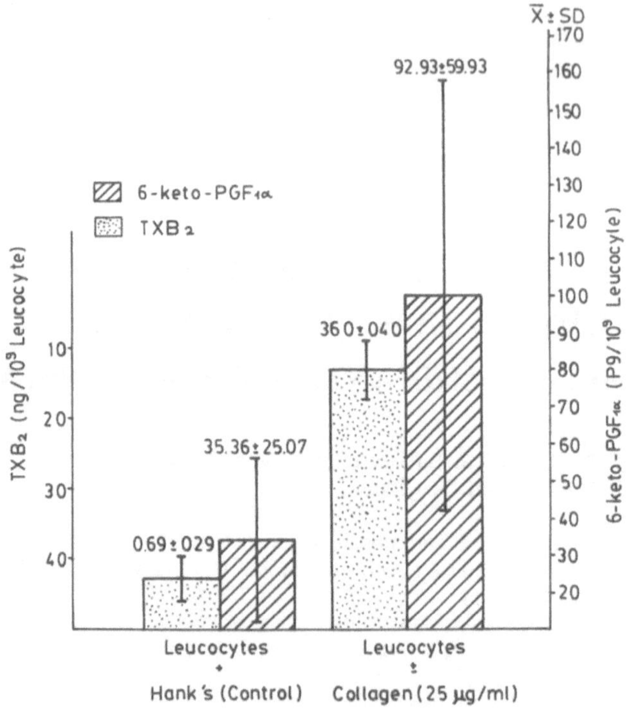

Fig. 1. Leucocytes isolated from healthy volunteers were shown to produce both TXA_2 and PGI_2.

ability of calcium ions to the cells. In this study as well in ours, the concentrations of defibrotide were higher than the therapeutic ranges.

Bilsel et al. (1) studied the effects of endothelial growth supplement and defibrotide on the human umbilical vein endothelial cell culture. Both of them caused endothelial cell proliferation in the culture, but endothelial growth supplement enhanced the proliferation of cells more profoundly. On the other hand, defibrotide stimulated the protein production more significantly. The remarkable increase of the protein production after incubation with defibrotide needs further clarification. Both findings are completely in accord with the findings observed in animals as well in human.

Animal experiments

Defibrotide has demonstrated strong antithrombotic effects in animal experiments (20, 24, 25, 30-33, 47, 55). In 1978 Prino et al. (30), and in 1981 and 1986 Niada et al. (31, 33) have shown that defibrotide has very strong antithrombotic effect on animal thrombosis models, has a profibrinolytic effect and causes an increase in PGI_2 formation in rabbits. These

Table 1. The Effect of Defibrotide in Isolated Leucocyte Suspension on the PGI_2 and TXA_2 Formation

	Control	Defibrotide (100 μg/ml)	Defibrotide (300 μg/ml)
$PGF_{1\alpha}$ pg/10^9 leucocyte	105.84 ± 68.35	33.27 ± 23.54	29.60 ± 20.21
TXB_2 ng/10^9 leucocyte	364.78 ± 182.46	304.07 ± 263.07	269.45 ± 87.05
X ± SD			

findings have been confirmed by other studies (20, 24, 25, 47). We also showed that defibrotide prevents experimental DIC in rabbits (37).

In one of our studies we already demonstrated that in the area of electrically mural thrombosis in dogs, the thickening of the vein wall with marked enlargement of the media and intima was observed. The thickening of the vessel wall was due mainly to the proliferation of endothelial and smooth muscle cells. The lumens were almost completely blocked by partially organized and retracted thrombi.

In dogs treated with defibrotide there were no or minimal histological changes of the vessel wall. The treated animals did show minimal morphological or ultrastructural alterations (SEM) (51-53).

Depending on these findings in dogs and on some observations in human we state in 1984 that "we can assume that there is a modifying effect of defibrotide on the impaired endothelial cell function (12, 13)." And now depending on the further studies we can state that defibrotide is a stimulating, supporting, preventing and modulating compound.

In 1988, in their laser-induced endothelial injury model in rats Weichert et al. (55) demonstrated that defibrotide intravenously and orally has dose-dependent antithrombotic effect.

In primates (*Macaca mulatta*) defibrotide increases PGI_2 production, inhibits TXB_2 generation, increases t-PA level progressively and causes a decrease in PAI (20, 24, 25). It has also been shown that jugular vein thrombosis model of rabbit defibrotide and heparin have synergic actions (20).

Defibrotide effectively inhibits platelet aggregate formation in the microarterial suture-line of end-to-end microarterial anostomosis in rats and also increases the reendothelization of anostomotic lien (34, 35). These findings fit to the platelet inhibition effect of the drug in human (9, 39, 44).

On the other hand Fumagelli et al. (21) and Sala et al. (39) studied antithrombotic effect of defibrotide morphologically and ultrastructurally applying "collagen coated nylon stread," which induced vena femoralis thrombosis in rats, and demonstrated that the drug decreases the size of thrombus significantly, inhibits the platelet formation of platelet aggregates and lesser amount of leucocytes found in the clot. It has been shown in guinea pigs and cats that defibrotide causes an increase of the level of PGI_2 in coronary vessels and so myocardial damage was less in ischaemic myocardium (26, 38). Finally the antithrombotic effect of the drug disappears by hepatic isolation in rabbits. According to this finding it is our opinion that it is possible that this compound converts to an active product in the liver (24, 25).

Clinical pharmacology in human

We are going to summarize the effect of defibrotide on different mechanisms as follows: a- profibrinolytic action; b- synergic action with heparin; c- the effect on protein C and S levels; d- the effect on PGI_2, MDA and TXA_2 levels; and e- the effects on platelets; such as differential counts of platelets on the foreign surface, the level of platelet c-AMP, platelet molecular markers, etc.

The effect of fibrinolytic activity

It has been shown that in different studies defibrotide activates fibrinolytic system in human (14, 22). This agent progressively corrects the low fibrinolytic activity and venous stasis test (cuff test) in vascular diseases (22, 23, 40, 48). Defibrotide causes an increase in the production and release of t-PA together with a decrease of PAI level. Cuff test values show a significant increase during defibrotide treatment in normal volunteers and in the patients with peripheral occlusive vascular diseases (POVO) (6, 7, 28) (Table 2). Similar findings were also demonstrated in primates (20, 24, 25). Defibrotide affecting the endothelial cells, results in an increase in t-PA and decrease in PAI, and causes an increase in global fibrinolytic activity which we detected using euglobulin lysis time and fibrin platelets methods. During defibrotide treatment plasma level of α_2-antiplasmin is also found decreased.

The synergic action of defibrotide with heparin

The synergism between defibrotide and heparin was detected first as a clinical observation (2) then this observation was confirmed by heparin loading test in our laboratory (41, 43, 46, 50) (Fig. 2).

Table 2. The Effect of Defibrotide on t-PA and PAI Plasma Levels in Normal Volunteers. Determination Has Been Done with Elisa Method Using Asserachrom t-PA and Stachrom PAI. Values Are ng/ml t-PA: Ag and PAI IU/ml. Defibrotide Was Given Intravenously as a Bolus 600 mg/daily

	Before	30 min	7th day
t-PA (N:14)	6.2±2.8	14.2±4.2 p<0.01	16.3±4.8 p<0.001
PAI (N:14)	7.4±3.2	5.1±2.6 p<0.1	3.8±1.8 p<0.005

As we mentioned before this synergism was shown with jugular vein thrombosis in rabbits (20). The relation with heparin and defibrotide was confirmed in animals (29) and in human studies (27). In our recent study beside synergic action with heparin we demonstrated that after defibrotide injections an anti-Xa activity appeared and increased in the plasma and when we applied defibrotide with LMW heparin (Cy 222) the anti-Xa activity increased more than heparin alone and this activity stayed longer time (42). (Table 3). These findings is one of the reason of high antithrombotic effect of defibrotide.

The effect of defibrotide on protein C and S levels

In our studies we found that defibrotide caused significant increase of the plasma levels of protein C and protein S (5, 45) (Table 4). These findings also explain why defibrotide has a very strong antithrombotic and preventive action on the thromboembolism. The increase of natural inhibitors (protein C and S) is one of the reasons of this effect. Protein C synthetized in the liver as one of the Vit-K dependent factors as well protein S. From this standpoint we may suggest that defibrotide also stimulates liver cells. As we mentioned before the isolation of liver in rabbits totally blocked the antithrombic actions

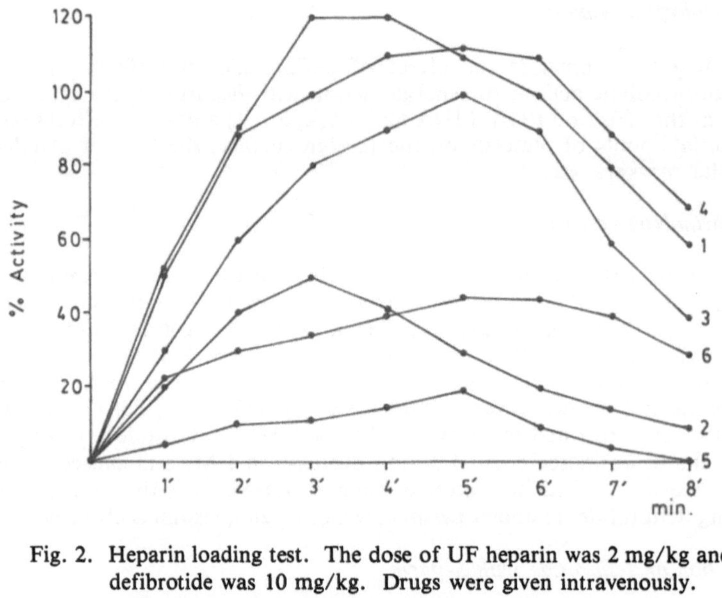

Fig. 2. Heparin loading test. The dose of UF heparin was 2 mg/kg and defibrotide was 10 mg/kg. Drugs were given intravenously.
1- before, 2- 20 min after heparin, 3- 60 min after heparin, 4- 30 min after defibrotide, 5- 20 min after heparin and defibrotide, 6- 60 min after heparin and defibrotide.

Table 3. **The Effect of Defibrotide on the Anti-Xa Activity of Plasma. Dilution Curve Was Done for Heptest with LMW Heparin (Cy 222)**

	Before	15'	30'	45'	60'	120'	180'	360'
Defibrotide γ/ml (N:13)	0	180.62	77.36	44.17	24.95	22.95	0	0
Heptest U/ml (N:11)	0	0.0177	0.0297	0.0329	0.0121	0.084	0.0058	0.0048

Table 4. **The Effect of Defibrotide on the Level of Protein C and Protein S. Determinations Were Done Using Asserachrom-Protein C and Asserachrom-Protein S Kits**

	Before	7th day	Significance
Protein C (N:14)	86.4 ± 11.4	122.6 ± 19.2	$p<0.005$
Protein S (N:14)	91.6 ± 12.6	104.4 ± 21.4	$p<0.05$

of defibrotide (24, 25) then we may also suggest that the liver has a role on the pharmacological action of defibrotide in *in vivo* conditions. Recently we showed (8) that defibrotide causes morphological changes in the rat liver. Ultrastructural evidence showed that defibrotide stimulates the liver cells (8). These results will be published separately with laboratory and electronmicroscopical data. Protein S synthetised in human endothelial cells beside liver, so the effect of defibrotide on this factor still needs further investigation (19).

The effects of defibrotide on PGI_2, TXA_2 and platelet c-AMP levels

In our laboratory we demonstrated in normal volunteers and the causes of POVD, that defibrotide caused a progressive increase in the production of PGI_2 with a simultaneous decrease in production of MDA and TXB_2 (9, 12, 13, 40, 46). Defibrotide is also caused an increase of platelet c-AMP level (Table 5). This also has been shown in primates (20, 24, 25, 47). We believe that the therapeutic action of compound is mostly due to the increase of PGI_2 production.

The effect of defibrotide on platelets

Defibrotide has an inhibitory effect on platelet functions, but we don't know if this

Table 5. **The Effect of Defibrotide on 6-Keto-$PGF_{1\alpha}$, TXB_2 and Platelet c-AMP Levels in Normal Subjects (N:14)**

	Before	30 min	3 hours	7th day
6-Keto-$PGF_{1\alpha}$ (Pg/ml)	54.2 ± 23.4	91.40 ± 37.1 $p<0.05$	—	101.60 ± 41.4 $p<0.005$
TXB_2 (ng/ml)	1.36 ± 0.91	0.98 ± 6.36 NS	0.62 ± 0.41 $p<0.025$	1.12 ± 0.54 NS
Platelet c-AMP (pmol/10^8 platelets)	12.90 ± 3.7	18.60 ± 4.1 $p<0.005$	—	15.30 ± 3.2 $p<0.05$

Table 6. The Differential Platelet Count on Formvar Membrane in 8 Cases of POVD Atherosclerotic Origin Before, 30 min After and on 7th Day of Defibrotide Treatment

	Round	Dendritic	Intermediate	Spread	Aggregate Single Small	Per Platelet Gross
Before	3	24	27	46	27	12
30 min. after	7	48	33	12	8	1
7th day	6	53	26	15	5	–

effect was direct one or through the increase of PGI_2 level. This point is still open for discussion.

In human defibrotide alters differential counts of platelets on foreign surfaces and also inhibits especially platelet aggregate formation and platelet adhesion (39, 44, 47) (Table 6).

Using thin section method of EM Ugur (39) showed that defibrotide inhibits platelet functions as follows: decreases adhesion, decreases surface connecting tubular systems, decreases dendritis formation, increases discoid forms, increases numbers of intracellular granules that are distributed in the cell separately. These findings also will be published with the other platelet alterations. These results conform to the findings about the inhibition of platelet aggregate formation on microarterial suture-line in rats (34, 35).

After defibrotide injections decrease of maximum amplitude of ADP and adrenalin induced aggregation curves and a prolongation of lag-period of collagen induced aggregation were observed (46). A significant increase of platelet c-AMP level was also demonstrated (9, 12, 40, 46) (Table 5).

During defibrotide treatment molecular markers of platelets decreased progressively such as ß-TG and PF4 (Table 7).

The results of clinical pharmacology of defibrotide have been summarized in different publications (15, 40, 43, 46). The findings which we discussed and summarized above are due to the results when the drug was given intravenously. The following table summarizes our results obtained with oral application in 15 normal volunteers and 17 patients with POVD (Table 8).

We would like to summarize our results in the scheme (Fig. 3). Defibrotide effects endothelial cell directly and/or through the liver.

Defibrotide has a multipotential effect so we may suggest that its effects are not related to a certain biochemical pathway but affect the cell itself.

All these findings lead us to assume that the mode of action of this drug is mainly through endothelial cell.

Due to all these effects the compound has been used in different diseases and we obtained encouraging results in DVT (18), POVD (40, 49) and stroke (3, 4).

We can name this therapy as "endothelium cell supporting therapy through functional modulation of endothelium."

Table 7. The Effect of Defibrotide on the Molecular Markers of Platelets. Determinations Were Done Using Asserachrom -BTG and Asserachrom PF4 Kits. Results Due to 16 Cases of POVD

	Before	30th min	24th hour	7th day
PF-4 IU/ml	10.2 ± 2.3	12.3 ± 2.06 NS	9.3 ± 1.9 $p < 0.10$	7.3 ± 1.96 $p < 0.01$
B-TG	36.7 ± 5.15	37.3 ± 4.55 NS	29.0 ± 4.45 $p < 0.01$	24.1 ± 2.96 $p < 0.001$

Table 8. The Effect of Per Oral Defibrotide (800 mg/daily) on the Levels of t-PA, PAI and 6-Keto-PGF$_{1\alpha}$ in Normal Volunteers and 17 Patients of POVD. Results Represent the Mean±SD Baseline and During Defibrotide Treatment. Student's Test, Baseline Versus of Defibrotide (1988)

	Before	7th day	30th day
t-PA Control (ng/ml)	6.21±2.41	9.80±2.9 $p<0.05$	17.20±4.3 $p<0.001$
POVD	11.42±3.4	12.52±3.1 NS	16.40±3.8 $p<0.01$
PAI POVD (IU/ml)	14.60±4.6	10.42±3.8 $p<0.01$	8.43±3.4 $p<0.005$
6-Keto-PGF$_{1\alpha}$ Control	64.2±16.2	86.60±21.4 $p<0.005$	91.30±36.4 $p<0.025$
POVD	44.60±22.2	57.30±19.8 $p<0.05$	78.50±28.8 $p<0.005$

REFERENCES

1. S. Bilsel, Y. Taga, A. S. Yalgin, K. Emerk, and O. N. Ulutin, Effect of defibrotide and endothelial cell growth supplement on human endothelial cells in culture. *Haematology Reviews,* **3**:29-34 (1989).
2. S. Balkuv-Ulutin (unpublished observation 1985).
3. S. Balkuv-Ulutin, E. Goldenberg, M. Bali, A. Kavukçu, and O. N. Ulutin, Clinical Results with Defibrotide in Patients with Acute Stroke. *Haemostasis,* **18**: Suppl. 2, 81 (1988).
4. S. Balkuv-Ulutin, M. Hacihanefioglu, and O. N. Ulutin, Clinical and laboratory results obtained with defibrotide in the cases of acute stroke (Abst. 1874). *XIIth Congress of ISTH,* Aug. 19-25, (1989) Tokyo.

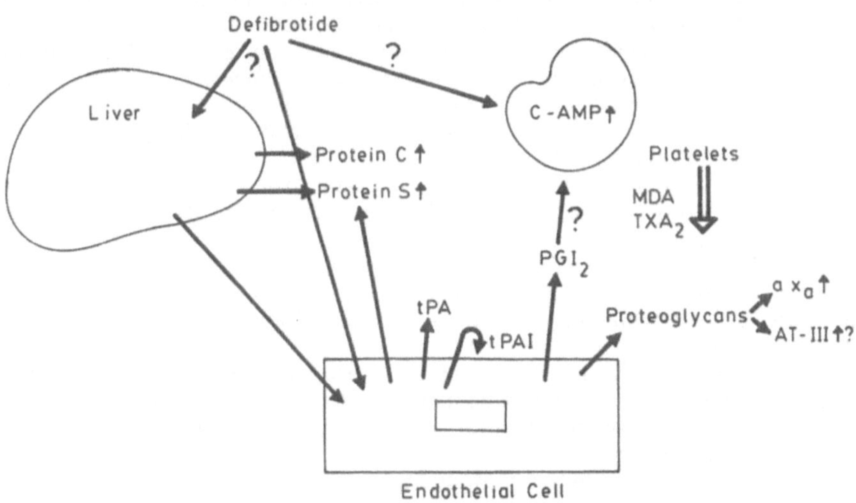

Fig. 3. Schema of the effect of defibrotide. Ulutin et al. (1981-1986).

5. S. Balkuv-Ulutin, T. Ulutin, G. Kaya, T. Yardimci, and O. N. Ulutin, The effect of defibrotide on protein C levels in human (preliminary results). *Thrombosis and Hemorrhagic Diseases,* ed. by O.N. Ulutin and H. Vinazzer, Gözlem Printing and Publishing Co., Istanbul (1986) pp. 409-410.
6. S. Balkuv-Ulutin, T. Ulutin, Y. Özsoy, G. Çizmeci, T. Yardimci, B. Ferhanoglu, and O. N. Ulutin, The effect of defibrotide on t-PA and PAI levels and also platelet function. *Marmara Med. J.,* **1:**48 (1988).
7. S. Balkuv-Ulutin, T. Ulutin, Y. Özsoy, B. Ferhanoglu, and O. N. Ulutin, The effects of defibrotide on tissue plasminogen activator and molecular markers of platelets (Abs. 316). *Xth Internat. Congress on Thrombosis,* May 23-27, 1988 Athens.
8. A. Çevikbas, O. Ulutin, U. Çevikbas, T. Yardimci, and T. Erbengi, Defibrotidin fare karacigerine etkisi. (An ultrastructural study). *Proc. of IXth National Congress of Electronmicroscopy,* May 29-31, (1989) p. 255-256.
9. G. Çizmeci: *In vivo* effect of defibrotide on platelet c-AMP and blood prostanoid levels. *Haemostasis,* **16:** Suppl. 1, 31-35, (1986).
10. G. Cizmeci, and O. N. Ulutin: In vitro effect of defibrotide on the production of PGI_2 and TXB_2 in leucocytes (Abs. 1464). *Thrombosis and Haemostasis,* **54:**210 (1985).
11. G. Çizmeci, and O. N. Ulutin: Prostanoid synthesis in isolated leucocytes and in-vitro effect of defibrotide. *Thrombosis and Hemorrhagic Disease,* ed. by O.N. Ulutin and H. Vinazzer, Gözlem Printing and Publishing Co., Istanbul (1986) pp. 199-204.
12. G. Çizmeci, and O. N. Ulutin: Defibrotide antithrombotic effect existing from modified endothelium function (Abst. 824). *XXth Congress ISH,* Sept. 1-7, (1984) Buenos Aires.
13. G. Çizmeci, and O. N. Ulutin: Corrective effect of defibrotide on altered endothelium cell function in atherosclerosis (Abst. 515). *Thromb. Haemost.,* **54:**87 (1985).
14. S. Coccheri, V. de Rosa, A. G. Dettori, D. Ponari, B. Bizzi, N. Ciaverella, and A. Isidori: Effect of fibrinolysis of a new antithrombotic agent: Fraction P (defibrotide) *Int. J. Clin. Pharmacol. Res.,* **11:**227-245 (1982).
15. U. Cornelli, and M. Nazzari: Defibrotide: An overview of clinical pharmacology and early clinical studies. *Seminars in Thrombosis and Hemostasis,* **14:**64-70 (1988).
16. T. Di Perri, F. Laghi-Pasini, P. L. Capecchi, L. Ceccatelli, A. L. Pasqui, and A. Orrico: Defibrotide *in vitro* inhibits neutrophil activation by a Ca** involving mechanism. *Int. J. Tiss. Reac.,* **IX (5):** 399-406 (1987).
17. T. Di Perri, and F. L. Pasini: Effect of defibrotide on polymorphonuclear leukocytes: Modulation of calcium entry and availability. *Seminars in Thrombosis and Haemostasis,* **14:**Suppl. 23-26 (1988).
18. T. Di Perri, A. Vittoria, G. L. Messa, and R. Cappelli, Defibrotide therapy for thrombophlebitis: Controlled clinical trial. *Haemostasis,* **16:** Suppl. 1, 42-47 (1986).
19. D. S. Fair, R. A. Marlar, and E. G. Levin, Human endothelial cells synthetize protein S. *Blood,* **67:**1168-1171 (1986).
20. J. Fareed, J. M. Walenga, D. A. Hoppenstead, A. Kumar, O. N. Ulutin, U. Cornelli, Pharmacologic profiling of defibrotide in experimental models. *Seminars in Thrombosis and Haemostasis,* **14:** Suppl. 1, 27-37 (1988).
21. G. Fumagalli, E. Angelaccio, N. Lambardo, and Francisco Clementi, Morphometric and ultrastructural study of experimental venous thrombosis. *Haemostasis,* **17:**361-370 (1987).
22. S. Ilhan-Berkel, S. Balkuv-Ulutin, and O. N. Ulutin, Fibrinolytic effect in man of a new extractive drug: defibrotide. VIIth Internat. Congr. on Fibrinolysis, Venice (1984) *Haemostasis,* **14:** Suppl. (1984).
23. N. Ilhan-Berkel, S. Balkuv-Ulutin, and O. N. Ulutin, The effect of defibrotide on the venous stasis test (Cuff test). VIIIth Internat. Congr. on Thrombosis, June 4-7, (1984) (Abst. 110).
24. A. Kumar, J. Fareed, W. H. Wehrmacher, D. Hoppensteadt, O. N. Ulutin, and J. M. Walenga, Endothelial function modulation and control of vascular or thrombotic disorders: Experimental results with a polydeoxyribonucleotide agent defibrotide. *Blood,* **68:** Suppl. 1, 356c (1986).
25. A. Kumar, J. Fareed, W. H. Wehrmacher, D. Hoppensteadt, O. Ulutin, and J. M. Walenga, Endothelial function modulation and control of vascular or thrombotic disorders: Experimental results with a polydeoxyribonucleotide agent defibrotide. *Thrombosis and Haemostasis,* **58:**350 (1987).
26. P. Löbel, and K. Schrör: Selective stimulation of coronary vascular PGI_2 but not of platelet thromboxane formation by defibrotide in the platelet perfused heart. *Naunyn-Schmiedeberg's Arch. Pharmacol.,* **331:**125-130 (1985).

27. E. M. Pogliani, M. Salvatore, C. Fausti, R. Girardello, and C. Marelli, Effects of a defibrotide-heparin combinations on some measures of haemostasis in healthy volunteers *J. Internat. Med. Res.*, **17**:36-40 (1989).
28. E. M. Pogliani, M. Salvatore, C. Faust, C. Marelli, Studio di bioequivalenza di due schemi posologici di defibrotide su parametri della fibrinolizi in soggetti valontari sani *Farmaci and Terapi*, **4**:1-5 (1987).
29. R. Porta, R. Pescador, M. Martovani, P. Alberico, R. Tettamanti, and G. Prino, Interference of defibrotide with heparin's anticoagulant activity and pharmacokinetics *6th Meeting of the Danubian League Against Thrombosis and Hemorrhagic Disorders*, May 31-June, **3**: (1989) p.46.
30. G. Prino, M. Mantovani, and R. Niada, Antithrombotic activity of a polydexynucleotidic-like substance (Fraction P). *Advance in coagulation fibrinolysis, platelet aggregation and atherosclerosis*, ed. by A. Stano C.E.P.I., Roma (1978), pp. 288-289.
31. R. Niada, M. Mantovani, G. Prino, R. Pescador, F. Berti, C. Onini, and G. C. Folco, Antithrombotic activity of a polydeoxyribanucleatidic substance extracted from mammalian organs: A possible link with prostacyclin. *Thrombosis Research*, **23**:233-246, (1981).
32. R. Niada, M. Mantovani, G. Prino, R. Pescador, R. Porta, F. Berti, G. C. Folco, C. Onini, and T. Vigano, PGI_2-generation and antithrombotic activity of orally administered defibrotide. *Pharmacol. Res. Commun.*, **14**:949-957 (1982).
33. R. Niada, R. Pescador, R. Porta, M. Mantovani, and G. Prino, Defibrotide is antithrombotic and thrombolytic against rabbit venous thrombosis. *Haemostasis*, **16**: Suppl. 3-8 (1986).
34. A. F. Özer, M. N. Pamir, T. Erbengi, and O. N. Ulutin, The early and late effect of defibrotide on the platelet aggregation and reendothelization on the microarteria suture-line. *XVth World Congr. of Internat. Union of Angiology*. Sept. 17-22, (1989) Rome.
35. A. F. Özer, M. N. Pamir, T. Erbengi, and O. N. Ulutin, The effect of defibrotide on thrombus formation in microarterial anostomosis. *Internat. World Congr. of Neurological Surgery*, Oct. 8-13 (1989), New Delhi.
36. C. Sala, E. Angelaccio, N. Lambardo, and G. Fumagalli, Morphometric and ultrastructural analysis of the antithrombotic activity of defibrotide. *Cellular blood components in haemostasis and thrombosis*, ed. by T. Barbui, S. Cartelazzo, P. Viero, S. Gorini and G. de Gaetano, John Libbey Eurotext, Paris, (1988) pp. 189-193.
37. A. Süer, S. Toplan, M. A. Özdemir, O. N. ve Ulutin, Defibrotid'in tavsanlarda DIC üzerine etkisi. (The effect of defibrotide on experimental DIC in rabbits). *Hematoloji*, **IX**:188-189, (1987).
38. C. Thiermann, P. Löbel, and K. Schrök, Uselfulness of defibrotide in protecting ischemic myocardium from early reperfusion damage. *Amer. J. Cardiology*, **56**:978-982, (1985).
39. M. S. Ugur: Defibrotide' in trombosit ultrastrukturune etkisi. (The effect of defibrotide on platelet ultrastructure). *Ph.D. thesis, Istanbul*, (1986).
40. O. N. Ulutin, Clinical effectiveness of defibrotide in vasocclusive disorders and its mode of action. *Seminars in Thrombosis and Haemostasis*, **14**: Suppl. 1, 58-63, (1988).
41. O. N. Ulutin, Clinical effectiveness and mode action of defibrotide. *Second Congr. Mediterranean Society of Therapy*, July 20-25 (1986), Istanbul, p. 98.
42. O. N. Ulutin, The clinical and pharmacological results of defibrotide during treatment of POVD. *15th World Congress of Internat. Union of Angiology*, Sept. 17-22, (1989), Rome.
43. O. N. Ulutin, S. Balkuv-Ulutin, G. Çizmeci, T. Ulutin, and B. Ferhanoglu, Some clinical pharmacological observations of the effect of defibrotide in human. *Acta Pharmaceutica Turcica*, **30**:57-62, (1988).
44. O. N. Ulutin, S. Balkuv-Ulutin, M. S. Ugur, T. Ulutin, T. Erbengi, B. Ferhanoglu, and T. Yardimci, Defibrotide and its effects on platelet functions. *6th Meeting of the Danubian League Against Thrombosis and Haemorrhagic Disorders*, May 31- June 3, (1989), Vienna, p. 48.
45. T. Ulutin, S. Balkuv-Ulutin, T. Yardimci, F. Demirkol, B. Ferhanoglu, and O. N. Ulutin, *XXth National Congress of Haematology*, Nov. 21-25, (1988) Ankara, Turkey p. 57.
46. O. N. Ulutin, G. Çizmeci, and S. Balkuv-Ulutin: Clinical pharmacology and mode of action of a new antithrombotic compound defibrotide. *Folia Haematol., Leipzig*, **115**:177-180, (1988).

47. O. N. Ulutin, G. Çizmeci, and M. S. Ugur, Clinical pharmacology of defibrotide and its effect on platelet function (Abs. 189). *Haemostasis,* **18:**, Suppl. 2, 143, (1988).
48. O. N. Ulutin, N. Ilhan-Berkel, and S. Belkuv-Ulutin, Some observations in man with a new extractive drug: defibrotide. *Progress in Fibrinolysis VII,* ed. by J. F. Davidson, M. B. Donati, S. Coccheri, Churcill-Livingstone, Edinburgh (1985), pp. 327-328.
49. O. N. Ulutin, N. Ilhan-Berkel, H. Tunali, M. Özer, S. Balkuv-Ulutin, and Ç. Önsel, Effects of defibrotide on peripheral obliterative vascular diseases. *Haemostasis,* **16:** Suppl. 1, 59-62, (1986).
50. O. N. Ulutin, and D. Sestakof, A study on hypercoagulability using heparin loading test. *Henry Ford Hosp. Med. Bull.* **8:**63-67, (1960).
51. O. N. Ulutin, H. Tunali, G. Girisken, S. Aytis, M. S. Ugur, and S. Balkuv-Ulutin, The effect of fibrinolytic on electrically induced experimental venous thrombosis and vessel wall in dogs. *XIIIth World Congr. of International Union of Angiology,* Sept. 11-15, (1983) Rochester, Minnesota.
52. O. N. Ulutin, H. Tunali, G. Girisken, S. S. Aytis, M. S. Ugur, and S. Balkuv-Ulutin, The effect of fraction P, on the electrically induced thrombus formation in dogs. (7th Internat. Congr. on Thrombosis, Valencia, Oct. 1982) *Haemostasis,* **12:**130, (1985).
53. O. N. Ulutin, H. Tunali, M. S. Ugur, S. Aytis, T. Erbengi, and S. Balkuv-Ulutin, Effect of defibrotide in electrically induced thrombosis in dogs. *Haemostasis,* **16:** Suppl. 1, 9-12, (1986).
54. W. Weichert, H. K. Breddin, and J. Staubesand, Application of a laser-induced endothelial injury model in the screening of antithrombotic drugs. *Seminars in Thrombosis and Hemostasis,* **14:**Suppl. 1, 108-114, (1988).

AUTHOR INDEX

Chen, Jia-Chyuan, 265
Institute of Life Science
National Tsing Hua University
Hsin-Chu, Taiwan, ROC

Cheng, S. M., 201
Biotechnology & Microbiology Division
Wyeth-Ayerst Research Inc.
P.O.Box 8299
Philadelphia, PA 19101, USA

Cheng, Xiang-Fei, 185
Dept. of Medical Biochemistry and Biophysics
University of Umea
S-90187 Umea, Sweden

Chiang, Benjamin N., 395
Division of Cardiology
Veterans General Hospital
No. 201, Shih-Pai Rd.,Sec. 2
Shih-Pai, Taipei, Taiwan, ROC

Chien, Shu, 409
Dept. of AMES-Bioengineering, R-012
University of California, San Diego
La Jolla, CA 92093, USA

Chung, Dominic W., 39
Dept. of Biochemistry
University of Washington
Seattle, WA 98195, USA

Cizmeci, G., 429
Haemostasis and Thrombosis Research Center
Cerrahpasa Medical School
Istanbul University
Istanbul, Turkey

Collen, Desire J., 333
Center for Thrombosis and Vascular Research
University of Leuven
B-3000 Leuven, Belgium

Colman, Robert W., 105
Thrombosis Research Center
Temple University
School of Medicine
3400 North Broad Street
Philadelphia, PA 19140, USA

Colman, Roberta F., 257
Dept. of Chemistry and Biochemistry
University of Delaware
Newark, DE 19716, USA

Cook, Jacquelynn J., 251
Dept. of Physiology and Thrombosis Research Center
Temple University
School of Medicine
3400, North Broad Street
Philadelphia, PA 19140, USA

Creighton, L.J., 419
National Institute for Biological Standards and Control
Blanche Lane, Potters Bar
Hertfordshire EN6 3QG, UK

Cruickshank, J.K., 145
MRC Epidemiology and Medical Care Unit
Northwick Park Hospital
Watford Road, Harrow
Middlesex, HA1 3UJ, UK

Davie, Earl W., 39
Dept. of Biochemistry
University of Washingtion
Seattle, WA 98195, USA

de Bono, David P., 377
Dept. of Cardiology
University of Leicester
Glenfield General Hospital
Leicester, LE3 9QP, UK

de Fouw, N.J., 235
Unilever Research Laboratorium
Vlaardingen, The Netherlands

DiBello, Patricia M., 73
Research Institute
The Cleveland Clinic Foundation
Cleveland, OH 44195, USA

Doolittle, Russell F., 25
Center for Molecular Genetics M-034
University of California, San Diego
La Jolla, CA 92093, USA

Eckardt, Annette, 251
Dept. of Physiology and Thrombosis Reserch Center
Temple University
School of Medicine
3400, North Broad Street
Philadelphia, PA 19140, USA

Fenoglio Jr., J., 313
Columbia University
College of Physicians and Surgeons
630 W. 168th St.
New York, NY 10032, USA

Fenton II, John W., 177
Wadswordd Center for Laboratories and Research
New York State Department of Health
Albany, NY, USA

Gaffney, P.J., 419
National Institute for Biological Standards
and Control
Blanche Lane, Potters Bar
Hertfordshire EN6 3QG, UK

Gascoine, P.S., 419
National Institute for Biological Standards
and Control
Blanche Lane, Potters Bar
Hertfordshire EN6 3QG, UK

Gold, H.K., 333
Massachusetts General Hospital
Harvard Medical School
Boston, MA, USA

Graor, Robert, 73
Research Institute
The Cleveland Clinic Foundation
Cleveland, OH 44195, USA

Grossman, Betty, 401
College of Physicians and Surgeons
Columbia University
630 W. 168th St.
New York, NY 10032, USA

Hamsten, A., 1
King Gustaf V Research Institute
Karolinska Hospital
S-10401 Stockholm, Sweden

Harris, Jeff E., 39
Dept. of Biochemistry
University of Washington
Seattle, WA 98195, USA

Hase, Sumihiro, 121
Dept. of Chemistry
Osaka University
College of Science
Osaka 565, Japan

Haverkate, Frits, 235
Gaubius Institute TNO
P.O Box 612
2300 AP Leiden
The Netherlands

Henschen, Agnes H., 49
Dept. of Molecular Biology and
Biochemistry
University of California, Irvine
Irvine, CA 92717, USA

Hessel, Birgit D., 1
Dept. of Blood Coagulation Reaserch
Karolinska Institute
P.O.Box 60 400
S-10401 Stockholm, Sweden

Houghten, Richard A., 133
Scripps Clinic and Research Foundation
BCR-8
10666 N. Torrey Pines Rd.
La Jolla, CA 92037, USA

Hu, S. J., 277
Dept. of Physiology
College of Medicine
National Cheng-Kung University
Tainan, Taiwan, ROC

Huang, Tur-Fu, 151
Institute of Pharmacology
College of Medicine
National Taiwan University
No. 1, Jen-Ai Rd., Sec. 1
Taipei 10018, Taiwan, ROC

Hung, Paul P., 201
Biotechnology & Microbiology Division
Wyeth-Ayerst Research Inc.
P.O.Box 8299
Philadelphia, PA 19101, USA

Ikenaka, Tokuji, 121
Dept. of Chemistry
Osaka University
College of Science
Osaka 565, Japan

Iwanaga, Sadaaki, 121
Dept. of Biology
Faculty of Science
Kyushu University 33
6-11-1, Hakozaki
Higashi-Ku
Fukuoka 812, Japan

James, H. L., 201
Washington University
St. Louis, MO 63110, USA

Jan, Kung-ming, 409
Division of Cardiology
Dept. of Medicine
College of Physicians and Surgeons
Columbia University
New York, NY 10032, USA

Jen, C. J., 277
Dept. of Physiology
College of Medicine
National Cheng-Kung University
Tainan 70102, Taiwan, ROC

Jung. F., 389
Dept. of Clinical Haemostaseology
and Transfusion Medicine
University of Saarland
D-6650 Homburg/Saar, FRG

Kahn, Nighat N., 271
Division of Cardiology
Montefiore Medical Center
Albert Einstein College of Medicine
111 E. 210th St.
Bronx, NY 10467, USA

Kalyan, N. K., 201
Biotechnology & Microbiology Division
Wyeth-Ayerst Research Inc.
P.O.Box 8299
Philadelphia, PA 19101, USA

Kaplan, Karan L., 313
Mount Sinai Services
City Hospital Center at Elmhurst
The City University of New York
79-01 Broadway, Elmhurst
New York, NY 11373, USA

Kawabata, Shun-Ichiro, 121
Dept. of Biology
Faculty of Science
Kyushu Univeristy 33
Fukuoka 812, Japan

Kisiel, Walter, 121
Blood System Foundation Laboratory
Dept. of Pathology
University of New Mexico
School of Medicine
Albuquerque, NM 87131, USA

Kudryk, Bohdan, 313, 401
The New York Blood Center
310 E. 67th St.
New York, NY 10021, USA

Kwaan, Hau C., 367
Dept. of Medicine
Northwestrn University
Medical School
Section of Hematology/Oncology
VA Lakeside Medical Center
Chicago, IL 60611-4494, USA

Lee, S. G., 201
Biotechnology & Microbiology Division
Wyeth-Ayerst Research Inc.
P.O.Box 8299
Philadelphia, PA 19101, USA

Lin, T. S., 277
Dept. of Electrical Engineering
National Cheng-Kung University
Tainan 70102, Taiwan, ROC

Liu, Chung Yuan, 319
Institute of Biomedical Sciences
Academia Sinica
Taipei 11529, Taiwan, ROC

Liu, H. Mei, 319
Department of Pathology
College of Medicine
National Cheng Kung University
Tainan 70102, Taiwan, ROC

Maekawa, Hisato, 63
Institute of Hematology
Jichi Medical School
Tochigi-Ken 329-04, Japan

Mao, C. W., 277
Dept. of Electrical Engineering
National Cheng-Kung University
Tainan 70102, Taiwan, ROC

Maraganore, John M., 177
Biogen Inc.
14 Cambridge Center
Cambridge, MA 02142, USA

Martin, J.C. 145
MRC Epidemiology and Medical Care Unit
Northwick Park Hospital
Watford Road, Harrow
Middlesex, HA1 3UJ, UK

Matsuda, Michio, 63
Institute of Hematology
Jichi Medical School
Minamikawachi-Machi
Tochigi-Ken, 329-04, Japan

Miller, George J., 145
MRC Epidemiology and Medical Care Unit
Northwick Park Hospital
Watford Road, Harrow
Middlesex, HA1 3UJ, UK

Mitropoulos, K.A., 145
MRC Epidemiology and Medical Care Unit
Northwick Park Hospital
Watford Road, Harrow
Middlesex, HA1 3UJ, UK

Miyashita, C., 389
Dept. of Clinical Haemostaseology and Transfusion Medicine
University of Saarland
D-6650 Homburg/Saar, FRG

Nachowiak, D., 201
Washington University
St. Louis, MO 63110, USA

Nichols, Allen, 409
Dept. of Medicine
College of Physicians and Surgeons

Columbia University
New York, NY 10032, USA

Nieuwenhuizen, Willem, 83
Gaubius Institute TNO
P.O.Box 612
2300 AP Leiden, The Netherlands

Niewiarowski, Stefan, 251
Dept. of Physiology
Temple University School of Medicine
3400 N. Broad Street
Philadelphia, PA 19140, USA

Nishimura, Hitoshi, 121
Dept. of Biology
Faculty of Science
Kyushu University 33
Fukuoka 812, Japan

Norton, Karin J., 251
Dept. of Physiology and Thrombosis
Research Center
Temple University School of Medicine
Philadelphia, PA 19140, USA

O'Brien, J. R., 287
Central Laboratory
Haematology Department
St. Mary's Hospital
Portsmouth, Hants., UK

Ogawa, Satoshi, 303
Dept. of Physiology and Cellular Biophysics
College of Physicians and Surgeons
Columbia University
630 W. 168th St.
New York, NY 10032, USA

Ohlsson, Per-Ingvar, 185
Dept. of Medical Biochemistry
and Biophysics
University of Umea
S-90187 Umea, Sweden

Ouyang, Chaoho, 151
Dept. of Pharmacology
College of Medicine
National Taiwan University
No. 1, Jen-Ai Rd., Sec.1
Taipei 10018, Taiwan, ROC

Owen, John, 401
Hematology/Oncology Division
Bowman Gray School of Medicine
Wake Forest University
300 S. Hawthornd Rd.
Winston-Salam, NC 27103, USA

Ozbek, C., 389
Dept. of Internal Medicine III
University of Saarland
D-6650 Homburg/Saar, FRG

Ozsoy, Y., 429
Haemostasis and Thrombosis Research
Center
Cerrahpasa Medical School
Istanbul University
Istanbul, Turkey

Pindur, G., 389
Dept. of Clinical Haemostaseology
and Transfusion Medicine
University of Saarland
D-6650 Homburg/Saar, FRG

Pirkle, Hubert, 165
Dept. of Pathology
College of Medicine
University of California
Irvine, CA 92717, USA

Power, Eric, 409
Dept. of Medicine
College of Physicians and Surgeons
Columbia University
New York, NY 10032, USA

Procyk, R., 1
The New York Blood Center
310 E. 67th St.
New York, NY 10021, USA

Rapaport, Samuel I., 97
Dept. of Medicine and Pathology
University of Callifornia,
San Diego Medical Center
225 Dickinson St. (H811K)
San Diego, CA 92103-9981, USA

Reinhart, Walter, 409
Department of Physiology and Cellular
Biophysics
College of Physicians and Surgeons
Columbia University
New York, NY 10032, USA

Reison, Dennis, 409
Dept. of Medicine
College of Physicians and Surgeons
Columbia University
New York, NY 10032, USA

Roberts, James R., 133
Scripps Clinic and Research Foundation
BCR-8
10666 North Torrey Pines Rd.
La Jolla, CA 92037, USA

Ross, Allan M., 361
George Washington University

2150 Pennsylvania Ave. NW, Rm 4-414
Washington D.C. 20037, USA

Ruggeri, Zaverio M., 133
Scripps Clinic and Research Foundation
BCR-8, 10666 North Torrey Pines Rd.
La Jolla, CA 92037, USA

Salmon, G. P., 287
Central Laboratory
Dept. of Haematology
St. Mary's Hospital
Portmouth, Hants, UK

Schieffer, H., 389
Dept. of Internal Medicine III
University of Saarland
D-6650 Homburg/Saar, FRG

Schielen, W.J.G., 83
The Laboratory of Organic Chemistry
The Catholic University
Nijmegen, The Netherlands

Schwartz, Allan, 409
Dept. of Medicine
College of Physicians and Surgeons
Columbia University
New York, NY 10032, USA

Schwerdt, H., 389
Dept. of Internal Medicine III
University of Saarland
D-6650 Homburg/Saar, FRG

Seegers, Walter H., 93
2857 Ptarmigan Dr. #5
Walnut Creek, CA 94595, USA

Sen, C., 389
Dept. of Internal Medicine III
University of Saarland
D-6650 Homburg/Saar, FRG

Shainoff, John R., 73
Research Institute FF4
Cleveland Clinic Foundation
9500 Euclid Ave.
Cleveland, OH 44195, USA

Shiao, Ming-Shi, 265
Institute of Life Science
National Tsing Hua University
Hsinchu, Taiwan, ROC

Shreeniwas, Revati, 303
Dept. of Physiology and
Cellular Biophysics
College of Physicians and Surgeons
Columbia University
New York, NY 10032, USA

Silveira, A., 1
Dept. of Blood Coagulation Research
Karolinska Institute
S-104 01 Stockholm, Sweden

Sinha, Asru Kumar, 271
Division of Cardiology
Montefiore Medical Center
Albert Einstein College of Medicine
111 E. 210th St.
Bronx, NY 10467, USA

Sobel, B. E., 201
Washington University
St. Louis, MO 63110, USA

Sobel, Joan H., 55, 401
Dept. of Medicine
College of Physicians & Surgeons
Columbia University
630 W. 168th St.
New York, NY 10032, USA

Stern, David M., 303
Dept. of Physiology and
Cellular Biophysics
College of Physisians and Surgeons
Columbia University
New York, NY 10032, USA

Takada, Akikazu, 209, 223
Dept. of Physiology
Hamamatsu University
School of Medicine
3600 Handa-cho, Hamamatsu-Shi
Shizuoka-Ken, 431-31, Japan

Takada, Yumiko, 209, 223
Dept. of Physiology
Hamamatsu University
School of Medicine
3600 Handa-cho
Hamamatsu-shi
Shizuoka-Ken, 431-31, Japan

Teng, Che-Ming, 151
Pharmacological Institute
College of Medicine
National Taiwan University
No 1, Jen-Ai Rd., Sec. 1,
Taipei 10018, Taiwan, ROC

Terukina, Shigeharu, 63
Institute of Hematology
Jichi Medical School
Tochigi-Ken 329-04, Japan

Tesser, G.I., 83
The Laboratory of Organic Chemistry
The Catholic University
Nijmegen, The Netherlands

Theodor, Ida, 165
Dept. of Pathology
College of Medicine
University of California
Irvine, CA 92717, USA

Tymkewycz, P.M., 419
National Institute for Biological Standards
and Control
Blanche Lane, Potters Bar
Hertfordshire EN6 3QG, UK

Ugur, M. S., 429
Haemostasis and Thrombosis
Research Center
Cerrahpasa Medical School
Istanbul University
Istanbul, Turkey

Ulutin, Orhan Nuri, 429
Haemostasis and Thrombosis
Research Center
Cerrahpasa Medical School
Istanbu University
Istanbul, Turkey

Ulutin, T., 429
Haemostasis and Thrombosis
Research Center
Cerrahpasa Medical School
Istanbul University
Istanbul, Turkey

Urano, Tetsumei, 209
Dept. of Physiology
Hamamatsu University
School of Medicine
3600 Handa-Cho
Hamamatsu-Shi
Shizuoka-ken, 431-31, Japan

Urbanic, David A., 73
Research Institute FF4
Cleveland Clinic Foundation
9500 Euclid Ave.
Cleveland, OH 44195, USA

Valenzuela, Rafael, 73
Research Institute FF4
Cleveland Clinic Foundation
9500 Euclid Ave.
Cleveland, OH 44195, USA

Voskuilen, M., 83
Gaubius Institute TNO
P.O.Box 612
2300 AP Leiden, The Netherlands

Wallen, Per B., 185
Dept. of Medical Biochemistry and
Biophysics
University of Umea
Umea, S-90187, Sweden

Wang, Chuen-Neu, 265
Institute of Life Science
National Tsing Hua University
Hsinchu, Taiwan, ROC

Wang, Cheng-Teh, 265
Institute of Life Science
National Tsing Hua University
Hsinchu, Taiwan, ROC

Wang, Danny Ling, 319
Institute of Biomedical Sciences
Academia Sinica
Taipei 11529, Taiwan, ROC

Wang, Shih-Pu, 395
Division of Cardiology
Veterans General Hospital
Shih-Pai, Taipei, Taiwan, ROC

Watson, Rita, 409
Dept. of Medicine
College of Physicians and Surgeons
Columbia University
New York, NY 10032, USA

Weinheimer, C. J., 201
Washington University
St. Louis, MO 63110, USA

Wenzel, Ernst, 389
Dept. of Clinical Haemostaseology
and Transfusion Medicine
University of Saarland
D-6650 Homburg/Saar, FRG

White, Harvey D., 383
Cardiovascular Research
Dept. of Cardiology
Green Lane Hospital
Private Bag Symonds Street
Auckland 1, New Zealand

Wilhelm, J., 201
Biotechnology & Microbiology Division
Wyeth-Ayerst Research Inc.
P.O.Box 8299
Philadelphia, PA 19101, USA

Wong, Patrick Y.-K., 245
Dept. of Physiology and Medicine
New York Medical College
Valhalla, NY 10595, USA

Wu, H. J., 277
Dept. of Physiology
College of Medicine
National Cheng-Kung University

No. 1, Ta Hsueh Rd.
Tainan 70102, Taiwan, ROC

Wu, Kenneth K., 297
Division of Hematology
Dept. of Medicine
University of Texas
Health Sciences Center
6431 Fannin
Houston, TX 77030, USA

Yamazumi, Kensuke, 63
Institute of Hematology
Jichi Medical School
Tochigi-Ken 329-04, Japan

Yonekawa, Osamu, 83
Gaubius Institute TNO
P.O.Box 612
2300 AP Leiden, The Netherlands

Yoshida, Nobuhiko, 63
Institute of Hematology
Jichi Medical School
Tochigi-Ken 329-04, Japan

Zacharski, L., 1
Veterans Administration Hospital
White River Junction, VT 05001, USA

SUBJECT INDEX